Spon's Mechanical and Electrical Services Price Book

2009

Spon's Mechanical and Electrical Services Price Book

Edited by

DAVIS LANGDON MOTT GREEN WALL
Engineering Services

2009

Fortieth edition

Taylor & Francis
Taylor & Francis Group

LONDON AND NEW YORK

First edition 1968
Fortieth edition published 2009
by Taylor & Francis
2 Park Square, Milton Park, Abingdon, Oxon OX14 4RN

Simultaneously published in the USA and Canada
by Taylor & Francis
270 Madison Avenue, New York, NY 10016

Taylor & Francis is an imprint of the Taylor & Francis Group, an informa business

© 2009 Taylor & Francis

The right of Davis Langdon Mott Green Wall to be identified as the Author of this Work has been asserted by them in accordance with the Copyright, Designs and Patents Act 1988

Printed and bound in Great Britain by
TJ International Ltd, Padstow, Cornwall

All rights reserved. No part of this book may be reprinted or reproduced or utilised in any form or by any electronic, mechanical, or other means, now known or hereafter invented, including photocopying and recording, or in any information storage or retrieval system, without permission in writing from the publishers.

The publisher makes no representation, express or implied, with regard to the accuracy of the information contained in this book and cannot accept any legal responsibility or liability for any errors or omissions that may be made.

Publisher's note
This book has been produced from camera-ready copy supplied by the authors.

British Library Cataloguing in Publication Data
A catalogue record for this book is available from the British Library

ISBN13: 978-0-415-46561-8
Ebook: 978-0-203-88673-1
ISSN: 0305-4543

Contents

Preface	vii
Special Acknowledgements	ix
Acknowledgements	xi

PART ONE: ENGINEERING FEATURES

Revisions to Part L of the Building Regulations	3
Renewable Energy Options	5
Grey Water	9
Ground Water Cooling	13
Fuel Cells	19
Biomass Energy	25
Getting the Connection	29
Typical Engineering Details	31

PART TWO: APPROXIMATE ESTIMATING

Directions	59
Cost Indices	60
RIBA Stage A Feasibility Costs	62
RIBA Stage C Elemental Rates	67
All-in-Rates	74
Elemental Costs	90

PART THREE: MATERIAL COSTS/MEASURED WORK PRICES

Mechanical Installations

Directions	119
R : Disposal Systems	
R10 : Rainwater Pipework/Gutters	125
R11 : Above Ground Drainage	145
S : Piped Supply Systems	
S10 : Cold Water	169
S11 : Hot Water	224
S32 : Natural Gas	228
S41 : Fuel Oil Storage/Distribution	232
S60 : Fire Hose Reels	233
S61 : Dry Risers	234
S63 : Sprinklers	235
S65 : Fire Hydrants	241
T : Mechanical/Cooling/Heating Systems	
T10 : Gas/Oil Fired Boilers	243
T13 : Packaged Steam Generators	257
T31 : Low Temperature Hot Water Heating	258
T33 : Steam Heating	339
T42 : Local Heating Units	343
T60 : Central Refrigeration Plant	344
T61 : Chilled Water	352
T70 : Local Cooling Units	358

U : Ventilation/Air Conditioning Systems
- U10 : General Ventilation — 360
- U14 : Ductwork: Fire Rated — 445
- U30 : Low Velocity Air Conditioning — 465
- U31 : VAV Air Conditioning — 468
- U41 : Fan Coil Air Conditioning — 469
- U70 : Air Curtains — 473

Electrical Installations

Directions — 477

V : Electrical Supply/Power/Lighting Systems
- V10 : Electrical Generation Plant — 481
- V11 : HV Supply — 483
- V20 : LV Distribution — 483
- V21 : General Lighting — 555
- V22 : General LV Power — 563
- V32 : Uninterruptible Power Supply — 570
- V40 : Emergency Lighting — 572

W : Communications/Security/Control
- W10 : Telecommunications — 577
- W20 : Radio/Television — 579
- W23 : Clocks — 581
- W30 : Data Transmission — 582
- W40 : Access Control — 588
- W41 : Security Detection and Alarm — 589
- W50 : Fire Detection and Alarm — 590
- W51 : Earthing and Bonding — 593
- W52 : Lightning Protection — 594
- W60 : Central Control/Building Management — 598

PART FOUR: RATES OF WAGES

Mechanical Installations

Rates of Wages — 603

Electrical Installations

Rates of Wages — 611

PART FIVE: DAYWORK

Heating and Ventilating Industry — 619
Electrical Industry — 622
Building Industry Plant Hire Costs — 625

Tables and Memoranda — 637

Index — 661

Preface

The Fortieth Edition of *SPON'S Mechanical and Electrical Services Price Book* continues to cover the widest range and depth of services, reflecting the many alternative systems and products that are commonly used in the industry as well as current industry trends.

In terms of current pricing levels, the continuing boom in the Chinese economy and it's consumption of raw materials has led to sharp increases in the price of steel and copper. Both steel and copper tube manufacturers are reporting monthly increases in prices, and whilst these have been incorporated at the time of going to press, readers are advised to check the currency of such prices before using them. Likewise, the sustained high cost of crude oil will impact on products that are oil based i.e. UPVC as well as transport costs. However, unlike last year, the picture is clearer concerning manufacturers and suppliers of plant and materials utilising steel and copper, with some increasing costs and lead times due to the increasing demand. Again, prices are current at the time of going to press, but readers are advised to check that such prices are still current before using them.

Before referring to prices or other information in the book, readers are advised to study the `Directions' which precede each section of the Materials Costs/Measured Work Prices. As before, no allowance has been made in any of the sections for Value Added Tax.

The order of the book reflects the order of the estimating process, from broad outline costs through to detailed unit rate items.

The approximate estimating section has been thoroughly reviewed to provide up to date key data in terms of square metre rates, all-in-rates for key elements and selected specialist activities and elemental analyses on a comprehensive range of building types.

The prime purpose of the Materials Costs/Measured Work Prices part is to provide industry average prices for mechanical and electrical services, giving a reasonably accurate indication of their likely cost. Supplementary information is included which will enable readers to make adjustments to suit their own requirements. It cannot be emphasised too strongly that it is not intended that these prices should be used in the preparation of an actual tender without adjustment for the circumstances of the particular project in terms of productivity, locality, project size and current market conditions. Adjustments should be made to standard rates for time, location, local conditions, site constraints and any other factor likely to affect the costs of a specific scheme. Readers are referred to the build up of the gang rates, where allowances are included for supervision, labour related insurances, and where the percentage allowances for overhead, profit and preliminaries are defined.

Readers are reminded of the service available on the Spon's website detailing significant changes to the published information. www.pricebooks.co.uk/updates

As with previous editions the Editors invite the views of readers, critical or otherwise, which might usefully be considered when preparing future editions of this work.

Whilst every effort is made to ensure the accuracy of the information given in this publication, neither the Editors nor Publishers in any way accept liability for loss of any kind resulting from the use made by any person of such information.

In conclusion, the Editors record their appreciation of the indispensable assistance received from the many individuals and organisations in compiling this book.

DAVIS LANGDON MOTT GREEN WALL
Engineering Services
MidCity Place
71 High Holborn
London WC1V 6QS

Telephone: 0207 061 7777
Facsimile: 0207 061 7009

SPON'S PRICEBOOKS 2009

Spon's Architects' and Builders' Price Book 2009

Davis Langdon

The most detailed, professionally relevant source of construction price information currently available anywhere.

In the 2009 edition the Preliminaries section has been re-written and based upon the 2005 JCT contract, and the new section on Sustainability has been re-written and considerably expanded. New Measured Works items include more economical, lower nickel, stainless steel rebars; Schüco curtain walling; brises soleil; "Ecologic" roofing tiles; Velfac 200 windows; and triple glazing. ... along with additional items of rainscreen cladding, damp proof courses, steel lintels, Patent Glazing items, and radiators and manhole covers.

Hb & electronic package*: 1000pp approx.
978-0-415-46555-7: **£145**
electronic package
(inc. sales tax where appropriate)
978-0-203-88676-2: only **£135**

Spon's Mechanical and Electrical Services Price Book 2009

Davis Langdon, Mott Green Wall

The only comprehensive and best annual services engineering price book available for the UK.

As well as an extensive overhaul of prices, updates to the 2009 edition include a new infrastructure section which includes typical connection scenarios and costs, extensive revisions - to the renewable energy sections, with descriptions of biomass boilers, ground source heap pumps and photovoltaics. Sprinklers and luminaries have been bought in-line with new regulations.

Hb & electronic package*: 640pp approx
978-0-415-46561-8: **£145**
electronic package
(inc. sales tax where appropriate)
978-0-203-88673-1: only **£135**

Spon's External Works and Landscape Price Book 2009

Davis Langdon

The only comprehensive source of information for detailed external works and landscape costs.

The 2009 edition gives more on costs of tender preparation and specialised surveys, under Preliminaries; and a number of new and heavily revised Measured Works items: repair and conservation of historic railings; new Aco drainage ranges; fin drains; concrete pumping and crushing costs; rainwater harvesting tanks; containerised tree planting; specialised mortars; park signage; and site clearance, notably the clearance of planting material and demolition of structures.

Hb & electronic package*: 424pp approx.
978-0-415-46559-5: **£115**
electronic package
(inc. sales tax where appropriate)
978-0-203-88674-8: only **£105**

Spon's Civil Engineering and Highway Works Price Book 2009

Davis Langdon

The 23rd edition, in its easy to read format, incorporates a general review throughout, and includes updates to the Capital Allowances and VAT and Construction sections to reflect the latest government legislation.

Hb & electronic package:* 800pp approx.
978-0-415-46557-1: **£155**
electronic package
(inc. sales tax where appropriate)
978-0-203-88675-5: only **£145**

---

*Receive our eBook free when you order any hard copy Spon 2009 Price Book, with free estimating software to help you produce tender documents, customise data, perform word searches and simple calculations.
Or buy just the ebook, with free estimating software.
Visit **www.pricebooks.co.uk**

To Order: Tel: +44 (0) 1235 400524 **Fax:** +44 (0) 1235 400525
or Post: Taylor and Francis Customer Services, Bookpoint Ltd, Unit T1, 200 Milton Park, Abingdon, Oxon, OX14 4TA UK
Email: book.orders@tandf.co.uk

For a complete listing of all our titles visit: www.tandf.co.uk

Taylor & Francis
Taylor & Francis Group

Special Acknowledgements

The Editors wish to record their appreciation of the special assistance given by the following organisations in the compilation of this edition.

Senior HARGREAVES
DUCTWORK SPECIALISTS

Lord Street, Bury, Lancashire. BL9 0RG
Tel: 0161 764 5082 • Fax: 0161 762 2336
E-Mail: sales@senior-hargreaves.co.uk

Comunica plc
The Hallmarks
146 Field End Road
Eastcote
Pinner
Middlesex
HA5 1RJ

Tel: 020 8429 9696
Fax: 020 8429 4982
Email: enquiries@comunicaplc.co.uk
www.comunicaplc.co.uk

Abbey

ABBEY THERMAL INSULATION LTD.
23-24, Riverside House,
Lower Southend Road, Wickford, Essex SS11 8BB
Telephone: 01268 572116- Facsimile: 01268 572117
E-mail: general@abbeythermal.com

AXIMA

Axima Building Services
Westmead, Farnborough
Bournemouth, Hants GU14 7LP
Tel: 01252 525500
Fax: 01252 378988
www.axima.eu.com

HOTCHKISS

Hampden Park Industrial Estate
Eastbourne
East Sussex
BN22 9AX
Tel : 01323 501234
Fax : 01323 508752
E-Mail : info@Hotchkiss.co.uk
www.Hotchkiss.co.uk

J. & M. Insulations Limited
For all aspects of Thermal Insulation & Fire Protection

257a Banbury Road,
Summertown, Oxford, OX2 7HN
Telephone: 01865 310220
Fax: 01865 512419
E-mail: jandminsualation@fsbdial.co.uk
www.jandminsulations.co.uk

Get more out of your Spon Price Book Library

Everything you need to price work quickly and accurately using your own custom rates together with your Spon Price libraries.

- **Resource Library** - Keep your costs up-to-date and in one place
- **Suppliers** - Allocate resources to individual suppliers
- **Price Lists** - Access to the latest product prices from the UK's largest merchant
- **Composite Resources** - A real time saver for mortar; labour gangs etc.
- **Price Book Options** - Access industry standard Price libraries from Spon
- **Item Library** - Maintain your own rate library (with full build ups)
- **Composite Items** - Save time by selecting a single item covering multiple tasks
- **BQ Preparation** - Deliver professional documents to clients and colleagues
- **Resource Requirements** - Specify exactly what you need to complete the job
- **Purchase Ordering -** Convert job requirements into enquiries and orders
- **Flexible** - Works for builders, contractors, sub-contractors and specialists
- **Price Excel® and CITE Bills** - Import, price and export bills supplied in these industry standard formats.
- **Output to Microsoft® Applications** - Word®, Excel®, Project®
- **Subcontractor Bills** - Generate bills from selected resources or tagged items.

Fast Estimate shares the same clear pricing processes as Spon Estimating, but with more powerful pricing tools to give you outstanding control over your tender documents.

Download your FREE trial copy:

www.estek.co.uk

Or call 01764 655331

Acknowledgements

The editors wish to record their appreciation of the assistance given by many individuals and organisations in the compilation of this edition.

Manufacturers, Distributors and Sub-Contractors who have contributed this year include:-

A C Plastics Industries Ltd
Armstrong Road
Daneshill East
Basingstoke RG24 8NU
GRP Water Storage Tanks
Tel: 01256 329334
Fax: 01256 817862
www.acplastiques.com

Alfa Laval Limited
Unit 1, 6 Wellheads Road
Farburn Industrial Estate
Dyce
Aberdeen AB21 7HG
Heat Exchangers
Tel : 01224 424300
Fax : 01224 725213
www.alfalaval.com

Aquilar Limited
Dial Post Court
Horsham Road
Rusper
West Sussex RH12 4QX
Leak Detection
Tel : 08707 940310
Fax : 08707 940320
www.aquilar.co.uk

Axima Building Services
80 Paul Street
London EC2A 4UD
Above Ground Drainage
Tel : (020) 7729 7634
Fax : (020) 7729 9756
www.axima-uk.com

Balmoral Tanks
Balmoral Park
Loirston
Aberdeen AB12 3GY
GRP Water Storage Tanks
Tel: 01224 859000
Fax: 01224 859123
www.balmoral-group.com

Braithwaite Engineers Ltd
Neptune Works
Uskway
Newport
South Wales NP9 2UY
Sectional Steel Water Storage Tanks
Tel: 01633 262141
Fax: 01633 250631
www.braithwaite.co.uk

Broadcrown Limited
Alliance Works
Airfield Industrial Estate
Hixon
Staffs ST18 0PF
Generators
Tel: 01889 272200
Fax: 01889 272220
www.broadcrown.co.uk

Caradon Stelrad Ideal Boilers
PO Box 103
National Avenue
Kingston-upon-Hall
North Humberside HU5 4JN
Boilers/Heating Products
Tel: 08708 400030
Fax: 08708 400059
www.rycroft.com

Carrier Air Conditioning
United Technologies House
Guildford Road
Leatherhead
Surrey KT22 9UT
Chilled Water Plant
Tel: 0870 6001100
Fax: 01372 220221
www.carrier.uk.com

Chloride Power Protection
Unit C, George Curl Way
Southampton SO18 2RY
Static UPS Systems
Tel: 023 8061 0311
Fax: 023 8061 0852
www.chloridepower.com

Communica
Chatteris Airfield
Near March
Cambridgeshire PE15 0EA
Telephone Cables
Tel : 01354 742340
www.communicaplc.co.uk

Cooper Lighting and Security
London Project Office
Suite 8, King Harold Court
Sun Street
Waltham Abbey
Essex
EN9 1ER
Emergency Lighting and Luminaires
Tel: 01302 303303
Fax: 01392 367155
www.cooper-ls.com

Danfoss Flowmetering Ltd
Magflo House
Ebley Road
Stonehouse
Glos GL10 2LU
Energy Meters
Tel: 01453 828891
Fax: 01453 853860
www.danfoss-randall.co.uk

Dewey Waters Limited
Cox's Green
Wrington
Bristol BS40 5QS
Tanks
Tel : 01934 862601
Fax : 01934 862604
www.deweywaters.co.uk

Diffusion
Benson Environmental Limited
47 Central Avenue
West Molesey
Surrey KT8 2QZ
Fan Coil Units
Tel: (020) 8783 0033
Fax: (020) 8783 0140
www.diffusionenv.com

Dunham-Bush Limited
8 Downley Road
Havant
Hampshire PO9 2JD
Convectors and Heaters
Tel : 02392 477700
Fax : 02392 450396
www.dunham-bush.com

EMS Radio Fire & Security Systems Limited
Technology House
Sea Street
Herne Bay
Kent CT6 8JZ
Security
Tel : 01227 369570
Fax : 01227 369679
www.emsgroup.co.uk

Engineering Appliances Ltd
Unit 11
Sunbury Cross Ind Est
Brooklands Close
Sunbury On Thames TW16 7DX
Expansion Joints, Air and Dirt Separators
Tel: 01932 788888
Fax: 01932 761263
e-mail: info@engineering-appliances.co.uk
www.engineeringappliances.com

FCS Ductwork Limited
3rd Floor
Thomas Telford House
1 Heron Quay
Canary Wharf
London E14 5JD
Fire Rated Ductwork
Tel: (020) 7987 7692
Fax: (020) 7537 5627
www.fcsgroup.co.uk

FKI Hawker Siddeley
Falcon Works
P O Box 7713
Meadow Lane
Loughborough
Leicestershire LE11 1ZF
HV Supply, Cables and HV Switchgear and Transformers
Tel: 01495 331024
Fax: 01495 331019
www.fkiswitchgear.com

Flakt Woods Limited
Tufnell Way
Colchester CO4 5AR
Air Handling Units
Tel : 01206 544122
Fax : 01206 574434

Hall Fire Protection Limited
186 Moorside Road
Swinton
Manchester M27 9HA
Fire Protection Equipment
Tel : 0161 793 4822
Fax : 0161 794 4950
www.hallfire.co.uk

Halton
5 Waterside Business Park
Witham
Essex CM8 3YQ
Chilled Beams
Tel : 01376 503040
Fax : 01376 503060
www.haltongroup.com

Hattersley, Newman, Hender Ltd
Burscough Road
Ormskirk
Lancashire L39 2XG
Valves
Tel: 01695 577199
Fax: 01695 578775
e-mail: uksales@hattersley-valves.co.uk
www.hattersley.com

Hitec Power Protection Limited
Unit B21a
Holly Farm Business Park
Honiley
Kenilworth
Warwickshire CV8 1NP
Uninterruptible Power Supply (Rotary/Diesel)
Tel: 01926 484535
Fax: 01926 484336
www.hitecups.co.uk

Honeywell CS Limited
Honeywell House
Anchor Boulevard
Crossways Business Park
Dartford
Kent DA2 6QH
Control Components
Tel : 01322 484800
Fax : 01322 484898
www.honeywell.com

Hoval Limited
Northgate
Newark
Notts NG24 1JN
Boilers
Tel : 01636 672711
Fax : 01636 673532
www.hoval.co.uk

HRS Hevac Ltd
10-12 Caxton Way
Watford Business Park
Watford
Herts WD18 8JY
Heat Exchangers
Tel: 01923 232335
Fax: 01923 230266
www.hrshevac.co.uk

Hudevad
Bridge House
Bridge Street
Walton on Thames
Radiators
Tel: 01932 247835
Fax: 01932 247694
www.hudevad.co.uk

Hydrotec (UK) Limited
Hydrotec House
5 Mannor Courtyard
Hughenden Avenue
High Wycombe HP13 5RE
Chemical Treatment
Tel : 01494 796040
Fax : 01494 796049
www.hydrotec.com

IAC
IEC House
Moorside Road
Winchester
Hampshire SO23 7US
Attenuators
Tel : 01962 873000
Fax : 01962 873102
www.industrialacoustics.com

IC Service & Maintenance Ltd
Unit K3 Temple Court
Knights Place
Knight Road
Strood
Kent ME2 2LT
Fire Detection & Alarm
Tel : 01634 290300
Fax : 01634 290700
www.icservice.biz

Ideal Boilers
P O Box 103
National Avenue
Kingston Upon Hull
East Yorkshire HU5 4JN
Boilers
Tel : 01482 492251
Fax : 01482 448858
www.idealboilers.com

K and W Fabrications Ltd
High Street
Handcross
Haywards Heath
West Sussex RH17 6BZ
Plastic Ductwork
Tel: 01444 401144
Fax: 01444 401188

Kampmann
Benson Environmental Limited
47 Central Avenue
West Molesey
Surrey KT8 2QZ
Trench Heating
Tel: (020) 8783 0033
Fax: (020) 8783 0140
www.diffusionenv.com

Kiddie Fire Protection Services
Enterprise House
Jasmine Grove
London SE20 8JW
Fire Protection Equipment
Tel : (020) 8659 7235
Fax : (020) 8659 7237
www.kfp.co.uk

Metcraft Ltd
Harwood Industrial Estate
Littlehampton
West Sussex BN17 7BB
Oil Storage Tanks
Tel: 01903 714226
Fax: 01903 723206
www.metcraft.co.uk

Osma Underfloor Heating
18 Apple Lane
Sowton Trade City
Exeter
Devon EX2 5GL
Underfloor Heating
Tel : 01392 444122
Fax : 01392 444135
www.osmaufh.co.uk

Pullen Pumps Limited
158 Beddington Lane
Croydon CR9 4PT
Pumps, Booster Sets
Tel: (020) 8684 9521
Fax: (020) 8689 8892
www.pullenpumps.co.uk

Reliance Hi-tech
Boundary House
Cricketfield Road
Uxbridge
Middlesex
UB8 1QG
Access Control and Security Detection and Alarm
Tel: 01895 205000
Fax: 01895 205100
www.reliancesecurity.co.uk

Rycroft
Duncombe Road
Bradford BD8 9TB
Storage Cylinders
Tel : 01274 490911
Fax : 01274 498580
www.rycroft.com

SF Limited
Pottington Business Park
Barnstaple
Devon EX31 1LZ
Flues
Tel : 01271 326633
Fax : 01271 334303

Simmtronic Limited
Waterside
Charlton Mead Lane
Hoddesdon
Hertfordshire EN11 0QR
Lighting Controls
Tel : 01992 456869
Fax : 01992 445132
www.simmtronic.com

Socomec Limited
Knowl Piece
Wilbury Way
Hitchin
Hertfordshire SG4 0TY
Automatic Transfer Switches
Tel : 01462 440033
Fax : 01462 431143
www.socomec.com

Spirax-Sarco Ltd
Charlton House
Cheltenham
Gloucestershire GL53 8ER
Traps and Valves
Tel: 01242 521361
Fax: 01242 573342
www.spiraxsarco.com

Tyco Limited
Unit 6 West Point Enterprize Park
Clarence Avenue
Trafford Park
Manchester M17 1QS
Fire Protection
Tel: 0161 875 0400
Fax: 0161 875 0491
www.tyco.com

Utile Engineering Company Ltd
Irthlingborough
Northants NN9 5UG
Gas Boosters
Tel: 01933 650216
Fax: 01933 652738
www.utileengineering.com

Vokes Ltd
Henley Park
Guildford
Surrey GU3 2AF
Air Filters
Tel: 01483 569971
Fax: 01483 235384
e-mail: vokes@btvinc.com
www.vokes.com

Waterloo Air Management
Mills Road
Aylesford
Kent ME20 7NB
Grilles & Diffusers
Tel: 01622 717861
Fax: 01622 710648
www.waterloo.co.uk

Whitecroft Lighting Limited
Burlington Street
Ashton-under-Lyne
Lancashire OL7 0AX
Lighting & Luminaires
Tel: 0870 5087087
Fax: 0870 5084210
www.whitecroftlighting.com

Woods of Colchester
Tufnell Way
Colchester
Essex CO4 5AR
Air Distribution, Fans, Anti-vibration mountings
Tel: 01206 544122
Fax: 01206 574434

eBooks – at www.eBookstore.tandf.co.uk

A library at your fingertips!

eBooks are electronic versions of printed books. You can store them on your PC/laptop or browse them online.

They have advantages for anyone needing rapid access to a wide variety of published, copyright information.

eBooks can help your research by enabling you to bookmark chapters, annotate text and use instant searches to find specific words or phrases. Several eBook files would fit on even a small laptop or PDA.

NEW: Save money by eSubscribing: cheap, online access to any eBook for as long as you need it.

Annual subscription packages

We now offer special low-cost bulk subscriptions to packages of eBooks in certain subject areas. These are available to libraries or to individuals.

For more information please contact webmaster.ebooks@tandf.co.uk

We're continually developing the eBook concept, so keep up to date by visiting the website.

www.eBookstore.tandf.co.uk

PART ONE

Engineering Features

In this, the Fortieth edition of *SPON'S Mechanical and Electrical Services Price Book,* a new section has been included for Engineering Features, dealing with current issues or technical advancements within the industry. These shall be complimented by cost models and/or itemised prices for items that form part of such.

The intention is that the book shall develop to provide more than just a schedule of prices to assist the user in the preparation and evaluation of costs.

- Revisions to Part L of the Building Regulations
- Renewable Energy Options
- Grey Water
- Ground Water Cooling
- Fuel Cells
- Biomass Energy
- Getting the Connection

ESSENTIAL READING FROM TAYLOR AND FRANCIS

Spon's Irish Construction Price Book

Third Edition

Franklin + Andrews

This new edition of *Spon's Irish Construction Price Book*, edited by Franklin + Andrews, is the only complete and up-to-date source of cost data for this important market.

- All the materials costs, labour rates, labour constants and cost per square metre are based on current conditions in Ireland
- Structured according to the new Agreed Rules of Measurement (second edition)
- 30 pages of Approximate Estimating Rates for quick pricing

This price book is an essential aid to profitable contracting for all those operating in Ireland's booming construction industry.

Franklin + Andrews, Construction Economists, have offices in 100 countries and in-depth experience and expertise in all sectors of the construction industry.

April 2008: 246x174 mm: 510 pages
Hb: 978-0-415-45637-1: **£135.00**

To Order: Tel: +44 (0) 1235 400524 **Fax:** +44 (0) 1235 400525
or Post: Taylor and Francis Customer Services,
Bookpoint Ltd, Unit T1, 200 Milton Park, Abingdon, Oxon, OX14 4TA UK
Email: book.orders@tandf.co.uk

For a complete listing of all our titles visit:
www.tandf.co.uk

Revisions to Part L of the Building Regulations

The following review of the effects of the revised Part L on non-residential buildings is based on updated material originally published in a Building Cost Model on Part L, issued in *Building* Magazine on 5 August 2005. The original article was produced in collaboration with Consultants, Arup.

Introduction

Part L (2006) came into force in April 2006 at a time when awareness of the threat posed by global warming has never been higher. The 2002 revisions signaled an important shift in emphasis from energy conservation to control of emissions of carbon and other greenhouse gases. Revised Part L 2006 is building on this foundation, reducing carbon emissions from new build further, and widening the scope of work required to improve performance in existing buildings. In raising standards, the approach adopted has been to give maximum flexibility to owners and designers in selecting means of compliance.

Changes in the Approach in Part L2

Part L2 deals with non-dwellings. The major headline change is the requirement to achieve savings in carbon emissions compared to the previous 2002 standard. For new build, air-conditioned and mechanically ventilated buildings, the reduction is 28%, whereas for naturally ventilated buildings the reduction is 23.5%.

In addition to the introduction of these challenging targets for carbon emissions reduction, the Revised Part L also involves some significant changes to practice that are having an impact on the work of consultants and contractors.

- The introduction of alternative assessment methodologies for new build and existing buildings. The new build method, based on a national standard assessment methodology is concerned with comparing the carbon emissions of the proposed development with those of a notional building, which complies with the previous 2002 Regulations. The approach for existing buildings is broadly intended to encourage proportional, technically feasible and economic improvements to building performance to ensure that alterations and improvements made to an existing building contribute to a reduction in carbon emissions.

- A move away from prescriptive technical guidance. Users are required to refer to technical details taken from a wide range of sources to identify how to meet the new standards. The new Part L only sets out the required benchmarks. .

- The adoption of a single national calculation methodology. This change, driven by the EPBD, has led to the adoption of a carbon emissions method for buildings other than dwellings. As a result, the simple elemental method of calculation used under previous versions of the regulations is no longer available.

- The Regulations include a provision to include the contribution of low or zero carbon (LZC) technologies, such as solar hot water heating, in the assessment. They also provide flexibility with regard to their application whilst the technologies mature and become more widely available.

The National Calculation Methodology

The new calculation methodology required under involves inputting all aspects of the building design into an assessment tool based on an agreed methodology. The National Calculation Tool, known as iSBEM (Simple Building Evaluation Method), developed by the BRE, is currently the most common application. Others such as CECM are becoming available. The comparative emissions calculation needs to be done fairly early on in the design process and then updated once the design is fixed. This will potentially make it more difficult for the design team to handle changes, and, given the resource requirements of recalculation, this could limit design iteration, although criteria such as perimeter heat gain can be set that will allow some design iterations without recalculation. Furthermore, contractors may be less likely to propose alternative design solutions as a result of the introduction of the new methodology, if they are subsequently required to demonstrate and achieve compliance.

Impact of the Regulations on design

The revised Part L is beginning to affect all aspects of the building design. The transition period was very short, and in effect, all buildings currently being designed will have to meet the new Standard. In addition to insulation levels, proportion and type of glazing, other factors that may be improved are:

- Shading to combat perimeter heat gains, e.g. external shading, double wall facades;
- Boiler efficiencies and chiller coefficients of performance;
- Building envelope air tightness;
- Lighting efficiency, lighting control;
- Fan and pump efficiencies, air and water pressure drops;
- Energy recovery.

A key characteristic of the revisions is, although the carbon emissions target has been cut by some 28% for cooled or mechanically ventilated buildings, none of the minimum acceptable standards to which the design must comply have been changed significantly from those set in 2002. It will continue to be possible to specify, for example, highly glazed buildings, albeit that significant improvement in other aspects of building performance will be necessary. Lighting specification and lamp efficiency are examples of areas where substantial reductions in both energy consumption and cooling loads could be made at a relatively low cost.

Indicative cost implications

Davis Langdon prepared a comparison of comparative office model solutions in Summer 2005 with Consulting Engineers, Arup, in order to show how the 2006 Regulations could be met by using either 40% glazing or different combinations of 100% glazing and solar shading. The cost implications are that using a low proportion of glazing, improvements to the glass specification, services plant efficiency and the introduction of simple heat recovery measures will actually result in a small reduction in capital costs resulting from the reduction in heating and cooling loads and plant sizing.

Option	Envelope solution	Saving/Extra £/m²	Saving/Extra %
40% glazing	High performance glass	(11.82)	(0.62)
100% glazing	High performance glass, fixed external louvres	61.29	3.19
100% glazing	Clear double glazing with low-e coating, external motorised slatted blinds	96.78	5.06
100% glazing	Double wall facade with externally ventilated cavity, with motorised slatted blinds at close centres	73.12	3.82

The costs in the comparison are at 3rd Quarter 2005 prices based on a central London location. The base building costs include the category A fit-out, preliminaries and contingencies. Costs of demolitions, external works, tenant fit-out, professional fees and VAT are excluded.

Due to the complexity of the changes and the recent introduction of the Regulations, it has not been possible to assess their impact on the rates within the 'Cost Models' and the 'Approximate Estimating' sections of the SPON'S Mechanical and Electrical Price Book 2007. These sections have not been adjusted to take into account the implications of works required to comply with the revised Part L Regulations. Users should note that the items and their rates in the 'Prices for Measured Works' section are compliant.

Renewable Energy Options

RENEWABLE ENERGY, TOGETHER WITH AN ANALYSIS OF WHERE THEY MAY BE INSTALLED TO BEST ADVANTAGE.

This article focuses on building-integrated options rather than large-scale utility solutions such as wind farms which will be addressed in the 2008 Edition.

The legislative background, imperatives and incentives

In recognition of likely causes and effects of global climate change, the Kyoto protocol was signed by the UK and other nations in 1992, with a commitment to reduce the emission of greenhouse gases relative to 1990 as the base year.
The first phase of European Union Emission Trading Scheme (EU ETS) covers the power sector and high-energy users such as oil refineries, metal processing, mineral and paper pulp industries. From 1st January 2005, all such companies in all 25 EU member states must limit their CO_2 emissions to allocated levels in line with Kyoto. The EU ETS principle is that participating organisations can:

- Meet the targets by reducing their own emissions, or
- Exceed the targets and sell or bank their excess emission allowances, or
- Fail to meet the targets and buy emission allowances from other participants.

These targets can only be met by either energy efficiency measures or making using more renewable energy instead of that derived from fossil fuels.
In the UK, the Utilities Act (2000) requires power suppliers to provide some electricity from renewables, starting at 3% in 2003 and rising to15% by 2015. In a similar way to EU ETS, generating companies receive and can trade Renewables Obligation Certificates (ROCs) for the qualifying electricity that they generate. For small renewable generators, Renewable Energy Guarantee of Origin Certificates (REGOs) has been introduced, in units of 1 kWh.

The initial focus is on Carbon Dioxide (CO_2), and the goals set by the UK government are:-

- 20% emission reduction by 2010 (and 10% of UK electricity from renewable sources)
- 60% emission reduction by 2050
- Real progress towards the 60% by 2020 (and 20% of UK electricity from renewable sources)

Four years after the introduction of ROCS, it was estimated that less than 3% of UK electricity was being generated from renewable sources. A 'step change' in policy was required, and the Office of the Deputy Prime Minister (ODPM) published 'Planning Policy Statement 22 (PPS 22): Renewable Energy' in order to promote renewable energy through the UK's regional and local planning authorities.

Local Planning & Building Regulations

More than 100 local authorities have already embraced PPS 22 and its Companion Guide published in December 2004, by adopting pro-renewables planning policies. Others are expected to follow; the typical requirement is for 10% of a site's electricity or heat to be derived from renewable sources, but at least one authority has already 'raised the bar' to 15%. The London Plan states that "The Mayor will and boroughs should require major developments to show how the development would generate a proportion of the site's electricity or heat from renewables".

Energy Performance Certificates, EPC

The EU Directive on the Energy Performance of Buildings (EPBD) requires that Energy Performance Certificates must be prominently placed on all buildings open to the public and commercial buildings built, sold or let from January 2006, thus enabling prospective purchasers and tenants to be more aware of a building's energy performance.

Assessing the carbon emissions associated with the operation of buildings is now an important part of the overall early design process for planning approval. Methods are set out in:

- Part L of the UK Building Regulations: (Conservation of fuel & power)
- BREEAM - (Building Research Establishment Environmental Assessment Method)
- Standard Assessment Procedure (SAP) for Energy Rating

On-site renewable energy sources are taken into account, but developers are not allowed to rely on any 'green tariff' as part of an assessment.

Technology Options and Applications

- **Wind Generators** - In a suitable location, wind energy can be an effective source of renewable power generation. Even without grant aid, an installed cost range of £2500 to £5000 per kW of generator capacity has been established over the past few years. The most common arrangement is a machine with three blades on a horizontal axis; all mounted on a tower or, increasingly for small generators in inner city areas, on top of a building. Average site wind speeds of 4 m/s can produce useful amounts of energy from a small generator up to say 3 kW, but larger generators require at least 7 m/s. A small increase in average site wind speed will produce a large

   Third party provision through an Energy Service Company (ESCo) can be successful for larger installations co-located with or close to the host building, especially in industrial settings where there may be less aesthetic or noise issues than inner city office or residential. The ESCo provides funding, installs and operates the plant and the client signs up for the renewable electrical energy at a fixed price for a period of time.

- **Building Integrated Photovoltaics (BIPV)** - Photovoltaic materials, commonly known as solar cells, generate direct current electrical power when exposed to light. Solar cells are constructed from semi-conducting materials that absorb solar radiation; electrons are displaced within the material, thus starting a flow of current through an external connected circuit. Conversion efficiency of solar energy to electrical power is improving with advances in technology and ranges from 7% to 18% under laboratory conditions. In practice, however, allowing for typical UK weather conditions, an installation of at least $7m^2$ of the latest high-efficiency hybrid modules is needed to produce 1000 watts peak (1 kWp), yielding perhaps 800 kWh in a year. Installed costs range from £300 to £450/m^2 for roof covering, and from £850 to £1300/m^2 for laminated glass.

- **Ground source heat pumps** - The ground temperature remains substantially constant throughout the year and heat can be extracted by circulating a fluid (normally water) through a system of pipes and into a heat exchanger. An electrically driven heat pump is then used to raise the fluid temperature via the compression cycle, and hot water is delivered to the building load as if from a normal boiler (albeit at a somewhat lower temperature than a normal boiler)

   Most ground heat systems consist of a cluster of pipes inserted into vertical holes typically 50 to 100 metres deep depending on space and ground type. Costs for the drilling operation vary according to location, site access and ground conditions. A geological investigation may be needed minimise the risk of failure and to improve cost certainty.

   Such systems can achieve a Coefficient of Performance (COP = heat output /electrical energy input) of between 3 and 4, achieving good savings of energy compared with conventional fossil fuels. Installed costs are in the range £800 to £1200/kW depending on system size and complexity.

- **Borehole cooling** - The constant ground temperature is well below ambient air temperature during the summer, so 'coolth' can be extracted and used to replace or, more likely commercially, to supplement conventional building cooling systems. Such borehole systems may be either 'open' – discharging ground water to river or sewer after passing it through a heat exchanger, or 'closed' – circulating a fluid (often water) through a heat exchanger and vertical pipes extending below the water table.

   Ground source heating and cooling systems are only 'partial renewable energy' because they rely on electrical power, mainly for pumping. Considerable carbon savings can be justifiably claimed, however, by avoiding the use of fossil fuel for heating and electrical power to drive conventional chillers. Indicative system costs are from £200 to £250/kW.

- **Solar Water Heating** - Simple flat-plate water-based collector panels have been used successfully on South-facing roofs over many years in the UK – especially by DIY enthusiasts prepared to devise their own simple control systems. The basic principle is to collect heat from the sun and circulate it to pre-heat space heating or domestic hot water, in either a separate tank or a twin coil hot water cylinder. Purpose-designed, evacuated tube collectors have been developed to increase performance and a typical 4 m^2 installed residential system has a cost range from £2500 to £4000 depending on pipe runs and complexity. Such a system could produce approximately 2000 kWh saving in energy use per year. Commercial systems are simply larger and slightly more complex but should achieve similar performance; low-density residential, retail and leisure developments with washrooms and showers may be suitable applications having adequate demand for hot water.

- **Biomass Boilers** - Wood chips or pellets derived from waste or farmed coppices or forests are available commercially and are considered carbon neutral, having absorbed carbon dioxide during growth. With a suitable fuel storage hopper and automatic screw drive and controls, biomass boilers can replace conventional boilers with little technical or aesthetic impact. They do, however, depend on a viable source of fuel, and there is a requirement for ash removal/disposal as well as periodic de-coking. In individual dwellings, space may be a problem because a biomass boiler does not integrate readily into a typical modern kitchen. Biomass boilers are available in a wide range of domestic and commercial sizes. For a large installation, they are more likely to form part of a modular system rather than to displace conventional boilers entirely. There is a cost premium for the biomass storage and feed system, and the cost of the fuel is currently comparable with other solid fuels. As an addition to a conventional system, installed costs could range from £200 to £250/kW.

- **Biomass Combined Heat & Power (CHP)** - Conventional CHP installations consist of either an internal combustion engine or a gas turbine driving an alternator, with maximum recovery of heat, particularly from the exhaust system. For best efficiency, there needs to be a convenient and constant requirement for the output heat energy, and the generated electricity should also be utilised locally, with any excess exported to the grid.

  Unless a source of fuel is available from landfill gas, or from a local biomass digester, then an on-site biomass to gas conversion plant would be needed to fuel the CHP engine. Considering the cost implications for biomass storage and handling as described for boilers, it appears that biomass CHP will only be viable in specific circumstances, with installed system costs in the order of £2500 to £3000/kW (electrical).

Investment 'yield' table for various renewable technologies

It can be quite difficult to compare renewable options in terms of how much energy they might save on a particular project, and how that translates into CO_2, especially if more than one option appears to be feasible. The table illustrates the potential saving per £100,000 of renewable investment i.e. £100,000 is the notional 'extra over' cost of introducing a proportion of renewable energy into the particular building service. Grant aid has been ignored in the table.

The photovoltaic options indicated include no allowance for the displacement of conventional building fabric. In practice, the 'yield per £100,000' may be higher, e.g. if PV is fully-integrated.

Renewable technology	Candidate buildings	Prerequisites	Potential barriers	Annual saving per £100,000 of capital cost	
				kWh	Kg CO_2
Tower-mounted wind generators	A	F	Environmental impact. Site space for large turbines	100,000	43,000 c.f. elec.
Building-mounted 'micro wind'	B	G	Environmental impact. Roof space for small turbines.	40,000	17,200 c.f. elec.
*Photovoltaic roof or panels	B	H	Available roof space	12,500	5,375 c.f. elec.
*Photovoltaic rain screen or glass	C	H	None	9,000	3,870 c.f. elec.
Passive solar water heating	D	J	None	50,000	9,500 c.f. gas
Ground source heat pump	B	K	Site space for pipes	40,000	7,600 c.f. gas
Borehole cooling	D	K	Site space for pipes	12,000	5,160 c.f. elec.
Biomass boilers	B	L	Environmental impact, & maintenance	100,000	19,000 c.f. gas
Biomass CHP	E	M	Environmental impact, & maintenance	28,000 + 63,000	12,000 c.f. electricity, + 12,000 c.f. gas

Key :

- A - Industrial, distribution centres
- B - Most types of building
- C - Prestige offices or retail
- D - Residential and commercial, hotels & leisure
- E - Industrial, Hotel, leisure, hospital
- F - Average site wind speed minimum 7 m/s
- G - Average site wind speed minimum 3.5 m/s
- H - Roughly south-facing, un-shaded
- J - Roughly south-facing, un-shaded - for hot water
- K - Feasible ground conditions
- L - Space and convenient source of fuel
- M - Space & convenient source of fuel - for summer heat

The following exclusions relate to all of the aforementioned indicative costs:

Inflation beyond second quarter 2005, maintenance charges, general builders work, main contractor's overheads, profit and attendance, main contract preliminaries, professional and prescribed fees, contingency and design reserve, grant aid, tax allowances, Value Added Tax. Price levels indicated are based on provincial locations.

Grey Water Recycling and Rainwater Harvesting

The potential for grey water recycling and rainwater harvesting for both domestic residential and for various types of commercial building, considering the circumstances in which the systems offer benefits, both as stand-alone installations and combined.

Water usage trends

Water usage in the UK has increased dramatically over last century or so and it is still accelerating. The current average per capita usage is estimated to be at least 150 litres per day, and the population is predicted to rise from 60 million now to 65 million by 2017 and to 75 million by 2031, with an attendant increase in loading on water supply and drainage infrastructures.

Even at the current levels of consumption, it is clear from recent experience that long, dry summers can expose the drier regions of the UK to water shortages and restrictions. The predicted effects of climate change include reduced summer rainfall, more extreme weather patterns, and an increase in the frequency of exceptionally warm dry summers. This is likely to result in a corresponding increase in demand to satisfy more irrigation of gardens, parks, additional usage of sports facilities and other open spaces, together with additional needs for agriculture. The net effect, therefore, at least in the drier regions of the UK, is for increased demand coincident with a reduction in water resource, thereby increasing the risk of shortages.

Water applications and re-use opportunities

Average domestic water utilisation can be summarised as follows, as a percentage of total usage (Source: - Three Valleys Water)

- Wash hand basin 8%
- Toilet 35%
- Dishwasher 4%
- Washing machine 12%
- Shower 5%
- Kitchen sink 15%
- Bath 15%
- External use 6%

Water for drinking and cooking makes up less than 20% of the total, and more than a third of the total is used for toilet flushing. The demand for garden watering, although still relatively small, is increasing year by year and coincides with summer shortages, thereby exacerbating the problem.

In many types of building it is feasible to collect rainwater from the roof area and to store it, after suitable filtration, in order to meet the demand for toilet flushing, cleaning, washing machines and outdoor use – thereby saving in many cases a third of the water demand. The other, often complementary recycling approach, is to collect and disinfect 'greywater' – the waste water from baths, showers and washbasins. Hotels, leisure centres, care homes and apartment blocks generate large volumes of waste water and therefore present a greater opportunity for recycling. With intelligent design, even offices can make worthwhile water savings by recycling greywater, not necessarily to flush all of the toilets in the building but perhaps just those in one or two primary cores, with the greywater plant and distribution pipework dimensioned accordingly.

Intuitively, rainwater harvesting and greywater recycling seem like 'the right things to do' and rainwater harvesting is already common practice in many counties in Northern continental Europe. In Germany, for example, some 60,000 to 80,000 systems are being installed every year – compared with perhaps 2,000 systems in the UK. Greywater recycling systems, which are less widespread than rainwater harvesting, have been developed over the last 20 years and the 'state of the art' is to use biological and UV (ultra-violet) disinfection rather than chemicals, and to reduce the associated energy use through advanced technology membrane micro filters.

In assessing the environmental credentials of new developments, the sustainable benefits are recognised by the Building Research Establishment Environmental Assessment Method (BREEAM) whereby additional points can be gained for efficient systems - those designed to achieve enough water savings to satisfy at least 50% of the relevant demand. Furthermore, rainwater harvesting systems from several manufacturers are included in the 'Energy Technology Product List' and thereby qualify for Enhanced Capital Allowances (ECAs). Claims are allowed not only for the equipment, but also to directly associated project costs including:

- Transportation - the cost of getting equipment to the site.

- Installation - cranage (to lift heavy equipment into place), project management costs and labour, plus any necessary modifications to the site or existing equipment.

- Professional Fees - if they are directly related to the acquisition and installation of the equipment.

Rainwater Harvesting

A typical domestic rainwater harvesting system can be installed at reasonable cost if properly designed and installed at the same time as building the house. The collection tank can either be buried or installed in a basement area. Rainwater enters the drainage system through sealed gullies and passes through a pre-filter to remove leaves and other debris before passing into the collection tank. A submersible pump, under the control of the monitoring and sensing panel, delivers recycled rainwater on demand. The non-potable distribution pipework to the washing machine, cleaner's tap, outside tap and toilets etc could be either a boosted system or configured for a header tank in the loft, with mains supply back-up, monitors and sensors located there instead of at the control panel.

Calculating the collection tank size brings into play the concept of system efficiency – relating the water volume saved to the annual demand. In favourable conditions – ample rainfall and large roof collection area – it would be possible in theory to achieve almost 100%. In practice, systems commonly achieve 50 to 70% efficiency, with enough storage to meet demand for typically one week, though this is subject to several variables. As well as reducing the demand for drinking quality mains supply water, rainwater harvesting tanks act as an effective storm water attenuator, thereby reducing the drainage burden and the risk of local flooding which is a benefit to the wider community. Many urban buildings are located where conditions are unfavourable for rainwater harvesting – low rainfall and small roof collection area. In these circumstances it may still be worth considering water savings through greywater recycling.

Greywater Recycling

In a greywater recycling system, waste water from baths, showers and washbasins is collected by conventional fittings and pipework, to enter a pre-treatment sedimentation tank which removes the larger dirt particles. This is followed by the aerobic treatment tank in which cleaning bacteria ensure that all bio-degradable substances are broken down. The water then passes onto a third tank, where an ultra-filtration membrane removes all particles larger than 0.00005 mm, (this includes viruses and bacteria) effectively disinfecting the recycled greywater. The clean water is then stored in the fourth tank from where it is pumped on demand under the control of monitors and sensors in the control panel. Recycled greywater may then be used for toilet and urinal flushing, for laundry and general cleaning, and for outdoor use such as vehicle washing and garden irrigation – a substantial water saving for premises such as hotels. If the tank becomes depleted, the distribution is switched automatically to the mains water back-up supply. If there is insufficient plant room space, then the tanks may be buried but with adequate arrangements for maintenance access. Overflow soakaways are recommended where feasible but are not an inherent part of the rainwater harvesting and greywater recycling systems.

Combined rainwater/greywater systems

Rainwater can be integrated into a greywater scheme with very little added complication other than increased tank size. In situations where adequate rainwater can be readily collected and diverted to pre-treatment, then heavy demands such as garden irrigation can be met more easily than with greywater alone. An additional benefit is that they reduce the risk of flooding by keeping collected storm water on site instead of passing it immediately into the drains.

Indicative system cost and payback considerations

Rainwater Harvesting Scenario: Office building in Leeds having a roof area of 2000 m^2 and accommodating 435 people over 3 floors. Local annual rainfall is 875 mm and the application is for toilet and urinal flushing. An underground collection tank of 25,000 litres has been specified to give 4.5 storage days.

Rainwater harvesting cost breakdown	Cost £
System tanks and filters and controls	18,000
Mains water back-up and distribution pump arrangement	4,000
Non-potable distribution pipework	1,000
Connections to drainage	1,000
Civil works and tank installation (assumption normal ground conditions)	7,000
System installation & commissioning	2,000
TOTAL COST	33,000

Rainwater Harvesting payback considerations

Annual water saving: 1100 m^3 @ average cost £2.00 / m^3 = £2200
Annual Maintenance and system energy cost = £700

Indicative payback period = 33,000/1,500 = 22 years

Greywater recycling scenario: Urban leisure hotel building with 200 bedrooms offers little opportunity for rainwater harvesting but has a greywater demand of up to 12000 litres per day for toilet and urinal flushing, plus a laundry. The greywater recycling plant is located in a basement plant room.

Greywater recycling cost breakdown	Cost £
System tanks and controls	34,000
Mains water back-up and distribution pump arrangement	4,000
Non-potable distribution pipework	4,000
Greywater waste collection pipework	4,500
Connections to drainage	1000
System installation & commissioning	5,000
TOTAL COST	52,500

Greywater recycling payback considerations

Annual water saving: 4260 m^3 @ average cost £2.00 / m^3 = £8,520
Annual Maintenance and system energy cost = £2,000

Indicative payback period = 52,500/6,520 = 8 years

Exclusions

- Site organisation and management costs other than specialist contractor's allowances
- Contingency/design reserve
- Main contractor's overhead and profit or management fee
- Professional fees
- Tax allowances
- Value Added Tax
- Inflation beyond third quarter 2007

Conclusions

There is a justified and growing interest in saving and recycling water by way of both greywater recycling and rainwater harvesting. The financial incentive at today's water cost is not great for small or inefficient systems but water costs are predicted to rise and demand to increase – not least as a result of population growth.

The payback periods for the above scenarios are not intended to compare to potential payback periods of rainwater harvesting against greywater recycling but rather to illustrate the importance of choosing 'horses for courses'. The office building with its relatively small roof area and limited demand for toilet flushing results in a fairly inefficient system. Burying the collection tank also adds a cost so that payback exceeds 20 years. Payback periods of less than 10 years are feasible for buildings with large roofs and a large demand for toilet flushing or for other uses. Therefore sports stadia, exhibition halls, supermarkets, schools and similar structures are likely to be suitable.

The hotel scenario is good application for greywater recycling. Many hotels and residential developments will generate more than enough greywater to meet the demand for toilet flushing etc. and in these circumstances large quantities of water can be saved and recycled with attractive payback periods.

ESSENTIAL READING FROM TAYLOR AND FRANCIS

The ZEDbook
Solutions for a Shrinking World

Bill Dunster

While Zero (fossil) Energy Development (ZED) is generally regarded as a positive innovation, in practice this option is not taken up or fully implemented. A planner, a councillor, a developer, a housing association representative, an architect, a contractor, a building component manufacturer and a government officer would all have different reasons for not espousing ZED standards for their new-build projects, but the result would be the same.

This book tackles the reasons why this happens, systematically dismantling the defence against its adoption. It explains the principles behind ZED fossil Energy Development, and provides a ZED toolkit of methods and case studies to enable construction professionals and policymakers to realize a ZED build as easily as a conventional scheme.

2007: 276x219: 276pp
Pb: 978-0-415-39199-3: **£40.00**

To Order: Tel: +44 (0) 1235 400524 **Fax:** +44 (0) 1235 400525
or Post: Taylor and Francis Customer Services,
Bookpoint Ltd, Unit T1, 200 Milton Park, Abingdon, Oxon, OX14 4TA UK
Email: book.orders@tandf.co.uk

For a complete listing of all our titles visit:
www.tandf.co.uk

Ground Water Cooling

THE USE OF GROUND WATER COOLING SYSTEMS, CONSIDERING THE TECHNICAL AND COST IMPLICATIONS OF THIS RENEWABLE ENERGY TECHNOLOGY.

The application of ground water cooling systems is quickly becoming an established technology in the UK with numerous installations having been completed for a wide range of building types, both new build and existing (refurbished).

Buildings in the UK are significant users of energy, accounting for 60% of UK carbon emissions in relation to their construction and occupation. The drivers for considering renewable technologies such as groundwater cooling are well documented and can briefly be summarised as follows:

- Government set targets – The Energy White Paper, published in 2003, setting a target of producing 10% of UK electricity from renewable sources by 2010 and the aspiration of doubling this by 2020.
- The proposed revision to the Building Regulations Part L 2006, in raising the overall energy efficiency of non domestic buildings, through the reduction in carbon emissions, by 27%.
- Local Government policy for sustainable development. In the case of London, major new developments (i.e. City of London schemes over 30,000m^2) are required to demonstrate how they will generate a proportion of the site's delivered energy requirements from on-site renewable sources where feasible. The GLA's expectation is that, overall, large developments will contribute 10% of their energy requirement using renewables, although the actual requirement will vary from site to site. Local authorities are also likely to set lower targets for buildings which fall below the GLA's renewables threshold.
- Company policies of building developers and end users to minimise detrimental impact to the environment.

The Ground as a Heat Source/Sink

The thermal capacity of the ground can provide an efficient means of tempering the internal climate of buildings. Whereas the annual swing in mean air temperature in the UK is around 20 K, the temperature of the ground is far more stable. At the modest depth of 2m, the swing in temperature reduces to 8 K, while at a depth of 50m the temperature of the ground is stable at 11-13°C. This stability and ambient temperature therefore makes groundwater a useful source of renewable energy for heating and cooling systems in buildings.

Furthermore, former industrial cities like Nottingham, Birmingham, Liverpool and London have a particular problem with rising ground water as they no longer need to abstract water from below ground for use in manufacturing. The use of groundwater for cooling is therefore encouraged by the Environment Agency in areas with rising groundwater as a means of combating this problem.

System Types

Ground water cooling systems may be defined as either open or closed loop.

Open loop systems

Open loop systems generally involve the direct abstraction and use of ground water, typically from aquifers (porous water bearing rock). Water is abstracted via one or more boreholes and passed through a heat exchanger and is returned via a separate borehole or boreholes, discharged to foul water drainage or released into a suitable available source such as a river. Typical ground water supply temperatures are in the range 6-10°C and typical re-injection temperatures 12-18°C (subject to the requirements of the abstraction licence).

Open loop systems fed by groundwater at 8°C, can typically cool water to 12°C on the secondary side of the heat exchanger to serve conventional cooling systems.

Open loop systems are thermally efficient but overtime can suffer from blockages caused by silt, and corrosion due to dissolved salts. As a result, additional cost may be incurred in having to provide filtration or water treatment, before the water can be used in the building.

Abstraction licence and discharge consent needs to be obtained for each installation, and this together with the maintenance and durability issues can significantly affect whole life operating costs, making this system less attractive.

Closed Loop Systems

Closed loop systems do not rely on the direct abstraction of water, but instead comprise a continuous pipework loop buried in the ground. Water circulates in the pipework and provides the means of heat transfer with the ground. Since ground water is not being directly used, closed loop systems therefore suffer fewer of the operational problems of open loop systems, being designed to be virtually maintenance free, but do not contribute to the control of groundwater levels.

There are two types of closed loop system:

Vertical Boreholes – Vertical loops are inserted as U tubes into pre-drilled boreholes, typically less than 150mm in diameter. These are backfilled with a high conductivity grout to seal the bore, prevent any cross contamination and to ensure good thermal conductivity between the pipe wall and surrounding ground. Vertical boreholes have the highest performance and means of heat rejection, but also have the highest cost due to associated drilling and excavation requirements.

As an alternative to having a separate borehole housing the pipe loop, it can also be integrated with the piling, where the loop is encased within the structural piles. This obviously saves on the costs of drilling and excavation since these would be carried out as part of the piling installation. The feasibility of this option would depend on marrying up the piling layout with the load requirement, and hence the number of loops, for the building.

Horizontal Loops – These are single (or pairs) of pipes laid in 2m deep trenches, which are backfilled with fine aggregate. These obviously require a greater physical area than vertical loops but are cheaper to install. As they are located closer to the surface where ground temperatures are less stable, efficiency is lower compared to open systems. Alternatively, coiled pipework can also be used where excavation is more straightforward and a large amount of land is available. Although performance may be reduced with this system as the pipe overlaps itself, it does represent a cost effective way of maximising the length of pipe installed and hence overall system capacity.

The Case for Heat Pumps

Instead of using the groundwater source directly in the building, referred to as passive cooling, when coupled to a reverse cycle heat pump, substantially increased cooling loads can be achieved.

Heat is extracted from the building and transferred by the heat pump into the water circulating through the loop. As it circulates, it gives up heat to the cooler earth, with the cooler water returning to the heat pump to pick up more heat. In heating mode the cycle is reversed, with the heat being extracted from the earth and being delivered to the HVAC system.

The use of heat pumps provides greater flexibility for heating and cooling applications within the building than passive systems. Ground source heat pumps are inherently more efficient than air source heat pumps, their energy requirement is therefore lower and their associated CO_2 emissions are also reduced, so they are well suited for connection to a groundwater source.

Closed loop systems can typically achieve outputs of 50W/m (of bore length), although this will vary with geology and borehole construction. When coupled to a reverse cycle heat pump, 1m of vertical borehole will typically deliver 140kWh of useful heating and 110kWh of cooling per annum, although this will depend on hours run and length of heating and cooling seasons.

Key factors Affecting Cost

- The cost is obviously dependant on the type of system used. Deciding on what system is best suited to a particular project is dependant on the peak cooling and heating loads of the building and its likely load profile. This in turn determines the performance required from the ground loop, in terms of area of coverage in the case of the horizontal looped system, and in the case of vertical boreholes, the depth and number or bores. The cost of the system is therefore a function of the building load.

- In the case of vertical boreholes, drilling costs are significant factor, as specific ground conditions can be variable, and there are potential problems in drilling through sand layers, pebble beds, gravels and clay, which may mean additional costs through having to drill additional holes or the provision of sleeving etc. The costs of excavation obviously make the vertical borehole solution significantly more expensive than the equivalent horizontal loop.

- The thermal efficiency of the building is also a factor. The higher load associated with a thermally inefficient building obviously results in the requirement for a greater number of boreholes or greater area of horizontal loop coverage, however in the case of boreholes the associated cost differential between a thermally inefficient building and a thermally efficient one is substantially greater than the equivalent increase in the cost of conventional plant. Reducing the energy consumption of the building is cheaper than producing the energy from renewables and the use of renewable energy only becomes cost effective, and indeed should only be considered, when a building is energy efficient.

- With open loop systems, the principal risk in terms of operation is that the user is not in control of the quantity or quality of the water being taken out of the ground, this being dependant on the local ground conditions. Reduced performance due to blockage (silting etc) may lead to the system not delivering the design duties whilst bacteriological contamination may lead to the expensive water treatment or the system being taken temporarily out of operation. In order to mitigate the above risk, it may be decided to provide additional means of heat rejection and heating by mechanical means as a back up to the borehole system, in the event of operational problems. This obviously carries a significant cost. If this additional plant were not provided, then there are space savings to be had over conventional systems due to the absence of heating, heat rejection and possibly refrigeration plant.

- Open loop systems may lend themselves particularly well to certain applications increasing their cost effectiveness, i.e. in the case of a leisure centre, the removal of heat from the air conditioned parts of the centre and the supply of fresh water to the swimming pool.

- In terms of the requirements for abstraction and disposal of the water for open loop systems, there are risks associated with the future availability and cost of the necessary licenses; particularly in areas of high forecast energy consumption, such as the South East of England, which needs to be borne in mind when selecting a suitable system.

Whilst open loop systems would suit certain applications or end user clients, for commercial buildings the risks associated with this system tend to mean that closed loop applications are the system of choice. When coupled to a reversible heat pump, the borehole acts simply as a heat sink or heat source so the problems associated with open loop systems do not arise.

Typical Costs

Table 1 gives details of the typical borehole cost to an existing site in Central London, using one 140m deep borehole working on the open loop principle, providing heat rejection for the 600kW of cooling provided to the building. The borehole passes through rubble, river gravel terraces, clay and finally chalk, and is lined above the chalk level to prevent the hole collapsing. The breakdown includes all costs associated with the provision of a working borehole up to the well head, including the manhole chamber and manhole. The costs of any plant or equipment from the well head are not included.

Heat is drawn out of the cooling circuit and the water is discharged into the Thames at an elevated temperature. In this instance, although the boreholes are more expensive than the dry air cooler alternative, the operating cost is significantly reduced as the system can operate at around three times the efficiency of conventional dry air coolers, so the payback period is a reasonable one. Additionally, the borehole system does not generate any noise, does not require rooftop space and does not require as much maintenance.

This is representative of a typical cost of providing a borehole for an open loop scheme within the London basin. There are obviously economies of scale to be had in drilling more than one well at the same time, with two wells saving approximately 10% of the comparative cost of two separate wells and four wells typically saving 15%.

Table 2 provides a summary of the typical range of costs that could expected for the different types of system based on current prices.

Table 1 : Breakdown of the Cost of a Typical Open Loop Borehole System

Description	Cost £
General Items	
• Mobilisation, Insurances, demobilisation on Completion	20,000
• Fencing around working area for the duration of drilling and testing	2,000
• Modifications to existing LV panel and installation of new power supplies for borehole installation	14,000
Trial Hole	
• Allowance for breakout access to nearest walkway (Existing borehole on site used for trial purposes, hence no drilling costs included)	3,000
Construct Borehole	
• Drilling, using temporary casing where required, permanent casing and grouting	31,000
Borehole Cap and Chamber	
• Cap borehole with PN16 flange, construct manhole chamber in roadway, rising main, header pipework, valves, flow meter	12,000
• Permanent pump	13,000

Samples

- Water Samples   300

Acidisation

Mobilisation, set up and removal of equipment for acidisation of borehole, carry out acidisation   11,500

Development and Test Pumping

- Mobilise pumping equipment and materials and remove on completion of testing   4,000
- Calibration test, pre-test monitoring, step testing   3,500
- Constant rate testing and monitoring   19,000
- Waste removal and disposal   3,000

Reinstatement

- Reinstatement and Making Good   1,500

Total   **144,300**

Free Updates

with three easy steps…

1. Register today on www.pricebooks.co.uk/updates

2. We'll alert you by email when new updates are posted on our website

3. Then go to www.pricebooks.co.uk/updates and download.

All four Spon Price Books – *Architects' and Builders'*, *Civil Engineering and Highway Works*, *External Works and Landscape* and *Mechanical and Electrical Services* – are supported by an updating service. Three updates are loaded on our website during the year, in November, February and May. Each gives details of changes in prices of materials, wage rates and other significant items, with regional price level adjustments for Northern Ireland, Scotland and Wales and regions of England. The updates terminate with the publication of the next annual edition.

As a purchaser of a Spon Price Book you are entitled to this updating service for the 2008 edition – free of charge. Simply register via the website www.pricebooks.co.uk/updates and we will send you an email when each update becomes available.

If you haven't got internet access we can supply the updates by an alternative means. Please write to us for details: Spon Price Book Updates, Taylor & Francis Marketing Department, 2 Park Square, Milton Park, Abingdon, Oxfordshire, OX14 4RN.

Find out more about Spon books
Visit www.tandfbuiltenvironment.com for more details.

New books from Spon

The following books can be ordered directly from Taylor & Francis or from your nearest good bookstore

Spon's Irish Construction Price Book 3rd edition, Franklin + Andrews
    hbk 978-0-415-45637-1
THE ZEDBOOK, B Dunster
    pbk 978-0-415-39199-3
Spon's Building Regulations Explained 8th edition, LDSA et al.
    hbk 978-0-415-43067-8
Spon's Estimating Costs Guide to Plumbing and Heating 4th edition, B Spain
    pbk 978-0-415-46905-0
Spon's Estimating Costs Guide to Electrical Works 4th edition, B Spain
    pbk 978-0-415-46904-3
Spon's Estimating Cost Guide to Minor Works 4th edition, B Spain
    pbk 978-0-415-46906-7
Ventilation Systems, H Awbi
    hbk 978-0-419-21700-8
Understanding the Building Regulations 4th edition, S Polley
    pbk 978-0-415-45272-4
Construction Delays, R Gibson
    hbk 978-0-415-34586-6
Building Acoustics, E Vigran
    hbk 978-0-415-42853-8
Ethics for the Built Environment, P Fewings
    hbk 978-0-415-42982-5, pbk 978-0-415-42983-2
Construction Contracts Q&A, D Chappell
    pbk 9780415375979
Project Management Demystified 3rd edition, G Reiss
    pbk 9780415421638
Building Services Engineering 5th edition, D Chadderton
    hbk 9780415413541, pbk 9780415413558
Heat and Mass Transfer in Buildings, K Moss
    hbk 9780415409070, pbk 9780415409087
Design of Electrical Services for Buildings 4th edition, B Rigby
    hbk 0415310822, pbk 0415310830
Outdoor Lighting Guide, Inst of Lighting Engineers
    hbk 0415370078
Acoustic Absorbers & Diffusers, T Cox et al.
    hbk 0415296498

Please send your order to:
Marketing Department, Taylor & Francis, 2 Park Square, Milton Park Abingdon, Oxfordshire, OX14 4RN.
or visit www.tandfbuiltenvironment.com for more details.

Table 2 Summary of the Range of Costs for Different Systems

System	Range			Notes
	Small – 4kWth	Medium – 50kWth	Large – 400kWth	
Heat pump (per unit)	£ 3,500 – 4,500	£ 30,000 – 40,000	£ 140,000 – 170,000	
Slinky pipe (per installation) including excavation	£ 3,000 – 4,000	£ 40,000 – 50,000	£ 360,000 – 390,000 [1]	[1] Based on 90 nr 50m lengths
Vertical, closed (per installation) using structural piles	N/A	£ 40,000 – 60,000	Not available	Based on 50 nr piles. Includes borehole cap and header pipework but excludes connection to pump room and heat pumps
Vertical, closed (per installation) including excavation	£ 2,000 – 3,000	£ 60,000 – 80,000	£ 360,000 – 390,000	Includes borehole cap and header pipework but excludes connection to pump room and heat pumps
Vertical, open (per installation) including excavation	£ 2,000 – 3,000	£ 45,000 – 65,000	£ 330,000 – 360,000	Excludes connection to pump room and heat exchangers

ESSENTIAL READING FROM TAYLOR AND FRANCIS

Spon's Building Regulations Explained

Eighth Edition

London District Surveyors Association and **John Stephenson**

This fully revised, essential reference takes into account all important aspects of building control including new legislation up to the start of 2008, covering major revisions to Parts A, B, C, F, J, L1A, L1B, L2A, L2B and P and revisions to Part E. Each chapter explains in clear terms the appropriate regulation and any other relevant legislation, before explaining the approved document.

Selected Contents:

1. The Development of Building Control
2. Control of Building Work
3. Application of Building Regulations to Inner London
4. Relaxation of Building Regulations
5. Exempt Buildings and Works
6. Notices and Plans
7. Approved Inspectors
8. Work Undertaken by Public Bodies
9. Approved Document to Support Regulation 7

Chapters 10-22. Approved Documents A to N 23. Other Approved Documents

2009: 297x210: 672pp
Hb: 978-0-415-43067-8 **£80.00**

To Order: Tel: +44 (0) 1235 400524 **Fax:** +44 (0) 1235 400525
or Post: Taylor and Francis Customer Services,
Bookpoint Ltd, Unit T1, 200 Milton Park, Abingdon, Oxon, OX14 4TA UK
Email: book.orders@tandf.co.uk

For a complete listing of all our titles visit:
www.tandf.co.uk

Fuel Cells

THE APPLICATION OF FUEL CELL TECHNOLOGY WITHIN BUILDINGS.

Fuel cells are electrochemical devices that convert the chemical energy in fuel into electrical energy directly, without combustion, with high electrical efficiency and low pollutant emissions. They represent a new type of power generation technology that offers modularity, efficient operation across a wide range of load conditions, and opportunities for integration into co-generation systems. With the publication of the energy white paper in February this year, the Government confirmed it's commitment to the development of fuel cells as a key technology in the UK's future energy system, as the move is made away from a carbon based economy.

There are currently very few fuel cells available commercially, and those that are available are not financially viable. Demand has therefore been limited to niche applications, where the end user is willing to pay the premium for what they consider to be the associated key benefits. Indeed, the UK currently has only one fuel cell in regular commercial operation. However, fuel cell technology has made significant progress in recent years, with prices predicted to approach those of the principal competition in the near future.

Fuel Cell Technology

A fuel cell is composed of an anode (a negative electrode that repels electrons), an electrolyte membrane in the centre, and a cathode (a positive electrode that attracts electrons). As hydrogen flows into the cell on the anode side, a platinum coating on the anode facilitates the separation of the hydrogen gas into electrons and protons. The electrolyte membrane only allows the protons to pass through to the cathode side of the fuel cell. The electrons cannot pass through this membrane and flow through an external circuit to form an electric current.

As oxygen flows into the fuel cell cathode, another platinum coating helps the oxygen, protons, and electrons combine to produce pure water and heat.

The voltage from a single cell is about 0.7 volts, just enough for a light bulb. However by stacking the cells, higher outputs are achieved, with the number of cells in the stack determining the total voltage, and the surface area of each cell determining the total current. Multiplying the two together yields the total electrical power generated.

In a fuel cell the conversion process from chemical energy to electricity is direct. In contrast, conventional energy conversion processes first transform chemical energy to heat through combustion and then convert heat to electricity through some form of power cycle (e.g. gas turbine or internal combustion engine) together with a generator.
The fuel cell is therefore not limited by the Carnot efficiency limits of an internal combustion engine in converting fuel to power, resulting in efficiencies 2 to3 times greater.

Fuel Cell Systems

In addition to the fuel cell itself, the system comprises the following sub-systems:

- A fuel processor – This allows the cell to operate with available hydrocarbon fuels, by cleaning the fuel and converting (or reforming) it as required.
- A power conditioner – This regulates the dc electricity output of the cell to meet the application, and to power the fuel cell auxiliary systems.
- An air management system – This delivers air at the required temperature, pressure and humidity to the fuel stack and fuel processor.
- A thermal management system – This heats or cools the various process streams entering and leaving the fuel cell and fuel processor, as required.
- A water management system – Pure water is required for fuel processing in all fuel cell systems, and for dehumidification in the PEMFC.

The overall electrical conversion efficiency of a fuel cell system (defined as the electrical power out divided by the chemical energy into the system, taking into account the individual efficiencies of the sub-systems) ranges from 35-55%. Taking into account the thermal energy available from the system, the overall or cogeneration efficiency is 75-90%.

Also, unlike most conventional generating systems (which operate most efficiently near full load, and then suffer declining efficiency as load decreases), fuel cell systems can maintain high efficiency at loads as low as 20% of full load.

Fuel cell systems also offer the following potential benefits:

- At operating temperature, they respond quickly to load changes, the limiting factor usually being the response time of the auxiliary systems.
- They are modular and can be built in a wide range of outputs. This also allows them to be located close to the point of electricity use, facilitating cogeneration systems.
- Noise levels are comparable with residential or light commercial air conditioning systems.
- Commercially available systems are designed to operate unattended and manufactured as packaged units.
- Since the fuel cell stack has no moving parts, other than the replacement of the stack at 3-5 year intervals there is little on-site maintenance. The maintenance requirements are well established for the auxiliary system plant.
- Fuel cell stacks fuelled by hydrogen produce only water, therefore the fuel processor is the primary source of emissions, and these are significantly lower than emissions from conventional combustion systems.
- Since fuel cell technology generates 50% more electricity than the conventional equivalent without directly burning any fuel, CO_2 emissions are significantly reduced in the production of the source fuel.
- Potentially zero carbon emissions when using hydrogen produced from renewable energy sources.
- The facilitation of embedded generation, where electricity is generated close to the point of use, minimising transmission losses.
- The fast response times of fuel cells offer potential for use in UPS systems, replacing batteries and standby generators.

Types of Fuel Cell

There are four main types of fuel cell technology that are applicable for building systems, classed in terms of the electrolyte they use. The chemical reactions involved in each cell are very different.

Phosphoric Acid Fuel Cells (PAFCs) are the dominant current technology for large stationary applications and have been available commercially for some time. The only working fuel cell installation in the UK, in Woking, uses a PAFC, rated at 200kW. There is less potential for PAFC unit cost reduction than for some other fuel cell systems, and this technology may be superseded in time by the other technologies.

The Solid Oxide Fuel Cell (SOFC) offers significant flexibility due to its large power range and wide fuel compatibility. SOFCs represent one of the most promising technologies for stationary applications. There are difficulties when operating at high temperatures with the stability of the materials, however, significant further development and cost reduction is anticipated with this type.

The relative complexity of Molten Carbonate Fuel Cells (MCFCs) has tended to limit developments to large scale stationary applications, although the technology is still very much in the development stages.

The quick start-up times and size range make Proton Exchange Membrane Fuel Cells (PEMFCs) suitable for small to medium sized stationary applications. They have a high power density and can vary output quickly, making them well suited for transport applications as well as UPS systems. The development efforts in the transport sector suggest there will continue to be substantial cost reductions over both the short and long term.

All four technologies remain the subject of extensive research and development programmes to reduce initial costs and improve reliability through improvements in materials, optimisation of operating conditions and advances in manufacturing. It is expected that all types will be commercially available in limited markets by 2006 and with mass market availability by 2010.

The market for fuel cells

The stationary applications market for fuel cells can be sectaries as follows:

- Distributed generation/CHP – For large scale applications, there are no drivers specifically advantageous to fuel cells, with economics (and specifically initial cost) therefore being the main consideration. So, until cost competitive and thoroughly proven and reliable fuel cells are available, their use is likely to be limited to niche applications such as environmentally sensitive areas from 2005. Wider commercialisation is likely closer to 2010, with high temperature cells (MCFC's and SOFC's) being most suitable, although PEMFC's may preferable in specific areas, i.e. where hydrogen is available.

- Domestic and small scale CHP – The drivers for the use of fuel cells in this emerging market are better value for customers than separate gas and electricity purchase, reduction in domestic CO_2 emissions, and potential reduction in electricity transmission costs. However, the barriers of resistance to distributed generation, high capital costs and competition from Stirling engines needs to be overcome. Commercialisation depends on cost reduction, and successful demonstration which is expected to begin in the next 2-3 years, leading to wider commercialisation before 2010. Systems based on SOFC's and PEMFC's are being developed for this application.

- Small generator sets and remote power – The drivers for the use of fuel cells are high reliability, low noise and low refuelling frequencies, which cannot be met by existing technologies. Since cost is often not the primary consideration, fuel cells will find early markets in this sector. Existing PEMFC systems are close to meeting the requirements in terms of cost, size and performance. Small SOFC's have potential in this market, but require further development.

Cost comparison

Table 1 provides an indication of the capital and operating costs of the different fuel cell types, together with comparative figures for the existing technologies. The projected figures for the fuel cell technologies are based on economies typically achieved through mass manufacture.

The projected costs for 2007 show the fuel cell technologies still being significantly more expensive than the existing technologies. To extend fuel cell application beyond niche markets, their cost needs to reduce significantly. The successful and wide-spread commercial application of fuel cells is dependant on the projected cost reductions indicated, with electricity generated from fuel cells being competitive with current centralised and distributed power generation.

It has been estimated that if the cost reductions are met, fuel cells could achieve up to 50% penetration of the global distributed energy market by 2020.

Typical Current Project Costs

Table 2 gives an indication of the typical cost breakdown to be expected for the installation of a fuel cell system in the UK. This is based on information from the manufacturer and from economic evaluation/feasibility studies, since with only one working system installed to date in the UK there is no available accurate cost data.

The unit is a standard commercially available PAFC, complete with fuel processor, fuel stack and power conditioning system. The parameters are as follows:

- Rating – 200kW/235kVA, 400V, 3ph
- Power generating efficiency – 40%
- Heat output – 204kW, 60°C hot water
- Fuel, consumption – Natural gas, 54m^3/hr
- External location
- Size – 5.5m x 3m x 3m (h)
- Weight – 20 tons
- Noise level at full load – 62dBA at 10m

The above illustrates the fact that the costs to supply, install and set to work a modestly sized fuel cell unit are prohibitively high, compared with incumbent generator technologies. Despite being the only commercially available unit, only 220 units have been sold worldwide, and so the full benefits of volume manufacture have not been realised, and a proportion of the costs associated with developing the unit is included within the unit cost.

Conclusions

Despite significant growth in recent years, fuel cells are still at a relatively early stage of commercial development, with prohibitively high capital costs preventing them from competing with the incumbent technology in the market place. However, costs are forecast to reduce significantly over the next five years as the technology moves from niche applications, and into mass production.

However, in order for these projected cost reductions to be achieved, customers need to be convinced that the end product is not only cost competitive but also thoroughly proven, and Government support represents a key part in achieving this.

The Governments of Canada, USA, Japan and Germany have all been active in supporting development of the fuel cell sector through integrated strategies, however the UK has been slow in this respect, and support has to date been small in comparison. It is clear that without Government intervention, fuel cell applications may struggle to reach the cost and performance requirements of the emerging fuel cell market.

Table 1 – Cost Comparison for Stationary Generation Equipment

Period	Fuel Cell Type			
	PEMFC	PAFC	MCFC	SOFC
Capital Costs (£/kW)	£	£	£	£
2004	2,600-6,500	2,000-3,400	2,000-5,000	5,000-10,000
2006	1,500	1,900	1,850	2,300
2010 (Projected)	700	1,500	950	1,000
2015 (Projected)	500	1,300	700	750
2020 (Projected)	300	1,100	550	550
Operating Costs (£/kW/h)				
2004	0.04-0.06	0.08	0.04-0.08	0.05
Maintenance Costs (£/kW/h)				
2006	0.003-0.01	0.003-0.01	0.003-0.01	0.003-0.01

Period	Conventional Systems		
	Internal Combustion - Generator	Micro Turbine - Generator	Gas Turbine
Capital Costs (£/kW)	£	£	£
2004	-	-	-
2006	200-820	450-820	450-570
2010 (Projected)	-	-	-
2015 (Projected)	-	-	-
2020 (Projected)	-	-	-
Operating Costs (£/kW/h)			
2004	-	-	-
Maintenance Costs (£/kW/h)			
2006	0.005-0.01	0.002-0.01	0.002-0.06

Table 2 – Breakdown of Typical Project Costs

Item	£
Fuel Cell Stack	225,000
Fuel Processor	85,000
Plant	113,600
Labour	113,400
Total	**567,000**
Delivery to Site (Outside USA)	20,000

Installation Costs (Site and Application dependant)

Standard (Generation Only)	50,000
Non Standard	80,000
CHP	100,000-200,000
Spares	10% of Capital Cost
Maintenance	16,000-26,000 pa
Total (£/kW)	**3,549-4,415**

Fuel Stack

Fuel Stack Replacement	150,000 every 4 years
Fuel Stack Refurbishment	120,000 every 4 years

Note – Excludes Incoming Gas Supply, BWIC, Main Contractor's Overheads ad Profit, attendance, Preliminaries and Professional Fees etc, VAT

ESSENTIAL READING FROM TAYLOR AND FRANCIS

Spon's Estimating Costs Guide to Plumbing and Heating

Unit Rates and Project Costs

Fourth Edition

Bryan Spain

Do you work on jobs between £50 and £50,000? - Then this book is for you.

All the cost data you need to keep your estimating accurate, competitive and profitable.

Specially written for contractors and small businesses carrying out small works, *Spon's Estimating Cost Guide to Plumbing and Heating* contains accurate information on thousands of rates each broken down to labour, material overheads and profit.

The first book to include typical project costs for:

- rainwater goods installations
- bathrooms
- external waste systems
- central heating systems
- hot and cold water systems.

July 2008: 216x138: 264pp
Pb: 978-0-415-46905-0: **£29.99**

To Order: Tel: +44 (0) 1235 400524 **Fax:** +44 (0) 1235 400525
or Post: Taylor and Francis Customer Services,
Bookpoint Ltd, Unit T1, 200 Milton Park, Abingdon, Oxon, OX14 4TA UK
Email: book.orders@tandf.co.uk

**For a complete listing of all our titles visit:
www.tandf.co.uk**

Biomass Energy

The potential for biomass energy systems, with regards to the adequacy of the fuel supply and the viability of various system types at different scales.

Biomass heating and combined heat and power (CHP) systems have become a major component of the low-carbon strategy for many projects, as they can provide a large renewable energy component at a relatively low initial cost. Work by the Carbon Trust has demonstrated that both large and small biomass systems were viable even before recent increases in gas and fuel oil prices, so it is no surprise that recent research by South Bank University into the renewables strategies to large London projects has found that 25% feature biomass or biofuel systems.

These proposals are not without risk, however. Although the technology is well established, few schemes are in operation in the UK and the long-term success depends more on the effectiveness of the local supply chain than the quality of the design and installation.

How the biomass market works

Biomass is defined as living or recently dead biological material that can be used as an energy source. Biomass is generally used to provide heat, generate electricity or drive CHP engines. The biomass family includes biofuels, which are being specified in city centre schemes, but which provide lower energy outputs and could transfer farmland away from food production.

In the UK, much of the focus in biomass development is on the better utilisation of waste materials such as timber and the use of set-aside land for low-intensity energy crops such as willow, rather than expansion of the biofuels sector. There are a variety of drivers behind the development of a biomass strategy. In addition to carbon neutrality, another policy goal is the promotion of the UK's energy security through the development of independent energy sources. A third objective is to address energy poverty, particularly for off-grid energy users, who are most vulnerable to the effects of high long-term costs of fuel oil and bottled gas.

Biomass' position in the zero-carbon hierarchy is a little ambiguous in that its production, transport and combustion all produce carbon emissions, albeit most is offset during a plant's growth cycle. The key to neutrality is that the growing and combustion cycles need to occur over a short period, so that combustion emissions are genuinely offset. Biomass strategy is also concerned about minimising waste and use of landfill, and the ash produced by combustion can be used as a fertiliser.

Dramatic increases in fossil fuel prices have swung considerations decisively in favour of technologies such as biomass. Research by the Carbon Trust has demonstrated that, with oil at $50 (£25) a barrel, rates of return of more than 10% could be achieved with both small and large heating installations. CHP and electricity-only schemes have more complex viability issues linked to renewable incentives, but with oil currently trading at over ($100) £50 a barrel and a plentiful supply of source material, it is argued that biomass input prices will not rise and so the sector should become increasingly competitive.

The main sources of biomass in the UK include:

- Forestry crops, including the waste products of tree surgery industry
- Industrial waste, particularly timber, paper and card: timber pallets account for 30% of this waste stream by weight
- Woody energy crops, particularly those grown through 'short rotation' methods such as willow coppicing
- Wastes and residues taken from food, agriculture and manufacturing

Biomass is an emerging UK energy sector. Most suppliers are small and there remains a high level of commercial risk associated with finding appropriate, reliable sources of biomass. This is particularly the case for larger-scale schemes such as those proposed for Greater London, which will have sourced biomass either from multiple UK suppliers or from overseas. Many have adopted biofuels as an alternative.

The UK's only large-scale biomass CHP in Slough has a throughput of 180,000 tonnes of biomass per year requiring the total production of more than 20 individual suppliers – not a recipe for easy management or product consistency. However, Carbon Trust research has identified significant potential capacity in waste wood (5 million to 6 million tonnes) and short-rotation coppicing, which could create the conditions for wider adoption of small and large-scale biomass.

Biomass technologies

A wide range of technologies have been developed for processing various forms of biomass, including anaerobic digesters and gasifiers. However, the main biomass technology is solid fuel combustion, as a heat source, CHP unit or energy source for electricity generation.

Solid fuel units use either wood chippings or wood pellets. Wood chippings are largely unprocessed and need few material inputs, other than seasoning, chipping and transport. Wood pellets are formed from compressed sawdust. As a result they have a lower moisture content than wood chippings and consistent dimensions, so are easier to handle but are about twice as expensive.

Solid fuel burners operate in the same way as other fossil fuel-based heat sources, with the following key differences:

- Biomass heat output can be controlled but not instantaneously, so systems cannot respond to rapid load changes. Solutions to provide more flexibility include provision of peak capacity from gas-fired systems, or the use of thermal stores that capture excess heat energy during off-peak periods, enabling extended operation of the biomass system itself
- Heat output cannot be throttled back by as much as gas-fired systems, so for heat-only installations it may be necessary to have an alternative summer system for water heating, such as a solar collector
- Biomass feedstock is bulky and needs a mechanised feed system as well as extensive storage
- Biomass systems are large, and the combustion unit, feed hopper and fuel store take up substantial floor area. A large unit with an output of 500kW has a footprint of 7.5m x 2m
- Biomass systems need maintenance related to fuel deliveries, combustion efficiency, ash removal, adding to the lifetime cost
- Fuel stores need to be physically isolated from the boiler and the rest of the building in order to minimise fire risk. The fuel store needs to be sized to provide for at least 100 hours of operation, which is approximately 100m² for a 500kW boiler. The space taken up by storage and delivery access may compromise other aspects of site planning
- Fire-protection measures include anti-blowback arrangements on conveyors and fire dampers, together with the specification of elements such as flues for higher operating temperatures
- Collocation of the fuel source and burner at ground level require larger, free-standing flues

As a result of these issues, which drive up initial costs, affect development efficiency and add to management overheads, take-up of biomass has initially been mostly at the small-scale, heat-only end of the market, based on locally sourced feedstock. In such systems the initial cost premium of the biomass boiler can be offset against long-term savings in fuel costs.

Sourcing biomass

Compared with solar or wind power installations, the initial costs of biomass systems are low, the technology is well established and energy output is dependable. As a result, the real challenge for successful operation of a biomass system is associated with the reliable sourcing of feedstock.

Heat-only systems themselves cost between £150 and £750 per Kw (excluding costs of storage), depending on scale and technology adopted. This compares with a typical cost of £50 to £300 per kW for a gas-fired boiler – which does not require further investment in fuel or thermal storage bunkers.

As a high proportion of lifetime cost is associated with the operation of a system, availability of good-quality, locally sourced feedstock is essential for long-term viability – particularly in areas where incentivisation through policies like the Merton rule is driving up demand.

Research funded by BioRegional in connection with medium-scale biomass systems in the South-east shows that considerable feedstock is already in the system but far more is required to respond to emerging requirements.

The researchers estimate that the existing biomass resource within 25km of London totals 330,000 tonnes a year, sourced from waste wood, energy crops and forestry (tree-surgery) byproducts. However, they calculate that a new 2,000-unit low-energy residential system would require 35,000 tonnes a year. This means that London's total biomass would have the potential to support the equivalent heating load of just 20,000 homes.

Fortunately, the scale of the UK's untapped resource is considerable, with 5 million to 6 million tonnes of waste wood going to landfill annually, and 680,000ha of set-aside land that could be used for energy crops without affecting agricultural output.

Based on these figures, it is estimated that 15% of the UK's building-related energy load could be supported without recourse to imported material. However, the supply chain is fragmented in terms of producers, processors and distributors – presenting potential biomass users with a range of complexities that gas users simply do not need to worry about. These include:

- Ensuring quality. Guaranteeing biomass quality is important for the assurance of performance and reliability. Variation in moisture content affects combustion, while inconsistent woodchip size or differences in sawdust content can result in malfunction. The presence of contaminants in waste wood causes problems too. High-profile schemes including the 180,000-tonne generator in Slough have had to shutdown because of variations in fuel quality. Use of pellets reduces to risk, but they are more expensive and require more energy for processing and transport
- Functioning markets. The scale of trade in biomass compares unfavourably with gas or oil, in that there are no standard contracts, fixed-price deals or opportunities for hedging which enable major users to manage their energy cost risk
- Security of supply. The potential for competing uses could lead to price inflation. Biofuels carry the greatest such risk, but many biomass streams have alternative uses. Lack of capacity in the marketplace is another security issue, with no mechanism to encourage strategic stockpiling for improved response to crop failure or fluctuations in demand
- Installation and maintenance infrastructure. The different technologies used in biomass systems creates maintenance requirements not yet met by a readily available pool of skilled system engineers

Optimum uses of biomass technology

Biomass is a high-grade, locally available source of energy that can be used at a range of scales to support domestic and commercial use. Following increases in fossil fuel prices, one of the main barriers to adoption is, now, the capability of the supply chain.

The Carbon Trust's biomass sector review, completed before the large energy price rises in 2007, drew the following key conclusions about the most effective application of the technologies:

- Returns on CHP and electricity-generating systems depend heavily on government incentives such as renewable obligations certificates. Under the present arrangements, large CHP systems provide the best returns
- Heat-only systems are very responsive to changes in fuel prices, with systems at all scales providing returns in excess of 10% when oil prices are at $50 (£25) a barrel
- Small-scale heat-only plants produce the best returns, because the cost of the displaced fuel (typically fuel oil) is more expensive
- Small-scale electricity and large-scale heat-only installations produce very poor returns
- There is little difference in the impact of fuel type in the returns generated by projects

The study also concluded that heat installations at all scales had the greater potential for carbon saving, based on a finite supply of biomass. This is because heat-generating processes have the greatest efficiency and, in the case of small-scale systems in isolated, off-grid dwellings, displace fuels such as oil that have the greatest carbon intensity. Ninety percent of the UK's existing biomass resource of 5.6m tonnes per annum could be used in displacing carbon-intensive off-grid heating, saving 2.5m tonnes of carbon emissions.

Small-scale systems are well established in Europe and the existing local supply chain suits the demand pattern. In addition, since the target market is in rural, off-grid locations, affected dwellings are less likely to suffer space constraints related to storage. As fuel costs continue to rise, the benefits of avoiding fuel poverty, combined with the effective reduction of carbon emissions from existing buildings, mean smaller systems are likely to offer the best mid-term use of the existing biomass supply base, with large-scale systems being developed as the supply chain matures and expands.

Large-scale systems also offer the opportunity to generate significant returns, but the barriers that developers or operators face are significant, particularly if there is an electricity supply component, which requires a supply agreement. However, while developers are required to delivery renewable energy on site, biomass in the form of biofuels, has the great attraction of being able to provide a scale of renewable energy generation that other systems such as ground source heating or photovoltaics simply cannot compete with.

Whether biomass plant should be used on commercial schemes in urban locations is potentially a policy issue. Sizing of both CHP and heat-only systems should be determined by the heat load, which for city-centre schemes may not be that large – affecting the potential for the CHP component. The costs of a district heating element on these schemes may also be prohibitively high, and considerations of biomass transport and storage also make it harder to get city-centre schemes to stack up.

It may be a more appropriate policy to encourage industrial users or large scale regenerators to take first call on the expanding biomass resource, rather than commercial schemes. The launch of a 45MW biomass power station in Scotland illustrates this trend. Data shows that 50% of the market potential for industrial applications of CHP could utilise 100% of the UK's available biomass resource. The issues that city-centre biomass schemes face in connection with storage, transport, emissions and supply chain management might be better addressed by industrial users or their energy suppliers in low-cost locations rather than by developers in prime city-centre sites.

Indicative costs

System	Indicative load kW	Capital cost (£/kWh)
Gas-fired boiler	50	85
	400	45
Biomass-fired boiler	50	500
	500	250
Biomass-fired CHP	1,000	450

Allowance for stand-alone boiler house and fuel store £30,000-60,000 for 50kWh system indicative costs exclude flues and plan room installation

Fuel costs

Wood chip : 1.5 to 2.5p/kWh
Wood pellet : 3 to 4.5p/kWh
Fuel oil : 4.3 to 4.8p/kWh
Natural gas : 2.5 to 3.2p/kWh
Bottled LPG : 6.8 to 7p/kWh

Getting the Connection

The provision of an electricity connection has both a physical and contractual element. Physically, it involves the design, planning and construction of electrical infrastructure (cables, switchgear, civil works), whilst contractually, it requires legal agreements to be drawn up and agreed (construction, connection, adoption).

Planning

The Planning Stage, typically RIBA Stage C, is where the site's developer (the Developer) should be formulating their plans for the scheme and, in doing so, consult the local distribution network operator's (the Host)[1] long-term development statement, which will identify potential connection point opportunities. This information is normally readily available from the Host for a small fee and should provide an early indication of whether the Host's network may need to be reinforced before the development can be connected.

As the Planning Stage progresses, the Developer should discuss its proposals with the Host. Relatively simple connections for single building supplies are straight forward, however schemes of a more complex nature may require some form of feasibility study to be carried out to assess connection options and provide indicative costs for the contestable[2] and non-contestable[3] work elements.

Design

The Design Stage, typically RIBA Stage D, is the point at which the Developer submits its formal connection request to the Host. It is important that this is completed in accordance with the specific procedures of the Host, since if it does not include all supporting information required by the Host's application process, there is likely to be a delay in the processing of a firm offer. The Host is expected to provide a documented application process to assist the applicant make a complete application.

The convention is to request a Section 16/16A connection[4] where the terms are standard and non-negotiable. The alternative is a Section 22 connection[2] offer, where the terms are fully negotiable. If the Section 22 route is chosen then caution is required, as disputes over Section 22 agreements cannot be referred to the regulator for determination after the connection agreement has been signed; a post contract dispute will have to be pursued as a civil action. It is always advised that a Section 16/16A offer is requested before considering the Section 22 option.

On receipt of the Host's firm offer[5], the Developer generally has 90 days to accept its terms and to undertake its own review of the offer to ensure it meets its requirements. If for any reason the Developer and Host are unable to reach agreement of the terms, it is recommended that the Developer seeks specialist advice. It should be noted that in extreme circumstances, it may be necessary to refer an issue to regulator for determination but, be warned, this can be a lengthy process, taking upto 16 weeks to conclude.

Competition

One of the key decisions affecting the way in which the connection process proceeds is whether the Developer wishes to introduce competition into the procurement process by appointing a third party to design and construct the contestable connection works. In these circumstances, the Developer can requisition a non contestable quotation from the Host or ask the third party to do so on its behalf. The Host is obliged to provide information within standardised time frames against which its performance is monitored by the industry regulator (Ofgem).

Contestable works are those that may be carried out either by the Host or by an approved contractor, on the Developer's behalf. This contrasts with non contestable works, which can only be carried out by the Host. Broadly speaking, the Host will make the connection to its network for the new supply and undertake any upstream network reinforcement works. All works downstream from the point of connection to the Host network into the site and to each building are contestable works. Therefore, the extent of the contestable works can vary significantly depending on

---

[1] *A distribution network operator (DNO) is a company that is responsible for the design, construction operation and maintenance of a public electricity distribution network. The host DNO is the electricity distribution network to which the development site will directly connect.*
[2] *Contestable is work in providing the connection that can be carried out by an accredited independent party.*
[3] *Non contestable is work that can only be carried by the Host*
[4] *The Electricity Act 1989.*
[5] *Unless the offer becomes interactive, i.e. another connection scheme is vying for network capacity. Where the connection request becomes interactive the Developer has 30 days to accept the offer. The Host will inform the Developer if the connection request is interactive.*

where the point of connection is designated by the Host. Alternatively, the Developer can request an independent licensed network operator (IDNO) to adopt assets constructed by the third party.

If the Developer decides to contract with a third party to construct the contestable works, it is the Developer's responsibility to ensure that the construction works meet the Host's network adoption requirements.

Network Reinforcement

Reinforcement works may be required to increase the capacity of the network to enable the connection to meet a site's projected demand. In terms of the capacity made available by reinforcement, the following possible scenarios arise:

- Where reinforcement works are necessary for sole use by the Developer, the Developer is charged the full cost of the works. In some circumstances the most economic method of reinforcement may introduce spare network capacity in excess of the Developer's requirements. In such circumstances the Developer can receive a rebate where this spare capacity is absorbed by subsequent developments. However, the possibility of a rebate is time constrained and the original development will only qualify for a rebate for up to 5 years after the connection is completed.

- Where reinforcement works are necessary but also cater for the Host's future network requirements, the Developer is charged for a proportion of the cost of the works in the form of a capital contribution. This is calculated using cost apportionment factors for security (the ratio of capacity requested to that which is made available) and fault level (the ratio of fault level contribution of connection to that which is made available).

The charges levied for reinforcement works are attributable to those reinforcement works undertaken at one voltage level above the connection voltage only. That is, if connection voltage is 400V then costs are chargeable for reinforcement works undertaken at the next highest voltage, which is usually 11kV, and not for works conducted at the next highest voltage, which is usually 33kV[6]. These deeper network reinforcement costs are recovered as part of the system charges built into the electricity supply tariffs.

Costs

The cost of the Developer's connection depends on the nature and extent of the works to be undertaken. The distance between the site and the Host's network, the size of the customer demand in relation to available capacity (and hence the potential need for reinforcement), customer-specific timescales and to an extent market value of raw material and labour, are all significant factors that will affect the cost.

As part of the firm offer received from the Host, the Developer is provided with a charging statement, which includes the charges to be levied for the following items:

- Assessment and Design - to identify and design the most appropriate point on the existing network for the connection.

- Design Approval – to ensure design of a connection meets the safety and operation requirements of the Host.

- Non-contestable Works/Reinforcement - to include circuits and plant forming part of the connection that can be undertaken by the Host only, including land rights issues and consents.

- Contestable Connection Works – to include circuits and plant forming part of the connection that can be undertaken by approved contractors or the Host.

- Inspection of Works – to ensure that works are being constructed in accordance with the design requirements of the Host.

- Commissioning of Works – to include circuit outages and testing that will ensure that connection is safe to be energised.

In some circumstances, in addition to the cost of the physical connection works there may be chargeable costs associated with operation, maintenance repair and replacement of the new or modified connection. These are known as O&M costs and are chargeable as capitalised up-front costs where the Developer requests a solution that is in excess of the minimum necessary to provide the connection. For example, where extra resilience is requested for a connection, over and above that which the Host is obligated to provide, then O&M costs can be levied but only for the extra resilience element of the connection, not the total cost of the connection.

---

[6] *Some DNOs use voltages of 6.6kV and 22kV.*

Typical Engineering Details

In addition to the Engineering Features, Typical Engineering Details are included. These are indicative schematics to assist in the compilation of costing exercises. The user should note that these are only examples and cannot be construed to reflect the design for each and every situation. They are merely provided to assist the user with gaining an understanding of the Engineering concepts and elements making up such.

ELECTRICAL

- Urban Network Mainly Underground
- Urban Network Mainly Underground with Reinforcement
- Urban Network Mainly Underground with Substation Reinforcement
- Typical Simple 11kV Network Connection For LV Intakes Up To 1000kVA
- Typical 11kV Network Connections For HV Intakes 1000kVA To 6000kVA
- Static UPS System - Simplified Single Line Schematic For a Single Module
- Typical Data Transmission (Structured Cabling)
- Typical Networked Lighting Control System
- Typical Standby Power System, Single Line Schematic
- Typical Fire Detection and Alarm Schematic
- Typical Block Diagram - Access Control System (ACS)
- Typical Block Diagram - Intruder Detection System (IDS)
- Typical Block Diagram - Digital CCTV

MECHANICAL

- BMS Controls For Low Pressure Hot Water (LPHW)
- BMS Controls For Primary Chillers and Chilled Water
- Fan Coil Unit System
- Displacement System
- Chilled Ceiling System (Passive System)
- Chilled Beam System (Passive System)
- Variable Air Volume (VAV)
- Variable Refrigerant Volume System (VRV)
- Alternative All Air System (FGU)
- Reverse cycle heat pump

Urban Network Mainly Underground

Details : Connection to small housing development 10 houses, 60m of LV cable from local 11/LV substation route in footpath and verge, 10m of service cable to each plot in verge

Supply Capacity : 200kVA

Connection Voltage : LV

Nr of Phases : 1Φ

Breakdown of Detailed Cost Information

	Labour	Plant	Materials	Overheads	Total
Cable	£ 228	£ 109	£ 1,119	£ 428	£ 1,884
Jointing	£ 1,536	£ 398	£ 539	£ 716	£ 3,189
Switchgear	£ -	£ -	£ -	£ -	£ -
Termination	£ 364	£ 94	£ 118	£ 167	£ 744
Transformer	£ -	£ -	£ -	£ -	£ -
Trench/Reinstate	£ 970	£ 596	£ 1,094	£ 1,102	£ 3,762
OHL LV	£ -	£ -	£ -	£ -	£ -
OHL HV	£ -	£ -	£ -	£ -	£ -
Other	£ -	£ -	£ -	£ -	£ -
Special / One-offs					£ -
Total Calculated Price	£ 3,099	£ 1,197	£ 2,871	£ 2,413	£ 9,579
Non-contestable Elements and Associated Charges					
Point of Connection Information Charges					£ 180
Design Approval Charges					£ 300
Final Connection Charges					£ 550
Inspection and Monitoring Charges					£ 100
Wayleave and Easement Charges					£ 100
General Business Overhead Recovery					£ 638
Total Non-contestable Elements and Associated Charges					£ 1,868
Grand Total Calculated Price					£ 11,447

Typical Engineering Details

Urban Network Mainly Underground with Reinforcement

Details : Connection to small housing development 10 houses, 60m of LV cable from local 11/LV substation route in footpath and verge, 10m of service cable to each plot in verge. Scheme includes reinforcement of LV distribution board

Supply Capacity : 200kVA

Connection Voltage : LV

Nr of Phases : 1Φ

Breakdown of Detailed Cost Information

	Labour	Plant	Materials	Overheads	Total
Cable	£ 228	£ 109	£ 1,119	£ 428	£ 1,884
Jointing	£ 1,536	£ 398	£ 539	£ 716	£ 3,189
Switchgear	£ 1,773	£ 574	£ 2,537	£ 1,425	£ 6,309
Termination	£ 364	£ 94	£ 118	£ 167	£ 744
Transformer	£ -	£ -	£ -	£ -	£ -
Trench/Reinstate	£ 970	£ 596	£ 1,094	£ 1,102	£ 3,762
OHL LV	£ -	£ -	£ -	£ -	£ -
OHL HV	£ -	£ -	£ -	£ -	£ -
Other	£ -	£ -	£ -	£ -	£ -
Special / One-offs					£ -
Total Calculated Price	£ 4,871	£ 1,771	£ 5,408	£ 3,838	£ 15,888
Non-contestable Elements and Associated Charges					
Point of Connection Information Charges					£ 180
Design Approval Charges					£ 300
Final Connection Charges					£ 550
Inspection and Monitoring Charges					£ 100
Wayleave and Easement Charges					£ 100
General Business Overhead Recovery					£ 1,010
Total Non-contestable Elements and Associated Charges					£ 2,240
Grand Total Calculated Price					£ 18,128

Urban Network Mainly Underground with Substation Reinforcement

Details : Connection to small housing development 10 houses, 60m of LV cable from local 11/LV substation route in footpath and verge, 10m of service cable to each plot in verge. Scheme includes reinforcement of LV distribution board and new substation and 20m of HV cable

Supply Capacity : 200kVA

Connection Voltage : LV

Nr of Phases : 1Φ

Breakdown of Detailed Cost Information

	Labour	Plant	Materials	Overheads	Total
Cable	£ 270	£ 129	£ 1,552	£ 573	£ 2,524
Jointing	£ 1,829	£ 474	£ 760	£ 888	£ 3,951
Switchgear	£ 5,428	£ 1,757	£ 15,037	£ 6,509	£ 28,731
Termination	£ 599	£ 155	£ 297	£ 305	£ 1,355
Transformer	£ 2,788	£ 902	£ 7,500	£ 3,277	£ 14,468
Trench/Reinstate	£ 1,506	£ 1,005	£ 1,870	£ 1,815	£ 6,197
OHL LV	£ -	£ -	£ -	£ -	£ -
OHL HV	£ -	£ -	£ -	£ -	£ -
Other	£ -	£ -	£ -	£ -	£ -
Special / One-offs					£ -
Total Calculated Price	£ 12,421	£ 4,422	£ 27,016	£ 13,367	£ 57,227
Non-contestable Elements and Associated Charges					
Point of Connection Information Charges					£ 180
Design Approval Charges					£ 300
Final Connection Charges					£ 550
Inspection and Monitoring Charges					£ 100
Wayleave and Easement Charges					£ 100
General Business Overhead Recovery					£ 3,449
Total Non-contestable Elements and Associated Charges					£ 4,679
Grand Total Calculated Price					£ 61,906

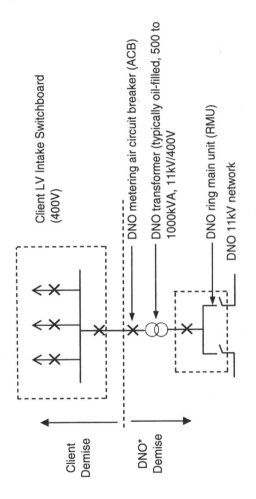

Typical Simple 11kV Network Connection For LV Intakes Up To 1000kVA

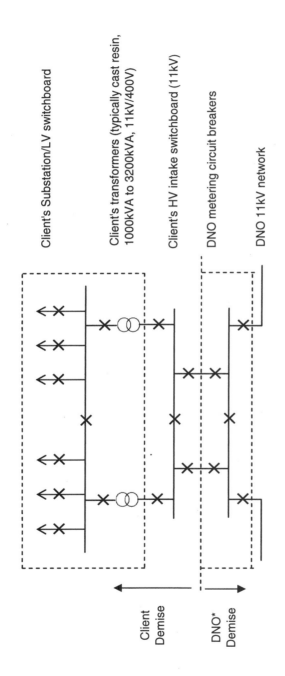

Typical 11kV Network Connections For HV Intakes 1000kVA To 6000kVA

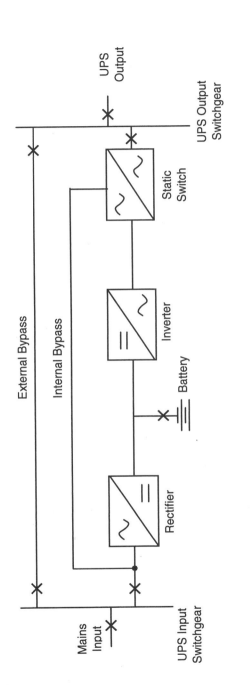

Static UPS System - Simplified Single Line Schematic For a Single Module

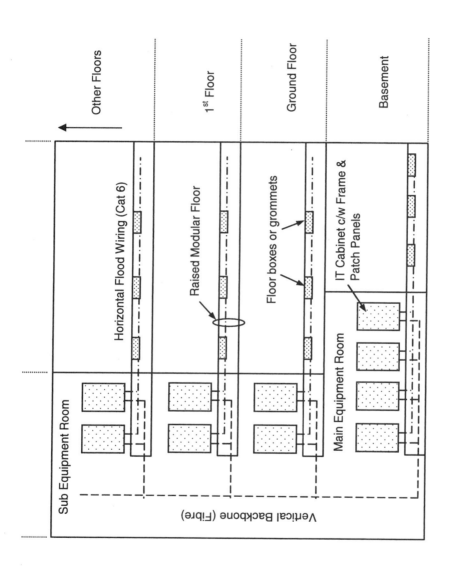

Typical Data Transmission (Structured Cabling)

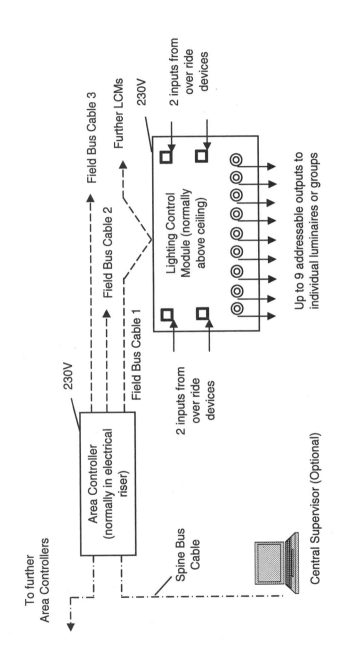

Typical Networked Lighting Control System

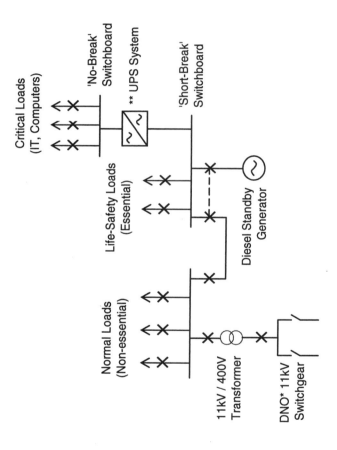

Typical Standby Power System, Single Line Schematic

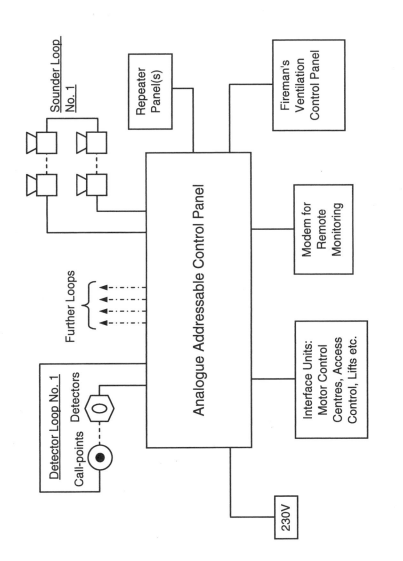

Typical Fire Detection and Alarm Schematic

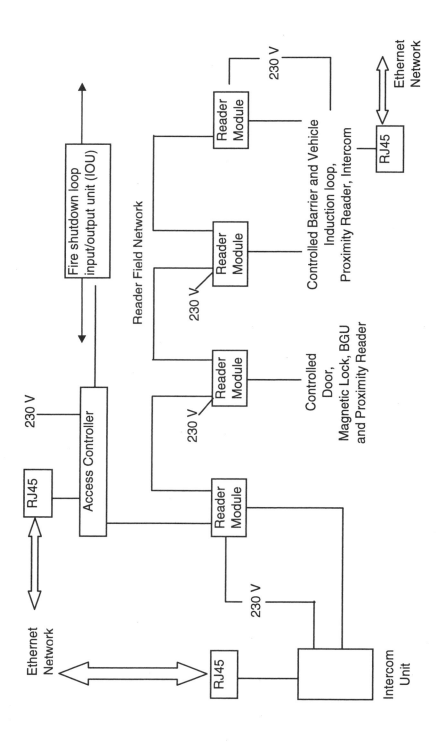

Typical Block Diagram - Access Control System (ACS)

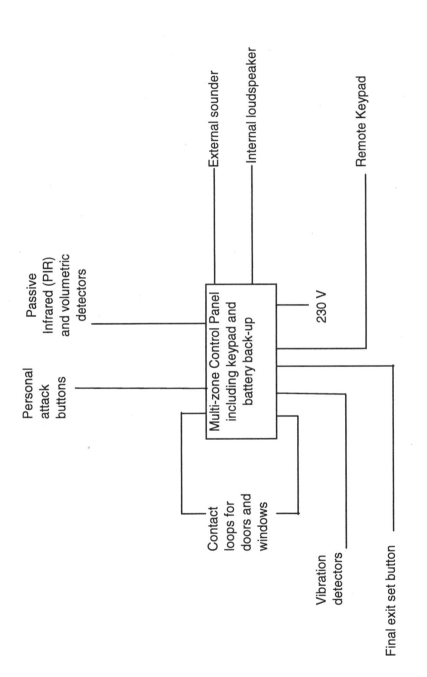

Typical block diagram - Intruder Detection System (IDS)

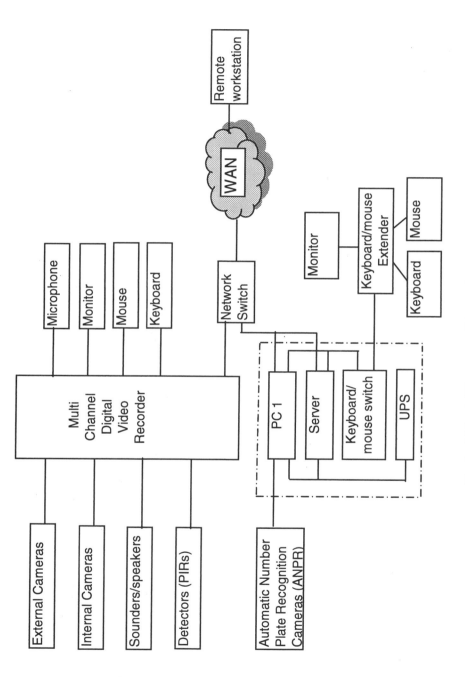

Typical Block Diagram - Digital CCTV

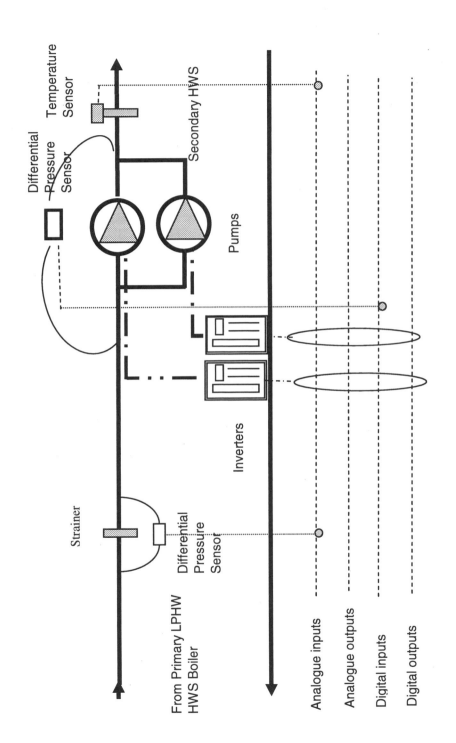

BMS Controls For Low Pressure Hot Water (LPHW)

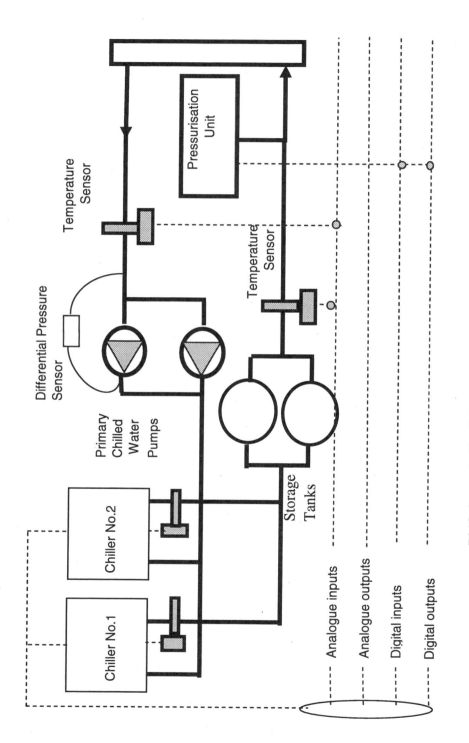

BMS Controls For Primary Chillers and Chilled Water

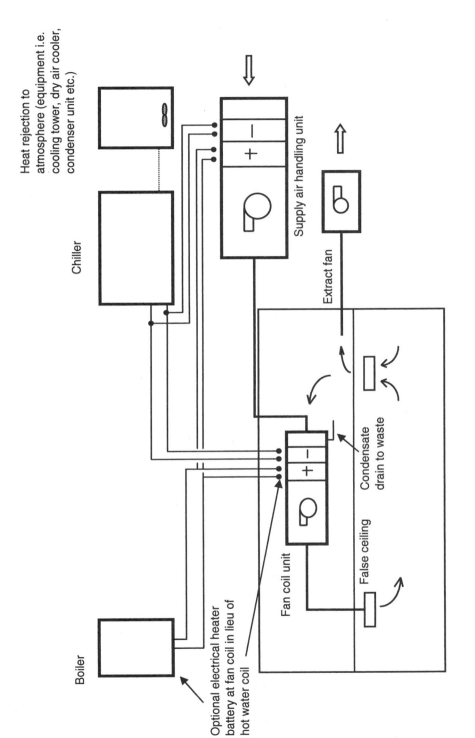

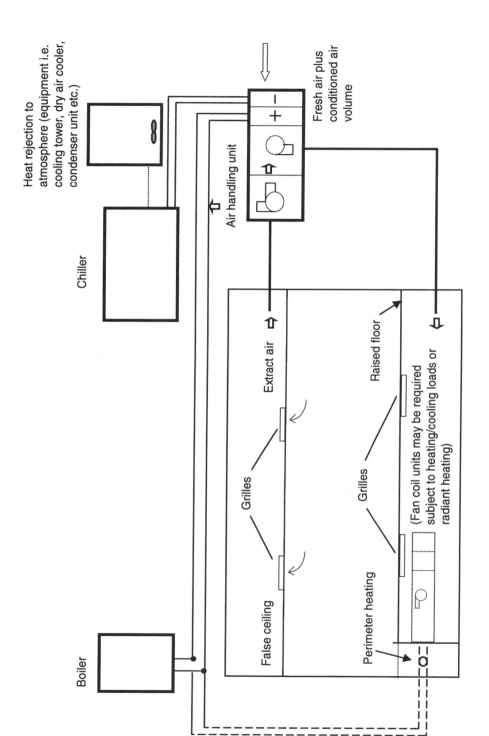

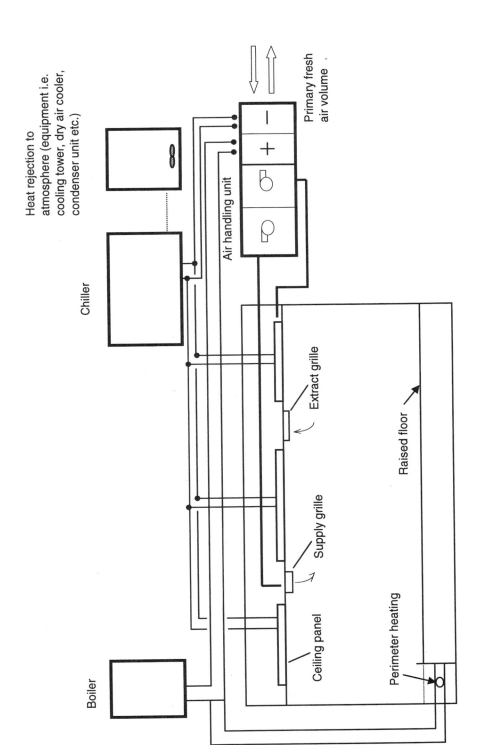

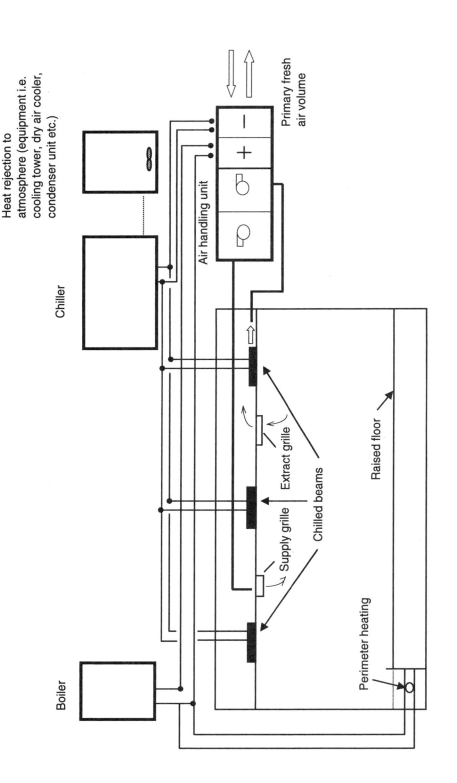

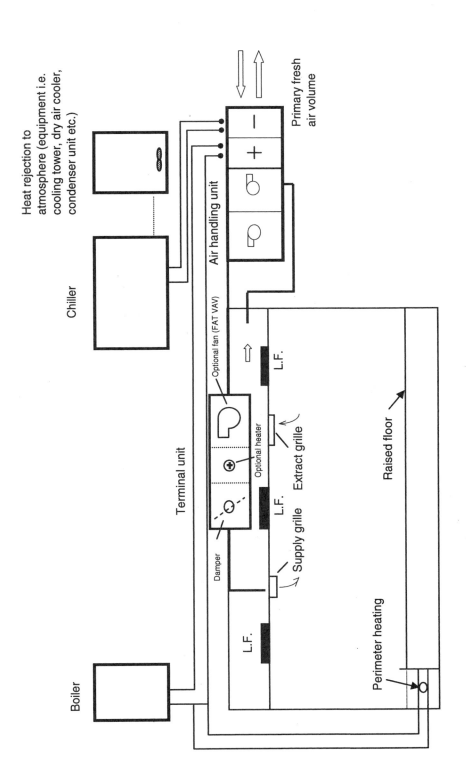

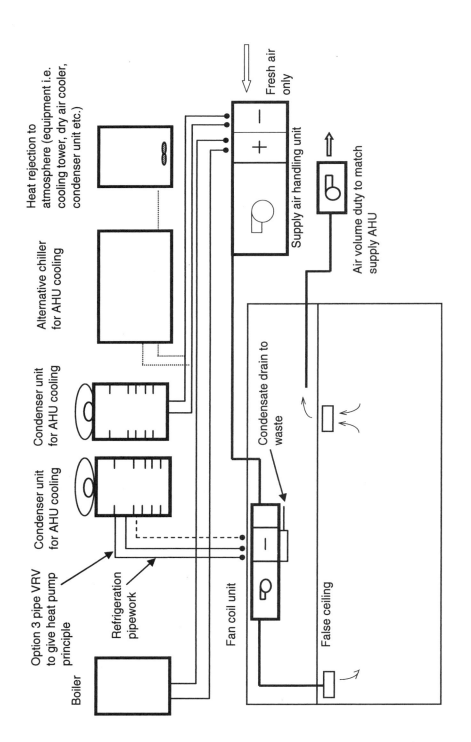

Variable Refrigerant Volume System (VRV)

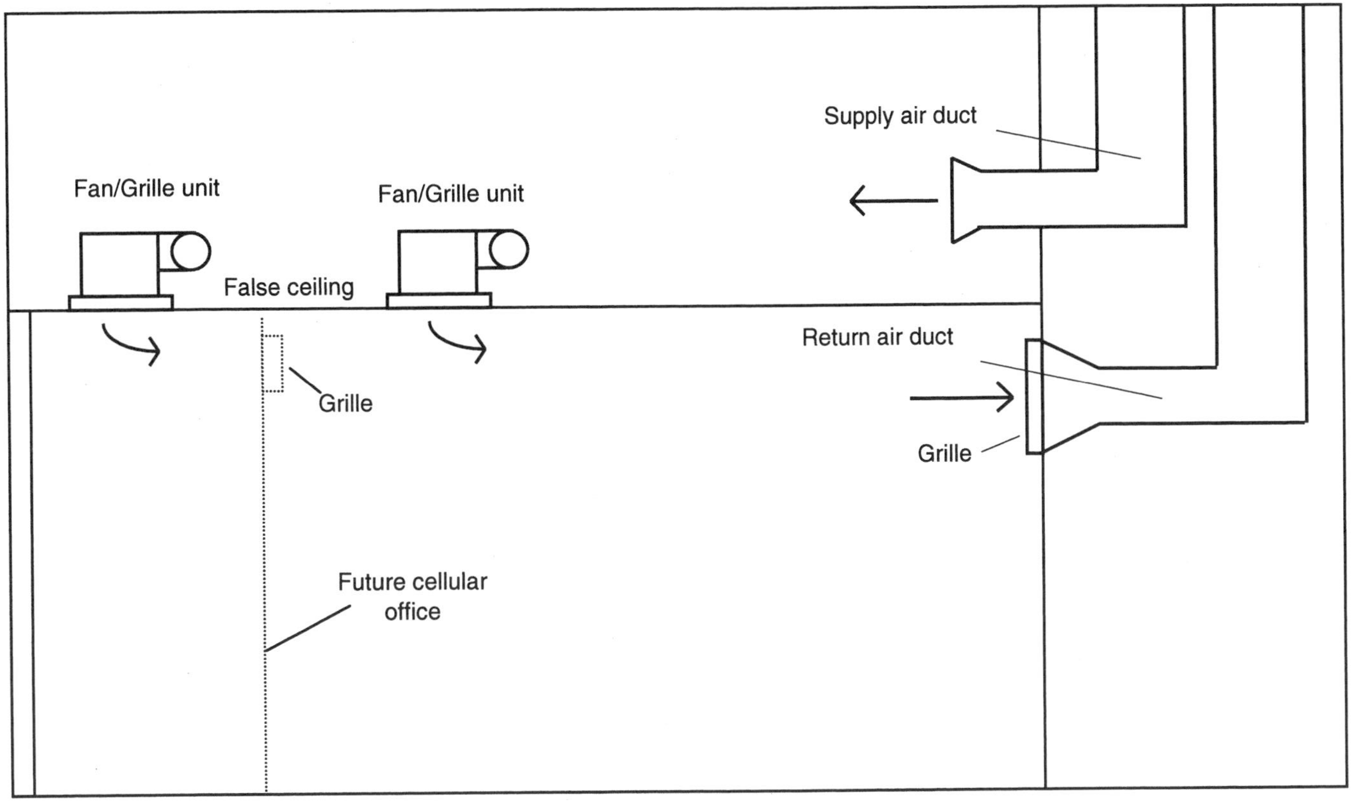

Supply air duct

Return air duct

Grille

Fan/Grille unit

Fan/Grille unit

False ceiling

Grille

Future cellular office

Alternative All Air System (FGU)

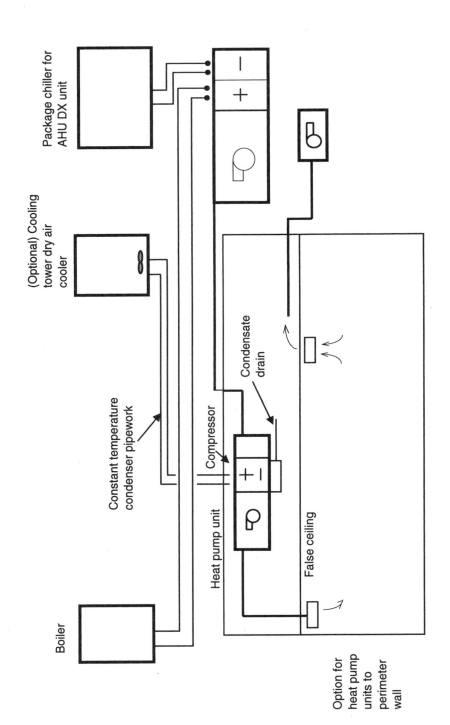

Reverse Cycle Heat Pump

ESSENTIAL READING FROM TAYLOR AND FRANCIS

Spon's Estimating Costs Guide to Electrical Works

Project Costs at a Glance

Fourth Edition

Bryan Spain

Specially written for contractors and small businesses carrying out small works, *Spon's Estimating Costs Guide to Electrical Works* provides accurate information on thousands of rates, each broken down to labour, material overheads and profit for residential, retail and light industrial premises. It is the first book to include typical project costs for new installations, stripping out, rewiring and upgrading for flats and houses.

In addition, vital information and advice is given on setting up and running a business, employing staff, tax, VAT and CIS4's.

For the cost of approximately two hours of your charge-out rate (or less), this book will help you to:

- produce estimates faster
- keep your estimates accurate and competitive
- run your business more effectively
- save time.

August 2008: 216x138: 304pp
Pb: 978-0-415-46904-3: **£29.99**

To Order: Tel: +44 (0) 1235 400524 **Fax:** +44 (0) 1235 400525
or Post: Taylor and Francis Customer Services,
Bookpoint Ltd, Unit T1, 200 Milton Park, Abingdon, Oxon, OX14 4TA UK
Email: book.orders@tandf.co.uk

**For a complete listing of all our titles visit:
www.tandf.co.uk**

ESSENTIAL READING FROM TAYLOR AND FRANCIS

Spon's Estimating Costs Guide to Minor Works,
Alterations and Repairs to Fire, Flood, Gale and Theft Damage

Fourth Edition

Bryan Spain

Specially written for contractors, quantity surveyors and clients carrying out small works, Spon's Estimating Costs Guide to Minor Works, Alterations and Repairs to Fire, Flood, Gale and Theft Damage contains accurate information on thousands of rates each broken down to labour, material overheads and profit.

Selected Contents: Introduction. Standard Method of Measurement/Trades Link. Part 1: Unit Rates. Part 2: Damage Repairs. Part 3: Approximate Estimating. Part 4: Plant and Tool Hire. Part 5: General Construction Data. Part 6: Business Matters

August 2008: 216x138: 320pp
Pb: 978-0-415-46906-7: **£29.99**

To Order: Tel: +44 (0) 1235 400524 **Fax:** +44 (0) 1235 400525
or Post: Taylor and Francis Customer Services,
Bookpoint Ltd, Unit T1, 200 Milton Park, Abingdon, Oxon, OX14 4TA UK
Email: book.orders@tandf.co.uk

For a complete listing of all our titles visit:
www.tandf.co.uk

PART TWO

Approximate Estimating

Directions, *page 59*
Cost Indices, *page 60*
RIBA Stage A Feasibility Costs, *page 62*
RIBA Stage C Elemental Rates, *page 67*
All-In-Rates, *page 74*
Elemental Costs, *page 90*

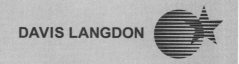

Constructing the best and most valued relationships in the industry

www.davislangdon.com

Offices in Europe & the Middle East, Africa, Asia Pacific, Australasia and the USA

Cost Management | Project Management
Banking Tax and Finance | Building Surveying | Engineering Services | Legal Support Group
Management Consultancy | Specifications and Design Management | VPR

SPON'S PRICEBOOKS 2009

Spon's Architects' and Builders' Price Book 2009
Davis Langdon

The most detailed, professionally relevant source of construction price information currently available anywhere.

In the 2009 edition the Preliminaries section has been re-written and based upon the 2005 JCT contract, and the new section on Sustainability has been re-written and considerably expanded. New Measured Works items include more economical, lower nickel, stainless steel rebars; Schüco curtain walling; brises soleil; "Ecologic" roofing tiles; Velfac 200 windows; and triple glazing. ... along with additional items of rainscreen cladding, damp proof courses, steel lintels, Patent Glazing items, and radiators and manhole covers.

Hb & electronic package*: 1000pp approx.
978-0-415-46555-7: **£145**
electronic package
(inc. sales tax where appropriate)
978-0-203-88676-2: only **£135**

Spon's Mechanical and Electrical Services Price Book 2009
Davis Langdon, Mott Green Wall

The only comprehensive and best annual services engineering price book available for the UK.

As well as an extensive overhaul of prices, updates to the 2009 edition include a new infrastructure section which includes typical connection scenarios and costs, extensive revisions - to the renewable energy sections, with descriptions of biomass boilers, ground source heap pumps and photovoltaics. Sprinklers and luminaries have been bought in-line with new regulations.

Hb & electronic package*: 640pp approx
978-0-415-46561-8: **£145**
electronic package
(inc. sales tax where appropriate)
978-0-203-88673-1: only **£135**

Spon's External Works and Landscape Price Book 2009
Davis Langdon

The only comprehensive source of information for detailed external works and landscape costs.

The 2009 edition gives more on costs of tender preparation and specialised surveys, under Preliminaries; and a number of new and heavily revised Measured Works items: repair and conservation of historic railings; new Aco drainage ranges; fin drains; concrete pumping and crushing costs; rainwater harvesting tanks; containerised tree planting; specialised mortars; park signage; and site clearance, notably the clearance of planting material and demolition of structures.

Hb & electronic package*: 424pp approx.
978-0-415-46559-5: **£115**
electronic package
(inc. sales tax where appropriate)
978-0-203-88674-8: only **£105**

Spon's Civil Engineering and Highway Works Price Book 2009
Davis Langdon

The 23rd edition, in its easy to read format, incorporates a general review throughout, and includes updates to the Capital Allowances and VAT and Construction sections to reflect the latest government legislation.

Hb & electronic package:* 800pp approx.
978-0-415-46557-1: **£155**
electronic package
(inc. sales tax where appropriate)
978-0-203-88675-5: only **£145**

---

*Receive our eBook free when you order any hard copy Spon 2009 Price Book, with free estimating software to help you produce tender documents, customise data, perform word searches and simple calculations.
Or buy just the ebook, with free estimating software.
Visit **www.pricebooks.co.uk**

To Order: Tel: +44 (0) 1235 400524 Fax: +44 (0) 1235 400525
or Post: Taylor and Francis Customer Services, Bookpoint Ltd, Unit T1, 200 Milton Park, Abingdon, Oxon, OX14 4TA UK
Email: book.orders@tandf.co.uk
For a complete listing of all our titles visit: www.tandf.co.uk

Approximate Estimating

DIRECTIONS

The prices shown in this section of the book are average prices on a fixed price basis for typical buildings tendered during the second quarter of 2008. Unless otherwise noted, they exclude external services and professional fees.

The information in this section has been arranged to follow more closely the order in which estimates may be developed, in accordance with the RIBA stages of work;

a) Cost Indices and Regional Variations – These provide information regarding the adjustments to be made to estimates taking into account current pricing levels for different locations in the UK.

b) Feasibility Costs – These provide a range of data (based on a rate per square metre) for all-in engineering costs, excluding lifts, associated with a wide variety of building types. These would typically be used at work stage A/B (feasibility) of a project.

c) Elemental Rates – The outline costs for offices have been developed further to provide rates for the alternative solutions for each of the services elements. These would typically be used at work stage C, outline proposal.

Where applicable, costs have been identified as Shell and Core and Fit Out to reflect projects where the choice of procurement has dictated that the project is divided into two distinctive contractual parts.

Such detail would typically be required at work stage D, detailed proposals.

d) All-in-Rates – These are provided for a number of items and complete parts of a system i.e. boiler plant, ductwork, pipework, electrical switchgear and small power distribution, together with lifts and escalators. Refer to the relevant section for further guidance notes.

e) Elemental Costs – These are provided for a diverse range of building types; offices, laboratory, shopping mall, airport terminal building, supermarket, performing arts centre, sports hall, luxury hotel, hospital and secondary school. Also included is a separate analysis of a building management system for an office block. In each case, a full analysis of engineering services costs is given to show the division between all elements and their relative costs to the total building area. A regional variation factor has been applied to bring these analyses to a common London base.

Prices should be applied to the total floor area of all storeys of the building under consideration. The area should be measured between the external walls without deduction for internal walls and staircases/lift shafts i.e. G.I.A (Gross Internal Area).

Although prices are reviewed in the light of recent tenders it has only been possible to provide a range of prices for each building type. This should serve to emphasise that these can only be average prices for typical requirements and that such prices can vary widely depending on variations in size, location, phasing, specification, site conditions, procurement route, programme, market conditions and net to gross area efficiencies. Rates per square metre should not therefore be used indiscriminately and each case needs to be assessed on its own merits.

The prices do not include for incidental builder's work nor for profit and attendance by a Main Contractor where the work is executed as a sub-contract: they do however include for preliminaries, profit and overheads for the services contractor. Capital contributions to statutory authorities and public undertakings and the cost of work carried out by them have been excluded.

Where services works are procured indirectly, i.e. ductwork via a mechanical Sub Contractor, the reader should make due allowance for the addition of a further level of profit etc.

COST INDICES

The following tables reflect the major changes in cost to contractors but do not necessarily reflect changes in tender levels. In addition to changes in labour and materials costs, tenders are affected by other factors such as the degree of competition in the particular industry, the area where the work is to be carried out, the availability of labour and the prevailing economic conditions. This has meant in recent years that, when there has been an abundance of work, tender levels have tended to increase at a greater rate than can be accounted for solely by increases in basic labour and material costs and, conversely, when there is a shortage of work this has tended to result in keener tenders. Allowances for these factors are impossible to assess on a general basis and can only be based on experience and knowledge of the particular circumstances.

In compiling the tables the cost of labour has been calculated on the basis of a notional gang as set out elsewhere in the book. The proportion of labour to materials has been assumed as follows:
Mechanical Services - 30:70, Electrical Services - 50:50, (1976 = 100)

Mechanical Services

Year	First Quarter	Second Quarter	Third Quarter	Fourth Quarter
1998	365	363	368	373
1999	368	363	370	384
2000	384	386	388	400
2001	401	401	405	411
2002	411	411	410	442
2003	443	446	447	456
2004	458	464	467	482
2005	486	487	492	508
2006	513	522	527	533
2007	535	541	546	556P
2008	555P	557F	559F	573F
2009	577F	579F	582F	598F

Electrical Services

Year	First Quarter	Second Quarter	Third Quarter	Fourth Quarter
1998	410	423	422	422
1999	433	432	431	446
2000	458	464	465	468
2001	485	484	484	487
2002	508	508	508	513
2003	530	533	533	541
2004	571	574	576	589
2005	607	608	607	615
2006	631	636	641	649
2007	665	666	668	676P
2008	693P	694F	695F	706F
2009	723F	724F	726F	737F

(P = Provisional)
(F = Forecast)

COST INDICES

Regional Variations

Prices throughout this Book apply to work in the London area (see Directions at the beginning of the Mechanical Installations and Electrical Installations sections). However, prices for mechanical and electrical services installations will of course vary from region to region, largely as a result of differing labour costs but also depending on the degree of accessibility, urbanisation and local market conditions.

The following table of regional factors is intended to provide readers with indicative adjustments that may be made to the prices in the Book for locations outside of London. The figures are of necessity averages for regions and further adjustments should be considered for city centre or very isolated locations, or other known local factors.

Region	Factor	Region	Factor
Greater London	1.00	Yorkshire & Humberside	0.97
South East	1.00	North West	0.97
South West	0.99	North East	0.97
East Midlands	0.97	Scotland	0.98
West Midlands	0.96	Wales	0.99
East Anglia	0.99	Northern Ireland	0.96

RIBA STAGE A FEASIBILITY COSTS

TYPICAL SQUARE METRE RATES FOR ENGINEERING SERVICES

The following examples indicate the range of rates within each building type for engineering services, excluding lifts etc, utilities services and professional fees. Based on Gross Internal Area (GIA).

Industrial Buildings $£/m^2$ GIA

Factories

Owner occupation: Includes for rainwater, soil/waste, LTHW heating via HL radiant heaters, BMS, LV installations, lighting, fire alarms, security, earthing.. 125

Owner occupation: Includes for rainwater, soil/waste, sprinklers, LTHW heating via HL radiant heaters, local air conditioning, BMS, HV/LV installations, lighting, fire alarms, security, earthing.................... 190

Warehouses

High bay for owner occupation: Includes for rainwater, soil/waste, LTHW heating via HL gas fired heaters, BMS, HV/LV installations, lighting, fire alarms, security, earthing....................................... 100

High bay for owner occupation: Includes for rainwater, soil/waste, sprinklers, LTHW heating via HL radiant heaters, local air conditioning, BMS, HV/LV installations, lighting, fire alarms, security, earthing 205

Distribution Centres

High bay for letting: Includes for rainwater, soil/waste, LTHW heating via HL gas fired heaters, BMS, HV/LV installations, lighting, fire alarms, security, earthing... 125

High bay for owner occupation: Includes for rainwater, soil/waste, sprinklers, LTHW heating via HL radiant heaters, local air conditioning, BMS, HV/LV installations, lighting, fire alarms, security...... 240

Office Buildings 5,000m² to 15,000m²

Offices for letting

Shell & Core and Cat A non air conditioned; Includes for rainwater, soil/waste, cold water, hot water via local electrical heaters, LTHW heating via radiator heaters, toilet extract, controls, LV installations, lighting, small power (landlords), fire alarms, earthing, security & IT wireways................................ 270

Shell & Core and Cat A non air conditioned; Includes for rainwater, soil/waste, cold water, hot water, LTHW heating via perimeter heaters, toilet extract, controls, LV installations, lighting, small power (landlords), fire alarms, earthing, security wireways, IT wireways.. 285

Shell & Core and Cat A air conditioned; Includes for rainwater, soil/waste, VRV 3 pipe heat pumps, toilet extract, BMS, LV installations, lighting, small power (landlords), fire alarms, earthing, security & IT wireways, ... 420

Shell & Core and Cat A air conditioned; Includes for rainwater, soil/waste, cold water, hot water via local electrical heaters, LTHW heating via perimeter heaters, 2 pipe, toilet extract, BMS, LV installations, lighting, small power (landlords), fire alarms, earthing, security & IT wireways.............. 438

Offices for owner occupation

Non air conditioned; Includes for rainwater, soil/waste, cold water, hot water via local electrical heaters, dry risers, LTHW heating via radiator heaters, toilet extract, controls, LV installations, lighting, small power (landlords), fire alarms, earthing, security, IT wireways... 306

Non air conditioned; Includes for rainwater, soil/waste, cold water, hot water, dry risers, LTHW heating via perimeter heaters, toilet extract, BMS, LV installations, lighting, small power (landlords), fire alarms, earthing, security, IT wireways 316

Approximate Estimating

RIBA STAGE A FEASIBILITY COSTS

TYPICAL SQUARE METRE RATES FOR ENGINEERING SERVICES

Office Buildings *continued*

Offices for owner occupation *continued* £/m² GIA

Air conditioned; Includes for rainwater, soil/waste, cold water, hot water via local electrical heaters, dry risers, LTHW heating via perimeter heating, 4 pipe air conditioning, toilet extract, kitchen extract, sprinkler protection, BMS, LV installations, life safety standby generators, lighting, small power, fire alarms, earthing, security, IT wireways small power, fire alarms L1/P1, earthing, security, and IT wireways 490

Health and Welfare Facilities

District general hospitals
Natural ventilation Includes for rainwater, soil/waste, cold water, hot water, dry risers, medical gases, LTHW heating via perimeter heating, toilet extract, kitchen extract, BMS, LV installations, standby generation, lighting, small power, fire alarms, earthing/lightning protection, nurse call systems, security, IT wireways... 460

Natural ventilation Includes for rainwater, soil/waste, cold water, hot water, dry risers, LTHW heating via perimeter heaters, localised VAV air conditioning, kitchen/toilet extract, BMS, LV installations, standby generation, lighting, small power, fire alarms, earthing/lightning protection, nurse call systems, security, IT wireways 650

Private hospitals
Air conditioned; Includes for rainwater, soil/waste, cold water, hot water, dry risers, medical gases, LTHW heating, 2 pipe air conditioning, toilet extract, kitchen extract, BMS, LV installations, standby generation, lighting, small power, fire alarms, earthing, nurse call systems, nurse call system, security, IT wireways 730

Air conditioned; Includes for rainwater, soil/waste, cold water, hot water, dry risers, medical gases, LTHW heating, 4 pipe air conditioning, kitchen/toilet extract, BMS, LV installations, standby generation, lighting, small power, fire alarms, earthing, nurse call system, security, IT wireways.......................... 790

Day care unit
Natural ventilation; Includes for rainwater, soil/waste, cold water, hot water via local electrical heaters, medical gases, LTHW heating, toilet extract, kitchen extract, BMS, LV installations, lighting, small power, fire alarms , earthing, security, IT wireways.. 490

Comfort cooled; Includes for rainwater, soil/waste, cold water, hot water, medical gases, LTHW heating, DX air conditioning, kitchen/toilet extract, BMS, LV installations, lighting, small power, fire alarms, earthing, security, IT wireways 530

Entertainment and Recreation Buildings

Non Performing
Natural Ventilation; Includes for rainwater, soil/waste, cold water, central hot water, dry risers, LTHW heating, toilet extract, kitchen extract, controls, LV installations, lighting, small power, fire alarms, earthing, security, IT wireways.. 350

Comfort cooled; Includes for rainwater, soil/waste, cold water, hot water via local electrical heaters sprinklers/dry risers, LTHW heating , DX air conditioning, kitchen/toilet extract, BMS, LV installations, lighting, small power, fire alarms, earthing, security, IT wireways 550

RIBA STAGE A FEASIBILITY COSTS

TYPICAL SQUARE METRE RATES FOR ENGINEERING SERVICES *continued*

Entertainment and Recreation Buildings *continued* £/m² GIA

Performing Arts (With Theatre)
Natural Ventilation; Includes for rainwater, soil/waste, cold water, central hot water, sprinklers/dry risers, LTHW heating, toilet extract, kitchen extract, controls, LV installations, lighting, small power, fire alarms, earthing, security, IT wireways 575

Comfort Cooled; Includes for rainwater, soil/waste, cold water, central hot water, sprinklers/dry risers, LTHW heating, DX air conditioning, kitchen/toilet extract, BMS, LV installations, lighting including enhanced dimming/scene setting, small power, fire alarms, earthing, security, IT wireways including for production, audio and video recording, EPOS system, ... 700

Sports Halls
Natural ventilation; Includes for rainwater, soil/waste, cold water, hot water gas fired heaters, LTHW heating, toilet extract, BMS, LV installations, lighting, small power, fire alarms, earthing, security 214

Comfort Cooled; Includes for rainwater, soil/waste, cold water, hot water via LTHW heat exchangers, LTHW heating, air conditioning via AHU's, toilet extract, BMS, LV installations, lighting, small power, fire alarms, earthing, security. 313

Multi Purpose Leisure Centre
Natural ventilation; Includes for rainwater, soil/waste, cold water, hot water gas fired heaters, LTHW heating, toilet extract, BMS, LV installations, lighting, small power, fire alarms, earthing, security, IT wireways 313

Comfort cooled; Includes for rainwater, soil/waste, cold water, hot water LTHW heat exchangers, LTHW heating, air conditioning via AHU, pool hall supply/extract, kitchen/toilet extract, BMS, LV installations, lighting, small power, fire alarms, earthing, security, IT wireways 424

Retail Buildings

Open Arcade
Natural ventilation; Includes for rainwater, soil/waste, cold water, hot water, sprinklers/dry risers, LTHW heating, toilet extract, smoke extract, BMS, LV installations, life safety standby generators, lighting, small power, fire alarms, public address, earthing/lightning protection, security, IT wireways 330

Enclosed Shopping Mall
Air conditioned; Includes for rainwater, soil/waste, cold water, hot water, sprinklers/dry risers, LTHW heating, air conditioning via AHU's, toilet extract, smoke extract, BMS, LV installations, life safety standby generators, lighting, small power, fire alarms, public address, earthing/lightning protection, CCTV/security, IT wireways people counting systems... 535

Department Stores
Air conditioned; Includes for sanitaryware, soil/waste, cold water, hot water, sprinklers/dry risers, LTHW heating, air conditioning via AHU's, toilet extract, smoke extract, BMS, LV installations, life safety standby generators, lighting, small power, fire alarms, public address, earthing, lightning protection, CCTV/security, IT installation wireways... 395

Approximate Estimating

RIBA STAGE A FEASIBILITY COSTS

TYPICAL SQUARE METRE RATES FOR ENGINEERING SERVICES *continued*

Retail Buildings *continued* £/m² GIA

Supermarkets

Air conditioned; Includes for rainwater, soil/waste, cold water, hot water, sprinklers, LTHW heating, air conditioning via AHU's, toilet extract, BMS, LV installations, lighting, small power, fire alarms, earthing, lightning protection, security, IT wireways, refrigeration.. 590

Educational Buildings

Secondary Schools (Academy)
Natural Ventilation; Includes for rainwater, soil/waste, cold water, central hot water, LTHW heating, toilet extract, BMS, LV installations, lighting, small power, fire alarms, earthing, security, IT wireways 330

Natural vent with comfort cooling to selected areas (BB93 compliant); Includes for rainwater, soil/waste, cold water, central hot water, LTHW heating, DX air conditioning, general supply/extract, toilet extract, BMS, LV installations, lighting, small power, fire alarms, earthing, security, IT wireways 410

Scientific Buildings

Educational Research
Comfort cooled; Includes for rainwater, soil/waste, cold water, central hot water, dry risers, compressed air, medical gases, LTHW heating, 4 pipe air conditioning, toilet extract, fume, BMS, LV installations, lighting, small power, fire alarms, earthing, lightning protection, security, IT wireways 800

Air conditioned; Includes for rainwater, soil/waste, laboratory waste, cold water, central hot water, specialist water, dry risers, compressed air, medical gases, steam, LTHW heating, VAV air conditioning, Comm's room cooling, toilet extract, fume extract, BMS, LV installations, UPS, standby generators, lighting, small power, fire alarms, earthing, lightning protection, security, IT wireways 1300

Commercial Research
Air conditioned; Includes for rainwater, soil/waste, laboratory waste, cold water, central hot water, specialist water, dry risers, compressed air, medical gases, steam, LTHW heating, VAV air conditioning, Comm room cooling, toilet extract, fume extract, BMS, LV installations, UPS, standby generators, lighting, small power, fire alarms, earthing, lightning protection, security, IT wireways 1300

RIBA STAGE A FEASIBILITY COSTS

TYPICAL SQUARE METRE RATES FOR ENGINEERING SERVICES *continued*

£/m² GIA

Hotels

1 to 3 Star; Includes for rainwater, soil/waste, cold water, hot water, dry risers, LTHW heating via radiators, toilet/bathroom extract, kitchen extract, BMS, LV installations, lighting, small power, fire alarms, earthing, lightning protection security, IT wireways .... 350 to 550

4 to 5 Star; Includes for rainwater, soil/waste, cold water, hot water, sprinklers, dry risers, 4 pipe air conditioning, kitchen extract, toilet/bathroom extract, BMS, LV installations, life safety standby generators, lighting, small power, fire alarms, earthing, security, IT wireways................................ 750 to 850

Approximate Estimating

RIBA STAGE C ELEMENTAL RATES

ELEMENTAL RATES FOR ALTERNATIVE ENGINEERING SERVICES SOLUTIONS

The following examples of building types indicate the range of rates for alternative design solutions for each of the engineering services elements based on Gross Internal Area for the Shell and Core and Net Internal Area for the Fit Out. Fit Out is assumed to be to Cat A standard.

Consideration should be made for the size of the building, which may affect the economies of scale for rates i.e. the larger the building the lower the rates

OFFICES MECHANICAL SERVICES	Shell & Core £/m² GIA.	Fit Out £/m² NIA.
Sanitaryware		
Building up to 3,000m²	8 to 10	-
Building over 3,000m² to 15,000m² (low rise)	6 to 9	-
Disposal installation		
Building up to 3,000m²	20 to 25	-
Building over 3,000m² to 15,000m²	10 to 15	-
Water installation		
Building up to 3,000m²	10 to 15	-
Building over 3,000m² to 15,000m²	10 to 15	-
LPHW Heating Installation; including gas installations		
Building up to 3,000m²	35 to 45	45 to 55
Building over 3,000m² to 15,000m²	30 to 35	35 to 45
Air Conditioning; including ventilation		
<u>Comfort Cooling:</u>		
2 pipe fan coil for building up to 3,000m²	55 to 70	95 to 105
2 pipe fan coil for building over 3,000m² to 15,000m²	50 to 65	85 to 95
2 pipe variable refrigerant volume (VRV) for building up to 3,000m²	40 to 50	65 to 75
<u>Full air conditioning:</u>		
4 Pipe fan coil for building up to 3,000m²	80 to 95	135 to 155
4 Pipe fan coil for building over 3,000m² to 15,000m²	70 to 85	115 to 135
3 pipe variable refrigerant volume for building up to 3,000m²	65 to 75	115 to 125
Ventilated (active) chilled beams for building over 3,000m² to 15,000m²	70 to 85	135 to 145
Chilled beam exposed services for building over 3,000m² to 15,000m²	70 to 85	225 to 245
Concealed passive chilled beams for building over 3,000m to 15,000m²	70 to 85	125 to 135
Chilled ceiling for building over 3,000m² to 15,000m²	70 to 85	215 to 225
Chilled ceiling/perimeter beams for building over 3,000m² to 15,000m²	70 to 85	235 to 255
Displacement for building over 3,000m² to 15,000m²	80 to 95	75 to 95

RIBA STAGE C ELEMENTAL RATES

ELEMENTAL RATES FOR ALTERNATIVE ENGINEERING SERVICES SOLUTIONS *continued*

OFFICES **MECHANICAL SERVICES** *continued*	Shell & Core £/m² GIA.	Fit Out £/m² NIA.
Ventilation systems		
Building up to 3,000m²	30 to 35	-
Building over 3,000m² to 15,000m²	25 to 30	-
Fire Protection over 3,000 m² to 15,000m²		
Dry risers	8 to 10	-
Sprinkler installation	20 to 30	25 to 30
BMS Controls: including MCC panels and control cabling		
Full air conditioning, fan coil/chilled ceiling	20 to 30	20 to 25
Full air conditioning, chilled beam	20 to 30	20 to 25

Approximate Estimating

RIBA STAGE C ELEMENTAL RATES

ELEMENTAL RATES FOR ALTERNATIVE ENGINEERING SERVICES SOLUTIONS *continued*

OFFICES ELECTRICAL SERVICES	Shell & Core £/m² GIA.	Fit Out £/m² NIA.
LV Installations		
Standby generators (life safety only)		
Buildings up to 3,000m².	20 to 25	-
Buildings over 3,000m² to 15,000m².	10 to 15	-
LV distribution		
Buildings up to 3,000m².	15 to 25	-
Buildings over 3,000m² to 15,000m².	20 to 30	-
Lighting Installations (including lighting controls and luminaries)		
Buildings up to 3,000m².	15 to 20	40 to 60
Buildings over 3,000m² to 15,000m².	15 to 20	40 to 60
Small Power		
Buildings up to 3,000m².	4 to 8	-
Buildings over 3,000m² to 15,000m².	4 to 6	-
Protective Installations		
Earthing		
Buildings up to 3,000m².	2 to 3	1 to 2
Buildings over 3,000m² to 15,000m².	2 to 3	1 to 2
Lightning Protection		
Buildings up to 3,000m².	3 to 4	-
Buildings over 3,000m² to 15,000m².	2 to 3	-
Communication Installations		
Fire Alarms (single stage)		
Buildings up to 3,000m².	6 to 10	8 to 10
Buildings over 3,000m² to 15,000m².	8 to 10	8 to 10
Fire Alarms (phased evacuation)		
Buildings over 3,000m² to 15,000m².	10 to 15	10 to 15
IT (Wireways only)		
Buildings up to 3,000m².	2 to 3	-
Buildings over 3,000m² to 15,000m².	2 to 3	-

RIBA STAGE C ELEMENTAL RATES

ELEMENTAL RATES FOR ALTERNATIVE ENGINEERING SERVICES SOLUTIONS *continued*

OFFICES ELECTRICAL SERVICES *continued*	Shell & Core £/m² G.I.A.	Fit Out £/m² N.I.A.
Security		
Buildings up to 3,000m²..	10 to 12	-
Buildings over 3,000m² to 15,000 m².....................................	8 to 10	-
Electrical Installations for Mechanical Plant		
Buildings up to 3,000m²..	5 to 8	-
Buildings over 3,000m² to 15,000 m².....................................	5 to 8	-

Approximate Estimating

RIBA STAGE C ELEMENTAL RATES

ELEMENTAL RATES FOR ALTERNATIVE ENGINEERING SERVICES SOLUTIONS *continued*

HOTELS
MECHANICAL SERVICES £/m² GIA.

Sanitaryware and above ground disposal installation

2 to 3 Star	20 to 30
4 to 5 Star	20 to 30

Water installation

2 to 3 Star	28 to 38
4 to 5 Star	40 to 50

LPHW Heating Installation; including gas installations

2 to 3 Star	30 to 45
4 to 5 Star	30 to 45

Air Conditioning; including ventilation

2 to 3 Star – 4 pipe Fan coil	200 to 240
4 to 5 Star – 4 pipe Fan coil	200 to 240
2 to 3 Star - 3 pipe variable refrigerant volume	130 to 150
4 to 5 Star - 3 pipe variable refrigerant volume	130 to 150

Fire Protection

2 to 3 Star - Dry risers	8 to 12
4 to 5 Star - Dry risers	8 to 12
2 to 3 Star - Sprinkler installation	5 to 10
4 to 5 Star - Sprinkler installation	5 to 10

BMS Controls: including MCC panels and control cabling

2 to 3 Star	10 to 12
4 to 5 Star	20 to 30

HOTELS
ELECTRICAL SERVICES

LV Installations

Standby generators (life safety only)

2 to 3 Star	10 to 20
4 to 5 Star	10 to 20

LV distribution

2 to 3 Star	25 to 35
4 to 5 Star	35 to 45

Approximate Estimating

RIBA STAGE C ELEMENTAL RATES

ELEMENTAL RATES FOR ALTERNATIVE ENGINEERING SERVICES SOLUTIONS *continued*

HOTELS
ELECTRICAL SERVICES *continued* £/m² GIA.

Lighting Installations
- 2 to 3 Star... 15 to 25
- 4 to 5 Star... 15 to 25

Small Power
- 2 to 3 Star... 5 to 10
- 4 to 5 Star... 10 to 15

Protective Installations

  Earthing

- 2 to 3 Star... 1 to 2
- 4 to 5 Star... 1 to 2

  Lightning Protection

- 2 to 3 Star... 1 to 2
- 4 to 5 Star... 1 to 2

Communication Installations

  Fire Alarms

- 2 to 3 Star... 10 to 15
- 4 to 5 Star... 10 to 15

  IT

- 2 to 3 Star... 10 to 20
- 4 to 5 Star... 10 to 20

  Security

- 2 to 3 Star... 15 to25
- 4 to 5 Star... 15 to25

Electrical Installations for Mechanical Plant

- 2 to 3 Star... 5 to 8
- 4 to 5 Star... 5 to 8

Approximate Estimating

RIBA STAGE C ELEMENTAL RATES

ELEMENTAL RATES FOR ALTERNATIVE ENGINEERING SERVICES SOLUTIONS *continued*

RESIDENTIAL MECHANICAL & ELECTRICAL SERVICES	Shell & Core £/m² GIA.	Fit Out £/m² NIA.
Sanitaryware and above ground disposal installation		
Affordable	0 to 5	25 to 35
Private	0 to 5	50 to 75
Disposal installation		
Affordable	10 to 15	5 to 15
Private	15 to 20	9 to 16
Water installation		
Affordable	20 to 35	30 to 65
Private	30 to 35	40 to 65
Heat Source		
Affordable	0 to 15	0 to 35
Private	10 to 15	0 to 35
Space Heating & Air Treatment		
Affordable	5 to 10	20 to 45
Private	40 to 60	150 to 250
Ventilation		
Affordable to façade	10 to 20	25 to 35
Private (whole house vent)	10 to 20	30 to 55
Electrical Installations		
Affordable	35 to 45	45 to 65
Private	45 to 60	80 to 110
Gas Installations		
Affordable	1 to 5	0 to 5
Private	5 to 10	0 to 20
Protective Installations		
Affordable	10 to 35	0 to 20
Private	10 to 40	0 to 20
Communication Installations		
Affordable	20 to 30	25 to 35
Private	30 to 45	35 to 60
Special Installations		
Affordable	5 to 10	5 to 10
Private	10 to 25	5 to 10

<u>Note</u>: *The range in cost differs due to the vast diversity in services strategies available. The lower end of the scale reflects all electric schemes (not always be possible due to Part L requirements) or local plant within apartment schemes, such as combi boilers, local ventilation etc. The high end of the scale is based on good quality apartments, which includes comfort cooling, sprinklers, home network installations, video entry, higher quality of sanitaryware and lighting.*

ALL-IN-RATES

ALL-IN-RATES FOR PRICING MECHANICAL APPROXIMATE QUANTITIES

	Cost per Point £
ABOVE GROUND DRAINAGE	
Soil and Waste	350 – 400
WATER INSTALLATIONS	
Cold Water	350 - 400
Hot Water	400 - 450

	Cost Per kW £
HEAT SOURCE	
Gas fired boilers including gas train and controls	30 - 35
Gas fired boilers including gas train, controls, flue, Plantroom pipework, valves and insulation, pumps and pressurisation unit	65 – 120

SPACE HEATING AND AIR TREATMENT

	Cost per kW £
CHILLED WATER	
Air cooled R134a refrigerant chiller including control panel, anti vibration mountings	120 - 140
Air cooled R134a refrigerant chiller including control panel, anti-vibration mountings, plantroom pipework, valves, insulation, pumps and pressurisation units	160 - 220
Water cooled R134a refrigerant chiller including control panel, anti vibration mountings	70 – 90
Water cooled R134a refrigerant chiller including control panel, anti-vibration mountings, plantroom pipework, valves, insulation, pumps and pressurisation units	110 - 190
Absorption steam medium chiller including control panel, anti-vibration mountings, plantroom pipework, valves, insulation, pumps and pressurisation units	170 - 270

	Cost per kW (heat rejection) £
HEAT REJECTION	
Open circuit, forced draft cooling tower	60 – 70
Closed circuit, forced draft cooling tower	75 - 85
Dry Air	60 – 70

Approximate Estimating

ALL-IN-RATES

ALL-IN-RATES FOR PRICING MECHANICAL APPROXIMATE QUANTITIES

SPACE HEATING AND AIR TREATMENT *continued*

PUMPS

Cost per kPa
£

Pumps including flexible connections, anti-vibration mountings	13 - 65
Pumps including flexible connections, anti-vibration mountings, plantroom pipework, valves, insulation and accessories	42 - 140

DUCTWORK

The rates below allow for ductwork and for all other labour and material in fabrication, fittings, supports and jointing to equipment, stop and capped ends, elbows, bends, diminishing and transition pieces, regular and reducing couplings, volume control dampers, branch diffuser and 'snap on' grille connections, ties, 'Ys', crossover spigots, etc., turning vanes, regulating dampers, access doors and openings, hand-holes, test holes and covers, blanking plates, flanges, stiffeners, tie rods and all supports and brackets fixed to structure.

Per m^2 of duct
£

Rectangular galvanised mild steel ductwork as HVCA DW 144 up to 1000mm longest side	50 – 56
Rectangular galvanised mild steel ductwork as HVCA DW 144 up to 2500mm longest side	56 – 67
Rectangular galvanised mild steel ductwork as HVCA DW 144 3000mm longest side and above	77 – 82
Circular galvanised mild steel ductwork as HVCA DW 144	56 – 78
Flat oval galvanised mild steel ductwork as HVCA DW 144 up to 545mm wide	56 – 62
Flat oval galvanised mild steel ductwork as HVCA DW 144 up to 880mm wide	64 – 76
Flat oval galvanised mild steel ductwork as HVCA DW 144 up to 1785mm wide	76 – 81

PACKAGED AIR HANDLING UNITS

Cost per m^3/s
£

Air handling unit including LPHW pre-heater coil, pre-filter panel, LPHW heater coils, chilled water coil, filter panels, inverter drive, motorised volume control dampers, sound attenuation, flexible connections to ductwork and all anti-vibration mountings.	6,500 – 8,500

EXTRACT FANS

Extract fan including inverter drive, sound attenuation, flexible connections to ductwork and all anti-vibration mountings	1,750 – 2,750

ALL-IN-RATES

ALL-IN-RATES FOR PRICING MECHANICAL APPROXIMATE QUANTITIES *continued*

PROTECTIVE INSTALLATIONS

SPRINKLER INSTALLATION £

Recommended maximum area coverage per sprinkler head:
  Extra light hazard, 21 m² of floor area
  Ordinary hazard, 12 m² of floor area
  Extra high hazard, 9 m² of floor area

Sprinkler equipment installation, pipework, valve sets, booster pumps and water storage..	50,000 - 75,000
Price per sprinkler head; including pipework, valves and supports.......................	180

PROTECTIVE INSTALLATIONS
HOSE REELS AND DRY RISERS

Wall mounted concealed hose reel with 36 metre hose including approximately 15 metres of pipework and isolating valve:

Price per hose reel..	1,750

100mm dry riser main including 2 way breeching valve and box,, 65mm landing valve, complete with padlock and leather strap and automatic air vent and drain valve.

Price per landing...	1,500

COMMUNICATIONS INSTALLATIONS

SECURITY

ACCESS CONTROL SYSTEMS

Door Mounted access control unit inclusive of door furniture, lock plus software. Including up to 50 meters of cable and termination. Including documentation testing and commissioning

Internal single leaf door...	1,100
Internal double door	1,250
External single leaf door	1,200
External Double leaf door..	1,350
Management control PC with printer software and commissioning up to 1000 users..	12,500

CCTV INSTALLATIONS

CCTV Equipment inclusive of 50 m of cable including testing and commissioning

Internal camera with Bracket	900
Internal camera with Housing...	950
Internal PTZ camera with Bracket...	1,750
External fixed camera with housing	1200
External PTZ camera dome..	2,750
External PTZ camera dome with power...	4,000

Approximate Estimating

ALL-IN-RATES

ALL-IN-RATES FOR PRICING MECHANICAL APPROXIMATE QUANTITIES

IT INSTALLATIONS

DATA CABLING

Cost per point
£

Complete channel link including, patch leads, cable, panels, testing and documentation (excludes cabinets and/or frames as well as containment).

Cat 5e up to 5,000 outlets	63.00
Cat 5e 5,000 to 15,000 outlets	57.75
Cat 6 up to 5,000 outlets	70.35
Cat 6 5,000 to 15,000 outlets	68.25

PIPEWORK

HOT AND COLD WATER *excludes insulation, valves and ancillaries etc*

Light gauge copper tube to EN1057 R250 (TX) formerly BS 2871 part 1 table X with joints as described including allowance for waste, fittings and supports assuming average runs with capillary joints up to 54mm and bronze welded thereafter

Cost per metre
£

Horizontal High Level Distribution

15mm	29.82
22mm	34.90
28mm	40.15
35mm	50.62
42mm	61.15
54mm	79.30
67mm	116.21
76mm	131.82
108mm	170.64

Risers

15mm	19.94
22mm	22.68
28mm	27.11
35mm	38.28
42mm	45.47
54mm	57.34
67mm	90.48
76mm	106.70
108mm	137.98

Toilet Areas etc at Low Level

15mm	67.92
22mm	73.50
28mm	93.41

ALL-IN-RATES

ALL-IN-RATES FOR PRICING MECHANICAL APPROXIMATE QUANTITIES *continued*

PIPEWORK *continued*

LTHW AND CHILLED WATER *excludes insulation, valves and ancillaries etc*

Black heavy weight mild steel tube to BS1387 with joints in the running length, allowance for waste, fittings and supports assuming average runs

	Cost per metre	
	LTHW £	Chilled Water £
Horizontal Distribution – Basements etc		
15mm	38.43	39.67
20mm	39.55	42.77
25mm	47.33	48.14
32mm	55.28	56.10
40mm	62.91	63.65
50mm	76.28	77.46
65mm	85.50	86.31
80mm	113.07	113.79
100mm	142.63	143.85
125mm	181.85	182.52
150mm	210.81	211.68
200mm	305.34	305.83
250mm	403.38	404.17
300mm	454.49	455.16
Risers		
15mm	23.53	24.38
20mm	25.97	26.65
25mm	30.26	30.96
32mm	35.82	36.52
40mm	39.92	40.45
50mm	49.51	50.33
65mm	66.63	67.16
80mm	86.40	86.87
100mm	108.45	109.25
125mm	148.91	149.42
150mm	174.16	174.94
200mm	228.61	228.97
250mm	306.64	307.23
300mm	324.59	325.09
On Floor Distribution		
15mm	36.68	37.83
20mm	40.75	41.58
25mm	45.89	46.74
32mm	53.64	54.50
40mm	61.07	61.85
50mm	74.58	75.92
65mm	-	81.01

Approximate Estimating

ALL-IN-RATES

ALL-IN-RATES FOR PRICING MECHANICAL APPROXIMATE QUANTITIES

PIPEWORK *continued*

LTHW AND CHILLED WATER *excludes insulation, valves and ancillaries etc (continued)*

Black heavy weight mild steel tube to BS1387 with joints in the running length, allowance for waste, fittings and supports assuming average runs

	Cost per metre	
	LTHW £	Chilled Water £
Plantroom Areas etc		
15mm	41.07	42.34
20mm	45.48	46.37
25mm	51.27	52.15
32mm	59.88	60.77
40mm	67.70	68.53
50mm	82.41	79.51
65mm	91.75	92.59
80mm	121.63	122.37
100mm	153.45	154.73
125mm	196.22	194.95
150mm	228.96	230.04
200mm	330.20	330.72
250mm	374.72	375.55
300mm	494.83	495.77

ALL-IN-RATES

ALL-IN-RATES FOR PRICING ELECTRICAL APPROXIMATE QUANTITIES

HV/LV INSTALLATIONS

The cost of HV/LV equipment will vary according to the electricity supplier's requirements, the duty required and the actual location of the site. For estimating purposes the items indicated below are typical of the equipment required in a HV substation incorporated into a building.

	Cost per Unit £
RING MAIN UNIT	
Ring Main Unit, 11kv including electrical terminations.................................	12,000 - 18,000

	Cost per KVA £
TRANSFORMERS	
Oil filled transformers, 11kv to 415v including electrical terminations	14 to 18
Cast Resin transformers, 11kv to 415v including electrical terminations......................	15 to 20

	Cost per Section £
HV SWITCHGEAR	
Cubicle section HV switchpanel, Form 4 type 6 including air circuit breakers, meters and electrical terminations..	20,000 – 25,000

	Cost per isolator £
LV SWITCHGEAR	
LV switchpanel, Form 3 including all isolators, fuses, meters and electrical terminations..	1,800 - 2,800
LV switchpanel, Form 4 type 5 including all isolators, fuses, meters and electrical terminations..	3,000 – 4,000

	£
EXTERNAL PACKAGED SUB-STATION	
Extra over cost for prefabricated packaged sub station housing excludes base and protective security fencing	20,000 – 25,000

	Cost per KVA £
STANDBY GENERATING SETS	
Diesel powered including control panel, flue, oil day tank and attenuation	
Approximate installed cost, LV	170 - 220
Approximate installed cost, HV	190 - 240

UNINTERRUPTIBLE POWER SUPPLY

Rotary UPS including control panel, automatic bypass, DC isolator and batteries for 30 minutes standby (excludes distribution)

Approximate installed cost (range 100KVA to 1000KVA)	250 - 350

Static UPS including control panel, automatic bypass, DC isolator and batteries for 30 minutes standby (excludes distribution)

Approximate installed cost (range 100KVA to 1000KVA)..	250 - 350

Approximate Estimating

ALL-IN-RATES

ALL-IN-RATES FOR PRICING ELECTRICAL APPROXIMATE QUANTITIES

SMALL POWER

Approximate prices for wiring of power points of length not exceeding 20m, including accessories, wireways but excluding distribution boards.

	Per Point £
13 amp Accessories	
Wired in PVC insulated twin and earth cable in ring main circuit	
Domestic properties..	55.00
Commercial properties..	75.00
Industrial properties..	75.00
Wired in PVC insulated twin and earth cable in radial circuit	
Domestic properties..	70.00
Commercial properties..	90.00
Industrial property..	90.00
Wired in LSF insulated single cable in ring main circuit	
Commercial properties..	90.00
Industrial property	90.00
Wired in LSF insulated single cable in radial circuit	
Commercial properties..	110.00
Industrial property..	110.00
45 amp wired in PVC insulated twin and earth cable	
Domestic properties..	110.00

Low voltage power circuits

Three phase four wire radial circuit feeding an individual load, wired in LSF insulated single cable including all wireways, isolator, *not exceeding 10 metres; in commercial properties.*

Cable size mm²	
1.5	180.00
2.5	195.00
4	210.00
6	230.00
10	265.00
16	290.00

Three phase four core radial circuit feeding an individual load item, wired in LSF/SWA/XLPE insulated cable including terminations, isolator; clipped to surface, *not exceeding 10 metres in commercial properties.*

Cable size mm²	
1.5	133.00
2.5	148.00
4	168.00
6	180.00
10	276.00
16	357.00

ALL-IN-RATES

ALL-IN-RATES FOR PRICING ELECTRICAL APPROXIMATE QUANTITIES *continued*

LIGHTING

Approximate prices for wiring of lighting points including rose, wireways but excluding distribution boards, luminaires and switches.

Per Point
£

Final Circuits
Wired in PVC insulated twin and earth cable
Domestic properties 40.00
Commercial properties 50.00
Industrial properties 50.00
Wired in LSF insulated single cable
Commercial properties 65.00
Industrial property 65.00

ELECTRICAL WORKS IN CONNECTION WITH MECHANICAL SERVICES

The cost of electrical connections to mechanical services equipment will vary depending on the type of building and complexity of the equipment. Therefore a rate of £ 5.00 per m² of gross floor area should be a useful guide to allow for power wiring, isolators and associated wireways.

FIRE ALARMS

Cost per point for two core MICC insulated wired system including all terminations, supports and wireways.

Call point 254.00
Smoke detector 222.00
Smoke/heat detector 254.00
Heat detector 245.00
Heat detector and sounder 215.00
Input/output/relay units 311.00
Alarm sounder 241.00
Alarm sounder/beacon 281.00
Speakers/voice sounders 282.00
Speakers/voice sounders (weatherproof) 297.00
Beacon/strobe 221.00
Beacon/strobe (weatherproof) 327.00
Door release units 324.00
Beam detector 869.00

Approximate Estimating

ALL-IN-RATES

ALL-IN-RATES FOR PRICING ELECTRICAL APPROXIMATE QUANTITIES

FIRE ALARMS *continued*

Cost per point for wireless system

	Per Point £
Call point	194.00
Smoke detector	193.00
Smoke/heat detector	232.00
Heat detector	202.00
Heat detector and sounder	406.00
Input/output/relay units	372.00
Alarm sounder	368.00
Alarm sounder/beacon	421.00
Speakers/voice sounders	382.00
Speakers/voice sounders (weatherproof)	461.00
Beacon/strobe	393.00
Beacon/strobe (weatherproof)	390.00
Door release units	362.00
Beam detector	1,028.00

For costs for zone control panel, battery chargers and batteries, see 'Prices for Measured Work' section.

EXTERNAL LIGHTING

Estate road lighting
Post type road lighting lantern 70 watt CDM-T 3000k complete with 5m high column with hinged lockable door, control gear and cut-out including 2.5 mm two core butyl cable internal wiring, interconnections and earthing fed by 16 mm² four core XLPE/SWA /LSF cable and terminations. Approximate installed price *per metre road length* (based on 300 metres run) including time switch but excluding builder's work in connection

    Columns erected at 30 m intervals .... £ 45.00 per m of road

Bollard lighting
Bollard lighting fitting 26 watt TC-D 3500k including control gear, all internal wiring, interconnections, earthing and 25 metres of 2.5 mm² three core XLPE/SWA/LSF cable

    Approximate installed price excluding builder's work in connection .... £ 938.00 *each*

Outdoor flood lighting
Wall mounted outdoor flood light fitting complete with tungsten halogen lamp, mounting bracket, wire guard and all internal wiring; fixed to brickwork or concrete and connected.

    Installed price 500 watt .... £88.00 - £154.00
    Installed price 1000 watt.. £122.00 - £176.00

Pedestal mounted outdoor floor light fitting complete with1000 watt MBF/U lamp, mounting bracket, control gear, contained in weatherproof steel box, all internal wiring, interconnections and earthing; fixed to brickwork or concrete and connected

    Approximate installed price excluding builder's work in connection .... £1,019.00 *each*

ALL-IN-RATES

ALL-IN-RATES FOR PRICING SPECIALIST APPROXIMATE QUANTITIES

LIFT INSTALLATIONS

The cost of lift installations will vary depending upon a variety of circumstances. The following prices assume a car height of 2.2 metres, manufacturers standard car finish, brushed stainless steel 2 panel centre opening doors to BSEN81 part 1 & 2 and Lift Regulations 1997.

Passenger Lifts – Machine Room Above	8 Person £	10 Person £	13 Person £	17 Person £	21 Person £	26 Person £
Electrically operated AC drive serving 2 levels with directional collective controls and a speed of 1.0 m/s	58,321	63,084	66,306	74,824	84,843	97,345
As above serving 4 levels and a speed of 1.0m/s	67,893	72,570	76,276	85,759	96,890	110,804
As above serving 6 levels and a speed of 1.0m/s	77,258	82,031	86,038	96,487	108,694	123,955
As above serving 8 levels and a speed of 1.0m/s	86,622	91,494	95,800	107,214	120,498	137,104
As above serving 10 levels and a speed of 1.0m/s	95,986	100,954	105,558	117,942	132,301	150,254
As above serving 12 levels and a speed of 1.0m/s	105,350	110,417	115,321	128,669	144,105	163,403
As above serving 14 levels and a speed of 1.0m/s	116,580	95,687	126,948	141,261	156,947	176,552
Add to above for:						
Increase speed from 1.0 to 1.6 m/s	4,141	3,925	4,194	4,183	4,183	4,183
Increase speed from 1.6m/s to 2.0m/s	878	878	1,167	1,167	1,474	1,474
Increase speed from 2.0m/s to 2.5m/s	2,040	2,040	2,492	2,492	2,934	2,934
Enhanced finish to car – Centre mirror, flat ceiling, carpet	2,816	3,076	2,992	3,512	4,083	4,797

ALL-IN-RATES

ALL-IN-RATES FOR PRICING SPECIALIST APPROXIMATE QUANTITIES

LIFT INSTALLATIONS *continued*

Passenger Lifts Machine Room Above (cont'd)	8 Person £	10 Person £	13 Person £	17 Person £	21 Person £	26 Person £
Bottom motor room	7,308	7,308	7,308	8,780	8,780	9,063
Fire fighting control	5,834	5,834	5,834	5,834	5,834	5,834
Glass back	2,492	2,889	3,445	4,192	4,192	4,192
Glass doors	20,053	20,053	22,004	22,658	22,658	22,658
Painting to entire pit	2,161	2,161	2,161	2,161	2,161	2,161
Dual seal shaft	4,233	4,233	4,233	5,085	5,085	5,085
Dust sealing machine room	816	816	1,360	1,360	1,360	1,360
Intercom to reception desk and security room	383	383	383	383	383	383
Heating, cooling and ventilation to machine room	849	849	849	849	849	849
Shaft lighting / small power	4,055	4,055	4,055	4,055	4,055	4,055
Motor room lighting / small power	1,360	1,360	1,529	1,677	1,677	1,677
Lifting beams	1,387	1,387	1,387	1,387	1,387	1,387
10mm Equipotential bonding of all entrance metalwork	918	918	918	918	918	918
Shaft secondary steelwork	6,147	6,288	6,571	6,719	6,719	6,719
Independent insurance inspection	1,980	1,980	1,980	1,980	1,980	1,980
12 Month warranty service	1,916	1,916	1,916	1,916	1,916	1,916

ALL-IN-RATES

ALL-IN-RATES FOR PRICING SPECIALIST APPROXIMATE QUANTITIES *continued*

LIFT INSTALLATIONS *continued*

Passenger Lifts – Machine room-less	8 Person £	10 Person £	13 Person £	17 Person £	21 Person £	26 Person £
Electrically operated AC drive serving 2 levels with directional collective controls and a speed of 1.0 m/s	52,687	57,779	61,619	74,696	81,650	90,151
As above serving 4 levels and a speed of 1.0m/s	61,609	66,203	70,901	84,676	91,974	101,563
As above serving 6 levels and a speed of 1.0m/s	70,564	75,199	80,062	94,531	102,214	112,768
As above serving 8 levels and a speed of 1.0m/s	79,475	84,233	89,260	104,427	112,349	123,972
As above serving 10 levels and a speed of 1.0m/s	88,437	93,314	98,505	114,370	122,623	135,180
As above serving 12 levels and a speed of 1.0m/s	97,454	102,451	107,808	124,246	132,974	144,117
As above serving 14 levels and a speed of 1.0m/s	109,161	111,452	118,693	135,694	145,785	160,000
Add to above for:						
Increase speed from 1.0 to 1.6 m/s	3,023	2,718	2,928	4,385	5,011	6,435
Enhanced finish to car – Centre mirror, flat ceiling, carpet	2,818	2,776	2,992	3,323	4,398	5,029
Fire fighting control	7,279	7,279	7,279	-	-	-
Painting to entire pit	806	806	806	806	806	806
Dual seal shaft	747	756	784	947	1,004	1,004
Shaft lighting / small power	4,055	4,055	4,055	-	-	-

ALL-IN-RATES

ALL-IN-RATES FOR PRICING SPECIALIST APPROXIMATE QUANTITIES

LIFT INSTALLATIONS *continued*

Passenger Lifts Machine Room-less (cont'd)	8 Person £	10 Person £	13 Person £	17 Person £	21 Person £	26 Person £
Add to above for:						
Intercom to reception desk and security room	382	382	382	382	382	382
Heating, cooling and ventilation to machine room	849	849	849	849	849	849
Lifting beams	906	906	906	1,265	1,437	1,331
10mm Equipotential bonding of all entrance metalwork	435	435	435	435	435	435
Shaft secondary steelwork	5,834	5,834	5,834	-	-	-
Independent insurance inspection	1,980	1,980	1,980	1,980	1,980	1,980
12 Month warranty service	840	870	881	907	933	933

Goods Lifts Machine Room Above	2000 kg £	2250 kg £	2500 kg £	3000 kg £
Electrically operated two speed serving 2 levels to take 1000 kg load, prime coated internal finish and a speed of 1.0 m/s	97,345	107,560	108,906	117,969
As above serving 4 levels and a speed of 1.0m/s	110,804	121,451	122,796	138,597
A As above serving 6 levels and a speed of 1.0m/s	123,954	135,058	136,377	152,458
As above serving 8 levels and a speed of 1.0m/s	137,104	148,613	149,958	169,702
As above serving 10 levels and a speed of 1.0m/s	150,254	162,197	163,539	186,946
As above serving 12 levels and a speed of 1.0m/s	163,406	175,773	177,119	204,189
As above serving 14 levels and a speed of 1.0m/s	176,552	189,355	190,699	221,434

ALL-IN-RATES

ALL-IN-RATES FOR PRICING SPECIALIST APPROXIMATE QUANTITIES *continued*

LIFT INSTALLATIONS *continued*

Goods Lifts Machine Room Above (cont'd)	2000 kg £	2250 kg £	2500 kg £	3000 kg £
Add to above for:				
Increased speed of travel from 1.0 to 1.6 metres per second	1,360	-	-	-
Enhanced finish to car – Centre mirror, flat ceiling, carpet	3,670	3,670	3,670	-
Bottom motor room	8,780	-	-	-
Painting to entire pit	660	1,397	1,397	1,397
Dual seal shaft	5,936	5,936	5,936	-
Intercom to reception desk and security room	383	383	383	383
Heating, cooling and ventilation to machine room	849	849	849	849
Lifting beams	1,265	1,265	1,265	1,265
10mm Equipotential bonding of all entrance metalwork	755	918	918	918
Independent insurance inspection	2,518	2,518	2,518	2,518
12 Month warranty service	690	1,264	1,264	-

Goods Lifts Machine Room-less	2000 kg £	2250 kg £	2500 kg £
Electrically operated two speed serving 2 levels to take 1000 kg load, prime coated internal finish and a speed of 1.0 m/s	87,980	94,101	100,229
As above serving 4 levels and a speed of 1.0m/s	99,392	105,723	112,057
As above serving 6 levels and a speed of 1.0m/s	110,597	110,597	123,883
As above serving 8 levels and a speed of 1.0m/s	121,802	128,767	135,711
As above serving 10 levels and a speed of 1.0m/s	133,009	140,275	147,542
As above serving 12 levels and a speed of 1.0m/s	144,211	151,790	159,367

Approximate Estimating

ALL-IN-RATES

ALL-IN-RATES FOR PRICING SPECIALIST APPROXIMATE QUANTITIES

LIFT INSTALLATIONS *continued*

Goods Lifts Machine Room-less (cont'd)	2000 kg £	2250 kg £	2500 kg £
As above serving 14 levels and a speed of 1.0m/s	157,830	164,513	171,195

Add to above for:

Increased speed of travel from 1.0 to 1.6 metres per second	5,018	8,498	-

Add to above for:

Enhanced finish to car – Centre mirror, flat ceiling, carpet	4,139	5,665	8,499

Add to above for:

Painting to entire pit	806	-	951
Dual seal shaft	742	-	-
Intercom to reception desk and security room	383	-	-
Heating, cooling and ventilation to machine room	425	849	849
Lifting beams	1,222	1,132	1,132
10mm Equipotential bonding of all entrance metalwork	435	-	-
Independent insurance inspection	1,980	1,980	1,980
12 Month warranty service	690	690	690

ESCALATOR INSTALLATIONS

	Each £
30Ø Pitch escalator with a rise of 3 to 6 metres with standard balustrades	
1000mm step width	97,997

Add to above for:

Balustrade Lighting	9,063
Skirting Lighting	9,800
Emergency stop button pedestals	3,070
Truss cladding - Stainless steel	27,984
Truss cladding - Spray painted steel	22,375

ELEMENTAL COSTS

AIRPORT TERMINAL BUILDING

New build airport terminal building, premium quality, located in the South East, handling both domestic and international flights with a gross internal floor area (GIA) of 25,000m². These costs exclude baggage handling, check-in systems, pre-check in and boarding security systems, vertical transportation and services to aircraft stands, with the heat source via district mains(excluded) and executed under landside access/logistics environment.

Cost Summary

El. Ref.	Element	Total Cost £	Cost/m² £
5A	Sanitaryware	75,000.00	3.00
5C	Disposal Installations		
	Rainwater	150,000.00	6.00
	Soil and waste	200,000.00	8.00
	Condensate	37,500.00	1.50
5D	Water Installations		
	Hot and cold water services	450,000.00	18.00
5F	Space Heating and Air Treatment		
	LTHW Heating system	2,000,000.00	80.00
	Chilled water system	1,875,000.00	75.00
	Supply and extract air conditioning system	5,500,000.00	220.00
	Allowance for services to communications rooms	250,000.00	10.00
5G	Ventilating Services		
	Mechanical ventilation to baggage handling and plantrooms	625,000.00	25.00
	Toilet extract ventilation	225,000.00	9.00
	Smoke extract installation	250,000.00	10.50
	Kitchen extract system	75,000.00	3.00
5H	Electrical Installation		
	HV/LV Switchgear	1,625,000.00	65.00
	Standby generator	1,000,000.00	40.00
	Mains and sub mains installation	1,125,000.00	45.00
	Small power installation	375,000.00	15.00
	Lighting and luminaires	2,450,000.00	98.00
	Emergency lighting installation	300,000.00	12.00
	Power to mechanical services	175,000.00	7.00
5I	Gas Installation	50,000.00	2.00
5K	Protective Installations		
	Lightning protection	75,000.00	3.00
	Earthing and bonding	100,000.00	4.00
	Sprinkler installation	875,000.00	35.00
	Dry riser and hosereel installations	175,000.00	7.00
	Fire suppression installation to communications room	75,000.00	3.00
	Carried forward	20,112,500.00	804.50

Approximate Estimating

ELEMENTAL COSTS

AIRPORT TERMINAL BUILDING

El. Ref.	Element	Total Cost £	Cost/m² £
	Brought forward	20,112,500.00	804.50
5L	Communications Installations		
	Fire and smoke detection and alarm system	750,000.00	30.00
	Voice/public address system	500,000.00	20.00
	Intruder detection	250,000.00	10.00
	Security, CCTV and access control	875,000.00	35.00
	Wireways for telephones, data and structured cable	375,000.00	15.00
	Structured cable installation	875,000.00	35.00
	Flight information display system	625,000.00	25.00
5M	Special Installations		
	BMS Installation	1,375,000.00	55.00
	Summary total.	23,762,500.00	1029.50

ELEMENTAL COSTS

SHOPPING MALL (TENANT'S FIT OUT EXCLUDED)

Natural ventilation shopping mall with approximately 33,000m² two storey retail area and a 13,000m² above ground covered car park, situated in a town centre in South East England

Cost Summary

El. Ref.	Element	Total Cost £	Cost/m² £
	RETAIL BUILDING 33,000m²		
5A	Sanitary appliances	26,000.00	0.79
5C	Disposal Installations		
	Rainwater	178,500.00	5.41
	Soil, waste and vent	210,000.00	6.36
5D	Water Installations		
	Cold water installation	200,000.00	6.06
	Hot water installation	189,000.00	5.73
5E	Heat Source	Included	
5F	Space Heating and Air Treatment		
	Condenser water system	1,050,000.00	31.82
	LTHW installation	140,000.00	4.24
	Air conditioning system	1,060,000.00	32.12
	Over-door heaters at entrances	31,500.00	0.95
5G	Ventilation Services		
	Public toilet ventilation	21,000.00	0.64
	Plant room ventilation	141,000.00	4.27
	Supply and extract systems to shop units	451,000.00	13.67
	Toilet extract systems to shop units	84,000.00	2.55
	Smoke ventilation system to Mall Area	340,000.00	10.30
	Service corridor ventilation	78,000.00	2.36
	Miscellaneous ventilation	577,000.00	17.48
5H	Electrical Installation		
	LV distribution	771,000.00	23.36
	Standby power	126,000.00	3.82
	General lighting	2,037,000.00	61.73
	External lighting	173,000.00	5.24
	Emergency lighting	388,000.00	11.76
	Small power	315,000.00	9.55
	Mechanical services power supplies	89,000.00	2.70
	General earthing	42,000.00	1.27
	UPS for security and CCTV	23,000.00	0.70
5I	Gas Installation		
	Gas supplies to boilers	18,000.00	0.55
	Gas supplies to Anchor (major) stores	11,500.00	0.35
5K	Protective Installations		
	Lightning protection	42,000.00	1.27
	Sprinkler installation	451,000.00	13.67
	Dry Risers	Excluded	
	Hosereel installation	Excluded	
	Carried forward	9,263,500.00	280.72

Approximate Estimating

ELEMENTAL COSTS

SHOPPING MALL (TENANT'S FIT OUT EXCLUDED)

El. Ref.	Element	Total Cost £	Cost/m² £
	Brought forward	9,263,500.00	280.72
5L	Communications Installations		
	Fire alarm installation..	262,000.00	7.94
	Public address/ voice alarm.....................................	173,000.00	5.24
	Security installation	378,000.00	11.45
	General containment	194,000.00	5.88
5M	Special Installations		
	BMS/Controls	577,000.00	17.48
	Summary total..	10,847,500.00	328.71
	CAR PARK - 13,000m²		
5C	Disposal Installations		
	Car park drainage..	78,000.00	6.00
5G	Ventilation Services		
	Car park ventilation (ducted system).......................	997,000.00	76.69
5H	Electrical Installation		
	LV distribution	152,000.00	11.69
	Standby power	Included	
	General lighting	378,000.00	29.08
	External lighting	Excluded	
	Emergency lighting	78,000.00	6.00
	Small power	152,000.00	11.69
	Mechanical services power supplies........................	52,000.00	4.00
	General earthing..	16,000.00	1.23
	Ramp frost protection	21,000.00	1.62
5K	Protective Installations		
	Sprinkler installation...	299,000.00	23.00
	Dry Riser and Hosereel Installation..........................	Excluded	
5L	Communications Installations		
	Fire alarm installation...	299,000.00	23.00
	Security installation	126,000.00	9.69
5M	Special Installations		
	BMS/Controls	79,000.00	6.08
	Entry/exit barriers, pay stations	73,500.00	5.65
	Summary total..	2,800,500.00	215.42

ELEMENTAL COSTS

OFFICE BUILDING

Speculative 14 storey office in Central London for single tenant occupancy with a gross floor area of 27,490m². A four pipe fan coil system, with roof mounted air cooled chillers, gas fired boilers.

Cost Summary

El. Ref.	Element	Total Cost £	Cost/m² £
	SHELL AND CORE 27,490m² GIA		
5A	Sanitaryware	178,960.00	6.51
5C	Disposal Installations		
	Rainwater/Soil and Waste	329,605.00	11.99
	Condensate	28,590.00	1.04
5D	Water Installations		
	Hot and cold water services	286,446.00	10.42
5E	Heat Source	Included in 5F	
5F	Space Heating and Air Treatment		
	LTHW Heating	328,231.00	11.94
	Chilled water	677,629.00	24.65
	Ductwork	1,680,189.00	61.12
5G	Ventilating Services		
	Toilet extract ventilation	77,247.00	2.81
	Basement extract	264,729.00	9.63
	Miscellaneous ventilation systems	195,179.00	7.10
5H	Electrical Installation		
	Generator	90,442.10	3.29
	HV/LV supply/distribution	1,447,623.00	52.66
	General lighting	711,716.10	25.89
	General power	97,589.50	3.55
	Electrical services for mechanical equipment	129,477.90	4.71
5I	Gas Installation	35,737.00	1.30
5K	Protection		
	Dry risers	53,606.00	1.95
	Sprinklers	481,075.00	17.50
	Earthing and bonding	54,980.00	2.00
	Lightning protection	54,980.00	2.00
5L	Communication Installation		
	Fire/Voice alarms	343,625.00	12.50
	Voice and data (wireways)	38,486.00	1.40
	Security (wireways)	32,988.00	1.20
	Disabled/refuge alarms	42,060.00	1.53
	Carried forward	7,661,191.00	278.69

Approximate Estimating

ELEMENTAL COSTS

OFFICE BUILDING

El. Ref.	Element	Total Cost £	Cost/m² £
	Brought forward..	7,661,191.00	278.69
5M	Special Installation Building management systems..	465,681.00	16.94
	Summary total (based on Gross Internal Area - GIA)..................	8,126,872.00	295.63

El. Ref.	Element	Total Cost £	Cost/m² £
	CATEGORY 'A' FIT OUT - 19,186m² NIA		
5C	Disposal Installations Condensate...	112,238.00	5.85

5F	Space Heating and Air Treatment
	    LTHW Heating..
	    Chilled water..
	    Ductwork...
		461,423.00
		620,667.00
		1,555,025.00
			24.15
			32.35
			81.05

5H	Electrical Installation
	    Lighting installation...
	    Electrical services in connection......................................
	    Tenant distribution board..
		1,432,427.00
		76,744.00
		86,529.00
			74.66
			4.00
			4.51

5K	Protection Sprinkler installation...	499,795.00	26.05
5L	Communication Installation Fire/Voice alarms...	265,918.00	13.86
5M	Special Installations Building management system...	374,894.00	19.54
	Summary total (based on Nett Internal Area – NIA)..................	5,485,660.00	285.92

ELEMENTAL COSTS

BUSINESS PARK

New build office in South East within the M25 part of a speculative business park consisting of two 3 storey existing buildings and 1 new build, fitted out to Category A specification. Four pipe system, external chiller, BMS controlled with all three buildings linked with an area of 7,500m² gross internal area and nett internal area of 6,000m².

Cost Summary

El. Ref.	Element	Total Cost £	Cost/m² £
	SHELL AND CORE – 7.500m² GIA		
5A	Sanitaryware	41,475.00	5.53
5C	Disposal Installations		
	Rainwater	23,100.00	3.08
	Soil and waste	63,840.00	8.51
5D	Water Installations		
	Cold water services	42,945.00	5.73
	Hot water services	13,090.00	1.75
5E	Heat Source	Included in 5F	
5F	Space Heating and Air Treatment		
	LTHW Heating; plantroom and risers	275,550.00	36.74
	Chilled water; plantroom and risers	296,560.00	39.54
5G	Ventilating Services		
	Toilet and miscellaneous ventilation	39,600.00	5.28
5H	Electrical Installation		
	LV supply/distribution	152,040.00	20.27
	General lighting	151,200.00	20.16
	General power	34,650.00	4.62
5I	Gas Installation	11,025.00	1.47
5K	Protective Installation		
	Earthing and bonding	13,200.00	1.76
	Lightning protection	15,000.00	2.00
5L	Communication Installation		
	Fire alarms	63,750.00	8.50
	Security (wireways)	12,600.00	1.68
	Data and voice (wireways)	12,600.00	1.68
5M	Special Installation		
	Building management systems	150,442.00	20.06
	Electrical services in connection	15,000.00	2.00
	Summary total (based on Gross Internal Area – GIA)	1,427,667.00	190.36

ELEMENTAL COSTS

BUSINESS PARK

El. Ref.	Element	Total Cost £	Cost/m² £
	CATEGORY 'A' FIT OUT – 6,000m² NIA		
5C	Disposal Installation		
	Condensate..	42,000.00	7.00
5F	Space Heating and Air Treatment		
	LTHW Heating..	124,080.00	20.68
	Chilled water..	151,690.00	25.28
	Supply and extract ductwork......................................	464,970.00	77.50
5H	Electrical Installation		
	Distribution boards...	18,000.00	3.00
	General lighting, recessed including lighting controls	277,200.00	46.20
5K	Protective Installation		
	Earthing and bonding..	6,000.00	1.00
5L	Communication Installation		
	Fire alarms...	30,000.00	5.00
5M	Special Installations		
	Building management systems....................................	72,000.00	12.00
	Electrical services in connection.................................	21,000.00	3.50
	Summary total (Based on Nett Internal Area - NIA).................	1,206,940.00	201.16

ELEMENTAL COSTS

PERFORMING ARTS CENTRE (LOW SPECIFICATION)

Performing Arts centre with a Gross Internal Area (GIA) of 6,000m², on a low specification for the theatre systems and with natural ventilation.

The development comprises dance studios and a theatre auditorium. The theatre would require all the necessary stage lighting, machinery and equipment installed in a modern professional theatre (these are excluded from the model, as assumed to be FF&E, but the containment and power wiring is included). Also not included are the staff call system, audio and video recording, EPOS ticket system, production recording and relay to TV screens and enhanced lighting including dimming – for such, refer to the High Specification Model.

Cost Summary

El. Ref.	Element	Total Cost £	Cost/m² £
5A	Sanitaryware	78,000.00	13.00
5C	Disposal Installations		
	Soil, Waste and Rainwater	102,000.00	17.00
5D	Water Installations		
	Cold water services	60,000.00	10.00
	Hot water services	60,000.00	10.00
5E	Heat Source	144,000.00	24.00
5F	Space Heating and Air Treatment		
	Heating with limited cooling	420,000.00	70.00
	DX Cooling to Comms and Amps rooms	36,000.00	6.00
5G	Ventilating Services		
	Ventilation and extract systems	726,000.00	121.00
5H	Electrical Installation		
	LV supply/distribution	330,000.00	55.00
	General lighting	522,000.00	87.00
	Small power	180,000.00	30.00
5I	Gas Installation	18,000.00	3.00
5K	Protection		
	Lighting protection	18,000.00	3.00
5L	Communication Installation		
	Fire alarms and detection	210,000.00	35.00
	Voice and Data (containment only)	42,000.00	7.00
	Security, Access, Control and Disabled alarms	186,000.00	31.00
5M	Special Installation		
	Building management systems	222,000.00	37.00
	Theatre systems includes for containment and power wiring	294,000.00	49.00
	Summary total	3,648,300.00	608.00

Approximate Estimating

ELEMENTAL COSTS

PERFORMING ARTS CENTRE (HIGH SPECIFICATION)

Performing Arts centres with a Gross Internal Area (GIA) of 6,000m², upon which this cost analysis has been based, on a high specification for the theatre systems and with cooling to the Auditorium.

The development comprises of dance studios and a theatre auditorium. The theatre would require all the necessary stage lighting, machinery and equipment installed in a modern professional theatre (these are excluded from the model, as assumed to be FF&E, but the containment and power wiring is included). Included are the staff call system/paging, audio and video recording, EPOS ticket system, production recording and relay to TV screens and enhanced lighting including dimming (not stage)

Cost Summary

El. Ref.	Element	Total Cost £	Cost/m² £
5A	Sanitaryware	84,000.00	14.00
5C	Disposal Installations		
	Soil, Waste and Rainwater	108,000.00	18.00
5D	Water Installations		
	Cold water services	60,000.00	10.00
	Hot water services	54,000.00	9.00
5E	Heat Source	144,000.00	24.00
5F	Space Heating and Air Treatment		
	Heating and ventilation	480,000.00	80.00
	Cooling to Auditorium with DX to Comms and Amps rooms	420,000.00	70.00
5G	Ventilating Services		
	Ventilation and extract systems to toilets, kitchen and workshop	840,000.00	140.00
5H	Electrical Installation		
	LV supply/distribution	360,000.00	60.00
	General lighting	660,000.00	110.00
	Small power	210,000.00	35.00
5I	Gas Installation	18,000.00	3.00
5K	Protection		
	Lighting protection	18,000.00	3.00
5L	Communication Installation		
	Fire alarms and detection	240,000.00	40.00
	Voice and Data complete installation (excluding active equipment)	120,000.00	20.00
	Security, Access, Control, Disabled alarms, Staff paging	210,000.00	35.00
5M	Special Installation		
	Building management systems	240,000.00	40.00
	Theatre systems includes for containment and power wiring	390,000.00	65.00
	Summary total	4,656,000.00	776.00

ELEMENTAL COSTS

SPORTS HALL

Single storey sports hall, located in the South East, with a gross internal area of 1,200m² (40m x 30m).

Cost Summary

El. Ref.	Element	Total Cost £	Cost/m² £
5A	Sanitaryware...	13,200.00	11.00
5C	Disposal Installations		
	Rainwater	4,800.00	4.00
	Soil and waste	8,400.00	7.00
5D	Water Installations		
	Hot and cold water services	20,400.00	17.00
5E	Heat Source		
	Boiler, flues, pumps and controls........................	15,600.00	13.00
5F	Space Heating and Air Treatment		
	Warm air heating to sports hall area...................	18,000.00	15.00
	Radiator heating to ancillary areas......................	28,800.00	24.00
5G	Ventilating Services		
	Ventilation to changing, fitness and sports hall areas.............	19,200.00	16.00
5H	Electrical Installations		
	Main switchgear and sub-mains...........................	15,600.00	13.00
	Small power...	13,200.00	11.00
	Lighting and luminaries to sports hall areas........	22,800.00	19.00
	Lighting and luminaries to ancillary areas...........	26,400.00	22.00
5I	Gas Installation...	Included in 5E	
5K	Protective Installations		
	Lightning protection...	4,800.00	4.00
5L	Communications Installations		
	Fire, smoke detection and alarm system, intruder detection....	14,400.00	12.00
	CCTV Installation...	18,000.00	15.00
	Public address and music systems......................	8,400.00	7.00
	Wireways for telephone and data........................	3,600.00	3.00
	Summary total..	256,80.00	214.00

Approximate Estimating

ELEMENTAL COSTS

HOTELS

200 Bedroom, four star hotel, situated in Central London, with a gross internal floor area of 16,500m².

The development comprises a ten storey building with large suites on each guest floor, together with banqueting, meeting rooms and leisure facilities.

Cost Summary

El. Ref.	Element	Total Cost £	Cost/m² £
5A	Sanitaryware	577,500.00	35.00
5C	Disposal Installations Rainwater, soil and waste	495,000.00	30.00
5D	Water Installations Hot and cold water services	742,500.00	45.00
5E	Heat Source Condensing boiler and pumps etc	198,000.00	12.00
5F	Space Heating and Air Treatment Air conditioning system; chillers, pumps, air handling units, ductwork, fan coil units etc; to guest rooms, public areas, meeting and banquet rooms	2,311,000.00	140.00
5G	Ventilating Services General toilet extract and ventilation to kitchens and bathrooms etc	742,500.00	45.00
5H	Electrical Installation HV/LV Installation, standby power, lighting, emergency lighting and small power to guest floors and public areas including earthing and lightning protection	3,053,000.00	185.00
5I	Gas Installation	41,250.00	2.50
5K	Protective Installations Dry risers and sprinkler installation	660,000.00	40.00
5L	Communications Installations Fire, smoke detection and alarm system/security CCTV Background music, AV wireways Telecommunications, data and T.V. wiring (no hotel management and head end equipment)	759,000.00 247,500.00 123,750.00	46.00 15.00 7.50
5M	Special Installations Building Management System	429,000.00	26.00
	Summary total	9,471,000.00	574.00

ELEMENTAL COSTS

STADIUM – NEW

A three storey stadium, located in Greater London with gross internal area of 85,000m² and incorporating 60,000 spectator seats

Cost Summary

El. Ref.	Element	Total Cost £	Cost/m² £
5A	Sanitaryware	935,000.00	11.00
5C	Disposal Installations		
	Rainwater	340,000.00	4.00
	Above ground drainage	1,020,000.00	12.00
5D	Water Installations		
	Hot and cold water	1,955,000.00	23.00
5E	Heat Source	850,000.00	10.00
5F	Space Heating and Air Treatment		
	Heating	595,000.00	7.00
	Cooling	1,360,000.00	16.00
5G	Ventilating Services		
	Ventilation	4,675,000.00	55.00
5H	Electrical Installation		
	HV/LV Supply	850,000.00	10.00
	LV Distribution	2,210,000.00	26.00
	General lighting	5,440,000.00	64.00
	Small power	1,360,000.00	16.00
	Earthing and bonding	105,000.00	1.00
	Power supply to mechanical equipment	115,000.00	1.00
	Pitch lighting	700,000.00	8.00
5I	Gas Installation	85,000.00	1.00
5K	Protective Installations		
	Lightning protection	115,000.00	1.00
	Hydrants	215,000.00	2.00
5L	Communications Installations		
	Wireways for data, TV, telecom and PA	680,000.00	8.00
	Public address	1,360,000.00	16.00
	Security	1,020,000.00	12.00
	Data voice installations	2,720,000.00	32.00
	Fire alarms	765,000.00	9.00
	Disabled/refuse alarm/call systems	255,000.00	3.00
5M	Special Installations		
	BMS/Controls	1,275,000.00	15.00
	Summary total	31,110,000.00	366.00

Cost per seat	£518.50	

Approximate Estimating

ELEMENTAL COSTS

PRIVATE HOSPITAL

New build project building. The works consist of a new 80 bed hospital of approximately 15,000m², eight storey with a plant room.

All heat is provided from existing steam boiler plant, medical gases are also served from existing plant. The project includes the provision of additional standby electrical generation to serve the wider site requirements.

This hospital has six operating theatres, ITU/HDU department, pathology facilities, diagnostic imaging, out patient facilities and physiotherapy.

Cost Summary

El. Ref.	Element	Total Cost £	Cost/m² £
5A	Sanitaryware	310,000.00	20.66
5C	Disposal Installations		
	Rainwater	40,000.00	2.66
	Soil and waste	359,000.00	23.93
	Specialist drainage (above ground)	22,000.00	1.46
5D	Water Installations		
	Hot and cold water services	787,000.00	52.46
5E	Heat Source	Included in 5F	
5F	Space Heating and Air Treatment		
	LPHW Heating	627,000.00	41.80
	Chilled Water	603,000.00	40.20
	Steam and condensate	364,000.00	24.26
5G	Ventilating Services		
	Ventilation, comfort cooling and air conditioning	2,041,000.00	136.06
5H	Electrical Installation		
	HV Distribution	36,000.00	2.40
	LV supply/distribution	424,000.00	28.26
	Standby Power	510,000.00	34.00
	UPS	352,000.00	23.45
	General lighting	610,000.00	40.66
	General power	607,000.00	40.46
	Emergency lighting	178,000.00	11.46
	Theatre lighting	217,000.00	14.46
	Specialist lighting	201,000.00	13.40
	External lighting	35,000.00	2.33
	Electrical supplies for mechanical equipment	183,000.00	12.20
5I	Gas Installation	50,000.00	3.33
	Oil Installations	103,000.00	6.86
5K	Protection		
	Dry risers	25,000.00	1.66
	Lightning Protection	7,000.00	0.46
	Carried forward	8,685,000.00	578.88

ELEMENTAL COSTS

PRIVATE HOSPITAL *continued*

El. Ref.	Element	Total Cost £	Cost/m² £
	Brought forward...	8,685,000.00	578.88
5L	Communication Installation		
	Fire alarms and detection...	275,000.00	18.33
	Voice and Data..	170,000.00	11.33
	Security and CCTV..	51,500.00	3.43
	Nurse call and cardiac alarm system.............................	253,500.00	16.90
	Personnel paging...	65,000.00	4.33
	Hospital radio (entertainment)..	335,500.00	22.37
5M	Special Installation		
	Building management systems.......................................	587,500.00	39.17
	Pneumatic tube conveying system.................................	41,500.00	2.77
	Group 1 Equipment...	565,500.00	37.70
	Summary total (based on gross floor area)............................	11,483,000.00	765.53

ELEMENTAL COSTS

SCHOOL

New build secondary school (Academy) located in Southern England, with a gross internal floor area of 10,000m².

The building comprises a three storey teaching block, including provision for music, drama, catering, sports hall, science laboratories, food technology, workshops and reception/admin (BB93 compliant). Excludes IT Cabling

Cost Summary

El. Ref.	Element	Total Cost £	Cost/m² £
5A	Sanitaryware		
	Toilet cores and changing facilities only............................	100,000.00	10.00
5C	Disposal Installations		
	Rainwater installations...	50,000.00	5.00
	Soil and waste...	120,000.00	12.00
5D	Water Installations		
	Potable hot and cold water services................................	200,000.00	20.00
	Non potable hot and cold water services to labs and art rooms	100,000.00	10.00
5E	Heat Source		
	Gas fired boiler installation..	150,000.00	15.00
5F	Space Heating and Air Treatment		
	LTHW Heating system (primary)......................................	400,000.00	40.00
	LTHW Heating system (secondary)...................................	100,000.00	10.00
	DX Cooling system to ICT server rooms...........................	50,000.00	5.00
	Mechanical supply and extract ventilation including DX type cooling to Music, Drama, Kitchen/Dining and Sports Hall......	600,000.00	60.00
5G	Ventilating Services		
	Toilet extract systems..	50,000.00	5.00
	Changing area extract systems.......................................	40,000.00	4.00
	Extract ventilation from design/food technology and science labs...	80,000.00	8.00
5H	Electrical Installation		
	Mains and sub-mains distribution.....................................	300,000.00	30.00
	Lighting and luminaries; including emergency fittings............	750,000.00	75.00
	Small power installation...	325,000.00	34.00
	Earthing and bonding...	20,000.00	2.00
5I	Gas Installation..	90,000.00	9.00
5K	Protective Installations		
	Lightning protection...	30,000.00	3.00
5L	Communications Installations		
	Containment for telephone, IT data, AV and security systems.	30,000.00	3.00
	Fire, smoke detection and alarm system............................	150,000.00	15.00
	Security installations including CCTV, access control and intruder alarm..	180,000.00	18.00
	Disabled toilet, refuge and induction loop systems...............	40,000.00	4.00
5M	Special Installations		
	Building Management system – To plant............................	220,000.00	22.00
	Building Management system – To opening vents/windows...	80,000.00	8.00
	Summary total...	4,270,000.00	427.00

ELEMENTAL COSTS

AFFORDABLE RESIDENTIAL DEVELOPMENT

A 12 storey, 50 apartment affordable residential development with a gross internal area of 3,400m² and a net internal area of 2,400m², situated within the London area. The development does not include a car park and is based on 71% efficiency.

Based on an individual radiator LTHW system within each apartment, with local gas combi boilers exhausting to building façade. Kitchens and bathrooms are also ventilated to the building façade, there are pendant light fittings, an audio entry system, telephone and satellite installation. Sanitaryware is of lower quality but a disabled refuge alarm is included. Full sprinkler installation installed throughout.

Cost Summary

El. Ref.	Element	Total Cost £	Cost/m² £
	SHELL & CORE		
5A	Sanitaryware..	6,800.00	2.00
5B	Services...	-	-
5C	Disposal Installations....................................	47,600.00	14.00
5D	Water Installations..	68,000.00	20.00
5E	Heat Source...	N/A	N/A
5F	Space Heating and Air Treatment.................	10,200.00	3.00
5G	Ventilating Services......................................	61,200.00	10.00
5H	Electrical Installation....................................	112,200.00	33.00
5I	Gas Installation..	17,000.00	5.00
5K	Protective Installations.................................	108,800.00	32.00
5L	Communications Installations.......................	85,000.00	25.00
5M	Special Installations.....................................	34,000.00	10.00
	Summary total (based on Gross Internal Area - GIA)...............	550,800.00	162.00

Approximate Estimating

ELEMENTAL COSTS

AFFORDABLE RESIDENTIAL DEVELOPMENT

Cost Summary

El. Ref.	Element	Total Cost £	Cost/m² £
	FITTING OUT		
5A	Sanitaryware..	60,000.00	25.00
5B	Services...	-	-
5C	Disposal Installations......................................	24,000.00	10.00
5D	Water Installations..	81,600.00	34.00
5E	Heat Source..	72,000.00	30.00
5F	Space Heating and Air Treatment...................	76,800.00	32.00
5G	Ventilating Services..	72,000.00	30.00
5H	Electrical Installation......................................	120,000.00	50.00
5I	Gas Installation...	28,800.00	12.00
5K	Protective Installations...................................	40,800.00	17.00
5L	Communications Installations.........................	72,000.00	30.00
5M	Special Installations..	16,800.00	7.00
	Summary total (Net Internal Area - NIA)...............	664,800.00	277.00

	Cost per apartment	£ 13,296.00

ELEMENTAL COSTS

PRIVATE RESIDENTIAL DEVELOPMENT

A 20 storey, 250 apartment private residential developments with a gross internal area of 22,750m² and a net internal area of 20,415m², situated within the London area. The development does not include a car park and is based on 90% efficiency.

Included is a central boiler and hot water installation with perimeter trench heating to each apartment and 30% LTHW radiators to supplement the 4 pipe fan coil unit installation. Central air cooled chillers system. No gas to apartments. Whole house ventilation system discharging to local façade, wet riser with full sprinkler installation. Video entry and TV and satellite installation. Flood wiring for apartment home automation and sound system.

Cost Summary

El. Ref.	Element	Total Cost £	Cost/m² £
	SHELL & CORE		
5A	Sanitaryware	22,750.00	1.00
5B	Services	-	-
5C	Disposal Installations	409,500.00	18.00
5D	Water Installations	568,750.00	25.00
5E	Heat Source	227,500.00	10.00
5F	Space Heating and Air Treatment	910,000.00	40.00
5G	Ventilating Services	182,000.00	8.00
5H	Electrical Installation	819,000.00	36.00
5I	Gas Installation	68,250.00	3.00
5K	Protective Installations	728,000.00	32.00
5L	Communications Installations	546,000.00	24.00
5M	Special Installations	409,500.00	18.00
	Summary total (based on Gross Internal Area - GIA)	4,891,250.00	215.00

Approximate Estimating

ELEMENTAL COSTS

PRIVATE RESIDENTIAL DEVELOPMENT

Cost Summary

El. Ref.	Element	Total Cost £	Cost/m² £
	FITTING OUT		
5A	Sanitaryware	1,122,825.00	55.00
5B	Services	-	-
5C	Disposal Installations	183,735.00	9.00
5D	Water Installations	816,600.00	40.00
5E	Heat Source	367,470.00	18.00
5F	Space Heating and Air Treatment	3,062,250.00	150.00
5G	Ventilating Services	714,525.00	35.00
5H	Electrical Installation	1,633,200.00	80.00
5I	Gas Installation	N/A	N/A
5K	Protective Installations	204,150.00	10.00
5L	Communications Installations	816,600.00	40.00
5M	Special Installations	61,245.00	3.00
	Summary total (based on Gross Internal Area - GIA)	8,982,600.00	440.00

	Cost per apartment	£ 35,930.00

ELEMENTAL COSTS

SUPERMARKET

Supermarket located in the South East with a total gross floor area of 4,000m², including a sales area of 2,350m². The building is on one level and incorporates a main sales, coffee shop, bakery, offices and amenities areas and warehouse.

Cost Summary

El. Ref.	Element	Total Cost £	Cost/m² £
5A	Sanitaryware...	8,312.00	2.08
5C	Disposal Installations		
	Soil and Waste..	11,777.00	2.94
5D	Water Installations		
	Hot and Cold water services..	41,142.00	10.29
5E	Heat Source...	45,877.00	11.47
5F	Space Heating and Air Treatment		
	Heating & ventilation with cooling via DX units................	94,331.00	23.58
5G	Ventilating Services		
	Supply and extract system..	61,978.00	15.49
5H	Electrical Installation		
	Panels / Boards...	108,946.00	27.24
	Containment..	7,902.00	1.98
	General lighting...	112,945.00	28.24
	Small power..	41,554.00	10.39
	Mechanical Services wiring...	8,044.00	2.01
5I	Gas Installation		
	Gas mains services to plantroom....................................	12,107.00	3.03
5K	Protection		
	Sprinklers..	131,495.00	32.87
	Lightning protection...	3,505.00	0.88
5L	Communication Installation		
	Fire alarms, detection and public address......................	29,387.00	7.34
	CCTV...	53,927.00	13.48
	Intruder alarm, detection and store security...................	79,903.00	19.98
	Telecom and structured cabling......................................	15,115.00	3.78
5M	Special Installations		
	BMS Installation..	31,313.00	7.83
	Refrigeration		
	Installation..	120,520.00	30.13
	Plant..	119,995.00	30.00
	Cold Store...	44,332.00	11.08
	Cabinets..	328,950.00	82.24
	Summary total	1,513,357.00	378.35

Approximate Estimating

ELEMENTAL COSTS

DISTRIBUTION CENTRE

Distribution centre located in London with a total gross floor area of 75,000m², including a refrigerated cold box of 17,500m².

The building is on one level and incorporates a office area, vehicle recovery unit, gate house and plantrooms

Cost Summary

El. Ref.	Element	Total Cost £	Cost/m² £
5C	Disposal Installations		
	Soil and Waste...	166,950.00	2.23
	Rainwater..	540,600.00	7.21
5D	Water Installations		
	Hot and Cold water services..	144,690.00	1.93
5F	Space Heating and Air Treatment		
	Heating with ventilation to offices, displacement system to main warehouse..	1,866,660.00	24.89
5G	Ventilating Services		
	Smoke extract system...	417,375.00	5.57
5H	Electrical Installation		
	Generator..	1,266,435.00	16.89
	Main HV installation..	1,462,800.00	19.50
	MV distribution..	906,300.00	12.08
	Lighting installation...	831,570.00	11.09
	Small power installation...	1,019,985.00	13.60
5I	Gas Installation		
	Gas mains services to plantroom.....................................	46,110.00	0.61
5K	Protection		
	Sprinklers including racking protection.............................	3,702,315.00	49.36
	Lightning protection...	8,745.00	0.12
5L	Communication Installation		
	Fire alarms, detection and public address.......................	942,075.00	12.56
	CCTV..	604,995.00	8.07
5M	Special Installations		
	BMS Installation	470,640.00	6.28
	Refrigeration Installation..	3,009,870.00	40.13
	Summary total...	17,408,115.00	232.12

ELEMENTAL COSTS

DATA CENTRE

New build data centre located in the London area/proximity to M25.

Net Technical Area (NTA) provided at 2,000m² with typically other areas of 250m² office space, 250m² ancillary space and 1,000m² internal plant. Total GIA 3,500m²

Power and cooling to Technical Space @ 1,000w/m². Cost/m² against Nett Technical Area - NTA.

Cost Summary

El. Ref.	Element	Total Cost £	Cost/m² £
5C	Disposal Installations		
	Soil & Waste	31,500.00	15.75
	Rainwater	21,000.00	10.50
	Condensate	21,000.00	10.50
5D	Water Installations		
	Hot and Cold water services	52,500.00	26.25
5F	Space Heating and Air Treatment		
	Chilled water plant with redundancy of N+1 to provide 1,000w/m² net technical space cooling	840,000.00	420.00
	Chilled water distribution to data centre to free standing cooling units and distribution to ancillary office and workshop/build areas	1,470,000.00	735.00
	Freestanding cooling units with redundancy of N+20% to technical space and switchrooms. Based on single coil cooling units 30% of units with humidification	682,500.00	341.25
	Floor grilles	157,500.00	78.75
5G	Ventilating Services		
	Supply and extract ventilation systems to data centre, switchrooms and ancillary spaces including dedicated gas extract and hot aisle extract system	924,000.00	462.00
5H	Electrical Installation		
	Main HV installations including transformers with redundant capacity of N+1	420,000.00	210.00
	LV distribution including cabling and busbar installations to provide full system – System dual supplies/redundancy including supplies to mechanical services, office and ancillary areas	1,260,000.00	630.00
	Generator Installation		
	Standby rated containerised generators with redundant capacity of N+1 and including synchronisation panel, 72 hours bulk fuel store and controls	1,785,000.00	892.50
	Uninterruptible Power Supplies		
	Static UPS to provide 2 x (N+1) system redundancy with 10 minute battery autonomy	1,260,000.00	630.00
	Carried Forward	8,924,500.00	4,462.50

Approximate Estimating

ELEMENTAL COSTS

DATA CENTRE

El. Ref.	Element	Total Cost £	Cost/m² £
	Brought Forward..	8,924,500.00	4,462.50
5H	Electrical Installation *continued*		
	LV Switchgear Incoming LV switchgear, UPS input and output boards for electrical and mechanical services systems........................	1,260,000.00	630.00
	Power Distribution Units (PDUs) PDUs to provide a redundant capacity of 2 Nr PDU to include static transfer switch and isolating transformer......................	1,092,000.00	546.00
	Cabinet Supplies A & B supply cables from PDUs to a BS4343 socket fixed under raised floor adjacent cabinet positions........................	588,000.00	294.00
	Lighting Installation Lighting to technical, workshop and plant areas including office/ancillary spaces and external areas...........................	210,000.00	105.00
5I	Gas Installation...	21,000.00	10.50
5K	Protective Installations Gaseous suppression to technical areas and switchrooms Lightning protection... Earthing and clean earth .... Leak detection	525,000.00 31,500.00 63,000.00 31,500.00	262.50 15.75 31.50 15.75
5L	Communications Installation Fire, smoke detection and alarm system .... Very early smoke detection alarm (VESDA) to technical areas.. CCTV installations.. Access control installations..	 273,000.00 346,500.00 294,000.00	 136.50 173.25 147.00
5M	Special Installations Building Management System.. PLC/Electrical monitoring system..	577,500.00 525,000.00	288.75 262.50
	Summary total (Based on Nett Technical Area - NTA)................	14,762,500.00	7,381.25

113

ELEMENTAL COSTS

BUILDING MANAGEMENT INSTALLATION

El. Ref.	Element	Total Cost £	Cost/Point £
	Category A Fit Out		
	Option 1 – 189 nr four pipe fan coil – 756 points		
1.0	**Field Equipment** Network devices Valves/actuators Sensing devices..	55,689.48	73.66
2.0	**Cabling** Power – from local isolator to DDC controller Control – from DDC controller to field equipment................	31,465.66	41.62
3.0	**Programming** Software – central facility Software – network devices Graphics..	17,418.55	23.04
4.0	**On site testing and commissioning** Equipment Programming/graphics Power and control cabling..................................	18,140.43	23.99
	Total Option 1 – Four pipe fan coil (On Point Basis)................	122,714.13	162.32
	Category A Fit Out		
	Option 2 – 189 nr two pipe fan coil system with electric heating – 756 points		
1.0	**Field Equipment** Network devices Valves/actuators/thyristors Sensing devices..	71,422.89	94.47
2.0	**Cabling** Power – from local isolator to DDC controller Control – from DDC controller to field equipment................	32,947.53	43.58
3.0	**Programming** Software – central facility Software – network devices Graphics..	17,418.55	23.04
4.0	**On site testing and commissioning** Equipment Programming/graphics Power and control cabling..................................	18,140.43	23.99
	Total Option 2 – Two pipe fan coil with electric heating (On Point Basis)..	139,929.40	185.09

Approximate Estimating

ELEMENTAL COSTS

BUILDING MANAGEMENT INSTALLATIONS

El: Ref.	Element	Total Cost £	Cost/Point £
	Category A Fit Out		
	Option 3 – 180 Nr Chilled Beams with perimeter heating – 567 points		
1.0	**Field Equipment** Network devices Valves/actuators Sensing devices…………………………………………………	50,445.78	88.96
2.0	**Cabling** Power – from local isolator to DDC controller Control – from DDC controller to field equipment……………….	33,463.81	59.01
3.0	**Programming** Software – central facility Software – network devices Graphics…………………………………………………………….	17,418.55	30.72
4.0	**On site testing and commissioning** Equipment Programming/graphics Power and control cabling……………………………………….	18,618.60	32.83
	Total Option 3 – Chilled beams with perimeter heating (On Point Basis) …………………………………………………….	119,946.75	211.54
	Combined Shell & Core and Fit Out		
1.0	**Option 1 – 4 pipe fan coil – 1196 points** Shell and core and Category A Fit out……………………………….	419,549.93	350.79
2.0	**Option 2 – 2 pipe fan coil with electrical heating – 1196 points** Shell and core and Category A Fit out……………………………	436,764.40	365.18
3.0	**Option 3 – Chilled beams with perimeter heating – 1007 points** Shell and core and Category A Fit out……………………………	416,469.90	413.57

ESSENTIAL READING FROM TAYLOR AND FRANCIS

Ventilation Systems
Design and Performance
Edited by **Hazim B. Awbi**

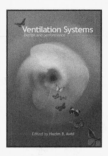

This comprehensive reference guide to ventilation systems provides up-to-date knowledge based on the experience of internationally-recognised experts to deal with current and future ventilation requirements in buildings. Presenting the most recent developments in ventilation research and its applications, this book covers the fundamentals as well as more advanced topics. With rigorous coverage for researchers and a practical edge for building professionals, Ventilation Systems is the one stop guide for the subject.

Selected Contents: 1. Airflow, Heat and Mass Transfer in Enclosures 2. Ventilation and Indoor Environmental Quality 3. Energy Implications of Indoor Environment Control 4. Modelling of Ventilation Airflow 5. Air Distribution – System Design 6. Characteristics of Mechanical Ventilation Systems 7. Characteristics of Natural and Hybrid Ventilation Systems 8. Measurement and Visualization of Air Movements

2007: 234x156: 464pp
Hb: 978-0-419-21700-8 **£95.00**

To Order: Tel: +44 (0) 1235 400524 Fax: +44 (0) 1235 400525,
or Post: Taylor and Francis Customer Services,
Bookpoint Ltd, Unit T1, 200 Milton Park, Abingdon, Oxon, OX14 4TA UK
Email: book.orders@tandf.co.uk

For a complete listing of all our titles visit:
www.tandf.co.uk

ESSENTIAL READING FROM TAYLOR AND FRANCIS

Understanding the Building Regulations
Fourth Edition

Simon Polley

Although the UK's Building Regulations are complex and lengthy, their essential features are relatively straightforward. Simon Polley brings his brief and practical guide up to date for Spon, in line with changes to Parts A, B, C, E, H, J, L and M, and with coverage of the new Part P.

Written by an industry practitioner, the book is well laid out, jargon-free, easy to use, and a simple alternative to the London District Surveyors Association's much larger and weightier The Building Regulations Explained 7th edition. It is an excellent introduction for students as well as for professionals who want something basic.

August 2008: 234x156: 296pp
Pb: 978-0-415-45272-4: **£18.99**

**To Order: Tel: +44 (0) 1235 400524 Fax: +44 (0) 1235 400525
or Post: Taylor and Francis Customer Services,
Bookpoint Ltd, Unit T1, 200 Milton Park, Abingdon, Oxon, OX14 4TA UK
Email: book.orders@tandf.co.uk**

**For a complete listing of all our titles visit:
www.tandf.co.uk**

ESSENTIAL READING FROM TAYLOR AND FRANCIS

Construction Delays
Extensions of Time and Prolongation Claims

Roger Gibson

Providing guidance on delay analysis, the author gives readers the information and practical details to be considered in formulating and resolving extension of time submissions and time-related prolongation claims. Useful guidance and recommended good practice is given on all the common delay analysis techniques. Worked examples of extension of time submissions and time-related prolongation claims are included.

Selected Contents:

1. Introduction
2. Programmes & Record Keeping
3. Contracts and Case Law
4. The 'Thorny Issues'
5. Extensions of Time
6. Prolongation Claims Summary

April 2008: 234x156: 374pp
Hb: 978-0-415-35486-6 **£70.00**

To Order: Tel: +44 (0) 1235 400524 **Fax:** +44 (0) 1235 400525
or Post: Taylor and Francis Customer Services,
Bookpoint Ltd, Unit T1, 200 Milton Park, Abingdon, Oxon, OX14 4TA UK
Email: book.orders@tandf.co.uk

For a complete listing of all our titles visit:
www.tandf.co.uk

PART THREE

Material Costs/ Measured Work Prices

Mechanical Installations

    R : Disposal Systems
        R10 : Rainwater Pipework/Gutters   125
        R11 : Above Ground Drainage   145

    S : Piped Supply Systems
        S10 : Cold Water   169
        S11 : Hot Water   224
        S32 : Natural Gas   228
        S41 : Fuel Oil Storage/Distribution   232
        S60 : Fire Hose Reels   233
        S61 : Dry Risers   234
        S63 : Sprinklers   235
        S65 : Fire Hydrants   241

    T : Mechanical/Cooling/Heating Systems
        T10 : Gas/Oil Fired Boilers   243
        T13 : Packaged Steam Generators   257
        T31 : Low Temperature Hot Water Heating   258
        T33 : Steam Heating   339
        T42 : Local Heating Units   343
        T60 : Central Refrigeration Plant   344
        T61 : Chilled Water   352
        T70 : Local Cooling Units   358

    U : Ventilation/Air Conditioning Systems
        U10 : General Ventilation   360
        U14 : Ductwork: Fire Rated   445
        U30 : Low Velocity Air Conditioning   465
        U31 : VAV Air Conditioning   468
        U41 : Fan Coil Air Conditioning   469
        U70 : Air Curtains   473

Electrical Installations

    V : Electrical Supply/Power/Lighting Systems
        V10 : Electrical Generation Plant   481
        V11 : HV Supply   483
        V20 : LV Distribution   488
        V21 : General Lighting   555
        V22 : General LV Power   563
        V32 : Uninterruptible Power Supply   570
        V40 : Emergency Lighting   572

Material Costs/Measured Work Prices

Electrical Installations *continued*

    W : Communications/Security/Control
        W10 : Telecommunications   577
        W20 : Radio/Television   579
        W23 : Clocks   581
        W30 : Data Transmission   582
        W40 : Access Control   588
        W41 : Security Detection and Alarm   589
        W50 : Fire Detection and Alarm   590
        W51 : Earthing and Bonding   593
        W52 : Lightning Protection   594
        W60 : Central Control/Building Management   598

Mechanical Installations

Material Costs/Measured Work Prices

DIRECTIONS

The following explanations are given for each of the column headings and letter codes.

Unit	Prices for each unit are given as singular (i.e. 1 metre, 1 nr) unless stated otherwise
Net price	Industry tender prices, plus nominal allowance for fixings (unless measured separately), waste and applicable trade discounts.
Material cost	Net price plus percentage allowance for overheads, profit and preliminaries.
Labour norms	In man-hours for each operation.
Labour cost	Labour constant multiplied by the appropriate all-in man-hour cost based on gang rate. (See also relevant Rates of Wages Section) plus percentage allowance for overheads, profit and preliminaries
Measured work Price (total rate)	Material cost plus Labour cost.

MATERIAL COSTS

The Material Costs given are based at Second Quarter 2008 but exclude any charges in respect of VAT. (At the time of going to press, the raw material cost of copper and steel was fluctuating but had not been reflected within factory exit prices for finished goods i.e. tube and fittings etc. Users of the book are advised to register on the SPON's website www.pricebooks.co.uk/updates to receive the free quarterly updates - alerts will then be provided by e-mail as changes arise).

MEASURED WORK PRICES

These prices are intended to apply to new work in the London area. The prices are for reasonable quantities of work and the user should make suitable adjustments if the quantities are especially small or especially large. Adjustments may also be required for locality (e.g. outside London - refer to cost indices in approximate estimating section for details of adjustment factors) and for the market conditions e.g. volume of work secured or being tendered) at the time of use.

MECHANICAL INSTALLATIONS

The labour rate has been based on average gang rates per man hour effective from 6 October 2008. To this rate has been added 12.5% and 7.5% to cover preliminary items, site and head office overheads together with 5% for profit, resulting in an inclusive rate of £27.10 per man hour. The rate has been calculated on a working year of 2,016 hours; a detailed build-up of the rate is given at the end of these directions.

DUCTWORK INSTALLATIONS

The labour rate basis is as per Mechanical above and to this rate has been added 25% plus 12.5% to cover shop, site and head office overheads and preliminary items together with 10% for profit, resulting in an inclusive rate of £31.62 per man hour. The rate has been calculated on a working year of 2,016 hours; a detailed build-up of the rate is given at the end of these directions.

In calculating the 'Measured Work Prices' the following assumptions have been made:
- (a) That the work is carried out as a sub-contract under the Standard Form of Building Contract.
- (b) That, unless otherwise stated, the work is being carried out in open areas at a height which would not require more than simple scaffolding.
- (c) That the building in which the work is being carried out is no more than six storey's high.

Where these assumptions are not valid, as for example where work is carried out in ducts and similar confined spaces or in multi-storey structures when additional time is needed to get to and from upper floors, then an appropriate adjustment must be made to the prices. Such adjustment will normally be to the labour element only. *Note : The rates do not include for any uplift applied if the ductwork package is procured via the Mechanical Sub Contractor*

Material Costs/Measured Work Prices – Mechanical Installations

DIRECTIONS

LABOUR RATE - MECHANICAL

The annual cost of a notional twelve man gang

		FOREMAN	SENIOR CRAFTSMAN (+2 Welding skill)	SENIOR CRAFTSMAN	CRAFTSMAN	INSTALLER	MATE (Over 18)	SUB TOTALS
		1 NR	1 NR	2 NR	4 NR	2 NR	2 NR	
Hourly Rate from 6 October 2008		14.71	12.66	12.16	11.16	10.11	8.52	
Working hours per annum per man		1,702.40	1,702.40	1,702.40	1,702.40	1,702.40	1,702.40	
x Hourly rate x nr of men = £ per annum		25,042.30	21,552.38	41,402.37	75,995.14	34,422.53	29,008.90	227,423.62
Overtime Rate		20.73	17.84	17.14	15.73	14.26	12.00	
Overtime hours per annum per man		313.60	313.60	313.60	313.60	313.60	313.60	
x Hourly rate x nr of men = £ per annum		6,500.93	5,594.62	10,750.21	19,731.71	8,943.87	7,526.40	59,047.74
Total		31,543.23	27,147.01	52,152.58	95,726.85	43,366.40	36,535.30	286,471.36
Incentive schemes	5.00%	1,577.16	1,357.35	2,607.63	4,786.34	2,168.32	1,826.76	14,323.57
Daily Travel Time Allowance (15-20 miles each way)		9.13	9.13	9.13	9.13	9.13	9.13	
Days per annum per man		224.00	224.00	224.00	224.00	224.00	224.00	
x nr of men = £ per annum		2,045.12	2,045.12	4,090.24	8,180.48	4,090.24	4,090.24	24,541.44
Daily Travel Fare (15-20 miles each way)		9.40	9.40	9.40	9.40	9.40	9.40	
Days per annum per man		224.00	224.00	224.00	224.00	224.00	224.00	
x nr of men = £ per annum		2,105.60	2,105.60	4,211.20	8,422.40	4,211.20	4,211.20	25,267.20
Employers Contributions to EasyBuild Stakeholder Pension (Death and accident cover is provided free):								
Number of weeks		52	52	52	52	52	52	
Total weekly £ contribution each		5.00	5.00	5.00	5.00	5.00	5.00	
£ Contributions/annum		260.00	260.00	520.00	1,040.00	520.00	520.00	3,120.00
National Insurance Contributions:								
Wkly gross pay (subject to NI) each		37,271.11	32,655.08	63,061.64	117,116.07	53,836.16	46,663.50	
% of NI Contributions		12.8	12.8	12.8	12.8	12.8	12.8	
£ Contributions/annum		3,884.68	3,293.82	6,299.84	11,446.75	5,118.98	4,200.88	34,244.94

DIRECTIONS

LABOUR RATE - MECHANICAL

The annual cost of a notional twelve man gang

	FOREMAN	SENIOR CRAFTSMAN (+2 Welding skill)	SENIOR CRAFTSMAN	CRAFTSMAN	INSTALLER	MATE (Over 18)	SUB TOTALS
	1 NR	1 NR	2 NR	4 NR	2 NR	2 NR	
Holiday Credit and Welfare Contributions:							
Number of weeks	52	52	52	52	52	52	
Total weekly £ contribution each	77.16	67.41	65.10	60.29	55.29	47.71	
x nr of men = £ contributions/annum	4,012.32	3,505.32	6,770.40	12,540.32	5,750.16	4,961.84	37,540.36
Holiday Top-up Funding including overtime	13.90	11.95	11.41	10.51	9.53	8.02	
Cost	722.80	621.40	1,186.64	2,186.08	991.12	834.08	6,542.12

	SUB-TOTAL		432,050.99
TRAINING (INCLUDING ANY TRADE REGISTRATIONS) – SAY		1.00%	4,320.51
SEVERANCE PAY AND SUNDRY COSTS – SAY		1.50%	6,545.57
EMPLOYER'S LIABILITY AND THIRD PARTY INSURANCE – SAY		2.00%	8,858.34
ANNUAL COST OF NOTIONAL GANG			451,775.41
MEN ACTUALLY WORKING = 10.5 THEREFORE ANNUAL COST PER PRODUCTIVE MAN			43,026.23
AVERAGE NR OF HOURS WORKED PER MAN = 2016 THEREFORE ALL IN MAN HOURS			21.34
PRELIMINARY ITEMS - SAY		7.50%	1.60
SITE AND HEAD OFFICE OVERHEADS - SAY		12.50%	2.87
PROFIT – SAY		5.00%	1.29
THEREFORE INCLUSIVE MAN HOUR RATE			27.10

Notes:

(1) The following assumptions have been made in the above calculations:-
    (a) The working week of 38 hours i.e. the normal working week as defined by the National Agreement.
    (b) The actual hours worked are five days of 9 hours each.
    (c) A working year of 2016 hours.
    (d) Five days in the year are lost through sickness or similar reason.
(2) The incentive scheme addition of 5% is intended to reflect bonus schemes typically in use.
(3) National insurance contributions are those effective from 6 April 2008. Calculation is based on employer making regular payment into the holiday pay scheme, allowing savings on NI.
(4) Weekly Holiday Credit/Welfare Stamp values are those effective from 6 October 2008.
(5) Rates are based from 6 October 2008.
(6) Overtime rates are based on Premium Rate 1.
(7) Easybuild Stakeholder Pension Contributions effective from April 2007.
(8) Fares (New Malden to Waterloo + Zone 1) Current at April 2008. Tel : 08457 484950

Material Costs/Measured Work Prices – Mechanical Installations

DIRECTIONS

LABOUR RATE - DUCTWORK

The annual cost of notional eight man gang

		FOREMAN 1 NR	SENIOR CRAFTSMAN 1 NR	CRAFTSMAN 4 NR	INSTALLER 2 NR	SUB TOTALS
Hourly Rate from 6 October 2008		14.71	12.16	11.16	10.11	
Working hours per annum per man		1,702.40	1,702.40	1,702.40	1,702.40	
x Hourly rate x nr of men = £ per annum		25,042.30	20,701.18	75,995.14	34,422.53	156,161.15
Overtime Rate		20.73	17.14	15.73	14.26	
Overtime hours per annum per man		313.60	313.60	313.60	313.60	
x hourly rate x nr of men = £ per		6,500.93	5,375.10	19,731.71	8,943.87	40.551.62
Total		31,543.23	26,076.29	95.726.85	43,366.40	196,712.77
Incentive schemes	5.00%	1,577.16	1,303.81	4,786.34	2,168.32	9,835.64
Daily Travel Time Allowance (15-20 miles each way)		9.13	9.13	9.13	9.13	
Days per annum per man		224	224	224	224	
x nr of men = £ per annum		2,045.12	2,045.12	8,140.48	4,090.24	16,360.96
Daily Travel Fare (15-20 miles each way)		9.40	9.40	9.40	9.40	
Days per annum per man		224	224	224	224	
x nr of men = £ per annum		2,105.60	2,105.60	8,422.40	4.211.20	16,844.80
Employers Contributions to EasyBuild Stakeholder Pension (Death cover is provided free)						
Number of weeks		52	52	52	52	
Total weekly £ contribution each		5.00	5.00	5.00	5.00	
£ Contributions/annum		260.00	260.00	1,040.00	520.00	2,080.00
National Insurance Contributions:						
Weekly gross pay (subject to NI) each		37,271.11	31,530.82	117,116.07	53,836.16	
% of NI Contributions		12.8	12.8	12.8	12.8	
£ Contributions/annum		3,884.68	3,149.92	11,446.75	5,118.98	23,600.32
Holiday Credit and Welfare contributions:						
Number of weeks		52	52	52	52	
Total weekly £ contribution each		77.16	65.10	60.29	55.29	
x nr of men = £ Contributions/annum		4,012.32	3,385.20	12,540.32	5,750.16	25,688.00

Material Costs/Measured Work Prices – Mechanical Installations

DIRECTIONS

LABOUR RATE - DUCTWORK

The annual cost of notional eight man gang

	FOREMAN 1 NR	SENIOR CRAFTSMAN 1 NR	CRAFTSMAN 4 NR	INSTALLER 2 NR	SUB TOTALS
Holiday Top-up Funding including overtime	13.90	11.41	10.51	9.53	
Cost	722.80	593.32	2,186.08	991.12	4,493.32
SUB-TOTAL					295,615.81
TRAINING (INCLUDING ANY TRADE REGISTRATIONS) - SAY				1.00%	2,956.16
SEVERANCE PAY AND SUNDRY COSTS - SAY				1.50%	4,478.58
EMPLOYER'S LIABILITY AND THIRD PARTY INSURANCE - SAY				2.00%	6,061.01
ANNUAL COST OF NOTIONAL GANG					309,111.56
MEN ACTUALLY WORKING = 7.5 — THEREFORE ANNUAL COST PER PRODUCTIVE MAN					41,214.87
AVERAGE NR OF HOURS WORKED PER MAN = 2016 — THEREFORE ALL IN MAN HOURS					20.44
PRELIMINARY ITEMS - SAY				12.50%	2.56
SITE AND HEAD OFFICE OVERHEADS - SAY				25.00%	5.75
PROFIT - SAY				10.00%	2.87
THEREFORE INCLUSIVE MAN HOUR RATE					31.62

Notes:

(1) The following assumptions have been made in the above calculations:-
    (a) The working week of 38 hours i.e. the normal working week as defined by the National Agreement.
    (b) The actual hours worked are five days of 9 hours each.
    (c) A working year of 2016 hours.
    (d) Five days in the year are lost through sickness or similar reason.
(2) The incentive scheme addition of 5% is intended to reflect bonus schemes typically in use.
(3) National insurance contributions are those effective from 6 April 2008. Calculation is based on employer making regular payment into the holiday pay scheme, allowing savings on NI.
(4) Weekly Holiday Credit/Welfare Stamp values are those effective from 1 October 2008.
(5) Rates are based from 6 October 2008.
(6) Fares (New Malden to Waterloo + Zone 1) current at April 2008. Tel : 08457 484950
(7) Easybuild Stakeholder Pension Contributions effective from April 2007.
(8) Overtime rates are based on Premium Rate 1.

ESSENTIAL READING FROM TAYLOR AND FRANCIS

Building Services Engineering

Fifth Edition

David V. Chadderton

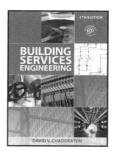

Updated and expanded, this core textbook introduces the range of building services found within modern buildings.

In this fifth edition coverage has been broadened as a response to the trend towards low energy mechanical services systems for the heating and cooling of buildings. New chapters have been included on mechanical transportation and on understanding units.

Now accompanied by a new instructor's resource, the book's style retains its focus on the fundamentals and it sets out applications in a general and easily understood manner. It is extensively illustrated with fully worked examples of all numerical problems and student-centred problems, complemented by full answers.

Suitable for distance learning and with a broad international applicability, *Building Services Engineering* provides for the higher education of building industry professionals, whether on higher certificate, higher diploma, undergraduate courses or graduate level conversion courses, across the building technology, architectural, surveying and services engineering disciplines.

2007: 246x174 mm: 448 pages
Hb: 978-0-415-41354-1: **£90.00**
Pb: 978-0-415-41355-8: **£29.99**

To Order: Tel: +44 (0) 1235 400524 **Fax:** +44 (0) 1235 400525
or Post: Taylor and Francis Customer Services,
Bookpoint Ltd, Unit T1, 200 Milton Park, Abingdon, Oxon, OX14 4TA UK
Email: book.orders@tandf.co.uk

**For a complete listing of all our titles visit:
www.tandf.co.uk**

R: DISPOSAL SYSTEMS

Item	Net Price £	Material £	Labour hours	Labour £	Unit	Total rate £
R10: RAINWATER PIPEWORK/GUTTERS						
PVC-U gutters: push fit joints; fixed with brackets to backgrounds; BS 4576 BS EN 607						
Half round gutter, with brackets measured separately						
75mm	1.46	1.87	0.69	18.70	m	**20.57**
100 mm	3.10	3.97	0.64	17.35	m	**21.32**
150mm	4.63	5.93	0.82	22.22	m	**28.16**
Brackets: including fixing to backgrounds. For minimum fixing distances, refer to the Tables and Memoranda at the rear of the book						
75mm; Fascia	0.66	0.85	0.15	4.07	nr	**4.91**
100mm; Jointing	1.70	2.18	0.16	4.34	nr	**6.51**
100mm; Support	0.73	0.94	0.16	4.34	nr	**5.27**
150mm; Fascia	1.15	1.47	0.16	4.34	nr	**5.81**
Bracket supports: including fixing to backgrounds. For minimum fixing distances, refer to the Tables and Memoranda at the rear of the book						
Side rafter	3.04	3.89	0.16	4.34	nr	**8.23**
Top rafter	3.04	3.89	0.16	4.34	nr	**8.23**
Rise and fall	2.94	3.77	0.16	4.34	nr	**8.10**
Extra over fittings half round PVC-U gutter						
Union						
75mm	1.11	1.42	0.19	5.15	nr	**6.57**
100mm	2.43	3.11	0.24	6.50	nr	**9.62**
150mm	3.38	4.55	0.28	7.59	nr	**12.13**
Rainwater pipe outlets						
Running: 75 x 53mm dia	2.56	3.28	0.12	3.25	nr	**6.53**
Running: 100 x 68mm dia	3.23	4.14	0.12	3.25	nr	**7.39**
Running: 150 x 110mm dia	6.55	8.39	0.12	3.25	nr	**11.64**
Stop end: 100 x 68mm dia	3.23	4.14	0.12	3.25	nr	**7.39**
Internal stop ends: short						
75mm	1.11	2.28	0.09	2.44	nr	**4.71**
100mm	1.29	2.46	0.09	2.44	nr	**4.89**
150mm	1.99	2.55	0.09	2.44	nr	**4.99**
External stop ends: short						
75mm	3.40	4.35	0.09	2.44	nr	**6.79**
100mm	3.40	5.64	0.09	2.44	nr	**8.08**
150mm	4.57	5.85	0.09	2.44	nr	**8.29**
Angles						
75mm; 45 °	3.02	3.87	0.20	5.42	nr	**9.29**
75mm; 90 °	3.02	3.87	0.20	5.42	nr	**9.29**
100mm; 90 °	3.19	4.29	0.20	5.42	nr	**9.71**
100mm; 120 °	3.53	4.52	0.20	5.42	nr	**9.94**
100mm; 135 °	3.53	4.52	0.20	5.42	nr	**9.94**
100mm; Prefabricated to special angle	15.05	19.28	0.23	6.23	nr	**25.51**
100mm; Prefabricated to raked angle	16.29	20.86	0.23	6.23	nr	**27.10**
150mm; 90 °	5.98	7.66	0.20	5.42	nr	**13.08**

R: DISPOSAL SYSTEMS

Item	Net Price £	Material £	Labour hours	Labour £	Unit	Total rate £
R10: RAINWATER PIPEWORK/GUTTERS (cont'd)						
Extra over fittings half round PVC-U gutter (cont'd)						
Gutter adaptors						
100mm; Stainless steel clip	1.71	2.19	0.16	4.34	nr	**6.53**
100mm; Cast iron spigot	4.69	6.01	0.23	6.23	nr	**12.24**
100mm; Cast iron socket	4.69	6.01	0.23	6.23	nr	**12.24**
100mm; Cast iron "ogee" spigot	4.75	6.08	0.23	6.23	nr	**12.32**
100mm; Cast iron "ogee" socket	4.75	6.08	0.23	6.23	nr	**12.32**
100mm; Half round to Square PVC-U	8.23	10.54	0.23	6.23	nr	**16.77**
100mm; Gutter overshoot guard	10.19	13.05	0.58	15.72	nr	**28.77**
Square gutter, with brackets measured separately						
120mm	2.51	3.21	0.82	22.22	m	**25.44**
Brackets: including fixing to backgrounds. For minimum fixing distances, refer to the Tables and Memoranda at the rear of the book						
Jointing	2.11	2.70	0.16	4.34	nr	**7.04**
Support	0.81	1.04	0.16	4.34	nr	**5.37**
Bracket support: including fixing to backgrounds. For minimum fixing distances, refer to the Tables and Memoranda at the rear of the book						
Side rafter	3.04	3.89	0.16	4.34	nr	**8.23**
Top rafter	3.04	3.89	0.16	4.34	nr	**8.23**
Rise and fall	4.75	6.08	0.16	4.34	nr	**10.42**
Extra over fittings square PVC-U gutter						
Rainwater pipe outlets						
Running: 62mm square	3.52	4.51	0.12	3.25	nr	**7.76**
Stop end: 62mm square	3.03	3.88	0.12	3.25	nr	**7.13**
Stop ends: short						
External	1.72	2.20	0.09	2.44	nr	**4.64**
Angles						
90°	3.57	4.80	0.20	5.42	nr	**10.22**
120°	13.18	17.72	0.20	5.42	nr	**23.15**
135°	3.57	4.80	0.20	5.42	nr	**10.22**
Prefabricated to special angle	15.44	19.78	0.23	6.23	nr	**26.01**
Prefabricated to raked angle	15.85	20.30	0.23	6.23	nr	**26.53**
Gutter adaptors						
Cast iron	8.22	10.53	0.23	6.23	nr	**16.76**
High capacity square gutter, with brackets measured separately						
137mm	6.43	8.24	0.82	22.22	m	**30.46**

R: DISPOSAL SYSTEMS

Item	Net Price £	Material £	Labour hours	Labour £	Unit	Total rate £
Brackets: including fixing to backgrounds. For minimum fixing distances, refer to the Tables and Memoranda at the rear of the book						
Jointing	5.83	7.84	0.16	4.34	nr	**12.18**
Support	2.43	3.27	0.16	4.34	nr	**7.60**
Overslung	2.26	3.04	0.16	4.34	nr	**7.38**
Bracket supports: including fixing to backgrounds. For minimum fixing distances, refer to the Tables and Memoranda at the rear of the book						
Side rafter	3.65	4.67	0.16	4.34	nr	**9.01**
Top rafter	3.65	4.67	0.16	4.34	nr	**9.01**
Rise and fall	5.17	6.62	0.16	4.34	nr	**10.96**
Extra over fittings high capacity square UPV-C						
Rainwater pipe outlets						
Running: 75mm square	10.23	13.10	0.12	3.25	nr	**16.35**
Running: 82mm dia	10.23	13.10	0.12	3.25	nr	**16.35**
Running: 110mm dia	8.96	11.48	0.12	3.25	nr	**14.73**
Screwed outlet adaptor						
75mm square pipe	5.38	6.89	0.23	6.23	nr	**13.12**
Stop ends: short						
External	3.40	4.35	0.09	2.44	nr	**6.79**
Angles						
90°	9.52	12.80	0.20	5.42	nr	**18.22**
135°	18.62	25.04	0.20	5.42	nr	**30.46**
Prefabricated to special angle	21.67	27.75	0.23	6.23	nr	**33.99**
Prefabricated to raked internal angle	37.72	48.31	0.23	6.23	nr	**54.55**
Prefabricated to raked external angle	37.72	48.31	0.23	6.23	nr	**54.55**
Deep elliptical gutter, with brackets measured separately						
137mm	3.05	3.91	0.82	22.22	m	**26.13**
Brackets: including fixing to backgrounds. For minimum fixing distances, refer to the Tables and Memoranda at the rear of the book						
Jointing	0.86	1.16	0.16	4.34	nr	**5.49**
Support	3.17	4.26	0.16	4.34	nr	**8.60**
Bracket support: including fixing to backgrounds. For minimum fixing distances, refer to the Tables and Memoranda at the rear of the book						
Side rafter	3.65	4.91	0.16	4.34	nr	**9.25**
Top rafter	3.65	4.91	0.16	4.34	nr	**9.25**
Rise and fall	5.16	6.94	0.16	4.34	nr	**11.28**

R: DISPOSAL SYSTEMS

Item	Net Price £	Material £	Labour hours	Labour £	Unit	Total rate £
R10: RAINWATER PIPEWORK/GUTTERS (cont'd)						
Extra over fittings deep elliptical PVC-U gutter						
Rainwater pipe outlets						
Running: 68mm dia	3.40	4.57	0.12	3.25	nr	7.82
Running: 82mm dia	3.40	4.35	0.12	3.25	nr	7.61
Stop end: 68mm dia	3.59	4.60	0.12	3.25	nr	7.85
Stop ends: short						
External	1.68	2.15	0.09	2.44	nr	4.59
Angles						
90°	3.74	5.03	0.20	5.42	nr	10.45
135°	3.74	5.03	0.20	5.42	nr	10.45
Prefabricated to special angle	10.65	14.32	0.23	6.23	nr	20.56
Gutter adaptors						
Stainless steel clip	2.47	3.16	0.16	4.34	nr	7.50
Marley deep flow	3.13	4.01	0.23	6.23	nr	10.24
Ogee profile PVC-U gutter, with brackets measured separately						
122mm	3.43	4.39	0.82	22.22	m	26.62
Brackets: including fixing to backgrounds. For minimum fixing distances, refer to the Tables and Memoranda at the rear of the book						
Jointing	2.69	3.45	0.16	4.34	nr	7.78
Support	1.09	1.40	0.16	4.34	nr	5.73
Overslung	1.09	1.40	0.16	4.34	nr	5.73
Extra over fittings Ogee profile PVC-U gutter						
Rainwater pipe outlets						
Running: 68mm dia	3.71	4.75	0.12	3.25	nr	8.00
Stop ends: short						
Internal/External: left or right hand	1.89	2.42	0.09	2.44	nr	4.86
Angles						
90°: internal or external	4.12	5.28	0.20	5.42	nr	10.70
135°: internal or external	4.12	5.28	0.20	5.42	nr	10.70
PVC-U rainwater pipe: dry push fit joints; fixed with brackets to backgrounds; BS 4576/ BS EN 607						
Pipe: circular, with brackets measured separately						
53mm	3.68	4.71	0.61	16.53	m	21.25
68mm	4.03	5.16	0.61	16.53	m	21.69
Pipe clip: including fixing to backgrounds. For minimum fixing distances, refer to the Tables and Memoranda at the rear of the book						
68mm	1.09	1.40	0.16	4.34	nr	5.73

R: DISPOSAL SYSTEMS

Item	Net Price £	Material £	Labour hours	Labour £	Unit	Total rate £
Pipe clip adjustable: including fixing to backgrounds. For minimum fixing distances, refer to the Tables and Memoranda at the rear of the book						
53mm	1.06	1.36	0.16	4.34	nr	**5.69**
68mm	2.36	3.02	0.16	4.34	nr	**7.36**
Pipe clip drive in: including fixing to backgrounds. For minimum fixing distances, refer to the Tables and Memoranda at the rear of the book						
68mm	2.68	3.43	0.16	4.34	nr	**7.77**
Extra over fittings circular pipework PVC-U						
Pipe coupler: PVC-U to PVC-U						
68mm	1.36	1.74	0.12	3.25	nr	**4.99**
Pipe coupler: PVC-U to Cast Iron						
68mm: to 3" cast iron	3.03	3.88	0.17	4.61	nr	**8.49**
68mm: to 3.3/4" cast iron	13.48	17.27	0.17	4.61	nr	**21.87**
Access pipe: single socket						
68mm	8.66	11.09	0.15	4.07	nr	**15.16**
Bend: short radius						
53mm: 67.5 °	1.88	2.41	0.20	5.42	nr	**7.83**
68mm: 92.5 °	2.36	3.02	0.20	5.42	nr	**8.44**
68mm: 112.5 °	1.97	2.52	0.20	5.42	nr	**7.94**
Bend: long radius						
68mm: 112 °	2.36	3.02	0.20	5.42	nr	**8.44**
Branch						
68mm: 92 °	10.94	14.71	0.23	6.23	nr	**20.95**
68mm: 112 °	10.99	14.78	0.23	6.23	nr	**21.01**
Double branch						
68mm: 112 °	22.76	29.15	0.24	6.50	nr	**35.66**
Shoe						
53mm	1.88	2.41	0.12	3.25	nr	**5.66**
68mm	1.60	2.05	0.12	3.25	nr	**5.30**
Rainwater head: including fixing to backgrounds						
68mm	8.75	11.21	0.29	7.86	nr	**19.07**
Pipe: square, with brackets measured separately						
62mm	3.02	3.87	0.45	12.20	m	**16.06**
75mm	5.18	6.63	0.45	12.20	m	**18.83**
Pipe clip: including fixing to backgrounds. For minimum fixing distances, refer to the Tables and Memoranda at the rear of the book						
62mm	1.01	1.29	0.16	4.34	nr	**5.63**
75mm	2.04	2.61	0.16	4.34	nr	**6.95**

R: DISPOSAL SYSTEMS

Item	Net Price £	Material £	Labour hours	Labour £	Unit	Total rate £
R10: RAINWATER PIPEWORK/GUTTERS (cont'd)						
Extra over fittings circular pipework PVC-U (cont'd)						
Pipe clip adjustable: including fixing to backgrounds. For minimum fixing distances, refer to the Tables and Memoranda at the rear of the book						
62mm	3.12	4.00	0.16	4.34	nr	8.33
Extra over fittings square pipework PVC-U						
Pipe coupler: PVC-U to PVC-U						
62mm	1.87	2.40	0.20	5.42	nr	7.82
75mm	2.43	3.11	0.20	5.42	nr	8.53
Square to circular adaptor: single socket						
62mm to 68mm	2.88	3.69	0.20	5.42	nr	9.11
Square to circular adaptor: single socket						
75mm to 62mm	3.66	4.69	0.20	5.42	nr	10.11
Access pipe						
62mm	11.91	15.25	0.16	4.34	nr	19.59
75mm	14.96	19.16	0.16	4.34	nr	23.50
Bends						
62mm: 92.5 °	2.54	3.25	0.20	5.42	nr	8.67
62mm: 112.5 °	2.09	2.68	0.20	5.42	nr	8.10
75mm: 112.5 °	4.87	6.24	0.20	5.42	nr	11.66
Bends: prefabricated special angle						
62mm	11.81	15.13	0.23	6.23	nr	21.36
75mm	16.50	21.13	0.23	6.23	nr	27.37
Offset						
62mm	3.82	4.89	0.20	5.42	nr	10.31
75mm	10.79	13.82	0.20	5.42	nr	19.24
Offset: prefabricated special angle						
62mm	11.86	15.19	0.23	6.23	nr	21.42
Shoe						
62mm	1.97	2.52	0.12	3.25	nr	5.78
75mm	2.51	3.21	0.12	3.25	nr	6.47
Branch						
62mm	6.10	7.81	0.23	6.23	nr	14.05
75mm	16.10	20.62	0.23	6.23	nr	26.85
Double branch						
62mm	21.42	27.43	0.24	6.50	nr	33.94
Rainwater head						
62mm	7.52	9.63	0.29	7.86	nr	17.49
75mm	28.76	36.84	3.45	93.46	nr	130.30

R: DISPOSAL SYSTEMS

Item	Net Price £	Material £	Labour hours	Labour £	Unit	Total rate £
PVC-U rainwater pipe: solvent welded joints; fixed with brackets to backgrounds; BS 4576/ BS EN 607						
Pipe: circular, with brackets measured separately						
82mm	8.08	10.35	0.35	9.49	m	**19.84**
Pipe clip: galvanised; including fixing to backgrounds. For minimum fixing distances, refer to the Tables and Memoranda at the rear of the book						
82mm	2.31	2.96	0.58	15.72	nr	**18.68**
Pipe clip: galvanised plastic coated; including fixing to backgrounds. For minimum fixing distances, refer to the Tables and Memoranda at the rear of the book						
82mm	3.21	4.11	0.58	15.72	nr	**19.83**
Pipe clip: PVC-U including fixing to backgrounds. For minimum fixing distances, refer to the Tables and Memoranda at the rear of the book						
82mm	1.73	2.22	0.58	15.72	nr	**17.94**
Pipe clip: PVC-U adjustable: including fixing to backgrounds. For minimum fixing distances, refer to the Tables and Memoranda at the rear of the book						
82mm	3.17	4.06	0.58	15.72	nr	**19.78**
Extra over fittings circular pipework PVC-U						
Pipe coupler: PVC-U to PVC-U						
82mm	3.39	4.34	0.21	5.69	nr	**10.03**
Access pipe						
82mm	18.35	23.50	0.23	6.23	nr	**29.74**
Bend						
82mm: 92, 112.5 and 135 °	7.56	9.68	0.29	7.86	nr	**17.54**
Shoe						
82mm	7.33	9.39	0.29	7.86	nr	**17.25**
110mm	9.22	11.81	0.32	8.67	nr	**20.48**
Branch						
82mm: 92, 112.5 and 135 °	11.79	15.10	0.35	9.49	nr	**24.59**
Rainwater head						
82mm	15.52	19.88	0.58	15.72	nr	**35.60**
110mm	14.16	18.14	0.58	15.72	nr	**33.86**
Roof outlets: 178 dia; Flat						
50mm	13.95	17.87	1.15	31.17	nr	**49.04**
82mm	13.95	17.87	1.15	31.17	nr	**49.04**
Roof outlets: 178mm dia; Domed						
50mm	13.95	17.87	1.15	31.17	nr	**49.04**
82mm	13.95	17.87	1.15	31.17	nr	**49.04**

R: DISPOSAL SYSTEMS

Item	Net Price £	Material £	Labour hours	Labour £	Unit	Total rate £
R10: RAINWATER PIPEWORK/GUTTERS (cont'd)						
Extra over fittings circular pipework PVC-U (cont'd)						
Roof outlets: 406mm dia; Flat						
82mm	28.60	36.63	1.15	31.17	nr	67.80
110mm	28.60	36.63	1.15	31.17	nr	67.80
Roof outlets: 406mm dia; Domed						
82mm	28.60	36.63	1.15	31.17	nr	67.80
110mm	28.60	36.63	1.15	31.17	nr	67.80
Roof outlets: 406mm dia; Inverted						
82mm	60.07	76.94	1.15	31.17	nr	108.11
110mm	60.07	76.94	1.15	31.17	nr	108.11
Roof outlets: 406mm dia; Vent Pipe						
82mm	41.35	52.96	1.15	31.17	nr	84.13
110mm	41.35	52.96	1.15	31.17	nr	84.13
Balcony outlets: screed						
82mm	24.34	31.17	1.15	31.17	nr	62.34
Balcony outlets: asphalt						
82mm	24.35	31.19	1.15	31.17	nr	62.36
Adaptors						
82mm x 62mm square pipe	1.90	2.43	0.21	5.69	nr	8.13
82mm x 68mm circular pipe	1.76	2.25	0.21	5.69	nr	7.95
For 110mm diameter pipework and fittings refer to R11: Above Ground Drainage						
Cast iron gutters: mastic and bolted joints; BS 460; fixed with brackets to backgrounds						
Half round gutter, with brackets measured separately						
100 mm	15.74	20.16	0.85	23.04	m	43.20
115 mm	17.23	22.07	0.97	26.29	m	48.36
125 mm	20.05	25.68	0.97	26.29	m	51.97
150 mm	33.43	42.82	1.12	30.36	m	73.17
Brackets; fixed to backgrounds. For minimum fixing distances, refer to the Tables and Memoranda at the rear of the book						
Fascia						
100 mm	1.79	2.29	0.16	4.34	nr	6.63
115 mm	1.79	2.29	0.16	4.34	nr	6.63
125 mm	1.79	2.29	0.16	4.34	nr	6.63
150 mm	2.25	2.88	0.16	4.34	nr	7.22
Rise and fall						
100 mm	2.66	3.41	0.39	10.57	nr	13.98
115 mm	2.66	3.41	0.39	10.57	nr	13.98
125 mm	3.19	4.09	0.39	10.57	nr	14.66
150 mm	3.74	4.79	0.39	10.57	nr	15.36

R: DISPOSAL SYSTEMS

Item	Net Price £	Material £	Labour hours	Labour £	Unit	Total rate £
Top rafter						
100 mm	1.71	2.19	0.16	4.34	nr	6.53
115 mm	1.71	2.19	0.16	4.34	nr	6.53
125 mm	1.77	2.27	0.16	4.34	nr	6.60
150 mm	2.66	3.41	0.16	4.34	nr	7.74
Side rafter						
100 mm	1.71	2.19	0.16	4.34	nr	6.53
115 mm	1.71	2.19	0.16	4.34	nr	6.53
125 mm	1.77	2.27	0.16	4.34	nr	6.60
150 mm	2.66	3.41	0.16	4.34	nr	7.74
Extra over fittings half round gutter cast iron BS 460						
Union						
100 mm	4.36	5.58	0.39	10.57	nr	16.15
115 mm	5.31	6.80	0.48	13.01	nr	19.81
125 mm	6.12	7.84	0.48	13.01	nr	20.85
150 mm	6.89	8.82	0.55	14.91	nr	23.73
Stop end; internal						
100 mm	2.22	2.84	0.12	3.25	nr	6.10
115 mm	2.87	3.68	0.15	4.07	nr	7.74
125 mm	2.87	3.68	0.15	4.07	nr	7.74
150 mm	3.83	4.91	0.20	5.42	nr	10.33
Stop end; external						
100 mm	2.22	2.84	0.12	3.25	nr	6.10
115 mm	2.87	3.68	0.15	4.07	nr	7.74
125 mm	2.87	3.68	0.15	4.07	nr	7.74
150 mm	3.83	4.91	0.20	5.42	nr	10.33
90 ° angle; single socket						
100 mm	6.58	8.43	0.39	10.57	nr	19.00
115 mm	6.79	8.70	0.43	11.65	nr	20.35
125 mm	8.00	10.25	0.43	11.65	nr	21.90
150 mm	14.62	18.73	0.50	13.55	nr	32.28
90 ° angle; double socket						
100 mm	7.99	10.23	0.39	10.57	nr	20.80
115 mm	8.49	10.87	0.43	11.65	nr	22.53
125 mm	10.97	14.05	0.43	11.65	nr	25.70
135 ° angle; single socket						
100 mm	6.12	7.84	0.39	10.57	nr	18.41
115 mm	6.79	8.70	0.43	11.65	nr	20.35
125 mm	10.02	12.83	0.43	11.65	nr	24.49
150 mm	13.38	17.14	0.50	13.55	nr	30.69
Running outlet						
65 mm outlet						
100 mm	6.42	8.22	0.39	10.57	nr	18.79
115 mm	7.00	8.97	0.43	11.65	nr	20.62
125 mm	8.00	10.25	0.43	11.65	nr	21.90

R: DISPOSAL SYSTEMS

Item	Net Price £	Material £	Labour hours	Labour £	Unit	Total rate £
R10: RAINWATER PIPEWORK/GUTTERS (cont'd)						
Extra over fittings half round gutter (cont'd)						
75 mm outlet						
100 mm	6.42	8.22	0.39	10.57	nr	18.79
115 mm	7.00	8.97	0.43	11.65	nr	20.62
125 mm	8.00	10.25	0.43	11.65	nr	21.90
150 mm	13.61	17.43	0.50	13.55	nr	30.98
100 mm outlet						
150 mm	13.61	13.61	0.50	13.55	nr	27.16
Stop end outlet; socket						
65 mm outlet						
100 mm	5.11	6.54	0.39	10.57	nr	17.12
115 mm	5.62	7.20	0.43	11.65	nr	18.85
75 mm outlet						
125 mm	7.14	9.14	0.43	11.65	nr	20.80
150 mm	13.61	17.43	0.50	13.55	nr	30.98
100mm outlet						
150 mm	13.61	17.43	0.50	13.55	nr	30.98
Stop end outlet; spigot						
65 mm outlet						
100 mm	5.11	6.54	0.39	10.57	nr	17.12
115 mm	5.62	7.20	0.43	11.65	nr	18.85
75 mm outlet						
125 mm	7.14	9.14	0.43	11.65	nr	20.80
150 mm	13.61	17.43	0.50	13.55	nr	30.98
100mm outlet						
150 mm	13.61	17.43	0.50	13.55	nr	30.98
Half round; 3 mm thick double beaded gutter, with brackets measured separately						
100 mm	9.59	12.28	0.85	23.04	m	35.32
115 mm	10.11	12.95	0.85	23.04	m	35.99
125 mm	11.33	14.51	0.97	26.29	m	40.80
Brackets; fixed to backgrounds. For minimum fixing distances, refer to the Tables and Memoranda at the rear of the book						
Fascia						
100 mm	4.04	5.17	0.16	4.34	nr	9.51
115 mm	4.04	5.17	0.16	4.34	nr	9.51
125 mm	4.94	6.33	0.16	4.34	nr	10.66

R: DISPOSAL SYSTEMS

Item	Net Price £	Material £	Labour hours	Labour £	Unit	Total rate £
Extra over fittings Half Round 3mm thick Gutter BS 460						
Union						
100 mm	4.34	5.56	0.38	10.30	nr	15.86
115 mm	5.28	6.76	0.38	10.30	nr	17.06
125 mm	6.04	7.74	0.43	11.65	nr	19.39
Stop end; internal						
100 mm	1.98	2.54	0.12	3.25	nr	5.79
115 mm	2.81	3.60	0.12	3.25	nr	6.85
125 mm	2.87	3.68	0.15	4.07	nr	7.74
Stop end; external						
100 mm	1.98	2.54	0.12	3.25	nr	5.79
115 mm	2.81	3.60	0.12	3.25	nr	6.85
125 mm	2.81	3.60	0.15	4.07	nr	7.66
90 ° angle; single socket						
100 mm	6.73	8.62	0.38	10.30	nr	18.92
115 mm	6.94	8.89	0.38	10.30	nr	19.19
125 mm	8.44	10.81	0.43	11.65	nr	22.46
135 ° angle; single socket						
100 mm	6.73	8.62	0.38	10.30	nr	18.92
115 mm	6.94	8.89	0.38	10.30	nr	19.19
125 mm	8.44	10.81	0.43	11.65	nr	22.46
Running outlet						
65 mm outlet						
100 mm	6.73	8.62	0.38	10.30	nr	18.92
115 mm	6.80	8.71	0.38	10.30	nr	19.01
125 mm	6.58	8.43	0.43	11.65	nr	20.08
75 mm outlet						
115 mm	6.94	8.89	0.38	10.30	nr	19.19
125 mm	8.28	10.60	0.43	11.65	nr	22.26
Stop end outlet; socket						
65 mm outlet						
100 mm	5.20	6.66	0.38	10.30	nr	16.96
115 mm	6.34	8.12	0.38	10.30	nr	18.42
125 mm	7.41	7.41	0.43	11.65	nr	19.06
75 mm outlet						
125 mm	7.10	9.09	0.43	11.65	nr	20.75
Stop end outlet; spigot						
65 mm outlet						
100 mm	5.20	6.66	0.38	10.30	nr	16.96
115 mm	6.34	8.12	0.38	10.30	nr	18.42
125 mm	7.24	9.27	0.43	11.65	nr	20.93
Deep half round gutter, with brackets measured separately						
100 x 75 mm	16.34	20.93	0.85	23.04	m	43.97
125 x 75 mm	21.15	27.09	0.97	26.29	m	53.38

R: DISPOSAL SYSTEMS

Item	Net Price £	Material £	Labour hours	Labour £	Unit	Total rate £
R10: RAINWATER PIPEWORK/GUTTERS (cont'd)						
Deep half round gutter (cont'd)						
Brackets; fixed to backgrounds. For minimum fixing distances, refer to the Tables and Memoranda at the rear of the book						
Fascia						
100 x 75 mm	6.25	8.00	0.16	4.34	nr	12.34
125 x 75 mm	13.33	17.07	0.16	4.34	nr	21.41
Extra over fittings Deep Half Round Gutter BS 460						
Union						
100 x 75 mm	7.14	9.14	0.38	10.30	nr	19.44
125 x 75 mm	7.71	9.87	0.43	11.65	nr	21.53
Stop end; internal						
100 x 75 mm	6.25	8.00	0.12	3.25	nr	11.26
125 x 75 mm	7.71	9.87	0.15	4.07	nr	13.94
Stop end; external						
100 x 75 mm	6.25	8.00	0.12	3.25	nr	11.26
125 x 75 mm	7.71	9.87	0.15	4.07	nr	13.94
90 ° angle; single socket						
100 x 75 mm	18.16	23.26	0.38	10.30	nr	33.56
125 x 75 mm	23.05	29.52	0.43	11.65	nr	41.18
135 ° angle; single socket						
100 x 75 mm	18.16	23.26	0.38	10.30	nr	33.56
125 x 75 mm	23.05	29.52	0.43	11.65	nr	41.18
Running outlet						
65 mm outlet						
100 x 75 mm	18.16	23.26	0.38	10.30	nr	33.56
125 x 75 mm	23.05	29.52	0.43	11.65	nr	41.18
75 mm outlet						
100 x 75 mm	13.62	17.44	0.38	10.30	nr	27.74
125 x 75 mm	16.40	21.00	0.43	11.65	nr	32.66
Stop end outlet; socket						
65 mm outlet						
100 x 75 mm	12.16	15.57	0.38	10.30	nr	25.87
75 mm outlet						
100 x 75 mm	12.16	15.57	0.38	10.30	nr	25.87
125 x 75 mm	15.54	19.90	0.43	11.65	nr	31.56
Stop end outlet; spigot						
65 mm outlet						
100 x 75 mm	12.16	15.57	0.38	10.30	nr	25.87
75 mm outlet						
100 x 75 mm	12.16	15.57	0.38	10.30	nr	25.87
125 x 75 mm	15.54	19.90	0.43	11.65	nr	31.56

R: DISPOSAL SYSTEMS

Item	Net Price £	Material £	Labour hours	Labour £	Unit	Total rate £
Ogee gutter, with brackets measured separately						
100 mm	10.75	13.77	0.85	23.04	m	36.81
115 mm	12.01	15.38	0.97	26.29	m	41.67
125 mm	12.74	16.32	0.97	26.29	m	42.61
Brackets; fixed to backgrounds. For minimum fixing distances, refer to the Tables and Memoranda at the rear of the book						
Fascia						
100 mm	3.82	4.89	0.16	4.34	nr	9.23
115 mm	4.11	5.26	0.16	4.34	nr	9.60
125 mm	4.78	6.12	0.16	4.34	nr	10.46
Extra over fittings Ogee Cast Iron Gutter BS 460						
Union						
100 mm	4.18	5.35	0.38	10.30	nr	15.65
115 mm	4.36	5.58	0.43	11.65	nr	17.24
125 mm	5.48	7.02	0.43	11.65	nr	18.67
Stop end; internal						
100 mm	2.05	2.63	0.12	3.25	nr	5.88
115 mm	2.71	3.47	0.15	4.07	nr	7.54
125 mm	2.71	3.47	0.15	4.07	nr	7.54
Stop end; external						
100 mm	2.05	2.63	0.12	3.25	nr	5.88
115 mm	2.71	3.47	0.15	4.07	nr	7.54
125 mm	2.71	3.47	0.15	4.07	nr	7.54
90 ° angle; internal						
100 mm	6.89	8.82	0.38	10.30	nr	19.12
115 mm	7.46	9.55	0.43	11.65	nr	21.21
125 mm	8.14	10.43	0.43	11.65	nr	22.08
90 ° angle; external						
100 mm	7.01	8.98	0.38	10.30	nr	19.28
115 mm	7.61	9.75	0.43	11.65	nr	21.40
125 mm	8.31	10.64	0.43	11.65	nr	22.30
135 ° angle; internal						
100 mm	6.89	8.82	0.38	10.30	nr	19.12
115 mm	7.46	9.55	0.43	11.65	nr	21.21
125 mm	8.14	10.43	0.43	11.65	nr	22.08
135 ° angle; external						
100 mm	6.89	8.82	0.38	10.30	nr	19.12
115 mm	7.46	9.55	0.43	11.65	nr	21.21
125 mm	8.31	10.64	0.43	11.65	nr	22.30
Running outlet						
65 mm outlet						
100 mm	7.53	9.64	0.38	10.30	nr	19.94
115 mm	8.10	10.37	0.43	11.65	nr	22.03
125 mm	8.88	11.37	0.43	11.65	nr	23.03

R: DISPOSAL SYSTEMS

Item	Net Price £	Material £	Labour hours	Labour £	Unit	Total rate £
R10: RAINWATER PIPEWORK/GUTTERS (cont'd)						
Extra over fittings Ogee Cast Iron Gutter (cont'd)						
75 mm outlet						
125 mm	8.88	11.37	0.43	11.65	nr	**23.03**
Stop end outlet; socket						
65 mm outlet						
100 mm	8.24	10.55	0.38	10.30	nr	**20.85**
115 mm	8.24	10.55	0.43	11.65	nr	**22.21**
125 mm	9.50	12.17	0.43	11.65	nr	**23.82**
75 mm outlet						
125 mm	10.58	13.55	0.43	11.65	nr	**25.21**
Stop end outlet; spigot						
65 mm outlet						
100 mm	8.24	10.55	0.38	10.30	nr	**20.85**
115 mm	8.24	10.55	0.43	11.65	nr	**22.21**
125 mm	9.50	12.17	0.43	11.65	nr	**23.82**
75 mm outlet						
125 mm	3.82	4.89	0.43	11.65	nr	**16.55**
Notts Ogee Gutter, with brackets measured separately						
115 mm	17.46	22.36	0.85	23.04	m	**45.40**
Brackets; fixed to backgrounds. For minimum fixing distances, refer to the Tables and Memoranda at the rear of the book						
Fascia						
115 mm	6.11	7.83	0.16	4.34	nr	**12.16**
Extra over fittings Notts Ogee Cast Iron Gutter BS 460						
Union						
115 mm	7.15	9.16	0.38	10.30	nr	**19.46**
Stop end; internal						
115 mm	6.11	7.83	0.16	4.34	nr	**12.16**
Stop end; external						
115 mm	6.11	7.83	0.16	4.34	nr	**12.16**
90° angle; internal						
115 mm	17.08	21.88	0.43	11.65	nr	**33.53**
90° angle; external						
115 mm	17.08	21.88	0.43	11.65	nr	**33.53**
135° angle; internal						
115 mm	17.08	21.88	0.43	11.65	nr	**33.53**

R: DISPOSAL SYSTEMS

Item	Net Price £	Material £	Labour hours	Labour £	Unit	Total rate £
135 ° angle; external						
115 mm	17.08	21.88	0.43	11.65	nr	**33.53**
Running outlet						
65 mm outlet						
115 mm	20.49	26.24	0.43	11.65	nr	**37.90**
75 mm outlet						
115 mm	17.10	21.90	0.43	11.65	nr	**33.56**
Stop end outlet; socket						
65 mm outlet						
115 mm	13.52	17.32	0.43	11.65	nr	**28.97**
Stop end outlet; spigot						
65 mm outlet						
115 mm	16.23	20.79	0.43	11.65	nr	**32.44**
No 46 moulded Gutter, with brackets measured separately						
100 x 75 mm	16.57	21.22	0.85	23.04	m	**44.26**
125 x 100 mm	23.84	30.53	0.97	26.29	m	**56.82**
Brackets; fixed to backgrounds. For minimum fixing distances, refer to the Tables and Memoranda at the rear of the book						
Fascia						
100 x 75 mm	3.38	4.33	0.16	4.34	nr	**8.67**
125 x 100 mm	3.38	4.33	0.16	4.34	nr	**8.67**
Extra over fittings						
Union						
100 x 75 mm	8.84	11.32	0.38	10.30	nr	**21.62**
125 x 100 mm	9.82	12.58	0.43	11.65	nr	**24.23**
Stop end; internal						
100 x 75 mm	6.38	8.17	0.12	3.25	nr	**11.42**
125 x 100 mm	8.27	10.59	0.15	4.07	nr	**14.66**
Stop end; external						
100 x 75 mm	6.38	8.17	0.12	3.25	nr	**11.42**
125 x 100 mm	8.27	10.59	0.15	4.07	nr	**14.66**
90 ° angle; internal						
100 x 75 mm	16.74	21.44	0.38	10.30	nr	**31.74**
125 x 100 mm	24.05	30.80	0.43	11.65	nr	**42.46**
90 ° angle; external						
100 x 75 mm	16.74	21.44	0.38	10.30	nr	**31.74**
125 x 100 mm	24.05	30.80	0.43	11.65	nr	**42.46**
135 ° angle; internal						
100 x 75 mm	16.74	21.44	0.38	10.30	nr	**31.74**
125 x 100 mm	24.05	30.80	0.43	11.65	nr	**42.46**

R: DISPOSAL SYSTEMS

Item	Net Price £	Material £	Labour hours	Labour £	Unit	Total rate £
R10: RAINWATER PIPEWORK/GUTTERS (cont'd)						
Extra over fittings (cont'd)						
135 ° angle; external						
100 x 75 mm	16.74	21.44	0.38	10.30	nr	31.74
125 x 100 mm	24.05	30.80	0.43	11.65	nr	42.46
Running outlet						
65 mm outlet						
100 x 75 mm	16.74	21.44	0.38	10.30	nr	31.74
125 x 100 mm	24.05	30.80	0.43	11.65	nr	42.46
75 mm outlet						
100 x 75 mm	16.74	21.44	0.38	10.30	nr	31.74
125 x 100 mm	24.05	30.80	0.43	11.65	nr	42.46
100 mm outlet						
100 x 75 mm	24.05	30.80	0.38	10.30	nr	41.10
125 x 100 mm	24.05	30.80	0.43	11.65	nr	42.46
100 x 75 mm outlet						
125 x 100 mm	24.05	30.80	0.43	11.65	nr	42.46
Stop end outlet; socket						
65 mm outlet						
100 x 75 mm	13.27	17.00	0.38	10.30	nr	27.30
75 mm outlet						
125 x 100 mm	16.38	20.98	0.43	11.65	nr	32.63
Stop end outlet; spigot						
65 mm outlet						
100 x 75 mm	13.27	17.00	0.38	10.30	nr	27.30
75 mm outlet						
125 x 100 mm	16.38	20.98	0.43	11.65	nr	32.63
Box gutter, with brackets measured separately						
100 x 75 mm	27.80	35.61	0.85	23.04	m	58.64
Brackets; fixed to backgrounds. For minimum fixing distances, refer to the Tables and Memoranda at the rear of the book						
Fascia						
100 x 75 mm	3.73	4.78	0.16	4.34	nr	9.11
Extra over fittings Box Cast Iron Gutter BS 460						
Union						
100 x 75 mm	4.75	6.08	0.38	10.30	nr	16.38
Stop end; external						
100 x 75 mm	3.66	4.69	0.12	3.25	nr	7.94
90 ° angle						
100 x 75 mm	12.64	16.19	0.38	10.30	nr	26.49

R: DISPOSAL SYSTEMS

Item	Net Price £	Material £	Labour hours	Labour £	Unit	Total rate £
135° angle						
100 x 75 mm	12.91	16.54	0.38	10.30	nr	**26.83**
Running outlet						
65 mm outlet						
100 x 75 mm	12.64	16.19	0.38	10.30	nr	**26.49**
75 mm outlet						
100 x 75 mm	12.64	16.19	0.38	10.30	nr	**26.49**
100 x 75 mm outlet						
100 x 75 mm	12.64	16.19	0.38	10.30	nr	**26.49**
Cast iron rainwater pipe; dry joints; BS 460; fixed to backgrounds						
Circular						
Plain socket pipe, with brackets measured separately						
65mm	16.49	21.12	0.69	18.70	m	**39.82**
75 mm	16.49	21.12	0.69	18.70	m	**39.82**
100 mm	22.51	28.83	0.69	18.70	m	**47.53**
Bracket; fixed to backgrounds. For minimum fixing distances, refer to the Tables and Memoranda at the rear of the book						
65mm	5.26	6.74	0.29	7.86	nr	**14.60**
75 mm	5.30	6.79	0.29	7.86	nr	**14.65**
100 mm	5.30	6.79	0.29	7.86	nr	**14.65**
Eared socket pipe, with wall spacers measured separately						
65mm	17.01	21.79	0.62	16.80	m	**38.59**
75 mm	17.01	21.79	0.62	16.80	m	**38.59**
100 mm	22.83	29.24	0.62	16.80	m	**46.04**
Wall spacer plate; eared pipework						
65mm	3.66	4.69	0.16	4.34	nr	**9.02**
75 mm	3.62	4.64	0.16	4.34	nr	**8.97**
100 mm	3.79	4.85	0.16	4.34	nr	**9.19**
Extra over fittings Circular Cast Iron Pipework BS 460						
Loose sockets						
Plain socket						
65mm	12.27	15.72	0.23	6.23	nr	**21.95**
75 mm	12.27	15.72	0.23	6.23	nr	**21.95**
100 mm	16.51	21.15	0.23	6.23	nr	**27.38**
Eared socket						
65mm	16.84	21.57	0.29	7.86	nr	**29.43**
75 mm	16.84	21.57	0.29	7.86	nr	**29.43**
100 mm	22.53	28.86	0.29	7.86	nr	**36.72**

R: DISPOSAL SYSTEMS

Item	Net Price £	Material £	Labour hours	Labour £	Unit	Total rate £
R10: RAINWATER PIPEWORK/GUTTERS (cont'd)						
Extra over fittings Circular (cont'd)						
Shoe; front projection						
Plain socket						
65mm	16.84	21.57	0.23	6.23	nr	27.80
75 mm	16.84	21.57	0.23	6.23	nr	27.80
100 mm	22.53	28.86	0.23	6.23	nr	35.09
Eared socket						
65mm	15.00	19.21	0.29	7.86	nr	27.07
75 mm	15.00	19.21	0.29	7.86	nr	27.07
100 mm	18.73	23.99	0.29	7.86	nr	31.85
Access Pipe						
65mm	23.40	29.97	0.23	6.23	nr	36.20
75 mm	24.58	31.48	0.23	6.23	nr	37.72
100 mm	42.93	54.98	0.23	6.23	nr	61.22
100 mm; eared	48.41	62.00	0.29	7.86	nr	69.86
Bends; any °						
65mm	9.18	11.76	0.23	6.23	nr	17.99
75 mm	11.17	14.31	0.23	6.23	nr	20.54
100 mm	22.51	28.83	0.23	6.23	nr	35.06
Branch						
92.5 °s						
65mm	17.72	22.70	0.29	7.86	nr	30.56
75 mm	19.53	25.01	0.29	7.86	nr	32.87
100 mm	23.22	29.74	0.29	7.86	nr	37.60
112.5 °s						
65mm	14.19	18.17	0.29	7.86	nr	26.03
75 mm	15.65	20.04	0.29	7.86	nr	27.90
135 °s						
65mm	17.72	22.70	0.29	7.86	nr	30.56
75 mm	19.53	25.01	0.29	7.86	nr	32.87
Offsets						
75 to 150 mm projection						
65mm	14.07	18.02	0.25	6.78	nr	24.80
75 mm	14.07	18.02	0.25	6.78	nr	24.80
100 mm	26.53	33.98	0.25	6.78	nr	40.76
225 mm projection						
65mm	16.38	20.98	0.25	6.78	nr	27.76
75 mm	16.38	20.98	0.25	6.78	nr	27.76
100 mm	32.15	41.18	0.25	6.78	nr	47.95
305 mm projection						
65mm	19.17	24.55	0.25	6.78	nr	31.33
75 mm	20.13	25.78	0.25	6.78	nr	32.56
100 mm	32.15	41.18	0.25	6.78	nr	47.95

R: DISPOSAL SYSTEMS

Item	Net Price £	Material £	Labour hours	Labour £	Unit	Total rate £
380 mm projection						
65mm	38.27	49.02	0.25	6.78	nr	55.79
75 mm	38.27	49.02	0.25	6.78	nr	55.79
100 mm	52.24	66.91	0.25	6.78	nr	73.68
455 mm projection						
65mm	44.80	57.38	0.25	6.78	nr	64.16
75 mm	44.80	57.38	0.25	6.78	nr	64.16
100 mm	63.50	81.33	0.25	6.78	nr	88.11
Rectangular						
Plain socket						
100 x 75 mm	66.87	85.65	1.04	28.19	m	113.83
Bracket; fixed to backgrounds. For minimum fixing distances, refer to the Tables and Memoranda at the rear of the book						
100 x 75mm; build in holdabat	23.28	29.82	0.35	9.49	nr	39.30
100 x 75mm; trefoil earband	18.21	23.32	0.29	7.86	nr	31.18
100 x 75mm; plain earband	17.60	22.54	0.29	7.86	nr	30.40
Eared Socket, with wall spacers measured separately						
100 x 75 mm	64.01	81.98	1.16	31.44	m	113.42
Wall spacer plate; eared pipework						
100 x 75	3.79	4.85	0.16	4.34	nr	9.19
Extra over fittings Rectangular Cast Iron Pipework BS 460						
Loose socket						
100 x 75 mm; plain	17.02	21.80	0.23	6.23	nr	28.03
100 x 75 mm; eared	31.78	40.70	0.29	7.86	nr	48.56
Shoe; front						
100 x 75 mm; plain	17.09	21.89	0.23	6.23	nr	28.12
100 x 75 mm; eared	28.31	36.26	0.29	7.86	nr	44.12
Shoe; side						
100 x 75 mm; plain	45.13	57.80	0.23	6.23	nr	64.04
100 x 75 mm; eared	57.39	73.50	0.29	7.86	nr	81.36
Bends; side; any °						
100 x 75 mm; plain	40.87	52.35	0.25	6.78	nr	59.12
100 x 75 mm; 135 °; plain	43.27	55.42	0.25	6.78	nr	62.20
Bends; side; any °						
100 x 75 mm; eared	51.09	65.44	0.25	6.78	nr	72.21
Bends; front; any °						
100 x 75 mm; plain	40.87	52.35	0.25	6.78	nr	59.12
100 x 75 mm; eared	51.09	65.44	0.25	6.78	nr	72.21
Offset; side;						
Plain socket						
75mm projection	55.80	71.47	0.25	6.78	nr	78.24
115mm projection	58.02	74.31	0.25	6.78	nr	81.09
225mm projection	72.27	92.56	0.25	6.78	nr	99.34
305mm projection	83.31	106.70	0.25	6.78	nr	113.48

R: DISPOSAL SYSTEMS

Item	Net Price £	Material £	Labour hours	Labour £	Unit	Total rate £
R10: RAINWATER PIPEWORK/GUTTERS (cont'd)						
Extra over fittings Rectangular (cont'd)						
Offset; front						
Plain socket						
75mm projection	55.80	71.47	0.25	6.78	nr	78.24
150mm projection	59.03	75.61	0.25	6.78	nr	82.38
225mm projection	72.27	92.56	0.25	6.78	nr	99.34
305mm projection	83.31	106.70	0.25	6.78	nr	113.48
Eared socket						
75mm projection	53.27	68.23	0.25	6.78	nr	75.00
150mm projection	59.03	75.61	0.25	6.78	nr	82.38
225mm projection	75.38	96.55	0.25	6.78	nr	103.32
305mm projection	80.57	103.19	0.25	6.78	nr	109.97
Offset; plinth						
115mm projection; plain	43.80	56.10	0.25	6.78	nr	62.87
115mm projection; eared	55.33	70.87	0.25	6.78	nr	77.64
Rainwater heads						
Flat hopper						
210 x 160 x 185 mm; 65 mm outlet	11.73	15.02	0.40	10.84	nr	25.87
210 x 160 x 185 mm; 75 mm outlet	13.33	17.07	0.40	10.84	nr	27.91
250 x 215 x 215 mm; 100 mm outlet	29.51	37.80	0.40	10.84	nr	48.64
Flat rectangular						
225 x 125 x 125 mm; 65 mm outlet	20.49	26.24	0.40	10.84	nr	37.08
225 x 125 x 125 mm; 75 mm outlet	20.49	26.24	0.40	10.84	nr	37.08
280 x 150 x 130 mm; 100 mm outlet	28.29	36.23	0.40	10.84	nr	47.07
Rectangular						
250 x 180 x 175mm; 75 mm outlet	29.41	37.67	0.40	10.84	nr	48.51
250 x 180 x 175mm; 100 mm outlet	31.00	39.70	0.40	10.84	nr	50.55
300 x 250 x 200mm; 65 mm outlet	55.11	70.58	0.40	10.84	nr	81.43
300 x 250 x 200mm; 75 mm outlet	55.11	70.58	0.40	10.84	nr	81.43
300 x 250 x 200mm; 100 mm outlet	55.11	70.58	0.40	10.84	nr	81.43
300 x 250 x 200mm; 100 x 75 mm outlet	55.11	70.58	0.40	10.84	nr	81.43
Castellated rectangular						
250 x 180 x 175mm; 65 mm outlet	55.11	70.58	0.40	10.84	nr	81.43

R: DISPOSAL SYSTEMS

Item	Net Price £	Material £	Labour hours	Labour £	Unit	Total rate £
R11: ABOVE GROUND DRAINAGE						
Pricing note: Degree angles are only indicated where material prices differ PVC-U overflow pipe; solvent welded joints; fixed with clips to backgrounds						
Pipe, with brackets measured separately						
19mm	2.14	2.74	0.21	5.69	m	**8.43**
Fixings						
Pipe clip: including fixing to backgrounds. For minimum fixing distances, refer to the Tables and Memoranda at the rear of the book						
19mm	0.36	0.46	0.18	4.88	nr	**5.34**
Extra over fittings overflow pipework PVC-U						
Straight coupler						
19mm	0.84	1.08	0.17	4.61	nr	**5.68**
Bend						
19mm: 91.25 °	1.00	1.28	0.17	4.61	nr	**5.89**
19mm: 135 °	1.02	1.31	0.17	4.61	nr	**5.91**
Tee						
19mm	1.09	1.40	0.18	4.88	nr	**6.27**
Reverse nut connector						
19mm	0.42	0.54	0.15	4.07	nr	**4.60**
BSP adaptor: solvent welded socket to threaded socket						
19mm x 3/4"	1.40	1.79	0.14	3.79	nr	**5.59**
Straight tank connector						
19mm	1.33	1.70	0.21	5.69	nr	**7.40**
32mm	2.34	3.00	0.28	7.59	nr	**10.59**
40mm	2.54	3.25	0.30	8.13	nr	**11.38**
Bent tank connector						
19mm	1.57	2.01	0.21	5.69	nr	**7.70**
Tundish						
19mm	22.57	28.91	0.38	10.30	nr	**39.21**
MuPVC waste pipe; solvent welded joints; fixed with clips to backgrounds; BS 5255						
Pipe, with brackets measured separately						
32mm	1.89	2.42	0.23	6.23	m	**8.65**
40mm	2.33	2.98	0.23	6.23	m	**9.22**
50mm	3.61	4.62	0.26	7.05	m	**11.67**

R: DISPOSAL SYSTEMS

Item	Net Price £	Material £	Labour hours	Labour £	Unit	Total rate £
R11: ABOVE GROUND DRAINAGE (cont'd)						
Fixings						
Pipe clip: including fixing to backgrounds. For minimum fixing distances, refer to the Tables and Memoranda at the rear of the book						
32mm	0.29	0.37	0.13	3.52	nr	3.89
40mm	0.38	0.49	0.13	3.52	nr	4.01
50mm	0.70	0.90	0.13	3.52	nr	4.42
Pipe clip: expansion: including fixing to backgrounds. For minimum fixing distances, refer to the Tables and Memoranda at the rear of the book						
32mm	0.32	0.41	0.13	3.52	nr	3.93
40mm	0.41	0.53	0.13	3.52	nr	4.05
50mm	1.12	1.43	0.13	3.52	nr	4.96
Pipe clip: metal; including fixing to backgrounds. For minimum fixing distances, refer to the Tables and Memoranda at the rear of the book						
32mm	1.31	1.68	0.13	3.52	nr	5.20
40mm	1.55	1.99	0.13	3.52	nr	5.51
50mm	1.96	2.51	0.13	3.52	nr	6.03
Extra over fittings waste pipework MuPVC						
Screwed access plug						
32mm	1.31	1.68	0.18	4.88	nr	6.56
40mm	1.38	1.77	0.18	4.88	nr	6.65
50mm	2.26	2.89	0.25	6.78	nr	9.67
Straight coupling						
32mm	0.86	1.10	0.27	7.32	nr	8.42
40mm	1.02	1.31	0.27	7.32	nr	8.62
50mm	1.57	2.01	0.27	7.32	nr	9.33
Expansion coupling						
32mm	1.51	1.93	0.27	7.32	nr	9.25
40mm	1.83	2.34	0.27	7.32	nr	9.66
50mm	2.47	3.16	0.27	7.32	nr	10.48
MuPVC to copper coupling						
32mm	1.39	1.78	0.27	7.32	nr	9.10
40mm	1.56	2.00	0.27	7.32	nr	9.32
50mm	2.09	2.68	0.27	7.32	nr	9.99
Spigot and socket coupling						
32mm	1.51	1.93	0.27	7.32	nr	9.25
40mm	1.83	2.34	0.27	7.32	nr	9.66
50mm	2.47	3.16	0.27	7.32	nr	10.48
Union						
32mm	3.60	4.61	0.28	7.59	nr	12.20
40mm	4.73	6.06	0.28	7.59	nr	13.65
50mm	7.38	9.45	0.28	7.59	nr	17.04

Material Costs/Prices for Measured Works – Mechanical Installations

R: DISPOSAL SYSTEMS

Item	Net Price £	Material £	Labour hours	Labour £	Unit	Total rate £
Reducer: socket						
32 x 19mm	1.08	1.38	0.27	7.32	nr	8.70
40 x 32mm	0.97	1.24	0.27	7.32	nr	8.56
50 x 32mm	1.66	2.13	0.27	7.32	nr	9.44
50 x 40mm	1.68	2.15	0.27	7.32	nr	9.47
Reducer: level invert						
40 x 32mm	1.48	1.90	0.27	7.32	nr	9.21
50 x 32mm	2.06	2.64	0.27	7.32	nr	9.96
50 x 40mm	1.96	2.51	0.27	7.32	nr	9.83
Swept bend						
32mm	1.35	1.73	0.27	7.32	nr	9.05
32mm: 165°	1.63	2.09	0.27	7.32	nr	9.41
40mm	1.50	1.92	0.27	7.32	nr	9.24
40mm: 165°	2.12	2.72	0.27	7.32	nr	10.03
50mm	2.49	3.19	0.30	8.13	nr	11.32
50mm: 165°	2.77	3.55	0.30	8.13	nr	11.68
Knuckle bend						
32mm	1.24	1.59	0.27	7.32	nr	8.91
40mm	1.36	1.74	0.27	7.32	nr	9.06
Spigot and socket bend						
32mm	1.53	1.96	0.27	7.32	nr	9.28
32mm: 150°	1.49	1.91	0.27	7.32	nr	9.23
40mm	1.70	2.18	0.27	7.32	nr	9.50
50mm	4.03	5.16	0.30	8.13	nr	13.29
Swept tee						
32mm: 91.25°	1.91	2.45	0.31	8.40	nr	10.85
32mm: 135°	2.26	2.89	0.31	8.40	nr	11.30
40mm: 91.25°	2.40	3.07	0.31	8.40	nr	11.48
40mm: 135°	2.85	3.65	0.31	8.40	nr	12.05
50mm	4.38	5.61	0.31	8.40	nr	14.01
Swept cross						
40mm: 91.25°	6.79	8.70	0.31	8.40	nr	17.10
50mm: 91.25°	7.80	9.99	0.43	11.65	nr	21.64
50mm: 135°	16.75	21.45	0.31	8.40	nr	29.86
Male iron adaptor						
32mm	1.38	1.77	0.28	7.59	nr	9.36
40mm	1.63	2.09	0.28	7.59	nr	9.68
Female iron adaptor						
32mm	1.38	1.77	0.28	7.59	nr	9.36
40mm	1.63	2.09	0.28	7.59	nr	9.68
50mm	2.34	3.00	0.31	8.40	nr	11.40
Reverse nut adaptor						
32mm	1.97	2.52	0.20	5.42	nr	7.94
40mm	1.97	2.52	0.20	5.42	nr	7.94
Automatic air admittance valve						
32mm	11.54	14.78	0.27	7.32	nr	22.10
40mm	11.54	14.78	0.28	7.59	nr	22.37
50mm	11.54	14.78	0.31	8.40	nr	23.18

R: DISPOSAL SYSTEMS

Item	Net Price £	Material £	Labour hours	Labour £	Unit	Total rate £
R11: ABOVE GROUND DRAINAGE (cont'd)						
Extra over fittings waste pipework MuPVC (cont'd)						
MuPVC to metal adaptor: including heat shrunk joint to metal						
50mm	4.73	6.06	0.38	10.30	nr	16.36
Caulking bush: including joint to metal						
32mm	2.06	2.64	0.31	8.40	nr	11.04
40mm	2.06	2.64	0.31	8.40	nr	11.04
50mm	2.06	2.64	0.32	8.67	nr	11.31
Weathering apron						
50mm	1.72	2.20	0.65	17.62	nr	19.82
Vent Cowl						
50mm	1.76	2.25	0.19	5.15	nr	7.40
ABS waste pipe; solvent welded joints; fixed with clips to backgrounds; BS 5255						
Pipe, with brackets measured separately						
32mm	1.29	1.65	0.23	6.23	m	7.89
40mm	1.55	1.99	0.23	6.23	m	8.22
50mm	2.04	2.61	0.26	7.05	m	9.66
Fixings						
Pipe clip: including fixing to backgrounds. For minimum fixing distances, refer to the Tables and Memoranda at the rear of the book						
32mm	0.28	0.36	0.17	4.61	nr	4.97
40mm	0.36	0.46	0.17	4.61	nr	5.07
50mm	0.68	0.87	0.17	4.61	nr	5.48
Pipe clip: expansion: including fixing to backgrounds. For minimum fixing distances, refer to the Tables and Memoranda at the rear of the book						
32mm	0.31	0.40	0.17	4.61	nr	5.00
40mm	0.38	0.49	0.17	4.61	nr	5.09
50mm	0.97	1.24	0.17	4.61	nr	5.85
Pipe clip: metal; including fixing to backgrounds. For minimum fixing distances, refer to the Tables and Memoranda at the rear of the book						
32mm	1.26	1.61	0.17	4.61	nr	6.22
40mm	1.50	1.92	0.17	4.61	nr	6.53
50mm	1.91	2.45	0.17	4.61	nr	7.05
Extra over fittings waste pipework ABS						
Screwed access plug						
32mm	0.64	0.82	0.18	4.88	nr	5.70
40mm	0.64	0.82	0.18	4.88	nr	5.70
50mm	1.47	1.88	0.25	6.78	nr	8.66

Material Costs/Prices for Measured Works – Mechanical Installations

R: DISPOSAL SYSTEMS

Item	Net Price £	Material £	Labour hours	Labour £	Unit	Total rate £
Straight coupling						
32mm	0.64	0.82	0.27	7.32	nr	**8.14**
40mm	0.64	0.82	0.27	7.32	nr	**8.14**
50mm	1.47	1.88	0.27	7.32	nr	**9.20**
Expansion coupling						
32mm	1.47	1.88	0.27	7.32	nr	**9.20**
40mm	1.47	1.88	0.27	7.32	nr	**9.20**
50mm	2.95	3.78	0.27	7.32	nr	**11.10**
ABS to Copper coupling						
32mm	1.47	1.88	0.27	7.32	nr	**9.20**
40mm	1.47	1.88	0.27	7.32	nr	**9.20**
50mm	2.95	3.78	0.27	7.32	nr	**11.10**
Reducer: socket						
40 x 32mm	0.64	0.82	0.27	7.32	nr	**8.14**
50 x 32mm	0.64	0.82	0.27	7.32	nr	**8.14**
50 x 40mm	1.47	1.88	0.27	7.32	nr	**9.20**
Swept bend						
32mm	0.64	0.82	0.27	7.32	nr	**8.14**
40mm	0.64	0.82	0.27	7.32	nr	**8.14**
50mm	1.47	1.88	0.30	8.13	nr	**10.01**
Knuckle bend						
32mm	0.64	0.82	0.27	7.32	nr	**8.14**
40mm	0.64	0.82	0.27	7.32	nr	**8.14**
Swept tee						
32mm	0.64	0.82	0.31	8.40	nr	**9.22**
40mm	0.64	0.82	0.31	8.40	nr	**9.22**
50mm	1.47	1.88	0.31	8.40	nr	**10.28**
Swept cross						
40mm	4.21	5.39	0.23	6.23	nr	**11.63**
50mm	5.40	6.92	0.43	11.65	nr	**18.57**
Male iron adaptor						
32mm	1.27	1.63	0.28	7.59	nr	**9.22**
40mm	1.55	1.99	0.28	7.59	nr	**9.57**
Female iron adaptor						
32mm	1.23	1.58	0.28	7.59	nr	**9.16**
40mm	1.55	1.99	0.28	7.59	nr	**9.57**
50mm	2.04	2.61	0.31	8.40	nr	**11.01**
Tank connectors						
32mm	2.39	3.06	0.29	7.86	nr	**10.92**
40mm	2.60	3.33	0.29	7.86	nr	**11.19**
Caulking bush: including joint to pipework						
50mm	2.39	3.06	0.50	13.55	nr	**16.61**

R: DISPOSAL SYSTEMS

Item	Net Price £	Material £	Labour hours	Labour £	Unit	Total rate £
R11: ABOVE GROUND DRAINAGE (cont'd)						
Fixings						
Polypropylene waste pipe; push fit joints; fixed with clips to backgrounds; BS 5254						
Pipe, with brackets measured separately						
32mm	1.04	1.33	0.21	5.69	m	7.02
40mm	1.19	1.52	0.21	5.69	m	7.22
50mm	1.89	2.42	0.38	10.30	m	12.72
Pipe clip: saddle; including fixing to backgrounds. For minimum fixing distances, refer to the Tables and Memoranda at the rear of the book						
32mm	0.22	0.28	0.17	4.61	nr	4.89
40mm	0.22	0.28	0.17	4.61	nr	4.89
Pipe clip: including fixing to backgrounds. For minimum fixing distances, refer to the Tables and Memoranda at the rear of the book						
50mm	0.55	0.70	0.17	4.61	nr	5.31
Extra over fittings waste pipework polypropylene						
Screwed access plug						
32mm	0.64	0.82	0.16	4.34	nr	5.16
40mm	0.64	0.82	0.16	4.34	nr	5.16
50mm	1.11	1.42	0.20	5.42	nr	6.84
Straight coupling						
32mm	0.64	0.82	0.19	5.15	nr	5.97
40mm	0.64	0.82	0.19	5.15	nr	5.97
50mm	1.11	1.42	0.20	5.42	nr	6.84
Universal waste pipe coupler						
32mm dia.	1.50	1.92	0.20	5.42	nr	7.34
40mm dia.	1.57	2.01	0.20	5.42	nr	7.43
Reducer						
40 x 32mm	1.75	2.24	0.19	5.15	nr	7.39
50 x 32mm	0.64	0.82	0.19	5.15	nr	5.97
50 x 40mm	1.11	1.42	0.20	5.42	nr	6.84
Swept bend						
32mm	0.64	0.82	0.19	5.15	nr	5.97
40mm	0.64	0.82	0.19	5.15	nr	5.97
50mm	1.11	1.42	0.20	5.42	nr	6.84
Knuckle bend						
32mm	0.64	0.82	0.19	5.15	nr	5.97
40mm	0.64	0.82	0.19	5.15	nr	5.97
50mm	1.11	1.42	0.20	5.42	nr	6.84
Spigot and socket bend						
32mm	0.64	0.82	0.19	5.15	nr	5.97
40mm	0.64	0.82	0.19	5.15	nr	5.97

Material Costs/Prices for Measured Works – Mechanical Installations

R: DISPOSAL SYSTEMS

Item	Net Price £	Material £	Labour hours	Labour £	Unit	Total rate £
Swept tee						
32mm	0.64	0.82	0.22	5.96	nr	6.78
40mm	0.64	0.82	0.22	5.96	nr	6.78
50mm	1.15	1.47	0.23	6.23	nr	7.71
Male iron adaptor						
32mm	1.21	1.55	0.13	3.52	nr	5.07
40mm	1.39	1.78	0.19	5.15	nr	6.93
50mm	1.75	2.24	0.15	4.07	nr	6.31
Tank connector						
32mm	0.64	0.82	0.24	6.50	nr	7.32
40mm	0.64	0.82	0.24	6.50	nr	7.32
50mm	1.04	1.33	0.35	9.49	nr	10.82
Polypropylene traps; including fixing to appliance and connection to pipework; BS 3943						
Tubular P trap; 75mm seal						
32mm dia.	3.45	4.42	0.20	5.42	nr	9.84
40mm dia.	3.98	5.10	0.20	5.42	nr	10.52
Tubular S trap; 75mm seal						
32mm dia.	4.36	5.58	0.20	5.42	nr	11.00
40mm dia.	5.12	6.56	0.20	5.42	nr	11.98
Running tubular P trap; 75mm seal						
32mm dia.	5.29	6.78	0.20	5.42	nr	12.20
40mm dia.	5.78	7.40	0.20	5.42	nr	12.82
Running tubular S trap; 75mm seal						
32mm dia.	6.36	8.15	0.20	5.42	nr	13.57
40mm dia.	6.86	8.79	0.20	5.42	nr	14.21
Spigot and socket bend; converter from P to S						
Trap						
32mm	1.35	1.73	0.20	5.42	nr	7.15
40mm	1.45	1.86	0.21	5.69	nr	7.55
Bottle P trap; 75mm seal						
32mm dia.	3.84	4.92	0.20	5.42	nr	10.34
40mm dia.	4.59	5.88	0.20	5.42	nr	11.30
Bottle S trap; 75mm seal						
32mm dia.	4.63	5.93	0.20	5.42	nr	11.35
40mm dia.	5.63	7.21	0.25	6.78	nr	13.99
Bottle P trap; resealing; 75mm seal						
32mm dia.	4.78	6.12	0.20	5.42	nr	11.54
40mm dia.	5.59	7.16	0.25	6.78	nr	13.94
Bottle S trap; resealing; 75mm seal						
32mm dia.	5.47	7.01	0.20	5.42	nr	12.43
40mm dia.	6.34	8.12	0.25	6.78	nr	14.90
Bath trap, low level; 38mm seal						
40mm dia.	4.79	6.13	0.25	6.78	nr	12.91

R: DISPOSAL SYSTEMS

Item	Net Price £	Material £	Labour hours	Labour £	Unit	Total rate £
R11: ABOVE GROUND DRAINAGE (cont'd)						
Spigot and socket bend (cont'd)						
Bath trap, low level; 38mm seal complete with overflow hose						
40mm dia.	7.41	9.49	0.25	6.78	nr	16.27
Bath trap; 75mm seal complete with overflow hose						
40mm dia.	7.38	9.45	0.25	6.78	nr	16.23
Bath trap; 75mm seal complete with overflow hose and overflow outlet						
40mm dia.	12.59	16.13	0.20	5.42	nr	21.55
Bath trap; 75mm seal complete with overflow hose, overflow outlet and ABS chrome waste						
40mm dia.	17.41	22.30	0.20	5.42	nr	27.72
Washing machine trap; 75mm seal including stand pipe						
40mm dia.	10.02	12.83	0.25	6.78	nr	19.61
Washing machine standpipe						
40mm dia.	5.23	6.70	0.25	6.78	nr	13.47
Plastic unslotted chrome plated basin/sink waste including plug						
32mm	6.30	8.07	0.34	9.22	nr	17.28
40mm	8.42	10.78	0.34	9.22	nr	20.00
Plastic slotted chrome plated basin/sink waste including plug						
32mm	4.99	6.39	0.34	9.22	nr	15.61
40mm	8.36	10.71	0.34	9.22	nr	19.92
Bath overflow outlet; plastic; white						
42mm	4.38	5.61	0.37	10.03	nr	15.64
Bath overflow outlet; plastic; chrome plated						
42mm	5.80	7.43	0.37	10.03	nr	17.46
Combined cistern and bath overflow outlet; plastic; white						
42mm	9.01	11.54	0.39	10.57	nr	22.11
Combined cistern and bath overflow outlet; plastic; chrome plated						
42mm	9.00	11.53	0.39	10.57	nr	22.10
Cistern overflow outlet; plastic; white						
42mm	7.25	9.29	0.15	4.07	nr	13.35
Cistern overflow outlet; plastic; chrome plated						
42mm	6.54	8.38	0.15	4.07	nr	12.44

R: DISPOSAL SYSTEMS

Item	Net Price £	Material £	Labour hours	Labour £	Unit	Total rate £
PVC-U soil and waste pipe; solvent welded joints; fixed with clips to backgrounds; BS 4514/ BS EN 607						
Pipe, with brackets measured separately						
82mm	7.04	9.02	0.35	9.49	m	**18.50**
110mm	7.09	9.08	0.41	11.11	m	**20.19**
160mm	19.20	24.59	0.51	13.82	m	**38.41**
Fixings						
Galvanised steel pipe clip: including fixing to backgrounds. For minimum fixing distances, refer to the Tables and Memoranda at the rear of the book						
82mm	2.36	3.02	0.18	4.88	nr	**7.90**
110mm	2.44	3.13	0.18	4.88	nr	**8.00**
160mm	5.91	7.57	0.18	4.88	nr	**12.45**
Plastic coated steel pipe clip: including fixing to backgrounds. For minimum fixing distances, refer to the Tables and Memoranda at the rear of the book						
82mm	3.25	4.16	0.18	4.88	nr	**9.04**
110mm	3.25	4.16	0.18	4.88	nr	**9.04**
160mm	5.68	7.27	0.18	4.88	nr	**12.15**
Plastic pipe clip: including fixing to backgrounds. For minimum fixing distances, refer to the Tables and Memoranda at the rear of the book						
82mm	3.25	4.16	0.18	4.88	nr	**9.04**
110mm	3.25	4.16	0.18	4.88	nr	**9.04**
Plastic coated steel pipe clip: adjustable; including fixing to backgrounds. For minimum fixing distances, refer to the Tables and Memoranda at the rear of the book						
82mm	3.23	4.14	0.20	5.42	nr	**9.56**
110mm	3.33	4.26	0.20	5.42	nr	**9.69**
Galvanised steel pipe clip: drive in; including fixing to backgrounds. For minimum fixing distances, refer to the Tables and Memoranda at the rear of the book						
110mm	5.09	6.52	0.22	5.96	nr	**12.48**
Extra over fittings solvent welded pipework PVC-U						
Straight coupling						
82mm	3.16	4.05	0.21	5.69	nr	**9.74**
110mm	3.24	4.15	0.22	5.96	nr	**10.11**
160mm	9.55	12.23	0.24	6.50	nr	**18.74**
Expansion coupling						
82mm	3.16	4.05	0.21	5.69	nr	**9.74**
110mm	3.82	4.89	0.22	5.96	nr	**10.86**
160mm	6.17	7.90	0.24	6.50	nr	**14.41**

R: DISPOSAL SYSTEMS

Item	Net Price £	Material £	Labour hours	Labour £	Unit	Total rate £
R11: ABOVE GROUND DRAINAGE (cont'd)						
Extra over fittings solvent welded (cont'd)						
Slip coupling; double ring socket						
82mm	9.40	12.04	0.21	5.69	nr	17.73
110mm	11.75	15.05	0.22	5.96	nr	21.01
160mm	28.87	36.98	0.24	6.50	nr	43.48
Puddle flanges						
110mm	90.18	115.50	0.45	12.20	nr	127.70
160mm	145.73	186.65	0.55	14.91	nr	201.56
Socket reducer						
82 to 50mm	4.58	5.87	0.18	4.88	nr	10.74
110 to 50mm	5.74	7.35	0.18	4.88	nr	12.23
110 to 82mm	5.91	7.57	0.22	5.96	nr	13.53
160 to 110mm	11.97	15.33	0.26	7.05	nr	22.38
Socket plugs						
82mm	4.48	5.74	0.15	4.07	nr	9.80
110mm	4.82	6.17	0.20	5.42	nr	11.59
160mm	9.96	12.76	0.27	7.32	nr	20.07
Access door; including cutting into pipe						
82mm	10.11	12.95	0.28	7.59	nr	20.54
110mm	10.11	12.95	0.34	9.22	nr	22.16
160mm	18.05	23.12	0.46	12.47	nr	35.59
Screwed access cap						
82mm	7.15	9.16	0.15	4.07	nr	13.22
110mm	8.42	10.78	0.20	5.42	nr	16.20
160mm	15.85	20.30	0.27	7.32	nr	27.62
Access pipe: spigot and socket						
110mm	12.45	15.95	0.22	5.96	nr	21.91
Access pipe: double socket						
110mm	12.45	15.95	0.22	5.96	nr	21.91
Swept bend						
82mm	7.59	9.72	0.29	7.86	nr	17.58
110mm	8.88	11.37	0.32	8.67	nr	20.05
160mm	22.12	28.33	0.49	13.28	nr	41.61
Bend; special angle						
82mm	14.68	18.80	0.29	7.86	nr	26.66
110mm	17.51	22.43	0.32	8.67	nr	31.10
160mm	29.55	37.85	0.49	13.28	nr	51.13
Spigot and socket bend						
82mm	7.35	9.41	0.26	7.05	nr	16.46
110mm	8.60	11.01	0.32	8.67	nr	19.69
110mm: 135 °	9.67	12.39	0.32	8.67	nr	21.06
160mm: 135 °	21.40	27.41	0.44	11.93	nr	39.33
Variable bend: single socket						
110mm	16.61	21.27	0.33	8.94	nr	30.22

R: DISPOSAL SYSTEMS

Item	Net Price £	Material £	Labour hours	Labour £	Unit	Total rate £
Variable bend: double socket						
110mm	16.64	21.31	0.33	8.94	nr	**30.26**
Access bend						
110mm	24.64	31.56	0.33	8.94	nr	**40.50**
Single branch: two bosses						
82mm	10.61	13.59	0.35	9.49	nr	**23.08**
82mm: 104 °	10.61	13.59	0.35	9.49	nr	**23.08**
110mm	11.17	14.31	0.42	11.38	nr	**25.69**
110mm: 135 °	12.25	15.69	0.42	11.38	nr	**27.07**
160mm	26.69	34.18	0.50	13.55	nr	**47.74**
160mm: 135 °	47.60	60.97	0.50	13.55	nr	**74.52**
Single branch; four bosses						
110mm	14.70	18.83	0.42	11.38	nr	**30.21**
Single access branch						
82mm	48.12	61.63	0.35	9.49	nr	**71.12**
110mm	28.22	36.14	0.42	11.38	nr	**47.53**
Unequal single branch						
160 x 160 x 110mm	28.16	36.07	0.50	13.55	nr	**49.62**
160 x 160 x 110mm: 135 °	30.27	38.77	0.50	13.55	nr	**52.32**
Double branch						
110mm	27.91	35.75	0.42	11.38	nr	**47.13**
110mm: 135 °	30.33	38.85	0.42	11.38	nr	**50.23**
Corner branch						
110mm	51.27	65.67	0.42	11.38	nr	**77.05**
Unequal double branch						
160 x 160 x 110mm	52.40	67.11	0.50	13.55	nr	**80.67**
Single boss pipe; single socket						
110 x 110 x 32mm	7.19	9.21	0.24	6.50	nr	**15.71**
110 x 110 x 40mm	7.19	9.21	0.24	6.50	nr	**15.71**
110 x 110 x 50mm	6.93	8.88	0.24	6.50	nr	**15.38**
Single boss pipe; triple socket						
110 x 110 x 40mm	4.95	6.34	0.24	6.50	nr	**12.84**
Waste boss; including cutting into pipe						
82 to 32mm	4.14	5.30	0.29	7.86	nr	**13.16**
82 to 40mm	4.14	5.30	0.29	7.86	nr	**13.16**
110 to 32mm	4.14	5.30	0.29	7.86	nr	**13.16**
110 to 40mm	4.14	5.30	0.29	7.86	nr	**13.16**
110 to 50mm	4.30	5.51	0.29	7.86	nr	**13.37**
160 to 32mm	5.86	7.51	0.30	8.13	nr	**15.64**
160 to 40mm	5.86	7.51	0.35	9.49	nr	**16.99**
160 to 50mm	5.86	7.51	0.40	10.84	nr	**18.35**
Self locking waste boss; including cutting into pipe						
110 to 32mm	5.54	7.10	0.30	8.13	nr	**15.23**
110 to 40mm	5.81	7.44	0.30	8.13	nr	**15.57**
110 to 50mm	6.56	8.40	0.30	8.13	nr	**16.53**

R: DISPOSAL SYSTEMS

Item	Net Price £	Material £	Labour hours	Labour £	Unit	Total rate £
R11: ABOVE GROUND DRAINAGE (cont'd)						
Extra over fittings solvent welded (cont'd)						
Adaptor saddle; including cutting to pipe						
82 to 32mm	2.53	3.24	0.29	7.86	nr	11.10
110 to 40mm	3.12	4.00	0.29	7.86	nr	11.86
160 to 50mm	5.65	7.24	0.29	7.86	nr	15.10
Branch boss adaptor						
32mm	1.64	2.10	0.26	7.05	nr	9.15
40 mm	1.64	2.10	0.26	7.05	nr	9.15
50 mm	2.33	2.98	0.26	7.05	nr	10.03
Branch boss adaptor bend						
32mm	2.48	3.18	0.26	7.05	nr	10.22
40 mm	2.70	3.46	0.26	7.05	nr	10.50
50 mm	2.65	3.39	0.26	7.05	nr	10.44
Automatic air admittance valve						
82 to 110mm	26.80	34.33	0.19	5.15	nr	39.47
PVC-U to metal adaptor: including heat shrunk joint to metal						
110mm	6.93	8.88	0.57	15.45	nr	24.32
Caulking bush: including joint to pipework						
82mm	7.10	9.09	0.46	12.47	nr	21.56
110mm	7.10	9.09	0.46	12.47	nr	21.56
Vent cowl						
82mm	2.14	2.74	0.13	3.52	nr	6.26
110mm	2.15	2.75	0.13	3.52	nr	6.28
160mm	5.65	7.24	0.13	3.52	nr	10.76
Weathering apron; to lead slates						
82mm	2.14	2.74	1.15	31.17	nr	33.91
110mm	2.44	3.13	1.15	31.17	nr	34.29
160mm	7.37	9.44	1.15	31.17	nr	40.61
Weathering apron; to asphalt						
82mm	9.03	11.57	1.10	29.81	nr	41.38
110mm	9.03	11.57	1.10	29.81	nr	41.38
Weathering slate; flat; 406 x 406mm						
82mm	25.44	32.58	1.04	28.19	nr	60.77
110mm	25.44	32.58	1.04	28.19	nr	60.77
Weathering slate; flat; 457 x 457mm						
82mm	26.08	33.40	1.04	28.19	nr	61.59
110mm	26.08	33.40	1.04	28.19	nr	61.59
Weathering slate; angled; 610 x 610mm						
82mm	35.23	45.12	1.04	28.19	nr	73.31
110mm	35.23	45.12	1.04	28.19	nr	73.31

R: DISPOSAL SYSTEMS

Item	Net Price £	Material £	Labour hours	Labour £	Unit	Total rate £
PVC-U soil and waste pipe; ring seal joints; fixed with clips to backgrounds; BS 4514/BS EN 607						
Pipe, with brackets measured separately						
82mm dia.	7.67	9.82	0.35	9.49	m	**19.31**
110mm dia.	7.32	9.38	0.41	11.11	m	**20.49**
160mm dia.	18.03	23.09	0.51	13.82	m	**36.92**
Fixings						
Galvanised steel pipe clip: including fixing to backgrounds. For minimum fixing distances, refer to the Tables and Memoranda at the rear of the book						
82mm	2.36	3.02	0.18	4.88	nr	**7.90**
110mm	2.44	3.13	0.18	4.88	nr	**8.00**
160mm	5.91	7.57	0.18	4.88	nr	**12.45**
Plastic coated steel pipe clip: including fixing to backgrounds. For minimum fixing distances, refer to the Tables and Memoranda at the rear of the book						
82mm	3.25	4.16	0.18	4.88	nr	**9.04**
110mm	3.25	4.16	0.18	4.88	nr	**9.04**
160mm	5.68	7.27	0.18	4.88	nr	**12.15**
Plastic pipe clip: including fixing to backgrounds. For minimum fixing distances, refer to the Tables and Memoranda at the rear of the book						
82mm	3.25	4.16	0.18	4.88	nr	**9.04**
110mm	3.25	4.16	0.18	4.88	nr	**9.04**
Plastic coated steel pipe clip: adjustable; including fixing to backgrounds. For minimum fixing distances, refer to the Tables and Memoranda at the rear of the book						
82mm	3.23	4.14	0.20	5.42	nr	**9.56**
110mm	3.33	4.26	0.20	5.42	nr	**9.69**
Galvanised steel pipe clip: drive in; including fixing to backgrounds. For minimum fixing distances, refer to the Tables and Memoranda at the rear of the book						
110mm	5.09	6.52	0.22	5.96	nr	**12.48**
Extra over fittings ring seal pipework PVC-U						
Straight coupling						
82mm	4.20	5.38	0.21	5.69	nr	**11.07**
110mm	4.48	5.74	0.22	5.96	nr	**11.70**
160mm	9.85	12.62	0.24	6.50	nr	**19.12**
Straight coupling; double socket						
82mm	4.25	5.44	0.21	5.69	nr	**11.14**
110mm	7.63	9.77	0.22	5.96	nr	**15.74**
160mm	15.77	20.20	0.24	6.50	nr	**26.70**

R: DISPOSAL SYSTEMS

Item	Net Price £	Material £	Labour hours	Labour £	Unit	Total rate £
R11: ABOVE GROUND DRAINAGE (cont'd)						
Extra over fittings ring seal pipework (cont'd)						
Reducer; socket						
82 to 50mm	6.27	8.03	0.15	4.07	nr	12.10
110 to 50mm	4.20	5.38	0.15	4.07	nr	9.44
110 to 82mm	8.54	10.94	0.19	5.15	nr	16.09
160 to 110	12.24	15.68	0.31	8.40	nr	24.08
Access Cap						
82mm	7.65	9.80	0.15	4.07	nr	13.86
110mm	7.99	10.23	0.17	4.61	nr	14.84
Access Cap; pressure plug						
160mm	22.19	28.42	0.33	8.94	nr	37.36
Access pipe						
82mm	14.81	18.97	0.22	5.96	nr	24.93
110mm	15.12	19.37	0.22	5.96	nr	25.33
160mm	35.96	46.06	0.24	6.50	nr	52.56
Bend						
82mm	8.68	11.12	0.29	7.86	nr	18.98
82mm; adjustable radius	14.07	18.02	0.29	7.86	nr	25.88
110mm	10.18	13.04	0.32	8.67	nr	21.71
110mm; adjustable radius	13.41	17.18	0.32	8.67	nr	25.85
160mm	25.30	32.40	0.49	13.28	nr	45.68
160mm; adjustable radius	26.81	34.34	0.49	13.28	nr	47.62
Bend; spigot and socket						
110mm	10.18	13.04	0.32	8.67	nr	21.71
Bend; offset						
82mm	8.49	10.87	0.21	5.69	nr	16.57
110mm	10.12	12.96	0.32	8.67	nr	21.63
160mm	27.37	35.06	-	-	nr	35.06
Bend; access						
110mm	21.18	27.13	0.33	8.94	nr	36.07
Single branch						
82mm	12.36	15.83	0.35	9.49	nr	25.32
110mm	15.56	19.93	0.42	11.38	nr	31.31
110mm; 45°	16.49	21.12	0.31	8.40	nr	29.52
160mm	45.68	58.51	0.50	13.55	nr	72.06
Single branch; access						
82mm	19.15	24.53	0.35	9.49	nr	34.01
110mm	27.55	35.29	0.42	11.38	nr	46.67
Unequal single branch						
160 x 160 x 110mm	30.27	38.77	0.50	13.55	nr	52.32
160 x 160 x 110mm; 45°	26.05	33.36	0.50	13.55	nr	46.92
Double branch; 4 bosses						
110mm	24.27	31.08	0.49	13.28	nr	44.37
Corner branch; 2 bosses						
110mm	57.51	73.66	0.49	13.28	nr	86.94

R: DISPOSAL SYSTEMS

Item	Net Price £	Material £	Labour hours	Labour £	Unit	Total rate £
Multibranch; 4 bosses						
110mm	46.08	59.02	0.52	14.09	nr	**73.11**
Boss Branch						
110 x 32mm	3.25	4.16	0.34	9.22	nr	**13.38**
110 x 40mm	3.25	4.16	0.34	9.22	nr	**13.38**
Strap on boss						
110 x 32mm	4.23	5.42	0.30	8.13	nr	**13.55**
110 x 40mm	4.23	5.42	0.30	8.13	nr	**13.55**
110 x 50mm	4.23	5.42	0.30	8.13	nr	**13.55**
Patch boss						
82 x 32mm	3.46	4.43	0.31	8.40	nr	**12.83**
82 x 40mm	3.46	4.43	0.31	8.40	nr	**12.83**
82 x 50mm	3.46	4.43	0.31	8.40	nr	**12.83**
Boss Pipe; collar 4 boss						
110mm	17.57	22.50	0.35	9.49	nr	**31.99**
Boss adaptor; rubber; push fit						
32mm	1.61	2.06	0.26	7.05	nr	**9.11**
40 mm	1.61	2.06	0.26	7.05	nr	**9.11**
50 mm	2.31	2.96	0.26	7.05	nr	**10.01**
WC connector; cap and seal; solvent socket						
110mm	4.00	5.12	0.23	6.23	nr	**11.36**
110mm; 90°	10.47	13.41	0.27	7.32	nr	**20.73**
Vent terminal						
82mm	2.58	3.30	0.13	3.52	nr	**6.83**
110mm	2.76	3.54	0.13	3.52	nr	**7.06**
160mm	8.82	11.30	0.13	3.52	nr	**14.82**
Weathering slate; inclined; 610 x 610mm						
82mm	29.41	37.67	1.04	28.19	nr	**65.86**
110mm	29.41	37.67	1.04	28.19	nr	**65.86**
Weathering slate; inclined; 450 x 450mm						
82mm	18.38	23.54	1.04	28.19	nr	**51.73**
110mm	18.38	23.54	1.04	28.19	nr	**51.73**
Weathering slate; flat; 400 x 400mm						
82mm	18.28	23.41	1.04	28.19	nr	**51.60**
110mm	18.28	23.41	1.04	28.19	nr	**51.60**
Air admittance valve						
82mm	40.38	51.72	0.19	5.15	nr	**56.87**
110mm	41.52	53.18	0.19	5.15	nr	**58.33**
Cast iron pipe; nitrile rubber gasket joint with continuity clip BS 416/6087; fixed vertically to backgrounds						
Pipe, with brackets and jointing couplings measured separately						
50 mm	15.66	20.06	0.25	6.78	m	**26.83**
75 mm	17.30	22.16	0.45	12.20	m	**34.35**
100 mm	20.62	26.41	0.60	16.27	m	**42.68**
150 mm	42.52	54.46	0.70	18.98	m	**73.44**

R: DISPOSAL SYSTEMS

Item	Net Price £	Material £	Labour hours	Labour £	Unit	Total rate £
R11: ABOVE GROUND DRAINAGE (cont'd)						
Fixings						
Brackets; fixed to backgrounds. For minimum fixing distances, refer to the Tables and Memoranda at the rear of the book						
50 mm	3.87	4.96	0.15	4.07	nr	9.02
75 mm	3.87	4.96	0.18	4.88	nr	9.84
100 mm	4.30	5.51	0.18	4.88	nr	10.39
150 mm	8.05	10.31	0.20	5.42	nr	15.73
Extra over fittings nitrile gasket cast iron pipework BS 416/6087, with jointing couplings measured separately						
Standard coupling						
50 mm	6.37	8.16	0.50	13.55	nr	21.71
75 mm	7.04	9.02	0.60	16.27	nr	25.28
100 mm	9.19	11.77	0.67	18.16	nr	29.94
150 mm	18.37	23.53	0.83	22.51	nr	46.04
Conversion coupling						
65 x 75 mm	7.45	9.54	0.60	16.27	nr	25.81
70 x 75 mm	7.45	9.54	0.60	16.27	nr	25.81
90 x 100 mm	7.45	9.54	0.67	18.16	nr	27.71
Access pipe; round door						
50 mm	26.49	33.93	0.41	11.11	nr	45.04
75 mm	27.16	34.79	0.46	12.47	nr	47.25
100 mm	28.41	36.39	0.67	18.19	nr	54.58
150 mm	44.58	57.10	0.83	22.51	nr	79.61
Access pipe; square door						
100 mm	55.97	71.69	0.67	18.19	nr	89.88
150 mm	85.68	109.74	0.83	22.51	nr	132.25
Taper reducer						
75 mm	14.74	18.88	0.60	16.27	nr	35.15
100 mm	19.60	25.10	0.67	18.16	nr	43.27
150 mm	38.19	48.91	0.83	22.51	nr	71.43
Blank cap						
50 mm	3.88	4.97	0.24	6.50	nr	11.47
75 mm	4.73	6.06	0.26	7.05	nr	13.11
100 mm	4.73	6.06	0.32	8.67	nr	14.73
150 mm	7.59	9.72	0.40	10.84	nr	20.56
Blank cap; 50 mm screwed tapping						
75 mm	9.58	12.27	0.26	7.05	nr	19.32
100 mm	11.29	14.46	0.32	8.67	nr	23.13
150 mm	12.14	15.55	0.40	10.84	nr	26.39
Universal connector						
50 x 56/48/40 mm	4.06	5.20	0.33	8.94	nr	14.14
Change piece; BS416						
100 mm	12.14	15.55	0.47	12.74	nr	28.29

Material Costs/Prices for Measured Works – Mechanical Installations

R: DISPOSAL SYSTEMS

Item	Net Price £	Material £	Labour hours	Labour £	Unit	**Total rate £**
WC connector						
100 mm	18.50	23.69	0.49	13.28	nr	**36.98**
Boss pipe; 2 " BSPT socket						
50 mm	23.25	29.78	0.58	15.72	nr	**45.50**
75 mm	23.25	29.78	0.65	17.62	nr	**47.40**
100 mm	27.78	35.58	0.79	21.41	nr	**56.99**
150 mm	45.32	58.05	0.86	23.31	nr	**81.35**
Boss pipe; 2 " BSPT socket; 135 °						
100 mm	33.46	42.86	0.79	21.41	nr	**64.27**
Boss pipe; 2 x 2 " BSPT socket; opposed						
75 mm	30.79	39.44	0.65	17.62	nr	**57.05**
100 mm	35.92	46.01	0.79	21.41	nr	**67.42**
Boss pipe; 2 x 2 " BSPT socket; in line						
100 mm	36.28	46.47	0.79	21.41	nr	**67.88**
Boss pipe; 2 x 2 " BSPT socket; 90 °						
100 mm	35.92	46.01	0.79	21.41	nr	**67.42**
Bend; short radius						
50 mm	11.22	14.37	0.50	13.55	nr	**27.92**
75 mm	11.22	14.37	0.60	16.27	nr	**30.64**
100 mm	15.56	19.93	0.67	18.16	nr	**38.09**
100 mm; 11 °	13.41	17.18	0.67	18.16	nr	**35.34**
100 mm; 67 °	15.56	19.93	0.67	18.16	nr	**38.09**
150 mm	26.00	33.30	0.83	22.51	nr	**55.81**
Access bend; short radius						
50 mm	27.66	35.43	0.50	13.55	nr	**48.98**
75 mm	27.66	35.43	0.60	16.27	nr	**51.69**
100 mm	32.87	42.10	0.67	18.16	nr	**60.26**
100 mm; 45 °	32.87	42.10	0.67	18.16	nr	**60.26**
150 mm	46.70	59.81	0.83	22.51	nr	**82.33**
150 mm; 45 °	46.70	59.81	0.83	22.51	nr	**82.33**
Long radius bend						
75 mm	21.22	27.18	0.60	16.27	nr	**43.45**
100 mm	25.18	32.25	0.67	18.16	nr	**50.42**
100 mm; 5 °	15.56	19.93	0.67	18.16	nr	**38.09**
150 mm	50.75	65.00	0.83	22.51	nr	**87.51**
150 mm; 22.5 °	44.48	56.97	0.83	22.51	nr	**79.48**
Access bend; long radius						
75 mm	37.62	48.18	0.60	16.27	nr	**64.45**
100 mm	42.53	54.47	0.67	18.16	nr	**72.64**
150 mm	74.92	95.96	0.83	22.51	nr	**118.47**
Long tail bend						
100 x 250 mm long	20.09	25.73	0.70	18.98	nr	**44.71**
100 x 815 mm long	48.88	62.60	0.70	18.98	nr	**81.59**
Offset						
75 mm projection						
75 mm	11.70	14.99	0.53	14.36	nr	**29.35**
100 mm	16.35	20.94	0.66	17.89	nr	**38.83**

R: DISPOSAL SYSTEMS

Item	Net Price £	Material £	Labour hours	Labour £	Unit	Total rate £
R11: ABOVE GROUND DRAINAGE (cont'd)						
Extra over fittings nitrile gasket (cont'd)						
115 mm projection						
75 mm	13.84	17.73	0.53	14.36	nr	**32.09**
100 mm	19.50	24.98	0.66	17.89	nr	**42.86**
150 mm projection						
75 mm	13.84	17.73	0.53	14.36	nr	**32.09**
100 mm	19.50	24.98	0.66	17.89	nr	**42.86**
225 mm projection						
100 mm	22.34	28.61	0.66	17.89	nr	**46.50**
300 mm projection						
100 mm	25.18	32.25	0.66	17.89	nr	**50.14**
Branch; equal and unequal						
50 mm	16.91	21.66	0.78	21.14	nr	**42.80**
75 mm	16.91	21.66	0.85	23.05	nr	**44.70**
100 mm	24.03	30.78	1.00	27.10	nr	**57.88**
150 mm	59.56	76.28	1.20	32.54	nr	**108.82**
150 x 100 mm; 87.5 °	45.58	58.38	1.21	32.66	nr	**91.04**
150 x 100 mm; 45 °	65.09	83.37	1.21	32.66	nr	**116.03**
Branch; 2" BSPT screwed socket						
100 mm	32.76	41.96	1.00	27.10	nr	**69.06**
Branch; long tail						
100 x 915 mm long	63.55	81.39	1.00	27.10	nr	**108.50**
Access branch; equal and unequal						
50 mm	33.36	42.73	0.78	21.14	nr	**63.87**
75 mm	33.36	42.73	0.85	23.05	nr	**65.77**
100 mm	41.38	53.00	1.02	27.66	nr	**80.66**
150 mm	81.80	104.77	1.20	32.54	nr	**137.31**
150 x 100 mm; 87.5 °	64.51	82.62	1.20	32.54	nr	**115.16**
150 x 100 mm; 45 °	80.27	102.81	1.20	32.54	nr	**135.35**
Parallel branch						
100 mm	25.18	32.25	1.00	27.10	nr	**59.35**
Double branch						
75 mm	28.41	36.39	0.95	25.75	nr	**62.14**
100 mm	29.73	38.08	1.30	35.25	nr	**73.32**
150 x 100 mm	83.72	107.23	1.56	42.28	nr	**149.51**
Double access branch						
100 mm	47.06	60.27	1.43	38.77	nr	**99.05**
Corner branch						
100 mm	39.94	51.15	1.30	35.25	nr	**86.40**
Puddle flange; grey epoxy coated						
100 mm	28.41	36.39	1.00	27.10	nr	**63.49**
Roof vent connector; asphalt						
75 mm	34.39	44.05	0.90	24.39	nr	**68.44**
100 mm	26.84	34.38	0.97	26.29	nr	**60.67**

R: DISPOSAL SYSTEMS

Item	Net Price £	Material £	Labour hours	Labour £	Unit	Total rate £
P trap						
100 mm	24.91	31.90	1.00	27.10	nr	59.01
P trap with access						
50 mm	38.13	48.84	0.77	20.87	nr	69.71
75 mm	38.13	48.84	0.90	24.39	nr	73.23
100 mm	42.26	54.13	1.16	31.44	nr	85.57
150 mm	70.71	90.56	1.77	47.97	nr	138.54
Bellmouth gully inlet						
100 mm	35.41	45.35	1.08	29.27	nr	74.62
Balcony gully inlet						
100 mm	76.27	97.69	1.08	29.27	nr	126.96
Roof outlet						
Flat grate						
75 mm	75.25	96.38	0.83	22.50	nr	118.88
100 mm	87.15	111.62	1.08	29.27	nr	140.89
Dome grate						
75 mm	75.95	97.28	0.83	22.50	nr	119.77
100 mm	88.57	113.44	1.08	29.27	nr	142.71
Top Hat						
100 mm	165.75	212.29	1.08	29.27	nr	241.56
Cast iron pipe; EPDM rubber gasket joint with continuity clip; BS EN877; fixed to backgrounds						
Pipe, with brackets and jointing couplings measured separately						
50 mm	11.43	15.08	0.25	6.78	m	21.85
70 mm	12.92	17.04	0.45	12.20	m	29.24
100 mm	15.21	20.07	0.60	16.26	m	36.33
125 mm	24.76	32.66	0.65	17.62	m	50.28
150 mm	29.99	39.56	0.70	18.97	m	58.54
200 mm	58.07	76.61	1.14	30.90	m	107.50
250 mm	68.90	90.89	1.25	33.88	m	124.77
300 mm	84.71	111.75	1.53	41.47	m	153.22
Fixings						
Brackets; fixed to backgrounds. For minimum fixing distances, refer to the Tables and Memoranda at the rear of the book						
Ductile iron						
50mm	3.67	4.70	0.10	2.71	nr	7.41
70 mm	3.67	4.70	0.10	2.71	nr	7.41
100 mm	4.03	5.16	0.15	4.07	nr	9.23
150 mm	7.63	9.77	0.20	5.42	nr	15.19
200 mm	29.07	37.23	0.25	6.78	nr	44.01
Mild steel; vertical						
125 mm	7.14	9.14	0.15	4.07	nr	13.21

R: DISPOSAL SYSTEMS

Item	Net Price £	Material £	Labour hours	Labour £	Unit	Total rate £
R11: ABOVE GROUND DRAINAGE (cont'd)						
Fixings (cont'd)						
Brackets; fixed to backgrounds (cont'd)						
Mild steel; stand off						
250 mm	15.55	19.92	0.25	6.78	nr	26.69
300 mm	17.11	21.91	0.25	6.78	nr	28.69
Stack support; rubber seal						
70 mm	8.49	10.87	0.74	20.06	nr	30.93
100 mm	11.25	14.41	0.88	23.85	nr	38.26
125 mm	12.47	15.97	1.00	27.10	nr	43.08
150 mm	16.86	21.59	1.19	32.25	nr	53.85
200 mm	34.60	44.32	1.31	35.51	nr	79.82
Wall spacer plate; cast iron (eared sockets)						
50 mm	3.20	4.10	0.10	2.71	nr	6.81
70 mm	3.20	4.10	0.10	2.71	nr	6.81
100 mm	3.20	4.10	0.10	2.71	nr	6.81
Extra over fittings EPDM rubber jointed cast iron pipework BS EN 877, with jointing couplings measured separately						
Coupling						
50 mm	3.92	5.02	0.50	13.55	nr	18.57
70 mm	4.30	5.51	0.60	16.27	nr	21.77
100 mm	5.62	7.20	0.67	18.16	nr	25.36
125 mm	6.96	8.91	0.75	20.34	nr	29.25
150mm	11.22	14.37	0.83	22.51	nr	36.88
200 mm	25.13	32.19	1.21	32.80	nr	64.98
250 mm	35.00	44.83	1.33	36.05	nr	80.88
300 mm	40.51	51.88	1.63	44.18	nr	96.06
Plain socket						
50 mm	9.39	12.03	0.25	6.78	nr	18.80
70 mm	9.39	12.03	0.25	6.78	nr	18.80
100 mm	10.70	13.70	0.25	6.78	nr	20.48
Eared socket						
50 mm	9.70	12.42	0.25	6.78	nr	19.20
70 mm	9.70	12.42	0.25	6.78	nr	19.20
100 mm	11.65	14.92	0.25	6.78	nr	21.70
Slip socket						
50 mm	12.07	15.46	0.25	6.78	nr	22.23
70 mm	12.07	15.46	0.25	6.78	nr	22.23
100 mm	14.18	18.16	0.25	6.78	nr	24.94
Stack support pipe						
70 mm	14.53	18.61	0.74	20.06	nr	38.67
100 mm	16.15	20.68	0.88	23.85	nr	44.54
125 mm	17.92	22.95	1.00	27.10	nr	50.06
150 mm	22.57	28.91	1.19	32.25	nr	61.16
200 mm	43.28	55.43	1.31	35.51	nr	90.94

Material Costs/Prices for Measured Works – Mechanical Installations

R: DISPOSAL SYSTEMS

Item	Net Price £	Material £	Labour hours	Labour £	Unit	Total rate £
Access pipe; round door						
50 mm	18.05	23.12	0.54	14.64	nr	37.75
70 mm	19.10	24.46	0.64	17.35	nr	41.81
100 mm	21.00	26.90	0.60	16.26	nr	43.16
150 mm	37.99	48.66	0.83	22.50	nr	71.15
Access pipe; square door						
100 mm	40.61	52.01	0.60	16.26	nr	68.28
125 mm	42.25	54.11	0.67	18.16	nr	72.27
150 mm	63.55	81.39	0.71	19.24	nr	100.64
200 mm	126.25	161.70	1.21	32.80	nr	194.49
250 mm	198.62	254.39	1.31	35.51	nr	289.90
300 mm	247.78	317.35	1.43	38.76	nr	356.11
Taper reducer						
70 mm	10.41	13.33	0.51	13.82	nr	27.16
100 mm	12.25	15.69	0.58	15.72	nr	31.41
125 mm	12.32	15.78	0.64	17.35	nr	33.13
150 mm	23.50	30.10	0.67	18.16	nr	48.26
200 mm	38.19	48.91	1.15	31.17	nr	80.08
250 mm	78.88	101.03	1.25	33.88	nr	134.91
300 mm	108.47	138.93	1.37	37.13	nr	176.06
Blank cap						
50 mm	2.73	3.50	0.24	6.50	nr	10.00
70 mm	2.87	3.68	0.26	7.05	nr	10.72
100 mm	3.35	4.29	0.32	8.67	nr	12.96
125 mm	4.73	6.06	0.35	9.49	nr	15.54
150 mm	4.83	6.19	0.40	10.84	nr	17.03
200 mm	21.39	27.40	0.60	16.26	nr	43.66
250 mm	43.68	55.94	0.65	17.62	nr	73.56
300 mm	57.20	73.26	0.72	19.51	nr	92.78
Blank cap; 50 mm screwed tapping						
70 mm	6.43	8.24	0.26	7.05	nr	15.28
100 mm	6.93	8.88	0.32	8.67	nr	17.55
150 mm	8.32	10.66	0.40	10.84	nr	21.50
Universal connector; EPDM rubber						
50 x 56/48/40 mm	4.06	5.20	0.30	8.13	nr	13.33
Blank end; push fit						
100 x 38/32 mm	5.87	7.52	0.39	10.57	nr	18.09
Boss pipe; 2 " BSPT socket						
50 mm	15.25	19.53	0.54	14.64	nr	34.17
75 mm	15.25	19.53	0.64	17.35	nr	36.88
100 mm	18.65	23.89	0.78	21.14	nr	45.03
150 mm	30.42	38.96	1.09	29.54	nr	68.50
Boss pipe; 2 x 2 " BSPT socket; opposed						
100 mm	24.10	30.87	0.78	21.14	nr	52.01
Boss pipe; 2 x 2 " BSPT socket; 90 °						
100 mm	24.10	30.87	0.78	21.14	nr	52.01
Manifold connector						
100 mm	37.19	47.63	0.78	21.14	nr	68.77

R: DISPOSAL SYSTEMS

Item	Net Price £	Material £	Labour hours	Labour £	Unit	Total rate £
R11: ABOVE GROUND DRAINAGE (cont'd)						
Extra over fittings EPDM rubber (cont'd)						
Bend; short radius						
50mm	6.77	8.67	0.50	13.55	nr	22.22
70mm	7.63	9.77	0.60	16.26	nr	26.03
100mm	9.03	11.57	0.67	18.16	nr	29.73
125 mm	16.00	20.49	0.78	21.14	nr	41.63
150mm	16.23	20.79	0.83	22.51	nr	43.30
200 mm; 45 °	48.26	61.81	1.21	32.80	nr	94.61
250 mm; 45 °	94.19	120.64	1.31	35.51	nr	156.14
300 mm; 45 °	132.47	169.67	1.43	38.76	nr	208.42
Access bend; short radius						
70 mm	14.78	18.93	0.64	17.35	nr	36.28
100mm	21.59	27.65	0.78	21.14	nr	48.79
150mm	33.56	42.98	0.83	22.51	nr	65.50
Bend; long radius bend						
100mm	21.39	27.40	0.67	18.16	nr	45.56
100 mm; 22 °	16.89	21.63	0.78	21.14	nr	42.77
150mm	63.93	81.88	0.83	22.51	nr	104.39
Access bend; long radius						
100mm	27.93	35.77	0.67	18.16	nr	53.94
Bend; long tail						
100 mm	15.83	20.27	0.78	21.14	nr	41.42
Bend; long tail double						
70 mm	24.51	31.39	0.64	17.35	nr	48.74
100 mm	27.04	34.63	0.78	21.14	nr	55.77
Bend; air pipe						
100 mm	28.97	37.10	0.78	21.14	nr	58.25
Offset						
75 mm projection						
100 mm	13.88	17.78	0.78	21.14	nr	38.92
130 mm projection						
50 mm	11.47	14.69	0.54	14.64	nr	29.33
70 mm	17.41	22.30	0.64	17.35	nr	39.64
100 mm	22.87	29.29	0.78	21.14	nr	50.43
125 mm	29.01	37.16	0.78	21.14	nr	58.30
Branch; equal and unequal						
50 mm	10.87	13.92	0.78	21.14	nr	35.06
70 mm	11.47	14.69	0.85	23.05	nr	37.74
100 mm	15.73	20.15	1.00	27.10	nr	47.25
125 mm	31.55	40.41	1.16	31.44	nr	71.85
150 mm	39.19	50.19	1.37	37.13	nr	87.33
200 mm	91.35	117.00	1.51	40.93	nr	157.93
250 mm	211.17	270.46	1.63	44.18	nr	314.64
300mm	322.39	412.91	1.77	47.97	nr	460.89

R: DISPOSAL SYSTEMS

Item	Net Price £	Material £	Labour hours	Labour £	Unit	Total rate £
Branch; radius; equal and unequal						
70 mm	13.99	17.92	0.79	21.41	nr	**39.33**
100 mm	16.15	20.68	0.96	26.02	nr	**46.70**
125 mm	39.56	50.67	1.16	31.44	nr	**82.11**
150 mm	44.41	56.88	1.37	37.13	nr	**94.01**
200 mm	123.00	157.54	1.51	40.93	nr	**198.46**
Branch; long tail						
100 mm	51.34	65.76	0.96	26.02	nr	**91.78**
Access branch; radius; equal and unequal						
70 mm	20.49	26.24	0.79	21.41	nr	**47.66**
100 mm	27.79	35.59	0.96	26.02	nr	**61.61**
150 mm	59.40	76.08	1.20	32.54	nr	**108.62**
Double branch; equal and unequal						
100mm	21.02	26.92	1.30	35.25	nr	**62.17**
100 mm; 69 °	23.14	29.64	1.30	35.25	nr	**64.88**
150 mm	67.78	86.81	1.37	37.13	nr	**123.94**
200 mm	124.14	159.00	1.51	40.93	nr	**199.92**
Double branch; radius; equal and unequal						
100 mm	19.99	25.60	1.30	35.25	nr	**60.85**
150 mm	71.90	92.09	1.37	37.13	nr	**129.22**
200 mm	124.14	159.00	1.51	40.93	nr	**199.92**
Corner branch						
100 mm	38.29	49.04	1.30	35.25	nr	**84.29**
Corner branch; long tail						
100 mm	60.32	77.26	1.30	35.25	nr	**112.50**
Roof vent connector; asphalt						
100 mm	25.34	32.46	0.78	21.14	nr	**53.60**
Roof vent connector; felt						
100 mm	57.21	73.27	0.78	21.14	nr	**94.41**
Movement connector						
100 mm	28.62	36.66	0.78	21.14	nr	**57.80**
150 mm	49.66	63.60	0.78	21.14	nr	**84.74**
Expansion plugs						
70 mm	13.15	16.84	0.32	8.67	nr	**25.52**
100 mm	14.20	18.19	0.39	10.57	nr	**28.76**
150 mm	21.24	27.20	0.55	14.91	nr	**42.11**
P trap						
100mm dia.	16.76	21.47	1.00	27.10	nr	**48.57**
P trap with access						
50 mm	25.58	32.76	0.54	14.64	nr	**47.40**
70 mm	25.58	32.76	0.64	17.35	nr	**50.11**
100mm	27.71	35.49	1.00	27.10	nr	**62.59**
150 mm	49.55	63.46	1.09	29.54	nr	**93.01**
Branch trap						
100 mm	61.24	78.44	1.17	31.71	nr	**110.15**

R: DISPOSAL SYSTEMS

Item	Net Price £	Material £	Labour hours	Labour £	Unit	Total rate £
R11: ABOVE GROUND DRAINAGE (cont'd)						
Extra over fittings EPDM rubber (cont'd)						
Balcony gully inlet						
100 mm	53.52	68.55	1.00	27.10	nr	**95.65**
Roof outlet						
Flat grate						
70 mm	74.12	94.93	1.00	27.10	nr	**122.04**
100 mm	87.12	111.58	1.00	27.10	nr	**138.69**
Dome grate						
70 mm	75.95	97.28	1.00	27.10	nr	**124.38**
100 mm	75.95	97.28	1.00	27.10	nr	**124.38**
Top Hat						
100 mm	172.09	220.41	1.00	27.10	nr	**247.51**
Floor drains; for cast iron pipework BS 416 and BS EN877						
Adjustable clamp plate body						
100 mm; 165mm nickel bronze grate and frame	59.28	75.93	0.50	13.55	nr	**89.48**
100 mm; 165mm nickel bronze rodding eye	69.68	89.25	0.50	13.55	nr	**102.80**
100 mm; 150 x 150mm nickel bronze grate and frame	63.64	81.51	0.50	13.55	nr	**95.06**
100 mm; 150 x 150 nickel bronze rodding eye	72.92	93.40	0.50	13.55	nr	**106.95**
Deck plate body						
100 mm; 165mm nickel bronze grate and frame	60.76	77.82	0.50	13.55	nr	**91.37**
100 mm; 165mm nickel bronze rodding eye	72.43	92.77	0.50	13.55	nr	**106.32**
100 mm; 150 x 150mm nickel bronze grate and frame	67.38	86.30	0.50	13.55	nr	**99.85**
100 mm; 150 x 150mm nickel bronze rodding eye	73.73	94.43	0.50	13.55	nr	**107.98**
Extra for						
100 mm; Screwed extension piece	22.47	28.78	0.30	8.13	nr	**36.91**
100 mm; Grating extension piece; screwed or spigot	18.72	23.98	0.30	8.13	nr	**32.11**
100 mm; Brewery trap	713.22	913.49	2.00	54.21	nr	**967.69**

S: PIPED SUPPLY SYSTEMS

Item	Net Price £	Material £	Labour hours	Labour £	Unit	Total rate £
S10 : COLD WATER						
Y10 - PIPELINES						
COPPER PIPEWORK						
Copper pipe; capillary or compression joints in the running length; EN1057 R250 (TX) formerly BS 2871 Table X						
Fixed vertically or at low level, with brackets measured separately						
12mm dia.	1.34	1.72	0.39	10.57	m	**12.29**
15mm dia.	1.46	1.87	0.40	10.84	m	**12.71**
22mm dia.	2.95	3.78	0.47	12.74	m	**16.52**
28mm dia.	3.79	4.85	0.51	13.82	m	**18.68**
35mm dia.	7.35	9.41	0.58	15.72	m	**25.13**
42mm dia.	9.18	11.76	0.66	17.89	m	**29.65**
54mm dia.	12.28	15.73	0.72	19.51	m	**35.24**
67mm dia.	16.12	20.65	0.75	20.33	m	**40.97**
76mm dia.	22.59	28.93	0.76	20.60	m	**49.53**
108mm dia.	32.42	41.52	0.78	21.14	m	**62.66**
133mm dia.	42.10	53.92	1.05	28.46	m	**82.38**
159mm dia.	64.35	82.42	1.15	31.17	m	**113.59**
Fixed horizontally at high level or suspended, with brackets measured separately						
12mm dia.	1.34	1.72	0.45	12.20	m	**13.91**
15mm dia.	1.46	1.87	0.46	12.47	m	**14.34**
22mm dia.	2.95	3.78	0.54	14.64	m	**18.41**
28mm dia.	3.79	9.71	0.59	15.99	m	**25.70**
35mm dia.	7.35	9.41	0.67	18.16	m	**27.57**
42mm dia.	9.18	11.76	0.76	20.60	m	**32.36**
54mm dia.	12.28	15.73	0.83	22.50	m	**38.22**
67mm dia.	16.12	20.65	0.86	23.31	m	**43.96**
76mm dia.	22.59	28.93	0.87	23.58	m	**52.51**
108mm dia.	32.42	41.52	0.90	24.39	m	**65.92**
133mm dia.	42.10	53.92	1.21	32.80	m	**86.72**
159mm dia.	64.35	82.42	1.32	35.78	m	**118.20**
Copper pipe; capillary or compression joints in the running length; EN1057 R250 (TY) formerly BS 2871 Table Y						
Fixed vertically or at low level with brackets measured separately (Refer to Copper Pipe Table X Section)						
12mm dia.	1.81	2.32	0.41	11.11	m	**13.43**
15mm dia.	2.64	3.38	0.43	11.65	m	**15.04**
22mm dia.	4.68	5.99	0.50	13.55	m	**19.55**
28mm dia.	6.13	7.85	0.54	14.64	m	**22.49**
35mm dia.	10.13	12.97	0.62	16.80	m	**29.78**
42mm dia.	12.52	16.04	0.71	19.24	m	**35.28**
54mm dia.	21.49	27.52	0.78	21.14	m	**48.66**
67mm dia.	27.93	35.77	0.82	22.22	m	**58.00**
76mm dia.	32.48	41.60	0.60	16.26	m	**57.86**
108mm dia.	56.86	72.83	0.88	23.85	m	**96.68**

S: PIPED SUPPLY SYSTEMS

Item	Net Price £	Material £	Labour hours	Labour £	Unit	Total rate £
S10 : COLD WATER (cont'd)						
COPPER PIPEWORK (cont'd)						
Copper pipe; capillary or compression joints in the running length; EN1057 R250 (TX) formerly BS 2871 Table X						
Plastic coated gas and cold water service pipe for corrosive environments, fixed vertically or at low level with brackets measured separately						
15mm dia. (white)	1.88	2.41	0.59	15.99	m	18.40
22mm dia. (white)	3.60	4.61	0.68	18.44	m	23.05
28mm dia. (white)	4.54	5.81	0.74	20.06	m	25.88
FIXINGS						
For copper pipework						
Saddle band						
6mm dia.	0.11	0.14	0.11	2.98	nr	3.12
8mm dia.	0.11	0.14	0.12	3.25	nr	3.39
10mm dia.	0.11	0.14	0.12	3.25	nr	3.39
12mm dia.	0.11	0.14	0.12	3.25	nr	3.39
15mm dia.	0.12	0.15	0.13	3.52	nr	3.68
22mm dia.	0.12	0.15	0.13	3.52	nr	3.68
28mm dia.	0.13	0.17	0.16	4.34	nr	4.50
35mm dia.	0.21	0.27	0.18	4.88	nr	5.15
42mm dia.	0.44	0.56	0.21	5.69	nr	6.26
54mm dia.	0.59	0.76	0.21	5.69	nr	6.45
Single spacing clip						
15mm dia.	0.13	0.17	0.14	3.79	nr	3.96
22mm dia.	0.13	0.17	0.15	4.07	nr	4.23
28mm dia.	0.31	0.40	0.17	4.61	nr	5.00
Two piece spacing clip						
8mm dia. Bottom	0.11	0.14	0.11	2.98	nr	3.12
8mm dia. Top	0.11	0.14	0.11	2.98	nr	3.12
12mm dia. Bottom	0.11	0.14	0.13	3.52	nr	3.66
12mm dia. Top	0.11	0.14	0.13	3.52	nr	3.66
15mm dia. Bottom	0.12	0.15	0.13	3.52	nr	3.68
15mm dia. Top	0.12	0.15	0.13	3.52	nr	3.68
22mm dia. Bottom	0.12	0.15	0.14	3.79	nr	3.95
22mm dia. Top	0.12	0.15	0.14	3.79	nr	3.95
28mm dia. Bottom	0.12	0.15	0.16	4.34	nr	4.49
28mm dia. Top	0.21	0.27	0.16	4.34	nr	4.61
35mm dia. Bottom	0.13	0.17	0.21	5.69	nr	5.86
35mm dia. Top	0.31	0.40	0.21	5.69	nr	6.09
Single pipe bracket						
15mm dia.	1.44	1.84	0.14	3.79	nr	5.64
22mm dia.	1.66	2.13	0.14	3.79	nr	5.92
28mm dia.	1.98	2.54	0.17	4.61	nr	7.14

S: PIPED SUPPLY SYSTEMS

Item	Net Price £	Material £	Labour hours	Labour £	Unit	Total rate £
Single pipe ring						
15mm dia.	2.19	2.80	0.26	7.05	nr	9.85
22mm dia.	2.35	3.01	0.26	7.05	nr	10.06
28mm dia.	2.79	3.57	0.31	8.40	nr	11.98
35mm dia.	2.94	3.77	0.32	8.67	nr	12.44
42mm dia.	3.29	4.21	0.32	8.67	nr	12.89
54mm dia.	3.92	5.02	0.34	9.22	nr	14.24
67mm dia.	9.20	11.78	0.35	9.49	nr	21.27
76mm dia.	11.61	14.87	0.42	11.38	nr	26.25
108mm dia.	17.98	23.03	0.42	11.38	nr	34.41
Double pipe ring						
15mm dia.	2.94	3.77	0.26	7.05	nr	10.81
22mm dia.	3.13	4.01	0.26	7.05	nr	11.06
28mm dia.	4.19	5.37	0.31	8.40	nr	13.77
35mm dia.	4.34	5.56	0.32	8.67	nr	14.23
42mm dia.	4.74	6.07	0.32	8.67	nr	14.74
54mm dia.	5.69	7.29	0.34	9.22	nr	16.50
67mm dia.	12.17	15.59	0.35	9.49	nr	25.07
76mm dia.	15.18	19.44	0.42	11.38	nr	30.83
108mm dia.	25.56	32.74	0.42	11.38	nr	44.12
Wall bracket						
15mm dia.	3.69	4.73	0.05	1.36	nr	6.08
22mm dia.	4.34	5.56	0.05	1.36	nr	6.91
28mm dia.	5.29	6.78	0.05	1.36	nr	8.13
35mm dia.	6.80	8.71	0.05	1.36	nr	10.06
42mm dia.	9.02	11.55	0.05	1.36	nr	12.91
54mm dia.	11.37	14.56	0.05	1.36	nr	15.92
Hospital bracket						
15mm dia.	3.13	4.01	0.26	7.05	nr	11.06
22mm dia.	3.29	4.21	0.26	7.05	nr	11.26
28mm dia.	4.03	5.16	0.31	8.40	nr	13.56
35mm dia.	4.34	5.56	0.32	8.67	nr	14.23
42mm dia.	6.10	7.81	0.32	8.67	nr	16.49
54mm dia.	8.28	10.60	0.34	9.22	nr	19.82
Screw on backplate, female						
All sizes 15mm to 54mm x 10mm	1.38	1.77	0.10	2.71	nr	4.48
Screw on backplate, male						
All sizes 15mm to 54mm x 10mm	1.23	1.58	0.10	2.71	nr	4.29
Pipe joist clips, single						
15mm dia.	0.64	0.82	0.08	2.17	nr	2.99
22mm dia.	0.64	0.82	0.08	2.17	nr	2.17
Pipe joist clips, double						
15mm dia.	0.90	1.15	0.08	2.17	nr	3.32
22mm dia.	0.90	1.15	0.08	2.17	nr	3.32
Extra Over channel sections for fabricated hangers and brackets						
Galvanised steel; including inserts, bolts, nuts, washers; fixed to backgrounds						
41 x 21mm	2.50	3.20	0.29	7.86	m	11.06
41 x 41mm	3.90	5.00	0.29	7.86	m	12.86

S: PIPED SUPPLY SYSTEMS

Item	Net Price £	Material £	Labour hours	Labour £	Unit	Total rate £
S10 : COLD WATER (cont'd)						
COPPER PIPEWORK (cont'd)						
Extra Over channel sections (cont'd)						
Threaded rods; metric thread; including nuts, washers, etc						
10mm dia x 600mm long for ring clips up to 54mm	1.31	1.68	0.18	4.88	nr	6.56
12mm dia x 600mm long for ring clips from 54mm	2.70	3.46	0.18	4.88	nr	8.34
Extra over copper pipes; capillary fittings; BS 864						
Stop end						
15mm dia.	1.02	1.31	0.13	3.52	nr	4.83
22mm dia.	1.91	2.45	0.14	3.79	nr	6.24
28mm dia.	3.41	4.37	0.17	4.61	nr	8.98
35mm dia.	7.53	9.64	0.19	5.15	nr	14.79
42mm dia.	12.97	16.61	0.22	5.96	nr	22.57
54mm dia.	18.11	23.20	0.23	6.23	nr	29.43
Straight coupling; copper to copper						
6mm dia.	0.86	1.10	0.23	6.23	nr	7.34
8mm dia.	0.88	1.13	0.23	6.23	nr	7.36
10mm dia.	0.45	0.58	0.23	6.23	nr	6.81
15mm dia.	0.14	0.18	0.23	6.23	nr	6.41
22mm dia.	0.36	0.46	0.26	7.05	nr	7.51
28mm dia.	0.95	1.22	0.30	8.13	nr	9.35
35mm dia.	3.08	3.94	0.34	9.22	nr	13.16
42mm dia.	5.14	6.58	0.38	10.30	nr	16.88
54mm dia.	9.48	12.14	0.42	11.38	nr	23.53
67mm dia.	28.16	36.07	0.53	14.36	nr	50.43
Adaptor coupling; imperial to metric						
1/2" x 15mm dia.	2.23	2.86	0.27	7.32	nr	10.17
3/4" x 22mm dia.	1.95	2.50	0.31	8.40	nr	10.90
1" x 28mm dia.	3.84	4.92	0.36	9.76	nr	14.68
1 1/4" x 35mm dia.	6.42	8.22	0.41	11.11	nr	19.34
1 1/2" x 42mm dia.	8.16	10.45	0.46	12.47	nr	22.92
Reducing coupling						
15 x 10mm dia.	1.89	2.42	0.23	6.23	nr	8.65
22 x 10mm dia.	2.76	3.54	0.26	7.05	nr	10.58
22 x 15mm dia.	1.85	2.37	0.27	7.32	nr	9.69
28 x 15mm dia.	4.23	5.42	0.28	7.59	nr	13.01
28 x 22mm dia.	2.59	3.32	0.30	8.13	nr	11.45
35 x 28mm dia.	6.11	7.83	0.34	9.22	nr	17.04
42 x 35mm dia.	8.96	11.48	0.38	10.30	nr	21.78
54 x 35mm dia.	15.71	20.12	0.42	11.38	nr	31.50
54 x 42mm dia.	17.19	22.02	0.42	11.38	nr	33.40
Straight female connector						
15mm x 1/2" dia.	2.33	2.98	0.27	7.32	nr	10.30
22mm x 3/4" dia.	3.38	4.33	0.31	8.40	nr	12.73
28mm x 1" dia.	6.37	8.16	0.36	9.76	nr	17.92
35mm x 1 1/4" dia.	11.01	14.10	0.41	11.11	nr	25.21

S: PIPED SUPPLY SYSTEMS

Item	Net Price £	Material £	Labour hours	Labour £	Unit	Total rate £
42mm x 1 1/2" dia.	14.30	18.32	0.46	12.47	nr	30.78
54mm x 2" dia.	22.68	29.05	0.52	14.09	nr	43.14
Straight male connector						
15mm x 1/2" dia.	1.98	2.54	0.27	7.32	nr	9.85
22mm x 3/4" dia.	3.54	4.53	0.31	8.40	nr	12.94
28mm x 1" dia.	5.71	7.31	0.36	9.76	nr	17.07
35mm x 1 1/4" dia.	10.04	12.86	0.41	11.11	nr	23.97
42mm x 1 1/2" dia.	12.93	16.56	0.46	12.47	nr	29.03
54mm x 2" dia.	19.64	25.15	0.52	14.09	nr	39.25
67mm x 2 1/2" dia.	31.36	40.17	0.63	17.08	nr	57.24
Female reducing connector						
15mm x 3/4" dia.	5.79	7.42	0.27	7.32	nr	14.73
Male reducing connector						
15mm x 3/4" dia.	5.17	6.62	0.27	7.32	nr	13.94
22mm x 1" dia.	7.88	10.09	0.31	8.40	nr	18.49
Flanged connector						
28mm dia.	36.97	47.35	0.36	9.76	nr	57.11
35mm dia.	46.80	59.94	0.41	11.11	nr	71.05
42mm dia.	55.94	71.65	0.46	12.47	nr	84.12
54mm dia.	84.57	108.32	0.52	14.09	nr	122.41
67mm dia.	104.42	133.74	0.61	16.53	nr	150.27
Tank connector						
15mm x 1/2" dia.	4.96	6.35	0.25	6.78	nr	13.13
22mm x 3/4" dia.	7.55	9.67	0.28	7.59	nr	17.26
28mm x 1" dia.	9.93	12.72	0.32	8.67	nr	21.39
35mm x 1 1/4" dia.	12.73	16.30	0.37	10.03	nr	26.33
42mm x 1 1/2" dia.	16.68	21.36	0.43	11.65	nr	33.02
54mm x 2" dia.	25.49	32.65	0.46	12.47	nr	45.11
Tank connector with long thread						
15mm x 1/2" dia.	6.42	8.22	0.30	8.13	nr	16.35
22mm x 3/4" dia.	9.14	11.71	0.33	8.94	nr	20.65
28mm x 1" dia.	11.29	14.46	0.39	10.57	nr	25.03
Reducer						
15 x 10mm dia.	0.63	0.81	0.23	6.23	nr	7.04
22 x 15mm dia.	0.62	0.79	0.26	7.05	nr	7.84
28 x 15mm dia.	2.09	2.68	0.28	7.59	nr	10.27
28 x 22mm dia.	1.60	2.05	0.30	8.13	nr	10.18
35 x 22mm dia.	5.93	7.60	0.34	9.22	nr	16.81
42 x 22mm dia.	10.70	13.70	0.36	9.76	nr	23.46
42 x 35mm dia.	8.28	10.60	0.38	10.30	nr	20.90
54 x 35mm dia.	17.38	22.26	0.40	10.84	nr	33.10
54 x 42mm dia.	14.99	19.20	0.42	11.38	nr	30.58
67 x 54mm dia.	20.38	26.10	0.53	14.36	nr	40.47
Adaptor; copper to female iron						
15mm x 1/2" dia.	4.01	5.14	0.27	7.32	nr	12.45
22mm x 3/4" dia.	6.12	7.84	0.31	8.40	nr	16.24
28mm x 1" dia.	8.61	11.03	0.36	9.76	nr	20.78
35mm x 1 1/4" dia.	15.58	19.95	0.41	11.11	nr	31.07
42mm x 1 1/2" dia.	19.63	25.14	0.46	12.47	nr	37.61
54mm x 2" dia.	23.64	30.28	0.52	14.09	nr	44.37

S: PIPED SUPPLY SYSTEMS

Item	Net Price £	Material £	Labour hours	Labour £	Unit	Total rate £
S10 : COLD WATER (cont'd)						
COPPER PIPEWORK (cont'd)						
Extra over for capillary fittings (cont'd)						
Adaptor; copper to male iron						
15mm x 1/2" dia.	4.10	5.25	0.27	7.32	nr	12.57
22mm x 3/4" dia.	5.23	6.70	0.31	8.40	nr	15.10
28mm x 1" dia.	8.74	11.19	0.36	9.76	nr	20.95
35mm x 1 1/4" dia.	12.73	16.30	0.41	11.11	nr	27.42
42mm x 1 1/2" dia.	17.60	22.54	0.46	12.47	nr	35.01
54mm x 2" dia.	23.64	30.28	0.52	14.09	nr	44.37
Union coupling						
15mm dia.	5.54	7.10	0.41	11.11	nr	18.21
22mm dia.	8.88	11.37	0.45	12.20	nr	23.57
28mm dia.	13.16	16.86	0.51	13.82	nr	30.68
35mm dia.	16.97	21.73	0.64	17.35	nr	39.08
42mm dia.	24.80	31.76	0.68	18.43	nr	50.19
54mm dia.	47.19	60.44	0.78	21.14	nr	81.58
67mm dia.	79.91	102.35	0.96	26.02	nr	128.37
Elbow						
15mm dia.	0.25	0.32	0.23	6.23	nr	6.55
22mm dia.	0.64	0.82	0.26	7.05	nr	7.87
28mm dia.	1.52	1.95	0.31	8.40	nr	10.35
35mm dia.	6.59	8.44	0.35	9.49	nr	17.93
42mm dia.	10.90	13.96	0.41	11.11	nr	25.07
54mm dia.	22.51	28.83	0.44	11.93	nr	40.76
67mm dia.	58.41	74.81	0.54	14.64	nr	89.45
Backplate elbow						
15mm dia.	4.16	5.33	0.51	13.82	nr	19.15
22mm dia.	8.93	11.44	0.54	14.64	nr	26.07
Overflow bend						
22mm dia.	12.59	16.13	0.26	7.05	nr	23.17
Return bend						
15mm dia.	6.23	7.98	0.23	6.23	nr	14.21
22mm dia.	12.25	15.69	0.26	7.05	nr	22.74
28mm dia.	15.65	20.04	0.31	8.40	nr	28.45
Obtuse elbow						
15mm dia.	0.79	1.01	0.23	6.23	nr	7.25
22mm dia.	1.66	2.13	0.26	7.05	nr	9.17
28mm dia.	3.20	4.10	0.31	8.40	nr	12.50
35mm dia.	9.95	12.74	0.36	9.76	nr	22.50
42mm dia.	17.71	22.68	0.41	11.11	nr	33.80
54mm dia.	32.04	41.04	0.44	11.93	nr	52.96
67mm dia.	58.13	74.45	0.54	14.64	nr	89.09
Straight tap connector						
15mm x 1/2" dia.	1.20	1.54	0.13	3.52	nr	5.06
22mm x 3/4" dia.	1.60	2.05	0.14	3.79	nr	5.84
Bent tap connector						
15mm x 1/2" dia.	1.58	2.02	0.13	3.52	nr	5.55
22mm x 3/4" dia.	4.85	6.21	0.14	3.79	nr	10.01

S: PIPED SUPPLY SYSTEMS

Item	Net Price £	Material £	Labour hours	Labour £	Unit	Total rate £
Bent male union connector						
15mm x 1/2" dia.	8.10	10.37	0.41	11.11	nr	21.49
22mm x 3/4" dia.	10.54	13.50	0.45	12.20	nr	25.70
28mm x 1" dia.	15.07	19.30	0.51	13.82	nr	33.12
35mm x 1 1/4" dia.	24.59	31.49	0.64	17.35	nr	48.84
42mm x 1 1/2" dia.	39.98	51.21	0.68	18.43	nr	69.64
54mm x 2" dia.	63.15	80.88	0.78	21.14	nr	102.02
Bent female union connector						
15mm dia.	8.10	10.37	0.41	11.11	nr	21.49
22mm x 3/4" dia.	10.54	13.50	0.45	12.20	nr	25.70
28mm x 1" dia.	15.07	19.30	0.51	13.82	nr	33.12
35mm x 1 1/4" dia.	24.59	31.49	0.64	17.35	nr	48.84
42mm x 1 1/2" dia.	39.98	51.21	0.68	18.43	nr	69.64
54mm x 2" dia.	63.15	80.88	0.78	21.14	nr	102.02
Straight union adaptor						
15mm x 3/4" dia.	3.47	4.44	0.41	11.11	nr	15.56
22mm x 1" dia.	4.91	6.29	0.45	12.20	nr	18.49
28mm x 1 1/4" dia.	7.95	10.18	0.51	13.82	nr	24.01
35mm x 1 1/2" dia.	12.25	15.69	0.64	17.35	nr	33.04
42mm x 2" dia.	15.46	19.80	0.68	18.43	nr	38.23
54mm x 2 1/2" dia.	23.87	30.57	0.78	21.14	nr	51.71
Straight male union connector						
15mm x 1/2" dia.	6.90	8.84	0.41	11.11	nr	19.95
22mm x 3/4" dia.	8.96	11.48	0.45	12.20	nr	23.67
28mm x 1" dia.	13.34	17.09	0.51	13.82	nr	30.91
35mm x 1 1/4" dia.	19.23	24.63	0.64	17.35	nr	41.98
42mm x 1 1/2" dia.	30.20	38.68	0.68	18.43	nr	57.11
54mm x 2" dia.	43.41	55.60	0.78	21.14	nr	76.74
Straight female union connector						
15mm x 1/2" dia.	6.90	8.84	0.41	11.11	nr	19.95
22mm x 3/4" dia.	8.96	11.48	0.45	12.20	nr	23.67
28mm x 1" dia.	13.34	17.09	0.51	13.82	nr	30.91
35mm x 1 1/4" dia.	19.23	24.63	0.64	17.35	nr	41.98
42mm x 1 1/2" dia.	30.21	38.69	0.68	18.43	nr	57.12
54mm x 2" dia.	43.41	55.60	0.78	21.14	nr	76.74
Male nipple						
3/4 x 1/2" dia.	3.41	4.37	0.28	7.59	nr	11.96
1 x 3/4" dia.	3.95	5.06	0.32	8.67	nr	13.73
1 1/4 x 1" dia.	5.39	6.90	0.37	10.03	nr	16.93
1 1/2 x 1 1/4" dia.	7.98	10.22	0.42	11.38	nr	21.60
2 x 1 1/2" dia.	16.33	20.92	0.46	12.47	nr	33.38
2 1/2 x 2" dia.	21.81	27.93	0.56	15.18	nr	43.11
Female nipple						
3/4 x 1/2" dia.	3.41	4.37	0.28	7.59	nr	11.96
1 x 3/4" dia.	3.95	5.06	0.32	8.67	nr	13.73
1 1/4 x 1" dia.	5.39	6.90	0.37	10.03	nr	16.93
1 1/2 x 1 1/4" dia.	7.98	10.22	0.42	11.38	nr	21.60
2 x 1 1/2" dia.	16.33	20.92	0.46	12.47	nr	33.38
2 1/2 x 2" dia.	21.81	27.93	0.56	15.18	nr	43.11

S: PIPED SUPPLY SYSTEMS

Item	Net Price £	Material £	Labour hours	Labour £	Unit	Total rate £
S10 : COLD WATER (cont'd)						
COPPER PIPEWORK (cont'd)						
Extra over for capillary fittings (cont'd)						
Equal tee						
10mm dia.	1.73	2.22	0.25	6.78	nr	8.99
15mm dia.	0.46	0.59	0.36	9.76	nr	10.35
22mm dia.	1.48	1.90	0.39	10.57	nr	12.47
28mm dia.	4.21	5.39	0.43	11.65	nr	17.05
35mm dia.	10.73	13.74	0.57	15.45	nr	29.19
42mm dia.	17.22	22.06	0.60	16.26	nr	38.32
54mm dia.	34.73	44.48	0.65	17.62	nr	62.10
67mm dia.	75.62	96.85	0.78	21.14	nr	117.99
Female tee, reducing branch FI						
15 x 15mm x 1/4" dia.	5.19	6.65	0.36	9.76	nr	16.40
22 x 22mm x 1/2" dia.	3.61	4.62	0.39	10.57	nr	15.19
28 x 28mm x 3/4" dia.	12.46	15.96	0.43	11.65	nr	27.61
35 x 35mm x 3/4" dia.	17.98	23.03	0.47	12.74	nr	35.77
42 x 42mm x 1/2" dia.	21.60	27.67	0.60	16.26	nr	43.93
Backplate tee						
15 x 15mm x 1/2" dia.	9.79	12.54	0.62	16.80	nr	29.34
Heater tee						
1/2 x 1/2" x 15mm dia.	8.85	11.34	0.36	9.76	nr	21.09
Union heater tee						
1/2 x 1/2" x 15mm dia.	12.62	16.16	0.36	9.76	nr	25.92
Sweep tee - equal						
15mm dia.	7.03	9.00	0.36	9.76	nr	18.76
22mm dia.	9.05	11.59	0.39	10.57	nr	22.16
28mm dia.	15.23	19.51	0.43	11.65	nr	31.16
35mm dia.	21.60	27.67	0.57	15.45	nr	43.11
42mm dia.	32.01	41.00	0.60	16.26	nr	57.26
54mm dia.	35.45	45.40	0.65	17.62	nr	63.02
67mm dia.	61.71	79.04	0.78	21.14	nr	100.18
Sweep tee - reducing						
22 x 22 x 15mm dia.	7.58	9.71	0.39	10.57	nr	20.28
28 x 28 x 22mm dia.	12.88	16.50	0.43	11.65	nr	28.15
35 x 35 x 22mm dia.	21.60	27.67	0.57	15.45	nr	43.11
Sweep tee - double						
15mm dia.	7.95	10.18	0.36	9.76	nr	19.94
22mm dia.	10.83	13.87	0.39	10.57	nr	24.44
28mm dia.	16.45	21.07	0.43	11.65	nr	32.72
Cross						
15mm dia.	10.52	13.47	0.48	13.01	nr	26.48
22mm dia.	11.75	15.05	0.53	14.36	nr	29.41
28mm dia.	16.86	21.59	0.61	16.53	nr	38.13

S: PIPED SUPPLY SYSTEMS

Item	Net Price £	Material £	Labour hours	Labour £	Unit	Total rate £
Extra over copper pipes; high duty capillary fittings; BS 864						
Stop end						
15mm dia.	5.19	6.65	0.16	4.34	nr	**10.98**
Straight coupling; copper to copper						
15mm dia.	2.38	3.05	0.27	7.32	nr	**10.37**
22mm dia.	3.81	4.88	0.32	8.67	nr	**13.55**
28mm dia.	5.40	6.92	0.37	10.03	nr	**16.94**
35mm dia.	9.51	12.18	0.43	11.65	nr	**23.83**
42mm dia.	10.40	13.32	0.50	13.55	nr	**26.87**
54mm dia.	15.30	19.60	0.54	14.64	nr	**34.23**
Reducing coupling						
15 x 12mm dia.	4.48	5.74	0.27	7.32	nr	**13.06**
22 x 15mm dia.	5.19	6.65	0.32	8.67	nr	**15.32**
28 x 22mm dia.	7.16	9.17	0.37	10.03	nr	**19.20**
Straight female connector						
15mm x 1/2" dia.	5.85	7.49	0.32	8.67	nr	**16.17**
22mm x 3/4" dia.	6.59	8.44	0.36	9.76	nr	**18.20**
28mm x 1" dia.	9.73	12.46	0.42	11.38	nr	**23.85**
Straight male connector						
15mm x 1/2" dia.	5.69	7.29	0.32	8.67	nr	**15.96**
22mm x 3/4" dia.	6.59	8.44	0.36	9.76	nr	**18.20**
28mm x 1" dia.	9.73	12.46	0.42	11.38	nr	**23.85**
42mm x 1 1/2" dia.	18.98	24.31	0.53	14.36	nr	**38.67**
54mm x 2" dia.	30.84	39.50	0.62	16.80	nr	**56.30**
Reducer						
15 x 12mm dia.	2.92	3.74	0.27	7.32	nr	**11.06**
22 x 15mm dia.	2.87	3.68	0.32	8.67	nr	**12.35**
28 x 22mm dia.	5.19	6.65	0.37	10.03	nr	**16.68**
35 x 28mm dia.	6.59	8.44	0.43	11.65	nr	**20.09**
42 x 35mm dia.	8.50	10.89	0.50	13.55	nr	**24.44**
54 x 42mm dia.	13.71	17.56	0.39	10.57	nr	**28.13**
Straight union adaptor						
15mm x 3/4" dia.	4.75	6.08	0.27	7.32	nr	**13.40**
22mm x 1" dia.	6.43	8.24	0.32	8.67	nr	**16.91**
28mm x 1 1/4" dia.	8.50	10.89	0.37	10.03	nr	**20.91**
35mm x 1 1/2" dia.	15.40	19.72	0.43	11.65	nr	**31.38**
42mm x 2" dia.	19.50	24.98	0.50	13.55	nr	**38.53**
Bent union adaptor						
15mm x 3/4" dia.	12.35	15.82	0.27	7.32	nr	**23.14**
22mm x 1" dia.	16.66	21.34	0.32	8.67	nr	**30.01**
28mm x 1 1/4" dia.	22.45	28.75	0.37	10.03	nr	**38.78**
Adaptor; male copper to FI						
15mm x 1/2" dia.	8.81	11.28	0.27	7.32	nr	**18.60**
22mm x 3/4" dia.	8.98	11.50	0.32	8.67	nr	**20.17**

S: PIPED SUPPLY SYSTEMS

Item	Net Price £	Material £	Labour hours	Labour £	Unit	Total rate £
S10 : COLD WATER (cont'd)						
COPPER PIPEWORK (cont'd)						
Extra over for capillary fittings (cont'd)						
Union coupling						
15mm dia.	10.71	13.72	0.54	14.64	nr	28.35
22mm dia.	13.71	17.56	0.60	16.26	nr	33.82
28mm dia.	19.03	24.37	0.68	18.43	nr	42.80
35mm dia.	33.20	42.52	0.83	22.50	nr	65.02
42mm dia.	39.11	50.09	0.89	24.12	nr	74.21
Elbow						
15mm dia.	6.90	8.84	0.27	7.32	nr	16.16
22mm dia.	7.36	9.43	0.32	8.67	nr	18.10
28mm dia.	10.94	14.01	0.37	10.03	nr	24.04
35mm dia.	17.10	21.90	0.43	11.65	nr	33.55
42mm dia.	21.31	27.29	0.50	13.55	nr	40.85
54mm dia.	37.06	47.47	0.52	14.09	nr	61.56
Return bend						
28mm dia.	22.45	28.75	0.37	10.03	nr	38.78
35mm dia.	26.07	33.39	0.43	11.65	nr	45.04
Bent male union connector						
15mm x 1/2" dia.	15.97	20.45	0.54	14.64	nr	35.09
22mm x 3/4" dia.	21.53	27.58	0.60	16.26	nr	43.84
28mm x 1" dia.	39.11	50.09	0.68	18.43	nr	68.52
Composite flange						
35mm dia.	35.64	45.65	0.38	10.30	nr	55.95
42mm dia.	46.22	59.20	0.41	11.11	nr	70.31
54mm dia.	58.86	75.39	0.43	11.66	nr	87.04
Equal tee						
15mm dia.	7.91	10.13	0.44	11.93	nr	22.06
22mm dia.	9.95	12.74	0.47	12.74	nr	25.48
28mm dia.	13.12	16.80	0.53	14.36	nr	31.17
35mm dia.	22.45	28.75	0.70	18.97	nr	47.73
42mm dia.	28.57	36.59	0.84	22.77	nr	59.36
54mm dia.	44.99	57.62	0.79	21.41	nr	79.03
Reducing tee						
15 x 12mm dia.	7.27	9.31	0.44	11.93	nr	21.24
22 x 15mm dia.	9.54	12.22	0.47	12.74	nr	24.96
28 x 22mm dia.	12.51	16.02	0.53	14.36	nr	30.39
35 x 28mm dia.	20.70	26.51	0.73	19.79	nr	46.30
42 x 28mm dia.	25.72	32.94	0.84	22.77	nr	55.71
54 x 28mm dia.	39.24	50.26	1.01	27.37	nr	77.63
Extra over copper pipes; compression fittings; BS 864						
Stop end						
15mm dia.	1.22	1.56	0.10	2.71	nr	4.27
22mm dia.	1.45	1.86	0.12	3.25	nr	5.11
28mm dia.	4.72	6.05	0.15	4.07	nr	10.11

S: PIPED SUPPLY SYSTEMS

Item	Net Price £	Material £	Labour hours	Labour £	Unit	Total rate £
Straight connector; copper to copper						
15mm dia.	0.79	1.01	0.18	4.88	nr	5.89
22mm dia.	1.35	1.73	0.21	5.69	nr	7.42
28mm dia.	4.25	5.44	0.24	6.50	nr	11.95
Straight connector; copper to imperial copper						
22mm dia.	3.99	5.11	0.21	5.69	nr	10.80
Male coupling; copper to MI (BSP)						
15mm dia.	0.72	0.92	0.19	5.15	nr	6.07
22mm dia.	1.14	1.46	0.23	6.23	nr	7.69
28mm dia.	2.52	3.23	0.26	7.05	nr	10.27
Male coupling with long thread and backnut						
15mm dia.	4.43	5.67	0.19	5.15	nr	10.82
22mm dia.	5.61	7.19	0.23	6.23	nr	13.42
Female coupling; copper to FI (BSP)						
15mm dia.	0.89	1.14	0.19	5.15	nr	6.29
22mm dia.	1.27	1.63	0.23	6.23	nr	7.86
28mm dia.	3.51	4.50	0.27	7.32	nr	11.81
Elbow						
15mm dia.	0.95	1.22	0.18	4.88	nr	6.10
22mm dia.	1.61	2.06	0.21	5.69	nr	7.75
28mm dia.	5.20	6.66	0.24	6.50	nr	13.16
Male elbow; copper to FI (BSP)						
15mm x 1/2" dia.	1.73	2.22	0.19	5.15	nr	7.37
22mm x 3/4" dia.	2.24	2.87	0.23	6.23	nr	9.10
28mm x 1" dia.	5.43	6.95	0.27	7.32	nr	14.27
Female elbow; copper to FI (BSP)						
15mm x 1/2" dia.	2.66	3.41	0.19	5.15	nr	8.56
22mm x 3/4" dia.	3.84	4.92	0.23	6.23	nr	11.15
28mm x 1" dia.	6.77	8.67	0.27	7.32	nr	15.99
Backplate elbow						
15mm x 1/2" dia.	3.84	4.92	0.50	13.55	nr	18.47
Tank coupling; long thread						
22mm dia.	4.43	5.67	0.46	12.47	nr	18.14
Tee equal						
15mm dia.	1.35	1.73	0.28	7.59	nr	9.32
22mm dia.	2.25	2.88	0.30	8.13	nr	11.01
28mm dia.	9.79	12.54	0.34	9.22	nr	21.75
Tee reducing						
22mm dia.	5.53	7.08	0.30	8.13	nr	15.21
Backplate tee						
15mm dia.	7.30	9.35	0.62	16.80	nr	26.15

S: PIPED SUPPLY SYSTEMS

Item	Net Price £	Material £	Labour hours	Labour £	Unit	Total rate £
S10 : COLD WATER (cont'd)						
COPPER PIPEWORK (cont'd)						
Extra over fittings; silver brazed welded joints						
Reducer						
76 x 67mm dia	33.33	42.69	1.40	37.94	nr	80.63
108 x 76mm dia	47.37	60.67	1.80	48.79	nr	109.46
133 x 108mm dia	72.82	93.27	2.20	59.63	nr	152.89
159 x 133mm dia	85.80	109.89	2.60	70.47	nr	180.36
90° elbow						
76mm dia	51.75	66.28	1.60	43.37	nr	109.65
108mm dia	86.63	110.95	2.00	54.21	nr	165.16
133mm dia	223.38	286.10	2.40	65.05	nr	351.15
159mm dia	259.47	332.33	2.80	75.89	nr	408.22
45° elbow						
76mm dia	58.59	75.04	1.60	43.37	nr	118.41
108mm dia	100.44	128.64	2.00	54.21	nr	182.85
133mm dia	201.15	257.63	2.40	65.05	nr	322.68
159mm dia	234.36	300.17	2.80	75.89	nr	376.06
Equal tee						
76mm dia	92.07	117.92	2.40	65.05	nr	182.97
108mm dia	125.55	160.80	3.00	81.31	nr	242.11
133mm dia	228.87	293.13	3.60	97.57	nr	390.71
159mm dia	284.58	364.49	4.20	113.83	nr	478.32
Extra over copper pipes; dezincification resistant compression fittings; BS 864						
Stop end						
15mm dia.	1.67	2.14	0.10	2.71	nr	4.85
22mm dia.	2.42	3.10	0.13	3.52	nr	6.62
28mm dia.	5.17	6.62	0.15	4.07	nr	10.69
35mm dia.	7.96	10.20	0.18	4.88	nr	15.07
42mm dia.	13.26	16.98	0.20	5.42	nr	22.40
Straight coupling; copper to copper						
15mm dia.	1.34	1.72	0.18	4.88	nr	6.59
22mm dia.	2.18	2.79	0.21	5.69	nr	8.48
28mm dia.	4.95	6.34	0.24	6.50	nr	12.84
35mm dia.	10.27	13.15	0.29	7.86	nr	21.01
42mm dia.	13.51	17.30	0.33	8.94	nr	26.25
54mm dia.	20.21	25.88	0.38	10.30	nr	36.18
Straight swivel connector; copper to imperial copper						
22mm dia.	4.80	6.15	0.20	5.42	nr	11.57
Male coupling; copper to MI (BSP)						
15mm x 1/2" dia.	1.19	1.52	0.19	5.15	nr	6.67
22mm x 3/4" dia.	1.81	2.32	0.23	6.23	nr	8.55
28mm x 1" dia.	3.51	4.50	0.26	7.05	nr	11.54
35mm x 1 1/4" dia.	7.81	10.00	0.32	8.67	nr	18.68
42mm x 1 1/2" dia.	11.72	15.01	0.37	10.03	nr	25.04
54mm x 2" dia.	17.30	22.16	0.57	15.45	nr	37.61

S: PIPED SUPPLY SYSTEMS

Item	Net Price £	Material £	Labour hours	Labour £	Unit	Total rate £
Male coupling with long thread and backnuts						
22mm dia.	6.77	8.67	0.23	6.23	nr	**14.90**
28mm dia.	7.50	9.61	0.24	6.50	nr	**16.11**
Female coupling; copper to FI (BSP)						
15mm x 1/2" dia.	1.44	1.84	0.19	5.15	nr	**6.99**
22mm x 3/4" dia.	2.10	2.69	0.23	6.23	nr	**8.92**
28mm x 1" dia.	4.54	5.81	0.27	7.32	nr	**13.13**
35mm x 1 1/4" dia.	9.38	12.01	0.32	8.67	nr	**20.69**
42mm x 1 1/2" dia.	12.61	16.15	0.37	10.03	nr	**26.18**
54mm x 2" dia.	18.50	23.69	0.42	11.38	nr	**35.08**
Elbow						
15mm dia.	1.61	2.06	0.18	4.88	nr	**6.94**
22mm dia.	2.57	3.29	0.21	5.69	nr	**8.98**
28mm dia.	6.39	8.18	0.24	6.50	nr	**14.69**
35mm dia.	13.87	17.76	0.29	7.86	nr	**25.62**
42mm dia.	18.78	24.05	0.33	8.94	nr	**33.00**
54mm dia.	32.32	41.40	0.38	10.30	nr	**51.69**
Male elbow; copper to MI (BSP)						
15mm x 1/2" dia.	2.79	3.57	0.19	5.15	nr	**8.72**
22mm x 3/4" dia.	3.12	4.00	0.23	6.23	nr	**10.23**
28mm x 1" dia.	5.86	7.51	0.27	7.32	nr	**14.82**
Female elbow; copper to FI (BSP)						
15mm x 1/2" dia.	2.99	3.83	0.19	5.15	nr	**8.98**
22mm x 3/4" dia.	4.30	5.51	0.23	6.23	nr	**11.74**
28mm x 1" dia.	7.14	9.14	0.27	7.32	nr	**16.46**
Backplate elbow						
15mm x 1/2" dia.	4.35	5.57	0.50	13.55	nr	**19.12**
Straight tap connector						
15mm dia.	2.47	3.16	0.13	3.52	nr	**6.69**
22mm dia.	5.36	6.87	0.15	4.07	nr	**10.93**
Tank coupling						
15mm dia.	3.40	4.35	0.19	5.15	nr	**9.50**
22mm dia.	3.77	4.83	0.23	6.23	nr	**11.06**
28mm dia.	7.95	10.18	0.27	7.32	nr	**17.50**
35mm dia.	13.78	17.65	0.32	8.67	nr	**26.32**
42mm dia.	22.40	28.69	0.37	10.03	nr	**38.72**
54mm dia.	28.85	36.95	0.31	8.40	nr	**45.35**
Tee equal						
15mm dia.	2.26	2.89	0.28	7.59	nr	**10.48**
22mm dia.	3.74	4.79	0.30	8.13	nr	**12.92**
28mm dia.	10.18	13.04	0.34	9.22	nr	**22.25**
35mm dia.	18.04	23.11	0.43	11.65	nr	**34.76**
42mm dia.	28.36	36.32	0.46	12.47	nr	**48.79**
54mm dia.	45.56	58.35	0.54	14.64	nr	**72.99**

S: PIPED SUPPLY SYSTEMS

Item	Net Price £	Material £	Labour hours	Labour £	Unit	Total rate £
S10 : COLD WATER (cont'd)						
COPPER PIPEWORK (cont'd)						
Extra over copper pipes; dezincification (cont'd)						
Tee reducing						
22mm dia.	5.97	7.65	0.30	8.13	nr	15.78
28mm dia.	9.83	12.59	0.34	9.22	nr	21.81
35mm dia.	17.63	22.58	0.43	11.65	nr	34.23
42mm dia.	27.25	34.90	0.46	12.47	nr	47.37
54mm dia.	45.56	58.35	0.54	14.64	nr	72.99
Extra over copper pipes; bronze one piece brazing flanges; metric, including jointing ring and bolts						
Bronze flange; PN6						
15mm dia.	13.33	17.07	0.27	7.32	nr	24.39
22mm dia.	15.82	20.26	0.32	8.67	nr	28.94
28mm dia.	18.15	23.25	0.36	9.76	nr	33.00
35mm dia.	25.55	32.72	0.47	12.74	nr	45.46
42mm dia.	31.26	40.04	0.54	14.64	nr	54.67
54mm dia.	44.51	57.01	0.63	17.08	nr	74.08
67mm dia.	51.36	65.78	0.77	20.87	nr	86.65
76mm dia.	58.45	74.86	0.93	25.21	nr	100.07
108mm dia.	78.54	100.59	1.14	30.90	nr	131.49
133mm dia.	95.35	122.12	1.41	38.22	nr	160.34
159mm dia.	135.13	173.07	1.74	47.16	nr	220.23
Bronze flange; PN10						
15mm dia.	17.42	22.31	0.27	7.32	nr	29.63
22mm dia.	20.29	25.99	0.32	8.67	nr	34.66
28mm dia.	20.41	26.14	0.38	10.30	nr	36.44
35mm dia.	27.87	35.70	0.47	12.74	nr	48.43
42mm dia.	33.26	42.60	0.54	14.64	nr	57.23
54mm dia.	47.32	60.61	0.63	17.08	nr	77.68
67mm dia.	51.36	65.78	0.77	20.87	nr	86.65
76mm dia.	65.00	83.25	0.93	25.21	nr	108.46
108mm dia.	96.25	123.28	1.14	30.90	nr	154.17
133mm dia.	110.63	141.69	1.41	38.22	nr	179.91
159mm dia.	169.22	216.74	1.74	47.16	nr	263.90
Bronze flange; PN16						
15mm dia.	17.42	22.31	0.27	7.32	nr	29.63
22mm dia.	20.29	25.99	0.32	8.67	nr	34.66
28mm dia.	21.05	26.96	0.38	10.30	nr	37.26
35mm dia.	27.87	35.70	0.47	12.74	nr	48.43
42mm dia.	33.26	42.60	0.54	14.64	nr	57.23
54mm dia.	49.99	64.03	0.63	17.08	nr	81.10
67mm dia.	60.87	77.96	0.77	20.87	nr	98.83
76mm dia.	77.35	99.07	0.93	25.21	nr	124.28
108mm dia.	99.40	127.31	1.14	30.90	nr	158.21
133mm dia.	164.47	210.65	1.41	38.22	nr	248.87
159mm dia.	204.83	262.34	1.74	47.16	nr	309.50

S: PIPED SUPPLY SYSTEMS

Item	Net Price £	Material £	Labour hours	Labour £	Unit	Total rate £
Extra Over copper pipes; bronze blank flanges; metric, including jointing ring and bolts						
Bronze blank flange; PN6						
15mm dia.	11.84	15.16	0.27	7.32	nr	**22.48**
22mm dia.	15.11	19.35	0.27	7.32	nr	**26.67**
28mm dia.	15.53	19.89	0.27	7.32	nr	**27.21**
35mm dia.	25.41	32.54	0.32	8.67	nr	**41.22**
42mm dia.	34.28	43.91	0.32	8.67	nr	**52.58**
54mm dia.	37.21	47.66	0.34	9.22	nr	**56.87**
67mm dia.	46.51	59.57	0.36	9.76	nr	**69.33**
76mm dia.	59.91	76.73	0.37	10.03	nr	**86.76**
108mm dia.	95.19	121.92	0.41	11.11	nr	**133.03**
133mm dia.	112.30	143.83	0.58	15.72	nr	**159.55**
159mm dia.	141.49	181.22	0.61	16.53	nr	**197.75**
Bronze blank flange; PN10						
15mm dia.	14.32	18.34	0.27	7.32	nr	**25.66**
22mm dia.	18.53	23.73	0.27	7.32	nr	**31.05**
28mm dia.	20.52	26.28	0.27	7.32	nr	**33.60**
35mm dia.	25.41	32.54	0.32	8.67	nr	**41.22**
42mm dia.	47.63	61.00	0.32	8.67	nr	**69.68**
54mm dia.	54.30	69.55	0.34	9.22	nr	**78.76**
67mm dia.	58.25	74.61	0.46	12.47	nr	**87.07**
76mm dia.	74.84	95.85	0.47	12.74	nr	**108.59**
108mm dia.	114.30	146.39	0.51	13.82	nr	**160.22**
133mm dia.	120.68	154.57	0.58	15.72	nr	**170.29**
159mm dia.	220.70	282.67	0.71	19.24	nr	**301.91**
Bronze blank flange; PN16						
15mm dia.	14.32	18.34	0.27	7.32	nr	**25.66**
22mm dia.	18.81	24.09	0.27	7.32	nr	**31.41**
28mm dia.	20.52	26.28	0.27	7.32	nr	**33.60**
35mm dia.	25.41	32.54	0.32	8.67	nr	**41.22**
42mm dia.	47.63	61.00	0.32	8.67	nr	**69.68**
54mm dia.	54.30	69.55	0.34	9.22	nr	**78.76**
67mm dia.	84.45	108.16	0.46	12.47	nr	**120.63**
76mm dia.	100.01	128.09	0.47	12.74	nr	**140.83**
108mm dia.	127.14	162.84	0.51	13.82	nr	**176.66**
133mm dia.	208.65	267.24	0.58	15.72	nr	**282.96**
159mm dia.	267.07	342.06	0.71	19.24	nr	**361.30**
Extra Over copper pipes; bronze screwed flanges; metric, including jointing ring and bolts						
Bronze screwed flange; 6 BSP						
15mm dia.	11.84	15.16	0.35	9.49	nr	**24.65**
22mm dia.	13.69	17.53	0.47	12.74	nr	**30.27**
28mm dia.	14.27	18.28	0.52	14.09	nr	**32.37**
35mm dia.	19.23	24.63	0.62	16.80	nr	**41.43**
42mm dia.	23.09	29.57	0.70	18.97	nr	**48.55**
54mm dia.	31.37	40.18	0.84	22.77	nr	**62.95**
67mm dia.	39.38	50.44	1.03	27.92	nr	**78.35**
76mm dia.	47.53	60.88	1.22	33.07	nr	**93.94**
108mm dia.	75.27	96.41	1.41	38.22	nr	**134.62**
133mm dia.	89.22	114.27	1.75	47.43	nr	**161.70**
159mm dia.	113.58	145.47	2.21	59.90	nr	**205.37**

S: PIPED SUPPLY SYSTEMS

Item	Net Price £	Material £	Labour hours	Labour £	Unit	Total rate £
S10 : COLD WATER (cont'd)						
COPPER PIPEWORK (cont'd)						
Extra Over copper pipes; bronze screwed (cont'd)						
Bronze screwed flange; 10 BSP						
15mm dia.	14.34	18.37	0.35	9.49	nr	27.85
22mm dia.	16.70	21.39	0.47	12.74	nr	34.13
28mm dia.	18.44	23.62	0.52	14.09	nr	37.71
35mm dia.	26.28	33.66	0.62	16.80	nr	50.46
42mm dia.	32.09	41.10	0.70	18.97	nr	60.07
54mm dia.	45.63	58.44	0.84	22.77	nr	81.21
67mm dia.	53.42	68.42	1.03	27.92	nr	96.34
76mm dia.	60.30	77.23	1.22	33.07	nr	110.30
108mm dia.	80.05	102.53	1.41	38.22	nr	140.74
133mm dia.	97.01	124.25	1.75	47.43	nr	171.68
159mm dia.	172.16	220.50	2.21	59.90	nr	280.40
Bronze screwed flange; 16 BSP						
15mm dia.	14.34	18.37	0.35	9.49	nr	27.85
22mm dia.	16.70	21.39	0.47	12.74	nr	34.13
28mm dia.	18.44	23.62	0.52	14.09	nr	37.71
35mm dia.	26.28	33.66	0.62	16.80	nr	50.46
42mm dia.	32.09	41.10	0.70	18.97	nr	60.07
54mm dia.	45.63	58.44	0.84	22.77	nr	81.21
67mm dia.	62.69	80.55	1.03	27.92	nr	108.47
76mm dia.	79.39	101.68	1.22	33.07	nr	134.75
108mm dia.	101.40	129.87	1.41	38.22	nr	168.09
133mm dia.	167.34	214.33	1.75	47.43	nr	261.76
159mm dia.	207.45	265.70	2.21	59.90	nr	325.60
Extra over copper pipes; labour						
Made bend						
15mm dia.	-	-	0.26	7.05	nr	7.05
22mm dia.	-	-	0.28	7.59	nr	7.59
28mm dia.	-	-	0.31	8.40	nr	8.40
35mm dia.	-	-	0.42	11.38	nr	11.38
42mm dia.	-	-	0.51	13.82	nr	13.82
54mm dia.	-	-	0.58	15.72	nr	15.72
67mm dia.	-	-	0.69	18.70	nr	18.70
76mm dia.	-	-	0.80	21.68	nr	21.68
Bronze butt weld						
15mm dia.	-	-	0.25	6.78	nr	6.78
22mm dia.	-	-	0.31	8.40	nr	8.40
28mm dia.	-	-	0.37	10.03	nr	10.03
35mm dia.	-	-	0.49	13.28	nr	13.28
42mm dia.	-	-	0.58	15.72	nr	15.72
54mm dia.	-	-	0.72	19.51	nr	19.51
67mm dia.	-	-	0.88	23.85	nr	23.85
76mm dia.	-	-	1.08	29.27	nr	29.27
108mm dia.	-	-	1.37	37.13	nr	37.13
133mm dia.	-	-	1.73	46.89	nr	46.89
159mm dia.	-	-	2.03	55.02	nr	55.02

S: PIPED SUPPLY SYSTEMS

Item	Net Price £	Material £	Labour hours	Labour £	Unit	Total rate £
PRESSFIT (copper fittings)						
Mechanical pressfit joints; butyl rubber O ring						
Coupler						
15mm dia	0.82	1.05	0.36	9.76	nr	10.81
22mm dia	1.30	1.67	0.36	9.76	nr	11.42
28mm dia	2.58	3.30	0.44	11.93	nr	15.23
35mm dia	3.26	4.18	0.44	11.93	nr	16.10
42mm dia	5.85	7.49	0.52	14.09	nr	21.59
54mm dia	7.47	9.57	0.60	16.26	nr	25.83
Stop end						
22mm dia	2.02	2.59	0.18	4.88	nr	7.47
28mm dia	3.18	4.07	0.22	5.96	nr	10.04
35mm dia	5.47	7.01	0.22	5.96	nr	12.97
42mm dia	8.23	10.54	0.26	7.05	nr	17.59
54mm dia	9.92	12.71	0.30	8.13	nr	20.84
Reducer						
22 x 15mm dia	0.95	1.22	0.36	9.76	nr	10.97
28 x 15mm dia	2.46	3.15	0.40	10.84	nr	13.99
28 x 22mm dia	2.55	3.27	0.40	10.84	nr	14.11
35 x 22mm dia	3.05	3.91	0.40	10.84	nr	14.75
35 x 28mm dia	3.38	4.33	0.44	11.93	nr	16.25
42 x 22mm dia	5.28	6.76	0.44	11.93	nr	18.69
42 x 28mm dia	5.03	6.44	0.48	13.01	nr	19.45
42 x 35mm dia	5.03	6.44	0.48	13.01	nr	19.45
54 x 35mm dia	6.81	8.72	0.52	14.09	nr	22.82
54 x 42mm dia	6.81	8.72	0.56	15.18	nr	23.90
90° elbow						
15mm dia	0.89	1.14	0.36	9.76	nr	10.90
22mm dia	1.49	1.91	0.36	9.76	nr	11.67
28mm dia	3.22	4.12	0.44	11.93	nr	16.05
35mm dia	6.44	8.25	0.44	11.93	nr	20.17
42mm dia	12.10	15.50	0.52	14.09	nr	29.59
54mm dia	16.78	21.49	0.60	16.26	nr	37.75
45° elbow						
15mm dia	1.07	1.37	0.36	9.76	nr	11.13
22mm dia	1.47	1.88	0.36	9.76	nr	11.64
28mm dia	4.42	5.66	0.44	11.93	nr	17.59
35mm dia	6.31	8.08	0.44	11.93	nr	20.01
42mm dia	10.49	13.44	0.52	14.09	nr	27.53
54mm dia	14.92	19.11	0.60	16.26	nr	35.37
Equal tee						
15mm dia	1.42	1.82	0.54	14.64	nr	16.45
22mm dia	2.58	3.30	0.54	14.64	nr	17.94
28mm dia	4.50	5.76	0.66	17.89	nr	23.65
35mm dia	7.79	9.98	0.66	17.89	nr	27.87
42mm dia	15.38	19.70	0.78	21.14	nr	40.84
54mm dia	19.18	24.57	0.90	24.39	nr	48.96

S: PIPED SUPPLY SYSTEMS

Item	Net Price £	Material £	Labour hours	Labour £	Unit	Total rate £
S10 : COLD WATER (cont'd)						
PRESSFIT (copper fittings) (cont'd)						
Mechanical pressfit joints; butyl rubber O ring (cont'd)						
Reducing tee						
22 x 15mm dia	2.10	2.69	0.54	14.64	nr	17.33
28 x 15mm dia	8.62	11.04	0.62	16.80	nr	27.84
28 x 22mm dia	5.32	6.81	0.62	16.80	nr	23.62
35 x 22mm dia	6.93	8.88	0.62	16.80	nr	25.68
35 x 28mm dia	7.71	9.87	0.62	16.80	nr	26.68
42 x 28mm dia	13.95	17.87	0.70	18.97	nr	36.84
42 x 35mm dia	13.95	17.87	0.70	18.97	nr	36.84
54 x 35mm dia	23.58	30.20	0.82	22.22	nr	52.43
54 x 42mm dia	23.58	30.20	0.82	22.22	nr	52.43
Male iron connector; BSP thread						
15mm dia	3.01	3.86	0.18	4.88	nr	8.73
22mm dia	4.32	5.53	0.18	4.88	nr	10.41
28mm dia	5.78	7.40	0.22	5.96	nr	13.37
35mm dia	10.45	13.38	0.22	5.96	nr	19.35
42mm dia	14.01	17.94	0.26	7.05	nr	24.99
54mm dia	27.02	34.61	0.30	8.13	nr	42.74
90° elbow; male iron BSP thread						
15mm dia	4.71	6.03	0.36	9.76	nr	15.79
22mm dia	7.37	9.44	0.36	9.76	nr	19.20
28mm dia	11.30	14.47	0.44	11.93	nr	26.40
35mm dia	14.69	18.81	0.44	11.93	nr	30.74
42mm dia	19.16	24.54	0.52	14.09	nr	38.63
54mm dia	28.00	35.86	0.60	16.26	nr	52.12
Female iron connector; BSP thread						
15mm dia	3.33	4.26	0.18	4.88	nr	9.14
22mm dia	4.39	5.62	0.18	4.88	nr	10.50
28mm dia	5.92	7.58	0.22	5.96	nr	13.55
35mm dia	11.57	14.82	0.22	5.96	nr	20.78
42mm dia	16.52	21.16	0.26	7.05	nr	28.21
54mm dia	28.33	36.28	0.30	8.13	nr	44.42
90° elbow; female iron BSP thread						
15mm dia	3.98	5.10	0.36	9.76	nr	14.85
22mm dia	5.85	7.49	0.36	9.76	nr	17.25
28mm dia	9.67	12.39	0.44	11.93	nr	24.31
35mm dia	12.49	16.00	0.44	11.93	nr	27.92
42mm dia	16.99	21.76	0.52	14.09	nr	35.85
54mm dia	25.00	32.02	0.60	16.26	nr	48.28

Material Costs/Prices for Measured Works – Mechanical Installations

S: PIPED SUPPLY SYSTEMS

Item	Net Price £	Material £	Labour hours	Labour £	Unit	Total rate £
STAINLESS STEEL PIPEWORK						
Stainless steel pipes; capillary or compression joints; BS 4127, vertical or at low level, with brackets measured separately						
Grade 304; satin finish						
15mm dia.	2.61	3.34	0.41	11.11	m	**14.46**
22mm dia.	3.66	4.69	0.51	13.83	m	**18.52**
28mm dia.	4.99	6.39	0.58	15.72	m	**22.11**
35mm dia.	7.53	9.64	0.65	17.62	m	**27.27**
42mm dia.	9.56	12.24	0.71	19.25	m	**31.49**
54mm dia.	13.32	17.06	0.80	21.68	m	**38.74**
Grade 316 satin finish						
15mm dia.	3.36	4.30	0.61	16.54	m	**20.84**
22mm dia.	6.27	8.03	0.76	20.61	m	**28.64**
28mm dia.	7.44	9.53	0.87	23.59	m	**33.12**
35mm dia.	13.50	17.29	0.98	26.57	m	**43.86**
42mm dia.	17.47	22.38	1.06	28.74	m	**51.12**
54mm dia.	20.36	26.08	1.16	31.44	m	**57.52**
FIXINGS						
For Stainless Steel pipework						
Single pipe ring						
15mm dia.	8.69	11.13	0.26	7.05	nr	**18.18**
22mm dia.	10.12	12.96	0.26	7.05	nr	**20.01**
28mm dia.	10.60	13.58	0.31	8.40	nr	**21.98**
35mm dia.	12.04	15.42	0.32	8.67	nr	**24.09**
42mm dia.	13.68	17.52	0.32	8.67	nr	**26.19**
54mm dia.	15.62	20.01	0.34	9.22	nr	**29.22**
Screw on backplate, female						
All sizes 15mm to 54mm dia.	7.83	10.03	0.10	2.71	nr	**12.74**
Screw on backplate, male						
All sizes 15mm to 54mm dia.	8.92	11.42	0.10	2.71	nr	**14.13**
Stainless steel threaded rods; metric thread; including nuts, washers etc						
10mm dia x 600mm long	9.33	11.95	0.18	4.88	nr	**16.83**
Extra over stainless steel pipes; capillary fittings						
Straight coupling						
15mm dia.	0.96	1.23	0.25	6.78	nr	**8.01**
22mm dia.	1.55	1.99	0.28	7.59	nr	**9.57**
28mm dia.	2.05	2.63	0.33	8.94	nr	**11.57**
35mm dia.	4.72	6.05	0.37	10.03	nr	**16.08**
42mm dia.	5.44	6.97	0.42	11.38	nr	**18.35**
54mm dia.	8.19	10.49	0.45	12.20	nr	**22.69**

S: PIPED SUPPLY SYSTEMS

Item	Net Price £	Material £	Labour hours	Labour £	Unit	Total rate £
S10 : COLD WATER (cont'd)						
STAINLESS STEEL PIPEWORK (cont'd)						
Extra over stainless steel pipes (cont'd)						
45° bend						
15mm dia.	5.15	6.60	0.25	6.78	nr	13.37
22mm dia.	6.77	8.67	0.30	8.04	nr	16.71
28mm dia.	8.31	10.64	0.33	8.94	nr	19.59
35mm dia.	9.85	12.62	0.37	10.03	nr	22.65
42mm dia.	12.68	16.24	0.42	11.38	nr	27.62
54mm dia.	15.67	20.07	0.45	12.20	nr	32.27
90° bend						
15mm dia.	2.66	3.41	0.28	7.59	nr	11.00
22mm dia.	3.59	4.60	0.28	7.59	nr	12.19
28mm dia.	5.07	6.49	0.33	8.94	nr	15.44
35mm dia.	12.36	15.83	0.37	10.03	nr	25.86
42mm dia.	17.02	21.80	0.42	11.38	nr	33.18
54mm dia.	23.07	29.55	0.45	12.20	nr	41.74
Reducer						
22 x 15mm dia.	6.14	7.86	0.28	7.59	nr	15.45
28 x 22mm dia.	6.85	8.77	0.33	8.94	nr	17.72
35 x 28mm dia.	8.38	10.73	0.37	10.03	nr	20.76
42 x 35mm dia.	9.03	11.57	0.42	11.38	nr	22.95
54 x 42mm dia.	26.79	34.31	0.48	13.03	nr	47.34
Tap connector						
15mm dia.	12.96	16.60	0.13	3.52	nr	20.12
22mm dia.	17.13	21.94	0.14	3.79	nr	25.73
28mm dia.	23.77	30.44	0.17	4.61	nr	35.05
Tank connector						
15mm dia.	16.74	21.44	0.13	3.52	nr	24.96
22mm dia.	24.91	31.90	0.13	3.52	nr	35.43
28mm dia	30.67	39.28	0.15	4.07	nr	43.35
35mm dia.	44.35	56.80	0.18	4.88	nr	61.68
42mm dia.	58.56	75.00	0.21	5.69	nr	80.69
54mm dia.	88.66	113.55	0.24	6.50	nr	120.06
Tee equal						
15mm dia.	4.77	6.11	0.37	10.03	nr	16.14
22mm dia.	5.94	7.61	0.40	10.84	nr	18.45
28mm dia.	7.18	9.20	0.45	12.20	nr	21.39
35mm dia.	17.24	22.08	0.59	16.00	nr	38.08
42mm dia.	21.27	27.24	0.62	16.81	nr	44.05
54mm dia.	42.96	55.02	0.67	18.16	nr	73.19
Unequal tee						
22 x 15mm dia.	9.68	12.40	0.37	10.03	nr	22.43
28 x 15mm dia.	10.90	13.96	0.45	12.20	nr	26.16
28 x 22mm dia.	10.90	13.96	0.45	12.20	nr	26.16
35 x 22mm dia.	19.04	24.39	0.59	16.00	nr	40.39
35 x 28mm dia.	19.04	24.39	0.59	16.00	nr	40.39
42 x 28mm dia.	23.41	29.98	0.62	16.81	nr	46.80
42 x 35mm dia.	23.41	29.98	0.62	16.81	nr	46.80
54 x 35mm dia.	48.47	62.08	0.67	18.16	nr	80.24
54 x 42mm dia.	48.47	62.08	0.67	18.16	nr	80.24

S: PIPED SUPPLY SYSTEMS

Item	Net Price £	Material £	Labour hours	Labour £	Unit	Total rate £
Union, conical seat						
15mm dia.	21.46	27.49	0.25	6.78	nr	34.26
22mm dia.	33.80	43.29	0.28	7.59	nr	50.88
28mm dia.	43.69	55.96	0.33	8.94	nr	64.90
35mm dia.	57.36	73.47	0.37	10.03	nr	83.50
42mm dia.	72.34	92.65	0.42	11.38	nr	104.04
54mm dia.	95.70	122.57	0.45	12.20	nr	134.77
Union, flat seat						
15mm dia.	22.41	28.70	0.25	6.78	nr	35.48
22mm dia.	34.90	44.70	0.28	7.59	nr	52.29
28mm dia.	45.11	57.78	0.33	8.94	nr	66.72
35mm dia.	58.94	75.49	0.37	10.03	nr	85.52
42mm dia.	74.25	95.10	0.42	11.38	nr	106.48
54mm dia.	99.57	127.53	0.45	12.20	nr	139.72
Extra over stainless steel pipes; compression fittings						
Straight coupling						
15mm dia.	18.65	23.89	0.18	4.88	nr	28.77
22mm dia.	35.54	45.52	0.22	5.96	nr	51.48
28mm dia.	47.84	61.27	0.25	6.78	nr	68.05
35mm dia.	73.86	94.60	0.30	8.13	nr	102.73
42mm dia.	86.21	110.42	0.40	10.84	nr	121.26
90° bend						
15mm dia.	23.52	30.12	0.18	4.88	nr	35.00
22mm dia.	46.71	59.83	0.22	5.96	nr	65.79
28mm dia.	63.71	81.60	0.25	6.78	nr	88.38
35mm dia.	129.00	165.22	0.33	8.94	nr	174.17
42mm dia.	188.52	241.45	0.35	9.49	nr	250.94
Reducer						
22 x 15mm dia.	33.85	43.35	0.28	7.59	nr	50.94
28 x 22mm dia.	46.36	59.38	0.28	7.59	nr	66.97
35 x 28mm dia.	67.74	86.76	0.30	8.13	nr	94.89
42 x 35mm dia.	90.12	115.42	0.37	10.03	nr	125.46
Stud coupling						
15mm dia.	19.39	24.83	0.42	11.38	nr	36.22
22mm dia.	32.80	42.01	0.25	6.78	nr	48.79
28mm dia.	45.53	58.31	0.25	6.78	nr	65.09
35mm dia.	72.93	93.41	0.37	10.03	nr	103.44
42mm dia.	86.21	110.42	0.42	11.38	nr	121.80
Equal tee						
15mm dia.	33.11	42.41	0.37	10.03	nr	52.44
22mm dia.	68.41	87.62	0.40	10.84	nr	98.46
28mm dia.	93.56	119.83	0.45	12.20	nr	132.03
35mm dia.	186.04	238.28	0.59	16.00	nr	254.28
42mm dia.	257.97	330.41	0.62	16.81	nr	347.22
Running tee						
15mm dia.	40.76	52.20	0.37	10.03	nr	62.24
22mm dia.	73.43	94.05	0.40	10.84	nr	104.89
28mm dia.	124.33	159.24	0.59	16.00	nr	175.24

S: PIPED SUPPLY SYSTEMS

Item	Net Price £	Material £	Labour hours	Labour £	Unit	Total rate £
S10 : COLD WATER (cont'd)						
PRESS FIT (stainless steel)						
Press fit jointing system; butyl rubber O ring mechanical joint						
Pipework						
15mm dia	3.25	4.16	0.46	12.47	m	16.63
22mm dia	5.12	6.56	0.48	13.01	m	19.57
28mm dia	6.52	8.35	0.52	14.09	m	22.44
35mm dia	9.78	12.53	0.56	15.18	m	27.70
42mm dia	11.75	15.05	0.58	15.72	m	30.77
54mm dia	14.80	18.96	0.66	17.89	m	36.84
FIXINGS						
For stainless steel pipes						
Refer to fixings for stainless steel pipes; capillary or compression joints; BS 4127						
Extra over stainless steel pipes; Press fit jointing system;						
Coupling						
15mm dia	2.04	2.61	0.36	9.76	nr	12.37
22mm dia	2.57	3.29	0.36	9.76	nr	13.05
28mm dia	2.90	3.71	0.44	11.93	nr	15.64
35mm dia	3.60	4.61	0.44	11.93	nr	16.54
42mm dia	4.92	6.30	0.52	14.09	nr	20.40
54mm dia	5.91	7.57	0.60	16.26	nr	23.83
Stop end						
22mm dia	2.27	2.91	0.18	4.88	nr	7.79
28mm dia	3.20	4.10	0.22	5.96	nr	10.06
35mm dia	3.70	4.74	0.22	5.96	nr	10.70
42mm dia	5.19	6.65	0.26	7.05	nr	13.69
54mm dia	6.01	7.70	0.30	8.13	nr	15.83
Reducer						
22 x 15mm dia	2.44	3.13	0.36	9.76	nr	12.88
28 x 15mm dia	2.76	3.54	0.40	10.84	nr	14.38
28 x 22mm dia	2.85	3.65	0.40	10.84	nr	14.49
35 x 22mm dia	3.48	4.46	0.40	10.84	nr	15.30
35 x 28mm dia	4.32	5.53	0.44	11.93	nr	17.46
42 x 35mm dia	4.55	5.83	0.48	13.01	nr	18.84
54 x 42mm dia	5.20	6.66	0.56	15.18	nr	21.84
90° bend						
15mm dia	2.92	3.74	0.36	9.76	nr	13.50
22mm dia	4.09	5.24	0.36	9.76	nr	15.00
28mm dia	5.16	6.61	0.44	11.93	nr	18.53
35mm dia	8.11	10.39	0.44	11.93	nr	22.31
42mm dia	13.56	17.37	0.52	14.09	nr	31.46
54mm dia	18.72	23.98	0.60	16.26	nr	40.24

S: PIPED SUPPLY SYSTEMS

Item	Net Price £	Material £	Labour hours	Labour £	Unit	Total rate £
45° bend						
15mm dia	3.97	5.08	0.36	9.76	nr	**14.84**
22mm dia	4.93	6.31	0.36	9.76	nr	**16.07**
28mm dia	5.74	7.35	0.44	11.93	nr	**19.28**
35mm dia	6.74	8.63	0.44	11.93	nr	**20.56**
42mm dia	10.85	13.90	0.52	14.09	nr	**27.99**
54mm dia	14.09	18.05	0.60	16.26	nr	**34.31**
Equal tee						
15mm dia	4.79	6.13	0.54	14.64	nr	**20.77**
22mm dia	5.88	7.53	0.54	14.64	nr	**22.17**
28mm dia	6.88	8.81	0.66	17.89	nr	**26.70**
35mm dia	8.71	11.16	0.66	17.89	nr	**29.04**
42mm dia	12.36	15.83	0.78	21.14	nr	**36.97**
54mm dia	14.79	18.94	0.90	24.39	nr	**43.34**
Reducing tee						
22 x 15mm dia	5.04	6.46	0.54	14.64	nr	**21.09**
28 x 15mm dia	6.10	7.81	0.62	16.80	nr	**24.62**
28 x 22mm dia	6.60	8.45	0.62	16.80	nr	**25.26**
35 x 22mm dia	7.83	10.03	0.62	16.80	nr	**26.83**
35 x 28mm dia	8.17	10.46	0.62	16.80	nr	**27.27**
42 x 28mm dia	11.61	14.87	0.70	18.97	nr	**33.84**
42 x 35mm dia	11.96	15.32	0.70	18.97	nr	**34.29**
54 x 35mm dia	13.51	17.30	0.82	22.22	nr	**39.53**
54 x 42mm dia	13.90	17.80	0.82	22.22	nr	**40.03**
FIXINGS						
For stainless steel pipes						
Refer to fixings for stainless steel pipes; capillary or compression joints; BS 4127						
MEDIUM DENSITY POLYETHYLENE - BLUE						
Note : MDPE is sized on Outside Diameter i.e. OD not ID						
Pipes for water distribution; laid underground; electrofusion joints in the running length; BS 6572						
Coiled service pipe						
20mm dia.	0.52	0.69	0.37	10.03	m	**10.72**
25mm dia.	0.60	0.79	0.41	11.11	m	**11.90**
32mm dia.	1.01	1.33	0.47	12.74	m	**14.07**
50mm dia.	2.42	3.19	0.53	14.37	m	**17.56**
63mm dia.	3.83	5.05	0.60	16.27	m	**21.32**
Mains service pipe						
90mm dia.	7.71	10.17	0.90	24.40	m	**34.57**
110mm dia	11.48	14.70	1.10	29.81	m	**44.52**
125mm dia.	14.63	19.30	1.20	32.54	m	**51.84**
160mm dia	25.27	32.37	1.48	40.11	m	**72.48**
180mm dia.	31.77	41.91	1.50	40.70	m	**82.61**
225mm dia	35.91	45.99	1.77	47.97	m	**93.97**
250mm dia.	44.19	58.30	1.75	47.47	m	**105.76**
315mm dia	70.15	89.85	1.90	51.50	m	**141.34**

S: PIPED SUPPLY SYSTEMS

Item	Net Price £	Material £	Labour hours	Labour £	Unit	Total rate £
S10 : COLD WATER (cont'd)						
MEDIUM DENSITY POLYETHYLENE - BLUE (cont'd)						
Extra over fittings; MDPE blue; electrofusion joints						
Coupler						
20mm dia	3.94	5.05	0.36	9.76	nr	**14.80**
25mm dia	3.94	5.05	0.40	10.84	nr	**15.89**
32mm dia	3.94	5.05	0.44	11.93	nr	**16.97**
40mm dia	3.94	5.05	0.48	13.01	nr	**18.06**
50mm dia	7.16	9.17	0.52	14.09	nr	**23.26**
63mm dia.	7.41	9.49	0.58	15.72	nr	**25.21**
90mm dia.	10.93	14.00	0.67	18.16	nr	**32.16**
110mm dia	17.63	22.58	0.74	20.06	nr	**42.64**
125mm dia.	19.84	25.41	0.83	22.50	nr	**47.91**
160mm dia	31.59	40.46	1.00	27.10	nr	**67.56**
180mm dia.	36.83	47.17	1.25	33.88	nr	**81.05**
225mm dia	59.14	75.75	1.35	36.59	nr	**112.34**
250mm dia.	86.35	110.60	1.50	40.66	nr	**151.25**
315mm dia	143.28	183.51	1.80	48.79	nr	**232.30**
Extra over fittings; MDPE blue; butt fused joints						
Cap						
25mm dia	7.82	10.02	0.20	5.42	nr	**15.44**
32mm dia	7.82	10.02	0.22	5.96	nr	**15.98**
40mm dia	13.19	16.89	0.24	6.50	nr	**23.40**
50mm dia	14.19	18.17	0.26	7.05	nr	**25.22**
63mm dia.	18.73	23.99	0.32	8.67	nr	**32.66**
90mm dia.	34.17	43.76	0.37	10.03	nr	**53.79**
110mm dia	53.55	68.59	0.40	10.84	nr	**79.43**
125mm dia.	65.40	83.76	0.46	12.47	nr	**96.23**
160mm dia	87.01	111.44	0.50	13.55	nr	**124.99**
180mm dia.	110.50	141.53	0.60	16.26	nr	**157.79**
225mm dia	138.07	176.84	0.68	18.43	nr	**195.27**
250mm dia	138.07	176.84	0.75	20.33	nr	**197.17**
315mm dia	307.17	393.42	0.90	24.39	nr	**417.81**
Reducer						
63 x 32mm dia	10.23	13.10	0.54	14.64	nr	**27.74**
63 x 50mm dia	12.03	15.41	0.60	16.26	nr	**31.67**
90 x 63mm dia.	12.03	15.41	0.67	18.16	nr	**33.57**
110 x 90mm dia	15.25	19.53	0.74	20.06	nr	**39.59**
125 x 90mm dia.	19.81	25.37	0.83	22.50	nr	**47.87**
125 x 110mm dia	30.58	39.17	1.00	27.10	nr	**66.27**
160 x 110mm dia	51.70	66.22	1.10	29.81	nr	**96.03**
180 x 125mm dia.	56.12	71.88	1.25	33.88	nr	**105.76**
225 x 160mm dia	79.34	101.62	1.40	37.94	nr	**139.56**
250 x 180mm dia	88.41	113.23	1.80	48.79	nr	**162.02**
315 x 250mm dia	144.83	185.50	2.40	65.05	nr	**250.55**

S: PIPED SUPPLY SYSTEMS

Item	Net Price £	Material £	Labour hours	Labour £	Unit	Total rate £
Bend; 45°						
50mm dia	19.10	24.46	0.50	13.55	nr	**38.01**
63mm dia.	19.10	24.46	0.58	15.72	nr	**40.18**
90mm dia.	29.48	37.76	0.67	18.16	nr	**55.92**
110mm dia	42.84	54.87	0.74	20.06	nr	**74.93**
125mm dia.	48.22	61.76	0.83	22.50	nr	**84.26**
160mm dia	89.55	114.69	1.00	27.10	nr	**141.80**
180mm dia.	109.39	140.11	1.25	33.88	nr	**173.99**
225mm dia	136.15	174.38	1.40	37.94	nr	**212.32**
250mm dia	167.89	215.03	1.80	48.79	nr	**263.82**
315mm dia	302.36	387.26	2.40	65.05	nr	**452.31**
Bend; 90°						
50mm dia	19.10	24.46	0.50	13.55	nr	**38.01**
63mm dia.	19.10	24.46	0.58	15.72	nr	**40.18**
90mm dia.	29.48	37.76	0.67	18.16	nr	**55.92**
110mm dia	36.71	47.02	0.74	20.06	nr	**67.07**
125mm dia.	48.22	61.76	0.83	22.50	nr	**84.26**
160mm dia	89.55	114.69	1.00	27.10	nr	**141.80**
180mm dia.	109.19	139.85	1.25	33.88	nr	**173.73**
225mm dia	173.32	221.99	1.40	37.94	nr	**259.93**
250mm dia	202.60	259.49	1.80	48.79	nr	**308.27**
315mm dia	411.13	526.57	2.40	65.05	nr	**591.62**
Equal tee						
50mm dia	17.36	22.23	0.70	18.97	nr	**41.21**
63mm dia.	18.73	23.99	0.75	20.33	nr	**44.32**
90mm dia.	34.17	43.76	0.87	23.58	nr	**67.34**
110mm dia	50.97	65.28	1.00	27.10	nr	**92.39**
125mm dia.	65.40	83.76	1.08	29.27	nr	**113.04**
160mm dia	107.74	137.99	1.35	36.59	nr	**174.58**
180mm dia.	110.50	141.53	1.63	44.18	nr	**185.71**
225mm dia	461.07	590.53	1.90	51.50	nr	**642.03**
250mm dia	560.46	717.83	2.70	73.18	nr	**791.01**
315mm dia	843.61	1080.49	3.60	97.57	nr	**1178.06**
Extra over plastic fittings, compression joints						
Straight connector						
20mm dia.	2.25	2.88	0.38	10.30	nr	**13.18**
25mm dia.	2.37	3.04	0.45	12.20	nr	**15.23**
32mm dia.	5.63	7.21	0.50	13.55	nr	**20.76**
50mm dia.	12.40	15.88	0.68	18.44	nr	**34.32**
63mm dia.	12.96	16.60	0.85	23.05	nr	**39.65**
Reducing connector						
25mm dia.	4.65	5.96	0.38	10.30	nr	**16.26**
32mm dia.	7.50	9.61	0.45	12.20	nr	**21.80**
50mm dia.	12.96	16.60	0.50	13.55	nr	**30.15**
63mm dia.	29.01	37.16	0.62	16.81	nr	**53.97**
Straight connector; polyethylene to MI						
20mm dia.	2.04	2.61	0.31	8.40	nr	**11.02**
25mm dia.	2.59	3.32	0.35	9.49	nr	**12.80**
32mm dia.	3.77	4.83	0.40	10.84	nr	**15.67**
50mm dia.	9.65	12.36	0.55	14.91	nr	**27.27**
63mm dia.	13.59	17.41	0.65	17.62	nr	**35.03**

S: PIPED SUPPLY SYSTEMS

Item	Net Price £	Material £	Labour hours	Labour £	Unit	Total rate £
S10 : COLD WATER (cont'd)						
MEDIUM DENSITY POLYETHYLENE - BLUE (cont'd)						
Extra over plastic fittings (cont'd)						
Straight connector; polyethylene to FI						
20mm dia.	2.76	3.54	0.31	8.40	nr	11.94
25mm dia.	2.98	3.82	0.35	9.49	nr	13.30
32mm dia.	3.56	4.56	0.40	10.84	nr	15.40
50mm dia.	11.32	14.50	0.55	14.91	nr	29.41
63mm dia.	15.87	20.33	0.75	20.33	nr	40.66
Elbow						
20mm dia.	3.02	3.87	0.38	10.30	nr	14.17
25mm dia.	4.45	5.70	0.45	12.20	nr	17.90
32mm dia.	6.49	8.31	0.50	13.55	nr	21.86
50mm dia.	13.89	17.79	0.68	18.44	nr	36.23
63mm dia.	20.48	26.23	0.80	21.68	nr	47.91
Elbow; polyethylene to MI						
25mm dia.	3.83	4.91	0.35	9.49	nr	14.39
Elbow; polyethylene to FI						
20mm dia.	2.74	3.51	0.31	8.40	nr	11.91
25mm dia.	3.72	4.76	0.35	9.49	nr	14.25
32mm dia.	5.58	7.15	0.42	11.38	nr	18.53
50mm dia.	13.25	16.97	0.50	13.55	nr	30.52
63mm dia.	17.38	22.26	0.55	14.91	nr	37.17
Tank coupling						
25mm dia.	5.75	7.36	0.42	11.38	nr	18.75
Equal tee						
20mm dia.	5.82	7.45	0.53	14.37	nr	21.82
25mm dia.	8.74	11.19	0.55	14.91	nr	26.10
32mm dia.	10.82	13.86	0.64	17.35	nr	31.21
50mm dia.	24.06	30.82	0.75	20.33	nr	51.15
63mm dia.	36.90	47.26	0.87	23.59	nr	70.85
Equal tee; FI branch						
20mm dia.	3.93	5.03	0.45	12.20	nr	17.23
25mm dia.	6.26	8.02	0.50	13.55	nr	21.57
32mm dia.	7.61	9.75	0.60	16.27	nr	26.01
50mm dia.	17.54	22.47	0.68	18.44	nr	40.90
63mm dia.	24.63	31.55	0.81	21.96	nr	53.51
Equal tee; MI branch						
25mm dia.	6.14	7.86	0.50	13.55	nr	21.42

S: PIPED SUPPLY SYSTEMS

Item	Net Price £	Material £	Labour hours	Labour £	Unit	Total rate £
ABS PIPEWORK						
Pipes; solvent welded joints in the running length, brackets measured separately						
Class C (9 bar pressure)						
1" dia.	2.35	3.01	0.30	8.13	m	**11.14**
1 1/4" dia.	3.95	5.06	0.33	8.94	m	**14.00**
1 1/2" dia.	5.02	6.43	0.36	9.76	m	**16.19**
2" dia.	6.76	8.66	0.39	10.57	m	**19.23**
3" dia.	13.93	17.84	0.46	12.47	m	**30.31**
4" dia.	22.96	29.41	0.53	14.36	m	**43.77**
6" dia.	45.40	58.15	0.76	20.60	m	**78.75**
8" dia.	77.96	99.85	0.97	26.29	m	**126.14**
Class E (15 bar pressure)						
1/2" dia.	1.79	2.29	0.24	6.50	m	**8.80**
3/4" dia.	2.76	3.54	0.27	7.32	m	**10.85**
1" dia.	3.64	4.66	0.30	8.13	m	**12.79**
1 1/4" dia.	5.43	6.95	0.33	8.94	m	**15.90**
1 1/2" dia.	7.16	9.17	0.36	9.76	m	**18.93**
2" dia.	8.97	11.49	0.39	10.57	m	**22.06**
3" dia.	18.03	23.09	0.49	13.28	m	**36.37**
4" dia.	29.00	37.14	0.57	15.45	m	**52.59**
Fixings						
Refer to steel pipes; galvanised iron. For minimum fixing dimensions, refer to the Tables and Memoranda at the rear of the book						
Extra over fittings; solvent welded joints						
Cap						
1/2" dia.	0.77	0.99	0.16	4.34	nr	**5.32**
3/4" dia.	0.89	1.14	0.19	5.15	nr	**6.29**
1" dia.	1.02	1.31	0.22	5.96	nr	**7.27**
1 1/4" dia.	1.70	2.18	0.25	6.78	nr	**8.95**
1 1/2" dia.	2.62	3.36	0.28	7.59	nr	**10.94**
2" dia.	3.32	4.25	0.31	8.40	nr	**12.65**
3" dia.	9.98	12.78	0.36	9.76	nr	**22.54**
4" dia.	15.26	19.54	0.44	11.93	nr	**31.47**
Elbow 90°						
1/2" dia.	1.07	1.37	0.29	7.86	nr	**9.23**
3/4" dia.	1.28	1.64	0.34	9.22	nr	**10.85**
1" dia.	1.79	2.29	0.40	10.84	nr	**13.13**
1 1/4" dia.	3.03	3.88	0.45	12.20	nr	**16.08**
1 1/2" dia.	3.94	5.05	0.51	13.82	nr	**18.87**
2" dia.	5.99	7.67	0.56	15.18	nr	**22.85**
3" dia.	17.21	22.04	0.65	17.62	nr	**39.66**
4" dia.	25.70	32.92	0.80	21.68	nr	**54.60**
6" dia.	103.46	132.51	1.21	32.80	nr	**165.31**
8" dia.	157.90	202.24	1.45	39.30	nr	**241.54**

S: PIPED SUPPLY SYSTEMS

Item	Net Price £	Material £	Labour hours	Labour £	Unit	Total rate £
S10 : COLD WATER (cont'd)						
ABS PIPEWORK (cont'd)						
Extra over fittings; solvent welded (cont'd)						
Elbow 45 °						
1/2" dia.	2.07	2.65	0.29	7.86	nr	**10.51**
3/4" dia.	2.10	2.69	0.34	9.22	nr	**11.90**
1" dia.	2.62	3.36	0.40	10.84	nr	**14.20**
1 1/4" dia.	3.83	4.91	0.45	12.20	nr	**17.10**
1 1/2" dia.	4.76	6.10	0.51	13.82	nr	**19.92**
2" dia.	6.61	8.47	0.56	15.18	nr	**23.64**
3" dia.	15.56	19.93	0.65	17.62	nr	**37.55**
4" dia.	32.26	41.32	0.80	21.68	nr	**63.00**
6" dia.	66.88	85.66	1.21	32.80	nr	**118.45**
8" dia.	143.92	184.33	1.45	39.30	nr	**223.63**
Reducing bush						
3/4" x 1/2" dia.	0.79	1.01	0.42	11.38	nr	**12.40**
1" x 1/2" dia.	1.02	1.31	0.45	12.20	nr	**13.50**
1" x 3/4" dia.	1.02	1.31	0.45	12.20	nr	**13.50**
1 1/4" x 1" dia.	1.37	1.75	0.48	13.01	nr	**14.76**
1 1/2" x 3/4" dia.	1.79	2.29	0.51	13.82	nr	**16.12**
1 1/2" x 1" dia.	1.79	2.29	0.51	13.82	nr	**16.12**
1 1/2" x 1 1/4" dia.	1.79	2.29	0.51	13.82	nr	**16.12**
2" x 1" dia.	2.35	3.01	0.56	15.18	nr	**18.19**
2" x 1 1/4" dia.	2.35	3.01	0.56	15.18	nr	**18.19**
2" x 1 1/2" dia.	2.35	3.01	0.56	15.18	nr	**18.19**
3" x 1 1/2" dia.	6.61	8.47	0.65	17.62	nr	**26.08**
3" x 2" dia.	6.61	8.47	0.65	17.62	nr	**26.08**
4" x 3" dia.	9.11	11.67	0.80	21.68	nr	**33.35**
6" x 4" dia.	28.07	35.95	1.21	32.80	nr	**68.75**
Union						
1/2" dia.	4.26	5.46	0.34	9.22	nr	**14.67**
3/4" dia.	4.59	5.88	0.39	10.57	nr	**16.45**
1" dia.	6.19	7.93	0.43	11.65	nr	**19.58**
1 1/4" dia.	7.59	9.72	0.50	13.55	nr	**23.27**
1 1/2" dia.	10.45	13.38	0.57	15.45	nr	**28.83**
2" dia.	13.63	17.46	0.62	16.80	nr	**34.26**
Sockets						
1/2" dia.	0.79	1.01	0.34	9.22	nr	**10.23**
3/4" dia.	0.89	1.14	0.39	10.57	nr	**11.71**
1" dia.	1.02	1.31	0.43	11.65	nr	**12.96**
1 1/4" dia.	1.79	2.29	0.50	13.55	nr	**15.84**
1 1/2" dia.	2.16	2.77	0.57	15.45	nr	**18.22**
2" dia.	3.03	3.88	0.62	16.80	nr	**20.69**
3" dia.	12.19	15.61	0.70	18.97	nr	**34.59**
4" dia.	17.30	22.16	0.70	18.97	nr	**41.13**
6" dia.	43.22	55.36	1.26	34.15	nr	**89.51**
8" dia.	86.33	110.57	1.55	42.01	nr	**152.58**

S: PIPED SUPPLY SYSTEMS

Item	Net Price £	Material £	Labour hours	Labour £	Unit	Total rate £
Barrel nipple						
1/2" dia.	1.49	1.91	0.34	9.22	nr	11.12
3/4" dia.	1.94	2.48	0.39	10.57	nr	13.06
1" dia.	2.51	3.21	0.43	11.65	nr	14.87
1 1/4" dia.	3.48	4.46	0.50	13.55	nr	18.01
1 1/2" dia.	4.10	5.25	0.57	15.45	nr	20.70
2" dia.	4.97	6.37	0.62	16.80	nr	23.17
3" dia.	13.21	16.92	0.70	18.97	nr	35.89
Tee, 90°						
1/2" dia.	1.22	1.56	0.41	11.11	nr	12.67
3/4" dia.	1.70	2.18	0.47	12.74	nr	14.92
1" dia.	2.35	3.01	0.55	14.91	nr	17.92
1 1/4" dia.	3.38	4.33	0.64	17.35	nr	21.68
1 1/2" dia.	4.97	6.37	0.71	19.24	nr	25.61
2" dia.	7.59	9.72	0.78	21.14	nr	30.86
3" dia.	22.13	28.34	0.91	24.66	nr	53.01
4" dia.	32.48	41.60	1.12	30.36	nr	71.96
6" dia.	113.50	145.37	1.69	45.80	nr	191.17
8" dia.	176.97	226.66	2.03	55.02	nr	281.68
Full face flange						
1/2" dia.	13.22	16.93	0.10	2.71	nr	19.64
3/4" dia.	13.53	17.33	0.13	3.52	nr	20.85
1" dia.	14.66	18.78	0.15	4.07	nr	22.84
1 1/4" dia.	16.29	20.86	0.18	4.88	nr	25.74
1 1/2" dia.	19.59	25.09	0.21	5.69	nr	30.78
2" dia.	26.52	33.97	0.29	7.86	nr	41.83
3" dia.	45.47	58.24	0.37	10.03	nr	68.27
4" dia.	59.59	76.32	0.41	11.11	nr	87.43

PVC-U PIPEWORK

Pipes; solvent welded joints in the running length, brackets Measured separately

Item	Net Price £	Material £	Labour hours	Labour £	Unit	Total rate £
Class C (9 bar pressure)						
2" dia.	7.49	9.59	0.41	11.11	m	20.71
3" dia.	14.35	18.38	0.47	12.74	m	31.12
4" dia.	25.46	32.61	0.50	13.55	m	46.16
6" dia.	55.09	70.56	1.76	47.70	m	118.26
Class D (12 bar pressure)						
1 1/4" dia.	4.36	5.58	0.41	11.11	m	16.70
1 1/2" dia.	6.00	7.68	0.42	11.38	m	19.07
2" dia.	9.30	11.91	0.45	12.20	m	24.11
3" dia.	19.92	25.51	0.48	13.01	m	38.52
4" dia.	33.35	42.71	0.53	14.36	m	57.08
6" dia.	61.88	79.26	0.58	15.72	m	94.98
Class E (15 bar pressure)						
1/2" dia.	2.13	2.73	0.38	10.30	m	13.03
3/4" dia.	3.05	3.91	0.40	10.84	m	14.75
1" dia.	3.55	4.55	0.41	11.11	m	15.66
1 1/4" dia.	5.21	6.67	0.41	11.11	m	17.79
1 1/2" dia.	6.78	8.68	0.42	11.38	m	20.07
2" dia.	10.59	13.56	0.45	12.20	m	25.76
3" dia.	22.94	29.38	0.47	12.74	m	42.12
4" dia.	37.66	48.23	0.50	13.55	m	61.79
6" dia.	81.58	104.49	0.53	14.36	m	118.85

S: PIPED SUPPLY SYSTEMS

Item	Net Price £	Material £	Labour hours	Labour £	Unit	Total rate £
S10 : COLD WATER (cont'd)						
PVC-U PIPEWORK (cont'd)						
Pipes; solvent welded joints (cont'd)						
Class 7						
1/2" dia.	3.78	4.84	0.32	8.67	m	**13.51**
3/4" dia.	5.29	6.78	0.33	8.94	m	**15.72**
1" dia.	8.07	10.34	0.40	10.84	m	**21.18**
1 1/4" dia.	11.09	14.20	0.40	10.84	m	**25.05**
1 1/2" dia.	13.73	17.59	0.41	11.11	m	**28.70**
2" dia.	22.82	29.23	0.43	11.65	m	**40.88**
Fixings						
Refer to steel pipes; galvanised iron. For minimum fixing dimensions, refer to the Tables and Memoranda at the rear of the book						
Extra over fittings; solvent welded joints						
End cap						
1/2" dia.	0.73	0.94	0.17	4.61	nr	**5.54**
3/4" dia.	0.85	1.09	0.19	5.15	nr	**6.24**
1" dia.	0.95	1.22	0.22	5.96	nr	**7.18**
1 1/4" dia.	1.49	1.91	0.25	6.78	nr	**8.68**
1 1/2" dia.	2.51	3.21	0.28	7.59	nr	**10.80**
2" dia.	3.07	3.93	0.31	8.40	nr	**12.33**
3" dia.	9.42	12.06	0.36	9.76	nr	**21.82**
4" dia.	14.54	18.62	0.44	11.93	nr	**30.55**
6" dia.	35.14	45.01	0.67	18.16	nr	**63.17**
Socket						
1/2" dia.	0.77	0.99	0.31	8.40	nr	**9.39**
3/4" dia.	0.85	1.09	0.35	9.49	nr	**10.57**
1" dia.	0.99	1.27	0.42	11.38	nr	**12.65**
1 1/4" dia.	1.79	2.29	0.45	12.20	nr	**14.49**
1 1/2" dia.	2.10	2.69	0.51	13.82	nr	**16.51**
2" dia.	2.98	3.82	0.56	15.18	nr	**18.99**
3" dia.	11.38	14.58	0.65	17.62	nr	**32.19**
4" dia.	16.50	21.13	0.80	21.68	nr	**42.82**
6" dia.	41.38	53.00	1.21	32.80	nr	**85.79**
Reducing socket						
3/4 x 1/2" dia.	0.90	1.15	0.31	8.40	nr	**9.55**
1 x 3/4" dia.	1.13	1.45	0.35	9.49	nr	**10.93**
1 1/4 x 1" dia.	2.16	2.77	0.42	11.38	nr	**14.15**
1 1/2 x 1 1/4" dia.	2.41	3.09	0.45	12.20	nr	**15.28**
2 x 1 1/2" dia.	3.64	4.66	0.51	13.82	nr	**18.48**
3 x 2" dia.	11.07	14.18	0.56	15.18	nr	**29.36**
4 x 3" dia.	16.39	20.99	0.65	17.62	nr	**38.61**
6 x 4" dia.	59.72	76.49	0.80	21.68	nr	**98.17**
8 x 6" dia.	92.50	118.47	1.21	32.80	nr	**151.27**

S: PIPED SUPPLY SYSTEMS

Item	Net Price £	Material £	Labour hours	Labour £	Unit	Total rate £
Elbow, 90°						
1/2" dia.	1.02	1.31	0.31	8.40	nr	9.71
3/4" dia.	1.22	1.56	0.35	9.49	nr	11.05
1" dia.	1.70	2.18	0.42	11.38	nr	13.56
1 1/4" dia.	2.98	3.82	0.45	12.20	nr	16.01
1 1/2" dia.	3.83	4.91	0.45	12.20	nr	17.10
2" dia.	5.68	7.27	0.56	15.18	nr	22.45
3" dia.	16.39	20.99	0.65	17.62	nr	38.61
4" dia.	24.68	31.61	0.80	21.68	nr	53.29
6" dia.	97.72	125.16	1.21	32.80	nr	157.95
Elbow 45°						
1/2" dia.	1.94	2.48	0.31	8.40	nr	10.89
3/4" dia.	2.07	2.65	0.35	9.49	nr	12.14
1" dia.	2.51	3.21	0.45	12.20	nr	15.41
1 1/4" dia.	3.59	4.60	0.45	12.20	nr	16.79
1 1/2" dia.	4.51	5.78	0.51	13.82	nr	19.60
2" dia.	6.35	8.13	0.56	15.18	nr	23.31
3" dia.	14.95	19.15	0.65	17.62	nr	36.77
4" dia.	30.73	39.36	0.80	21.68	nr	61.04
6" dia.	63.40	81.20	1.21	32.80	nr	114.00
Bend 90° (long radius)						
3" dia.	45.69	58.52	0.65	17.62	nr	76.14
4" dia.	92.29	118.20	0.80	21.68	nr	139.89
6" dia.	202.81	259.76	1.21	32.80	nr	292.55
Bend 45° (long radius)						
1 1/2" dia.	10.85	13.90	0.51	13.82	nr	27.72
2" dia.	17.73	22.71	0.56	15.18	nr	37.89
3" dia.	37.89	48.53	0.65	17.62	nr	66.15
4" dia.	73.74	94.45	0.80	21.68	nr	116.13
Socket union						
1/2" dia.	3.94	5.05	0.34	9.22	nr	14.26
3/4" dia.	4.51	5.78	0.39	10.57	nr	16.35
1" dia.	5.85	7.49	0.45	12.20	nr	19.69
1 1/4" dia.	7.28	9.32	0.50	13.55	nr	22.88
1 1/2" dia.	9.98	12.78	0.57	15.45	nr	28.23
2" dia.	12.90	16.52	0.62	16.80	nr	33.33
3" dia.	48.04	61.53	0.70	18.97	nr	80.50
4" dia.	65.04	83.30	0.89	24.12	nr	107.42
Saddle plain						
2" x 1 1/4" dia.	10.14	12.99	0.42	11.38	nr	24.37
3" x 1 1/2" dia.	14.24	18.24	0.48	13.01	nr	31.25
4" x 2" dia.	16.05	20.56	0.68	18.43	nr	38.99
6" x 2" dia.	18.84	24.13	0.91	24.66	nr	48.79
Straight tank connector						
1/2" dia.	2.62	3.36	0.13	3.52	nr	6.88
3/4" dia.	2.96	3.79	0.14	3.79	nr	7.59
1" dia.	6.29	8.06	0.14	3.79	nr	11.85
1 1/4" dia.	15.98	20.47	0.16	4.34	nr	24.80
1 1/2" dia.	17.52	22.44	0.18	4.88	nr	27.32
2" dia.	20.99	26.88	0.24	6.50	nr	33.39
3" dia.	21.52	27.56	0.29	7.86	nr	35.42

S: PIPED SUPPLY SYSTEMS

Item	Net Price £	Material £	Labour hours	Labour £	Unit	Total rate £
S10 : COLD WATER (cont'd)						
PVC-U PIPEWORK (cont'd)						
Extra over fittings; solvent welded (cont'd)						
Equal tee						
1/2" dia.	1.18	1.51	0.44	11.93	nr	13.44
3/4" dia.	1.49	1.91	0.48	13.01	nr	14.92
1" dia.	2.25	2.88	0.54	14.64	nr	17.52
1 1/4" dia.	3.18	4.07	0.70	18.97	nr	23.05
1 1/2" dia.	4.59	5.88	0.74	20.06	nr	25.94
2" dia.	7.28	9.32	0.80	21.68	nr	31.01
3" dia.	21.10	27.02	1.04	28.19	nr	55.21
4" dia.	30.94	39.63	1.28	34.69	nr	74.32
6" dia.	107.76	138.02	1.93	52.31	nr	190.33
PVC - C						
Pipes; solvent welded in the running length, brackets measured separately						
Pipe; 3m long; PN25						
16 x 2.0mm	2.08	2.66	0.20	5.42	m	8.08
20 x 2.3mm	3.15	4.03	0.20	5.42	m	9.46
25 x 2.8mm	4.08	5.23	0.20	5.42	m	10.65
32 x 3.6mm	6.06	7.76	0.20	5.42	m	13.18
Pipe; 5m long; PN25						
40 x 4.5mm	7.39	9.46	0.20	5.42	m	14.89
50 x 5.6mm	11.13	14.26	0.20	5.42	m	19.68
63 x 7.0mm	17.19	22.02	0.20	5.42	m	27.44
Fixings						
Refer to steel pipes; galvanised iron. For minimum fixing dimensions, refer to the Tables and Memoranda at the rear of the book						
Extra over fittings; solvent welded joints						
Straight coupling; PN25						
16mm	0.32	0.41	0.20	5.42	nr	5.83
20mm	0.45	0.58	0.20	5.42	nr	6.00
25mm	0.56	0.72	0.20	5.42	nr	6.14
32mm	1.73	2.22	0.20	5.42	nr	7.64
40mm	2.22	2.84	0.20	5.42	nr	8.26
50mm	2.98	3.82	0.20	5.42	nr	9.24
63mm	5.26	6.74	0.20	5.42	nr	12.16
Elbow; 90°; PN25						
16mm	0.52	0.67	0.20	5.42	nr	6.09
20mm	0.80	1.02	0.20	5.42	nr	6.45
25mm	0.99	1.27	0.20	5.42	nr	6.69
32mm	2.07	2.65	0.20	5.42	nr	8.07
40mm	3.19	4.09	0.20	5.42	nr	9.51
50mm	4.42	5.66	0.20	5.42	nr	11.08
63mm	7.55	9.67	0.20	5.42	nr	15.09

S: PIPED SUPPLY SYSTEMS

Item	Net Price £	Material £	Labour hours	Labour £	Unit	Total rate £
Elbow; 45°; PN25						
20mm	0:80	1.02	0.20	5.42	nr	6.45
25mm	0.99	1.27	0.20	5.42	nr	6.69
32mm	2.07	2.65	0.20	5.42	nr	8.07
40mm	3.19	4.09	0.20	5.42	nr	9.51
50mm	4.42	5.66	0.20	5.42	nr	11.08
63mm	7.55	9.67	0.20	5.42	nr	15.09
Reducer fitting; single stage reduction						
20/16mm	0.56	0.72	0.20	5.42	nr	6.14
25/20mm	0.68	0.87	0.20	5.42	nr	6.29
32/25mm	1.36	1.74	0.20	5.42	nr	7.16
40/32mm	1.79	2.29	0.20	5.42	nr	7.71
50/40mm	2.07	2.65	0.20	5.42	nr	8.07
63/50mm	3.14	4.02	0.20	5.42	nr	9.44
Equal tee; 90°; PN25						
16mm	0.87	1.11	0.20	5.42	nr	6.54
20mm	1.19	1.52	0.20	5.42	nr	6.94
25mm	1.51	1.93	0.20	5.42	nr	7.35
32mm	2.46	3.15	0.20	5.42	nr	8.57
40mm	4.25	5.44	0.20	5.42	nr	10.86
50mm	6.36	8.15	0.20	5.42	nr	13.57
63mm	10.73	13.74	0.20	5.42	nr	19.16
Cap; PN25						
20mm	0.60	0.77	0.20	5.42	nr	6.19
25mm	0.80	1.02	0.20	5.42	nr	6.45
32mm	1.16	1.49	0.20	5.42	nr	6.91
40mm	1.59	2.04	0.20	5.42	nr	7.46
50mm	2.22	2.84	0.20	5.42	nr	8.26
63mm	3.54	4.53	0.20	5.42	nr	9.95
SCREWED STEEL PIPEWORK						
Galvanised steel pipes; screwed and socketed joints; BS 1387: 1985						
Galvanised; medium, fixed vertically, with brackets measured separately, screwed joints are within the running length, but any flanges are additional						
10mm dia.	2.88	3.69	0.51	13.82	m	17.51
15mm dia.	2.83	3.62	0.52	14.09	m	17.72
20mm dia.	2.97	3.80	0.55	14.91	m	18.71
25mm dia.	4.16	5.33	0.60	16.26	m	21.59
32mm dia.	5.19	6.65	0.67	18.16	m	24.81
40mm dia.	6.06	7.76	0.75	20.33	m	28.09
50mm dia.	8.56	10.96	0.85	23.04	m	34.00
65mm dia.	11.99	15.36	0.93	25.21	m	40.56
80mm dia.	15.53	19.89	1.07	29.00	m	48.89
100mm dia.	22.70	29.07	1.46	39.57	m	68.64
125mm dia.	37.38	47.88	1.72	46.62	m	94.49
150mm dia.	43.29	55.45	1.96	53.12	m	108.57

S: PIPED SUPPLY SYSTEMS

Item	Net Price £	Material £	Labour hours	Labour £	Unit	Total rate £
S10 : COLD WATER (cont'd)						
SCREWED STEEL PIPEWORK (cont'd)						
Galvanised steel pipes; screwed (cont'd)						
Galvanised; heavy, fixed vertically, with brackets measured separately, screwed joints are within the running length, but any flanges are additional						
15mm dia.	3.20	4.10	0.52	14.09	m	18.19
20mm dia.	3.51	4.50	0.55	14.91	m	19.40
25mm dia.	5.00	6.40	0.60	16.26	m	22.67
32mm dia.	6.25	8.00	0.67	18.16	m	26.16
40mm dia.	7.31	9.36	0.75	20.33	m	29.69
50mm dia.	10.20	13.06	0.85	23.04	m	36.10
65mm dia.	14.23	18.23	0.93	25.21	m	43.43
80mm dia.	18.11	23.20	1.07	29.00	m	52.20
100mm dia.	26.01	33.31	1.46	39.57	m	72.88
125mm dia.	40.54	51.92	1.72	46.62	m	98.54
150mm dia.	45.71	58.54	1.96	53.12	m	111.67
Galvanised; medium, fixed horizontally or suspended at high level, with brackets measured separately, screwed joints are within the running length, but any flanges are additional						
10mm dia.	2.88	3.69	0.51	13.82	m	17.51
15mm dia.	2.83	3.62	0.52	14.09	m	17.72
20mm dia.	2.97	3.80	0.55	14.91	m	18.71
25mm dia.	4.16	5.33	0.60	16.26	m	21.59
32mm dia.	5.19	6.65	0.67	18.16	m	24.81
40mm dia.	6.06	7.76	0.75	20.33	m	28.09
50mm dia.	8.56	10.96	0.85	23.04	m	34.00
65mm dia.	11.99	15.36	0.93	25.21	m	40.56
80mm dia.	15.53	19.89	1.07	29.00	m	48.89
100mm dia.	22.70	29.07	1.46	39.57	m	68.64
125mm dia.	37.38	47.88	1.72	46.62	m	94.49
150mm dia.	43.29	55.45	1.96	53.12	m	108.57
Galvanised; heavy, fixed horizontally or suspended at high level, with brackets measured separately, screwed joints are within the running length, but any flanges are additional						
15mm dia.	3.20	4.10	0.52	14.09	m	18.19
20mm dia.	3.51	4.50	0.55	14.91	m	19.40
25mm dia.	5.00	6.40	0.60	16.26	m	22.67
32mm dia.	6.25	8.00	0.67	18.16	m	26.16
40mm dia.	7.31	9.36	0.75	20.33	m	29.69
50mm dia.	10.20	13.06	0.85	23.04	m	36.10
65mm dia.	14.23	18.23	0.93	25.21	m	43.43
80mm dia.	18.11	23.20	1.07	29.00	m	52.20
100mm dia.	26.01	33.31	1.46	39.57	m	72.88
125mm dia.	40.54	51.92	1.72	46.62	m	98.54
150mm dia.	45.71	58.54	1.96	53.12	m	111.67

S: PIPED SUPPLY SYSTEMS

Item	Net Price £	Material £	Labour hours	Labour £	Unit	Total rate £
FIXINGS						
For steel pipes; galvanised iron. For minimum fixing dimensions, refer to the Tables and Memoranda at the rear of the book						
Single pipe bracket, screw on, galvanised iron; screwed to wood						
15mm dia	0.51	0.65	0.14	3.79	nr	4.45
20mm dia	0.57	0.73	0.14	3.79	nr	4.52
25mm dia	0.67	0.86	0.17	4.61	nr	5.47
32mm dia	0.91	1.17	0.19	5.15	nr	6.32
40mm dia	1.21	1.55	0.22	5.96	nr	7.51
50mm dia	1.60	2.05	0.22	5.96	nr	8.01
65mm dia	2.12	2.72	0.28	7.59	nr	10.30
80mm dia	2.91	3.73	0.32	8.67	nr	12.40
100mm dia	4.24	5.43	0.35	9.49	nr	14.92
Single pipe bracket, screw on, galvanised iron; plugged and screwed						
15mm dia	0.51	0.65	0.25	6.78	nr	7.43
20mm dia	0.57	0.73	0.25	6.78	nr	7.51
25mm dia	0.67	0.86	0.30	8.13	nr	8.99
32mm dia	0.91	1.17	0.32	8.67	nr	9.84
40mm dia	1.21	1.55	0.32	8.67	nr	10.22
50mm dia	1.60	2.05	0.32	8.67	nr	10.72
65mm dia	2.12	2.72	0.35	9.49	nr	12.20
80mm dia	2.91	3.73	0.42	11.38	nr	15.11
100mm dia	4.24	5.43	0.42	11.38	nr	16.81
Single pipe bracket for building in, galvanised iron						
15mm dia	1.68	2.15	0.10	2.71	nr	4.86
20mm dia	1.68	2.15	0.11	2.98	nr	5.13
25mm dia	1.68	2.15	0.12	3.25	nr	5.40
32mm dia	1.90	2.43	0.14	3.79	nr	6.23
40mm dia	1.92	2.46	0.15	4.07	nr	6.52
50mm dia	1.99	2.55	0.16	4.34	nr	6.89
Pipe ring, single socket, galvanised iron						
15mm dia.	0.54	0.69	0.10	2.71	nr	3.40
20mm dia.	0.61	0.78	0.11	2.98	nr	3.76
25mm dia	0.67	0.86	0.12	3.25	nr	4.11
32mm dia	0.70	0.90	0.15	4.07	nr	4.96
40mm dia	0.91	1.17	0.15	4.07	nr	5.23
50mm dia	1.15	1.47	0.16	4.34	nr	5.81
65mm dia	1.67	2.14	0.30	8.13	nr	10.27
80mm dia	2.00	2.56	0.35	9.49	nr	12.05
100mm dia	3.03	3.88	0.40	10.84	nr	14.72
125mm dia	16.09	20.61	0.60	16.27	nr	36.88
150mm dia	17.99	23.04	0.77	20.87	nr	43.91
Pipe ring, double socket, galvanised iron						
15mm dia	1.16	1.49	0.10	2.71	nr	4.20
20mm dia	1.33	1.70	0.11	2.98	nr	4.68
25mm dia	1.50	1.92	0.12	3.25	nr	5.17
32mm dia	1.78	2.28	0.14	3.79	nr	6.07
40mm dia	2.06	2.64	0.15	4.07	nr	6.70
50mm dia	1.45	1.86	0.16	4.34	nr	6.19

S: PIPED SUPPLY SYSTEMS

Item	Net Price £	Material £	Labour hours	Labour £	Unit	Total rate £
S10 : COLD WATER (cont'd)						
SCREWED STEEL PIPEWORK (cont'd)						
For steel pipes; galvanised iron (cont'd)						
Screw on backplate (Male), galvanised iron; plugged and screwed						
All sizes 15mm to 50mm x M12	0.43	0.55	0.10	2.71	nr	**3.26**
Screw on backplate (Female), galvanised iron; plugged and screwed						
All sizes 15mm to 50mm x M12	0.43	0.55	0.10	2.71	nr	**3.26**
Extra Over channel sections for fabricated hangers and brackets						
Galvanised steel; including inserts, bolts, nuts, washers; fixed to backgrounds						
41 x 21mm	2.50	3.20	0.29	7.86	m	**11.06**
41 x 41mm	3.90	5.00	0.29	7.86	m	**12.86**
Threaded rods; metric thread; including nuts, washers, etc						
10mm dia x 600mm long	1.31	1.68	0.18	4.88	nr	**6.56**
12mm dia x 600mm long	2.70	3.46	0.18	4.88	nr	**8.34**
Extra over steel flanges, screwed and drilled; metric; BS 4504						
Screwed flanges; PN6						
15mm dia.	10.83	13.87	0.35	9.49	nr	**23.36**
20mm dia.	11.25	14.41	0.47	12.74	nr	**27.15**
25mm dia.	11.36	14.55	0.53	14.36	nr	**28.91**
32mm dia.	11.37	14.56	0.62	16.80	nr	**31.37**
40mm dia.	11.37	14.56	0.70	18.97	nr	**33.53**
50mm dia.	12.26	15.70	0.84	22.77	nr	**38.47**
65mm dia.	16.75	21.45	1.03	27.92	nr	**49.37**
80mm dia.	20.22	25.90	1.23	33.34	nr	**59.23**
100mm dia.	24.24	31.05	1.41	38.22	nr	**69.26**
125mm dia.	51.28	65.68	1.77	47.97	nr	**113.65**
150mm dia.	51.28	65.68	2.21	59.90	nr	**125.58**
Screwed flanges; PN16						
15mm dia.	13.84	17.73	0.35	9.49	nr	**27.21**
20mm dia.	13.94	17.85	0.47	12.74	nr	**30.59**
25mm dia.	14.05	18.00	0.53	14.36	nr	**32.36**
32mm dia.	15.66	20.06	0.62	16.80	nr	**36.86**
40mm dia.	15.66	20.06	0.70	18.97	nr	**39.03**
50mm dia.	16.54	21.18	0.84	22.77	nr	**43.95**
65mm dia.	20.69	26.50	1.03	27.92	nr	**54.42**
80mm dia.	24.35	31.19	1.23	33.34	nr	**64.52**
100mm dia.	29.18	37.37	1.41	38.22	nr	**75.59**
125mm dia.	53.17	68.10	1.77	47.97	nr	**116.07**
150mm dia.	56.05	71.79	2.21	59.90	nr	**131.69**

S: PIPED SUPPLY SYSTEMS

Item	Net Price £	Material £	Labour hours	Labour £	Unit	Total rate £
Extra over steel flanges, screwed and drilled; imperial; BS 10						
Screwed flanges; table E						
1/2" dia.	17.23	22.07	0.35	9.49	nr	31.55
3/4" dia.	17.65	22.61	0.47	12.74	nr	35.34
1" dia.	17.76	22.75	0.53	14.36	nr	37.11
1 1/4" dia.	17.77	22.76	0.62	16.80	nr	39.56
1 1/2" dia.	17.77	22.76	0.70	18.97	nr	41.73
2" dia.	17.99	23.04	0.84	22.77	nr	45.81
2 1/2" dia.	21.37	27.37	1.03	27.92	nr	55.29
3" dia.	24.86	31.84	1.23	33.34	nr	65.18
4" dia.	33.39	42.77	1.41	38.22	nr	80.98
5" dia.	69.10	88.50	1.77	47.97	nr	136.48
Extra over steel flange connections						
Bolted connection between pair of flanges; including gasket, bolts, nuts and washers						
50mm dia.	31.04	39.76	0.53	14.36	nr	54.12
65mm dia.	39.10	50.08	0.53	14.36	nr	64.44
80mm dia.	44.85	57.44	0.53	14.36	nr	71.81
100mm dia.	53.53	68.56	0.53	14.36	nr	82.93
125mm dia.	101.11	129.50	0.61	16.53	nr	146.03
150mm dia.	103.99	133.19	0.90	24.39	nr	157.58
Extra over heavy steel tubular fittings; BS 1387						
Long screw connection with socket and backnut						
15mm dia.	5.29	6.78	0.63	17.08	nr	23.85
20mm dia.	5.42	6.94	0.84	22.77	nr	29.71
25mm dia.	7.09	9.08	0.95	25.75	nr	34.83
32mm dia.	9.27	11.87	1.11	30.08	nr	41.96
40mm dia.	11.28	14.45	1.28	34.69	nr	49.14
50mm dia.	16.59	21.25	1.53	41.47	nr	62.72
65mm dia.	37.68	48.26	1.87	50.68	nr	98.94
80mm dia.	47.92	61.38	2.21	59.90	nr	121.27
100mm dia.	119.03	152.45	3.05	82.67	nr	235.12
Running nipple						
15mm dia.	1.12	1.43	0.50	13.55	nr	14.99
20mm dia.	1.39	1.78	0.68	18.43	nr	20.21
25mm dia.	1.50	1.92	0.77	20.87	nr	22.79
32mm dia.	2.41	3.09	0.90	24.39	nr	27.48
40mm dia.	3.25	4.16	1.03	27.92	nr	32.08
50mm dia.	4.95	6.34	1.23	33.34	nr	39.68
65mm dia.	10.65	13.64	1.50	40.66	nr	54.30
80mm dia.	16.60	21.26	1.78	48.24	nr	69.51
100mm dia.	26.01	33.31	2.38	64.51	nr	97.82

S: PIPED SUPPLY SYSTEMS

Item	Net Price £	Material £	Labour hours	Labour £	Unit	Total rate £
S10 : COLD WATER (cont'd)						
SCREWED STEEL PIPEWORK (cont'd)						
Extra over heavy steel tubular fittings (cont'd)						
Barrel nipple						
15mm dia.	0.89	1.14	0.50	13.55	nr	14.69
20mm dia.	0.99	1.27	0.68	18.43	nr	19.70
25mm dia.	1.41	1.81	0.77	20.87	nr	22.68
32mm dia.	1.79	2.29	0.90	24.39	nr	26.69
40mm dia.	2.22	2.84	1.03	27.92	nr	30.76
50mm dia.	3.14	4.02	1.23	33.34	nr	37.36
65mm dia.	5.71	7.31	1.50	40.66	nr	47.97
80mm dia.	7.95	10.18	1.78	48.24	nr	58.43
100mm dia.	14.40	18.44	2.38	64.51	nr	82.95
125mm dia.	26.73	34.24	2.87	77.79	nr	112.02
150mm dia.	42.10	53.92	3.39	91.88	nr	145.80
Close taper nipple						
15mm dia.	1.32	1.69	0.50	13.55	nr	15.24
20mm dia.	1.71	2.19	0.68	18.43	nr	20.62
25mm dia.	2.25	2.88	0.77	20.87	nr	23.75
32mm dia.	3.36	4.30	0.90	24.39	nr	28.70
40mm dia.	4.17	5.34	1.03	27.92	nr	33.26
50mm dia.	6.40	8.20	1.23	33.34	nr	41.53
65mm dia.	10.11	12.95	1.50	40.66	nr	53.60
80mm dia.	16.56	21.21	1.78	48.24	nr	69.45
100mm dia.	31.49	40.33	2.38	64.51	nr	104.84
90° bend with socket						
15mm dia.	3.60	4.61	0.64	17.35	nr	21.96
20mm dia.	4.84	6.20	0.85	23.04	nr	29.24
25mm dia.	7.42	9.50	0.97	26.29	nr	35.79
32mm dia.	10.65	13.64	1.12	30.36	nr	44.00
40mm dia.	13.01	16.66	1.29	34.96	nr	51.63
50mm dia.	20.23	25.91	1.55	42.01	nr	67.92
65mm dia.	40.84	52.31	1.89	51.23	nr	103.53
80mm dia.	60.71	77.76	2.24	60.71	nr	138.47
100mm dia.	107.58	137.79	3.09	83.75	nr	221.54
125mm dia.	264.84	339.20	3.92	106.25	nr	445.45
150mm dia.	398.27	510.10	4.74	128.47	nr	638.57
Extra over heavy steel fittings; BS 1740						
Plug						
15mm dia.	0.92	1.18	0.28	7.59	nr	8.77
20mm dia.	1.43	1.83	0.38	10.30	nr	12.13
25mm dia.	2.52	3.23	0.44	11.93	nr	15.15
32mm dia.	3.90	5.00	0.51	13.82	nr	18.82
40mm dia.	4.31	5.52	0.59	15.99	nr	21.51
50mm dia.	6.16	7.89	0.70	18.97	nr	26.86
65mm dia.	14.73	18.87	0.85	23.04	nr	41.90
80mm dia.	27.57	35.31	1.00	27.10	nr	62.41
100mm dia.	52.94	67.81	1.44	39.03	nr	106.83

S: PIPED SUPPLY SYSTEMS

Item	Net Price £	Material £	Labour hours	Labour £	Unit	Total rate £
Socket						
15mm dia.	0.85	1.09	0.64	17.35	nr	18.43
20mm dia.	0.96	1.23	0.85	23.04	nr	24.27
25mm dia.	1.36	1.74	0.97	26.29	nr	28.03
32mm dia.	1.97	2.52	1.12	30.36	nr	32.88
40mm dia.	2.39	3.06	1.29	34.96	nr	38.02
50mm dia.	3.69	4.73	1.55	42.01	nr	46.74
65mm dia.	7.31	9.36	1.89	51.23	nr	60.59
80mm dia.	9.45	12.10	2.24	60.71	nr	72.82
100mm dia.	17.78	22.77	3.09	83.75	nr	106.52
150mm dia.	42.45	54.37	4.74	128.47	nr	182.84
Elbow, female/female						
15mm dia.	4.88	6.25	0.64	17.35	nr	23.60
20mm dia.	6.37	8.16	0.85	23.04	nr	31.20
25mm dia.	8.64	11.07	0.97	26.29	nr	37.36
32mm dia.	16.08	20.60	1.12	30.36	nr	50.95
40mm dia.	19.18	24.57	1.29	34.96	nr	59.53
50mm dia.	31.44	40.27	1.55	42.01	nr	82.28
65mm dia.	76.82	98.39	1.89	51.23	nr	149.62
80mm dia.	91.60	117.32	2.24	60.71	nr	178.03
100mm dia.	158.63	203.17	3.09	83.75	nr	286.92
Equal tee						
15mm dia.	6.07	7.77	0.91	24.66	nr	32.44
20mm dia.	7.05	9.03	1.22	33.07	nr	42.10
25mm dia.	10.39	13.31	1.40	37.94	nr	51.25
32mm dia.	21.46	27.49	1.62	43.91	nr	71.39
40mm dia.	23.36	29.92	1.86	50.41	nr	80.33
50mm dia.	37.99	48.66	2.21	59.90	nr	108.56
65mm dia.	91.86	117.65	2.72	73.72	nr	191.37
80mm dia.	98.58	126.26	3.21	87.00	nr	213.26
100mm dia.	158.65	203.20	4.44	120.34	nr	323.54
Extra over malleable iron fittings; BS 143						
Cap						
15mm dia.	0.77	0.99	0.32	8.67	nr	9.66
20mm dia.	0.89	1.14	0.43	11.65	nr	12.79
25mm dia.	1.18	1.51	0.49	13.28	nr	14.79
32mm dia.	1.85	2.37	0.58	15.72	nr	18.09
40mm dia.	2.36	3.02	0.66	17.89	nr	20.91
50mm dia.	4.41	5.65	0.78	21.14	nr	26.79
65mm dia.	7.56	9.68	0.96	26.02	nr	35.70
80mm dia.	8.57	10.98	1.13	30.63	nr	41.60
100mm dia.	18.76	24.03	1.70	46.08	nr	70.10
Plain plug, hollow						
15mm dia.	0.60	0.77	0.28	7.59	nr	8.36
20mm dia.	0.76	0.97	0.38	10.30	nr	11.27
25mm dia.	0.93	1.19	0.44	11.93	nr	13.12
32mm dia.	1.47	1.88	0.51	13.82	nr	15.71
40mm dia.	2.28	2.92	0.59	15.99	nr	18.91
50mm dia.	3.19	4.09	0.70	18.97	nr	23.06
65mm dia.	5.57	7.13	0.85	23.04	nr	30.17
80mm dia.	8.32	10.66	1.00	27.10	nr	37.76
100mm dia.	15.32	19.62	1.44	39.03	nr	58.65

S: PIPED SUPPLY SYSTEMS

Item	Net Price £	Material £	Labour hours	Labour £	Unit	Total rate £
S10 : COLD WATER (cont'd)						
SCREWED STEEL PIPEWORK (cont'd)						
Extra over malleable iron fittings (cont'd)						
Plain plug, solid						
15mm dia.	1.67	2.14	0.29	7.86	nr	**10.00**
20mm dia.	1.79	2.29	0.38	10.30	nr	**12.59**
25mm dia.	2.37	3.04	0.44	11.93	nr	**14.96**
32mm dia.	3.32	4.25	0.51	13.82	nr	**18.08**
40mm dia.	4.48	5.74	0.59	15.99	nr	**21.73**
50mm dia.	5.88	7.53	0.70	18.97	nr	**26.50**
Elbow, male/female						
15mm dia.	0.88	1.13	0.64	17.35	nr	**18.47**
20mm dia.	1.17	1.50	0.85	23.04	nr	**24.54**
25mm dia.	1.95	2.50	0.97	26.29	nr	**28.79**
32mm dia.	4.04	5.17	1.12	30.36	nr	**35.53**
40mm dia.	5.57	7.13	1.29	34.96	nr	**42.10**
50mm dia.	7.17	9.18	1.55	42.01	nr	**51.19**
65mm dia.	16.00	20.49	1.89	51.23	nr	**71.72**
80mm dia.	21.88	28.02	2.24	60.71	nr	**88.74**
100mm dia.	38.26	49.00	3.09	83.75	nr	**132.75**
Elbow						
15mm dia.	0.78	1.00	0.64	17.35	nr	**18.35**
20mm dia.	1.07	1.37	0.85	23.04	nr	**24.41**
25mm dia.	1.66	2.13	0.97	26.29	nr	**28.42**
32mm dia.	3.16	4.05	1.12	30.36	nr	**34.40**
40mm dia.	4.73	6.06	1.29	34.96	nr	**41.02**
50mm dia.	5.54	7.10	1.55	42.01	nr	**49.11**
65mm dia.	12.36	15.83	1.89	51.23	nr	**67.06**
80mm dia.	18.15	23.25	2.24	60.71	nr	**83.96**
100mm dia.	31.18	39.94	3.09	83.75	nr	**123.68**
125mm dia.	75.02	96.08	4.44	120.34	nr	**216.42**
150mm dia.	139.68	178.90	5.79	156.93	nr	**335.83**
45° elbow						
15mm dia.	2.02	2.59	0.64	17.35	nr	**19.93**
20mm dia.	2.49	3.19	0.85	23.04	nr	**26.23**
25mm dia.	3.42	4.38	0.97	26.29	nr	**30.67**
32mm dia.	7.28	9.32	1.12	30.36	nr	**39.68**
40mm dia.	8.57	10.98	1.29	34.96	nr	**45.94**
50mm dia.	11.75	15.05	1.55	42.01	nr	**57.06**
65mm dia.	16.51	21.15	1.89	51.23	nr	**72.37**
80mm dia.	24.82	31.79	2.24	60.71	nr	**92.50**
100mm dia.	47.95	61.41	3.09	83.75	nr	**145.16**
150mm dia.	128.35	164.39	5.79	156.93	nr	**321.32**
Bend, male/female						
15mm dia.	1.54	1.97	0.64	17.35	nr	**19.32**
20mm dia.	2.54	3.25	0.85	23.04	nr	**26.29**
25mm dia.	3.56	4.56	0.97	26.29	nr	**30.85**
32mm dia.	5.50	7.04	1.12	30.36	nr	**37.40**
40mm dia.	8.06	10.32	1.29	34.96	nr	**45.29**
50mm dia.	15.15	19.40	1.55	42.01	nr	**61.41**
65mm dia.	23.19	29.70	1.89	51.23	nr	**80.93**
80mm dia.	31.45	40.28	2.24	60.71	nr	**100.99**
100mm dia.	77.90	99.77	3.09	83.75	nr	**183.52**

Material Costs/Prices for Measured Works – Mechanical Installations

S: PIPED SUPPLY SYSTEMS

Item	Net Price £	Material £	Labour hours	Labour £	Unit	Total rate £
Bend, male						
15mm dia.	3.53	4.52	0.64	17.35	nr	21.87
20mm dia.	3.96	5.07	0.85	23.04	nr	28.11
25mm dia.	5.82	7.45	0.97	26.29	nr	33.74
32mm dia.	11.76	15.06	1.12	30.36	nr	45.42
40mm dia.	16.50	21.13	1.29	34.96	nr	56.10
50mm dia.	22.06	28.25	1.55	42.01	nr	70.26
Bend, female						
15mm dia.	1.58	2.02	0.64	17.35	nr	19.37
20mm dia.	2.26	2.89	0.85	23.04	nr	25.93
25mm dia.	3.17	4.06	0.97	26.29	nr	30.35
32mm dia.	5.64	7.22	1.12	30.36	nr	37.58
40mm dia.	6.72	8.61	1.29	34.96	nr	43.57
50mm dia.	10.59	13.56	1.55	42.01	nr	55.57
65mm dia.	23.19	29.70	1.89	51.23	nr	80.93
80mm dia.	34.39	44.05	2.24	60.71	nr	104.76
100mm dia.	72.15	92.41	3.09	83.75	nr	176.16
125mm dia.	192.46	246.50	4.44	120.34	nr	366.84
150mm dia.	281.72	360.82	5.79	156.93	nr	517.75
Return bend						
15mm dia.	7.12	9.12	0.64	17.35	nr	26.47
20mm dia.	11.52	14.75	0.85	23.04	nr	37.79
25mm dia.	14.37	18.41	0.97	26.29	nr	44.70
32mm dia.	20.71	26.53	1.12	30.36	nr	56.88
40mm dia.	24.68	31.61	1.29	34.96	nr	66.57
50mm dia.	37.66	48.23	1.55	42.01	nr	90.25
Equal socket, parallel thread						
15mm dia.	0.83	1.06	0.64	17.35	nr	18.41
20mm dia.	0.99	1.27	0.85	23.04	nr	24.31
25mm dia.	1.33	1.70	0.97	26.29	nr	27.99
32mm dia.	2.51	3.21	1.12	30.36	nr	33.57
40mm dia.	3.58	4.59	1.29	34.96	nr	39.55
50mm dia.	5.14	6.58	1.55	42.01	nr	48.59
65mm dia.	9.01	11.54	1.89	51.23	nr	62.77
80mm dia.	12.38	15.86	2.24	60.71	nr	76.57
100mm dia.	21.00	26.90	3.09	83.75	nr	110.65
Concentric reducing socket						
20 x 15mm dia.	1.21	1.55	0.76	20.60	nr	22.15
25 x 15mm dia.	1.58	2.02	0.86	23.31	nr	25.33
25 x 20mm dia.	1.49	1.91	0.86	23.31	nr	25.22
32 x 25mm dia.	2.62	3.36	1.01	27.37	nr	30.73
40 x 25mm dia.	3.46	4.43	1.16	31.44	nr	35.87
40 x 32mm dia.	3.84	4.92	1.16	31.44	nr	36.36
50 x 25mm dia.	6.65	8.52	1.38	37.40	nr	45.92
50 x 40mm dia.	5.37	6.88	1.38	37.40	nr	44.28
65 x 50mm dia.	9.38	12.01	1.69	45.80	nr	57.82
80 x 50mm dia.	11.69	14.97	2.00	54.21	nr	69.18
100 x 50mm dia.	23.32	29.87	2.75	74.53	nr	104.40
100 x 80mm dia.	21.63	27.70	2.75	74.53	nr	102.24
150 x 100mm dia.	57.08	73.11	4.10	111.12	nr	184.23

S: PIPED SUPPLY SYSTEMS

Item	Net Price £	Material £	Labour hours	Labour £	Unit	Total rate £
S10 : COLD WATER (cont'd)						
SCREWED STEEL PIPEWORK (cont'd)						
Extra over malleable iron fittings (cont'd)						
Eccentric reducing socket						
20 x 15mm dia.	2.41	3.09	0.76	20.60	nr	23.69
25 x 15mm dia.	6.87	8.80	0.86	23.31	nr	32.11
25 x 20mm dia.	7.80	9.99	0.86	23.31	nr	33.30
32 x 25mm dia.	9.20	11.78	1.01	27.37	nr	39.16
40 x 25mm dia.	10.55	13.51	1.16	31.44	nr	44.95
40 x 32mm dia.	5.70	7.30	1.16	31.44	nr	38.74
50 x 25mm dia.	6.84	8.76	1.38	37.40	nr	46.16
50 x 40mm dia.	6.84	8.76	1.38	37.40	nr	46.16
65 x 50mm dia.	11.69	14.97	1.69	45.80	nr	60.78
80 x 50mm dia.	19.01	24.35	2.00	54.21	nr	78.55
Hexagon bush						
20 x 15mm dia.	0.68	0.87	0.37	10.03	nr	10.90
25 x 15mm dia.	0.93	1.19	0.43	11.65	nr	12.85
25 x 20mm dia.	0.88	1.13	0.43	11.65	nr	12.78
32 x 25mm dia.	1.08	1.38	0.51	13.82	nr	15.21
40 x 25mm dia.	1.61	2.06	0.58	15.72	nr	17.78
40 x 32mm dia.	1.71	2.19	0.58	15.72	nr	17.91
50 x 25mm dia.	3.42	4.38	0.71	19.24	nr	23.62
50 x 40mm dia.	3.19	4.09	0.71	19.24	nr	23.33
65 x 50mm dia.	5.87	7.52	0.84	22.77	nr	30.29
80 x 50mm dia.	8.86	11.35	1.00	27.10	nr	38.45
100 x 50mm dia.	20.51	26.27	1.52	41.20	nr	67.47
100 x 80mm dia.	17.06	21.85	1.52	41.20	nr	63.05
150 x 100mm dia.	54.02	69.19	2.48	67.22	nr	136.41
Hexagon nipple						
15mm dia.	0.73	0.94	0.28	7.59	nr	8.52
20mm dia.	0.83	1.06	0.38	10.30	nr	11.36
25mm dia.	1.17	1.50	0.44	11.93	nr	13.42
32mm dia.	2.28	2.92	0.51	13.82	nr	16.74
40mm dia.	2.62	3.36	0.59	15.99	nr	19.35
50mm dia.	4.78	6.12	0.70	18.97	nr	25.09
65mm dia.	8.00	10.25	0.85	23.04	nr	33.28
80mm dia.	11.07	14.18	1.00	27.10	nr	41.28
100mm dia.	19.93	25.53	1.44	39.03	nr	64.56
150mm dia.	55.52	71.11	2.32	62.88	nr	133.99
Union, male/female						
15mm dia.	3.68	4.71	0.64	17.35	nr	22.06
20mm dia.	4.50	5.76	0.85	23.04	nr	28.80
25mm dia.	5.24	6.71	0.97	26.29	nr	33.00
32mm dia.	8.44	10.81	1.12	30.36	nr	41.17
40mm dia.	10.81	13.85	1.29	34.96	nr	48.81
50mm dia.	17.01	21.79	1.55	42.01	nr	63.80
65mm dia.	33.51	42.92	1.89	51.23	nr	94.14

S: PIPED SUPPLY SYSTEMS

Item	Net Price £	Material £	Labour hours	Labour £	Unit	Total rate £
Union, female						
15mm dia.	8.38	10.73	0.64	17.35	nr	28.08
20mm dia.	9.96	12.76	0.85	23.04	nr	35.79
25mm dia.	12.24	15.68	0.97	26.29	nr	41.97
32mm dia.	18.73	23.99	1.12	30.36	nr	54.35
40mm dia.	22.64	29.00	1.29	34.96	nr	63.96
50mm dia.	26.76	34.27	1.55	42.01	nr	76.28
65mm dia.	60.02	76.87	1.89	51.23	nr	128.10
80mm dia.	98.03	125.56	2.24	60.71	nr	186.27
100mm dia.	141.17	180.81	3.09	83.75	nr	264.56
Union elbow, male/female						
15mm dia.	4.89	6.26	0.64	17.35	nr	23.61
20mm dia.	6.13	7.85	0.85	23.04	nr	30.89
25mm dia.	8.60	11.01	0.97	26.29	nr	37.31
Twin elbow						
15mm dia.	4.70	6.02	0.91	24.66	nr	30.68
20mm dia.	5.20	6.66	1.22	33.07	nr	39.73
25mm dia.	8.42	10.78	1.39	37.67	nr	48.46
32mm dia.	15.40	19.72	1.62	43.91	nr	63.63
40mm dia.	19.51	24.99	1.86	50.41	nr	75.40
50mm dia.	25.07	32.11	2.21	59.90	nr	92.01
65mm dia.	40.51	51.88	2.72	73.72	nr	125.61
80mm dia.	69.02	88.40	3.21	87.00	nr	175.40
Equal tee						
15mm dia.	1.07	1.37	0.91	24.66	nr	26.03
20mm dia.	1.57	2.01	1.22	33.07	nr	35.08
25mm dia.	2.24	2.87	1.39	37.67	nr	40.54
32mm dia.	4.33	5.55	1.62	43.91	nr	49.45
40mm dia.	5.93	7.60	1.86	50.41	nr	58.01
50mm dia.	8.54	10.94	2.21	59.90	nr	70.84
65mm dia.	20.00	25.62	2.72	73.72	nr	99.34
80mm dia.	23.32	29.87	3.21	87.00	nr	116.87
100mm dia.	42.26	54.13	4.44	120.34	nr	174.47
125mm dia.	103.66	132.77	5.38	145.82	nr	278.58
150mm dia.	165.18	211.56	6.31	171.02	nr	382.58
Tee reducing on branch						
20 x 15mm dia.	1.60	2.05	1.22	33.07	nr	35.12
25 x 15mm dia.	2.18	2.79	1.39	37.67	nr	40.47
25 x 20mm dia.	2.49	3.19	1.39	37.67	nr	40.86
32 x 25mm dia.	4.40	5.64	1.62	43.91	nr	49.54
40 x 25mm dia.	5.57	7.13	1.86	50.41	nr	57.55
40 x 32mm dia.	8.18	10.48	1.86	50.41	nr	60.89
50 x 25mm dia.	7.41	9.49	2.21	59.90	nr	69.39
50 x 40mm dia.	11.51	14.74	2.21	59.90	nr	74.64
65 x 50mm dia.	17.75	22.73	2.72	73.72	nr	96.46
80 x 50mm dia.	24.01	30.75	3.21	87.00	nr	117.75
100 x 50mm dia.	35.01	44.84	4.44	120.34	nr	165.18
100 x 80mm dia.	54.02	69.19	4.44	120.34	nr	189.53
150 x 100mm dia.	121.67	155.83	6.31	171.02	nr	326.86

S: PIPED SUPPLY SYSTEMS

Item	Net Price £	Material £	Labour hours	Labour £	Unit	Total rate £
S10 : COLD WATER (cont'd)						
SCREWED STEEL PIPEWORK (cont'd)						
Extra over malleable iron fittings (cont'd)						
Equal pitcher tee						
15mm dia.	3.71	4.75	0.91	24.66	nr	**29.42**
20mm dia.	4.58	5.87	1.22	33.07	nr	**38.93**
25mm dia.	6.87	8.80	1.39	37.67	nr	**46.47**
32mm dia.	9.71	12.44	1.62	43.91	nr	**56.34**
40mm dia.	15.03	19.25	1.86	50.41	nr	**69.66**
50mm dia.	21.10	27.02	2.21	59.90	nr	**86.92**
65mm dia.	30.02	38.45	2.72	73.72	nr	**112.17**
80mm dia.	41.26	52.85	3.21	87.00	nr	**139.85**
100mm dia.	92.84	118.91	4.44	120.34	nr	**239.25**
Cross						
15mm dia.	3.08	3.94	1.00	27.10	nr	**31.05**
20mm dia.	4.83	6.19	1.33	36.05	nr	**42.23**
25mm dia.	6.13	7.85	1.51	40.93	nr	**48.78**
32mm dia.	8.31	10.64	1.77	47.97	nr	**58.62**
40mm dia.	11.19	14.33	2.02	54.75	nr	**69.08**
50mm dia.	17.39	22.27	2.42	65.59	nr	**87.86**
65mm dia.	24.82	31.79	2.97	80.50	nr	**112.29**
80mm dia.	33.01	42.28	3.50	94.86	nr	**137.14**
100mm dia.	60.02	76.87	4.84	131.18	nr	**208.05**

S: PIPED SUPPLY SYSTEMS

Item	Net Price £	Material £	Labour hours	Labour £	Unit	Total rate £
Y11 - PIPELINE ANCILLARIES						
VALVES						
Regulators						
Gunmetal; self-acting two port thermostat; single seat; screwed; normally closed; with adjustable or fixed bleed device						
25mm dia.	360.63	461.89	1.46	39.57	nr	**501.46**
32mm dia.	371.04	475.22	1.45	39.30	nr	**514.52**
40mm dia.	396.60	507.96	1.55	42.02	nr	**549.98**
50mm dia.	477.42	611.47	1.68	45.53	nr	**657.01**
Self acting temperature regulator for storage calorifier; integral sensing element and pocket; screwed ends						
15mm dia.	354.82	454.45	1.32	35.78	nr	**490.23**
25mm dia.	389.47	498.83	1.52	41.20	nr	**540.03**
32mm dia.	502.95	644.17	1.79	48.52	nr	**692.69**
40mm dia.	615.26	788.02	1.99	53.94	nr	**841.95**
50mm dia.	719.20	921.14	2.26	61.25	nr	**982.40**
Self acting temperature regulator for storage calorifier; integral sensing element and pocket; flanged ends; bolted connection						
15mm dia.	520.88	667.14	0.61	16.53	nr	**683.67**
25mm dia.	596.14	763.53	0.72	19.51	nr	**783.04**
32mm dia.	751.45	962.45	0.94	25.48	nr	**987.93**
40mm dia.	890.04	1139.95	1.03	27.92	nr	**1167.87**
50mm dia.	1033.40	1323.57	1.18	31.98	nr	**1355.55**
Chrome plated thermostatic mixing valves including non-return valves and inlet swivel connections with strainers; copper compression fittings						
15mm dia.	57.49	73.63	0.69	18.70	nr	**92.33**
Chrome plated thermostatic mixing valves including non-return valves and inlet swivel connections with angle pattern combined isolating valves and strainers; copper compression fittings						
15mm dia.	94.34	120.83	0.69	18.70	nr	**139.53**
Gunmetal thermostatic mixing valves including non-return valves and inlet swivel connections with strainers; copper compression fittings						
15mm dia.	119.31	152.81	0.69	18.70	nr	**171.51**
Gunmetal thermostatic mixing valves including non-return valves and inlet swivel connections with angle pattern combined isolating valves and strainers; copper compression fittings						
15mm dia.	125.44	160.66	0.69	18.70	nr	**179.36**

S: PIPED SUPPLY SYSTEMS

Item	Net Price £	Material £	Labour hours	Labour £	Unit	Total rate £
S10 : COLD WATER (cont'd)						
VALVES (cont'd)						
Ball float valves						
Bronze, equilibrium; copper float; working pressure cold services up to 16 bar; flanged ends; BS 4504 Table 16/21; bolted connections						
25mm dia.	309.44	396.33	1.04	28.19	nr	424.52
32mm dia.	429.29	549.83	1.22	33.07	nr	582.90
40mm dia.	582.22	745.70	1.38	37.40	nr	783.10
50mm dia.	938.03	1201.42	1.66	44.99	nr	1246.41
65mm dia.	981.17	1256.67	1.93	52.31	nr	1308.98
80mm dia.	1212.97	1553.56	2.16	58.54	nr	1612.10
Heavy, equilibrium; with long tail and backnut; copper float; screwed for iron						
25mm dia.	169.98	217.71	1.58	42.82	nr	260.53
32mm dia.	264.11	338.27	1.78	48.24	nr	386.51
40mm dia.	286.82	367.36	1.90	51.50	nr	418.85
50mm dia.	469.43	601.24	2.65	71.82	nr	673.07
Brass, ball valve; BS 1212; copper float; screwed						
15mm dia.	9.62	12.32	0.25	6.78	nr	19.10
22mm dia	16.21	20.76	0.29	7.86	nr	28.62
28mm dia	61.03	78.17	0.35	9.49	nr	87.65
Gate valves						
DZR copper alloy wedge non-rising stem; capillary joint to copper						
15mm dia.	12.25	15.69	0.84	22.77	nr	38.46
22mm dia.	15.04	19.26	1.01	27.37	nr	46.64
28mm dia.	20.33	26.04	1.19	32.25	nr	58.29
35mm dia.	36.48	46.72	1.38	37.40	nr	84.13
42mm dia.	61.99	79.40	1.62	43.91	nr	123.30
54mm dia.	86.84	111.22	1.94	52.58	nr	163.80
Cocks; capillary joints to copper						
Stopcock; brass head with gun metal body						
15mm dia.	3.15	4.03	0.45	12.20	nr	16.23
22mm dia.	6.07	7.77	0.46	12.47	nr	20.24
28mm dia.	17.26	22.11	0.54	14.64	nr	36.74
Lockshield stop cocks; brass head with gun metal body						
15mm dia.	8.22	10.53	0.45	12.20	nr	22.72
22mm dia.	11.81	15.13	0.46	12.47	nr	27.59
28mm dia.	20.92	26.79	0.54	14.64	nr	41.43
DZR stopcock; brass head with gun metal body						
15mm dia.	8.16	10.45	0.45	12.20	nr	22.65
22mm dia.	14.15	18.12	0.46	12.47	nr	30.59
28mm dia.	23.57	30.19	0.54	14.64	nr	44.82

S: PIPED SUPPLY SYSTEMS

Item	Net Price £	Material £	Labour hours	Labour £	Unit	Total rate £
Gunmetal stopcock						
35mm dia.	36.97	47.35	0.69	18.70	nr	**66.05**
42mm dia.	49.08	62.86	0.71	19.24	nr	**82.10**
54mm dia.	73.32	93.91	0.81	21.95	nr	**115.86**
Double union stopcock						
15mm dia.	15.74	20.16	0.60	16.26	nr	**36.42**
22mm dia.	19.36	24.80	0.60	16.26	nr	**41.06**
28mm dia.	35.79	45.84	0.69	18.70	nr	**64.54**
Double union DZR stopcock						
15mm dia.	17.11	21.91	0.60	16.26	nr	**38.18**
22mm dia.	21.04	26.95	0.61	16.53	nr	**43.48**
28mm dia.	38.91	49.84	0.69	18.70	nr	**68.54**
Double union gun metal stopcock						
35mm dia.	64.82	83.02	0.63	17.08	nr	**100.10**
42mm dia.	88.95	113.93	0.67	18.16	nr	**132.09**
54mm dia.	139.93	179.22	0.85	23.04	nr	**202.26**
Double union stopcock with easy clean cover						
15mm dia.	18.29	23.43	0.60	16.26	nr	**39.69**
22mm dia.	22.82	29.23	0.61	16.53	nr	**45.76**
28mm dia.	42.43	54.34	0.69	18.70	nr	**73.05**
Combined stopcock and drain						
15mm dia.	18.03	23.09	0.67	18.16	nr	**41.25**
22mm dia.	22.14	28.36	0.68	18.43	nr	**46.79**
Combined DZR stopcock and drain						
15mm dia.	23.82	30.51	0.67	18.16	nr	**48.67**
Gate valve						
DZR copper alloy wedge non-rising stem; compression joint to copper						
15mm dia.	12.25	15.69	0.84	22.77	nr	**38.46**
22mm dia.	15.04	19.26	1.01	27.37	nr	**46.64**
28mm dia.	20.33	26.04	1.19	32.25	nr	**58.29**
35mm dia.	36.48	46.72	1.38	37.40	nr	**84.13**
42mm dia.	61.99	79.40	1.62	43.91	nr	**123.30**
54mm dia.	86.84	111.22	1.94	52.58	nr	**163.80**
Cocks; compression joints to copper						
Stopcock; brass head gun metal body						
15mm dia.	3.95	5.06	0.42	11.38	nr	**16.44**
22mm dia.	6.93	8.88	0.42	11.38	nr	**20.26**
28mm dia.	18.07	23.14	0.45	12.20	nr	**35.34**
Lockshield stopcock; brass head gun metal body						
15mm dia.	8.57	10.98	0.42	11.38	nr	**22.36**
22mm dia.	12.09	15.48	0.42	11.38	nr	**26.87**
28mm dia.	25.66	32.87	0.45	12.20	nr	**45.06**

S: PIPED SUPPLY SYSTEMS

Item	Net Price £	Material £	Labour hours	Labour £	Unit	Total rate £
S10 : COLD WATER (cont'd)						
VALVES (cont'd)						
Cocks; compression joints to copper (cont'd)						
DZR Stopcock						
15mm dia.	9.45	12.10	0.38	10.30	nr	22.40
22mm dia.	15.50	19.85	0.39	10.57	nr	30.42
28mm dia.	25.67	32.88	0.40	10.84	nr	43.72
35mm dia.	49.66	63.60	0.52	14.09	nr	77.70
42mm dia.	70.44	90.22	0.54	14.64	nr	104.85
54mm dia.	96.09	123.07	0.63	17.08	nr	140.15
DZR Lockshield stopcock						
15mm dia.	11.64	14.91	0.38	10.30	nr	25.21
22mm dia.	18.24	23.36	0.39	10.57	nr	33.93
Combined stop/draincock						
15mm dia.	14.58	18.67	0.22	5.96	nr	24.64
22mm dia.	18.82	24.10	0.45	12.20	nr	36.30
DZR Combined stop/draincock						
15mm dia.	18.28	23.41	0.41	11.11	nr	34.53
22mm dia.	25.04	32.07	0.42	11.38	nr	43.45
Stopcock to polyethylene						
15mm dia.	9.29	11.90	0.38	10.30	nr	22.20
20mm dia.	14.42	18.47	0.39	10.57	nr	29.04
25mm dia.	17.67	22.63	0.40	10.84	nr	33.47
Draw off coupling						
15mm dia.	5.72	7.33	0.38	10.30	nr	17.63
DZR Draw off coupling						
15mm dia.	7.90	10.12	0.38	10.30	nr	20.42
22mm dia.	9.14	11.71	0.39	10.57	nr	22.28
Draw off elbow						
15mm dia.	5.85	7.49	0.38	10.30	nr	17.79
22mm dia.	6.72	8.61	0.39	10.57	nr	19.18
Lockshield drain cock						
15mm dia.	11.98	15.34	0.41	11.11	nr	26.46
Check valves						
DZR copper alloy and bronze, WRC approved cartridge double check valve; BS 6282; working pressure cold services up to 10 bar at 65°C; screwed ends						
32mm dia.	92.00	117.83	1.38	37.40	nr	155.24
40mm dia.	136.76	175.16	1.62	43.91	nr	219.07
50mm dia.	161.02	206.23	1.94	52.58	nr	258.81

S: PIPED SUPPLY SYSTEMS

Item	Net Price £	Material £	Labour hours	Labour £	Unit	Total rate £
Y20 - PUMPS						
Packaged cold water pressure booster set; fully automatic; 3 phase supply; includes fixing in position; electrical work elsewhere						
Pressure booster set						
0.75 l/s @ 30m head	3578.47	4583.27	9.38	254.23	nr	**4837.50**
1.5 l/s @ 30m head	4249.43	5442.63	9.38	254.23	nr	**5696.86**
3 l/s @ 30m head	5191.97	6649.82	10.38	281.33	nr	**6931.16**
6 l/s @ 30m head	11725.88	15018.39	10.38	281.33	nr	**15299.72**
12 l/s @ 30m head	14649.35	18762.74	12.38	335.54	nr	**19098.28**
0.75 l/s @ 50m head	4073.71	5217.57	9.38	254.23	nr	**5471.80**
1.5 l/s @ 50m head	5191.97	6649.82	9.38	254.23	nr	**6904.05**
3 l/s @ 50m head	5735.14	7345.51	10.38	281.33	nr	**7626.84**
6 l/s @ 50m head	13096.56	16773.94	10.38	281.33	nr	**17055.28**
12 l/s @ 50m head	16087.13	20604.24	12.38	335.54	nr	**20939.78**
0.75 l/s @ 70m head	4457.11	5708.62	9.38	254.23	nr	**5962.85**
1.5 l/s @ 70m head	5731.95	7341.42	9.38	254.23	nr	**7595.66**
3 l/s @ 70m head	6102.57	7816.11	10.38	281.33	nr	**8097.44**
6 l/s @ 70m head	14297.90	18312.61	10.38	281.33	nr	**18593.94**
12 l/s @ 70m head	17381.12	22261.56	12.38	335.54	nr	**22597.11**
Automatic sump pump for clear and drainage water; single stage centrifugal pump, pressure tight electric motor; single phase supply; includes fixing in position; electrical work elsewhere						
Single pump						
1 l/s @ 2.68m total head	178.66	228.83	3.50	94.86	nr	**323.69**
1 l/s @ 4.68m total head	196.66	251.88	3.50	94.86	nr	**346.74**
1 l/s @ 6.68m total head	259.64	332.54	3.50	94.86	nr	**427.41**
2 l/s @ 4.38m total head	259.64	332.54	4.00	108.41	nr	**440.96**
2 l/s @ 6.38m total head	259.64	332.54	4.00	108.41	nr	**440.96**
2 l/s @ 8.38m total head	334.19	428.03	4.00	108.41	nr	**536.44**
3 l/s @ 3.7m total head	259.64	332.54	4.50	121.97	nr	**454.51**
3 l/s @ 5.7m total head	334.19	428.03	4.50	121.97	nr	**549.99**
4 l/s @ 2.9m total head	259.64	332.54	5.00	135.52	nr	**468.06**
4 l/s @ 4.9m total head	334.19	428.03	5.00	135.52	nr	**563.54**
4 l/s @ 6.9m total head	926.74	1186.96	5.00	135.52	nr	**1322.48**
Extra for high level alarm box with single float switch, local alarm and volt free contacts for remote alarm.	271.95	348.31	-	-	nr	**348.31**
Duty/standby pump unit						
1 l/s @ 2.68m total head	339.33	434.61	5.00	135.52	nr	**570.13**
1 l/s @ 4.68m total head	372.76	477.43	5.00	135.52	nr	**612.94**
1 l/s @ 6.68 total head	503.86	645.34	5.00	135.52	nr	**780.86**
2 l/s @ 4.38m total head	503.86	645.34	5.50	149.07	nr	**794.41**
2 l/s @ 6.38m total head	503.86	645.34	5.50	149.07	nr	**794.41**
2 l/s @ 8.38m total head	645.25	826.43	5.50	149.07	nr	**975.50**
3 l/s @ 3.7m total head	503.86	645.34	6.00	162.62	nr	**807.96**
3 l/s @ 5.7m total head	645.25	826.43	6.00	162.62	nr	**989.05**
4 l/s @ 2.9m total head	503.86	645.34	6.50	176.17	nr	**821.51**
4 l/s @ 4.9m total head	645.25	826.43	6.50	176.17	nr	**1002.60**
4 /s @ 6.9m total head	1714.66	2196.12	7.00	189.72	nr	**2385.84**
Extra for 4nr float switches to give pump on, off and high level alarm	284.13	363.91	-	-	nr	**363.91**
Extra for dual pump control panel, internal wall mounted IP54, including volt free contacts	1353.00	1732.91	4.00	108.41	nr	**1841.32**

S: PIPED SUPPLY SYSTEMS

Item	Net Price £	Material £	Labour hours	Labour £	Unit	Total rate £
S10 : COLD WATER (cont'd)						
Y21 - TANKS						
Cisterns; fibreglass; complete with ball valve, fixing plate and fitted covers						
Rectangular						
70 litres capacity	200.64	287.11	1.33	36.05	nr	323.15
110 litres capacity	208.56	298.44	1.40	37.94	nr	336.38
170 litres capacity	220.44	315.44	1.61	43.64	nr	359.08
280 litres capacity	359.04	513.77	1.61	43.64	nr	557.41
420 litres capacity	376.20	538.32	1.99	53.94	nr	592.26
710 litres capacity	620.40	887.76	3.31	89.71	nr	977.48
840 litres capacity	753.72	1078.54	3.60	97.57	nr	1176.11
1590 litres capacity	1054.68	1509.20	13.32	361.02	nr	1870.22
2275 litres capacity	1318.68	1886.97	20.18	546.95	nr	2433.92
3365 litres capacity	1642.08	2349.74	24.50	664.04	nr	3013.77
4545 litres capacity	1965.48	2812.51	29.91	810.67	nr	3623.17
Cisterns; polypropylene; complete with ball valve, fixing plate and cover; includes placing in position						
Rectangular						
18 litres capacity	9.67	12.28	1.00	27.10	nr	39.38
68 litres capacity	26.52	33.68	1.00	27.10	nr	60.78
91 litres capacity	27.04	34.34	1.00	27.10	nr	61.44
114 litres capacity	35.93	45.63	1.00	27.10	nr	72.73
182 litres capacity	64.24	81.57	1.00	27.10	nr	108.68
227 litres capacity	64.76	82.23	1.00	27.10	nr	109.34
Circular						
114 litres capacity	30.20	38.35	1.00	27.10	nr	65.45
227 litres capacity	45.82	58.18	1.00	27.10	nr	85.29
318 litres capacity	122.96	156.14	1.00	27.10	nr	183.24
455 litres capacity	139.74	177.45	1.00	27.10	nr	204.55
Steel sectional water storage tank; hot pressed steel tank to BS 1564 TYPE 1; 5mm plate; pre-insulated and complete with all connections and fittings to comply with BSEN 13280; 2001 and WRAS water supply (water fittings) regulations 1999; externally flanged base and sides; cost of erection (on prepared base) is included within the net price, labour cost allows for offloading and positioning materials						
Note - Prices are based on the most economical tank size for each volume, and the cost will vary with differing tank dimensions, for the same volume						
Volume, size						
4,900 litres, 3.66m x 1.22m x 1,22m (h)	5256.12	6567.52	6.00	162.62	nr	6730.14
20,300 litres, 3.66m x 2.4m x 2.4m (h)	11196.37	13989.86	12.00	325.24	nr	14315.11
52,000 litres, 6.1m x 3.6m x 2.4m (h)	19963.67	24944.61	19.00	514.97	nr	25459.57
94,000 litres, 7.3m x 3.6m x 3.6m (h)	30034.02	37527.51	28.00	758.90	nr	38286.41
140,000 litres, 9.7m x 6.1m x 2.44m (h)	39224.57	49011.10	28.00	758.90	nr	49770.00

S: PIPED SUPPLY SYSTEMS

Item	Net Price £	Material £	Labour hours	Labour £	Unit	Total rate £
GRP sectional water storage tank; pre-insulated and complete with all connections and fittings to comply with BSEN 13280; 2001 and WRAS water supply (water fittings) regulations 1999; externally flanged base and sides; cost of erection (on prepared base) is included within the net price, labour cost allows for offloading and positioning materials						
Note - Prices are based on the most economical tank size for each volume, and the cost will vary with differing tank dimensions, for the same volume						
Volume, size						
4,500 litres, 3m x 1m x 1.5m (h)	3695.46	4617.48	5.00	135.52	nr	**4752.99**
10,000 litres, 2.5m x 2m x 2m (h)	4912.16	6137.74	7.00	189.72	nr	**6327.47**
20,000 litres, 4m x 2.5m x 2m (h)	7065.63	8828.50	10.00	271.04	nr	**9099.54**
30,000 litres 5m x 3m x 2m (h)	8816.92	11016.74	12.00	325.24	nr	**11341.98**
40,000 litres, 5m x 4m x 2m (h)	10331.16	12908.78	12.00	325.24	nr	**13234.03**
50,000 litres, 5m x 4m x 2.5m (h)	12736.80	15914.63	14.00	379.45	nr	**16294.08**
60,000 litres, 6m x 4m x 2.5m (h)	14674.68	18336.01	16.00	433.66	nr	**18769.67**
70,000 litres, 7m x 4m x 2.5m (h)	16260.79	20317.86	16.00	433.66	nr	**20751.51**
80,000 litres, 8m x 4m x 2.5m (h)	18482.37	23093.72	16.00	433.66	nr	**23527.38**
90,000 litres, 6m x 5m x 3m (h)	18835.39	23534.82	16.00	433.66	nr	**23968.48**
105,000 litres, 7m x 5m x 3m (h)	20819.92	26014.49	24.00	650.48	nr	**26664.97**
120,000 litres, 8m x 5m x 3m (h)	21968.53	27449.68	24.00	650.48	nr	**28100.16**
135,000 litres, 9m x 6m x 2.5m (h)	25321.04	31638.64	24.00	650.48	nr	**32289.12**
144,000 litres, 8m x 6m x 3m (h)	24451.09	30551.64	24.00	650.48	nr	**31202.12**

S: PIPED SUPPLY SYSTEMS

Item	Net Price £	Material £	Labour hours	Labour £	Unit	Total rate £
S10 : COLD WATER (cont'd)						
Y25 - CLEANING AND CHEMICAL TREATMENT						
Electromagnetic water conditioner, complete with control box; maximum inlet pressure 16 bar; electrical work elsewhere						
Connection size, nominal flow rate at 50mbar						
20mm dia, 0.3l/s	1382.50	1770.69	1.25	33.88	nr	1804.57
25mm dia, 0.6l/s	1912.50	2449.51	1.45	39.30	nr	2488.81
32mm dia, 1.2l/s	2762.50	3538.18	1.55	42.01	nr	3580.19
40mm dia, 1.7l/s	3382.50	4332.27	1.65	44.72	nr	4376.99
50mm dia, 3.4l/s	4519.50	5788.53	1.75	47.43	nr	5835.96
65mm dia, 5.2l/s	4909.50	6288.04	1.90	51.50	nr	6339.54
100mm dia, 30.5l/s	16220.00	20774.41	3.00	81.31	nr	20855.72
Ultraviolet water sterilising unit, complete with control unit; UV lamp housed in quartz tube; unit complete with UV intensity sensor, flushing and discharge valve and facilities for remote alarm; electrical work elsewhere						
Maximum flow rate (@ 250J/m2 exposure), connection size						
0.82l/s, 40mm dia	2192.50	2808.13	1.98	53.66	nr	2861.80
1.28l/s, 40mm dia	2517.50	3224.39	1.98	53.66	nr	3278.05
2.00l/s, 40mm dia	3292.50	4217.00	1.98	53.66	nr	4270.67
4.14l/s, 50mm dia	3617.50	4633.26	2.10	56.92	nr	4690.18
1.28l/s, 40mm dia	3592.50	4601.24	1.98	53.66	nr	4654.90
2.00l/s, 40mm dia	4992.50	6394.34	1.98	53.66	nr	6448.01
4.14l/s, 50mm dia	5717.50	7322.92	2.10	56.92	nr	7379.83
7.4l/s, 80mm dia	9585.00	12276.37	3.60	97.57	nr	12373.94
16.8l/s 100mm dia	11955.00	15311.84	3.60	97.57	nr	15409.42
32.3l/s 100mm dia	13430.00	17201.01	3.60	97.57	nr	17298.58
Base exchange water softener complete with resin tank, brine tank and consumption data monitoring facilities						
Capacities of softeners are based on 300ppm hardness and quoted in m³ of softened water produced. Design flow rates are recommended for continuous use						
Simplex configuration						
Design flow rate, min-max softened water produced						
1l/s, 5.8m³-11.2m³	1580.00	2023.65	8.00	216.83	nr	2240.48
1.3l/s, 11.7m³-21.4m³	2160.00	2766.51	8.00	216.83	nr	2983.33
1.3l/s, 15.5m³-28.5m³	2460.00	3150.74	10.00	271.04	nr	3421.78
1.6l/s, 23.3m³-42.7m³	3125.00	4002.47	10.00	271.04	nr	4273.50
1.6l/s, 38.8m³-71.2m³	3520.00	4508.38	12.00	325.24	nr	4833.62
1.9l/s, 11.7m³-21.4m³	3550.00	4546.80	12.00	325.24	nr	4872.05
3.2l/s, 19.4m³-35.6m³	3810.00	4879.81	12.00	325.24	nr	5205.05
4.4l/s, 31m³-57m³	4490.00	5750.75	15.00	406.55	nr	6157.30
5.1l/s, 46.6m³-85.4m³	6690.00	8568.49	15.00	406.55	nr	8975.04
5.1l/s, 77.7m³-142.4m³	8180.00	10476.86	18.00	487.86	nr	10964.73

S: PIPED SUPPLY SYSTEMS

Item	Net Price £	Material £	Labour hours	Labour £	Unit	Total rate £
Duplex configuration Design flow rate, min-max softened water produced						
2l/s, 5.8m^3-22.4m^3	2630.00	3368.48	12.00	325.24	nr	**3693.72**
2.6l/s, 11.7m^3-42.8m^3	3570.00	4572.42	12.00	325.24	nr	**4897.66**
2.6l/s, 15.5m^3-57m^3	3740.00	4790.15	15.00	406.55	nr	**5196.71**
3.2l/s, 23.3m^3-85.4m^3	4830.00	6186.22	15.00	406.55	nr	**6592.77**
3.2l/s, 38.8m^3-142.4m^3	5770.00	7390.16	18.00	487.86	nr	**7878.02**
3.8l/s, 11.7m^3-42.8m^3	6220.00	7966.51	18.00	487.86	nr	**8454.38**
6.4l/s, 19.4m^3-71.2m^3	6690.00	8568.49	18.00	487.86	nr	**9056.35**
8.8l/s, 31.1m^3-114m^3	7670.00	9823.66	23.00	623.38	nr	**10447.04**
10.2l/s, 46.6m^3-170.8m^3	12290.00	15740.91	23.00	623.38	nr	**16364.29**
10.2l/s, 77.7m^3-284.8m^3	15240.00	19519.24	27.00	731.79	nr	**20251.03**
Triplex configuration Design flow rate, min-max softened water produced						
3l/s, 5.8m^3-33.6m^3	3655.00	4681.29	15.00	406.55	nr	**5087.84**
3.9l/s, 11.7m^3-64.2m^3	5180.00	6634.49	15.00	406.55	nr	**7041.04**
3.9l/s, 15.5m^3-85.5m^3	5470.00	7005.92	18.00	487.86	nr	**7493.78**
4.8l/s, 23.3m^3-128.1m^3	7120.00	9119.22	18.00	487.86	nr	**9607.09**
4.8l/s, 38.8m^3-213.6m^3	8180.00	10476.86	22.00	596.28	nr	**11073.14**
5.7l/s, 11.7m^3-64.2m^3	9160.00	11732.04	22.00	596.28	nr	**12328.31**
9.6l/s, 19.4m^3-106.8m^3	9930.00	12718.24	22.00	596.28	nr	**13314.52**
13.2l/s, 31.1m^3-171.0m^3	11390.00	14588.20	27.00	731.79	nr	**15319.99**
15.3l/s, 46.6m^3-256.2m^3	18320.00	23464.07	27.00	731.79	nr	**24195.87**
15.3l/s, 77.7m^3-427.2m^3	22890.00	29317.28	32.00	867.31	nr	**30184.60**

S: PIPED SUPPLY SYSTEMS

Item	Net Price £	Material £	Labour hours	Labour £	Unit	Total rate £
S10 : COLD WATER (cont'd)						
Y50 -THERMAL INSULATION						
Flexible closed cell walled insulation; Class 1/ Class O; adhesive joints; including around fittings						
6mm wall thickness						
15mm diameter	0.68	0.89	0.15	3.30	m	**4.19**
22mm diameter	0.80	1.05	0.15	3.30	m	**4.35**
28mm diameter	1.02	1.33	0.15	3.30	m	**4.63**
9mm wall thickness						
15mm diameter	0.72	0.94	0.15	3.30	m	**4.24**
22mm diameter	0.87	1.14	0.15	3.30	m	**4.44**
28mm diameter	0.96	1.25	0.15	3.30	m	**4.55**
35mm diameter	1.11	1.46	0.15	3.30	m	**4.75**
42mm diameter	1.29	1.69	0.15	3.30	m	**4.99**
54mm diameter	1.86	2.43	0.15	3.30	m	**5.73**
13mm wall thickness						
15mm diameter	0.93	1.22	0.15	3.30	m	**4.52**
22mm diameter	1.14	1.49	0.15	3.30	m	**4.78**
28mm diameter	1.39	1.81	0.15	3.30	m	**5.11**
35mm diameter	1.50	1.96	0.15	3.30	m	**5.26**
42mm diameter	1.79	2.34	0.15	3.30	m	**5.63**
54mm diameter	2.22	2.90	0.15	3.30	m	**6.20**
67mm diameter	3.50	4.58	0.15	3.30	m	**7.87**
76mm diameter	4.06	5.31	0.15	3.30	m	**8.60**
108mm diameter	5.94	7.77	0.15	3.30	m	**11.07**
19mm wall thickness						
15mm diameter	1.56	2.04	0.15	3.30	m	**5.34**
22mm diameter	1.90	2.49	0.15	3.30	m	**5.78**
28mm diameter	2.61	3.41	0.15	3.30	m	**6.71**
35mm diameter	3.04	3.98	0.15	3.30	m	**7.27**
42mm diameter	3.60	4.71	0.15	3.30	m	**8.00**
54mm diameter	4.58	5.99	0.15	3.30	m	**9.29**
67mm diameter	5.48	7.16	0.15	3.30	m	**10.46**
76mm diameter	6.30	8.24	0.22	4.84	m	**13.08**
108mm diameter	9.29	12.15	0.22	4.84	m	**16.98**
25mm wall thickness						
15mm diameter	3.10	4.05	0.15	3.30	m	**7.35**
22mm diameter	3.42	4.47	0.15	3.30	m	**7.77**
28mm diameter	3.88	5.07	0.15	3.30	m	**8.37**
35mm diameter	4.31	5.64	0.15	3.30	m	**8.93**
42mm diameter	4.59	6.01	0.15	3.30	m	**9.30**
54mm diameter	5.42	7.09	0.15	3.30	m	**10.39**
67mm diameter	6.61	8.65	0.15	3.30	m	**11.95**
76mm diameter	7.68	10.04	0.22	4.84	m	**14.88**

S: PIPED SUPPLY SYSTEMS

Item	Net Price £	Material £	Labour hours	Labour £	Unit	Total rate £
32mm wall thickness						
15mm diameter	3.93	5.13	0.15	3.30	m	**8.43**
22mm diameter	4.31	5.64	0.15	3.30	m	**8.93**
28mm diameter	5.01	6.55	0.15	3.30	m	**9.85**
35mm diameter	5.08	6.64	0.15	3.30	m	**9.94**
42mm diameter	5.91	7.73	0.15	3.30	m	**11.02**
54mm diameter	7.44	9.73	0.15	3.30	m	**13.03**
76mm diameter	11.13	14.56	0.22	4.84	m	**19.40**

Note : For mineral fibre sectional insulation; bright class O foil faced; bright class O foil taped joints; 19mm aluminium bands rates, refer to section T31 - Low Temperature Hot Water Heating, Y50 Thermal Insulation

Note : For mineral fibre sectional insulation; bright class O foil faced; bright class O foil taped joints; 22 swg plain/embossed aluminium cladding; pop riveted rates, refer to section T31 - Low Temperature Hot Water Heating, Y50 Thermal Insulation

Note : For mineral fibre sectional insulation; bright class O foil faced; bright class O foil taped joints; 0.8mm polyisobutylene sheeting; welded joints rates, refer to section T31 - Low Temperature Hot Water Heating, Y50 Thermal Insulation

S: PIPED SUPPLY SYSTEMS

Item	Net Price £	Material £	Labour hours	Labour £	Unit	Total rate £
S11 - HOT WATER						
Y10 - PIPELINES						
Note : For pipework prices refer to section S10 - Cold Water						
Y11 - PIPELINE ANCILLARIES						
Note : For prices for ancillaries refer to section S10 - Cold Water						
Y23 - STORAGE						
CYLINDERS/CALORIFIERS						
CYLINDERS						
Insulated copper storage cylinders; BS 699; includes placing in position						
Grade 3 (maximum 10m working head)						
BS size 6; 115 litres capacity; 400mm dia.; 1050mm high	114.36	146.47	1.50	40.70	nr	187.17
BS size 7; 120 litres capacity; 450mm dia.; 900mm high	135.21	173.18	2.00	54.21	nr	227.38
BS size 8; 144 litres capacity; 450mm dia.; 1050mm high	143.73	184.09	2.80	75.92	nr	260.01
Grade 4 (maximum 6m working head)						
BS size 2; 96 litres capacity; 400mm dia.; 900mm high	88.54	113.40	1.50	40.70	nr	154.10
BS size 7; 120 litres capacity; 450mm dia.; 900mm high	93.05	119.18	1.50	40.70	nr	159.87
BS size 8; 144 litres capacity; 450mm dia.; 1050mm high	111.08	142.27	1.50	40.70	nr	182.97
BS size 9; 166 litres capacity; 450mm dia.; 1200mm high	162.42	208.03	1.50	40.70	nr	248.72
Storage cylinders; brazed copper construction; to BS 699; screwed bosses; includes placing in position						
Tested to 2.2 bar, 15m maximum head						
144 litres	503.38	644.72	3.00	81.39	nr	726.12
160 litres	569.05	728.83	3.00	81.39	nr	810.23
200 litres	586.55	751.25	3.76	101.89	nr	853.14
255 litres	667.52	854.95	3.76	101.89	nr	956.85
290 litres	901.70	1154.89	3.76	101.89	nr	1256.78
370 litres	1033.03	1323.09	4.50	122.09	nr	1445.18
450 litres	1407.20	1802.33	5.00	135.52	nr	1937.85
Tested to 2.55 bar, 17m maximum head						
550 litres	1523.72	1951.57	5.00	135.52	nr	2087.08
700 litres	1781.37	2281.56	6.02	163.27	nr	2444.84
800 litres	2060.38	2638.91	6.54	177.15	nr	2816.06
900 litres	2229.54	2855.57	8.00	216.83	nr	3072.40
1000 litres	2352.79	3013.43	8.00	216.83	nr	3230.26
1250 litres	2576.84	3300.39	13.16	356.63	nr	3657.02

S: PIPED SUPPLY SYSTEMS

Item	Net Price £	Material £	Labour hours	Labour £	Unit	Total rate £
1500 litres	3921.28	5022.34	15.15	410.66	nr	**5432.99**
2000 litres	4705.88	6027.24	17.24	467.30	nr	**6494.55**
3000 litres	6610.15	8466.21	24.39	661.06	nr	**9127.27**
Indirect cylinders; copper; bolted top; up to 5 tappings for connections; BS 1586; includes placing in position						
Grade 3, tested to 1.45 bar, 10m maximum head						
74 litres capacity	229.30	293.69	1.50	40.70	nr	**334.38**
96 litres capacity	233.45	299.00	1.50	40.70	nr	**339.70**
114 litres capacity	239.70	307.01	1.50	40.70	nr	**347.70**
117 litres capacity	248.79	318.65	2.00	54.21	nr	**372.85**
140 litres capacity	256.37	328.36	2.50	67.76	nr	**396.11**
162 litres capacity	358.51	459.18	3.00	81.39	nr	**540.57**
190 litres capacity	391.88	501.92	3.51	95.10	nr	**597.02**
245 litres capacity	458.58	587.34	3.80	103.06	nr	**690.40**
280 litres capacity	812.91	1041.17	4.00	108.41	nr	**1149.58**
360 litres capacity	879.62	1126.61	4.50	122.09	nr	**1248.70**
440 litres capacity	1021.36	1308.15	4.50	122.09	nr	**1430.24**
Grade 2, tested to 2.2 bar, 15m maximum head						
117 litres capacity	331.41	424.47	2.00	54.21	nr	**478.67**
140 litres capacity	360.61	461.87	2.50	67.76	nr	**529.62**
162 litres capacity	412.71	528.59	2.80	75.92	nr	**604.51**
190 litres capacity	479.44	614.06	3.00	81.39	nr	**695.45**
245 litres capacity	579.48	742.19	4.00	108.41	nr	**850.61**
280 litres capacity	925.47	1185.33	4.00	108.41	nr	**1293.75**
360 litres capacity	1021.36	1308.15	4.50	122.09	nr	**1430.24**
440 litres capacity	1208.94	1548.40	4.50	122.09	nr	**1670.49**
Grade 1, tested 3.65 bar, 25m maximum head						
190 litres capacity	712.88	913.05	3.00	81.39	nr	**994.44**
245 litres capacity	810.81	1038.48	3.00	81.39	nr	**1119.87**
280 litres capacity	1146.43	1468.34	4.00	108.41	nr	**1576.75**
360 litres capacity	1450.77	1858.13	4.50	122.09	nr	**1980.22**
440 litres capacity	1761.31	2255.87	4.50	122.09	nr	**2377.96**
Indirect cylinders, including manhole; BS 853						
Grade 3, tested to 1.5 bar, 10m maximum head						
550 litres capacity	1482.60	1898.90	5.21	141.16	nr	**2040.06**
700 litres capacity	1641.46	2102.37	6.02	163.27	nr	**2265.64**
800 litres capacity	1906.21	2441.45	6.54	177.15	nr	**2618.60**
1000 litres capacity	2382.78	3051.84	7.04	190.87	nr	**3242.71**
1500 litres capacity	2753.42	3526.55	10.00	271.04	nr	**3797.59**
2000 litres capacity	3812.45	4882.95	16.13	437.15	nr	**5320.10**
Grade 2, tested to 2.55 bar, 15m maximum head						
550 litres capacity	1643.74	2105.29	5.21	141.16	nr	**2246.45**
700 litres capacity	2057.25	2634.91	6.02	163.27	nr	**2798.18**
800 litres capacity	2170.97	2780.56	6.54	177.15	nr	**2957.70**
1000 litres capacity	2687.87	3442.60	7.04	190.87	nr	**3633.47**
1500 litres capacity	3308.15	4237.05	10.00	271.04	nr	**4508.08**
2000 litres capacity	4135.20	5296.32	16.13	437.15	nr	**5733.48**

S: PIPED SUPPLY SYSTEMS

Item	Net Price £	Material £	Labour hours	Labour £	Unit	Total rate £
S11 - HOT WATER (cont'd)						
CYLINDERS (cont'd)						
Indirect cylinders, including manhole (cont'd)						
Grade 1, tested to 4 bar, 25m maximum head						
550 litres capacity	1912.52	2449.54	5.21	141.16	nr	2590.70
700 litres capacity	2170.97	2780.56	6.02	163.27	nr	2943.83
800 litres capacity	2326.05	2979.18	6.54	177.15	nr	3156.33
1000 litres capacity	3101.41	3972.25	7.04	190.87	nr	4163.13
1500 litres capacity	3721.65	4766.65	10.00	271.04	nr	5037.69
2000 litres capacity	4548.70	5825.93	16.13	437.15	nr	6263.08
Storage calorifiers; copper; heater battery capable of raising temperature of contents from 10°C to 65°C in one hour; static head not exceeding 1.35 bar; BS 853; includes fixing in position on cradles or legs						
Horizontal; primary LPHW at 82°C/71°C						
400 litres capacity	2122.41	2718.36	7.04	190.87	nr	2909.23
1000 litres capacity	3395.84	4349.36	8.00	216.83	nr	4566.19
2000 litres capacity	6791.72	8698.77	14.08	381.74	nr	9080.51
3000 litres capacity	8383.51	10737.52	25.00	677.59	nr	11415.10
4000 litres capacity	10187.55	13048.11	40.00	1084.14	nr	14132.25
4500 litres capacity	11480.92	14704.65	50.00	1355.17	nr	16059.82
Vertical; primary LPHW at 82°C/71°C						
400 litres capacity	2391.96	3063.60	7.04	190.87	nr	3254.47
1000 litres capacity	3844.24	4923.66	8.00	216.83	nr	5140.49
2000 litres capacity	7322.32	9378.35	14.08	381.74	nr	9760.09
3000 litres capacity	9152.90	11722.94	25.00	677.59	nr	12400.53
4000 litres capacity	11227.52	14380.10	40.00	1084.14	nr	15464.24
4500 litres capacity	12692.00	16255.79	50.00	1355.17	nr	17610.96
Storage calorifiers; galvanised mild steel; heater battery capable of raising temperature of contents from 10°C to 65°C in one hour; static head not exceeding 1.35 bar; BS 853; includes fixing in position on cradles or legs						
Horizontal; primary LPHW at 82°C/71°C						
400 litres capacity	2122.41	2718.36	7.04	190.87	nr	2909.23
1000 litres capacity	3395.84	4349.36	8.00	216.83	nr	4566.19
2000 litres capacity	6791.72	8698.77	14.08	381.74	nr	9080.51
3000 litres capacity	8383.51	10737.52	25.00	677.59	nr	11415.10
4000 litres capacity	10187.55	13048.11	40.00	1084.14	nr	14132.25
4500 litres capacity	11480.92	14704.65	50.00	1355.17	nr	16059.82
Vertical; primary LPHW at 82°C/71°C						
400 litres capacity	2391.96	3063.60	7.04	190.87	nr	3254.47
1000 litres capacity	3844.24	4923.66	8.00	216.83	nr	5140.49
2000 litres capacity	7322.32	9378.35	14.08	381.74	nr	9760.09
3000 litres capacity	9152.90	11722.94	25.00	677.59	nr	12400.53
4000 litres capacity	11227.52	14380.10	40.00	1084.14	nr	15464.24
4500 litres capacity	12692.00	16255.79	50.00	1355.17	nr	17610.96

S: PIPED SUPPLY SYSTEMS

Item	Net Price £	Material £	Labour hours	Labour £	Unit	Total rate £
Indirect cylinders; mild steel, welded throughout, galvanised; with bolted connections; includes placing in position						
3.2mm plate						
136 litres capacity	819.47	1049.57	2.50	67.76	nr	1117.33
159 litres capacity	882.52	1130.32	2.80	75.92	nr	1206.24
182 litres capacity	1084.26	1388.71	3.00	81.39	nr	1470.10
227 litres capacity	1336.38	1711.62	3.00	81.39	nr	1793.01
273 litres capacity	1558.89	1996.61	4.00	108.41	nr	2105.02
364 litres capacity	1815.46	2325.22	4.50	122.09	nr	2447.31
455 litres capacity	1875.15	2401.67	5.00	135.52	nr	2537.19
683 litres capacity	3034.56	3886.63	6.02	163.27	nr	4049.91
910 litres capacity	3524.40	4514.02	7.04	190.87	nr	4704.89
Y50 - INSULATION						
Refer to sections S10 - Cold Water and T31 – Low Temperature Hot Water Heating for details						
Local Electric Hot Water Heaters						
Unvented multi-point water heater; providing hot water for one or more outlets: Used with conventional taps or mixers: Factory fitted temperature and pressure relief valve: Externally adjustable thermostat: Elemental 'on' indicator: Fitted with 1 metre of 3 core cable: Electrical supply and connection excluded						
5 litre capacity, 2.2kW rating	169.29	216.82	1.50	40.66	nr	257.48
10 litre capacity, 2.2kW rating	188.80	241.81	1.50	40.66	nr	282.47
15 litre capacity, 2.2kW rating	209.80	268.71	1.50	40.66	nr	309.37
30 litre capacity, 3kW rating	373.92	478.91	2.00	54.21	nr	533.12
50 litre capacity, 3kW rating	393.67	504.21	2.00	54.21	nr	558.42
80 litre capacity, 3kW rating	774.62	992.13	2.00	54.21	nr	1046.33
100 litre capacity, 3kW rating	831.60	1065.11	2.00	54.21	nr	1119.31
Accessories						
Pressure reducing valve and expansion kit	141.41	181.12	2.00	54.21	nr	235.32
Thermostatic blending valve	75.25	96.38	1.00	27.10	nr	123.48

S: PIPED SUPPLY SYSTEMS

Item	Net Price £	Material £	Labour hours	Labour £	Unit	Total rate £
S32 : NATURAL GAS						
Y10 - PIPELINES						
MEDIUM DENSITY POLYETHELENE - YELLOW						
Pipe; laid underground; electrofusion joints in the running length; BS 6572; BGT PL2 standards						
Coiled service pipe						
20mm dia.	0.84	1.11	0.37	10.03	m	11.14
25mm dia.	1.10	1.45	0.41	11.11	m	12.56
32mm dia.	1.81	2.39	0.47	12.74	m	15.13
63mm dia.	6.90	9.10	0.60	16.26	m	25.36
90mm dia.	9.06	11.95	0.90	24.40	m	36.35
Mains service pipe						
63mm dia.	6.90	8.84	0.60	16.26	m	25.10
90mm dia.	9.06	11.60	0.90	24.39	m	36.00
125mm dia.	17.50	22.41	1.20	32.52	m	54.94
180mm dia.	36.13	46.27	1.50	40.66	m	86.93
250mm dia.	66.51	85.19	1.75	47.43	m	132.62
Extra over fittings, electrofusion joints						
Straight connector						
32mm dia.	5.41	6.93	0.47	12.74	nr	19.67
63mm dia.	10.17	13.03	0.58	15.72	nr	28.75
90mm dia.	15.00	19.21	0.67	18.16	nr	37.37
125mm dia.	28.08	35.96	0.83	22.50	nr	58.46
180mm dia.	50.57	64.77	1.25	33.88	nr	98.65
Reducing connector						
90 x 63mm dia.	20.94	26.82	0.67	18.16	nr	44.98
125 x 90mm dia.	41.99	53.78	0.83	22.50	nr	76.28
180 x 125mm dia.	77.05	98.68	1.25	33.88	nr	132.56
Bend; 45°						
90mm dia.	40.47	51.83	0.67	18.16	nr	69.99
125mm dia.	66.20	84.79	0.83	22.50	nr	107.28
180mm dia.	150.18	192.35	1.25	33.88	nr	226.23
Bend; 90°						
63mm dia.	26.22	33.58	0.58	15.72	nr	49.30
90mm dia.	40.47	51.83	0.67	18.16	nr	69.99
125mm dia.	66.20	84.79	0.83	22.50	nr	107.28
180mm dia.	150.18	192.35	1.25	33.88	nr	226.23
Extra over malleable iron fittings, compression joints						
Straight connector						
20mm dia.	7.72	9.89	0.38	10.30	nr	20.19
25mm dia.	8.42	10.78	0.45	12.20	nr	22.98
32mm dia.	9.44	12.09	0.50	13.55	nr	25.64
63mm dia.	18.96	24.28	0.85	23.05	nr	47.33

Material Costs/Prices for Measured Works – Mechanical Installations

S: PIPED SUPPLY SYSTEMS

Item	Net Price £	Material £	Labour hours	Labour £	Unit	Total rate £
Straight connector; polyethylene to MI						
20mm dia.	6.56	8.40	0.31	8.40	nr	16.81
25mm dia.	7.15	9.16	0.35	9.49	nr	18.64
32mm dia.	7.99	10.23	0.40	10.84	nr	21.07
63mm dia.	13.39	17.15	0.65	17.62	nr	34.77
Straight connector; polyethylene to FI						
20mm dia.	6.56	8.40	0.31	8.40	nr	16.81
25mm dia.	7.99	10.23	0.35	9.49	nr	19.72
32mm dia.	7.99	10.23	0.40	10.84	nr	21.07
63mm dia.	13.39	17.15	0.75	20.33	nr	37.48
Elbow						
20mm dia.	10.04	12.86	0.38	10.30	nr	23.16
25mm dia.	10.95	14.02	0.45	12.20	nr	26.22
32mm dia.	12.27	15.72	0.50	13.55	nr	29.27
63mm dia.	24.65	31.57	0.80	21.68	nr	53.25
Equal tee						
20mm dia.	13.39	17.15	0.53	14.37	nr	31.52
25mm dia.	15.65	20.04	0.55	14.91	nr	34.95
32mm dia.	19.71	25.24	0.64	17.35	nr	42.60
SCREWED STEEL						
For prices for steel pipework refer to section T31 - Low Temperature Hot Water Heating						
PIPE IN PIPE						
Note - for pipe in pipe, a sleeve size two pipe sizes bigger than actual pipe size has been allowed. All rates refer to actual pipe size						
Black steel pipes						
Screwed and socketed joints; BS 1387: 1985 up to 50mm pipe size. Butt welded joints; BS 1387: 1985 65mm pipe size and above						
Pipe						
25mm	8.58	10.99	1.73	46.89	m	57.88
32mm	11.23	14.38	1.95	52.85	m	67.24
40mm	12.86	16.47	2.16	58.54	m	75.01
50mm	16.75	21.45	2.44	66.13	m	87.59
65mm	21.89	28.04	2.95	79.96	m	107.99
80mm	28.38	36.35	3.42	92.69	m	129.04
100mm	35.54	45.52	4.00	108.41	m	153.93

S: PIPED SUPPLY SYSTEMS

Item	Net Price £	Material £	Labour hours	Labour £	Unit	Total rate £
S32 : NATURAL GAS (cont'd)						
PIPE IN PIPE (cont'd)						
Extra over black steel pipes - Screwed pipework; black malleable iron fittings; BS 143. Welded pipework; butt welded steel fittings; BS 1965						
Bend, 90 °						
25mm	4.72	6.05	2.91	78.87	m	84.92
32mm	7.22	9.25	3.45	93.51	m	102.75
40mm	7.90	10.12	5.34	144.73	m	154.85
50mm	9.50	12.17	6.53	176.99	m	189.15
65mm	11.69	14.97	8.84	239.59	m	254.57
80mm	18.80	24.08	10.73	290.82	m	314.90
100mm	26.03	33.34	12.76	345.84	m	379.18
Bend, 45 °						
25mm	5.81	7.44	2.91	78.87	m	86.31
32mm	8.95	11.46	3.45	93.51	m	104.97
40mm	9.09	11.64	5.34	144.73	m	156.38
50mm	10.48	13.42	6.53	176.99	m	190.41
65mm	11.77	15.07	8.84	239.59	m	254.67
80mm	19.59	25.09	10.73	290.82	m	315.91
100mm	25.14	32.20	12.76	345.84	m	378.04
Equal tee						
25mm	9.92	12.71	4.18	113.29	m	126.00
32mm	14.08	18.03	4.94	133.89	m	151.92
40mm	28.80	36.89	7.28	197.31	m	234.20
50mm	32.30	41.37	8.48	229.84	m	271.21
65mm	50.67	64.90	11.47	310.88	m	375.77
80mm	88.47	113.31	14.23	385.68	m	498.99
100mm	100.65	128.91	17.92	485.69	m	614.61
Copper pipe; capillary or compression joints in the running length; EN1057 R250 (TX) formerly BS 2871 Table X						
Plastic coated gas service pipe for corrosive environments, fixed vertically or at low level with brackets measured separately						
15mm dia. (yellow)	1.86	2.38	0.85	23.05	m	25.43
22mm dia. (yellow)	3.56	4.56	0.96	26.04	m	30.60
28mm dia. (yellow)	4.52	5.79	1.06	28.73	m	34.52
Copper pipe; capillary or compression joints in the running length; EN1057 R250 (TY) formerly BS 2871 Table Y						
Plastic coated gas and cold water service pipe for corrosive environments, fixed vertically or at low level with brackets measured separately (Refer to Copper Pipe Table X Section)						
15mm dia. (yellow)	2.89	3.70	0.61	16.53	m	20.23
22mm dia. (yellow)	4.56	5.84	0.69	18.70	m	24.54

S: PIPED SUPPLY SYSTEMS

Item	Net Price £	Material £	Labour hours	Labour £	Unit	Total rate £
FIXINGS						
Refer to Section S10 Cold Water						
Extra over copper pipes; capillary fittings; BS 864						
Refer to Section S10 Cold Water						
GAS BOOSTERS						
Complete skid mounted gas booster set, including AV mounts, flexible connections, low pressure switch, control panel and NRV (for run/standby unit); 3 phase supply; in accordance with IGE/UP/2; includes delivery, offloading and positioning						
Single unit						
Flow, pressure range						
0-200 m³/hour, 0.1-2.6 kPa	1239.91	1588.06	10.00	271.04	nr	1859.10
0-200 m³/hour, 0.1-4.0 kPa	1414.77	1812.02	10.00	271.04	nr	2083.06
0-200 m³/hour, 0.1-7 kPa	1646.39	2108.68	10.00	271.04	nr	2379.71
0-200 m³/hour, 0.1-9.5 kPa	1714.52	2195.94	10.00	271.04	nr	2466.98
0-200 m³/hour 0.1-11.0 kPa	1915.50	2453.35	10.00	271.04	nr	2724.39
0-400 m³/hour, 0.1-4.0 kPa	1562.37	2001.07	10.00	271.04	nr	2272.10
0-1000 m³/hour, 0.1-7.4 kPa	2164.15	2771.82	10.00	271.04	nr	3042.86
50-1000 m³/hour, 0.1-16.0 kPa	4043.32	5178.64	20.00	542.07	nr	5720.71
50-1000 m³/hour, 0.1-24.5 kPa	4571.29	5854.86	20.00	542.07	nr	6396.93
50-1000 m³/hour, 0.1-31.0 kPa	5199.20	6659.08	20.00	542.07	nr	7201.15
50-1000 m³/hour, 0.1-41.0 kPa	5738.54	7349.86	20.00	542.07	nr	7891.93
50-1000 m³/hour, 0.1-51.0 kPa	5954.26	7626.16	20.00	542.07	nr	8168.23
100-1800 m³/hour, 3.5-23.5 kPa	6109.83	7825.41	20.00	542.07	nr	8367.48
100-1800 m³/hour, 4.5-27.0 kPa	6590.12	8440.56	20.00	542.07	nr	8982.63
100-1800 m³/hour, 6.0-32.5 kPa	7298.64	9348.03	20.00	542.07	nr	9890.10
100-1800 m³/hour, 7.2-39.0 kPa	8396.61	10754.29	20.00	542.07	nr	11296.36
100-1800 m³/hour, 9.0-42.0 kPa	8825.81	11304.01	20.00	542.07	nr	11846.08
Run/Standby unit						
Flow, pressure range						
0-200 m³/hour, 0.1-2.6 kPa	7081.77	9070.26	16.00	433.66	nr	9503.92
0-200 m³/hour, 0.1-4.0 kPa	7255.49	9292.76	16.00	433.66	nr	9726.42
0-200 m³/hour, 0.1-7 kPa	7423.53	9507.98	16.00	433.66	nr	9941.64
0-200 m³/hour, 0.1-9.5 kPa	7607.48	9743.58	16.00	433.66	nr	10177.24
0-200 m³/hour 0.1-11.0 kPa	7733.51	9905.00	16.00	433.66	nr	10338.66
0-400 m³/hour, 0.1-4.0 kPa	8528.32	10922.99	16.00	433.66	nr	11356.64
0-1000 m³/hour, 0.1-7.4 kPa	10870.74	13923.14	25.00	677.59	nr	14600.72
50-1000 m³/hour, 0.1-16.0 kPa	14899.29	19082.86	25.00	677.59	nr	19760.45
50-1000 m³/hour, 0.1-24.5 kPa	16844.31	21574.02	25.00	677.59	nr	22251.61
50-1000 m³/hour, 0.1-31.0 kPa	18784.78	24059.36	25.00	677.59	nr	24736.95
50-1000 m³/hour, 0.1-41.0 kPa	20730.93	26551.97	25.00	677.59	nr	27229.56
50-1000 m³/hour, 0.1-51.0 kPa	20486.65	26239.10	25.00	677.59	nr	26916.68
100-1800 m³/hour, 3.5-23.5 kPa	18292.00	23428.21	25.00	677.59	nr	24105.80
100-1800 m³/hour, 4.5-27.0 kPa	19729.47	25269.31	25.00	677.59	nr	25946.90
100-1800 m³/hour, 6.0-32.5 kPa	20696.87	26508.34	25.00	677.59	nr	27185.93
100-1800 m³/hour, 7.2-39.0 kPa	25136.45	32194.51	25.00	677.59	nr	32872.10
100-1800 m³/hour, 9.0-42.0 kPa	26421.78	33840.75	25.00	677.59	nr	34518.34

S: PIPED SUPPLY SYSTEMS

Item	Net Price £	Material £	Labour hours	Labour £	Unit	Total rate £
S41 : FUEL OIL STORAGE/DISTRIBUTION						
Y10 - PIPELINES						
For pipework prices refer to Section T31 – Low Temperature Hot Water Heating						
Y21 – TANKS						
Fuel storage tanks; mild steel; with all necessary screwed bosses; oil resistant joint rings; includes placing in position						
Rectangular						
1360 litres (300 gallon) capacity; 2mm plate	329.09	421.50	12.03	326.06	nr	747.55
2730 litres (600 gallon) capacity; 2.5mm plate	439.75	563.23	18.60	504.13	nr	1067.35
4550 litres (1000 gallon) capacity; 3mm plate	921.95	1180.82	25.00	677.59	nr	1858.41
Fuel storage tanks; 5mm plate mild steel to BS 799 type J; complete with raised neck manhole with bolted cover, screwed connections, vent and fill connections, drain valve, gauge and overfill alarm; includes placing in position; excludes pumps and control panel						
Nominal capacity, size						
5,600 litres, 2.5m x 1.5m x 1.5m high	2151.88	2756.11	20.00	542.07	nr	3298.18
Extra for bund unit (internal use)	1237.50	1584.98	30.00	813.11	nr	2398.08
Extra for external use with bund (watertight)	783.75	1003.82	2.00	54.21	nr	1058.03
10,200 litres, 3.05m x 1.83m x 1.83m high	2674.38	3425.32	30.00	813.11	nr	4238.42
Extra for bund unit (internal use)	1801.25	2307.02	40.00	1084.14	nr	3391.16
Extra for external use with bund (watertight)	962.50	1232.76	2.00	54.21	nr	1286.97
15,000 litres, 3.75m x 2m x 2m high	3382.50	4332.27	40.00	1084.14	nr	5416.41
Extra for bund unit (internal use)	2420.00	3099.51	55.00	1490.69	nr	4590.20
Extra for external use with bund (watertight)	1127.50	1444.09	2.00	54.21	nr	1498.30
20,000 litres, 4m x 2.5m x 2m high	4461.88	5714.73	50.00	1355.17	nr	7069.91
Extra for bund unit (internal use)	2935.63	3759.93	65.00	1761.73	nr	5521.65
Extra for external use with bund (watertight)	1333.75	1708.25	2.00	54.21	nr	1762.46
Extra for BMS output (all tank sizes)	467.50	598.77	-	-	nr	598.77
Fuel storage tanks; plastic; with all necessary screwed bosses; oil resistant joint rings; includes placing in position						
Cylindrical; horizontal						
1250 litres (285 gallon) capacity	266.26	341.02	3.73	101.10	nr	442.12
1350 litres (300 gallon) capacity	244.62	313.31	4.30	116.55	nr	429.85
2500 litres (550 gallon) capacity	418.90	536.52	4.88	132.27	nr	668.79
Cylindrical; vertical						
1365 litres (300 gallon) capacity	168.89	216.31	3.73	101.10	nr	317.41
2600 litres (570 gallon) capacity	257.04	329.21	4.88	132.27	nr	461.48
3635 litres (800 gallon) capacity	400.73	513.25	4.88	132.27	nr	645.52
5455 litres (1200 gallon) capacity	583.77	747.69	5.95	161.27	nr	908.95
Bunded tanks						
1135 litres (250 gallon) capacity	541.41	693.43	4.30	116.55	nr	809.98
1590 litres (350 gallon) capacity	644.09	824.94	4.88	132.27	nr	957.21
2500 litres (550 gallon) capacity	765.59	980.56	5.95	161.27	nr	1141.83
5000 litres (1100 gallon) capacity	1544.88	1978.67	6.53	176.99	nr	2155.65

Material Costs/Prices for Measured Works – Mechanical Installations

S: PIPED SUPPLY SYSTEMS

Item	Net Price £	Material £	Labour hours	Labour £	Unit	Total rate £
S60 : FIRE HOSE REELS						
Y10 - PIPELINES						
For pipework prices refer to section S10 - Cold water						
Y11 - PIPELINE ANCILLARIES						
For prices for ancillaries refer to section S10 - Cold water						
Hose reels; automatic; connection to 25mm screwed joint; reel with 30.5 metres, 19mm rubber hose; suitable for working pressure up to 7 bar						
Reels						
Non-swing pattern	185.22	231.08	3.75	98.62	nr	**329.70**
Recessed non-swing pattern	237.54	296.35	3.75	98.62	nr	**394.98**
Swinging pattern	247.34	308.58	3.75	98.62	nr	**407.20**
Recessed swinging pattern	255.17	318.35	3.75	98.62	nr	**416.97**
Hose reels; manual; connection to 25mm screwed joint; reel with 30.5 metres, 19mm rubber hose; suitable for working pressure up to 7 bar						
Reels						
Non-swing pattern	186.55	232.74	3.25	85.47	nr	**318.21**
Recessed non-swing pattern	218.97	273.19	3.25	85.47	nr	**358.66**
Swinging pattern	236.29	294.79	3.25	85.47	nr	**380.27**
Recessed swinging pattern	244.79	305.40	3.25	85.47	nr	**390.87**

S: PIPED SUPPLY SYSTEMS

Item	Net Price £	Material £	Labour hours	Labour £	Unit	Total rate £
S61 : DRY RISERS						
Y10 - PIPELINES						
For pipework prices refer to section S10 - Cold water						
Y11 - PIPELINE ANCILLARIES						
VALVES (BS 5041, parts 2 and 3)						
Bronze/gunmetal inlet breeching for pumping in with 65mm dia. instantaneous male coupling; with cap, chain and 25mm drain valve						
Double inlet with back pressure valve, flanged to steel	167.80	209.35	1.75	46.02	nr	**255.37**
Quadruple inlet with back pressure valve, flanged to steel	375.71	468.73	1.75	46.02	nr	**514.76**
Bronze/gunmetal gate type outlet valve with 65mm dia. instantaneous female coupling; cap and chain; wheel head secured by padlock and leather strap						
Flanged to BS 4504 PN6 (bolted connection to counter flanges measured separately)	134.57	167.89	1.75	46.02	nr	**213.91**
Bronze/gunmetal landing type outlet valve, with 65mm dia. instantaneous female coupling; cap and chain; wheelhead secured by padlock and leather strap; bolted connections to counter flanges measured separately						
Horizontal, flanged to BS 4504 PN6	146.56	182.85	1.50	39.45	nr	**222.30**
Oblique, flanged to BS 4504 PN6	146.56	182.85	1.50	39.45	nr	**222.30**
Air valve, screwed joints to steel						
25mm dia.	29.13	36.34	0.55	14.46	nr	**50.81**
INLET BOXES (BS 5041, part 5)						
Steel dry riser inlet box with hinged wire glazed door suitably lettered (fixing by others)						
610 x 460 x 325mm; double inlet	178.59	222.81	3.00	78.90	nr	**301.71**
610 x 610 x 356mm; quadruple inlet	333.30	415.82	3.00	78.90	nr	**494.72**
OUTLET BOXES (BS 5041, part 5)						
Steel dry riser outlet box with hinged wire glazed door suitably lettered (fixing by others)						
610 x 460 x 325; single outlet	174.40	217.58	3.00	78.90	nr	**296.48**

S: PIPED SUPPLY SYSTEMS

Item	Net Price £	Material £	Labour hours	Labour £	Unit	Total rate £
S63 : SPRINKLERS						
Y10 - PIPELINES						
Prefabricated black steel pipework; screwed joints, including all couplings, unions and the like to BS 1387:1985; includes fixing to backgrounds, with brackets measured separately						
Heavy weight						
25mm dia	4.11	5.13	0.47	12.36	m	17.49
32mm dia	5.15	6.43	0.53	13.94	m	20.36
40mm dia	6.00	7.49	0.58	15.25	m	22.74
50mm dia	8.38	10.45	0.63	16.57	m	27.02
FIXINGS						
For steel pipes; black malleable iron. For minimum fixing distances, refer to the Tables and Memoranda at the rear of the book						
Pipe ring, single socket, black malleable iron						
25mm dia	0.49	0.63	0.12	3.25	nr	3.88
32mm dia	0.52	0.67	0.14	3.79	nr	4.46
40mm dia	0.68	0.87	0.15	4.07	nr	4.94
50mm dia	0.85	1.09	0.16	4.34	nr	5.43
Extra Over channel sections for fabricated hangers and brackets						
Galvanised steel; including inserts, bolts, nuts, washers; fixed to backgrounds						
41 x 21mm	2.50	3.20	0.29	7.86	m	11.06
41 x 41mm	3.90	5.00	0.29	7.86	m	12.86
Threaded rods; metric thread; including nuts, washers, etc						
12mm dia x 600mm long	2.70	3.46	0.18	4.88	nr	8.34
Extra over for black malleable iron fittings; BS 143						
Plain plug, solid						
25mm dia	1.05	1.31	0.40	10.52	nr	11.83
32mm dia	1.12	1.40	0.44	11.57	nr	12.97
40mm dia	1.50	1.87	0.48	12.62	nr	14.50
50mm dia	2.00	2.50	0.56	14.73	nr	17.22
Concentric reducing socket						
32mm dia	1.65	2.06	0.48	12.62	nr	14.68
40mm dia	2.15	2.68	0.55	14.46	nr	17.15
50mm dia	4.02	5.02	0.60	15.78	nr	20.79
Elbow; 90 ° female/female						
25mm dia	1.05	1.31	0.44	11.57	nr	12.88
32mm dia	1.77	2.21	0.53	13.94	nr	16.15
40mm dia	2.98	3.72	0.60	15.78	nr	19.50
50mm dia	3.49	4.35	0.65	17.09	nr	21.45

S: PIPED SUPPLY SYSTEMS

Item	Net Price £	Material £	Labour hours	Labour £	Unit	Total rate £
S63 : SPRINKLERS (cont'd)						
Extra over for black malleable iron (cont'd)						
Tee						
25mm dia equal	1.41	1.76	0.51	13.41	nr	15.17
32mm dia reducing to 25mm dia	3.31	4.13	0.54	14.20	nr	18.33
40mm dia	5.73	7.15	0.65	17.09	nr	24.24
50mm dia	6.51	8.12	0.78	20.51	nr	28.64
Cross tee						
25mm dia equal	5.05	6.30	1.16	30.51	nr	36.81
32mm dia	7.32	9.13	1.40	36.82	nr	45.95
40mm dia	9.43	11.76	1.60	42.08	nr	53.84
50mm dia	15.30	19.09	1.68	44.18	nr	63.27
Prefabricated black steel pipework; welded joints, including all couplings, unions and the like to BS 1387:1985; fixing to backgrounds						
Heavy weight						
65mm dia	9.32	11.63	0.65	17.09	m	28.72
80mm dia	11.86	14.80	0.70	18.41	m	33.21
100mm dia	16.54	20.64	0.85	22.35	m	42.99
150mm dia	25.96	32.39	1.15	30.24	m	62.63
FIXINGS						
For steel pipes; black malleable iron. For minimum fixing distances, refer to the Tables and Memoranda at the rear of the book						
Pipe ring, single socket, black malleable iron						
65mm dia	1.24	1.59	0.30	8.13	nr	9.72
80mm dia	1.48	1.90	0.35	9.49	nr	11.38
100mm dia	2.25	2.88	0.40	10.84	nr	13.72
150mm dia	5.09	6.52	0.77	20.87	nr	27.39
Extra over fittings						
Reducer (one size down)						
65mm dia	7.79	9.72	2.70	71.01	nr	80.73
80mm dia	7.91	9.87	2.86	75.22	nr	85.08
100mm dia	8.98	11.20	3.22	84.68	nr	95.89
150mm dia	21.32	26.60	4.20	110.46	nr	137.06
Elbow; 90°						
65mm dia	5.85	7.30	3.06	80.48	nr	87.77
80mm dia	6.69	8.35	3.40	89.42	nr	97.76
100mm dia	10.29	12.84	3.70	97.31	nr	110.15
150mm dia	26.62	33.21	5.20	136.76	nr	169.97
Branch bend						
65mm dia	5.85	7.30	3.60	94.68	nr	101.98
80mm dia	6.69	8.35	3.80	99.94	nr	108.28
100mm dia	10.29	12.84	5.10	134.13	nr	146.96
150mm dia	26.62	33.21	7.50	197.24	nr	230.46

S: PIPED SUPPLY SYSTEMS

Item	Net Price £	Material £	Labour hours	Labour £	Unit	Total rate £
Prefabricated black steel pipe; victaulic joints; including all couplings and the like to BS 1387: 1985; fixing to backgrounds						
Heavy weight						
65mm dia	9.74	12.15	0.70	18.41	m	**30.56**
80mm dia	12.65	15.78	0.78	20.51	m	**36.30**
100mm dia	18.44	23.01	0.93	24.46	m	**47.46**
150mm dia	36.72	45.81	1.25	32.87	m	**78.69**
FIXINGS						
For fixings refer to For steel pipes; black malleable iron						
Extra over fittings						
Coupling						
65mm dia	8.00	9.98	0.26	6.84	nr	**16.82**
80mm dia	8.58	10.70	0.26	6.84	nr	**17.54**
100mm dia	11.24	14.02	0.32	8.42	nr	**22.44**
150mm dia	19.79	24.69	0.35	9.20	nr	**33.89**
Reducer						
65mm dia	8.65	10.79	0.48	12.62	nr	**23.42**
80mm dia	10.45	13.04	0.43	11.31	nr	**24.35**
100mm dia	22.63	28.23	0.46	12.10	nr	**40.33**
150mm dia	24.41	30.45	0.45	11.83	nr	**42.29**
Elbow; any °						
65mm dia	10.38	12.95	0.56	14.73	nr	**27.68**
80mm dia	10.60	13.22	0.63	16.57	nr	**29.79**
100mm dia	14.20	17.72	0.71	18.67	nr	**36.39**
150mm dia	30.39	37.91	0.80	21.04	nr	**58.95**
Equal tee						
65mm dia	18.71	23.34	0.74	19.46	nr	**42.80**
80mm dia	19.82	24.73	0.83	21.83	nr	**46.56**
100mm dia	22.17	27.66	0.94	24.72	nr	**52.38**
150mm dia	54.49	67.98	1.05	27.61	nr	**95.60**

S: PIPED SUPPLY SYSTEMS

Item	Net Price £	Material £	Labour hours	Labour £	Unit	Total rate £
S63 : SPRINKLERS (cont'd)						
Y11 - PIPELINE ANCILLARIES						
SPRINKLER HEADS						
Sprinkler heads; brass body; frangible glass bulb; manufactured to standard operating temperature of 57-141°C; quick response; RTI<50						
Conventional pattern; 15mm dia.	3.50	4.37	0.15	3.94	nr	8.31
Sidewall pattern; 15mm dia.	5.14	6.41	0.15	3.94	nr	10.36
Conventional pattern; 15mm dia.; satin chrome plated	4.10	5.12	0.15	3.94	nr	9.06
Sidewall pattern; 15mm dia.; satin chrome plated	5.34	6.66	0.15	3.94	nr	10.61
Fully concealed; fusible link; 15mm dia	12.82	15.99	0.15	3.94	nr	19.94
VALVES						
Wet system alarm valves; including internal non-return valve; working pressure up to 12.5 bar; BS4504 PN16 flanged ends; bolted connections						
100mm dia.	1109.38	1384.06	25.00	657.48	nr	2041.54
150mm dia.	1355.13	1690.65	25.00	657.48	nr	2348.13
Wet system by-pass alarm valves; including internal non-return valve; working pressure up to 12.5 bar; BS4504 PN16 flanged ends; bolted connections						
100mm dia.	1955.53	2439.71	25.00	657.48	nr	3097.19
150mm dia.	2472.45	3084.61	25.00	657.48	nr	3742.10
Alternate system wet/dry alarm station; including butterfly valve, wet alarm valve, dry pipe differential pressure valve and pressure gauges; working pressure up to 10.5 bar; BS4505 PN16 flanged ends; bolted connections						
100mm dia.	2423.58	3023.64	40.00	1051.97	nr	4075.62
150mm dia.	2824.21	3523.47	40.00	1051.97	nr	4575.44
Alternate system wet/dry alarm station; including electrically operated butterfly valve, water supply accelerator set, wet alarm valve, dry pipe differential pressure valve and pressure gauges; working pressure up to 10.5 bar; BS4505 PN16 flanged ends; bolted connections						
100mm dia.	2762.58	3446.58	45.00	1183.47	nr	4630.05
150mm dia.	3167.22	3951.40	45.00	1183.47	nr	5134.87
ALARM/GONGS						
Water operated motor alarm and gong; stainless steel and aluminium body and gong; screwed connections						
Connection to sprinkler system and drain pipework	338.16	421.89	6.00	157.80	nr	579.68

S: PIPED SUPPLY SYSTEMS

Item	Net Price £	Material £	Labour hours	Labour £	Unit	Total rate £
Y21 - WATER TANKS						
Note - Prices are based on the most economical tank size for each volume, and the cost will vary with differing tank dimensions, for the same volume						
Steel sectional sprinkler tank; ordinary hazard, life safety classification; two compartment tank, complete with all fittings and accessories to comply with LPCB type A requirements; cost of erection (on prepared supports) is included within net price, labour cost allows for offloading and positioning of materials						
Volume, size						
70m^3, 6.1m x 4.88m x 2.44m (h)	23296.57	29109.06	24.00	650.48	nr	**29759.55**
105m^3, 7.3m x 6.1m x 2.44m (h)	27337.20	34157.83	28.00	758.90	nr	**34916.73**
168m^3, 9.76m x 4.88mx 3.66m (h)	33775.83	42202.90	28.00	758.90	nr	**42961.80**
211m^3, 9.76m x 6.1m x 3.66m (h)	37288.90	46592.48	32.00	867.31	nr	**47459.79**
Steel sectional sprinkler tank; ordinary hazard, property protection classification; single compartment tank, complete with all fittings and accessories to comply with LPCB type A requirements; cost of erection (on prepared supports) is included within net price, labour cost allows for offloading and positioning of materials						
Volume, size						
70m^3, 6.1m x 4.88m x 2.44m (h)	16426.30	20524.66	24.00	650.48	nr	**21175.15**
105m^3, 7.3m x 6.1m x 2.44m (h)	20293.08	25356.20	28.00	758.90	nr	**26115.10**
168m^3, 9.76m x 4.88m x 3.66m (h)	26222.13	32764.55	28.00	758.90	nr	**33523.45**
211m^3, 9.76m x 6.1m x 3.66m (h)	29423.46	36764.61	32.00	867.31	nr	**37631.93**
GRP sectional sprinkler tank; ordinary hazard, life safety classification; two compartment tank, complete with all fittings and accessories to comply with LPCB type A requirements; cost of erection (on prepared supports) is included within net price, labour cost allows for offloading and positioning of materials						
Volume, size						
55m^3, 6m x 4m x 3m (h)	25996.96	32483.20	14.00	379.45	nr	**32862.65**
70m^3, 6m x 5m x 3m (h)	29066.49	36318.58	16.00	433.66	nr	**36752.24**
80m^3, 8m x 4m x 3m (h)	30123.28	37639.04	16.00	433.66	nr	**38072.69**
105m^3, 10m x 4m x 3m (h)	34086.84	42591.51	24.00	650.48	nr	**43241.99**
125m^3, 10m x 5m x 3m (h)	38442.40	48033.78	24.00	650.48	nr	**48684.26**
140m^3, 8m x 7m x 3m (h)	41244.86	51535.45	24.00	650.48	nr	**52185.94**
135m^3, 9m x 6m x 3m (h)	40034.48	50023.08	24.00	650.48	nr	**50673.57**
160m^3, 13m x 5m x 3m (h)	45282.92	56581.01	24.00	650.48	nr	**57231.49**
185m^3, 12m x 6m x 3m (h)	47900.84	59852.10	24.00	650.48	nr	**60502.58**

S: PIPED SUPPLY SYSTEMS

Item	Net Price £	Material £	Labour hours	Labour £	Unit	Total rate £
S63 : SPRINKLERS (cont'd)						
GRP sectional sprinkler tank; ordinary hazard, property protection classification; single compartment tank, complete with all fittings and accessories to comply with LPCB type A requirements; cost of erection (on prepared supports) is within net price, labour cost allows for offloading and positioning of materials						
Volume, size						
55m^3, 6m x 4m x 3m (h)	20000.05	24990.06	14.00	379.45	nr	**25369.51**
70m^3, 6m x 5m x 3m (h)	22595.05	28232.51	16.00	433.66	nr	**28666.17**
80m^3, 8m x 4m x 3m (h)	24117.19	30134.43	16.00	433.66	nr	**30568.08**
105m^3, 10m x 4m x 3m (h)	28253.83	35303.16	24.00	650.48	nr	**35953.64**
125m^3, 10m x 5m x 3m (h)	31949.18	39920.50	24.00	650.48	nr	**40570.98**
140m^3, 8m x 7m x 3m (h)	33818.63	42256.38	24.00	650.48	nr	**42906.86**
135m^3, 9m x 6m x 3m (h)	33074.75	41326.90	24.00	650.48	nr	**41977.38**
160m^3, 13m x 5m x 3m (h)	38958.19	48678.26	24.00	650.48	nr	**49328.74**
185m^3, 12m x 6m x 3m (h)	40926.22	51137.31	24.00	650.48	nr	**51787.80**

S: PIPED SUPPLY SYSTEMS

Item	Net Price £	Material £	Labour hours	Labour £	Unit	Total rate £
S65 : FIRE HYDRANTS						
EXTINGUISHERS						
Fire extinguishers; hand held; BS 5423; placed in position						
Water type; cartridge operated; for Class A fires						
Water type, 9 litres capacity; 55gm CO2 cartridge; Class A fires (fire rating 13A)	78.24	97.61	1.00	26.30	nr	**123.91**
Foam type, 9 litres capacity; 75gm CO2 cartridge; Class A & B fires (fire rating 13A:183B)	92.78	115.75	1.00	26.30	nr	**142.05**
Dry powder type; cartridge operated; for Class A, B & C fires and electrical equipment fires						
Dry powder type, 1kg capacity; 12gm CO2 cartridge; Class A, B & C fires (fire rating 5A:34B)	33.09	41.28	1.00	26.30	nr	**67.58**
Dry powder type, 2kg capacity; 28gm CO2 cartridge; Class A, B & C fires (fire rating 13A:55B)	44.55	55.58	1.00	26.30	nr	**81.88**
Dry powder type, 4kg capacity; 90gm CO2 cartridge; Class A, B & C fires (fire rating 21A:183B)	79.36	99.01	1.00	26.30	nr	**125.31**
Dry powder type, 9kg capacity; 190gm CO2 cartridge; Class A, B & C fires (fire rating 43A:233B)	106.33	132.66	1.00	26.30	nr	**158.96**
Dry powder type; stored pressure type; for Class A, B & C fires and electrical equipment fires						
Dry powder type, 1kg capacity; Class A, B & C fires (fire rating 5A:34B)	30.48	38.03	1.00	26.30	nr	**64.33**
Dry powder type, 2kg capacity; Class A, B & C fires (fire rating 13A:55B)	36.18	45.14	1.00	26.30	nr	**71.44**
Dry powder type, 4kg capacity; Class A, B & C fires (fire rating 21A:183B)	67.58	84.31	1.00	26.30	nr	**110.61**
Dry powder type, 9kg capacity; Class A, B & C fires (fire rating 43A:233B)	87.56	109.24	1.00	26.30	nr	**135.54**
Carbon dioxide type; for Class B fires and electrical equipment fires						
CO2 type with hose and horn, 2kg capacity, Class B fires (fire rating 34B)	83.70	104.42	1.00	26.30	nr	**130.72**
CO2 type with hose and horn, 5kg capacity, Class B fires (fire rating 55B)	131.16	163.63	1.00	26.30	nr	**189.93**
Glass fibre blanket, in GRP container						
1100 x 1100mm	26.56	33.14	0.50	13.15	nr	**46.29**
1200 x 1200mm	29.11	36.32	0.50	13.15	nr	**49.47**
1800 x 1200mm	38.83	48.44	0.50	13.15	nr	**61.59**
HYDRANTS						
Fire hydrants; bolted connections						
Underground hydrants, complete with frost plug to BS 750						
sluice valve pattern type 1	218.45	272.54	4.50	118.35	nr	**390.88**
screw down pattern type 2	158.67	197.96	4.50	118.35	nr	**316.30**

S: PIPED SUPPLY SYSTEMS

Item	Net Price £	Material £	Labour hours	Labour £	Unit	Total rate £
S65 : FIRE HYDRANTS (cont'd)						
HYDRANTS (cont'd)						
Fire hydrants; bolted connections (cont'd)						
Stand pipe for underground hydrant; screwed base; light alloy						
Single outlet	125.64	156.75	1.00	26.30	nr	**183.05**
Double outlet	183.83	229.35	1.00	26.30	nr	**255.64**
64mm diameter bronze/gunmetal outlet valves						
Oblique flanged landing valve	137.52	171.57	1.00	26.30	nr	**197.87**
Oblique screwed landing valve	137.52	171.57	1.00	26.30	nr	**197.87**
Cast iron surface box; fixing by others						
400 x 200 x 100mm	120.41	150.22	1.00	26.30	nr	**176.52**
500 x 200 x 150mm	162.24	202.41	1.00	26.30	nr	**228.71**
Frost Plug	29.22	36.45	0.25	6.57	nr	**43.03**

T: MECHANICAL/COOLING/HEATING SYSTEMS

Item	Net Price £	Material £	Labour hours	Labour £	Unit	Total rate £
T10 : GAS/OIL FIRED BOILERS						
DOMESTIC						
Domestic water boilers; stove enamelled casing; electric controls; placing in position; assembling and connecting; electrical work elsewhere						
Gas fired; floor standing; connected to conventional flue						
9 to 12 kW	534.95	685.16	8.59	232.82	nr	917.98
12 to 15 kW	562.77	720.79	8.59	232.82	nr	953.61
15 to 18 kW	599.41	767.72	8.88	240.68	nr	1008.40
18 to 21 kW	696.38	891.92	9.92	268.87	nr	1160.78
21 to 23 kW	772.01	988.78	10.66	288.92	nr	1277.71
23 to 29 kW	1001.29	1282.44	11.81	320.09	nr	1602.53
29 to 37 kW	1184.28	1516.81	11.81	320.09	nr	1836.91
37 to 41 kW	1231.60	1577.42	12.68	343.67	nr	1921.09
Gas fired; wall hung; connected to conventional flue						
9 to 12 kW	470.17	602.19	8.59	232.82	nr	835.01
12 to 15 kW	607.02	777.47	8.59	232.82	nr	1010.28
13 to 18 kW	770.36	986.67	8.59	232.82	nr	1219.49
Gas fired; floor standing; connected to balanced flue						
9 to 12 kW	668.25	855.89	9.16	248.27	nr	1104.16
12 to 15 kW	695.28	890.51	10.95	296.78	nr	1187.29
15 to 18 kW	748.41	958.56	11.98	324.70	nr	1283.26
18 to 21 kW	881.93	1129.57	12.78	346.38	nr	1475.95
21 to 23 kW	1017.19	1302.81	12.78	346.38	nr	1649.19
23 to 29 kW	1296.52	1660.57	15.45	418.75	nr	2079.32
29 to 37 kW	2492.44	3192.29	17.65	478.38	nr	3670.67
Gas fired; wall hung; connected to balanced flue						
6 to 9 kW	504.55	646.22	9.16	248.27	nr	894.49
9 to 12 kW	577.37	739.49	9.16	248.27	nr	987.76
12 to 15kW	653.79	837.37	9.45	256.13	nr	1093.50
15 to 18kW	783.93	1004.05	9.74	263.99	nr	1268.04
18 to 22kW	832.11	1065.76	9.74	263.99	nr	1329.75
Gas fired; wall hung; connected to fan flue (including flue kit)						
6 to 9kW	577.06	739.09	9.16	248.27	nr	987.36
9 to 12kW	643.08	823.65	9.16	248.27	nr	1071.92
12 to 15kW	698.09	894.11	10.95	296.78	nr	1190.89
15 to 18kW	751.95	963.09	11.98	324.70	nr	1287.79
18 to 23kW	979.76	1254.87	12.78	346.38	nr	1601.25
23 to 29kW	1268.77	1625.03	15.45	418.75	nr	2043.78
29 to 35kW	1557.69	1995.07	17.65	478.38	nr	2473.45
Oil fired; floor standing; connected to conventional flue						
12 to 15	1001.94	1283.27	10.38	281.33	nr	1564.61
15 to 19	1045.18	1338.66	12.20	330.66	nr	1669.32
21 to 25	1192.85	1527.79	14.30	387.58	nr	1915.37
26 to 32	1307.80	1675.02	15.80	428.24	nr	2103.25
35 to 50	1476.55	1891.15	20.46	554.54	nr	2445.69

T: MECHANICAL/COOLING/HEATING SYSTEMS

Item	Net Price £	Material £	Labour hours	Labour £	Unit	Total rate £
T10 : GAS/OIL FIRED BOILERS (cont'd)						
DOMESTIC (cont'd)						
Domestic water boilers; stove enamelled (cont'd)						
Fire place mounted natural gas fire and back boiler; cast iron water boiler; electric control box; fire output 3kW with wood surround						
10.50kW	292.30	374.37	8.88	240.68	nr	615.05
FORCED DRAFT						
Commercial cast iron sectional floor standing boilers; pressure jet burner; including controls, enamelled jacket, insulation, assembly and commissioning; electrical work elsewhere						
Gas fired (on/off type), connected to conventional flue						
16-26 kW; 3 sections; 125mm dia. flue	2126.41	2723.48	8.00	216.83	nr	2940.31
26-33 kW; 4 sections; 125mm dia. flue	2267.13	2903.72	8.00	216.83	nr	3120.55
33-40 kW; 5 sections; 125mm dia. flue	2564.21	3284.21	8.00	216.83	nr	3501.04
35-50 kW; 3 sections; 153mm dia. flue	2814.37	3604.62	8.00	216.83	nr	3821.45
50-65 kW; 4 sections; 153mm dia. flue	3033.27	3884.98	8.00	216.83	nr	4101.81
65-80 kW; 5 sections; 153mm dia. flue	3299.07	4225.42	8.00	216.83	nr	4442.24
80-100 kW; 6 sections; 180mm dia. flue	3833.80	4910.29	8.00	216.83	nr	5127.12
100-120 kW; 7 sections; 180mm dia. flue	4846.97	6207.95	8.00	216.83	nr	6424.78
105-140 kW; 5 sections; 180mm dia. flue	5628.74	7209.23	8.00	216.83	nr	7426.06
140-180 kW; 6 sections; 180mm dia. flue	6410.51	8210.52	8.00	216.83	nr	8427.35
180-230 kW; 7 sections; 200mm dia. flue	7192.28	9211.80	8.00	216.83	nr	9428.63
230-280 kW; 8 sections; 200mm dia. flue	7661.35	9812.58	8.00	216.83	nr	10029.41
280-330 kW; 9 sections; 200mm dia. flue	8599.47	11014.12	8.00	216.83	nr	11230.94
Gas fired (high/low type), connected to conventional flue						
105-140 kW; 5 sections; 180mm dia. flue	6723.22	8611.03	8.00	216.83	nr	8827.86
140-180 kW; 6 sections; 180mm dia. flue	7223.56	9251.86	8.00	216.83	nr	9468.69
180-230 kW; 7 sections; 200mm dia. flue	8192.95	10493.45	8.00	216.83	nr	10710.28
230-280 kW; 8 sections; 200mm dia. flue	8755.82	11214.37	8.00	216.83	nr	11431.19
280-330 kW; 9 sections; 200mm dia. flue	9381.24	12015.40	8.00	216.83	nr	12232.23
300-390 kW; 8 sections; 250mm dia. flue	11257.49	14418.48	12.00	325.24	nr	14743.72
390-450 kW; 9 sections; 250mm dia. flue	12398.87	15880.35	12.00	325.24	nr	16205.59
450-540 kW; 10 sections; 250mm dia. flue	13602.80	17422.33	12.00	325.24	nr	17747.57
540-600 kW; 11 sections; 300mm dia. flue	13915.51	17822.85	12.00	325.24	nr	18148.09
600-670 kW; 12 sections; 300mm dia. flue	17082.24	23108.27	12.00	325.24	nr	23433.51
670-720 kW; 13 sections; 300mm dia. flue	17394.95	23508.79	12.00	325.24	nr	23834.03
720-780 kW; 14 sections; 300mm dia. flue	17629.48	23809.17	12.00	325.24	nr	24134.41
754-812 kW; 14 sections; 400mm dia. flue	17785.83	24009.42	12.00	325.24	nr	24334.66
812-870 kW; 15 sections; 400mm dia. flue	22476.45	30017.12	12.00	325.24	nr	30342.36
870-928 kW; 16 sections; 400mm dia. flue	24352.70	32420.20	12.00	325.24	nr	32745.44
928-986 kW; 17 sections; 400mm dia. flue	24978.12	33221.23	12.00	325.24	nr	33546.48
986-1044 kW; 18 sections; 400mm dia. flue	26072.60	34623.03	12.00	325.24	nr	34948.27
1044-1102 kW; 19 sections; 400mm dia. flue	26698.01	35424.05	12.00	325.24	nr	35749.29
1102-1160 kW; 20 sections; 400mm dia. flue	27385.97	36305.18	12.00	325.24	nr	36630.43
1160-1218 kW; 21 sections; exceeding 400mm dia. flue	28886.97	38227.65	12.00	325.24	nr	38552.89

T: MECHANICAL/COOLING/HEATING SYSTEMS

Item	Net Price £	Material £	Labour hours	Labour £	Unit	Total rate £
1218-1276 kW; 22 sections; exceeding 400mm dia. flue	30367.80	40301.06	12.00	325.24	nr	**40626.30**
1276-1334 kW; 23 sections; exceeding 400mm dia. flue	31446.65	41682.84	12.00	325.24	nr	**42008.08**
1334-1392 kW; 24 sections; exceeding 400mm dia. flue	32853.83	43485.14	12.00	325.24	nr	**43810.38**
1392-1450 kW; 25 sections; exceeding 400mm dia. flue	33791.96	44686.69	12.00	325.24	nr	**45011.93**
Oil fired (on/off type), connected to conventional flue						
16-26 kW; 3 sections; 125mm dia. flue	1563.54	2002.57	8.00	216.83	nr	**2219.39**
26-33 kW; 4 sections; 125mm dia. flue	1719.89	2202.82	8.00	216.83	nr	**2419.65**
33-40 kW; 5 sections; 125mm dia. flue	1876.25	2403.08	8.00	216.83	nr	**2619.91**
35-50 kW; 3 sections; 153mm dia. flue	2204.59	2823.62	8.00	216.83	nr	**3040.44**
50-65 kW; 4 sections; 153mm dia. flue	2314.04	2963.80	8.00	216.83	nr	**3180.63**
65-80 kW; 5 sections; 153mm dia. flue	2407.86	3083.96	8.00	216.83	nr	**3300.79**
80-100 kW; 6 sections; 180mm dia. flue	3596.14	4605.90	8.00	216.83	nr	**4822.73**
100-120 kW; 7 sections; 180mm dia. flue	4065.20	5206.67	8.00	216.83	nr	**5423.50**
105-140 kW; 5 sections; 180mm dia. flue	4690.62	6007.70	8.00	216.83	nr	**6224.53**
140-180 kW; 6 sections; 180mm dia. flue	5316.04	6808.73	8.00	216.83	nr	**7025.56**
180-230 kW; 7 sections; 200mm dia. flue	6566.87	8410.78	8.00	216.83	nr	**8627.61**
230-280 kW; 8 sections; 200mm dia. flue	7192.28	9211.80	8.00	216.83	nr	**9428.63**
280-330 kW; 9 sections; 200mm dia. flue	7974.05	10213.08	8.00	216.83	nr	**10429.91**
Oil fired (high/low type), connected to conventional flue						
105-140 kW; 5 sections; 180mm dia. flue	3768.13	4826.18	8.00	216.83	nr	**5043.01**
140-180 kW; 6 sections; 180mm dia. flue	4065.20	5206.67	8.00	216.83	nr	**5423.50**
180-230 kW; 7 sections; 200mm dia. flue	4534.27	5807.45	8.00	216.83	nr	**6024.28**
230-280 kW; 8 sections; 200mm dia. flue	5347.31	6848.78	8.00	216.83	nr	**7065.61**
280-330 kW; 9 sections; 200mm dia. flue	8443.12	10813.86	8.00	216.83	nr	**11030.69**
300-390 kW; 8 sections; 250mm dia. flue	9381.24	12015.40	12.00	325.24	nr	**12340.64**
390-450 kW; 9 sections; 250mm dia. flue	10084.83	12916.55	12.00	325.24	nr	**13241.79**
450-540 kW; 10 sections; 250mm dia. flue	11413.84	14618.73	12.00	325.24	nr	**14943.97**
540-600 kW; 11 sections; 300mm dia. flue	11992.35	15359.68	12.00	325.24	nr	**15684.92**
600-670 kW; 12 sections; 300mm dia. flue	12821.03	16421.05	12.00	325.24	nr	**16746.29**
670-720 kW; 13 sections; 300mm dia. flue	14540.92	18623.86	12.00	325.24	nr	**18949.11**
720-780 kW; 14 sections; 300mm dia. flue	14775.45	18924.25	12.00	325.24	nr	**19249.49**
754-812 kW; 14 sections; 400mm dia. flue	14978.71	19184.58	12.00	325.24	nr	**19509.82**
812-870 kW; 15 sections; 400mm dia. flue	20013.31	25632.85	12.00	325.24	nr	**25958.09**
870-928 kW; 16 sections; 400mm dia. flue	20779.45	26614.11	12.00	325.24	nr	**26939.35**
928-986 kW; 17 sections; 400mm dia. flue	21498.67	27535.28	12.00	325.24	nr	**27860.52**
986-1044 kW; 18 sections; 400mm dia. flue	22358.62	28636.70	12.00	325.24	nr	**28961.94**
1044-1102 kW; 19 sections; 400mm dia. flue	23453.10	30038.50	12.00	325.24	nr	**30363.74**
1102-1160 kW; 20 sections; 400 mm dia. flue	24766.48	31720.66	12.00	325.24	nr	**32045.90**
1160-1218 kW; 21 sections; exceeding 400mm dia. flue	26142.39	33482.91	12.00	325.24	nr	**33808.15**
1218-1276 kW; 22 sections; exceeding 400mm dia. flue	27440.13	35145.04	12.00	325.24	nr	**35470.29**
1276-1334 kW; 23 sections; exceeding 400mm dia. flue	28409.52	36386.63	12.00	325.24	nr	**36711.87**
1334-1392 kW; 24 sections; exceeding 400mm dia. flue	30074.69	38519.36	12.00	325.24	nr	**38844.60**
1392-1450 kW; 25 sections; exceeding 400mm dia. flue	31176.99	39931.18	12.00	325.24	nr	**40256.42**

T: MECHANICAL/COOLING/HEATING SYSTEMS

Item	Net Price £	Material £	Labour hours	Labour £	Unit	Total rate £
T10 : GAS/OIL FIRED BOILERS (cont'd)						
FORCED DRAFT (cont'd)						
Commercial steel shell floor standing boilers; pressure jet burner; including controls, enamelled jacket, insulation, placing in position and commissioning; electrical work elsewhere						
Gas fired (on/off type), connected to conventional flue						
130-190kW	5605.97	7180.07	8.00	216.83	nr	7396.90
200-250kW	6349.79	8132.75	8.00	216.83	nr	8349.58
280-360kW	7605.18	9740.64	8.00	216.83	nr	9957.47
375-500kW	9268.91	11871.53	8.00	216.83	nr	12088.36
Gas fired (high/low type), connected to conventional flue						
130-190kW	6827.96	8745.18	8.00	216.83	nr	8962.01
200-250kW	7292.47	9340.12	8.00	216.83	nr	9556.95
280-360kW	8716.36	11163.83	8.00	216.83	nr	11380.65
375-500kW	9724.31	12454.80	8.00	216.83	nr	12671.63
580-730kW	12274.55	15721.12	10.00	271.04	nr	15992.16
655-820kW	12525.02	16041.92	10.00	271.04	nr	16312.96
830-1040kW	12569.04	16098.30	12.00	325.24	nr	16423.54
1070-1400kW	16623.62	21291.37	12.00	325.24	nr	21616.61
1420-1850kW	20700.97	26513.60	12.00	325.24	nr	26838.84
1850-2350kW	23512.30	30114.32	14.00	379.45	nr	30493.77
2300-3000kW	27742.97	35532.92	14.00	379.45	nr	35912.37
2800-3500kW	35979.64	46082.36	14.00	379.45	nr	46461.81
Oil fired (on/off type), connected to conventional flue						
130-190kW	5132.36	6573.48	8.00	216.83	nr	6790.30
200-250kW	5604.46	7178.14	8.00	216.83	nr	7394.96
Oil fired (high/low type), connected to conventional flue						
130-190kW	5586.24	7154.80	8.00	216.83	nr	7371.63
200-250kW	6058.34	7759.46	8.00	216.83	nr	7976.29
280-360kW	6984.32	8945.45	8.00	216.83	nr	9162.28
375-500kW	8350.52	10695.26	8.00	216.83	nr	10912.09
580-730kW	9288.64	11896.80	10.00	271.04	nr	12167.83
655-820kW	9537.59	12215.65	10.00	271.04	nr	12486.68
830-1040kW	11034.34	14132.67	12.00	325.24	nr	14457.91
1070-1400kW	13482.88	17268.74	12.00	325.24	nr	17593.98
1420-1850kW	16326.09	20910.29	12.00	325.24	nr	21235.53
1850-2350kW	20907.41	26778.00	14.00	379.45	nr	27157.45
2300-3000kW	24421.58	31278.92	14.00	379.45	nr	31658.36
2800-3500kW	32658.25	41828.36	14.00	379.45	nr	42207.81

T: MECHANICAL/COOLING/HEATING SYSTEMS

Item	Net Price £	Material £	Labour hours	Labour £	Unit	Total rate £
ATMOSPHERIC						
Commercial cast iron sectional floor standing boilers; atmospheric; including controls, enamelled jacket, insulation, assembly and commissioning; electrical work elsewhere						
Gas (on/off type), connected to conventional flue						
30-40kW	2003.14	2565.60	8.00	216.83	nr	2782.43
40-50kW	2196.13	2812.78	8.00	216.83	nr	3029.61
50-60kW	2333.76	2989.06	8.00	216.83	nr	3205.88
60-70kW	2547.69	3263.06	8.00	216.83	nr	3479.88
70-80kW	2700.28	3458.49	8.00	216.83	nr	3675.32
80-90kW	3155.06	4040.97	8.00	216.83	nr	4257.80
90-100kW	3282.22	4203.83	8.00	216.83	nr	4420.66
100-110kW	3478.20	4454.84	8.00	216.83	nr	4671.67
110-120kW	3678.66	4711.59	8.00	216.83	nr	4928.42
Gas (high/low type), connected to conventional flue						
30-40kW	2273.92	2912.41	8.00	216.83	nr	3129.24
40-50kW	2411.55	3088.69	8.00	216.83	nr	3305.52
50-60kW	2592.57	3320.54	8.00	216.83	nr	3537.37
60-70kW	2827.44	3621.36	8.00	216.83	nr	3838.18
70-80kW	2993.50	3834.04	8.00	216.83	nr	4050.87
80-90kW	3327.10	4261.32	8.00	216.83	nr	4478.14
90-100kW	3508.12	4493.16	8.00	216.83	nr	4709.99
100-110kW	3714.57	4757.58	8.00	216.83	nr	4974.41
110-120kW	3946.45	5054.57	8.00	216.83	nr	5271.40
120-140kW	5919.67	7581.85	8.00	216.83	nr	7798.68
140-160kW	6069.27	7773.46	8.00	216.83	nr	7990.29
160-180kW	6172.50	7905.68	8.00	216.83	nr	8122.50
180-200kW	6749.95	8645.27	8.00	216.83	nr	8862.10
200-220kW	7041.67	9018.90	10.00	271.04	nr	9289.94
220-260kW	7652.04	9800.66	10.00	271.04	nr	10071.69
260-300kW	8344.69	10687.80	10.00	271.04	nr	10958.83
300-340kW	8899.70	11398.65	12.00	325.24	nr	11723.89
CONDENSING						
Low Nox wall mounted condensing boiler with high efficiency modulating pre-mix burner: Aluminium heat exchanger: Placing in position						
Maximum output						
35kW	1672.00	2141.48	11.00	298.14	nr	2439.62
45kW	1728.10	2213.33	11.00	298.14	nr	2511.47
60kW	1952.50	2500.74	11.00	298.14	nr	2798.88
80kW	2349.60	3009.34	11.00	298.14	nr	3307.48

T: MECHANICAL/COOLING/HEATING SYSTEMS

Item	Net Price £	Material £	Labour hours	Labour £	Unit	Total rate £
T10 : GAS/OIL FIRED BOILERS (cont'd)						
CONDENSING (cont'd)						
Low Nox floor standing condensing boiler with high efficiency modulating pre-mix burner: Stainless steel heat exchanger: Including controls: Placing in position						
Maximum output						
50kW	2747.80	3519.35	8.00	216.83	nr	3736.18
60kW	2937.00	3761.68	8.00	216.83	nr	3978.51
80kW	3206.50	4106.85	8.00	216.83	nr	4323.68
100kW	3687.20	4722.53	8.00	216.83	nr	4939.36
125kW	4582.60	5869.35	10.00	271.04	nr	6140.38
150kW	5071.00	6494.89	10.00	271.04	nr	6765.92
200kW	7010.30	8978.72	10.00	271.04	nr	9249.76
250kW	7674.70	9829.68	10.00	271.04	nr	10100.71
300kW	9716.30	12444.54	10.00	271.04	nr	12715.57
350kW	10586.40	13558.96	10.00	271.04	nr	13829.99
400kW	11388.30	14586.02	10.00	271.04	nr	14857.06
450kW	11578.60	14829.76	10.00	271.04	nr	15100.79
500kW	11768.90	15073.49	10.00	271.04	nr	15344.52
650kW	12375.00	15849.78	10.00	271.04	nr	16120.81
700kW	19769.20	25320.19	10.00	271.04	nr	25591.23
800kW	21140.90	27077.05	12.00	325.24	nr	27402.30
900kW	21869.10	28009.72	12.00	325.24	nr	28334.97
1000kW	22599.50	28945.21	12.00	325.24	nr	29270.46
1300kW	23974.50	30706.30	12.00	325.24	nr	31031.54
FLUE SYSTEMS						
Flues; suitable for domestic, medium sized industrial and commercial oil and gas appliances; stainless steel, twin wall, insulated; for use internally or externally						
Straight length; 120mm long; including one locking band						
127mm dia.	39.92	51.13	0.49	13.28	nr	64.41
152mm dia.	44.74	57.30	0.51	13.82	nr	71.13
175mm dia.	51.93	66.51	0.54	14.64	nr	81.15
203mm dia.	59.09	75.68	0.58	15.72	nr	91.40
254mm dia.	69.88	89.50	0.70	18.97	nr	108.47
304mm dia.	87.54	112.12	0.74	20.06	nr	132.18
355mm dia.	125.36	160.56	0.80	21.68	nr	182.24
Straight length; 300mm long; including one locking band						
127mm dia.	61.59	78.88	0.52	14.09	nr	92.98
152mm dia.	69.63	89.18	0.52	14.09	nr	103.28
178mm dia.	80.00	102.46	0.55	14.91	nr	117.37
203mm dia.	90.54	115.96	0.64	17.35	nr	133.31
254mm dia.	100.41	128.60	0.79	21.41	nr	150.02
304mm dia.	120.46	154.28	0.86	23.31	nr	177.59
355mm dia.	132.18	169.29	0.94	25.48	nr	194.77
400mm dia.	141.45	181.17	1.03	27.92	nr	209.08
450mm dia.	161.79	207.22	1.03	27.92	nr	235.14

T: MECHANICAL/COOLING/HEATING SYSTEMS

Item	Net Price £	Material £	Labour hours	Labour £	Unit	Total rate £
500mm dia.	173.53	222.26	1.10	29.81	nr	252.07
550mm dia.	191.50	245.27	1.10	29.81	nr	275.09
600mm dia.	211.31	270.64	1.10	29.81	nr	300.46
Straight length; 500mm long; including one locking band						
127mm dia.	72.39	92.72	0.55	14.91	nr	107.62
152mm dia.	80.70	103.36	0.55	14.91	nr	118.27
178mm dia.	90.89	116.41	0.63	17.08	nr	133.49
203mm dia.	106.24	136.07	0.63	17.08	nr	153.15
254mm dia.	123.39	158.04	0.86	23.31	nr	181.35
304mm dia.	147.83	189.34	0.95	25.75	nr	215.09
355mm dia.	166.01	212.62	1.03	27.92	nr	240.54
400mm dia.	181.28	232.18	1.12	30.36	nr	262.54
450mm dia.	209.35	268.13	1.12	30.36	nr	298.49
500mm dia.	225.58	288.92	1.19	32.25	nr	321.17
550mm dia.	248.66	318.48	1.19	32.25	nr	350.73
600mm dia.	257.58	329.91	1.19	32.25	nr	362.16
Straight length; 1000mm long; including one locking band						
127mm dia.	129.40	165.73	0.62	16.80	nr	182.54
152mm dia.	144.22	184.72	0.68	18.43	nr	203.15
178mm dia.	162.27	207.83	0.74	20.06	nr	227.89
203mm dia.	191.13	244.80	0.80	21.68	nr	266.48
254mm dia.	216.61	277.43	0.87	23.58	nr	301.01
304mm dia.	250.03	320.24	1.06	28.73	nr	348.97
355mm dia.	286.91	367.47	1.16	31.44	nr	398.91
400mm dia.	307.26	393.54	1.26	34.15	nr	427.69
450mm dia.	324.21	415.24	1.26	34.15	nr	449.40
500mm dia.	351.84	450.63	1.33	36.05	nr	486.68
550mm dia.	386.95	495.60	1.33	36.05	nr	531.65
600mm dia.	406.51	520.65	1.33	36.05	nr	556.70
Adjustable length; boiler removal; internal use only; including one locking band						
127mm dia.	59.31	75.96	0.52	14.09	nr	90.06
152mm dia.	67.03	85.85	0.55	14.91	nr	100.76
178mm dia.	75.96	97.29	0.59	15.99	nr	113.28
203mm dia.	87.11	111.57	0.64	17.35	nr	128.92
254mm dia.	129.16	165.43	0.79	21.41	nr	186.84
304mm dia.	154.70	198.14	0.86	23.31	nr	221.45
355mm dia.	173.49	222.20	0.99	26.83	nr	249.04
400mm dia.	303.58	388.82	0.91	24.66	nr	413.49
450mm dia.	324.96	416.21	0.91	24.66	nr	440.87
500mm dia.	354.42	453.94	0.99	26.83	nr	480.77
550mm dia.	386.65	495.22	0.99	26.83	nr	522.05
600mm dia.	404.76	518.41	0.99	26.83	nr	545.25
Inspection length; 500mm long; including one locking band						
127mm dia.	152.46	195.27	0.55	14.91	nr	210.18
152mm dia.	157.87	202.20	0.55	14.91	nr	217.11
178mm dia.	166.28	212.97	0.63	17.08	nr	230.04
203mm dia.	176.08	225.52	0.63	17.08	nr	242.60
254mm dia.	227.66	291.58	0.86	23.31	nr	314.89
304mm dia.	246.73	316.01	0.95	25.75	nr	341.76
355mm dia.	278.59	356.82	1.03	27.92	nr	384.73

T: MECHANICAL/COOLING/HEATING SYSTEMS

Item	Net Price £	Material £	Labour hours	Labour £	Unit	Total rate £
T10 : GAS/OIL FIRED BOILERS (cont'd)						
FLUE SYSTEMS (cont'd)						
Flues; suitable for domestic, medium (cont'd)						
Inspection length; 500mm long (cont'd)						
400mm dia.	446.96	572.46	1.12	30.36	nr	602.82
450mm dia.	459.12	588.04	1.12	30.36	nr	618.39
500mm dia.	503.79	645.25	1.19	32.25	nr	677.50
550mm dia.	526.81	674.73	1.19	32.25	nr	706.99
600mm dia.	535.76	686.20	1.19	32.25	nr	718.45
Adapters						
127mm dia.	12.75	16.33	0.49	13.28	nr	29.61
152mm dia.	13.95	17.87	0.51	13.82	nr	31.69
178mm dia.	14.92	19.11	0.54	14.64	nr	33.75
203mm dia.	16.98	21.75	0.58	15.72	nr	37.47
254mm dia.	18.69	23.94	0.70	18.97	nr	42.91
304mm dia.	23.31	29.86	0.74	20.06	nr	49.91
355mm dia.	28.31	36.26	0.80	21.68	nr	57.94
400mm dia.	31.76	40.68	0.89	24.12	nr	64.80
450mm dia.	33.92	43.44	0.89	24.12	nr	67.57
500mm dia.	36.00	46.11	0.96	26.02	nr	72.13
550mm dia.	41.87	53.63	0.96	26.02	nr	79.65
600mm dia.	50.30	64.42	0.96	26.02	nr	90.44
Fittings for flue system						
90° insulated tee; including two locking bands						
127mm dia.	148.23	189.85	1.89	51.23	nr	241.08
152mm dia.	171.08	219.12	2.04	55.29	nr	274.41
178mm dia.	186.98	239.48	2.39	64.78	nr	304.26
203mm dia.	218.93	280.40	2.56	69.39	nr	349.79
254mm dia.	221.47	283.66	2.95	79.96	nr	363.61
304mm dia.	275.81	353.25	3.41	92.42	nr	445.68
355mm dia.	351.63	450.36	3.77	102.18	nr	552.54
400mm dia.	463.34	593.44	4.25	115.19	nr	708.63
450mm dia.	482.93	618.53	4.76	129.01	nr	747.54
500mm dia.	546.48	699.93	5.12	138.77	nr	838.70
550mm dia.	587.42	752.36	5.61	152.05	nr	904.41
600mm dia.	611.22	782.84	5.98	162.08	nr	944.92
135° insulated tee; including two locking bands						
127mm dia.	192.25	246.23	1.89	51.23	nr	297.46
152mm dia.	207.87	266.24	2.04	55.29	nr	321.53
178mm dia.	227.27	291.09	2.39	64.78	nr	355.86
203mm dia.	288.43	369.42	2.56	69.39	nr	438.80
254mm dia.	327.73	419.75	2.95	79.96	nr	499.71
304mm dia.	246.93	316.27	3.41	92.42	nr	408.69
355mm dia.	485.88	622.31	3.77	102.18	nr	724.49
400mm dia.	639.60	819.19	4.25	115.19	nr	934.38
450mm dia.	692.40	886.82	4.76	129.01	nr	1015.83
500mm dia.	806.64	1033.14	5.12	138.77	nr	1171.91
550mm dia.	827.61	1059.99	5.61	152.05	nr	1212.05
600mm dia.	873.46	1118.72	5.98	162.08	nr	1280.80

T: MECHANICAL/COOLING/HEATING SYSTEMS

Item	Net Price £	Material £	Labour hours	Labour £	Unit	Total rate £
Wall sleeve; for 135° tee through wall						
127mm dia.	19.68	25.21	1.89	51.23	nr	76.43
152mm dia.	26.40	33.81	2.04	55.29	nr	89.10
178mm dia.	27.65	35.41	2.39	64.78	nr	100.19
203mm dia.	31.10	39.83	2.56	69.39	nr	109.22
254mm dia.	34.69	44.43	2.95	79.96	nr	124.39
304mm dia.	40.43	51.78	3.41	92.42	nr	144.21
355mm dia.	44.91	57.52	3.77	102.18	nr	159.70
15° insulated elbow; including two locking bands						
127mm dia.	102.87	131.75	1.57	42.55	nr	174.31
152mm dia.	114.32	146.42	1.79	48.52	nr	194.94
178mm dia.	122.27	156.60	2.05	55.56	nr	212.16
203mm dia.	129.62	166.02	2.33	63.15	nr	229.17
254mm dia.	133.50	170.99	2.45	66.40	nr	237.39
304mm dia.	168.78	216.17	3.43	92.97	nr	309.14
355mm dia.	225.80	289.20	4.71	127.66	nr	416.86
30° insulated elbow; including two locking bands						
127mm dia.	102.87	131.75	1.44	39.03	nr	170.78
152mm dia.	114.32	146.42	1.62	43.91	nr	190.33
178mm dia.	122.27	156.60	1.89	51.23	nr	207.83
203mm dia.	129.62	166.02	2.17	58.81	nr	224.83
254mm dia.	133.50	170.99	2.16	58.54	nr	229.53
304mm dia.	168.78	216.17	2.74	74.26	nr	290.44
355mm dia.	225.37	288.65	3.17	85.92	nr	374.57
400mm dia.	225.80	289.20	3.53	95.68	nr	384.88
450mm dia.	261.69	335.17	3.88	105.16	nr	440.33
500mm dia.	274.35	351.38	4.24	114.92	nr	466.30
550mm dia.	294.44	377.12	4.61	124.95	nr	502.06
600mm dia.	321.06	411.21	4.96	134.43	nr	545.64
45° insulated elbow; including two locking bands						
127mm dia.	102.87	131.75	1.44	39.03	nr	170.78
152mm dia.	114.32	146.42	1.51	40.93	nr	187.35
178mm dia.	122.27	156.60	1.58	42.82	nr	199.43
203mm dia.	129.62	166.02	1.66	44.99	nr	211.01
254mm dia.	133.50	170.99	1.72	46.62	nr	217.60
304mm dia.	168.78	216.17	1.80	48.79	nr	264.96
355mm dia.	225.80	289.20	1.94	52.58	nr	341.78
400mm dia.	287.94	368.79	2.01	54.48	nr	423.27
450mm dia.	301.61	386.30	2.09	56.65	nr	442.95
500mm dia.	323.52	414.36	2.16	58.54	nr	472.90
550mm dia.	352.85	451.93	2.23	60.44	nr	512.37
600mm dia.	361.79	463.38	2.30	62.34	nr	525.72
Flue supports						
Wall support, galvanised; including plate and brackets						
127mm dia.	62.45	79.99	2.24	60.71	nr	140.70
152mm dia.	68.66	87.94	2.44	66.13	nr	154.07
178mm dia.	75.80	97.08	2.52	68.30	nr	165.38
203mm dia.	78.93	101.09	2.77	75.08	nr	176.17
254mm dia.	94.13	120.56	2.98	80.77	nr	201.33
304mm dia.	106.43	136.31	3.46	93.78	nr	230.09
355mm dia.	142.46	182.46	4.08	110.58	nr	293.04

T: MECHANICAL/COOLING/HEATING SYSTEMS

Item	Net Price £	Material £	Labour hours	Labour £	Unit	Total rate £
T10 : GAS/OIL FIRED BOILERS (cont'd)						
FLUE SYSTEMS (cont'd)						
Flue supports (cont'd)						
Wall support, galvanised; including plate (cont'd)						
400mm dia.; with 300mm support length and collar	365.93	468.68	4.80	130.10	nr	**598.78**
450mm dia.; with 300mm support length and collar	390.14	499.69	5.62	152.32	nr	**652.01**
500mm dia.; with 300mm support length and collar	426.22	545.90	6.24	169.13	nr	**715.02**
550mm dia.; with 300mm support length and collar	466.31	597.25	6.97	188.91	nr	**786.16**
600mm dia.; with 300mm support length and collar	492.27	630.49	7.49	203.01	nr	**833.50**
Ceiling/floor support						
127mm dia.	19.33	24.76	1.86	50.41	nr	**75.17**
152mm dia.	21.49	27.52	2.14	58.00	nr	**85.53**
178mm dia.	25.73	32.95	1.93	52.31	nr	**85.26**
203mm dia.	38.74	49.62	2.74	74.26	nr	**123.88**
254mm dia.	44.20	56.61	3.21	87.00	nr	**143.61**
304mm dia.	50.51	64.69	3.68	99.74	nr	**164.43**
355mm dia.	60.20	77.10	4.28	116.00	nr	**193.11**
400mm dia.	305.47	391.24	4.86	131.72	nr	**522.97**
450mm dia.	321.60	411.90	5.46	147.99	nr	**559.89**
500mm dia.	339.48	434.80	6.04	163.71	nr	**598.51**
550mm dia.	371.65	476.01	6.65	180.24	nr	**656.24**
600mm dia.	377.99	484.13	7.24	196.23	nr	**680.36**
Ceiling/floor firestop spacer						
127mm dia.	3.76	4.82	0.66	17.89	nr	**22.70**
152mm dia.	4.19	5.37	0.69	18.70	nr	**24.07**
178mm dia.	4.73	6.06	0.70	18.97	nr	**25.03**
203mm dia.	5.61	7.19	0.87	23.58	nr	**30.77**
254mm dia.	5.79	7.42	0.91	24.66	nr	**32.08**
304mm dia.	6.96	8.91	0.95	25.75	nr	**34.66**
355mm dia.	12.68	16.24	0.99	26.83	nr	**43.07**
Wall band; internal or external use						
127mm dia.	22.77	29.16	1.03	27.92	nr	**57.08**
152mm dia.	23.77	30.44	1.07	29.00	nr	**59.45**
178mm dia.	24.71	31.65	1.11	30.08	nr	**61.73**
203mm dia.	26.09	33.42	1.18	31.98	nr	**65.40**
254mm dia.	27.15	34.77	1.30	35.23	nr	**70.01**
304mm dia.	29.33	37.57	1.45	39.30	nr	**76.87**
355mm dia.	31.22	39.99	1.65	44.72	nr	**84.71**
400mm dia.	38.71	49.58	1.85	50.14	nr	**99.72**
450mm dia.	41.25	52.83	2.39	64.78	nr	**117.61**
500mm dia.	49.58	63.50	2.25	60.98	nr	**124.48**
550mm dia.	51.93	66.51	2.45	66.40	nr	**132.91**
600mm dia.	54.90	70.32	2.66	72.10	nr	**142.41**

T: MECHANICAL/COOLING/HEATING SYSTEMS

Item	Net Price £	Material £	Labour hours	Labour £	Unit	Total rate £
Flashings and terminals						
Insulated top stub; including one locking band						
127mm dia.	56.09	71.84	1.49	40.38	nr	112.22
152mm dia.	63.42	81.23	1.90	51.50	nr	132.72
178mm dia.	68.28	87.45	1.92	52.04	nr	139.49
203mm dia.	72.66	93.06	2.20	59.63	nr	152.69
254mm dia.	77.14	98.80	2.49	67.49	nr	166.29
304mm dia.	105.46	135.07	2.79	75.62	nr	210.69
355mm dia.	139.22	178.31	3.19	86.46	nr	264.77
400mm dia.	134.23	171.92	3.59	97.30	nr	269.22
450mm dia.	145.83	186.78	3.97	107.60	nr	294.38
500mm dia.	168.70	216.07	4.38	118.71	nr	334.78
550mm dia.	176.52	226.09	4.78	129.55	nr	355.64
600mm dia.	182.69	233.99	5.17	140.13	nr	374.11
Rain cap; including one locking band						
127mm dia.	29.98	38.40	1.49	40.38	nr	78.78
152mm dia.	31.33	40.13	1.54	41.74	nr	81.87
178mm dia.	34.50	44.19	1.72	46.62	nr	90.81
203mm dia.	41.26	52.85	2.00	54.21	nr	107.05
254mm dia.	54.25	69.48	2.49	67.49	nr	136.97
304mm dia.	37.14	47.57	2.80	75.89	nr	123.46
355mm dia.	97.99	125.50	3.19	86.46	nr	211.96
400mm dia.	97.76	125.21	3.45	93.51	nr	218.72
450mm dia.	106.32	136.17	3.97	107.60	nr	243.77
500mm dia.	114.96	147.24	4.38	118.71	nr	265.95
550mm dia.	123.35	157.99	4.78	129.55	nr	287.54
600mm dia.	131.88	168.91	5.17	140.13	nr	309.04
Round top; including one locking band						
127mm dia	57.32	73.41	1.49	40.38	nr	113.80
152mm dia	62.50	80.05	1.65	44.72	nr	124.77
178mm dia	71.26	91.27	1.92	52.04	nr	143.31
203mm dia	83.86	107.41	2.20	59.63	nr	167.03
254mm dia	99.14	126.98	2.49	67.49	nr	194.47
304mm dia	130.22	166.78	2.80	75.89	nr	242.67
355mm dia	173.72	222.50	3.19	86.46	nr	308.96
Coping cap; including one locking band						
127mm dia.	32.10	41.11	1.49	40.38	nr	81.50
152mm dia.	33.63	43.07	1.65	44.72	nr	87.79
178mm dia.	37.00	47.39	1.92	52.04	nr	99.43
203mm dia.	44.40	56.87	2.20	59.63	nr	116.49
254mm dia.	54.25	69.48	2.49	67.49	nr	136.97
304mm dia.	73.14	93.68	2.79	75.62	nr	169.30
355mm dia.	97.99	125.50	3.19	86.46	nr	211.96
Storm collar						
127mm dia.	6.09	7.80	0.52	14.09	nr	21.89
152mm dia.	6.51	8.34	0.55	14.91	nr	23.24
178mm dia.	7.22	9.25	0.57	15.45	nr	24.70
203mm dia.	7.56	9.68	0.66	17.89	nr	27.57
254mm dia.	9.47	12.13	0.66	17.89	nr	30.02
304mm dia.	9.87	12.64	0.72	19.51	nr	32.16
355mm dia.	10.54	13.50	0.77	20.87	nr	34.37
400mm dia.	27.96	35.81	0.82	22.22	nr	58.04
450mm dia.	30.78	39.42	0.87	23.58	nr	63.00

T: MECHANICAL/COOLING/HEATING SYSTEMS

Item	Net Price £	Material £	Labour hours	Labour £	Unit	Total rate £
T10 : GAS/OIL FIRED BOILERS (cont'd)						
FLUE SYSTEMS (cont'd)						
Flashings and terminals (cont'd)						
Storm collar (cont'd)						
500mm dia.	33.56	42.98	0.92	24.94	nr	67.92
550mm dia.	36.37	46.58	0.98	26.56	nr	73.14
600mm dia.	39.16	50.16	1.03	27.92	nr	78.07
Flat flashing; including storm collar and sealant						
127mm dia.	34.18	43.78	1.49	40.38	nr	84.16
152mm dia.	35.32	45.24	1.65	44.72	nr	89.96
178mm dia.	37.06	47.47	1.92	52.04	nr	99.50
203mm dia.	40.58	51.97	2.20	59.63	nr	111.60
254mm dia.	55.46	71.03	2.49	67.49	nr	138.52
304mm dia.	66.26	84.87	2.80	75.89	nr	160.75
355mm dia.	104.42	133.74	3.20	86.73	nr	220.47
400mm dia.	145.67	186.57	3.59	97.30	nr	283.87
450mm dia.	167.34	214.33	3.97	107.60	nr	321.93
500mm dia.	181.29	232.19	4.38	118.71	nr	350.91
550mm dia.	192.49	246.54	4.78	129.55	nr	376.09
600mm dia.	199.44	255.44	5.17	140.13	nr	395.57
5°-30° rigid adjustable flashing; including storm collar and sealant						
127mm dia.	58.21	74.55	1.49	40.38	nr	114.94
152mm dia.	61.34	78.56	1.65	44.72	nr	123.28
178mm dia.	65.23	83.55	1.92	52.04	nr	135.58
203mm dia.	68.59	87.85	2.20	59.63	nr	147.48
254mm dia.	72.20	92.47	2.49	67.49	nr	159.96
304mm dia.	89.35	114.44	2.80	75.89	nr	190.33
355mm dia.	101.62	130.15	3.19	86.46	nr	216.61
400mm dia.	327.62	419.61	3.59	97.30	nr	516.91
450mm dia.	382.34	489.70	3.97	107.60	nr	597.30
500mm dia.	408.18	522.79	4.38	118.71	nr	641.51
550mm dia.	427.13	547.06	4.77	129.28	nr	676.35
600mm dia.	464.72	595.21	5.17	140.13	nr	735.33
Domestic and small commercial; twin walled gas vent system suitable for gas fired appliances; domestic gas boilers; small commercial boilers with internal or external flues						
152mm long						
100mm dia.	6.23	7.98	0.52	14.09	nr	22.07
125mm dia.	7.66	9.81	0.52	14.09	nr	23.90
150mm dia.	8.29	10.62	0.52	14.09	nr	24.71
305mm long						
100mm dia.	9.45	12.10	0.52	14.09	nr	26.20
125mm dia.	11.09	14.20	0.52	14.09	nr	28.30
150mm dia.	13.15	16.84	0.52	14.09	nr	30.94
457mm long						
100mm dia.	10.46	13.40	0.55	14.91	nr	28.30
125mm dia.	11.75	15.05	0.55	14.91	nr	29.96
150mm dia.	14.54	18.62	0.55	14.91	nr	33.53

T: MECHANICAL/COOLING/HEATING SYSTEMS

Item	Net Price £	Material £	Labour hours	Labour £	Unit	Total rate £
914mm long						
100mm dia.	18.68	23.93	0.62	16.80	nr	**40.73**
125mm dia.	21.77	27.88	0.62	16.80	nr	**44.69**
150mm dia.	24.96	31.97	0.62	16.80	nr	**48.77**
1524mm long						
100mm dia.	26.98	34.56	0.82	22.22	nr	**56.78**
125mm dia.	33.20	42.52	0.84	22.77	nr	**65.29**
150mm dia.	35.65	45.66	0.84	22.77	nr	**68.43**
Adjustable length 305mm long						
100mm dia.	11.96	15.32	0.56	15.18	nr	**30.50**
125mm dia.	13.43	17.20	0.56	15.18	nr	**32.38**
150mm dia.	16.93	21.68	0.56	15.18	nr	**36.86**
Adjustable length 457mm long						
100mm dia.	16.12	20.65	0.56	15.18	nr	**35.82**
125mm dia.	19.55	25.04	0.56	15.18	nr	**40.22**
150mm dia.	21.75	27.86	0.56	15.18	nr	**43.04**
Adjustable elbow 0°-90°						
100mm dia.	13.64	17.47	0.48	13.01	nr	**30.48**
125mm dia.	16.12	20.65	0.48	13.01	nr	**33.66**
150mm dia.	20.20	25.87	0.48	13.01	nr	**38.88**
Draughthood connector						
100mm dia.	4.21	5.39	0.48	13.01	nr	**18.40**
125mm dia.	4.74	6.07	0.48	13.01	nr	**19.08**
150mm dia.	5.15	6.60	0.48	13.01	nr	**19.61**
Adaptor						
100mm dia.	10.21	13.08	0.48	13.01	nr	**26.09**
125mm dia.	10.43	13.36	0.48	13.01	nr	**26.37**
150mm dia.	10.65	13.64	0.48	13.01	nr	**26.65**
Support plate						
100mm dia.	7.37	9.44	0.48	13.01	nr	**22.45**
125mm dia.	7.84	10.04	0.48	13.01	nr	**23.05**
150mm dia.	8.39	10.75	0.48	13.01	nr	**23.76**
Wall band						
100mm dia.	6.66	8.53	0.48	13.01	nr	**21.54**
125mm dia.	7.10	9.09	0.48	13.01	nr	**22.10**
150mm dia.	9.00	11.53	0.48	13.01	nr	**24.54**
Firestop						
100mm dia.	2.89	3.70	0.48	13.01	nr	**16.71**
125mm dia.	2.89	3.70	0.48	13.01	nr	**16.71**
150mm dia.	3.31	4.24	0.48	13.01	nr	**17.25**
Flat flashing						
125mm dia.	19.64	25.15	0.55	14.91	nr	**40.06**
150mm dia.	27.31	34.98	0.55	14.91	nr	**49.89**
Adjustable flashing 5°-30°						
100mm dia.	51.17	65.54	0.55	14.91	nr	**80.44**
125mm dia.	79.89	102.32	0.55	14.91	nr	**117.23**

T: MECHANICAL/COOLING/HEATING SYSTEMS

Item	Net Price £	Material £	Labour hours	Labour £	Unit	Total rate £
T10 : GAS/OIL FIRED BOILERS (cont'd)						
FLUE SYSTEMS (cont'd)						
Domestic and small commercial (cont'd)						
Storm collar						
100mm dia.	4.08	5.23	0.55	14.91	nr	**20.13**
125mm dia.	4.17	5.34	0.55	14.91	nr	**20.25**
150mm dia.	4.28	5.48	0.55	14.91	nr	**20.39**
Gas vent terminal						
100mm dia.	15.57	19.94	0.55	14.91	nr	**34.85**
125mm dia.	17.11	21.91	0.55	14.91	nr	**36.82**
150mm dia.	21.95	28.11	0.55	14.91	nr	**43.02**
Twin wall galvanised steel flue box, 125mm dia.; fitted for gas fire, where no chimney exists						
Free standing	98.95	126.73	2.15	58.27	nr	**185.01**
Recess	98.95	126.73	2.15	58.27	nr	**185.01**
Back boiler	72.90	93.37	2.40	65.05	nr	**158.42**

T: MECHANICAL/COOLING/HEATING SYSTEMS

Item	Net Price £	Material £	Labour hours	Labour £	Unit	Total rate £
T13 : PACKAGED STEAM GENERATORS						
Packaged steam boilers; boiler mountings centrifugal water feed pump; insulation; and sheet steel wrap around casing; plastic coated						
Gas fired						
293 kW rating	17554.39	22483.49	86.45	2343.10	nr	**24826.59**
1465 kW rating	37443.07	47956.72	148.22	4017.28	nr	**51974.00**
2930 kW rating	53769.61	68867.57	207.50	5623.98	nr	**74491.55**
Oil fired						
293 kW rating	15921.74	20392.41	86.45	2343.10	nr	**22735.50**
1465 kW rating	34636.53	44362.12	148.22	4017.28	nr	**48379.40**
2930 kW rating	52420.31	67139.40	207.50	5623.98	nr	**72763.38**

T: MECHANICAL/COOLING/HEATING SYSTEMS

Item	Net Price £	Material £	Labour hours	Labour £	Unit	Total rate £
T31 : LOW TEMPERATURE HOT WATER HEATING						
Y10 - PIPELINES						
SCREWED STEEL						
Black steel pipes; screwed and socketed joints; BS 1387: 1985. Fixed vertically, brackets measured separately. Screwed joints are within the running length, but any flanges are additional						
Medium weight						
10mm dia	1.85	2.37	0.37	10.03	m	12.40
15mm dia	1.99	2.55	0.37	10.03	m	12.58
20mm dia	2.28	2.92	0.37	10.03	m	12.95
25mm dia	3.27	4.19	0.41	11.11	m	15.30
32mm dia	4.09	5.24	0.48	13.01	m	18.25
40mm dia	4.75	6.08	0.52	14.09	m	20.18
50mm dia	6.73	8.62	0.62	16.80	m	25.42
65mm dia	9.43	12.08	0.65	17.62	m	29.70
80mm dia	12.25	15.69	1.10	29.81	m	45.50
100mm dia	17.93	22.96	1.31	35.51	m	58.47
125mm dia	28.91	37.03	1.66	44.99	m	82.02
150mm dia	34.98	44.80	1.88	50.95	m	95.76
Heavy weight						
15mm dia	2.44	3.13	0.37	10.03	m	13.15
20mm dia	2.79	3.57	0.37	10.03	m	13.60
25mm dia	4.06	5.20	0.41	11.11	m	16.31
32mm dia	5.09	6.52	0.48	13.01	m	19.53
40mm dia	5.93	7.60	0.52	14.09	m	21.69
50mm dia	8.28	10.60	0.62	16.80	m	27.41
65mm dia	11.56	14.81	0.64	17.35	m	32.15
80mm dia	14.74	18.88	1.10	29.81	m	48.69
100mm dia	21.17	27.11	1.31	35.51	m	62.62
125mm dia	30.48	39.04	1.66	44.99	m	84.03
150mm dia	37.01	47.40	1.88	50.95	m	98.36
200mm dia	49.08	62.86	2.99	81.04	m	143.90
250mm dia	63.38	81.18	3.49	94.59	m	175.77
300mm dia	104.78	134.20	3.91	105.97	m	240.18
Black steel pipes; screwed and socketed joints; BS 1387: 1985. Fixed at high level or suspended, brackets measured separately. Screwed joints are within the running length, but any flanges are additional						
Medium weight						
10mm dia	1.85	2.37	0.58	15.72	m	18.09
15mm dia	1.99	2.55	0.58	15.72	m	18.27
20mm dia	2.28	2.92	0.58	15.72	m	18.64
25mm dia	3.27	4.19	0.60	16.26	m	20.45
32mm dia	4.09	5.24	0.68	18.43	m	23.67
40mm dia	4.75	6.08	0.73	19.79	m	25.87
50mm dia	6.73	8.62	0.85	23.04	m	31.66
65mm dia	9.43	12.08	0.88	23.85	m	35.93

T: MECHANICAL/COOLING/HEATING SYSTEMS

Item	Net Price £	Material £	Labour hours	Labour £	Unit	Total rate £
80mm dia	12.25	15.69	1.45	39.30	m	54.99
100mm dia	17.93	22.96	1.74	47.16	m	70.12
125mm dia	28.91	37.03	2.21	59.90	m	96.93
150mm dia	34.98	44.80	2.50	67.76	m	112.56
Heavy weight						
15mm dia	2.44	3.13	0.58	15.72	m	18.85
20mm dia	2.79	3.57	0.58	15.72	m	19.29
25mm dia	4.06	5.20	0.60	16.26	m	21.46
32mm dia	5.09	6.52	0.68	18.43	m	24.95
40mm dia	5.93	7.60	0.73	19.79	m	27.38
50mm dia	8.28	10.60	0.85	23.04	m	33.64
65mm dia	11.56	14.81	0.88	23.85	m	38.66
80mm dia	14.74	18.88	1.45	39.30	m	58.18
100mm dia	21.17	27.11	1.74	47.16	m	74.27
125mm dia	30.48	39.04	2.21	59.90	m	98.94
150mm dia	37.01	47.40	2.50	67.76	m	115.16
200mm dia	49.08	62.86	2.99	81.04	m	143.90
250mm dia	63.38	81.18	3.49	94.59	m	175.77
300mm dia	104.78	134.20	3.91	105.97	m	240.18
FIXINGS						
For steel pipes; black malleable iron. For minimum fixing distances, refer to the Tables and Memoranda to the rear of the book						
Single pipe bracket, screw on, black malleable iron; screwed to wood						
15mm dia	0.38	0.49	0.14	3.79	nr	4.28
20mm dia	0.43	0.55	0.14	3.79	nr	4.35
25mm dia	0.49	0.63	0.17	4.61	nr	5.24
32mm dia	0.68	0.87	0.19	5.15	nr	6.02
40mm dia	0.90	1.15	0.22	5.96	nr	7.12
50mm dia	1.20	1.54	0.22	5.96	nr	7.50
65mm dia	1.58	2.02	0.28	7.59	nr	9.61
80mm dia	2.16	2.77	0.32	8.67	nr	11.44
100mm dia	3.15	4.03	0.35	9.49	nr	13.52
Single pipe bracket, screw on, black malleable iron; plugged and screwed						
15mm dia	0.38	0.49	0.25	6.78	nr	7.26
20mm dia	0.43	0.55	0.25	6.78	nr	7.33
25mm dia	0.49	0.63	0.30	8.13	nr	8.76
32mm dia	0.68	0.87	0.32	8.67	nr	9.54
40mm dia	0.90	1.15	0.32	8.67	nr	9.83
50mm dia	1.20	1.54	0.32	8.67	nr	10.21
65mm dia	1.58	2.02	0.35	9.49	nr	11.51
80mm dia	2.16	2.77	0.42	11.38	nr	14.15
100mm dia	3.15	4.03	0.42	11.38	nr	15.42
Single pipe bracket for building in, black malleable iron						
15mm dia	1.12	1.43	0.10	2.71	nr	4.14
20mm dia	1.12	1.43	0.11	2.98	nr	4.42
25mm dia	1.12	1.43	0.12	3.25	nr	4.69
32mm dia	1.27	1.63	0.14	3.79	nr	5.42
40mm dia	1.28	1.64	0.15	4.07	nr	5.70
50mm dia	1.33	1.70	0.16	4.34	nr	6.04

T: MECHANICAL/COOLING/HEATING SYSTEMS

Item	Net Price £	Material £	Labour hours	Labour £	Unit	Total rate £
T31 : LOW TEMPERATURE HOT WATER HEATING (cont'd)						
SCREWED STEEL (cont'd)						
Fixings; steel pipes; black malleable iron (cont'd)						
Pipe ring, single socket, black malleable iron						
15mm dia	0.40	0.51	0.10	2.71	nr	3.22
20mm dia	0.45	0.58	0.11	2.98	nr	3.56
25mm dia	0.49	0.63	0.12	3.25	nr	3.88
32mm dia	0.52	0.67	0.14	3.79	nr	4.46
40mm dia	0.68	0.87	0.15	4.07	nr	4.94
50mm dia	0.85	1.09	0.16	4.34	nr	5.43
65mm dia	1.24	1.59	0.30	8.13	nr	9.72
80mm dia	1.48	1.90	0.35	9.49	nr	11.38
100mm dia	2.25	2.88	0.40	10.84	nr	13.72
125mm dia	4.54	5.81	0.60	16.27	nr	22.08
150mm dia	5.09	6.52	0.77	20.87	nr	27.39
200mm dia	14.16	18.14	0.90	24.39	nr	42.53
250mm dia	17.70	22.67	1.10	29.81	nr	52.48
300mm dia	21.24	27.20	1.25	33.88	nr	61.08
350mm dia	24.79	31.75	1.50	40.66	nr	72.41
400mm dia	28.33	36.28	1.75	47.43	nr	83.72
Pipe ring, double socket, black malleable iron						
15mm dia	0.47	0.60	0.10	2.71	nr	3.31
20mm dia	0.54	0.69	0.11	2.98	nr	3.67
25mm dia	0.61	0.78	0.12	3.25	nr	4.03
32mm dia	0.72	0.92	0.14	3.79	nr	4.72
40mm dia	0.84	1.08	0.15	4.07	nr	5.14
50mm dia	0.95	1.22	0.16	4.34	nr	5.55
Screw on backplate (Male), black malleable iron; plugged and screwed						
M12	0.32	0.41	0.10	2.71	nr	3.12
Screw on backplate (Female), black malleable iron; plugged and screwed						
M12	0.32	0.41	0.10	2.71	nr	3.12
Extra Over channel sections for fabricated hangers and brackets						
Galvanised steel; including inserts, bolts, nuts, washers; fixed to backgrounds						
41 x 21mm	2.50	3.20	0.29	7.86	m	11.06
41 x 41mm	3.90	5.00	0.29	7.86	m	12.86
Threaded rods; metric thread; including nuts, washers, etc						
12mm dia x 600mm long	2.70	3.46	0.18	4.88	nr	8.34

T: MECHANICAL/COOLING/HEATING SYSTEMS

Item	Net Price £	Material £	Labour hours	Labour £	Unit	Total rate £
Pipe roller and chair						
Roller and chair; black malleable						
Up to 50mm dia	2.93	3.75	0.20	5.42	nr	9.17
65mm dia	2.93	3.75	0.20	5.42	nr	9.17
80mm dia	4.19	5.37	0.20	5.42	nr	10.79
100mm dia	4.50	5.76	0.20	5.42	nr	11.18
125mm dia	4.92	6.30	0.20	5.42	nr	11.72
150mm dia	5.46	6.99	0.30	8.13	nr	15.12
175mm dia	13.69	17.53	0.30	8.13	nr	25.67
200mm dia	13.69	17.53	0.30	8.13	nr	25.67
250mm dia	19.56	25.05	0.30	8.13	nr	33.18
300mm dia	20.54	26.31	0.30	8.13	nr	34.44
Roller bracket; black malleable						
25mm dia	1.78	2.28	0.20	5.42	nr	7.70
32mm dia	1.87	2.40	0.20	5.42	nr	7.82
40mm dia	2.00	2.56	0.20	5.42	nr	7.98
50mm dia	2.12	2.72	0.20	5.42	nr	8.14
65mm dia	2.78	3.56	0.20	5.42	nr	8.98
80mm dia	4.01	5.14	0.20	5.42	nr	10.56
100mm dia	4.45	5.70	0.20	5.42	nr	11.12
125mm dia	7.35	9.41	0.20	5.42	nr	14.83
150mm dia	7.35	9.41	0.30	8.13	nr	17.54
175mm dia	16.38	20.98	0.30	8.13	nr	29.11
200mm dia	16.38	20.98	0.30	8.13	nr	29.11
250mm dia	21.77	27.88	0.30	8.13	nr	36.01
300mm dia	26.98	34.56	0.30	8.13	nr	42.69
350mm dia	43.37	55.55	0.30	8.13	nr	63.68
400mm dia	49.57	63.49	0.30	8.13	nr	71.62
Extra over black steel screwed pipes; black steel flanges, screwed and drilled; metric; BS 4504						
Screwed flanges; PN6						
15mm dia	5.74	7.35	0.35	9.49	nr	16.84
20mm dia	6.10	7.81	0.47	12.74	nr	20.55
25mm dia	6.27	8.03	0.53	14.36	nr	22.40
32mm dia	6.28	8.04	0.62	16.80	nr	24.85
40mm dia	6.28	8.04	0.70	18.97	nr	27.02
50mm dia	6.83	8.75	0.84	22.77	nr	31.51
65mm dia	9.63	12.33	1.03	27.92	nr	40.25
80mm dia	11.87	15.20	1.23	33.34	nr	48.54
100mm dia	14.30	18.32	1.41	38.22	nr	56.53
125mm dia	29.44	37.71	1.77	47.97	nr	85.68
150mm dia	29.44	37.71	2.21	59.90	nr	97.61
Screwed flanges; PN16						
15mm dia	7.36	9.43	0.35	9.49	nr	18.91
20mm dia	7.46	9.55	0.47	12.74	nr	22.29
25mm dia	7.57	9.70	0.53	14.36	nr	24.06
32mm dia	8.74	11.19	0.62	16.80	nr	28.00
40mm dia	8.74	11.19	0.70	18.97	nr	30.17
50mm dia	9.29	11.90	0.84	22.77	nr	34.67
65mm dia	11.48	14.70	1.03	27.92	nr	42.62
80mm dia	14.10	18.06	1.23	33.34	nr	51.40
100mm dia	17.01	21.79	1.41	38.22	nr	60.00
125mm dia	29.20	37.40	1.77	47.97	nr	85.37
150mm dia	32.08	41.09	2.21	59.90	nr	100.99

T: MECHANICAL/COOLING/HEATING SYSTEMS

Item	Net Price £	Material £	Labour hours	Labour £	Unit	Total rate £
T31 : LOW TEMPERATURE HOT WATER HEATING (cont'd)						
SCREWED STEEL (cont'd)						
Extra over black steel screwed pipes; black steel flanges, screwed and drilled; imperial; BS 10						
Screwed flanges; Table E						
1/2" dia	10.55	13.51	0.35	9.49	nr	23.00
3/4" dia	10.65	13.64	0.47	12.74	nr	26.38
1" dia	10.76	13.78	0.53	14.36	nr	28.15
1 1/4" dia	11.49	14.72	0.62	16.80	nr	31.52
1 1/2" dia	11.49	14.72	0.70	18.97	nr	33.69
2" dia	11.71	15.00	0.84	22.77	nr	37.77
2 1/2" dia	13.79	17.66	1.03	27.92	nr	45.58
3" dia	16.90	21.65	1.23	33.34	nr	54.98
4" dia	22.47	28.78	1.41	38.22	nr	67.00
5" dia	22.87	29.29	1.77	47.97	nr	77.26
6" dia	45.50	58.28	2.21	59.90	nr	118.17
Extra over black steel screwed pipes; black steel flange connections						
Bolted connection between pair of flanges; including gasket, bolts, nuts and washers						
50mm dia	12.48	15.98	0.53	14.36	nr	30.35
65mm dia	14.28	18.29	0.53	14.36	nr	32.65
80mm dia	18.25	23.37	0.53	14.36	nr	37.74
100mm dia	22.79	29.19	0.61	16.53	nr	45.72
125mm dia	31.63	40.51	0.61	16.53	nr	57.04
150mm dia	34.13	43.71	0.90	24.39	nr	68.11
Extra over black steel screwed pipes; black heavy steel tubular fittings; BS 1387						
Long screw connection with socket and backnut						
15mm dia	3.38	4.33	0.63	17.08	nr	21.40
20mm dia	4.26	5.46	0.84	22.77	nr	28.22
25mm dia	5.59	7.16	0.95	25.75	nr	32.91
32mm dia	7.32	9.38	1.11	30.08	nr	39.46
40mm dia	8.95	11.46	1.28	34.69	nr	46.16
50mm dia	13.13	16.82	1.53	41.47	nr	58.29
65mm dia	29.95	38.36	1.87	50.68	nr	89.04
80mm dia	41.17	52.73	2.21	59.90	nr	112.63
100mm dia	66.54	85.22	3.05	82.67	nr	167.89
Running nipple						
15mm dia	0.86	1.10	0.50	13.55	nr	14.65
20mm dia	1.09	1.40	0.68	18.43	nr	19.83
25mm dia	1.34	1.72	0.77	20.87	nr	22.59
32mm dia	1.87	2.40	0.90	24.39	nr	26.79
40mm dia	2.53	3.24	1.03	27.92	nr	31.16
50mm dia	3.85	4.93	1.23	33.34	nr	38.27
65mm dia	8.28	10.60	1.50	40.66	nr	51.26
80mm dia	12.91	16.54	1.78	48.24	nr	64.78
100mm dia	20.23	25.91	2.38	64.51	nr	90.42

T: MECHANICAL/COOLING/HEATING SYSTEMS

Item	Net Price £	Material £	Labour hours	Labour £	Unit	Total rate £
Barrel nipple						
15mm dia	0.49	0.63	0.50	13.55	nr	**14.18**
20mm dia	0.78	1.00	0.68	18.43	nr	**19.43**
25mm dia	1.03	1.32	0.77	20.87	nr	**22.19**
32mm dia	1.54	1.97	0.90	24.39	nr	**26.37**
40mm dia	1.90	2.43	1.03	27.92	nr	**30.35**
50mm dia	2.70	3.46	1.23	33.34	nr	**36.80**
65mm dia	5.78	7.40	1.50	40.66	nr	**48.06**
80mm dia	8.07	10.34	1.78	48.24	nr	**58.58**
100mm dia	14.61	18.71	2.38	64.51	nr	**83.22**
125mm dia	27.12	34.73	2.87	77.79	nr	**112.52**
150mm dia	42.74	54.74	3.39	91.88	nr	**146.62**
Close taper nipple						
15mm dia	1.03	1.32	0.50	13.55	nr	**14.87**
20mm dia	1.34	1.72	0.68	18.43	nr	**20.15**
25mm dia	1.74	2.23	0.77	20.87	nr	**23.10**
32mm dia	2.62	3.36	0.90	24.39	nr	**27.75**
40mm dia	3.24	4.15	1.03	27.92	nr	**32.07**
50mm dia	4.98	6.38	1.23	33.34	nr	**39.72**
65mm dia	9.67	12.39	1.50	40.66	nr	**53.04**
80mm dia	11.93	15.28	1.78	48.24	nr	**63.52**
100mm dia	24.49	31.37	2.38	64.51	nr	**95.87**
Extra over black steel screwed pipes; black malleable iron fittings; BS 143						
Cap						
15mm dia	0.50	0.64	0.32	8.67	nr	**9.31**
20mm dia	0.58	0.74	0.43	11.65	nr	**12.40**
25mm dia	0.73	0.94	0.49	13.28	nr	**14.22**
32mm dia	1.18	1.51	0.58	15.72	nr	**17.23**
40mm dia	1.40	1.79	0.66	17.89	nr	**19.68**
50mm dia	2.93	3.75	0.78	21.14	nr	**24.89**
65mm dia	4.88	6.25	0.96	26.02	nr	**32.27**
80mm dia	5.52	7.07	1.13	30.63	nr	**37.70**
100mm dia	12.08	15.47	1.70	46.08	nr	**61.55**
Plain plug, hollow						
15mm dia	0.35	0.45	0.28	7.59	nr	**8.04**
20mm dia	0.45	0.58	0.38	10.30	nr	**10.88**
25mm dia	0.61	0.78	0.44	11.93	nr	**12.71**
32mm dia	0.80	1.02	0.51	13.82	nr	**14.85**
40mm dia	1.50	1.92	0.59	15.99	nr	**17.91**
50mm dia	2.10	2.69	0.70	18.97	nr	**21.66**
65mm dia	3.43	4.39	0.85	23.04	nr	**27.43**
80mm dia	5.36	6.87	1.00	27.10	nr	**33.97**
100mm dia	9.87	12.64	1.44	39.03	nr	**51.67**
Plain plug, solid						
15mm dia	1.10	1.41	0.28	7.59	nr	**9.00**
20mm dia	1.06	1.36	0.38	10.30	nr	**11.66**
25mm dia	1.57	2.01	0.44	11.93	nr	**13.94**
32mm dia	2.11	2.70	0.51	13.82	nr	**16.53**
40mm dia	2.98	3.82	0.59	15.99	nr	**19.81**
50mm dia	3.92	5.02	0.70	18.97	nr	**23.99**

T: MECHANICAL/COOLING/HEATING SYSTEMS

Item	Net Price £	Material £	Labour hours	Labour £	Unit	Total rate £
T31 : LOW TEMPERATURE HOT WATER HEATING (cont'd)						
SCREWED STEEL (cont'd)						
Extra for black malleable fittings (cont'd)						
90° Elbow, male/female						
15mm dia	0.58	0.74	0.64	17.35	nr	18.09
20mm dia	0.78	1.00	0.85	23.04	nr	24.04
25mm dia	1.29	1.65	0.97	26.29	nr	27.94
32mm dia	2.21	2.83	1.12	30.36	nr	33.19
40mm dia	3.71	4.75	1.29	34.96	nr	39.72
50mm dia	4.77	6.11	1.55	42.01	nr	48.12
65mm dia	10.31	13.20	1.89	51.23	nr	64.43
80mm dia	14.09	18.05	2.24	60.71	nr	78.76
100mm dia	24.64	31.56	3.09	83.75	nr	115.31
90° Elbow						
15mm dia	0.52	0.67	0.64	17.35	nr	18.01
20mm dia	0.71	0.91	0.85	23.04	nr	23.95
25mm dia	1.10	1.41	0.97	26.29	nr	27.70
32mm dia	1.87	2.40	1.12	30.36	nr	32.75
40mm dia	3.15	4.03	1.29	34.96	nr	39.00
50mm dia	3.69	4.73	1.55	42.01	nr	46.74
65mm dia	7.96	10.20	1.89	51.23	nr	61.42
80mm dia	11.70	14.99	2.24	60.71	nr	75.70
100mm dia	22.56	28.89	3.09	83.75	nr	112.64
125mm dia	48.33	61.90	4.44	120.34	nr	182.24
150mm dia	89.96	115.22	5.79	156.93	nr	272.15
45° Elbow						
15mm dia	1.23	1.58	0.64	17.35	nr	18.92
20mm dia	1.51	1.93	0.85	23.04	nr	24.97
25mm dia	2.26	2.89	0.97	26.29	nr	29.19
32mm dia	4.32	5.53	1.12	30.36	nr	35.89
40mm dia	5.30	6.79	1.29	34.96	nr	41.75
50mm dia	7.27	9.31	1.55	42.01	nr	51.32
65mm dia	10.64	13.63	1.89	51.23	nr	64.85
80mm dia	15.99	20.48	2.24	60.71	nr	81.19
100mm dia	30.89	39.56	3.09	83.75	nr	123.31
150mm dia	86.27	110.49	5.79	156.93	nr	267.42
90° Bend, male/female						
15mm dia	1.01	1.29	0.64	17.35	nr	18.64
20mm dia	1.48	1.90	0.85	23.04	nr	24.93
25mm dia	2.18	2.79	0.97	26.29	nr	29.08
32mm dia	3.25	4.16	1.12	30.36	nr	34.52
40mm dia	5.13	6.57	1.29	34.96	nr	41.53
50mm dia	8.97	11.49	1.55	42.01	nr	53.50
65mm dia	14.94	19.14	1.89	51.23	nr	70.36
80mm dia	20.25	25.94	2.24	60.71	nr	86.65
100mm dia	50.18	64.27	3.09	83.75	nr	148.02

T: MECHANICAL/COOLING/HEATING SYSTEMS

Item	Net Price £	Material £	Labour hours	Labour £	Unit	Total rate £
90° Bend, male						
15mm dia	2.33	2.98	0.64	17.35	nr	20.33
20mm dia	2.62	3.36	0.85	23.04	nr	26.39
25mm dia	3.84	4.92	0.97	26.29	nr	31.21
32mm dia	7.82	10.02	1.12	30.36	nr	40.37
40mm dia	10.97	14.05	1.29	34.96	nr	49.01
50mm dia	14.67	18.79	1.55	42.01	nr	60.80
90° Bend, female						
15mm dia	0.93	1.19	0.64	17.35	nr	18.54
20mm dia	1.32	1.69	0.85	23.04	nr	24.73
25mm dia	1.87	2.40	0.97	26.29	nr	28.69
32mm dia	3.32	4.25	1.12	30.36	nr	34.61
40mm dia	4.43	5.67	1.29	34.96	nr	40.64
50mm dia	6.23	7.98	1.55	42.01	nr	49.99
65mm dia	13.31	17.05	1.89	51.23	nr	68.27
80mm dia	21.20	27.15	2.24	60.71	nr	87.86
100mm dia	46.48	59.53	3.09	83.75	nr	143.28
125mm dia	123.96	158.77	4.44	120.34	nr	279.11
150mm dia	189.35	242.52	5.79	156.93	nr	399.45
Return bend						
15mm dia	4.70	6.02	0.64	17.35	nr	23.37
20mm dia	7.60	9.73	0.85	23.04	nr	32.77
25mm dia	9.48	12.14	0.97	26.29	nr	38.43
32mm dia	13.78	17.65	1.12	30.36	nr	48.01
40mm dia	16.41	21.02	1.29	34.96	nr	55.98
50mm dia	25.05	32.08	1.55	42.01	nr	74.09
Equal socket, parallel thread						
15mm dia	0.54	0.69	0.64	17.35	nr	18.04
20mm dia	0.66	0.85	0.85	23.04	nr	23.88
25mm dia	0.87	1.11	0.97	26.29	nr	27.40
32mm dia	1.55	1.99	1.12	30.36	nr	32.34
40mm dia	2.29	2.93	1.29	34.96	nr	37.90
50mm dia	3.42	4.38	1.55	42.01	nr	46.39
65mm dia	5.80	7.43	1.89	51.23	nr	58.65
80mm dia	7.97	10.21	2.24	60.71	nr	70.92
100mm dia	13.53	17.33	3.09	83.75	nr	101.08
Concentric reducing socket						
20 x 15mm dia	0.79	1.01	0.76	20.60	nr	21.61
25 x 15mm dia	0.93	1.19	0.85	23.04	nr	24.23
25 x 20mm dia	0.99	1.27	0.86	23.31	nr	24.58
32 x 25mm dia	1.75	2.24	1.01	27.37	nr	29.62
40 x 25mm dia	2.19	2.80	1.16	31.44	nr	34.25
40 x 32mm dia	2.27	2.91	1.16	31.44	nr	34.35
50 x 25mm dia	4.24	5.43	1.38	37.40	nr	42.83
50 x 40mm dia	3.18	4.07	1.38	37.40	nr	41.48
65 x 50mm dia	5.78	7.40	1.69	45.80	nr	53.21
80 x 50mm dia	7.53	9.64	2.00	54.21	nr	63.85
100 x 50mm dia	15.02	19.24	2.75	74.53	nr	93.77
100 x 80mm dia	13.93	17.84	2.75	74.53	nr	92.38
150 x 100mm dia	36.77	47.09	4.10	111.12	nr	158.22

T: MECHANICAL/COOLING/HEATING SYSTEMS

Item	Net Price £	Material £	Labour hours	Labour £	Unit	Total rate £
T31 : LOW TEMPERATURE HOT WATER HEATING (cont'd)						
SCREWED STEEL (cont'd)						
Extra for black malleable fittings (cont'd)						
Eccentric reducing socket						
20 x 15mm dia	1.41	1.81	0.73	19.79	nr	21.59
25 x 15mm dia	4.04	5.17	0.85	23.04	nr	28.21
25 x 20mm dia	4.59	5.88	0.85	23.04	nr	28.92
32 x 25mm dia	5.45	6.98	1.01	27.37	nr	34.35
40 x 25mm dia	6.72	8.61	1.16	31.44	nr	40.05
40 x 32mm dia	3.37	4.32	1.16	31.44	nr	35.76
50 x 25mm dia	4.36	5.58	1.38	37.40	nr	42.99
50 x 40mm dia	4.05	5.19	1.38	37.40	nr	42.59
65 x 50mm dia	7.53	9.64	1.69	45.80	nr	55.45
80 x 50mm dia	12.24	15.68	2.00	54.21	nr	69.88
Hexagon bush						
20 x 15mm dia	0.45	0.58	0.37	10.03	nr	10.60
25 x 15mm dia	0.55	0.70	0.43	11.65	nr	12.36
25 x 20mm dia	0.58	0.74	0.43	11.65	nr	12.40
32 x 25mm dia	0.70	0.90	0.51	13.82	nr	14.72
40 x 25mm dia	1.10	1.41	0.58	15.72	nr	17.13
40 x 32mm dia	1.01	1.29	0.58	15.72	nr	17.01
50 x 25mm dia	2.27	2.91	0.71	19.24	nr	22.15
50 x 40mm dia	2.10	2.69	0.71	19.24	nr	21.93
65 x 50mm dia	3.51	4.50	0.85	23.04	nr	27.53
80 x 50mm dia	5.71	7.31	1.00	27.10	nr	34.42
100 x 50mm dia	13.20	16.91	1.52	41.20	nr	58.10
100 x 80mm dia	10.99	14.08	1.52	41.20	nr	55.27
150 x 100mm dia	34.79	44.56	2.57	69.66	nr	114.21
Hexagon nipple						
15mm dia	0.48	0.61	0.28	7.59	nr	8.20
20mm dia	0.55	0.70	0.38	10.30	nr	11.00
25mm dia	0.78	1.00	0.44	11.93	nr	12.92
32mm dia	1.50	1.92	0.51	13.82	nr	15.74
40mm dia	1.75	2.24	0.59	15.99	nr	18.23
50mm dia	3.16	4.05	0.70	18.97	nr	23.02
65mm dia	5.15	6.60	0.85	23.04	nr	29.63
80mm dia	7.45	9.54	1.00	27.10	nr	36.65
100mm dia	12.64	16.19	1.44	39.03	nr	55.22
150mm dia	37.31	47.79	2.32	62.88	nr	110.67
Union, male/female						
15mm dia	2.35	3.01	0.64	17.35	nr	20.36
20mm dia	2.89	3.70	0.85	23.04	nr	26.74
25mm dia	3.61	4.62	0.97	26.29	nr	30.91
32mm dia	5.61	7.19	1.12	30.36	nr	37.54
40mm dia	7.18	9.20	1.29	34.96	nr	44.16
50mm dia	11.31	14.49	1.55	42.01	nr	56.50
65mm dia	21.59	27.65	1.89	51.23	nr	78.88
80mm dia	29.68	38.01	2.24	60.71	nr	98.73

T: MECHANICAL/COOLING/HEATING SYSTEMS

Item	Net Price £	Material £	Labour hours	Labour £	Unit	Total rate £
Union, female						
15mm dia	5.48	7.02	0.64	17.35	nr	24.36
20mm dia	6.06	7.76	0.85	23.04	nr	30.80
25mm dia	8.00	10.25	0.97	26.29	nr	36.54
32mm dia	11.48	14.70	1.12	30.36	nr	45.06
40mm dia	13.89	17.79	1.29	34.96	nr	52.75
50mm dia	17.63	22.58	1.55	42.01	nr	64.59
65mm dia	38.66	49.52	1.89	51.23	nr	100.74
80mm dia	63.14	80.87	2.24	60.71	nr	141.58
100mm dia	90.93	116.46	3.09	83.75	nr	200.21
Union elbow, male/female						
15mm dia	9.40	12.04	0.55	14.91	nr	26.95
20mm dia	12.66	16.21	0.85	23.04	nr	39.25
25mm dia	15.27	19.56	0.97	26.29	nr	45.85
Twin elbow						
15mm dia	2.97	3.80	0.91	24.66	nr	28.47
20mm dia	3.28	4.20	1.22	33.07	nr	37.27
25mm dia	5.32	6.81	1.39	37.67	nr	44.49
32mm dia	10.24	13.12	1.62	43.91	nr	57.02
40mm dia	12.97	16.61	1.86	50.41	nr	67.02
50mm dia	16.67	21.35	2.21	59.90	nr	81.25
65mm dia	26.09	33.42	2.72	73.72	nr	107.14
80mm dia	44.46	56.94	3.21	87.00	nr	143.95
Equal tee						
15mm dia	0.71	0.91	0.91	24.66	nr	25.57
20mm dia	1.03	1.32	1.22	33.07	nr	34.39
25mm dia	1.48	1.90	1.39	37.67	nr	39.57
32mm dia	2.55	3.27	1.62	43.91	nr	47.17
40mm dia	3.91	5.01	1.86	50.41	nr	55.42
50mm dia	5.63	7.21	2.21	59.90	nr	67.11
65mm dia	12.34	15.80	2.72	73.72	nr	89.53
80mm dia	15.02	19.24	3.21	87.00	nr	106.24
100mm dia	27.21	34.85	4.44	120.34	nr	155.19
125mm dia	69.67	89.23	5.38	145.82	nr	235.05
150mm dia	111.02	142.19	6.31	171.02	nr	313.22
Tee reducing on branch						
20 x 15mm dia	0.93	1.19	1.22	33.07	nr	34.26
25 x 15mm dia	1.29	1.65	1.39	37.67	nr	39.33
25 x 20mm dia	1.53	1.96	1.39	37.67	nr	39.63
32 x 25mm dia	2.51	3.21	1.62	43.91	nr	47.12
40 x 25mm dia	3.31	4.24	1.86	50.41	nr	54.65
40 x 32mm dia	4.33	5.55	1.86	50.41	nr	55.96
50 x 25mm dia	4.92	6.30	2.21	59.90	nr	66.20
50 x 40mm dia	7.33	9.39	2.21	59.90	nr	69.29
65 x 50mm dia	11.44	14.65	2.72	73.72	nr	88.37
80 x 50mm dia	15.46	19.80	3.21	87.00	nr	106.80
100 x 50mm dia	22.55	28.88	4.44	120.34	nr	149.22
100 x 80mm dia	34.79	44.56	4.44	120.34	nr	164.90
150 x 100mm dia	81.77	104.73	6.31	171.02	nr	275.75

T: MECHANICAL/COOLING/HEATING SYSTEMS

Item	Net Price £	Material £	Labour hours	Labour £	Unit	Total rate £
T31 : LOW TEMPERATURE HOT WATER HEATING (cont'd)						
SCREWED STEEL (cont'd)						
Extra for black malleable fittings (cont'd)						
Equal pitcher tee						
15mm dia	2.16	2.77	0.91	24.66	nr	27.43
20mm dia	2.69	3.45	1.22	33.07	nr	36.51
25mm dia	4.04	5.17	1.39	37.67	nr	42.85
32mm dia	5.76	7.38	1.62	43.91	nr	51.29
40mm dia	8.90	11.40	1.86	50.41	nr	61.81
50mm dia	12.50	16.01	2.21	59.90	nr	75.91
65mm dia	19.33	24.76	2.72	73.72	nr	98.48
80mm dia	26.58	34.04	3.21	87.00	nr	121.05
100mm dia	59.80	76.59	4.44	120.34	nr	196.93
Cross						
15mm dia	2.12	2.72	1.00	27.10	nr	29.82
20mm dia	3.18	4.07	1.33	36.05	nr	40.12
25mm dia	4.04	5.17	1.51	40.93	nr	46.10
32mm dia	5.53	7.08	1.77	47.97	nr	55.06
40mm dia	7.12	9.12	2.02	54.75	nr	63.87
50mm dia	11.57	14.82	2.42	65.59	nr	80.41
65mm dia	15.99	20.48	2.97	80.50	nr	100.98
80mm dia	21.26	27.23	3.50	94.86	nr	122.09
100mm dia	38.66	49.52	4.84	131.18	nr	180.70
BLACK WELDED STEEL						
Black steel pipes; butt welded joints; BS 1387: 1985; including protective painting. Fixed Vertical with brackets measured separately (Refer to Screwed Steel Section). Welded butt joints are within the running length, but any flanges are additional						
Medium weight						
10mm dia	1.77	2.27	0.37	10.03	m	12.30
15mm dia	1.88	2.41	0.37	10.03	m	12.44
20mm dia	2.15	2.75	0.37	10.03	m	12.78
25mm dia	3.09	3.96	0.41	11.11	m	15.07
32mm dia	3.82	4.89	0.48	13.01	m	17.90
40mm dia	4.44	5.69	0.52	14.09	m	19.78
50mm dia	6.25	8.00	0.62	16.80	m	24.81
65mm dia	8.48	10.86	0.64	17.35	m	28.21
80mm dia	11.02	14.11	1.10	29.81	m	43.93
100mm dia	15.61	19.99	1.31	35.51	m	55.50
125mm dia	23.72	30.38	1.66	44.99	m	75.37
150mm dia	27.55	35.29	1.88	50.95	m	86.24

T: MECHANICAL/COOLING/HEATING SYSTEMS

Item	Net Price £	Material £	Labour hours	Labour £	Unit	Total rate £
Heavy weight						
15mm dia	2.33	2.98	0.37	10.03	m	13.01
20mm dia	2.66	3.41	0.37	10.03	m	13.44
25mm dia	3.88	4.97	0.41	11.11	m	16.08
32mm dia	4.82	6.17	0.48	13.01	m	19.18
40mm dia	5.62	7.20	0.52	14.09	m	21.29
50mm dia	7.80	9.99	0.62	16.80	m	26.79
65mm dia	10.61	13.59	0.64	17.35	m	30.94
80mm dia	13.51	17.30	1.10	29.81	m	47.12
100mm dia	18.85	24.14	1.31	35.51	m	59.65
125mm dia	25.29	32.39	1.66	44.99	m	77.38
150mm dia	29.58	37.89	1.88	50.95	m	88.84
Black steel pipes; butt welded joints; BS 1387: 1985; including protective painting. Fixed at High Level or Suspended with brackets measured separately (Refer to Screwed Steel Section). Welded butt joints are within the running length, but any flanges are additional						
Medium weight						
10mm dia	1.77	2.27	0.58	15.72	m	17.99
15mm dia	1.88	2.41	0.58	15.72	m	18.13
20mm dia	2.15	2.75	0.58	15.72	m	18.47
25mm dia	3.09	3.96	0.60	16.26	m	20.22
32mm dia	3.82	4.89	0.68	18.43	m	23.32
40mm dia	4.44	5.69	0.73	19.79	m	25.47
50mm dia	6.25	8.00	0.85	23.04	m	31.04
65mm dia	8.48	10.86	0.88	23.85	m	34.71
80mm dia	11.02	14.11	1.45	39.30	m	53.41
100mm dia	15.61	19.99	1.74	47.16	m	67.15
125mm dia	23.72	30.38	2.21	59.90	m	90.28
150mm dia	27.55	35.29	2.50	67.76	m	103.04
Heavy weight						
15mm dia	2.33	2.98	0.58	15.72	m	18.70
20mm dia	2.66	3.41	0.58	15.72	m	19.13
25mm dia	3.88	4.97	0.60	16.26	m	21.23
32mm dia	4.82	6.17	0.68	18.43	m	24.60
40mm dia	5.62	7.20	0.73	19.79	m	26.98
50mm dia	7.80	9.99	0.85	23.04	m	33.03
65mm dia	10.61	13.59	0.88	23.85	m	37.44
80mm dia	13.51	17.30	1.45	39.30	m	56.60
100mm dia	18.85	24.14	1.74	47.16	m	71.30
125mm dia	25.29	32.39	2.21	59.90	m	92.29
150mm dia	29.58	37.89	2.50	67.76	m	105.64

T: MECHANICAL/COOLING/HEATING SYSTEMS

Item	Net Price £	Material £	Labour hours	Labour £	Unit	Total rate £
T31 : LOW TEMPERATURE HOT WATER HEATING (cont'd)						
BLACK WELDED STEEL (cont'd)						
FIXINGS						
Refer to steel pipes; black malleable iron. For minimum fixing distances, refer to the Tables and Memoranda to the rear of the book						
Extra over black steel butt welded pipes; black steel flanges, welded and drilled; metric; BS 4504						
Welded flanges; PN6						
15mm dia	2.77	3.55	0.59	15.99	nr	19.54
20mm dia	3.19	4.09	0.69	18.70	nr	22.79
25mm dia	3.30	4.23	0.84	22.77	nr	26.99
32mm dia	4.06	5.20	1.00	27.10	nr	32.30
40mm dia	4.06	5.20	1.11	30.08	nr	35.28
50mm dia	4.40	5.64	1.37	37.13	nr	42.77
65mm dia	5.17	6.62	1.54	41.74	nr	48.36
80mm dia	8.21	10.52	1.67	45.26	nr	55.78
100mm dia	9.98	12.78	2.22	60.17	nr	72.95
125mm dia	15.84	20.29	2.61	70.74	nr	91.03
150mm dia	15.84	20.29	2.99	81.04	nr	101.33
Welded flanges; PN16						
15mm dia	3.68	4.71	0.59	15.99	nr	20.70
20mm dia	3.78	4.84	0.69	18.70	nr	23.54
25mm dia	3.89	4.98	0.84	22.77	nr	27.75
32mm dia	5.87	7.52	1.00	27.10	nr	34.62
40mm dia	5.87	7.52	1.11	30.08	nr	37.60
50mm dia	7.26	9.30	1.37	37.13	nr	46.43
65mm dia	8.28	10.60	1.54	41.74	nr	52.34
80mm dia	11.05	14.15	1.67	45.26	nr	59.42
100mm dia	13.81	17.69	2.22	60.17	nr	77.86
125mm dia	18.43	23.61	2.61	70.74	nr	94.35
150mm dia	21.12	27.05	2.99	81.04	nr	108.09
Blank flanges, slip on for welding; PN6						
15mm dia	1.14	1.46	0.48	13.01	nr	14.47
20mm dia	1.14	1.46	0.55	14.91	nr	16.37
25mm dia	1.14	1.46	0.64	17.35	nr	18.81
32mm dia	1.90	2.43	0.76	20.60	nr	23.03
40mm dia	1.92	2.46	0.84	22.77	nr	25.23
50mm dia	2.11	2.70	1.01	27.37	nr	30.08
65mm dia	4.20	5.38	1.30	35.23	nr	40.61
80mm dia	4.26	5.46	1.41	38.22	nr	43.67
100mm dia	4.50	5.76	1.78	48.24	nr	54.01
125mm dia	7.34	9.40	2.06	55.83	nr	65.23
150mm dia	8.05	10.31	2.35	63.69	nr	74.00

T: MECHANICAL/COOLING/HEATING SYSTEMS

Item	Net Price £	Material £	Labour hours	Labour £	Unit	Total rate £
Blank flanges, slip on for welding; PN16						
15mm dia	1.05	1.34	0.48	13.01	nr	14.35
20mm dia	1.19	1.52	0.55	14.91	nr	16.43
25mm dia	1.30	1.67	0.64	17.35	nr	19.01
32mm dia	2.49	3.19	0.76	20.60	nr	23.79
40mm dia	2.73	3.50	0.84	22.77	nr	26.26
50mm dia	3.15	4.03	1.01	27.37	nr	31.41
65mm dia	4.13	5.29	1.30	35.23	nr	40.52
80mm dia	4.92	6.30	1.41	38.22	nr	44.52
100mm dia	6.37	8.16	1.78	48.24	nr	56.40
125mm dia	9.48	12.14	2.06	55.83	nr	67.98
150mm dia	11.13	14.26	2.35	63.69	nr	77.95
Extra over black steel butt welded pipes; black steel flanges, welding and drilled; imperial; BS 10						
Welded flanges; Table E						
15mm dia	5.73	7.34	0.59	15.99	nr	23.33
20mm dia	5.83	7.47	0.69	18.70	nr	26.17
25mm dia	5.94	7.61	0.84	22.77	nr	30.37
32mm dia	6.67	8.54	1.00	27.10	nr	35.65
40mm dia	6.67	8.54	1.11	30.08	nr	38.63
50mm dia	6.89	8.82	1.37	37.13	nr	45.96
65mm dia	8.28	10.60	1.54	41.74	nr	52.34
80mm dia	9.96	12.76	1.67	45.26	nr	58.02
100mm dia	13.94	17.85	2.22	60.17	nr	78.02
125mm dia	25.18	32.25	2.61	70.74	nr	102.99
150mm dia	30.32	38.83	2.99	81.04	nr	119.87
Blank flanges, slip on for welding; Table E						
15mm dia	3.78	4.84	0.48	13.01	nr	17.85
20mm dia	3.78	4.84	0.55	14.91	nr	19.75
25mm dia	3.78	4.84	0.64	17.35	nr	22.19
32mm dia	4.77	6.11	0.76	20.60	nr	26.71
40mm dia	5.19	6.65	0.84	22.77	nr	29.41
50mm dia	5.73	7.34	1.01	27.37	nr	34.71
65mm dia	6.60	8.45	1.30	35.23	nr	43.69
80mm dia	7.80	9.99	1.41	38.22	nr	48.21
100mm dia	11.34	14.52	1.78	48.24	nr	62.77
125mm dia	18.09	23.17	2.06	55.83	nr	79.00
150mm dia	25.92	33.20	2.35	63.69	nr	96.89
Extra over black steel butt welded pipes; black steel flange connections						
Bolted connection between pair of flanges; including gasket, bolts, nuts and washers						
50mm dia	12.48	15.98	0.50	13.55	nr	29.54
65mm dia	14.28	18.29	0.50	13.55	nr	31.84
80mm dia	18.25	23.37	0.50	13.55	nr	36.93
100mm dia	22.79	29.19	0.50	13.55	nr	42.74
125mm dia	31.63	40.51	0.50	13.55	nr	54.06
150mm dia	34.13	43.71	0.88	23.85	nr	67.56

T: MECHANICAL/COOLING/HEATING SYSTEMS

Item	Net Price £	Material £	Labour hours	Labour £	Unit	Total rate £
T31 : LOW TEMPERATURE HOT WATER HEATING (cont'd)						
BLACK WELDED STEEL (cont'd)						
Extra over fittings; BS 1965; butt welded						
Cap						
25mm dia	10.45	13.38	0.47	12.74	nr	26.12
32mm dia	10.45	13.38	0.59	15.99	nr	29.38
40mm dia	10.45	13.38	0.70	18.97	nr	32.36
50mm dia	12.29	15.74	0.99	26.83	nr	42.57
65mm dia	14.43	18.48	1.35	36.59	nr	55.07
80mm dia	14.69	18.81	1.66	44.99	nr	63.81
100mm dia	19.21	24.60	2.23	60.44	nr	85.04
125mm dia	27.02	34.61	3.03	82.12	nr	116.73
150mm dia	31.25	40.02	3.79	102.72	nr	142.75
Concentric reducer						
20 x 15mm dia	5.83	7.47	0.69	18.70	nr	26.17
25 x 15mm dia	5.62	7.20	0.87	23.58	nr	30.78
25 x 20mm dia	7.26	9.30	0.87	23.58	nr	32.88
32 x 25mm dia	7.75	9.93	1.08	29.27	nr	39.20
40 x 25mm dia	10.09	12.92	1.38	37.40	nr	50.33
40 x 32mm dia	6.91	8.85	1.38	37.40	nr	46.25
50 x 25mm dia	9.33	11.95	1.82	49.33	nr	61.28
50 x 40mm dia	6.80	8.71	1.82	49.33	nr	58.04
65 x 50mm dia	9.38	12.01	2.52	68.30	nr	80.31
80 x 50mm dia	9.55	12.23	3.24	87.82	nr	100.05
100 x 50mm dia	15.81	20.25	4.08	110.58	nr	130.83
100 x 80mm dia	10.88	13.94	4.08	110.58	nr	124.52
125 x 80mm dia	23.76	30.43	4.71	127.66	nr	158.09
150 x 100mm dia	25.69	32.90	5.33	144.46	nr	177.37
Eccentric reducer						
20 x 15mm dia	8.65	11.08	0.69	18.70	nr	29.78
25 x 15mm dia	11.33	14.51	0.87	23.58	nr	38.09
25 x 20mm dia	9.51	12.18	0.87	23.58	nr	35.76
32 x 25mm dia	10.52	13.47	1.08	29.27	nr	42.75
40 x 25mm dia	12.98	16.62	1.38	37.40	nr	54.03
40 x 32mm dia	12.40	15.88	1.38	37.40	nr	53.28
50 x 25mm dia	16.20	20.75	1.82	49.33	nr	70.08
50 x 40mm dia	10.66	13.65	1.82	49.33	nr	62.98
65 x 50mm dia	12.58	16.11	2.52	68.30	nr	84.41
80 x 50mm dia	15.39	19.71	3.24	87.82	nr	107.53
100 x 50mm dia	26.07	33.39	4.08	110.58	nr	143.97
100 x 80mm dia	19.24	24.64	4.08	110.58	nr	135.22
125 x 80mm dia	51.39	65.82	4.71	127.66	nr	193.48
150 x 100mm dia	38.92	49.85	5.33	144.46	nr	194.31
45° elbow, long radius						
15mm dia	2.19	2.80	0.56	15.18	nr	17.98
20mm dia	2.21	2.83	0.75	20.33	nr	23.16
25mm dia	2.86	3.66	0.93	25.21	nr	28.87
32mm dia	4.16	5.33	1.17	31.71	nr	37.04
40mm dia	3.84	4.92	1.46	39.57	nr	44.49
50mm dia	5.27	6.75	1.97	53.39	nr	60.14
65mm dia	6.85	8.77	2.70	73.18	nr	81.95
80mm dia	6.55	8.39	3.32	89.98	nr	98.37

T: MECHANICAL/COOLING/HEATING SYSTEMS

Item	Net Price £	Material £	Labour hours	Labour £	Unit	Total rate £
100mm dia	10.08	12.91	4.09	110.85	nr	123.76
125mm dia	19.92	25.51	4.94	133.89	nr	159.40
150mm dia	26.07	33.39	5.78	156.66	nr	190.05
90° elbow, long radius						
15mm dia	2.26	2.89	0.56	15.18	nr	18.07
20mm dia	2.21	2.83	0.75	20.33	nr	23.16
25mm dia	2.91	3.73	0.93	25.21	nr	28.93
32mm dia	4.17	5.34	1.17	31.71	nr	37.05
40mm dia	3.84	4.92	1.46	39.57	nr	44.49
50mm dia	5.27	6.75	1.97	53.39	nr	60.14
65mm dia	6.85	8.77	2.70	73.18	nr	81.95
80mm dia	7.70	9.86	3.32	89.98	nr	99.85
100mm dia	11.86	15.19	4.09	110.85	nr	126.04
125mm dia	22.75	29.14	4.94	133.89	nr	163.03
150mm dia	30.68	39.29	5.78	156.66	nr	195.95
Branch Bend (based on branch and pipe sizes being the same)						
15mm dia	9.00	11.53	0.85	23.04	nr	34.57
20mm dia	8.92	11.42	0.85	23.04	nr	34.46
25mm dia	9.01	11.54	1.02	27.65	nr	39.19
32mm dia	8.51	10.90	1.11	30.08	nr	40.98
40mm dia	8.40	10.76	1.36	36.86	nr	47.62
50mm dia	8.16	10.45	1.70	46.08	nr	56.53
65mm dia	12.07	15.46	1.78	48.24	nr	63.70
80mm dia	19.44	24.90	1.82	49.33	nr	74.23
100mm dia	26.82	34.35	1.87	50.68	nr	85.03
125mm dia	49.23	63.05	2.21	59.90	nr	122.95
150mm dia	75.72	96.98	2.65	71.82	nr	168.81
Equal tee						
15mm dia	21.34	27.33	0.82	22.22	nr	49.56
20mm dia	21.34	27.33	1.10	29.81	nr	57.15
25mm dia	21.34	27.33	1.35	36.59	nr	63.92
32mm dia	21.34	27.33	1.63	44.18	nr	71.51
40mm dia	21.34	27.33	2.14	58.00	nr	85.33
50mm dia	22.46	28.77	3.02	81.85	nr	110.62
65mm dia	32.77	41.97	3.61	97.84	nr	139.82
80mm dia	32.82	42.04	4.18	113.29	nr	155.33
100mm dia	40.84	52.31	5.24	142.02	nr	194.33
125mm dia	93.48	119.73	6.70	181.59	nr	301.32
150mm dia	102.11	130.78	8.45	229.02	nr	359.81
Extra over black steel butt welded pipes; labour						
Made bend						
15mm dia	-	-	0.42	11.38	nr	11.38
20mm dia	-	-	0.42	11.38	nr	11.38
25mm dia	-	-	0.50	13.55	nr	13.55
32mm dia	-	-	0.62	16.80	nr	16.80
40mm dia	-	-	0.74	20.06	nr	20.06
50mm dia	-	-	0.89	24.12	nr	24.12
65mm dia	-	-	1.05	28.46	nr	28.46
80mm dia	-	-	1.13	30.63	nr	30.63
100mm dia	-	-	2.90	78.60	nr	78.60
125mm dia	-	-	3.56	96.49	nr	96.49
150mm dia	-	-	4.18	113.29	nr	113.29

T: MECHANICAL/COOLING/HEATING SYSTEMS

Item	Net Price £	Material £	Labour hours	Labour £	Unit	Total rate £
T31 : LOW TEMPERATURE HOT WATER HEATING (cont'd)						
BLACK WELDED STEEL (cont'd)						
Extra over black steel butt welded pipes; labour (cont'd)						
Splay cut end						
15mm dia	-	-	0.14	3.79	nr	3.79
20mm dia	-	-	0.16	4.34	nr	4.34
25mm dia	-	-	0.18	4.88	nr	4.88
32mm dia	-	-	0.25	6.78	nr	6.78
40mm dia	-	-	0.27	7.32	nr	7.32
50mm dia	-	-	0.31	8.40	nr	8.40
65mm dia	-	-	0.35	9.49	nr	9.49
80mm dia	-	-	0.40	10.84	nr	10.84
100mm dia	-	-	0.48	13.01	nr	13.01
125mm dia	-	-	0.56	15.18	nr	15.18
150mm dia	-	-	0.64	17.35	nr	17.35
Screwed joint to fitting						
15mm dia	-	-	0.30	8.13	nr	8.13
20mm dia	-	-	0.40	10.84	nr	10.84
25mm dia	-	-	0.46	12.47	nr	12.47
32mm dia	-	-	0.53	14.36	nr	14.36
40mm dia	-	-	0.61	16.53	nr	16.53
50mm dia	-	-	0.73	19.79	nr	19.79
65mm dia	-	-	0.89	24.12	nr	24.12
80mm dia	-	-	1.05	28.46	nr	28.46
100mm dia	-	-	1.46	39.57	nr	39.57
125mm dia	-	-	2.10	56.92	nr	56.92
150mm dia	-	-	2.73	73.99	nr	73.99
Straight butt weld						
15mm dia	-	-	0.31	8.40	nr	8.40
20mm dia	-	-	0.42	11.38	nr	11.38
25mm dia	-	-	0.52	14.09	nr	14.09
32mm dia	-	-	0.69	18.70	nr	18.70
40mm dia	-	-	0.83	22.50	nr	22.50
50mm dia	-	-	1.22	33.07	nr	33.07
65mm dia	-	-	1.57	42.55	nr	42.55
80mm dia	-	-	1.95	52.85	nr	52.85
100mm dia	-	-	2.38	64.51	nr	64.51
125mm dia	-	-	2.83	76.70	nr	76.70
150mm dia	-	-	3.27	88.63	nr	88.63

T: MECHANICAL/COOLING/HEATING SYSTEMS

Item	Net Price £	Material £	Labour hours	Labour £	Unit	Total rate £
CARBON WELDED STEEL						
Hot finished seamless carbon steel pipe; BS 806 and BS 3601; wall thickness to BS 3600; butt welded joints; including protective painting, fixed vertically or at low level, brackets measured separately (Refer to Screwed Pipework Section). Welded butt joints are within the running length, but any flanges are additional						
Pipework						
200mm dia	59.34	76.00	2.04	55.29	m	131.29
250mm dia	78.93	101.09	2.59	70.20	m	171.29
300mm dia	87.25	111.75	2.99	81.04	m	192.79
350mm dia	128.80	164.97	3.52	95.40	m	260.37
400mm dia	286.16	366.51	4.08	110.58	m	477.09
Hot finished seamless carbon steel pipe; BS 806 and BS 3601; wall thickness to BS 3600; butt welded joints; including protective painting, fixed at high level or suspended, brackets measured separately (Refer to Screwed Pipework Section). Welded butt joints are within the running length, but any flanges are additional						
Pipework						
200mm dia	63.30	81.07	3.70	100.28	m	181.36
250mm dia	78.93	101.09	4.73	128.20	m	229.29
300mm dia	87.25	111.75	5.65	153.13	m	264.88
350mm dia	128.50	164.97	6.68	181.05	m	346.02
400mm dia	286.16	366.51	7.70	208.70	m	575.21
FIXINGS						
Refer to steel pipes; black malleable iron. For minimum fixing distances, refer to the Tables and Memoranda to the rear of the book						
Extra over fittings; BS 1965 part 1; butt welded						
Cap						
200mm dia	43.87	56.19	3.70	100.28	nr	156.47
250mm dia	84.71	108.50	4.73	128.20	nr	236.70
300mm dia	55.07	70.53	5.65	153.13	nr	223.67
350mm dia	69.63	89.18	6.68	181.05	nr	270.23
400mm dia	77.85	99.71	7.70	208.70	nr	308.41

T: MECHANICAL/COOLING/HEATING SYSTEMS

Item	Net Price £	Material £	Labour hours	Labour £	Unit	Total rate £
T31 : LOW TEMPERATURE HOT WATER HEATING (cont'd)						
CARBON WELDED STEEL (cont'd)						
Extra over fittings; BS 1965 part 1; butt welded (cont'd)						
Concentric reducer						
200mm x 150mm dia	48.23	61.77	7.27	197.04	nr	258.81
250mm x 150mm dia	69.93	89.57	9.05	245.29	nr	334.85
250mm x 200mm dia	42.80	54.82	9.10	246.64	nr	301.46
300mm x 150mm dia	146.80	188.02	10.75	291.36	nr	479.38
300mm x 200mm dia	83.65	107.14	10.75	291.36	nr	398.50
300mm x 250mm dia	74.37	95.25	11.15	302.20	nr	397.46
350mm x 200mm dia	101.14	129.54	12.50	338.79	nr	468.33
350mm x 250mm dia	110.26	141.22	12.70	344.21	nr	485.43
350mm x 300mm dia	120.32	154.10	13.00	352.35	nr	506.45
400mm x 250mm dia	130.68	167.37	14.46	391.92	nr	559.29
400mm x 300mm dia	136.24	174.49	14.51	393.27	nr	567.77
400mm x 350mm dia	138.22	177.03	15.16	410.89	nr	587.92
Eccentric reducer						
200mm x 150mm dia	87.09	111.54	7.27	197.04	nr	308.59
250mm x 150mm dia	116.05	148.64	9.05	245.29	nr	393.92
250mm x 200mm dia	75.94	97.26	9.10	246.64	nr	343.91
300mm x 150mm dia	168.46	215.76	10.75	291.36	nr	507.12
300mm x 200mm dia	160.58	205.67	10.75	291.36	nr	497.03
300mm x 250mm dia	129.25	165.54	11.15	302.20	nr	467.75
350mm x 200mm dia	132.66	169.91	12.50	338.79	nr	508.70
350mm x 250mm dia	140.27	179.66	12.70	344.21	nr	523.87
350mm x 300mm dia	111.67	143.03	13.00	352.35	nr	495.37
400mm x 250mm dia	145.26	186.05	14.46	391.92	nr	577.96
400mm x 300mm dia	167.95	215.11	14.51	393.27	nr	608.38
400mm x 350mm dia	148.99	190.82	15.16	410.89	nr	601.71
45° elbow						
200mm dia	46.12	59.07	7.75	210.05	nr	269.12
250mm dia	86.63	110.95	10.05	272.39	nr	383.35
300mm dia	127.72	163.58	12.20	330.66	nr	494.25
350mm dia	93.28	119.47	14.65	397.07	nr	516.54
400mm dia	121.00	154.98	17.12	464.01	nr	618.99
90° elbow						
200mm dia	53.00	67.88	7.75	210.05	nr	277.93
250mm dia	102.94	131.84	10.05	272.39	nr	404.23
300mm dia	150.27	192.46	12.20	330.66	nr	523.13
350mm dia	116.61	149.35	14.65	397.07	nr	546.42
400mm dia	151.25	193.72	17.12	464.01	nr	657.73
Equal tee						
200mm dia	143.36	183.61	11.25	304.91	nr	488.53
250mm dia	245.81	314.83	14.53	393.81	nr	708.64
300mm dia	139.85	179.12	17.55	475.67	nr	654.78
350mm dia	146.38	187.48	20.98	568.63	nr	756.11
400mm dia	165.24	211.64	24.38	660.78	nr	872.42

T: MECHANICAL/COOLING/HEATING SYSTEMS

Item	Net Price £	Material £	Labour hours	Labour £	Unit	Total rate £
Extra over black steel butt welded pipes; labour						
Straight butt weld						
200mm dia	-	-	4.08	110.58	nr	110.58
250mm dia	-	-	5.20	140.94	nr	140.94
300mm dia	-	-	6.22	168.58	nr	168.58
350mm dia	-	-	7.33	198.67	nr	198.67
400mm dia	-	-	8.41	227.94	nr	227.94
Branch weld						
100mm dia	-	-	3.46	93.78	nr	93.78
125mm dia	-	-	4.23	114.65	nr	114.65
150mm dia	-	-	5.00	135.52	nr	135.52
Extra over black steel butt welded pipes; black steel flanges, welding and drilled; metric; BS 4504						
Welded flanges; PN16						
200mm dia	37.39	47.89	4.10	111.12	nr	159.01
250mm dia	56.61	72.51	5.33	144.46	nr	216.97
300mm dia	74.72	95.70	6.40	173.46	nr	269.16
350mm dia	160.98	206.18	7.43	201.38	nr	407.56
400mm dia	186.71	239.14	8.45	229.02	nr	468.16
Welded flanges; PN25						
200mm dia	120.85	154.78	4.10	111.12	nr	265.91
250mm dia	144.86	185.54	5.33	144.46	nr	330.00
300mm dia	195.90	250.91	6.40	173.46	nr	424.37
Blank flanges, slip on for welding; PN16						
200mm dia	18.27	23.40	2.70	73.18	nr	96.58
250mm dia	26.13	33.47	3.48	94.32	nr	127.79
300mm dia	36.47	46.71	4.20	113.83	nr	160.55
350mm dia	66.41	85.06	4.78	129.55	nr	214.61
400mm dia	91.32	116.96	5.35	145.00	nr	261.97
Blank flanges, slip on for welding; PN25						
200mm dia	139.24	178.34	2.70	73.18	nr	251.52
250mm dia	207.03	265.16	3.48	94.32	nr	359.48
300mm dia	285.53	-365.70	4.20	113.83	nr	479.54
Extra over black steel butt welded pipes; black steel flange connections						
Bolted connection between pair of flanges; including gasket, bolts, nuts and washers						
200mm dia	60.30	77.23	3.83	103.81	nr	181.04
250mm dia	93.33	119.54	4.93	133.62	nr	253.16
300mm dia	127.24	162.97	5.90	159.91	nr	322.88

T: MECHANICAL/COOLING/HEATING SYSTEMS

Item	Net Price £	Material £	Labour hours	Labour £	Unit	Total rate £
T31 : LOW TEMPERATURE HOT WATER HEATING (cont'd)						
PRESS FIT						
Press fit jointing system; operating temperature -20°C to +120°C; operating pressure 16 bar; butyl rubber 'O' ring mechanical joint. With brackets measured separately (Refer to Screwed Steel Section)						
Carbon steel						
Pipework						
15mm dia	1.07	1.37	0.46	12.47	m	**13.84**
20mm dia	1.66	2.13	0.48	13.01	m	**15.14**
25mm dia	4.76	6.10	0.52	14.09	m	**20.19**
32mm dia	6.22	7.97	0.56	15.18	m	**23.14**
40mm dia	8.50	10.89	0.58	15.72	m	**26.61**
50mm dia	10.97	14.05	0.66	17.89	m	**31.94**
Extra over for Carbon Steel pressfit fittings						
Coupling						
15mm dia	0.73	0.94	0.36	9.76	nr	**10.69**
20mm dia	0.88	1.13	0.36	9.76	nr	**10.88**
25mm dia	1.11	1.42	0.44	11.93	nr	**13.35**
32mm dia	1.86	2.38	0.44	11.93	nr	**14.31**
40mm dia	2.49	3.19	0.52	14.09	nr	**17.28**
50mm dia	2.94	3.77	0.60	16.26	nr	**20.03**
Reducer						
20 x 15mm dia	0.67	0.86	0.36	9.76	nr	**10.62**
25 x 15mm dia	0.88	1.13	0.40	10.84	nr	**11.97**
25 x 20mm dia	0.93	1.19	0.40	10.84	nr	**12.03**
32 x 20mm dia	1.03	1.32	0.40	10.84	nr	**12.16**
32 x 25mm dia	1.10	1.41	0.44	11.93	nr	**13.33**
40 x 32mm dia	2.39	3.06	0.48	13.01	nr	**16.07**
50 x 20mm dia	6.89	8.82	0.48	13.01	nr	**21.83**
50 x 25mm dia	6.95	8.90	0.52	14.09	nr	**23.00**
50 x 40mm dia	7.30	9.35	0.56	15.18	nr	**24.53**
90° Elbow						
15mm dia	1.05	1.34	0.36	9.76	nr	**11.10**
20mm dia	1.38	1.77	0.36	9.76	nr	**11.52**
25mm dia	1.88	2.41	0.44	11.93	nr	**14.33**
32mm dia	4.71	6.03	0.44	11.93	nr	**17.96**
40mm dia	7.54	9.66	0.52	14.09	nr	**23.75**
50mm dia	9.01	11.54	0.60	16.26	nr	**27.80**
45° Elbow						
15mm dia	1.25	1.60	0.36	9.76	nr	**11.36**
20mm dia	1.39	1.78	0.36	9.76	nr	**11.54**
25mm dia	1.90	2.43	0.44	11.93	nr	**14.36**
32mm dia	3.73	4.78	0.44	11.93	nr	**16.70**
40mm dia	4.68	5.99	0.52	14.09	nr	**20.09**
50mm dia	5.29	6.78	0.60	16.26	nr	**23.04**

T: MECHANICAL/COOLING/HEATING SYSTEMS

Item	Net Price £	Material £	Labour hours	Labour £	Unit	Total rate £
Equal tee						
15mm dia	2.01	2.57	0.54	14.64	nr	**17.21**
20mm dia	2.32	2.97	0.54	14.64	nr	**17.61**
25mm dia	3.11	3.98	0.66	17.89	nr	**21.87**
32mm dia	4.84	6.20	0.66	17.89	nr	**24.09**
40mm dia	7.15	9.16	0.78	21.14	nr	**30.30**
50mm dia	8.58	10.99	0.90	24.39	nr	**35.38**
Reducing tee						
20 x 15mm dia	2.28	2.92	0.54	14.64	nr	**17.56**
25 x 15mm dia	3.08	3.94	0.62	16.80	nr	**20.75**
25 x 20mm dia	3.34	4.28	0.62	16.80	nr	**21.08**
32 x 15mm dia	4.52	5.79	0.62	16.80	nr	**22.59**
32 x 20mm dia	4.88	6.25	0.62	16.80	nr	**23.05**
32 x 25mm dia	4.95	6.34	0.62	16.80	nr	**23.14**
40 x 20mm dia	6.53	8.36	0.70	18.97	nr	**27.34**
40 x 25mm dia	6.77	8.67	0.70	18.97	nr	**27.64**
40 x 32mm dia	6.62	8.48	0.70	18.97	nr	**27.45**
50 x 20mm dia	7.79	9.98	0.82	22.22	nr	**32.20**
50 x 25mm dia	7.94	10.17	0.82	22.22	nr	**32.39**
50 x 32mm dia	8.17	10.46	0.82	22.22	nr	**32.69**
50 x 40mm dia	8.57	10.98	0.82	22.22	nr	**33.20**
MECHANICAL GROOVED						
Mechanical grooved jointing system; working temperature not exceeding 82 °s C BS 5750; pipework complete with grooved joints; painted finish. With brackets measured separately (Refer to Screwed Steel Section)						
Grooved Joints						
65 mm	11.94	15.29	0.58	15.72	m	**31.01**
80 mm	15.08	19.31	0.68	18.43	m	**37.74**
100 mm	20.90	26.77	0.79	21.41	m	**48.18**
125 mm	28.16	36.07	1.02	27.65	m	**63.71**
150mm	33.20	42.52	1.15	31.17	m	**73.69**
Extra over mechanical grooved system fittings						
Couplings						
65mm	7.61	9.75	0.41	11.11	nr	**20.86**
80mm	7.90	10.12	0.41	11.11	nr	**21.23**
100mm	9.79	12.54	0.66	17.89	nr	**30.43**
125mm	18.43	23.61	0.68	18.43	nr	**42.04**
150mm	15.19	19.46	0.80	21.68	nr	**41.14**
Concentric reducers (one size down)						
80mm	26.62	34.09	0.59	15.99	nr	**50.09**
100mm	31.08	39.81	0.71	19.24	nr	**59.05**
125mm	54.22	69.44	0.85	23.04	nr	**92.48**
150mm	55.21	70.71	0.98	26.56	nr	**97.27**
Short radius elbow; 90°						
65mm	26.77	34.29	0.53	14.36	nr	**48.65**
80mm	29.84	38.22	0.61	16.53	nr	**54.75**
100mm	39.43	50.50	0.80	21.68	nr	**72.18**
125mm	58.99	75.55	0.90	24.39	nr	**99.95**
150mm	75.23	96.35	0.94	25.48	nr	**121.83**

T: MECHANICAL/COOLING/HEATING SYSTEMS

Item	Net Price £	Material £	Labour hours	Labour £	Unit	Total rate £
T31 : LOW TEMPERATURE HOT WATER HEATING (cont'd)						
MECHANICAL GROOVED (cont'd)						
Extra over mechanical grooved (cont'd)						
Short radius elbow; 45°						
65mm	25.16	32.22	0.53	14.36	nr	**46.59**
80mm	29.19	37.39	0.61	16.53	nr	**53.92**
100mm	37.51	48.04	0.80	21.68	nr	**69.73**
125mm	56.41	72.25	0.90	24.39	nr	**96.64**
150mm	67.43	86.36	0.94	25.48	nr	**111.84**
Equal tee						
65mm	43.51	55.73	0.83	22.50	nr	**78.22**
80mm	48.97	62.72	0.93	25.21	nr	**87.93**
100mm	60.06	76.92	1.18	31.98	nr	**108.91**
125mm	113.92	145.91	1.37	37.13	nr	**183.04**
150mm	122.42	156.79	1.43	38.76	nr	**195.55**
PLASTIC PIPEWORK						
Polypropylene PP-R 80 pipe, mechanically stabilised by fibre compound mixture in middle layer; suitable for continuous working temperatures of 0-90 °s C; thermally fused joints in the running length						
Pipe; 4m long; PN 20						
20mm dia	1.58	2.02	0.35	9.49	m	**11.51**
25mm dia	2.38	3.05	0.39	10.57	m	**13.62**
32mm dia	2.71	3.47	0.43	11.65	m	**15.13**
40mm dia	3.62	4.64	0.47	12.74	m	**17.38**
50mm dia	5.27	6.75	0.51	13.82	m	**20.57**
63mm dia	8.69	11.13	0.52	14.09	m	**25.22**
75mm dia	11.24	14.40	0.60	16.26	m	**30.66**
90mm dia	17.33	22.20	0.69	18.70	m	**40.90**
110mm dia	26.06	33.38	0.69	18.70	m	**52.08**
125mm dia	27.92	35.76	0.85	23.04	m	**58.80**
FIXINGS						
Refer to steel pipes; black malleable iron. For minimum fixing distances, refer to the Tables and Memoranda to the rear of the book						
Extra over fittings; thermally fused joints						
Overbridge bow						
20mm dia	1.05	1.34	0.51	13.82	nr	**15.17**
25mm dia	1.93	2.47	0.56	15.18	nr	**17.65**
32mm dia	3.85	4.93	0.65	17.62	nr	**22.55**

T: MECHANICAL/COOLING/HEATING SYSTEMS

Item	Net Price £	Material £	Labour hours	Labour £	Unit	Total rate £
Elbow 90°						
20mm dia	0.36	0.46	0.44	11.93	nr	**12.39**
25mm dia	0.47	0.60	0.52	14.09	nr	**14.70**
32mm dia	0.68	0.87	0.59	15.99	nr	**16.86**
40mm dia	1.06	1.36	0.66	17.89	nr	**19.25**
50mm dia	2.28	2.92	0.73	19.79	nr	**22.71**
63mm dia	3.49	4.47	0.85	23.04	nr	**27.51**
75mm dia	7.72	9.89	0.85	23.04	nr	**32.93**
90mm dia	14.25	18.25	1.04	28.19	nr	**46.44**
110mm dia	20.27	25.96	1.04	28.19	nr	**54.15**
125mm dia	31.22	39.99	1.30	35.23	nr	**75.22**
Long bend 90°						
20mm dia	1.88	2.41	0.48	13.01	nr	**15.42**
25mm dia	1.96	2.51	0.57	15.45	nr	**17.96**
32mm dia	2.30	2.95	0.65	17.62	nr	**20.56**
40mm dia	4.21	5.39	0.73	19.79	nr	**25.18**
Elbow 90°, female/male						
20mm dia	0.36	0.46	0.44	11.93	nr	**12.39**
25mm dia	0.47	0.60	0.52	14.09	nr	**14.70**
32mm dia	0.68	0.87	0.59	15.99	nr	**16.86**
Elbow 45°						
20mm dia	0.36	0.46	0.44	11.93	nr	**12.39**
25mm dia	0.47	0.60	0.52	14.09	nr	**14.70**
32mm dia	0.68	0.87	0.59	15.99	nr	**16.86**
40mm dia	1.06	1.36	0.66	17.89	nr	**19.25**
50mm dia	2.28	2.92	0.73	19.79	nr	**22.71**
63mm dia	3.49	4.47	0.85	23.04	nr	**27.51**
75mm dia	7.72	9.89	0.85	23.04	nr	**32.93**
90mm dia	14.25	18.25	1.04	28.19	nr	**46.44**
110mm dia	20.27	25.96	1.04	28.19	nr	**54.15**
125mm dia	31.22	39.99	1.30	35.23	nr	**75.22**
Elbow 45°, female/male						
20mm dia	0.36	0.46	0.44	11.93	nr	**12.39**
25mm dia	0.47	0.60	0.52	14.09	nr	**14.70**
32mm dia	0.68	0.87	0.59	15.99	nr	**16.86**
T-Piece 90°						
20mm dia	0.49	0.63	0.61	16.53	nr	**17.16**
25mm dia	0.67	0.86	0.72	19.51	nr	**20.37**
32mm dia	0.87	1.11	0.83	22.50	nr	**23.61**
40mm dia	1.33	1.70	0.92	24.94	nr	**26.64**
50mm dia	3.79	4.85	1.01	27.37	nr	**32.23**
63mm dia	5.44	6.97	1.11	30.08	nr	**37.05**
75mm dia	9.06	11.60	1.18	31.98	nr	**43.59**
90mm dia	16.68	21.36	1.46	39.57	nr	**60.93**
110mm dia	26.03	33.34	1.46	39.57	nr	**72.91**
125mm dia	34.58	44.29	1.82	49.33	nr	**93.62**

T: MECHANICAL/COOLING/HEATING SYSTEMS

Item	Net Price £	Material £	Labour hours	Labour £	Unit	Total rate £
T31 : LOW TEMPERATURE HOT WATER HEATING (cont'd)						
PLASTIC PIPEWORK (cont'd)						
Extra over fittings; thermally fused (cont'd)						
T-Piece 90° reducing						
25 x 20 x 25mm	0.68	0.87	0.72	19.51	nr	20.39
32 x 20 x 32mm	0.87	1.11	0.83	22.50	nr	23.61
32 x 25 x 32mm	0.87	1.11	0.83	22.50	nr	23.61
40 x 20 x 40mm	1.33	1.70	0.92	24.94	nr	26.64
40 x 25 x 40mm	1.33	1.70	0.92	24.94	nr	26.64
40 x 32 x 40mm	1.33	1.70	0.92	24.94	nr	26.64
50 x 25 x 50mm	3.79	4.85	1.01	27.37	nr	32.23
50 x 32 x 50mm	3.79	4.85	1.01	27.37	nr	32.23
50 x 40 x 50mm	3.79	4.85	1.01	27.37	nr	32.23
63 x 20 x 63mm	5.12	6.56	1.11	30.08	nr	36.64
63 x 25 x 63mm	5.12	6.56	1.11	30.08	nr	36.64
63 x 32 x 63mm	5.12	6.56	1.11	30.08	nr	36.64
63 x 40 x 63mm	5.12	6.56	1.01	27.37	nr	33.93
63 x 50 x 63mm	5.12	6.56	1.01	27.37	nr	33.93
75 x 20 x 75mm	8.32	10.66	1.18	31.98	nr	42.64
75 x 25 x 75mm	8.32	10.66	1.18	31.98	nr	42.64
75 x 32 x 75mm	8.32	10.66	1.18	31.98	nr	42.64
75 x 40 x 75mm	8.32	10.66	1.18	31.98	nr	42.64
75 x 50 x 75mm	8.32	10.66	1.18	31.98	nr	42.64
75 x 63 x 75mm	8.32	10.66	1.18	31.98	nr	42.64
25 x 20 x 20mm	0.68	0.87	0.72	19.51	nr	20.39
90 x 63 x 90mm	16.68	21.36	1.46	39.57	nr	60.93
110 x 75 x 110mm	26.03	33.34	1.46	39.57	nr	72.91
110 x 90 x 110mm	26.03	33.34	1.46	39.57	nr	72.91
125 x 90 x 125mm	30.96	39.65	1.82	49.33	nr	88.98
125 x 110 x 125mm	31.59	40.46	1.82	49.33	nr	89.79
Reducer						
25 x 20mm	0.39	0.50	0.59	15.99	nr	16.49
32 x 20mm	0.50	0.64	0.62	16.80	nr	17.44
32 x 25mm	0.50	0.64	0.62	16.80	nr	17.44
40 x 20mm	0.80	1.02	0.66	17.89	nr	18.91
40 x 25mm	0.80	1.02	0.66	17.89	nr	18.91
40 x 32mm	0.80	1.02	0.66	17.89	nr	18.91
50 x 20mm	1.28	1.64	0.73	19.79	nr	21.42
50 x 25mm	1.28	1.64	0.73	19.79	nr	21.42
50 x 32mm	1.28	1.64	0.73	19.79	nr	21.42
50 x 40mm	1.28	1.64	0.73	19.79	nr	21.42
63 x 40mm	2.58	3.30	0.78	21.14	nr	24.45
63 x 25mm	2.58	3.30	0.78	21.14	nr	24.45
63 x 32mm	2.58	3.30	0.78	21.14	nr	24.45
63 x 50mm	2.58	3.30	0.78	21.14	nr	24.45
75 x 50mm	2.88	3.69	0.85	23.04	nr	26.73
75 x 63mm	2.88	3.69	0.85	23.04	nr	26.73
90 x 63mm	6.44	8.25	1.04	28.19	nr	36.44
90 x 75mm	6.44	8.25	1.04	28.19	nr	36.44
110 x 90mm	10.40	13.32	1.17	31.71	nr	45.03
125 x 110mm	16.26	20.83	1.43	38.76	nr	59.58

T: MECHANICAL/COOLING/HEATING SYSTEMS

Item	Net Price £	Material £	Labour hours	Labour £	Unit	Total rate £
Socket						
20mm dia	0.35	0.45	0.51	13.82	nr	14.27
25mm dia	0.39	0.50	0.56	15.18	nr	15.68
32mm dia	0.50	0.64	0.65	17.62	nr	18.26
40mm dia	0.62	0.79	0.74	20.06	nr	20.85
50mm dia	1.28	1.64	0.81	21.95	nr	23.59
63mm dia	2.58	3.30	0.86	23.31	nr	26.61
75mm dia	2.88	3.69	0.91	24.66	nr	28.35
90mm dia	7.44	9.53	0.91	24.66	nr	34.19
110mm dia	12.64	16.19	0.91	24.66	nr	40.85
125mm dia	17.61	22.55	1.30	35.23	nr	57.79
End Cap						
20mm dia	0.54	0.69	0.25	6.78	nr	7.47
25mm dia	0.67	0.86	0.29	7.86	nr	8.72
32mm dia	0.82	1.05	0.33	8.94	nr	9.99
40mm dia	1.31	1.68	0.36	9.76	nr	11.44
50mm dia	1.82	2.33	0.40	10.84	nr	13.17
63mm dia	3.06	3.92	0.44	11.93	nr	15.84
75mm dia	4.41	5.65	0.47	12.74	nr	18.39
90mm dia	9.99	12.80	0.57	15.45	nr	28.24
110mm dia	12.01	15.38	0.57	15.45	nr	30.83
125mm dia	18.30	23.44	0.85	23.04	nr	46.48
Stub flange with gasket						
32mm dia	12.14	15.55	0.23	6.23	nr	21.78
40mm dia	15.27	19.56	0.27	7.32	nr	26.88
50mm dia	18.47	23.66	0.38	10.30	nr	33.96
63mm dia	22.18	28.41	0.43	11.65	nr	40.06
75mm dia	26.03	33.34	0.48	13.01	nr	46.35
90mm dia	35.22	45.11	0.53	14.36	nr	59.47
110mm dia	49.32	63.17	0.53	14.36	nr	77.53
125mm dia	70.69	90.54	0.75	20.33	nr	110.87
Weld in saddle with female thread						
40 - 1/2"	0.88	1.13	0.36	9.76	nr	10.88
50 - 1/2"	0.88	1.13	0.36	9.76	nr	10.88
63 - 1/2"	0.88	1.13	0.40	10.84	nr	11.97
75 - 1/2"	0.88	1.13	0.40	10.84	nr	11.97
90 - 1/2"	0.88	1.13	0.46	12.47	nr	13.59
110 - 1/2"	0.88	1.13	0.46	12.47	nr	13.59
Weld in saddle with male thread						
50 - 1/2"	0.88	1.13	0.36	9.76	nr	10.88
63 - 1/2"	0.88	1.13	0.40	10.84	nr	11.97
75 - 1/2"	0.88	1.13	0.40	10.84	nr	11.97
90 - 1/2"	0.88	1.13	0.46	12.47	nr	13.59
110 - 1/2"	0.88	1.13	0.46	12.47	nr	13.59
Transition piece, round with female thread						
20 x 1/2"	2.04	2.61	0.29	7.86	nr	10.47
20 x 3/4"	2.69	3.45	0.29	7.86	nr	11.31
25 x 1/2"	2.04	2.61	0.33	8.94	nr	11.56
25 x 3/4"	2.69	3.45	0.33	8.94	nr	12.39

T: MECHANICAL/COOLING/HEATING SYSTEMS

Item	Net Price £	Material £	Labour hours	Labour £	Unit	Total rate £
T31 : LOW TEMPERATURE HOT WATER HEATING (cont'd)						
PLASTIC PIPEWORK (cont'd)						
Extra over fittings; thermally fused (cont'd)						
Transition piece, hexagon with female thread						
32 x 1"	7.60	9.73	0.36	9.76	nr	19.49
40 x 1 1/4"	12.02	15.40	0.36	9.76	nr	25.15
50 x 1 1/2"	13.95	17.87	0.36	9.76	nr	27.62
63 x 2"	21.62	27.69	0.40	10.84	nr	38.53
75 x 2"	22.55	28.88	0.40	10.84	nr	39.72
125 x 5"	128.04	163.99	0.51	13.82	nr	177.82
Stop valve for surface assembly						
20mm dia	8.39	10.75	0.25	6.78	nr	17.52
25mm dia	8.39	10.75	0.29	7.86	nr	18.61
32mm dia	15.80	20.24	0.33	8.94	nr	29.18
Ball valve						
20mm dia	29.54	37.83	0.25	6.78	nr	44.61
25mm dia	31.65	40.54	0.29	7.86	nr	48.40
32mm dia	38.03	48.71	0.33	8.94	nr	57.65
40mm dia	48.54	62.17	0.36	9.76	nr	71.93
50mm dia	66.64	85.35	0.40	10.84	nr	96.19
63mm dia	75.13	96.23	0.44	11.93	nr	108.15
Floor or ceiling cover plates						
Plastic						
15mm dia	0.32	0.41	0.16	4.34	nr	4.75
20mm dia	0.32	0.41	0.22	5.96	nr	6.37
25mm dia	0.34	0.44	0.22	5.96	nr	6.40
32mm dia	0.38	0.49	0.24	6.50	nr	6.99
40mm dia	0.84	1.08	0.26	7.05	nr	8.12
50mm dia	0.93	1.19	0.26	7.05	nr	8.24
Chromium plated						
15mm dia	1.95	2.50	0.16	4.34	nr	6.83
20mm dia	2.08	2.66	0.17	4.61	nr	7.27
25mm dia	2.16	2.77	0.21	5.69	nr	8.46
32mm dia	2.21	2.83	0.22	5.96	nr	8.79
40mm dia	2.49	3.19	0.26	7.05	nr	10.24
50mm dia	2.98	3.82	0.26	7.05	nr	10.86

T: MECHANICAL/COOLING/HEATING SYSTEMS

Item	Net Price £	Material £	Labour hours	Labour £	Unit	Total rate £
Y11 - PIPELINE ANCILLARIES						
EXPANSION JOINTS						
Axial movement bellows expansion joints; stainless steel						
Screwed ends for steel pipework; up to 6 bar G at 100°C						
15mm dia	48.00	61.48	0.68	18.43	nr	**79.91**
20mm dia	57.00	73.00	0.81	21.95	nr	**94.96**
25mm dia	57.00	73.00	0.93	25.21	nr	**98.21**
32mm dia	63.00	80.69	1.06	28.73	nr	**109.42**
40mm dia	80.00	102.46	1.16	31.44	nr	**133.90**
50mm dia	111.00	142.17	1.19	32.25	nr	**174.42**
Screwed ends for steel pipework; aluminium and steel outer sleeves; up to 16 bar G at 120°C						
20mm dia	71.00	90.94	1.32	35.78	nr	**126.71**
25mm dia	77.00	98.62	1.52	41.20	nr	**139.82**
32mm dia	110.00	140.89	1.80	48.79	nr	**189.67**
40mm dia	119.00	152.41	2.03	55.02	nr	**207.43**
50mm dia	143.00	183.15	2.26	61.25	nr	**244.41**
Flanged ends for steel pipework; aluminium and steel outer sleeves; up to 16 bar G at 120°C						
20mm dia	84.00	107.59	0.53	14.36	nr	**121.95**
25mm dia	90.00	115.27	0.64	17.35	nr	**132.62**
32mm dia	124.00	158.82	0.74	20.06	nr	**178.87**
40mm dia	134.00	171.63	0.82	22.22	nr	**193.85**
50mm dia	159.00	203.65	0.89	24.12	nr	**227.77**
Flanged ends for steel pipework; up to 16 bar G at 120°C						
65mm dia	147.00	188.28	1.10	29.81	nr	**218.09**
80mm dia	158.00	202.36	1.31	35.51	nr	**237.87**
100mm dia	181.00	231.82	1.78	48.24	nr	**280.07**
150mm dia	277.00	354.78	3.08	83.48	nr	**438.26**
Screwed ends for non-ferrous pipework; up to 6 bar G at 100°C						
20mm dia	88.00	112.71	0.72	19.51	nr	**132.22**
25mm dia	90.00	115.27	0.84	22.77	nr	**138.04**
32mm dia	101.00	129.36	1.02	27.65	nr	**157.01**
40mm dia	118.00	151.13	1.11	30.08	nr	**181.22**
50mm dia	157.00	201.08	1.18	31.98	nr	**233.07**
Flanged ends for steel, copper or non-ferrous pipework; up to 16 bar G at 120°C						
65mm dia	189.00	242.07	0.87	23.58	nr	**265.65**
80mm dia	195.00	249.75	0.95	25.75	nr	**275.50**
100mm dia	225.00	288.18	1.15	31.17	nr	**319.35**
150mm dia	369.00	472.61	1.36	36.86	nr	**509.47**

T: MECHANICAL/COOLING/HEATING SYSTEMS

Item	Net Price £	Material £	Labour hours	Labour £	Unit	Total rate £
T31 : LOW TEMPERATURE HOT WATER HEATING (cont'd)						
EXPANSION JOINTS (cont'd)						
Angular movement bellows expansion joints; stainless steel						
Flanged ends for steel pipework; up to 16 bar G at 120°C						
50mm dia	410.00	525.12	0.71	19.24	nr	544.37
65mm dia	428.00	548.18	0.83	22.50	nr	570.67
80mm dia	450.00	576.36	0.91	24.66	nr	601.02
100mm dia	512.00	655.76	0.97	26.29	nr	682.05
125mm dia	554.00	709.56	1.16	31.44	nr	741.00
150mm dia	586.00	750.54	1.18	31.98	nr	782.52
VALVES						
Isolating valves						
Bronze gate valve; non-rising stem; BS 5154, series B, PN 32; working pressure up to 14 bar for saturated steam, 32 bar from -10°C to 100°C; screwed ends to steel						
15mm dia	24.30	31.12	1.11	30.08	nr	61.21
20mm dia	31.32	40.11	1.28	34.69	nr	74.81
25mm dia	42.45	54.37	1.49	40.38	nr	94.75
32mm dia	62.05	79.47	1.88	50.95	nr	130.43
40mm dia	80.19	102.71	2.31	62.61	nr	165.32
50mm dia	115.90	148.44	2.80	75.89	nr	224.33
Bronze gate valve; non-rising stem; BS 5154, series B, PN 20; working pressure up to 9 bar for saturated steam, 20 bar from -10°C to 100°C; screwed ends to steel						
15mm dia	16.07	20.58	0.84	22.77	nr	43.35
20mm dia	22.80	29.20	1.01	27.37	nr	56.58
25mm dia	29.86	38.24	1.19	32.25	nr	70.50
32mm dia	42.59	54.55	1.38	37.40	nr	91.95
40mm dia	58.91	75.45	1.62	43.91	nr	119.36
50mm dia	85.51	109.52	1.94	52.58	nr	162.10
Bronze gate valve; non-rising stem; BS 5154, series B, PN 16; working pressure up to 7 bar for saturated steam, 16 bar from -10°C to 100°C; BS4504 flanged ends; bolted connections						
15mm dia	-	-	2.81	76.16	nr	76.16
20mm dia	95.97	122.92	1.24	33.61	nr	156.53
25mm dia	125.86	161.20	1.31	35.51	nr	196.71
32mm dia	163.99	210.04	1.43	38.76	nr	248.79
40mm dia	196.45	251.61	1.53	41.47	nr	293.08
50mm dia	273.95	350.87	1.63	44.18	nr	395.05
65mm dia	422.08	540.60	1.71	46.35	nr	586.94
80mm dia	596.94	764.55	1.88	50.95	nr	815.51
100mm dia	1058.69	1355.96	2.03	55.02	nr	1410.98

T: MECHANICAL/COOLING/HEATING SYSTEMS

Item	Net Price £	Material £	Labour hours	Labour £	Unit	Total rate £
Cast iron gate valve; bronze trim; non rising stem; BS 5150, PN6; working pressure 6 bar from -10°C to 120°C; BS4504 flanged ends; bolted connections						
50mm dia	171.68	219.89	1.85	50.14	nr	270.03
65mm dia	171.68	219.89	2.00	54.21	nr	274.09
80mm dia	198.41	254.12	2.27	61.52	nr	315.65
100mm dia	262.16	335.77	2.76	74.81	nr	410.58
125mm dia	367.33	470.47	6.05	163.98	nr	634.45
150mm dia	441.90	565.98	8.03	217.64	nr	783.62
200mm dia	805.12	1031.19	9.17	248.54	nr	1279.73
250mm dia	1220.42	1563.10	10.72	290.55	nr	1853.65
300mm dia	1447.51	1853.96	11.75	318.47	nr	2172.42
Cast iron gate valve; bronze trim; non rising stem; BS 5150, PN10; working pressure up to 8.4 bar for saturated steam, 10 bar from -10°C to 120°C; BS4504 flanged ends; bolted connections						
50mm dia	184.94	236.87	1.85	50.14	nr	287.01
65mm dia	184.94	236.87	2.00	54.21	nr	291.08
80mm dia	214.42	274.63	2.27	61.52	nr	336.15
100mm dia	284.53	364.42	2.76	74.81	nr	439.23
125mm dia	401.39	514.10	6.05	163.98	nr	678.07
150mm dia	462.37	592.20	8.03	217.64	nr	809.84
200mm dia	1223.64	1567.23	9.17	248.54	nr	1815.76
250mm dia	1557.36	1994.65	10.72	290.55	nr	2285.20
300mm dia	2069.28	2650.31	11.75	318.47	nr	2968.78
350mm dia	3381.81	4331.39	12.67	343.40	nr	4674.79
Cast iron gate valve; bronze trim; non rising stem; BS 5163 series A, PN16; working pressure for cold water services up to 16 bar; BS4504 flanged ends; bolted connections						
50mm dia	349.06	447.07	1.85	50.14	nr	497.21
65mm dia	362.55	464.35	2.00	54.21	nr	518.56
80mm dia	372.96	477.68	2.27	61.52	nr	539.21
100mm dia	484.30	620.29	2.76	74.81	nr	695.09
125mm dia	631.51	808.83	6.05	163.98	nr	972.81
150mm dia	761.60	975.45	8.03	217.64	nr	1193.09
Ball valves						
Malleable iron body; lever operated stainless steel ball and stem; working pressure up to 12 bar; flanged ends to BS 4504 16/11; bolted connections						
40mm dia	150.19	192.36	1.54	41.74	nr	234.10
50mm dia	188.95	242.01	1.64	44.45	nr	286.45
80mm dia	316.00	404.73	1.92	52.04	nr	456.77
100mm dia	584.09	748.10	2.80	75.89	nr	823.99
150mm dia	794.60	1017.72	12.05	326.60	nr	1344.31
Malleable iron body; lever operated stainless steel ball and stem; working pressure up to 16 bar; screwed ends to steel						
20mm dia	33.04	42.32	1.34	36.33	nr	78.65
25mm dia	33.98	43.52	1.40	37.96	nr	81.48
32mm dia	46.80	59.94	1.46	39.63	nr	99.57
40mm dia	46.80	59.94	1.54	41.76	nr	101.70
50mm dia	55.97	71.69	1.64	44.50	nr	116.19

T: MECHANICAL/COOLING/HEATING SYSTEMS

Item	Net Price £	Material £	Labour hours	Labour £	Unit	Total rate £
T31 : LOW TEMPERATURE HOT WATER HEATING (cont'd)						
VALVES (cont'd)						
Ball valves (cont'd)						
Carbon steel body; lever operated stainless steel ball and stem; Class 150; working pressure up to 19 bar; screwed ends to steel						
15mm dia	24.27	31.08	0.84	22.78	nr	53.86
20mm dia	25.48	32.63	1.14	30.91	nr	63.54
25mm dia	29.17	37.36	1.30	35.25	nr	72.61
Globe valves						
Bronze; rising stem; renewable disc; BS 5154 series B, PN32; working pressure up to 14 bar for saturated steam, 32 bar from -10°C to 100°C; screwed ends to steel						
15mm dia	26.54	33.99	0.77	20.87	nr	54.86
20mm dia	37.10	47.52	1.03	27.92	nr	75.43
25mm dia	52.06	66.68	1.19	32.25	nr	98.93
32mm dia	72.93	93.41	1.38	37.40	nr	130.81
40mm dia	101.34	129.80	1.62	43.91	nr	173.70
50mm dia	155.56	199.24	1.61	43.64	nr	242.88
Bronze; needle valve; rising stem; BS 5154, series B, PN32; working pressure up to 14 bar for saturated steam, 32 bar from -10°C to 100°C; screwed ends to steel						
15mm dia	27.30	34.97	1.07	29.00	nr	63.97
20mm dia	46.22	59.20	1.18	31.98	nr	91.18
25mm dia	65.17	83.47	1.27	34.42	nr	117.89
32mm dia	135.93	174.10	1.35	36.59	nr	210.69
40mm dia	214.14	274.27	1.47	39.84	nr	314.11
50mm dia	271.22	347.38	1.61	43.64	nr	391.01
Bronze; rising stem; renewable disc; BS 5154, series B, PN16; working pressure up to 7 bar for saturated steam, 16 bar from -10°C to 100°C; BS4504 flanged ends; bolted connections						
15mm dia	72.58	92.96	1.16	31.44	nr	124.40
20mm dia	83.94	107.51	1.26	34.15	nr	141.66
25mm dia	147.21	188.55	1.38	37.40	nr	225.95
32mm dia	185.83	238.01	1.47	39.84	nr	277.85
40mm dia	211.23	270.54	1.56	42.28	nr	312.82
50mm dia	303.07	388.17	1.71	46.35	nr	434.52

T: MECHANICAL/COOLING/HEATING SYSTEMS

Item	Net Price £	Material £	Labour hours	Labour £	Unit	Total rate £
Bronze; rising stem; renewable disc; BS 2060, class 250; working pressure up to 24 bar for saturated steam, 38 bar from -10°C to 100°C; flanged ends (BS 10 table H); bolted connections						
15mm dia	185.17	237.16	1.16	31.44	nr	268.60
20mm dia	215.20	275.63	1.26	34.15	nr	309.78
25mm dia	295.72	378.76	1.38	37.40	nr	416.16
32mm dia	386.32	494.79	1.47	39.84	nr	534.64
40mm dia	462.61	592.51	1.56	42.28	nr	634.79
50mm dia	719.35	921.34	1.71	46.35	nr	967.68
65mm dia	860.28	1101.84	1.88	50.95	nr	1152.79
80mm dia	2083.29	2668.26	2.03	55.02	nr	2723.28
Check valves						
Bronze; swing pattern; BS 5154 series B, PN 25; working pressure up to 10.5 bar for saturated steam, 25 bar from -10°C to 100°C; screwed ends to steel						
15mm dia	16.90	21.65	0.77	20.87	nr	42.52
20mm dia	20.08	25.72	1.03	27.92	nr	53.63
25mm dia	27.83	35.64	1.19	32.25	nr	67.90
32mm dia	47.15	60.39	1.38	37.40	nr	97.79
40mm dia	58.67	75.14	1.62	43.91	nr	119.05
50mm dia	89.97	115.23	1.94	52.58	nr	167.81
65mm dia	183.65	235.22	2.45	66.40	nr	301.62
80mm dia	259.68	332.60	2.83	76.70	nr	409.30
Bronze; vertical lift pattern; BS 5154 series B, PN32; working pressure up to 14 bar for saturated steam, 32 bar from -10°C to 100°C; screwed ends to steel						
15mm dia	31.31	40.10	0.96	26.02	nr	66.12
20mm dia	35.95	46.04	1.07	29.00	nr	75.05
25mm dia	39.75	50.91	1.17	31.71	nr	82.62
32mm dia	49.62	63.55	1.33	36.05	nr	99.60
40mm dia	66.18	84.76	1.41	38.22	nr	122.98
50mm dia	101.78	130.36	1.55	42.01	nr	172.37
65mm dia	313.95	402.10	1.80	48.79	nr	450.89
80mm dia	469.35	601.14	1.99	53.94	nr	655.07
Bronze; oblique swing pattern; BS 5154 series A, PN32; working pressure up to 14 bar for saturated steam, 32 bar from -10°C to 120°C; screwed connections to steel						
15mm dia	48.66	62.32	0.96	26.02	nr	88.34
20mm dia	53.44	68.45	1.07	29.00	nr	97.45
25mm dia	81.43	104.29	1.17	31.71	nr	136.01
32mm dia	117.68	150.72	1.33	36.05	nr	186.77
40mm dia	130.41	167.03	1.41	38.22	nr	205.24
50mm dia	187.00	239.51	1.55	42.01	nr	281.52

T: MECHANICAL/COOLING/HEATING SYSTEMS

Item	Net Price £	Material £	Labour hours	Labour £	Unit	Total rate £
T31 : LOW TEMPERATURE HOT WATER HEATING (cont'd)						
VALVES (cont'd)						
Check valves (cont'd)						
Cast iron; swing pattern; BS 5153 PN6; working pressure up to 6 bar from -10°C to 120°C; BS 4504 flanged ends; bolted connections						
50mm dia	214.93	275.28	1.86	50.41	nr	325.69
65mm dia	235.33	301.41	2.00	54.21	nr	355.62
80mm dia	264.88	339.26	2.56	69.39	nr	408.64
100mm dia	339.32	434.60	2.76	74.81	nr	509.40
125mm dia	506.22	648.36	6.05	163.98	nr	812.34
150mm dia	569.18	729.00	8.11	219.81	nr	948.81
200mm dia	1258.73	1612.17	9.26	250.98	nr	1863.15
250mm dia	-	-	10.72	290.55	nr	290.55
300mm dia	-	-	11.75	318.47	nr	318.47
Cast iron; horizontal lift pattern; BS 5153 PN16; working pressure up to 13 bar for saturated steam, 16 bar from -10°C to 120°C; BS 4504 flanged ends; bolted connections						
50mm dia	194.09	248.59	1.86	50.41	nr	299.00
65mm dia	194.09	248.59	2.00	54.21	nr	302.80
80mm dia	240.30	307.77	2.56	69.39	nr	377.16
100mm dia	279.84	358.42	2.96	80.23	nr	438.64
125mm dia	417.51	534.74	7.76	210.32	nr	745.07
150mm dia	469.41	601.22	10.50	284.59	nr	885.80
Cast iron; semi lugged butterfly valve; BS5155 PN16; working pressure 16 bar from -10°C to 120°C; BS 4504 flanged ends; bolted connections						
50mm dia	62.08	79.51	2.20	59.63	nr	139.14
65mm dia	62.08	79.51	2.31	62.61	nr	142.12
80mm dia	74.51	95.43	2.88	78.06	nr	173.49
100mm dia	112.13	143.62	3.11	84.29	nr	227.91
125mm dia	163.99	210.04	5.02	136.06	nr	346.10
150mm dia	190.29	243.72	6.98	189.18	nr	432.90
200mm dia	332.32	425.63	8.25	223.60	nr	649.24
250mm dia	663.45	849.74	10.47	283.77	nr	1133.51
300mm dia	841.56	1077.86	11.48	311.15	nr	1389.01
Commissioning valves						
Bronze commissioning set; metering station; double regulating valve; BS5154 PN20 Series B; working pressure 20 bar from -10°C to 100°C; screwed ends to steel						
15mm dia	40.61	52.01	1.08	29.27	nr	81.28
20mm dia	65.42	83.79	1.46	39.57	nr	123.36
25mm dia	80.08	102.57	1.68	45.53	nr	148.10
32mm dia	106.26	136.10	1.95	52.85	nr	188.95
40mm dia	157.92	202.26	2.27	61.52	nr	263.79
50mm dia	226.73	290.39	2.73	73.99	nr	364.39

T: MECHANICAL/COOLING/HEATING SYSTEMS

Item	Net Price £	Material £	Labour hours	Labour £	Unit	Total rate £
Cast iron commissioning set; metering station; double regulating valve; BS5152 PN16; working pressure 16 bar from -10°C to 90°C; flanged ends (BS 4504, Part 1, Table 16); bolted connections						
65mm dia	372.93	477.64	1.80	48.79	nr	526.43
80mm dia	448.18	574.02	2.56	69.39	nr	643.41
100mm dia	606.64	776.98	2.30	62.34	nr	839.32
125mm dia	987.42	1264.68	2.44	66.13	nr	1330.81
150mm dia	1239.07	1586.99	2.90	78.60	nr	1665.59
200mm dia	3616.89	4632.48	8.26	223.87	nr	4856.35
250mm dia	5214.07	6678.13	10.49	284.32	nr	6962.44
300mm dia	8215.81	10522.73	11.49	311.42	nr	10834.15
Cast iron variable orifice double regulating valve; orifice valve; BS5152 PN16; working pressure 16 bar from -10° to 90°C; flanged ends (BS 4504, Part 1, Table 16); bolted connections						
65mm dia	343.72	440.23	2.00	54.21	nr	494.44
80mm dia	428.58	548.92	2.56	69.39	nr	618.31
100mm dia	570.87	731.16	2.96	80.23	nr	811.39
125mm dia	835.70	1070.36	7.76	210.32	nr	1280.68
150mm dia	1037.42	1328.72	10.50	284.59	nr	1613.30
200mm dia	2685.79	3439.93	8.26	223.87	nr	3663.81
250mm dia	4279.28	5480.86	10.49	284.32	nr	5765.17
300mm dia	6815.31	8728.98	11.49	311.42	nr	9040.40
Cast iron globe valve with double regulating feature; BS5152 PN16; working pressure 16 bar from -10°C to 120°C; flanged ends (BS 4504, Part 1, Table 16); bolted connections						
65mm dia	264.39	338.63	2.00	54.21	nr	392.84
80mm dia	324.38	415.46	2.56	69.39	nr	484.85
100mm dia	455.62	583.55	2.96	80.23	nr	663.78
125mm dia	699.27	895.62	7.76	210.32	nr	1105.94
150mm dia	902.90	1156.43	10.50	284.59	nr	1441.01
200mm dia	2719.06	3482.54	8.26	223.87	nr	3706.42
250mm dia	4216.35	5400.26	10.49	284.32	nr	5684.57
300mm dia	6671.55	8544.85	11.49	311.42	nr	8856.27
Bronze autoflow commissioning valve; PN25; working pressure 25 bar up to 100°C; screwed ends to steel						
15mm dia	88.22	112.99	0.82	22.22	nr	135.22
20mm dia	91.34	116.99	1.08	29.27	nr	146.26
25mm dia	134.67	172.48	1.27	34.42	nr	206.91
32mm dia	173.84	222.65	1.50	40.66	nr	263.31
40mm dia	217.94	279.14	1.76	47.70	nr	326.84
50mm dia	275.03	352.26	2.13	57.73	nr	409.99
Ductile iron autoflow commissioning valves; PN16; working pressure 16 bar from -10°C to 120°C; for ANSI 150 flanged ends						
65mm dia	586.22	750.82	2.31	62.61	nr	813.43
80mm dia	648.13	830.12	2.88	78.06	nr	908.18
100mm dia	1010.74	1294.55	3.11	84.29	nr	1378.84
150mm dia	1705.63	2184.55	6.98	189.18	nr	2373.74
200mm dia	2461.63	3152.83	8.26	223.87	nr	3376.71
250mm dia	3418.84	4378.82	10.49	284.32	nr	4663.13
300mm dia	4366.39	5592.43	11.49	311.42	nr	5903.85

T: MECHANICAL/COOLING/HEATING SYSTEMS

Item	Net Price £	Material £	Labour hours	Labour £	Unit	Total rate £
T31 : LOW TEMPERATURE HOT WATER HEATING (cont'd)						
VALVES (cont'd)						
Strainers						
Bronze strainer; Y type; PN32 ; working pressure 32 bar from -10°C to 100°C; screwed ends to steel						
15mm dia	18.46	23.64	0.82	22.22	nr	**45.87**
20mm dia	23.48	30.07	1.08	29.27	nr	**59.34**
25mm dia	33.08	42.37	1.27	34.42	nr	**76.79**
32mm dia	55.63	71.25	1.50	40.66	nr	**111.91**
40mm dia	75.57	96.79	1.76	47.70	nr	**144.49**
50mm dia	126.56	162.10	2.13	57.73	nr	**219.83**
Cast iron strainer; Y type; PN16; working pressure 16 bar from -10°C to 120°C; BS 4504 flanged ends						
65mm dia	105.77	135.47	2.31	62.61	nr	**198.08**
80mm dia	122.52	156.92	2.88	78.06	nr	**234.98**
100mm dia	181.40	232.34	3.11	84.29	nr	**316.63**
125mm dia	374.50	479.66	5.02	136.06	nr	**615.72**
150mm dia	485.19	621.43	6.98	189.18	nr	**810.61**
200mm dia	792.94	1015.59	8.26	223.87	nr	**1239.46**
250mm dia	1198.54	1535.08	10.49	284.32	nr	**1819.39**
300mm dia	2011.01	2575.68	11.49	311.42	nr	**2887.10**
Regulators						
Gunmetal; self-acting two port thermostatic regulator; single seat; screwed ends; complete with sensing element, 2m long capillary tube						
15mm dia	470.40	602.48	1.37	37.13	nr	**639.62**
20mm dia	482.53	618.02	1.24	33.61	nr	**651.63**
25mm dia	498.28	638.19	1.34	36.32	nr	**674.51**
Gunmetal; self-acting two port thermostatic regulator; double seat; flanged ends (BS 4504 PN25); with sensing element, 2m long capillary tube; steel body						
65mm dia	1744.46	2234.29	1.23	33.32	nr	**2267.61**
80mm dia	2058.98	2637.12	1.62	43.89	nr	**2681.01**

T: MECHANICAL/COOLING/HEATING SYSTEMS

Item	Net Price £	Material £	Labour hours	Labour £	Unit	Total rate £
Control valves; electrically operated (electrical work elsewhere)						
Cast iron; butterfly type; two position electrically controlled 240V motor and linkage mechanism; for low pressure hot water; maximum pressure 6 bar at 120°C; flanged ends						
25mm dia	369.31	473.01	1.47	39.84	nr	**512.85**
32mm dia	381.60	488.75	1.52	41.20	nr	**529.95**
40mm dia	533.22	682.94	1.61	43.64	nr	**726.58**
50mm dia	548.38	702.36	1.71	46.35	nr	**748.71**
65mm dia	562.11	719.94	2.51	68.03	nr	**787.97**
80mm dia	584.09	748.10	2.69	72.91	nr	**821.00**
100mm dia	612.95	785.06	2.81	76.16	nr	**861.22**
125mm dia	678.92	869.55	2.94	79.68	nr	**949.24**
150mm dia	735.72	942.30	3.33	90.25	nr	**1032.56**
200mm dia	912.54	1168.77	3.67	99.47	nr	**1268.24**
Cast iron; three way 240V motorized; for low pressure hot water; maximum pressure 6 bar 120°C; flanged ends, drilled (BS 10, Table F)						
25mm dia	412.76	528.66	1.99	53.94	nr	**582.59**
40mm dia	432.92	554.48	2.13	57.73	nr	**612.21**
50mm dia	427.42	547.44	3.21	87.00	nr	**634.44**
65mm dia	477.57	611.67	3.23	87.54	nr	**699.21**
80mm dia	535.12	685.38	3.50	94.86	nr	**780.24**
Two port normally closed motorised valve; electric actuator; spring return; domestic usage						
22mm dia	58.52	74.95	1.18	31.98	nr	**106.93**
28mm dia	79.32	101.59	1.35	36.59	nr	**138.18**
Two port on/off motorised valve; electric actuator; spring return; domestic usage						
22mm dia	58.52	74.95	1.18	31.98	nr	**106.93**
Three port motorised valve; electric actuator; spring return; domestic usage						
22mm dia	85.39	109.37	1.18	31.98	nr	**141.35**
Safety and relief valves						
Bronze relief valve; spring type; side outlet; working pressure up to 20.7 bar at 120°C; screwed ends to steel						
15mm dia	81.95	104.96	0.26	7.05	nr	**112.01**
20mm dia	110.61	141.67	0.36	9.76	nr	**151.43**
Bronze relief valve; spring type; side outlet; working pressure up to 17.2 bar at 120°C; screwed ends to steel						
25mm dia	132.28	169.42	0.38	10.30	nr	**179.72**
32mm dia	201.40	257.95	0.48	13.01	nr	**270.96**

T: MECHANICAL/COOLING/HEATING SYSTEMS

Item	Net Price £	Material £	Labour hours	Labour £	Unit	Total rate £
T31 : LOW TEMPERATURE HOT WATER HEATING (cont'd)						
VALVES (cont'd)						
Safety and relief valves (cont'd)						
Bronze relief valve; spring type; side outlet; working pressure up to 13.8 bar at 120°C; screwed ends to steel						
40mm dia	175.89	225.28	0.64	17.35	nr	242.62
50mm dia	267.52	342.64	0.76	20.60	nr	363.24
65mm dia	424.76	544.03	0.94	25.48	nr	569.51
80mm dia	557.28	713.76	1.10	29.81	nr	743.57
Cocks; screwed joints to steel						
Bronze gland cock; complete with malleable iron lever; working pressure up to 10 bar at 100°C; screwed ends to steel						
15mm dia	26.46	33.89	0.77	20.87	nr	54.76
20mm dia	38.56	49.39	1.03	27.92	nr	77.30
25mm dia	55.37	70.92	1.19	32.25	nr	103.17
32mm dia	200.08	256.26	1.38	37.40	nr	293.66
40mm dia	284.10	363.87	1.62	43.91	nr	407.78
50mm dia	405.57	519.45	1.94	52.58	nr	572.03
Bronze three-way plug cock; complete with malleable iron lever; working pressure up to 10 bar at 100°C; screwed ends to steel						
15mm dia	47.24	60.50	0.77	20.87	nr	81.37
20mm dia	54.68	70.03	1.03	27.92	nr	97.95
25mm dia	76.44	97.90	1.19	32.25	nr	130.16
32mm dia	108.41	138.85	1.38	37.40	nr	176.25
40mm dia	130.71	167.41	1.62	43.91	nr	211.32
Air vents; including regulating, adjusting and testing						
Automatic air vent; maximum pressure up to 7 bar at 93°C; screwed ends to steel						
15mm dia	89.97	115.23	0.80	21.68	nr	136.92
Automatic air vent; maximum pressure up to 7 bar at 93°C; lockhead isolating valve; screwed ends to steel						
15mm dia	98.53	126.20	0.83	22.50	nr	148.69
Automatic air vent; maximum pressure up to 17 bar at 200°C; flanged ends (BS10, Table H); bolted connections to counter flange (measured separately)						
15mm dia	399.90	512.19	0.83	22.50	nr	534.68

T: MECHANICAL/COOLING/HEATING SYSTEMS

Item	Net Price £	Material £	Labour hours	Labour £	Unit	Total rate £
Radiator valves						
Bronze; wheelhead or lockshield; chromium plated finish; screwed joints to steel						
Straight						
15mm dia	28.43	36.41	0.59	15.99	nr	**52.40**
20mm dia	36.96	47.34	0.73	19.79	nr	**67.12**
25mm dia	46.20	59.17	0.85	23.04	nr	**82.21**
Angled						
15mm dia	20.21	25.88	0.59	15.99	nr	**41.88**
20mm dia	26.63	34.11	0.73	19.79	nr	**53.89**
25mm dia	34.34	43.98	0.85	23.04	nr	**67.02**
Bronze; wheelhead or lockshield; chromium plated finish; compression joints to copper						
Straight						
15mm dia	30.55	39.13	0.59	15.99	nr	**55.12**
20mm dia	38.12	48.82	0.73	19.79	nr	**68.61**
25mm dia	47.91	61.36	0.85	23.04	nr	**84.40**
Angled						
15mm dia	20.21	25.88	0.59	15.99	nr	**41.88**
20mm dia	28.05	35.93	0.73	19.79	nr	**55.71**
25mm dia	35.75	45.79	0.85	23.04	nr	**68.83**
Twin entry						
8mm dia	31.42	40.64	0.23	6.23	nr	**46.88**
10mm dia	35.66	46.13	0.23	6.23	nr	**52.36**
Bronze; thermostatic head; chromium plated finish; compression joints to copper						
Straight						
15mm dia	25.81	33.06	0.59	15.99	nr	**49.05**
20mm dia	30.11	38.56	0.73	19.79	nr	**58.35**
Angled						
15mm dia	27.19	34.82	0.59	15.99	nr	**50.82**
20mm dia	35.25	45.15	0.73	19.79	nr	**64.93**
GAUGES						
Thermometers and pressure gauges						
Dial thermometer; coated steel case and dial; glass window; brass pocket; BS 5235; pocket length 100mm; screwed end						
Back/bottom entry						
100mm dia face	57.68	73.88	0.81	21.96	nr	**95.84**
150mm dia face	65.79	84.26	0.81	21.96	nr	**106.23**
Dial pressure/altitude gauge; bronze bourdon tube type; coated steel case and dial; glass window BS 1780; screwed end						
100mm dia face	33.05	42.33	0.81	21.96	nr	**64.29**
150mm dia face	35.89	45.97	0.81	21.96	nr	**67.93**

T: MECHANICAL/COOLING/HEATING SYSTEMS

Item	Net Price £	Material £	Labour hours	Labour £	Unit	Total rate £
T31 : LOW TEMPERATURE HOT WATER HEATING (cont'd)						
EQUIPMENT						
PRESSURISATION UNITS						
LTHW pressurisation unit complete with expansion vessel(s), interconnecting pipework and all necessary isolating and drain valves; includes placing in position; electrical work elsewhere. Selection based on a final working pressure of 4 bar, a 3m static head and system operating temperatures of 82/71 °s C						
System volume						
2,400 litres	1653.75	2118.11	15.00	406.55	nr	2524.66
6,000 - 20,000 litres	1813.23	2322.37	22.00	596.28	nr	2918.64
25,000 litres	2251.23	2883.35	22.00	596.28	nr	3479.63
DIRT SEPARATORS						
Dirt separator; maximum operating temperature and pressure of 110°C and 10 bar; fitted with drain valve						
Bore size, flow rate (at 1.0m/s velocity); threaded connections						
32mm dia, 3.7m³/h	63.00	80.69	2.29	62.07	nr	142.76
40mm dia, 5.0m³/h	75.00	96.06	2.45	66.40	nr	162.46
Bore size, flow rate (at 1.5m/s velocity); flanged connections to PN16						
50mm dia, 13.0m³/h	424.00	543.05	3.00	81.31	nr	624.37
65mm dia, 21.0m³/h	442.00	566.11	3.00	81.31	nr	647.42
80mm dia, 29.0m³/h	622.00	796.65	3.84	104.08	nr	900.73
100mm dia, 49.0m³/h	658.00	842.76	4.44	120.34	nr	963.10
125mm dia, 74.0m³/h	1261.00	1615.08	11.64	315.48	nr	1930.56
150mm dia, 109.0m³/h	1315.00	1684.24	15.75	426.88	nr	2111.12
200mm dia, 181.0m³/h	1982.00	2538.53	15.75	426.88	nr	2965.41
250mm dia, 288.0m³/h	3513.00	4499.42	15.75	426.88	nr	4926.30
300mm dia, 407.0m³/h	5135.00	6576.86	17.24	467.26	nr	7044.12
MICROBUBBLE DEAERATORS						
Microbubble deaerator; maximum operating temperature and pressure of 110°C and 10 bar; fitted with drain valve						
Bore size, flow rate (at 1.0m/s velocity); threaded connections						
32mm dia, 3.7m³/h	58.00	74.29	2.29	62.07	nr	136.35
40mm dia, 5.0m³/h	69.00	88.37	2.45	66.40	nr	154.78

T: MECHANICAL/COOLING/HEATING SYSTEMS

Item	Net Price £	Material £	Labour hours	Labour £	Unit	Total rate £
Bore size, flow rate (at 1.5m/s velocity); flanged connections to PN16						
50mm dia, 13.0m³/h	424.00	543.05	3.00	81.31	nr	624.37
65mm dia, 21.0m³/h	442.00	566.11	3.00	81.31	nr	647.42
80mm dia, 29.0m³/h	622.00	796.65	3.84	104.08	nr	900.73
100mm dia, 49.0m³/h	658.00	842.76	4.44	120.34	nr	963.10
125mm dia, 74.0m³/h	1261.00	1615.08	11.64	315.48	nr	1930.56
150mm dia, 109.0m³/h	1315.00	1684.24	15.75	426.88	nr	2111.12
200mm dia, 181.0m³/h	1982.00	2538.53	15.75	426.88	nr	2965.41
250mm dia, 288.0m³/h	3513.00	4499.42	15.75	426.88	nr	4926.30
300mm dia, 407.0m³/h	5135.00	6576.86	17.24	467.26	nr	7044.12
COMBINED MICROBUBBLE DEAERATORS AND DIRT SEPARATORS						
Combined deaerator and dirt separators; maximum operating temperature and pressure of 110°C and 10 bar; fitted with drain valve						
Bore size, flow rate (at 1.5m/s velocity); threaded connections						
25mm dia, 2.0m³/h	80.00	102.46	2.75	74.53	nr	177.00
Bore size, flow rate (at 1.5m/s velocity); flanged connections to PN16						
50mm dia, 13.0m³/h	541.00	692.91	3.60	97.57	nr	790.48
65mm dia, 21.0m³/h	559.00	715.96	3.60	97.57	nr	813.53
80mm dia, 29.0m³/h	757.00	969.56	4.61	124.95	nr	1094.51
100mm dia, 49.0m³/h	793.00	1015.67	5.33	144.46	nr	1160.13
125mm dia, 74.0m³/h	1513.00	1937.84	13.97	378.64	nr	2316.47
150mm dia, 109.0m³/h	1568.00	2008.28	18.90	512.26	nr	2520.53
200mm dia, 181.0m³/h	2432.00	3114.88	18.90	512.26	nr	3627.14
250mm dia, 288.0m³/h	4288.00	5492.03	18.90	512.26	nr	6004.28
300mm dia, 407.0m³/h	6576.00	8422.48	20.69	560.77	nr	8983.25

T: MECHANICAL/COOLING/HEATING SYSTEMS

Item	Net Price £	Material £	Labour hours	Labour £	Unit	Total rate £
T31 : LOW TEMPERATURE HOT WATER HEATING (cont'd)						
Y20 - PUMPS						
Centrifugal heating and chilled water pump; belt drive; 3 phase, 1450 rpm motor; max. pressure 1000kN/m²; max. temperature 125°C; bed plate; coupling guard; bolted connections; supply only mating flanges; includes fixing on prepared concrete base; electrical work elsewhere						
40mm pump size; 4.0 l/s at 70 kPa max head; 0.25kW max motor rating	1147.53	1469.74	7.59	205.72	nr	**1675.46**
40mm pump size; 4.0 l/s at 130 kPa max head; 1.5kW max motor rating	1436.12	1839.37	8.09	219.27	nr	**2058.64**
50mm pump size; 8.5 l/s at 90 kPa max head; 2.2kW max motor rating	1481.04	1896.90	8.67	234.99	nr	**2131.89**
50mm pump size; 8.5 l/s at 190 kPa max head; 3kW max motor rating	2574.12	3296.91	11.20	303.56	nr	**3600.47**
50mm pump size; 8.5 l/s at 215 kPa max head; 4 kW max motor rating	1682.51	2154.94	11.70	317.11	nr	**2472.05**
65mm pump size; 14.0 l/s at 90 kPa max head; 3kW max motor rating	1475.60	1889.93	11.70	317.11	nr	**2207.04**
65mm pump size; 14.0 l/s at 160 kPa max head; 4 kW max motor rating	1636.23	2095.67	11.70	317.11	nr	**2412.78**
80mm pump size; 14.5 l/s at 210 kPa max head; 5.5 kW max motor rating	2510.15	3214.97	11.70	317.11	nr	**3532.09**
80mm pump size; 22.0 l/s at 130 kPa max head; 5.5 kW max motor rating	2510.15	3214.97	13.64	369.69	nr	**3584.67**
80mm pump size; 22.0 l/s at 200 kPa max head; 7.5 kW max motor rating	2628.57	3366.65	13.64	369.69	nr	**3736.34**
100mm pump size; 22.0 l/s at 250 kPa max head; 11kW max motor rating	3688.99	4724.82	13.64	369.69	nr	**5094.51**
100mm pump size; 30.0 l/s at 100 kPa max head; 4.0 kW max motor rating	2065.02	2644.86	19.15	519.03	nr	**3163.89**
100mm pump size; 36.0 l/s at 250 kPa max head; 15.0kW max motor rating	3906.79	5003.78	19.15	519.03	nr	**5522.81**
100mm pump size; 36.0 l/s at 550 kPa max head; 30.0 kW max motor rating	5302.07	6790.84	19.15	519.03	nr	**7309.87**
Centrifugal heating and chilled water pump; TWIN HEAD BELT DRIVE; 3 phase, 1450 rpm motor; max. pressure 1000kN/m²; max. temperature 125°C; bed plate; coupling guard; bolted connections; supply only mating flanges; includes fixing on prepared concrete base; electrical work elsewhere						
40mm pump size; 4.0 l/s at 70 kPa max head; 0.75kW max motor rating	2458.42	3148.72	7.59	205.72	nr	**3354.44**
40mm pump size; 4.0 l/s at 130 kPa max head; 1.5kW max motor rating	2959.35	3790.31	8.09	219.27	nr	**4009.57**
50mm pump size; 8.5 l/s at 90 kPa max head; 2.2kW max motor rating	3023.34	3872.26	8.67	234.99	nr	**4107.25**
50mm pump size; 8.5 l/s at 190 kPa max head; 4kW max motor rating	3484.80	4463.30	11.20	303.56	nr	**4766.86**

T: MECHANICAL/COOLING/HEATING SYSTEMS

Item	Net Price £	Material £	Labour hours	Labour £	Unit	Total rate £
65mm pump size; 8.5 l/s at 215 kPa max head; 4 kW max motor rating	3634.54	4655.08	11.70	317.11	nr	**4972.19**
65mm pump size; 14.0 l/s at 90 kPa max head; 3kW max motor rating	3254.75	4168.65	11.70	317.11	nr	**4485.76**
65mm pump size; 14.0 l/s at 160 kPa max head; 4 kW max motor rating	3634.54	4655.08	11.70	317.11	nr	**4972.19**
80mm pump size; 14.5 l/s at 210 kPa max head; 7.5 kW max motor rating	5459.97	6993.07	13.64	369.69	nr	**7362.77**
Centrifugal heating and chilled water pump; CLOSE COUPLED; 3 phase, 1450 rpm motor; max. pressure 1000kN/m²; max. temperature 110°C; bed plate; coupling guard; bolted connections; supply only mating flanges; includes fixing on prepared concrete base; electrical work elsewhere						
40mm pump size; 4.0 l/s at 23 kPa max head; 0.55kW max motor rating	690.15	883.94	7.31	198.13	nr	**1082.06**
50mm pump size; 4.0 l/s at 75 kPa max head; 0.75 kW max motor rating	765.03	979.84	7.31	198.13	nr	**1177.97**
50mm pump size; 7.0 l/s at 65 kPa max head; 0.75 kW max motor rating	765.03	979.84	8.01	217.10	nr	**1196.94**
65mm pump size; 10.0 l/s at 33 kPa max head; 0.75kW max motor rating	867.12	1110.60	8.01	217.10	nr	**1327.70**
50mm pump size; 4.0 l/s at 120 kPa max head; 1.5 kW max motor rating	1044.08	1337.25	8.01	217.10	nr	**1554.35**
80mm pump size; 16.0 l/s at 80 kPa max head; 2.2 kW max motor rating	1351.72	1731.27	12.35	334.73	nr	**2066.00**
80mm pump size; 16.0 l/s at 120 kPa max head; 4.0 kW max motor rating	1317.69	1687.68	12.35	334.73	nr	**2022.41**
100mm pump size; 28.0 l/s at 40 kPa max head; 2.2 kW max motor rating	1335.39	1710.35	17.86	484.07	nr	**2194.42**
100mm pump size; 28.0 l/s at 90 kPa max head; 4.0 kW max motor rating	1474.23	1888.18	17.86	484.07	nr	**2372.25**
125mm pump size; 40.0 l/s at 50 kPa max head; 3.0 kW max motor rating	1406.17	1801.01	25.85	700.63	nr	**2501.63**
125mm pump size; 40.0 l/s at 120 kPa max head; 7.5 kW max motor rating	1735.59	2222.93	25.85	700.63	nr	**2923.55**
150mm pump size; 70.0 l/s at 75 kPa max head; 11 kW max motor rating	2550.98	3267.27	30.43	824.76	nr	**4092.03**
150mm pump size; 70.0 l/s at 120 kPa max head; 15.0 kW max motor rating	2726.58	3492.18	30.43	824.76	nr	**4316.94**
150mm pump size; 70.0 l/s at 150 kPa max head; 15.0 kW max motor rating	2726.58	3492.18	30.43	824.76	nr	**4316.94**

T: MECHANICAL/COOLING/HEATING SYSTEMS

Item	Net Price £	Material £	Labour hours	Labour £	Unit	Total rate £
T31 : LOW TEMPERATURE HOT WATER HEATING (cont'd)						
Y20 – PUMPS (cont'd)						
Centrifugal heating & chilled water pump; close coupled; 3 phase, VARIABLE SPEED motor; max. system pressure 1000 kN/m²; max. temperature 110°C; bed plate; coupling guard; bolted connections; supply only mating flanges; includes fixing on prepared concrete base; electrical work elsewhere.						
40mm pump size; 4.0 l/s at 23 kPa max head; 0.55kW max motor rating	1245.54	1595.28	7.31	198.13	nr	**1793.40**
40mm pump size; 4.0 l/s at 75 kPa max head; 0.75kW max motor rating	1374.87	1760.92	7.31	198.13	nr	**1959.05**
50mm pump size; 7.0 l/s at 65 kPa max head; 1.5kW max motor rating	1686.59	2160.17	8.01	217.10	nr	**2377.27**
50mm pump size; 10.0 l/s at 33 kPa max head; 1.5kW max motor rating	1686.59	2160.17	8.01	217.10	nr	**2377.27**
50mm pump size; 4.0 l/s at 120 kPa max head; 1.5kW max motor rating	1686.59	2160.17	8.01	217.10	nr	**2377.27**
80mm pump size; 16.0 l/s at 80 kPa max head; 2.2kW max motor rating	2198.42	2815.71	12.35	334.73	nr	**3150.44**
80mm pump size; 16.0 l/s at 120 kPa max head; 3.0kW max motor rating	2371.30	3037.14	12.35	334.73	nr	**3371.87**
100mm pump size; 28.0 l/s at 40 kPa max head; 2.2kW max motor rating	2252.87	2885.45	17.86	484.07	nr	**3369.52**
100mm pump size; 28.0 l/s at 90 kPa max head; 4.0kW max motor rating	2687.11	3441.62	17.86	484.07	nr	**3925.69**
125mm pump size; 40.0 l/s at 50 kPa max head; 3.0kW max motor rating	2579.57	3303.89	25.85	700.63	nr	**4004.51**
125mm pump size; 40.0 l/s at 120 kPa max head; 7.5kW max motor rating	3630.45	4649.84	25.85	700.63	nr	**5350.47**
150mm pump size; 70.0 l/s at 75 kPa max head; 7.5kW max motor rating	4196.73	5375.13	30.43	824.76	nr	**6199.89**
Glandless domestic heating pump; for low pressure domestic hot water heating systems; 240 volt; 50Hz electric motor; max working pressure 1000N/m² and max temperature of 130°C; includes fixing in position; electrical work elsewhere						
1" BSP unions - 2 speed	124.15	159.01	1.58	42.82	nr	**201.83**
1.25" BSP unions - 3 speed	182.95	234.32	1.58	42.82	nr	**277.14**
Glandless pumps; for hot water secondary supply; silent running; 3 phase; max pressure 1000kN/m²; max temperature 130°C ; bolted connections; supply only mating flanges; including fixing in position; electrical elsewhere						
1" BSP unions - 3 speed	240.94	308.59	1.58	42.82	nr	**351.42**

T: MECHANICAL/COOLING/HEATING SYSTEMS

Item	Net Price £	Material £	Labour hours	Labour £	Unit	Total rate £
Pipeline mounted circulator; for heating and chilled water; silent running; 3 phase; 1450 rpm motor; max pressure 1000 kN/m²; max temperature 120°C; bolted connections; supply only mating flanges; includes fixing in position; electrical elsewhere						
32mm pump size; 2.0 l/s at 17 kPa max head; 0.2kW max motor rating	481.89	617.20	6.44	174.55	nr	**791.75**
50mm pump size; 3.0 l/s at 20 kPa max head; 0.2kW max motor rating	479.16	613.70	6.86	185.93	nr	**799.63**
65mm pump size; 5.0 l/s at 30 kPa max head; 0.37 kW max motor rating	762.30	976.35	7.48	202.73	nr	**1179.08**
65mm pump size; 8.0 l/s at 37 kPa max head; 0.75 kW max motor rating	853.50	1093.15	7.48	202.73	nr	**1295.89**
80mm pump size; 12.0 l/s at 42 kPa max head; 1.1 kW max motor rating	992.35	1270.99	8.01	217.10	nr	**1488.09**
100mm pump size; 25.0 l/s at 37 kPa max head; 2.2 kW max motor rating	1414.34	1811.47	9.11	246.91	nr	**2058.39**
Dual pipeline mounted circulator; for heating & chilled water; silent running; 3 phase; 1450 rpm motor; max pressure 1000 kN/m²; max temperature 120°C; bolted connections; supply only mating flanges; includes fixing in position; electrical work elsewhere						
40mm pump size; 2.0 l/s at 17 kPa max head; 0.8kW max motor rating	888.90	1138.49	7.88	213.58	nr	**1352.07**
50mm pump size; 3.0 l/s at 20 kPa max head; 0.2 kW max motor rating	891.62	1141.98	8.01	217.10	nr	**1359.08**
65mm pump size; 5.0 l/s at 30 kPa max head; 0.37 kW max motor rating	1447.01	1853.32	9.20	249.35	nr	**2102.67**
65mm pump size; 8.0 l/s at 37 kPa max head; 0.75 kW max motor rating	1672.98	2142.74	9.20	249.35	nr	**2392.09**
100mm pump size; 12.0 l/s at 42 kPa max head; 1.1 kW max motor rating	1892.14	2423.43	9.45	256.13	nr	**2679.56**
Glandless accelerator pumps; for low and medium pressure heating services; silent running; 3 phase; 1450 rpm motor; max pressure 1000 kN/m²; max temperature 130°C; bolted connections; supply only mating flanges; includes fixing in position; electrical work elsewhere						
40mm pump size; 4.0 l/s at 15 kPa max head; 0.35kW max motor rating	393.75	504.31	6.94	188.10	nr	**692.41**
50mm pump size; 6.0 l/s at 20 kPa max head; 0.45kW max motor rating	419.63	537.45	7.35	199.21	nr	**736.66**
80mm pump size; 13.0 l/s at 28 kPa max head; 0.58kW max motor rating	826.88	1059.05	7.76	210.32	nr	**1269.38**

T: MECHANICAL/COOLING/HEATING SYSTEMS

Item	Net Price £	Material £	Labour hours	Labour £	Unit	Total rate £
T31 : LOW TEMPERATURE HOT WATER HEATING (cont'd)						
Y22 - HEAT EXCHANGERS						
Plate heat exchanger; for use in LTHW systems; painted carbon steel frame; stainless steel plates, nitrile rubber gaskets; design pressure of 10 bar and operating temperature of 110/135°C						
Primary side; 80°C in, 69°C out; secondary side; 82°C in, 71°C out						
107 KW, 2.38 l/s	1910.00	2446.31	10.00	271.04	nr	2717.34
245 kW, 5.46 l/s	2935.00	3759.12	10.00	271.04	nr	4030.15
287 kW, 6.38 l/s	3258.00	4172.81	10.00	271.04	nr	4443.85
328 kW, 7.31 l/s	3555.00	4553.21	10.00	271.04	nr	4824.24
364 kW, 8.11 l/s	4219.00	5403.65	10.00	271.04	nr	5674.69
403 kW, 8.96 l/s	4429.00	5672.62	10.00	271.04	nr	5943.65
453 kW, 10.09 l/s	4608.00	5901.88	10.00	271.04	nr	6172.92
490 kW, 10.89 l/s	4777.00	6118.33	10.00	271.04	nr	6389.37
1000 kW, 21.7 l/s	3554.00	4551.93	12.00	325.24	nr	4877.17
1500 kW, 32.6 l/s	4528.00	5799.42	12.00	325.24	nr	6124.66
2000 kW, 43.4 l/s	5432.00	6957.25	15.00	406.55	nr	7363.80
2500 kW, 54.3 l/s	6677.00	8551.83	15.00	406.55	nr	8958.39
Note - For temperature conditions different to those above, the cost of the units can vary significantly, and so manufacturers advice should be sought.						

T: MECHANICAL/COOLING/HEATING SYSTEMS

Item	Net Price £	Material £	Labour hours	Labour £	Unit	Total rate £
Y23 - CALORIFIERS						
Non-storage calorifiers; mild steel; heater battery duty 82°C/71°C to BS 853, maximum test on shell 11.55 bar, tubes 26.25 bar						
Horizontal or vertical; primary water at 116°C on, 90°C off						
40 kW capacity	518.94	664.65	3.00	81.31	nr	745.96
88 kW capacity	604.24	773.90	5.00	135.52	nr	909.42
176 kW capacity	936.94	1200.02	7.04	190.87	nr	1390.89
293 kW capacity	1391.90	1782.73	9.01	244.18	nr	2026.91
586 kW capacity	2045.90	2620.37	22.22	602.30	nr	3222.67
879 kW capacity	2667.20	3416.12	28.57	774.38	nr	4190.51
1465 kW capacity	4216.92	5400.99	50.00	1355.17	nr	6756.16
2000 kW capacity	5829.18	7465.96	60.00	1626.21	nr	9092.17
HEAT EMITTERS						
Perimeter convector heating; metal casing with standard finish; aluminium extruded grille; including backplates						
Top/sloping/flat front outlet						
60 x 200mm	31.85	40.79	2.00	54.21	m	95.00
60 x 300mm	34.64	44.37	2.00	54.21	m	98.57
60 x 450mm	42.94	55.00	2.00	54.21	m	109.20
60 x 525mm	45.71	58.54	2.00	54.21	m	112.75
60 x 600mm	49.86	63.86	2.00	54.21	m	118.07
90 x 260mm	34.64	44.37	2.00	54.21	m	98.57
90 x 300mm	36.01	46.12	2.00	54.21	m	100.33
90 x 450mm	44.34	56.79	2.00	54.21	m	111.00
90 x 525mm	48.49	62.11	2.00	54.21	m	116.31
90 x 600mm	51.26	65.65	2.00	54.21	m	119.86
Extra over for dampers						
Damper	15.23	19.51	0.25	6.78	nr	26.28
Extra over for fittings						
60mm End caps	12.19	15.61	0.25	6.78	nr	22.39
90mm End caps	19.18	24.57	0.25	6.78	nr	31.34
60mm Corners	25.91	33.19	0.25	6.78	nr	39.96
90mm Corners	38.10	48.80	0.25	6.78	nr	55.57
Radiant Strip Heaters						
Suitable for connection to hot water system; aluminium sheet panels with steel pipe clamped to upper surface; including insulation, sliding brackets, cover plates, end closures; weld or screwed BSP ends						
One pipe						
1500mm long	64.41	82.50	3.11	84.29	nr	166.79
3000mm long	101.88	130.49	3.11	84.29	nr	214.78
4500mm long	138.19	176.99	3.11	84.29	nr	261.28
6000mm long	190.50	243.99	3.11	84.29	nr	328.28

T: MECHANICAL/COOLING/HEATING SYSTEMS

Item	Net Price £	Material £	Labour hours	Labour £	Unit	Total rate £
T31 : LOW TEMPERATURE HOT WATER HEATING (cont'd)						
HEAT EMITTERS (cont'd)						
Radiant Strip Heaters (cont'd)						
Suitable for connection to hot water (cont'd)						
Two pipe						
1500mm long	120.25	154.01	4.15	112.48	nr	266.49
3000mm long	189.94	243.27	4.15	112.48	nr	355.75
4500mm long	259.38	332.21	4.15	112.48	nr	444.69
6000mm long	349.45	447.57	4.15	112.48	nr	560.05
Pressed steel panel type radiators; fixed with and including brackets; taking down once for decoration; refixing						
300mm high; single panel						
500mm length	17.58	22.52	2.03	55.02	nr	77.54
1000mm length	35.14	45.01	2.03	55.02	nr	100.03
1500mm length	46.04	58.97	2.03	55.02	nr	113.99
2000mm length	51.65	66.15	2.47	66.95	nr	133.10
2500mm length	57.25	73.33	2.97	80.50	nr	153.82
3000mm length	68.51	87.75	3.22	87.27	nr	175.02
300mm high; double panel; convector						
500mm length	33.80	43.29	2.13	57.73	nr	101.02
1000mm length	67.65	86.65	2.13	57.73	nr	144.38
1500mm length	101.46	129.95	2.13	57.73	nr	187.68
2000mm length	135.29	173.28	2.57	69.66	nr	242.93
2500mm length	169.09	216.57	3.07	83.21	nr	299.78
3000mm length	202.93	259.91	3.31	89.71	nr	349.62
450mm high; single panel						
500mm length	16.40	21.00	2.08	56.38	nr	77.38
1000mm length	32.83	42.05	2.08	56.38	nr	98.42
1600mm length	52.52	67.27	2.53	68.57	nr	135.84
2000mm length	65.66	84.10	2.97	80.50	nr	164.59
2400mm length	78.77	100.89	3.47	94.05	nr	194.94
3000mm length	98.47	126.12	3.82	103.54	nr	229.65
450mm high; double panel; convector						
500mm length	30.06	38.50	2.18	59.09	nr	97.59
1000mm length	60.11	76.99	2.18	59.09	nr	136.07
1600mm length	110.09	141.00	2.63	71.28	nr	212.28
2000mm length	185.08	237.05	3.06	82.94	nr	319.99
2400mm length	222.11	284.48	3.37	91.34	nr	375.82
3000mm length	277.63	355.59	3.92	106.25	nr	461.83
600mm high; single panel						
500mm length	21.98	28.15	2.18	59.09	nr	87.24
1000mm length	43.97	56.32	2.43	65.86	nr	122.18
1600mm length	70.34	90.09	3.13	84.83	nr	174.92
2000mm length	87.92	112.61	3.77	102.18	nr	214.79
2400mm length	105.52	135.15	4.07	110.31	nr	245.46
3000mm length	131.90	168.94	5.11	138.50	nr	307.44

T: MECHANICAL/COOLING/HEATING SYSTEMS

Item	Net Price £	Material £	Labour hours	Labour £	Unit	Total rate £
600mm high; double panel; convector						
500mm length	37.85	48.48	2.28	61.80	nr	110.27
1000mm length	75.69	96.94	2.28	61.80	nr	158.74
1600mm length	138.64	177.57	3.23	87.54	nr	265.11
2000mm length	233.07	298.51	3.87	104.89	nr	403.40
2400mm length	279.68	358.21	4.17	113.02	nr	471.23
3000mm length	349.59	447.75	5.24	142.02	nr	589.77
700mm high; single panel						
500mm length	25.72	32.94	2.23	60.44	nr	93.38
1000mm length	51.42	65.86	2.83	76.70	nr	142.56
1600mm length	82.28	105.38	3.73	101.10	nr	206.48
2000mm length	102.86	131.74	4.46	120.88	nr	252.62
2400mm length	123.45	158.11	4.48	121.42	nr	279.54
3000mm length	154.28	197.60	5.24	142.02	nr	339.62
700mm high; double panel; convector						
500mm length	49.19	63.00	2.33	63.15	nr	126.15
1000mm length	132.34	169.50	3.08	83.48	nr	252.98
1600mm length	211.73	271.18	3.83	103.81	nr	374.99
2000mm length	264.66	338.97	4.17	113.02	nr	452.00
2400mm length	317.61	406.79	4.37	118.44	nr	525.23
3000mm length	397.00	508.47	4.82	130.64	nr	639.11
Flat panel type steel radiators; fixed with and including brackets; taking down once for decoration; refixing						
300mm high; single panel (44mm deep)						
500mm length	64.07	82.06	2.03	55.02	nr	137.08
1000mm length	106.97	137.01	2.03	55.02	nr	192.03
1500mm length	149.87	191.95	2.03	55.02	nr	246.97
2000mm length	192.77	246.90	2.47	66.95	nr	313.84
2400mm length	227.09	290.85	2.97	80.50	nr	371.35
3000mm length	278.57	356.79	3.22	87.27	nr	444.06
300mm high; double panel (100mm deep)						
500mm length	160.75	205.89	2.03	55.02	nr	260.91
1000mm length	275.95	353.43	2.03	55.02	nr	408.45
1500mm length	391.15	500.98	2.03	55.02	nr	556.00
2000mm length	506.35	648.53	2.47	66.95	nr	715.47
2400mm length	598.51	766.57	2.97	80.50	nr	847.06
3000mm length	736.75	943.62	3.22	87.27	nr	1030.90
500mm high; single panel (44mm deep)						
500mm length	79.84	102.26	2.13	57.73	nr	159.99
1000mm length	137.44	176.03	2.13	57.73	nr	233.76
1500mm length	195.04	249.81	2.13	57.73	nr	307.54
2000mm length	252.64	323.58	2.57	69.66	nr	393.23
2400mm length	298.72	382.60	3.07	83.21	nr	465.81
3000mm length	367.84	471.13	3.31	89.71	nr	560.84
500mm high; double panel (100mm deep)						
500mm length	200.48	256.77	2.08	56.38	nr	313.15
1000mm length	353.18	452.35	2.08	56.38	nr	508.72
1500mm length	505.88	647.93	2.53	68.57	nr	716.50
2000mm length	658.58	843.50	2.97	80.50	nr	924.00
2400mm length	780.74	999.96	3.47	94.05	nr	1094.01
3000mm length	963.98	1234.66	3.82	103.54	nr	1338.19

T: MECHANICAL/COOLING/HEATING SYSTEMS

Item	Net Price £	Material £	Labour hours	Labour £	Unit	Total rate £
T31 : LOW TEMPERATURE HOT WATER HEATING (cont'd)						
HEAT EMITTERS (cont'd)						
Flat panel type steel radiators (cont'd)						
600mm high; single panel (44mm deep)						
500mm length	86.94	111.35	2.18	59.09	nr	170.44
1000mm length	151.29	193.77	2.18	59.09	nr	252.86
1500mm length	215.64	469.96	2.63	71.28	nr	541.24
2000mm length	279.99	358.61	3.06	82.94	nr	441.55
2400mm length	331.47	424.54	3.37	91.34	nr	515.88
3000mm length	408.69	523.45	3.92	106.25	nr	629.69
600mm high; double panel (100mm deep)						
500mm length	219.88	281.62	2.18	59.09	nr	340.71
1000mm length	391.18	501.02	2.43	65.86	nr	566.88
1500mm length	562.48	720.42	3.13	84.83	nr	805.25
2000mm length	733.78	939.82	3.77	102.18	nr	1042.00
2400mm length	870.82	1115.34	4.07	110.31	nr	1225.65
3000mm length	1076.38	1378.62	5.11	138.50	nr	1517.12
700mm high; single panel (44mm deep)						
500mm length	95.93	122.87	2.28	61.80	nr	184.66
1000mm length	167.03	213.93	2.28	61.80	nr	275.73
1500mm length	238.13	304.99	3.23	87.54	nr	392.54
2000mm length	309.23	396.06	3.87	104.89	nr	500.95
2400mm length	366.11	468.91	4.17	113.02	nr	581.93
3000mm length	451.43	578.19	5.24	142.02	nr	720.21
700mm high; double panel (100mm deep)						
500mm length	249.01	318.93	2.23	60.44	nr	379.37
1000mm length	444.64	569.49	2.83	76.70	nr	646.19
1500mm length	640.21	819.97	3.73	101.10	nr	921.07
2000mm length	835.81	1070.50	4.46	120.88	nr	1191.38
2400mm length	992.29	1270.92	4.48	121.42	nr	1392.34
3000mm length	1227.01	1571.54	5.24	142.02	nr	1713.56
Fan convector; sheet metal casing with lockable access panel; centrifugal fan; air filter; LPHW heating coil; extruded aluminium grilles; 3 speed; includes fixing in position; electrical work elsewhere						
Free standing flat top, 695mm high, medium speed rating						
Entering air temperature, 18°C						
695mm long, 1 row 1.94 kW, 75 l/sec	633.08	810.84	2.73	73.99	nr	884.84
695mm long, 2 row 2.64 kW, 75 l/sec	633.08	810.84	2.73	73.99	nr	884.84
895mm long, 1 row 4.02 kW, 150 l/sec	713.38	913.69	2.73	73.99	nr	987.68
895mm long, 2 row 5.62 kW, 150 l/sec	713.38	913.69	2.73	73.99	nr	987.68
1195mm long, 1 row 6.58 kW, 250 l/sec	812.19	1040.24	3.00	81.31	nr	1121.56
1195mm long, 2 row 9.27 kW, 250 l/sec	812.19	1040.24	3.00	81.31	nr	1121.56
1495mm long, 1 row 9.04 kW, 340 l/sec	906.39	1160.90	3.26	88.36	nr	1249.25
1495mm long, 2 row 12.73 kW, 340 l/sec	906.39	1160.90	3.26	88.36	nr	1249.25

T: MECHANICAL/COOLING/HEATING SYSTEMS

Item	Net Price £	Material £	Labour hours	Labour £	Unit	Total rate £
Free standing flat top, 695mm high, medium speed rating, c/w floor plinth						
695mm long, 1 row 1.94 kW, 75 l/sec	664.73	851.38	2.73	73.99	nr	**925.37**
695mm long, 2 row 2.64 kW, 75 l/sec	664.73	851.38	2.73	73.99	nr	**925.37**
895mm long, 1 row 4.02 kW, 150 l/sec	749.04	959.36	2.73	73.99	nr	**1033.36**
895mm long, 2 row 5.62 kW, 150 l/sec	749.04	959.36	2.73	73.99	nr	**1033.36**
1195mm long, 1 row 6.58 kW, 250 l/sec	852.81	1092.27	3.00	81.31	nr	**1173.58**
1195mm long, 2 row 9.27 kW, 250 l/sec	852.81	1092.27	3.00	81.31	nr	**1173.58**
1495mm long, 1 row 9.04 kW, 340 l/sec	951.70	1218.93	3.26	88.36	nr	**1307.29**
1495mm long, 2 row 12.73 kW, 340 l/sec	951.70	1218.93	3.26	88.36	nr	**1307.29**
Free standing sloping top, 695mm high, medium speed rating, c/w floor plinth						
695mm long, 1 row 1.94 kW, 75 l/sec	687.89	881.04	2.73	73.99	nr	**955.04**
695mm long, 2 row 2.64 kW, 75 l/sec	687.89	881.04	2.73	73.99	nr	**955.04**
895mm long, 1 row 4.02 kW, 150 l/sec	772.20	989.03	2.73	73.99	nr	**1063.02**
895mm long, 2 row 5.62 kW, 150 l/sec	772.20	989.03	2.73	73.99	nr	**1063.02**
1195mm long, 1 row 6.58 kW, 250 l/sec	875.97	1121.93	3.00	81.31	nr	**1203.24**
1195mm long, 2 row 9.27 kW, 250 l/sec	875.97	1121.93	3.00	81.31	nr	**1203.24**
1495mm long, 1 row 9.04 kW, 340 l/sec	974.86	1248.59	3.26	88.36	nr	**1336.95**
1495mm long, 2 row 12.73 kW, 340 l/sec	974.86	1248.59	3.26	88.36	nr	**1336.95**
Wall mounted high level sloping discharge						
695mm long, 1 row 1.94 kW, 75 l/sec	694.84	889.94	2.73	73.99	nr	**963.94**
695mm long, 2 row 2.64 kW, 75 l/sec	694.84	889.94	2.73	73.99	nr	**963.94**
895mm long, 1 row 4.02 kW, 150 l/sec	714.93	915.68	2.73	73.99	nr	**989.67**
895mm long, 2 row 5.62 kW, 150 l/sec	714.93	915.68	2.73	73.99	nr	**989.67**
1195mm long, 1 row 6.58 kW, 250 l/sec	870.08	1114.39	3.00	81.31	nr	**1195.70**
1195mm long, 2 row 9.27 kW, 250 l/sec	870.08	1114.39	3.00	81.31	nr	**1195.70**
1495mm long, 1 row 9.04 kW, 340 l/sec	943.43	1208.34	3.26	88.36	nr	**1296.69**
1495mm long, 2 row 12.73 kW, 340 l/sec	943.43	1208.34	3.26	88.36	nr	**1296.69**
Ceiling mounted sloping inlet/outlet 665mm wide						
895mm long, 1 row 4.02 kW, 150 l/sec	785.94	1006.62	4.15	112.48	nr	**1119.10**
895mm long, 2 row 5.62 kW, 150 l/sec	785.94	1006.62	4.15	112.48	nr	**1119.10**
1195mm long, 1 row 6.58 kW, 250 l/sec	883.23	1131.23	4.15	112.48	nr	**1243.71**
1195mm long, 2 row 9.27 kW, 250 l/sec	883.23	1131.23	4.15	112.48	nr	**1243.71**
1495mm long, 1 row 9.04 kW, 340 l/sec	969.69	1241.97	4.15	112.48	nr	**1354.45**
1495mm long, 2 row 12.73 kW, 340 l/sec	969.69	1241.97	4.15	112.48	nr	**1354.45**
Free standing unit, extended height 1700/1900/2100mm						
895mm long, 1 row 4.02 kW, 150 l/sec	909.78	1165.24	3.11	84.29	nr	**1249.53**
895mm long, 2 row 5.62 kW, 150 l/sec	909.48	1164.85	3.11	84.29	nr	**1249.14**
1195mm long, 1 row 6.58 kW, 250 l/sec	1062.34	1360.63	3.11	84.29	nr	**1444.93**
1195mm long, 2 row 9.27 kW, 250 l/sec	1062.34	1360.63	3.11	84.29	nr	**1444.93**
1495mm long, 1 row 9.04 kW, 340 l/sec	1168.89	1497.10	3.11	84.29	nr	**1581.39**
1495mm long, 2 row 12.73 kW, 340 l/sec	1168.89	1497.10	3.11	84.29	nr	**1581.39**

T: MECHANICAL/COOLING/HEATING SYSTEMS

Item	Net Price £	Material £	Labour hours	Labour £	Unit	Total rate £
T31 : LOW TEMPERATURE HOT WATER HEATING (cont'd)						
HEAT EMITTERS (cont'd)						
LTHW trench heating; water temperatures 90°C/70°C; room air temperature 20°C; convector with copper tubes and aluminium fins within steel duct; Includes fixing within floor screed; electrical work elsewhere						
Natural convection type						
Normal capacity, 182mm width, complete with linear, natural anodised aluminium grille (grille also costed separately below)						
92mm deep						
1250mm long, 234 W output	222.34	284.77	2.00	54.21	nr	338.98
2250mm long, 471 W output	355.74	455.63	4.00	108.41	nr	564.04
3250mm long, 709 W output	486.61	623.25	5.00	135.52	nr	758.76
4250mm long, 946 W output	621.28	795.73	7.00	189.72	nr	985.45
5000mm long, 1124 W output	848.69	1086.99	8.00	216.83	nr	1303.82
120mm deep						
1250mm long, 294 W output	259.18	331.96	2.00	54.21	nr	386.16
2250mm long, 471 W output	407.84	522.36	4.00	108.41	nr	630.77
3250mm long, 891 W output	555.21	711.11	5.00	135.52	nr	846.62
4250mm long, 1190 W output	702.59	899.87	7.00	189.72	nr	1089.59
5000mm long, 1414 W output	946.53	1212.31	8.00	216.83	nr	1429.13
150mm deep						
1250mm long, 329 W output	265.54	340.10	2.00	54.21	nr	394.31
2250mm long, 664 W output	419.27	537.00	4.00	108.41	nr	645.41
3250mm long, 998 W output	570.46	730.64	5.00	135.52	nr	866.16
4250mm long, 1333 W output	725.46	929.16	7.00	189.72	nr	1118.89
5000mm long, 1584 W output	973.20	1246.46	8.00	216.83	nr	1463.29
200mm deep						
1250mm long, 396 W output	276.97	354.74	2.00	54.21	nr	408.95
2250mm long, 799 W output	506.94	649.28	4.00	108.41	nr	757.70
3250mm long, 1201 W output	603.49	772.94	5.00	135.52	nr	908.46
4250mm long, 1603 W output	772.46	989.36	7.00	189.72	nr	1179.08
5000mm long, 1905 W output	1026.56	1314.81	8.00	216.83	nr	1531.64

T: MECHANICAL/COOLING/HEATING SYSTEMS

Item	Net Price £	Material £	Labour hours	Labour £	Unit	Total rate £
Fan assisted type (outputs assume fan at 50%)						
Normal capacity, 182mm width, complete with Natural anodised aluminium grille						
112mm deep						
1250mm long, 437 W output	586.97	751.79	2.00	54.21	nr	805.99
2250mm long, 1019 W output	710.22	909.64	4.00	108.41	nr	1018.06
3250mm long, 1488 W output	834.72	1069.10	5.00	135.52	nr	1204.62
4250mm long, 1845 W output	957.96	1226.95	7.00	189.72	nr	1416.67
5000mm long, 2038 W output	1081.20	1384.79	8.00	216.83	nr	1601.62
Linear grille anodised aluminium, 170mm width (if supplied as a separate item)	138.49	177.38	-	-	m	177.38
Roll up grille, natural anodised aluminium	138.49	177.38	-	-	m	177.38
Thermostatic valve with remote regulator (c/w valve body)	57.18	73.24	4.00	108.41	nr	181.65
Fan speed controller	27.95	35.80	2.00	54.21	nr	90.01
Note - as an alternative to thermostatic control, the system can be controlled via two port valves. Refer to valve section in T31 for valve prices						
LTHW underfloor heating; water flow and return temperatures of 60°C and 70°C; pipework at 300mm centres; pipe fixings; flow and return manifolds and zone actuators; wiring block; insulation; includes fixing in position; excludes secondary pump, mixing valve, zone thermostats and floor finishes ; electrical work elsewhere						
Note - All rates are expressed on a m² basis, for the following example areas						
Screeded floor with 15-25mm stone/marble finish (producing 80-100W/m²)						
250m² area (single zone)	17.89	22.91	0.14	3.79	m²	26.71
1000m² area (single zone)	17.42	22.31	0.12	3.25	m²	25.56
5000m² area (multi-zone)	17.38	22.26	0.10	2.71	m²	24.97
Screeded floor with 10mm carpet tile (producing 80-100W/m²)						
250m² area (single zone)	17.89	22.91	0.14	3.79	m²	26.71
1000m² area (single zone)	17.42	22.31	0.12	3.25	m²	25.56
5000m² area (multi-zone)	17.38	22.26	0.10	2.71	m²	24.97
Floating timber floor with 20mm timber finish (producing 70-80Wm²)						
250m² are (single zone)	22.54	28.87	0.14	3.79	m²	32.66
1000m² area(single zone)	24.16	30.94	0.12	3.25	m²	34.20
5000m² area (multi-zone)	23.08	29.56	0.10	2.71	m²	32.27

T: MECHANICAL/COOLING/HEATING SYSTEMS

Item	Net Price £	Material £	Labour hours	Labour £	Unit	Total rate £
T31 : LOW TEMPERATURE HOT WATER HEATING (cont'd)						
HEAT EMITTERS (cont'd)						
LTHW underfloor heating; water flow (cont'd)						
Floating timber floor with 10mm carpet tile (producing 70-80W/m²)						
250m² are (single zone)	29.33	37.57	0.16	4.34	m²	41.90
1000m² area (single zone)	27.20	34.84	0.12	3.25	m²	38.09
5000m² area (multi-zone)	26.02	33.33	0.10	2.71	m²	36.04
PIPE FREEZING						
Freeze isolation of carbon steel or copper pipelines containing static water, either side of work location, freeze duration not exceeding 4 hours assuming that flow and return circuits are treated concurrently and activities undertaken during normal working hours						
Up to 4 freezes						
50mm dia	380.26	487.03	1.00	27.10	nr	514.14
65mm dia	380.26	487.03	1.00	27.10	nr	514.14
80mm dia	434.72	556.78	1.00	27.10	nr	583.89
100mm dia	489.19	626.55	1.00	27.10	nr	653.65
150mm dia	813.98	1042.54	1.00	27.10	nr	1069.64
200mm dia	1247.69	1598.03	1.00	27.10	nr	1625.13
ENERGY METERS						
Ultrasonic						
Energy meter for measuring energy use in LTHW systems; includes ultrasonic flow meter (with sensor and signal converter), energy calculator, pair of temperature sensors with brass pockets, and 3m of interconnecting cable; includes fixing in position; electrical work elsewhere						
Pipe size (flanged connections to PN16); maximum flow rate						
50mm, 36m³/hr	1122.84	1438.12	1.80	48.79	nr	1486.91
65mm, 60m³/hr	1237.71	1585.25	2.32	62.88	nr	1648.13
80mm, 100m³/hr	1387.38	1776.94	2.56	69.39	nr	1846.33
125mm, 250m³/hr	1607.41	2058.75	3.60	97.57	nr	2156.33
150mm, 360m³/hr	1744.93	2234.89	4.80	130.10	nr	2364.99
200mm, 600m³/hr	1949.60	2497.03	6.24	169.13	nr	2666.15
250mm, 1000m³/hr	2249.72	2881.42	9.60	260.19	nr	3141.61
300mm, 1500m³/hr	2644.50	3387.05	10.80	292.72	nr	3679.77
350mm, 2000m³/hr	3186.50	4081.24	13.20	357.77	nr	4439.00
400mm, 2500m³/hr	3647.61	4671.82	15.60	422.81	nr	5094.64
500mm, 3000m³/hr	4139.45	5301.77	24.00	650.48	nr	5952.25
600mm, 3500m³/hr	4650.72	5956.60	28.00	758.90	nr	6715.49

T: MECHANICAL/COOLING/HEATING SYSTEMS

Item	Net Price £	Material £	Labour hours	Labour £	Unit	Total rate £
Y53 - CONTROL COMPONENTS - MECHANICAL						
Room thermostats; light and medium duty; installed and connected						
Range 3°C to 27°C; 240 Volt						
1 amp; on/off type	24.63	31.55	0.30	8.13	nr	**39.68**
Range 0°C to +15°C; 240 Volt						
6 amp; frost thermostat	16.61	21.27	0.30	8.13	nr	**29.41**
Range 3°C to 27°C; 250 Volt						
2 amp; changeover type; dead zone	41.20	52.77	0.30	8.13	nr	**60.90**
2 amp; changeover type	19.11	24.48	0.30	8.13	nr	**32.61**
2 amp; changeover type; concealed setting	24.10	30.87	0.30	8.13	nr	**39.00**
6 amp; on/off type	14.88	19.06	0.30	8.13	nr	**27.19**
6 amp; temperature set-back	30.79	39.44	0.30	8.13	nr	**47.57**
16 amp; on/off type	22.91	29.34	0.30	8.13	nr	**37.47**
16 amp; on/off type; concealed setting	24.98	31.99	0.30	8.13	nr	**40.13**
20 amp; on/off type; concealed setting	21.28	27.26	0.30	8.13	nr	**35.39**
20 amp; indicated "off" position	26.83	34.36	0.30	8.13	nr	**42.49**
20 amp; manual; double pole on/off and neon indicator	48.84	62.55	0.30	8.13	nr	**70.68**
20 amp; indicated "off" position	31.96	40.93	0.30	8.13	nr	**49.07**
Range 10°C to 40°C; 240 Volt						
20 amp; changeover contacts	29.12	37.30	0.30	8.13	nr	**45.43**
2 amp; 'heating-cooling' switch	62.99	80.68	0.30	8.13	nr	**88.81**
Surface thermostats						
Cylinder thermostat						
6 amp; changeover type; with cable	15.93	20.91	0.25	6.78	nr	**27.69**
Electrical thermostats; installed and connected						
Range 5°C to 30°C; 230 Volt standard port single time						
10 amp with sensor	23.88	30.59	0.30	8.13	nr	**38.72**
Range 5°C to 30°C; 230 Volt standard port double time						
10 amp with sensor	27.06	34.66	0.30	8.13	nr	**42.79**
10 amp with sensor and on/off switch	41.28	52.87	0.30	8.13	nr	**61.00**
Radiator thermostats						
Angled valve body; thermostatic head; built in sensor						
15mm; liquid filled	14.89	19.07	0.84	22.78	nr	**41.85**
15mm; wax filled	14.89	19.07	0.84	22.78	nr	**41.85**

T: MECHANICAL/COOLING/HEATING SYSTEMS

Item	Net Price £	Material £	Labour hours	Labour £	Unit	Total rate £
T31 : LOW TEMPERATURE HOT WATER HEATING (cont'd)						
Immersion thermostats; stem type; domestic water boilers; fitted; electrical work elsewhere						
Temperature range 0°C to 40°C						
Non standard; 280mm stem	9.06	11.60	0.25	6.78	nr	18.38
Temperature range 18°C to 88°C						
13 amp; 178mm stem	6.02	7.71	0.25	6.78	nr	14.49
20 amp; 178mm stem	9.35	11.98	0.25	6.78	nr	18.75
Non standard; pocket clip; 280mm stem	8.68	11.12	0.25	6.78	nr	17.89
Temperature range 40°C to 80°C						
13 amp; 178mm stem	3.30	4.23	0.25	6.78	nr	11.00
20 amp; 178mm stem	6.25	8.00	0.25	6.78	nr	14.78
Non standard; pocket clip; 280mm stem	9.72	12.45	0.25	6.78	nr	19.23
13 amp; 457mm stem	3.91	5.01	0.25	6.78	nr	11.78
20 amp; 457mm stem	6.84	8.76	0.25	6.78	nr	15.54
Temperature range 50°C to 100°C						
Non standard; 1780mm stem	8.18	10.48	0.25	6.78	nr	17.25
Non standard; 280mm stem	8.53	10.93	0.25	6.78	nr	17.70
Pockets for thermostats						
For 178mm stem	10.70	13.70	0.25	6.78	nr	20.48
For 280mm stem	10.60	13.58	0.25	6.78	nr	20.35
Immersion thermostats; stem type; industrial installations; fitted; electrical work elsewhere						
Temperature range 5°C to 105°C						
For 305mm stem	134.46	172.22	0.50	13.55	nr	185.77

T: MECHANICAL/COOLING/HEATING SYSTEMS

Item	Net Price £	Material £	Labour hours	Labour £	Unit	Total rate £
Y50 -THERMAL INSULATION						
For flexible closed cell insulation see Section S10 - Cold Water						
Mineral fibre sectional insulation; bright class O foil faced; bright class O foil taped joints; 19mm aluminium bands						
Concealed pipework						
20mm thick						
15mm diameter	2.64	3.45	0.15	3.30	m	6.75
20mm diameter	2.81	3.68	0.15	3.30	m	6.98
25mm diameter	3.02	3.96	0.15	3.30	m	7.25
32mm diameter	3.37	4.41	0.15	3.30	m	7.71
40mm diameter	3.61	4.72	0.15	3.30	m	8.01
50mm diameter	4.12	5.38	0.15	3.30	m	8.68
Extra over for fittings concealed insulation						
Flange/union						
15mm diameter	1.31	1.72	0.13	2.86	nr	4.58
20mm diameter	1.41	1.85	0.13	2.86	nr	4.71
25mm diameter	1.51	1.98	0.13	2.86	nr	4.84
32mm diameter	1.69	2.20	0.13	2.86	nr	5.06
40mm diameter	1.80	2.35	0.13	2.86	nr	5.21
50mm diameter	2.06	2.69	0.13	2.86	nr	5.55
Valves						
15mm diameter	2.64	3.45	0.15	3.30	nr	6.75
20mm diameter	3.02	3.96	0.15	3.30	nr	7.25
25mm diameter	3.02	3.96	0.15	3.30	nr	7.25
32mm diameter	3.37	4.41	0.15	3.30	nr	7.71
40mm diameter	3.61	4.72	0.15	3.30	nr	8.01
50mm diameter	4.12	5.38	0.15	3.30	nr	8.68
Expansion bellows						
15mm diameter	5.27	6.89	0.22	4.84	nr	11.73
20mm diameter	5.63	7.36	0.22	4.84	nr	12.20
25mm diameter	6.05	7.91	0.22	4.84	nr	12.75
32mm diameter	6.73	8.80	0.22	4.84	nr	13.64
40mm diameter	7.21	9.43	0.22	4.84	nr	14.27
50mm diameter	8.24	10.78	0.22	4.84	nr	15.62
25mm thick						
15mm diameter	2.90	3.79	0.15	3.30	m	7.09
20mm diameter	3.12	4.08	0.15	3.30	m	7.38
25mm diameter	3.50	4.57	0.15	3.30	m	7.87
32mm diameter	3.81	4.98	0.15	3.30	m	8.27
40mm diameter	4.08	5.33	0.15	3.30	m	8.63
50mm diameter	4.67	6.11	0.15	3.30	m	9.41
65mm diameter	5.33	6.97	0.15	3.30	m	10.27
80mm diameter	5.85	7.65	0.22	4.84	m	12.49

T: MECHANICAL/COOLING/HEATING SYSTEMS

Item	Net Price £	Material £	Labour hours	Labour £	Unit	Total rate £
T31 : LOW TEMPERATURE HOT WATER HEATING (cont'd)						
Concealed pipework (cont'd)						
25mm thick (cont'd)						
100mm diameter	7.71	10.08	0.22	4.84	m	14.92
125mm diameter	8.91	11.65	0.22	4.84	m	16.49
150mm diameter	10.62	13.89	0.22	4.84	m	18.73
200mm diameter	15.01	19.63	0.25	5.50	m	25.13
250mm diameter	17.97	23.50	0.25	5.50	m	29.00
300mm diameter	19.21	25.13	0.25	5.50	m	30.62
Extra over for fittings concealed insulation						
Flange/union						
15mm diameter	1.45	1.90	0.13	2.86	nr	4.75
20mm diameter	1.56	2.04	0.13	2.86	nr	4.90
25mm diameter	1.75	2.29	0.13	2.86	nr	5.14
32mm diameter	1.91	2.50	0.13	2.86	nr	5.35
40mm diameter	2.03	2.66	0.13	2.86	nr	5.52
50mm diameter	2.33	3.05	0.13	2.86	nr	5.91
65mm diameter	2.67	3.49	0.13	2.86	nr	6.34
80mm diameter	2.93	3.83	0.18	3.96	nr	7.78
100mm diameter	3.86	5.04	0.18	3.96	nr	9.00
125mm diameter	4.46	5.84	0.18	3.96	nr	9.79
150mm diameter	5.31	6.94	0.18	3.96	nr	10.89
200mm diameter	7.50	9.81	0.22	4.84	nr	14.64
250mm diameter	8.99	11.75	0.22	4.84	nr	16.59
300mm diameter	9.61	12.56	0.22	4.84	nr	17.40
Valves						
15mm diameter	2.90	3.79	0.15	3.30	nr	7.09
20mm diameter	3.12	4.08	0.15	3.30	nr	7.38
25mm diameter	3.50	4.57	0.15	3.30	nr	7.87
32mm diameter	3.81	4.98	0.15	3.30	nr	8.27
40mm diameter	4.08	5.33	0.15	3.30	nr	8.63
50mm diameter	4.67	6.11	0.15	3.30	nr	9.41
65mm diameter	5.33	6.97	0.15	3.30	nr	10.27
80mm diameter	5.85	7.65	0.20	4.40	nr	12.05
100mm diameter	7.71	10.08	0.20	4.40	nr	14.48
125mm diameter	8.91	11.65	0.20	4.40	nr	16.05
150mm diameter	10.62	13.89	0.20	4.40	nr	18.29
200mm diameter	15.01	19.63	0.25	5.50	nr	25.13
250mm diameter	17.97	23.50	0.25	5.50	nr	29.00
300mm diameter	19.21	25.13	0.25	5.50	nr	30.62
Expansion bellows						
15mm diameter	5.80	7.59	0.22	4.84	nr	12.42
20mm diameter	6.26	8.19	0.22	4.84	nr	13.02
25mm diameter	6.98	9.13	0.22	4.84	nr	13.96
32mm diameter	7.61	9.95	0.22	4.84	nr	14.79
40mm diameter	8.16	10.67	0.22	4.84	nr	15.50
50mm diameter	9.35	12.22	0.22	4.84	nr	17.06
65mm diameter	10.67	13.96	0.22	4.84	nr	18.79
80mm diameter	11.69	15.29	0.29	6.37	nr	21.66
100mm diameter	15.41	20.15	0.29	6.37	nr	26.52
125mm diameter	17.83	23.31	0.29	6.37	nr	29.68
150mm diameter	21.25	27.78	0.29	6.37	nr	34.16

Material Costs/Prices for Measured Works – Mechanical Installations

T: MECHANICAL/COOLING/HEATING SYSTEMS

Item	Net Price £	Material £	Labour hours	Labour £	Unit	Total rate £
200mm diameter	30.01	39.24	0.36	7.91	nr	47.16
250mm diameter	35.94	46.99	0.36	7.91	nr	54.91
300mm diameter	38.43	50.25	0.36	7.91	nr	58.16
30mm thick						
15mm diameter	3.77	4.93	0.15	3.30	m	8.23
20mm diameter	4.04	5.28	0.15	3.30	m	8.58
25mm diameter	4.28	5.59	0.15	3.30	m	8.89
32mm diameter	4.66	6.09	0.15	3.30	m	9.39
40mm diameter	4.93	6.45	0.15	3.30	m	9.75
50mm diameter	5.64	7.38	0.15	3.30	m	10.67
65mm diameter	6.38	8.35	0.15	3.30	m	11.65
80mm diameter	6.97	9.11	0.22	4.84	m	13.95
100mm diameter	9.01	11.78	0.22	4.84	m	16.62
125mm diameter	10.38	13.57	0.22	4.84	m	18.40
150mm diameter	12.19	15.93	0.22	4.84	m	20.77
200mm diameter	17.04	22.29	0.25	5.50	m	27.78
250mm diameter	20.27	26.50	0.25	5.50	m	32.00
300mm diameter	21.48	28.09	0.25	5.50	m	33.59
350mm diameter	23.60	30.86	0.25	5.50	m	36.36
Extra over for fittings concealed insulation						
Flange/union						
15mm diameter	1.88	2.46	0.13	2.86	nr	5.32
20mm diameter	2.02	2.64	0.13	2.86	nr	5.50
25mm diameter	2.13	2.79	0.13	2.86	nr	5.65
32mm diameter	2.33	3.05	0.13	2.86	nr	5.91
40mm diameter	2.47	3.23	0.13	2.86	nr	6.08
50mm diameter	2.83	3.70	0.13	2.86	nr	6.55
65mm diameter	3.20	4.18	0.13	2.86	nr	7.04
80mm diameter	3.48	4.55	0.18	3.96	nr	8.51
100mm diameter	4.51	5.90	0.18	3.96	nr	9.86
125mm diameter	5.18	6.78	0.18	3.96	nr	10.73
150mm diameter	6.10	7.98	0.18	3.96	nr	11.93
200mm diameter	8.52	11.14	0.22	4.84	nr	15.97
250mm diameter	10.14	13.26	0.22	4.84	nr	18.10
300mm diameter	10.75	14.05	0.22	4.84	nr	18.89
350mm diameter	11.80	15.43	0.22	4.84	nr	20.27
Valves						
15mm diameter	4.04	5.28	0.15	3.30	nr	8.58
20mm diameter	4.04	5.28	0.15	3.30	nr	8.58
25mm diameter	4.28	5.59	0.15	3.30	nr	8.89
32mm diameter	4.66	6.09	0.15	3.30	nr	9.39
40mm diameter	4.93	6.45	0.15	3.30	nr	9.75
50mm diameter	5.64	7.38	0.15	3.30	nr	10.67
65mm diameter	6.38	8.35	0.15	3.30	nr	11.65
80mm diameter	6.97	9.11	0.20	4.40	nr	13.51
100mm diameter	9.01	11.78	0.20	4.40	nr	16.18
125mm diameter	10.38	13.57	0.20	4.40	nr	17.96
150mm diameter	12.19	15.93	0.20	4.40	nr	20.33
200mm diameter	17.04	22.29	0.25	5.50	nr	27.78
250mm diameter	20.27	26.50	0.25	5.50	nr	32.00
300mm diameter	21.48	28.09	0.25	5.50	nr	33.59
350mm diameter	23.60	30.86	0.25	5.50	nr	36.36

T: MECHANICAL/COOLING/HEATING SYSTEMS

Item	Net Price £	Material £	Labour hours	Labour £	Unit	Total rate £
T31 : LOW TEMPERATURE HOT WATER HEATING (cont'd)						
Concealed pipework (cont'd)						
30mm thick (cont'd)						
Extra over for fittings concealed insulation (cont'd)						
Expansion bellows						
15mm diameter	7.55	9.87	0.22	4.84	nr	14.71
20mm diameter	8.07	10.55	0.22	4.84	nr	15.39
25mm diameter	8.54	11.17	0.22	4.84	nr	16.00
32mm diameter	9.33	12.21	0.22	4.84	nr	17.04
40mm diameter	9.87	12.90	0.22	4.84	nr	17.74
50mm diameter	11.29	14.77	0.22	4.84	nr	19.60
65mm diameter	12.77	16.70	0.22	4.84	nr	21.53
80mm diameter	13.93	18.22	0.29	6.37	nr	24.59
100mm diameter	18.04	23.59	0.29	6.37	nr	29.96
125mm diameter	20.74	27.12	0.29	6.37	nr	33.49
150mm diameter	24.37	31.87	0.29	6.37	nr	38.24
200mm diameter	34.08	44.56	0.36	7.91	nr	52.47
250mm diameter	40.53	53.01	0.36	7.91	nr	60.92
300mm diameter	42.98	56.20	0.36	7.91	nr	64.11
350mm diameter	47.19	61.71	0.36	7.91	nr	69.62
40mm thick						
15mm diameter	4.83	6.32	0.15	3.30	m	9.62
20mm diameter	4.98	6.52	0.15	3.30	m	9.81
25mm diameter	5.36	7.00	0.15	3.30	m	10.30
32mm diameter	5.71	7.47	0.15	3.30	m	10.77
40mm diameter	6.00	7.85	0.15	3.30	m	11.14
50mm diameter	6.79	8.88	0.15	3.30	m	12.18
65mm diameter	7.62	9.97	0.15	3.30	m	13.27
80mm diameter	8.29	10.84	0.22	4.84	m	15.68
100mm diameter	10.76	14.07	0.22	4.84	m	18.91
125mm diameter	12.16	15.90	0.22	4.84	m	20.74
150mm diameter	14.19	18.56	0.22	4.84	m	23.40
200mm diameter	19.61	25.64	0.25	5.50	m	31.14
250mm diameter	22.98	30.05	0.25	5.50	m	35.55
300mm diameter	24.57	32.13	0.25	5.50	m	37.62
350mm diameter	27.17	35.53	0.25	5.50	m	41.03
400mm diameter	30.28	39.60	0.25	5.50	m	45.10
Extra over for fittings concealed insulation						
Flange/union						
15mm diameter	2.42	3.16	0.13	2.86	nr	6.02
20mm diameter	2.49	3.26	0.13	2.86	nr	6.12
25mm diameter	2.68	3.50	0.13	2.86	nr	6.36
32mm diameter	2.85	3.73	0.13	2.86	nr	6.59
40mm diameter	3.00	3.92	0.13	2.86	nr	6.78
50mm diameter	3.40	4.44	0.13	2.86	nr	7.30
65mm diameter	3.81	4.98	0.13	2.86	nr	7.83
80mm diameter	4.15	5.43	0.18	3.96	nr	9.39
100mm diameter	5.38	7.04	0.18	3.96	nr	10.99
125mm diameter	6.07	7.94	0.18	3.96	nr	11.90
150mm diameter	7.10	9.29	0.18	3.96	nr	13.24

T: MECHANICAL/COOLING/HEATING SYSTEMS

Item	Net Price £	Material £	Labour hours	Labour £	Unit	Total rate £
200mm diameter	9.81	12.82	0.22	4.84	nr	17.66
250mm diameter	11.49	15.03	0.22	4.84	nr	19.86
300mm diameter	12.28	16.06	0.22	4.84	nr	20.90
350mm diameter	13.59	17.77	0.22	4.84	nr	22.60
400mm diameter	15.15	19.81	0.22	4.84	nr	24.64
Valves						
15mm diameter	4.83	6.32	0.15	3.30	nr	9.62
20mm diameter	4.98	6.52	0.15	3.30	nr	9.81
25mm diameter	5.36	7.00	0.15	3.30	nr	10.30
32mm diameter	5.71	7.47	0.15	3.30	nr	10.77
40mm diameter	6.00	7.85	0.15	3.30	nr	11.14
50mm diameter	6.79	8.88	0.15	3.30	nr	12.18
65mm diameter	7.62	9.97	0.15	3.30	nr	13.27
80mm diameter	8.29	10.84	0.20	4.40	nr	15.24
100mm diameter	10.76	14.07	0.20	4.40	nr	18.47
125mm diameter	12.16	15.90	0.20	4.40	nr	20.30
150mm diameter	14.19	18.56	0.20	4.40	nr	22.96
200mm diameter	19.61	25.64	0.25	5.50	nr	31.14
250mm diameter	22.98	30.05	0.25	5.50	nr	35.55
300mm diameter	24.57	32.13	0.25	5.50	nr	37.62
350mm diameter	27.17	35.53	0.25	5.50	nr	41.03
400mm diameter	30.28	39.60	0.25	5.50	nr	45.10
Expansion bellows						
15mm diameter	9.68	12.66	0.22	4.84	nr	17.50
20mm diameter	9.95	13.02	0.22	4.84	nr	17.85
25mm diameter	10.72	14.02	0.22	4.84	nr	18.86
32mm diameter	11.42	14.93	0.22	4.84	nr	19.77
40mm diameter	12.01	15.71	0.22	4.84	nr	20.54
50mm diameter	13.59	17.77	0.22	4.84	nr	22.60
65mm diameter	15.23	19.92	0.22	4.84	nr	24.76
80mm diameter	16.59	21.69	0.29	6.37	nr	28.06
100mm diameter	21.51	28.12	0.29	6.37	nr	34.50
125mm diameter	24.31	31.79	0.29	6.37	nr	38.16
150mm diameter	28.39	37.12	0.29	6.37	nr	43.50
200mm diameter	39.22	51.29	0.36	7.91	nr	59.20
250mm diameter	45.96	60.11	0.36	7.91	nr	68.02
300mm diameter	49.14	64.26	0.36	7.91	nr	72.17
350mm diameter	27.17	35.53	0.36	7.91	nr	43.45
400mm diameter	60.57	79.20	0.36	7.91	nr	87.11
50mm thick						
15mm diameter	6.72	8.79	0.15	3.30	m	12.08
20mm diameter	7.07	9.24	0.15	3.30	m	12.54
25mm diameter	7.51	9.82	0.15	3.30	m	13.12
32mm diameter	7.85	10.26	0.15	3.30	m	13.56
40mm diameter	8.26	10.80	0.15	3.30	m	14.09
50mm diameter	9.26	12.11	0.15	3.30	m	15.41
65mm diameter	10.12	13.23	0.15	3.30	m	16.52
80mm diameter	10.83	14.17	0.22	4.84	m	19.00
100mm diameter	13.85	18.11	0.22	4.84	m	22.94
125mm diameter	15.54	20.33	0.22	4.84	m	25.16
150mm diameter	17.94	23.46	0.22	4.84	m	28.29
200mm diameter	24.49	32.03	0.25	5.50	m	37.53
250mm diameter	28.29	36.99	0.25	5.50	m	42.49
300mm diameter	29.97	39.20	0.25	5.50	m	44.69
350mm diameter	33.08	43.26	0.25	5.50	m	48.76
400mm diameter	36.69	47.98	0.25	5.50	m	53.48

T: MECHANICAL/COOLING/HEATING SYSTEMS

Item	Net Price £	Material £	Labour hours	Labour £	Unit	Total rate £
T31 : LOW TEMPERATURE HOT WATER HEATING (cont'd)						
Concealed pipework (cont'd)						
50mm thick (cont'd)						
Extra over for fittings concealed insulation						
Flange/union						
15mm diameter	3.36	4.39	0.13	2.86	nr	7.25
20mm diameter	3.53	4.62	0.13	2.86	nr	7.48
25mm diameter	3.76	4.91	0.13	2.86	nr	7.77
32mm diameter	3.93	5.14	0.13	2.86	nr	8.00
40mm diameter	4.13	5.40	0.13	2.86	nr	8.26
50mm diameter	4.62	6.05	0.13	2.86	nr	8.90
65mm diameter	5.06	6.61	0.13	2.86	nr	9.47
80mm diameter	5.42	7.08	0.18	3.96	nr	11.04
100mm diameter	6.93	9.06	0.18	3.96	nr	13.02
125mm diameter	7.77	10.16	0.18	3.96	nr	14.12
150mm diameter	8.97	11.74	0.18	3.96	nr	15.69
200mm diameter	12.25	16.02	0.22	4.84	nr	20.85
250mm diameter	14.14	18.50	0.22	4.84	nr	23.33
300mm diameter	14.99	19.60	0.22	4.84	nr	24.43
350mm diameter	16.55	21.64	0.22	4.84	nr	26.48
400mm diameter	18.35	23.99	0.22	4.84	nr	28.83
Valves						
15mm diameter	6.72	8.79	0.15	3.30	nr	12.08
20mm diameter	7.07	9.24	0.15	3.30	nr	12.54
25mm diameter	7.51	9.82	0.15	3.30	nr	13.12
32mm diameter	7.85	10.26	0.15	3.30	nr	13.56
40mm diameter	8.26	10.80	0.15	3.30	nr	14.09
50mm diameter	9.26	12.11	0.15	3.30	nr	15.41
65mm diameter	10.12	13.23	0.15	3.30	nr	16.52
80mm diameter	10.83	14.17	0.20	4.40	nr	18.56
100mm diameter	13.85	18.11	0.20	4.40	nr	22.50
125mm diameter	15.54	20.33	0.20	4.40	nr	24.72
150mm diameter	17.94	23.46	0.20	4.40	nr	27.85
200mm diameter	24.49	32.03	0.25	5.50	nr	37.53
250mm diameter	28.29	36.99	0.25	5.50	nr	42.49
300mm diameter	29.97	39.20	0.25	5.50	nr	44.69
350mm diameter	33.08	43.26	0.25	5.50	nr	48.76
400mm diameter	36.69	47.98	0.25	5.50	nr	53.48
Expansion bellows						
15mm diameter	13.42	17.56	0.22	4.84	nr	22.39
20mm diameter	14.13	18.48	0.22	4.84	nr	23.32
25mm diameter	15.02	19.65	0.22	4.84	nr	24.48
32mm diameter	15.71	20.54	0.22	4.84	nr	25.37
40mm diameter	16.51	21.59	0.22	4.84	nr	26.43
50mm diameter	18.51	24.20	0.22	4.84	nr	29.04
65mm diameter	20.23	26.45	0.22	4.84	nr	31.29
80mm diameter	21.67	28.33	0.29	6.37	nr	34.71
100mm diameter	27.69	36.21	0.29	6.37	nr	42.59
125mm diameter	31.08	40.64	0.29	6.37	nr	47.01
150mm diameter	35.89	46.93	0.29	6.37	nr	53.30
200mm diameter	49.00	64.08	0.36	7.91	nr	71.99
250mm diameter	56.56	73.97	0.36	7.91	nr	81.88
300mm diameter	59.96	78.41	0.36	7.91	nr	86.32

T: MECHANICAL/COOLING/HEATING SYSTEMS

Item	Net Price £	Material £	Labour hours	Labour £	Unit	Total rate £
350mm diameter	66.17	86.53	0.36	7.91	nr	94.44
400mm diameter	73.37	95.95	0.36	7.91	nr	103.86
Mineral fibre sectional insulation; bright class O foil faced; bright class O foil taped joints; 22 swg plain/embossed aluminium cladding; pop riveted						
Plantroom pipework						
20mm thick						
15mm diameter	4.30	5.62	0.44	9.67	m	15.29
20mm diameter	4.55	5.95	0.44	9.67	m	15.62
25mm diameter	4.87	6.37	0.44	9.67	m	16.04
32mm diameter	5.28	6.90	0.44	9.67	m	16.58
40mm diameter	5.56	7.27	0.44	9.67	m	16.94
50mm diameter	6.19	8.10	0.44	9.67	m	17.77
Extra over for fittings plantroom insulation						
Flange/union						
15mm diameter	4.81	6.29	0.58	12.75	nr	19.04
20mm diameter	5.10	6.67	0.58	12.75	nr	19.42
25mm diameter	5.47	7.16	0.58	12.75	nr	19.91
32mm diameter	5.98	7.81	0.58	12.75	nr	20.56
40mm diameter	6.31	8.25	0.58	12.75	nr	21.00
50mm diameter	7.09	9.27	0.58	12.75	nr	22.02
Bends						
15mm diameter	2.36	3.09	0.44	9.67	nr	12.76
20mm diameter	2.51	3.28	0.44	9.67	nr	12.95
25mm diameter	2.69	3.52	0.44	9.67	nr	13.19
32mm diameter	2.90	3.80	0.44	9.67	nr	13.47
40mm diameter	3.05	3.99	0.44	9.67	nr	13.66
50mm diameter	3.41	4.46	0.44	9.67	nr	14.13
Tees						
15mm diameter	1.42	1.85	0.44	9.67	nr	11.52
20mm diameter	1.50	1.96	0.44	9.67	nr	11.63
25mm diameter	1.61	2.10	0.44	9.67	nr	11.77
32mm diameter	1.74	2.28	0.44	9.67	nr	11.95
40mm diameter	1.84	2.40	0.44	9.67	nr	12.07
50mm diameter	2.04	2.67	0.44	9.67	nr	12.34
Valves						
15mm diameter	2.64	3.45	0.78	17.15	nr	20.60
20mm diameter	3.02	3.96	0.78	17.15	nr	21.10
25mm diameter	3.02	3.96	0.78	17.15	nr	21.10
32mm diameter	3.37	4.41	0.78	17.15	nr	21.55
40mm diameter	3.61	4.72	0.78	17.15	nr	21.86
50mm diameter	4.12	5.38	0.78	17.15	nr	22.53
Pumps						
15mm diameter	14.14	18.49	2.34	51.44	nr	69.92
20mm diameter	15.00	19.62	2.34	51.44	nr	71.05
25mm diameter	16.10	21.06	2.34	51.44	nr	72.50
32mm diameter	17.57	22.97	2.34	51.44	nr	74.41
40mm diameter	18.56	24.28	2.34	51.44	nr	75.71
50mm diameter	20.87	27.29	2.34	51.44	nr	78.73

T: MECHANICAL/COOLING/HEATING SYSTEMS

Item	Net Price £	Material £	Labour hours	Labour £	Unit	Total rate £
T31 : LOW TEMPERATURE HOT WATER HEATING (cont'd)						
Plantroom pipework (cont'd)						
20mm thick (cont'd)						
Extra over for fittings plantroom insulation (cont'd)						
Expansion bellows						
15mm diameter	11.30	14.78	1.05	23.08	nr	37.86
20mm diameter	12.00	15.69	1.05	23.08	nr	38.77
25mm diameter	12.89	16.85	1.05	23.08	nr	39.93
32mm diameter	14.05	18.38	1.05	23.08	nr	41.46
40mm diameter	14.84	19.41	1.05	23.08	nr	42.49
50mm diameter	16.69	21.83	1.05	23.08	nr	44.91
25mm thick						
15mm diameter	4.72	6.17	0.44	9.67	m	15.84
20mm diameter	5.10	6.67	0.44	9.67	m	16.34
25mm diameter	5.60	7.33	0.44	9.67	m	17.00
32mm diameter	5.96	7.80	0.44	9.67	m	17.47
40mm diameter	6.41	8.38	0.44	9.67	m	18.05
50mm diameter	7.18	9.38	0.44	9.67	m	19.06
65mm diameter	8.11	10.61	0.44	9.67	m	20.28
80mm diameter	8.80	11.50	0.52	11.43	m	22.93
100mm diameter	10.92	14.28	0.52	11.43	m	25.71
125mm diameter	12.68	16.59	0.52	11.43	m	28.02
150mm diameter	14.76	19.30	0.52	11.43	m	30.73
200mm diameter	20.16	26.36	0.60	13.19	m	39.55
250mm diameter	23.95	31.32	0.60	13.19	m	44.51
300mm diameter	26.82	35.07	0.60	13.19	m	48.26
Extra over for fittings plantroom insulation						
Flange/union						
15mm diameter	5.28	6.90	0.58	12.75	nr	19.65
20mm diameter	5.71	7.47	0.58	12.75	nr	20.22
25mm diameter	6.30	8.24	0.58	12.75	nr	20.99
32mm diameter	6.76	8.83	0.58	12.75	nr	21.58
40mm diameter	7.26	9.49	0.58	12.75	nr	22.24
50mm diameter	8.18	10.70	0.58	12.75	nr	23.45
65mm diameter	9.28	12.13	0.58	12.75	nr	24.88
80mm diameter	10.09	13.20	0.67	14.73	nr	27.92
100mm diameter	12.77	16.70	0.67	14.73	nr	31.42
125mm diameter	14.83	19.40	0.67	14.73	nr	34.12
150mm diameter	17.38	22.72	0.67	14.73	nr	37.45
200mm diameter	24.01	31.40	0.87	19.12	nr	50.52
250mm diameter	28.61	37.41	0.87	19.12	nr	56.53
300mm diameter	31.61	41.33	0.87	19.12	nr	60.46

T: MECHANICAL/COOLING/HEATING SYSTEMS

Item	Net Price £	Material £	Labour hours	Labour £	Unit	Total rate £
Bends						
15mm diameter	2.59	3.39	0.44	9.67	nr	**13.06**
20mm diameter	2.81	3.67	0.44	9.67	nr	**13.34**
25mm diameter	3.08	4.03	0.44	9.67	nr	**13.70**
32mm diameter	3.29	4.30	0.44	9.67	nr	**13.97**
40mm diameter	3.53	4.61	0.44	9.67	nr	**14.29**
50mm diameter	3.95	5.16	0.44	9.67	nr	**14.83**
65mm diameter	4.46	5.84	0.44	9.67	nr	**15.51**
80mm diameter	4.84	6.32	0.52	11.43	nr	**17.75**
100mm diameter	6.00	7.85	0.52	11.43	nr	**19.28**
125mm diameter	6.98	9.13	0.52	11.43	nr	**20.56**
150mm diameter	8.11	10.61	0.52	11.43	nr	**22.04**
200mm diameter	11.09	14.50	0.60	13.19	nr	**27.69**
250mm diameter	13.18	17.23	0.60	13.19	nr	**30.42**
300mm diameter	14.75	19.29	0.60	13.19	nr	**32.47**
Tees						
15mm diameter	1.56	2.04	0.44	9.67	nr	**11.71**
20mm diameter	1.68	2.20	0.44	9.67	nr	**11.87**
25mm diameter	1.85	2.42	0.44	9.67	nr	**12.09**
32mm diameter	1.97	2.57	0.44	9.67	nr	**12.25**
40mm diameter	2.11	2.76	0.44	9.67	nr	**12.43**
50mm diameter	2.36	3.09	0.44	9.67	nr	**12.76**
65mm diameter	2.68	3.50	0.44	9.67	nr	**13.17**
80mm diameter	2.90	3.80	0.52	11.43	nr	**15.23**
100mm diameter	3.60	4.71	0.52	11.43	nr	**16.14**
125mm diameter	4.19	5.48	0.52	11.43	nr	**16.91**
150mm diameter	4.87	6.37	0.52	11.43	nr	**17.80**
200mm diameter	6.65	8.69	0.60	13.19	nr	**21.88**
250mm diameter	7.91	10.34	0.60	13.19	nr	**23.53**
300mm diameter	8.86	11.58	0.60	13.19	nr	**24.77**
Valves						
15mm diameter	8.39	10.97	0.78	17.15	nr	**28.11**
20mm diameter	9.07	11.86	0.78	17.15	nr	**29.01**
25mm diameter	10.01	13.09	0.78	17.15	nr	**30.23**
32mm diameter	10.74	14.04	0.78	17.15	nr	**31.19**
40mm diameter	11.52	15.06	0.78	17.15	nr	**32.21**
50mm diameter	12.98	16.98	0.78	17.15	nr	**34.12**
65mm diameter	14.74	19.27	0.78	17.15	nr	**36.42**
80mm diameter	16.03	20.96	0.92	20.22	nr	**41.19**
100mm diameter	20.28	26.52	0.92	20.22	nr	**46.74**
125mm diameter	23.56	30.80	0.92	20.22	nr	**51.03**
150mm diameter	27.60	36.09	0.92	20.22	nr	**56.31**
200mm diameter	38.14	49.87	1.12	24.62	nr	**74.49**
250mm diameter	45.44	59.43	1.12	24.62	nr	**84.05**
300mm diameter	50.20	65.64	1.12	24.62	nr	**90.26**
Pumps						
15mm diameter	15.53	20.31	2.34	51.44	nr	**71.74**
20mm diameter	16.79	21.95	2.34	51.44	nr	**73.39**
25mm diameter	18.54	24.24	2.34	51.44	nr	**75.68**
32mm diameter	19.88	26.00	2.34	51.44	nr	**77.44**
40mm diameter	21.34	27.90	2.34	51.44	nr	**79.34**
50mm diameter	24.06	31.46	2.34	51.44	nr	**82.90**
65mm diameter	27.29	35.68	2.34	51.44	nr	**87.12**
80mm diameter	29.69	38.82	2.76	60.67	nr	**99.49**
100mm diameter	37.55	49.10	2.76	60.67	nr	**109.77**
125mm diameter	43.61	57.02	2.76	60.67	nr	**117.69**

T: MECHANICAL/COOLING/HEATING SYSTEMS

Item	Net Price £	Material £	Labour hours	Labour £	Unit	Total rate £
T31 : LOW TEMPERATURE HOT WATER HEATING (cont'd)						
Plantroom pipework (cont'd)						
25mm thick (cont'd)						
Extra over for fittings plantroom insulation (cont'd)						
Pumps (cont'd)						
150mm diameter	51.11	66.83	2.76	60.67	nr	127.50
200mm diameter	70.63	92.36	3.36	73.86	nr	166.22
250mm diameter	84.14	110.03	3.36	73.86	nr	183.89
300mm diameter	92.95	121.55	3.36	73.86	nr	195.41
Expansion bellows						
15mm diameter	12.42	16.24	1.05	23.08	nr	39.32
20mm diameter	13.43	17.56	1.05	23.08	nr	40.64
25mm diameter	14.83	19.40	1.05	23.08	nr	42.48
32mm diameter	15.91	20.81	1.05	23.08	nr	43.89
40mm diameter	17.08	22.33	1.05	23.08	nr	45.41
50mm diameter	19.25	25.17	1.05	23.08	nr	48.25
65mm diameter	21.83	28.54	1.05	23.08	nr	51.62
80mm diameter	23.75	31.05	1.26	27.70	nr	58.75
100mm diameter	30.04	39.28	1.26	27.70	nr	66.97
125mm diameter	34.88	45.62	1.26	27.70	nr	73.31
150mm diameter	40.88	53.46	1.26	27.70	nr	81.16
200mm diameter	56.51	73.89	1.53	33.63	nr	107.53
250mm diameter	67.32	88.03	1.53	33.63	nr	121.66
300mm diameter	74.36	97.24	1.53	33.63	nr	130.88
30mm thick						
15mm diameter	5.82	7.61	0.44	9.67	m	17.28
20mm diameter	6.16	8.05	0.44	9.67	m	17.72
25mm diameter	6.52	8.52	0.44	9.67	m	18.19
32mm diameter	7.14	9.34	0.44	9.67	m	19.01
40mm diameter	7.44	9.73	0.44	9.67	m	19.40
50mm diameter	8.26	10.80	0.44	9.67	m	20.47
65mm diameter	9.31	12.18	0.44	9.67	m	21.85
80mm diameter	10.18	13.31	0.52	11.43	m	24.74
100mm diameter	12.50	16.35	0.52	11.43	m	27.78
125mm diameter	14.29	18.69	0.52	11.43	m	30.12
150mm diameter	16.66	21.78	0.52	11.43	m	33.21
200mm diameter	22.38	29.27	0.60	13.19	m	42.45
250mm diameter	26.48	34.63	0.60	13.19	m	47.82
300mm diameter	29.15	38.12	0.60	13.19	m	51.31
350mm diameter	32.46	42.45	0.60	13.19	m	55.64

T: MECHANICAL/COOLING/HEATING SYSTEMS

Item	Net Price £	Material £	Labour hours	Labour £	Unit	Total rate £
Extra over for fittings plantroom insulation						
Flange/union						
15mm diameter	6.62	8.66	0.58	12.75	nr	**21.41**
20mm diameter	7.03	9.20	0.58	12.75	nr	**21.95**
25mm diameter	7.45	9.74	0.58	12.75	nr	**22.49**
32mm diameter	8.16	10.67	0.58	12.75	nr	**23.42**
40mm diameter	8.53	11.16	0.58	12.75	nr	**23.91**
50mm diameter	9.56	12.51	0.58	12.75	nr	**25.26**
65mm diameter	10.80	14.12	0.58	12.75	nr	**26.87**
80mm diameter	11.80	15.43	0.67	14.73	nr	**30.15**
100mm diameter	14.74	19.27	0.67	14.73	nr	**34.00**
125mm diameter	16.87	22.06	0.67	14.73	nr	**36.79**
150mm diameter	19.73	25.80	0.67	14.73	nr	**40.53**
200mm diameter	26.87	35.13	0.87	19.12	nr	**54.26**
250mm diameter	31.86	41.66	0.87	19.12	nr	**60.79**
300mm diameter	34.66	45.32	0.87	19.12	nr	**64.44**
350mm diameter	38.45	50.28	0.87	19.12	nr	**69.40**
Bends						
15mm diameter	3.20	4.19	0.44	9.67	nr	**13.86**
20mm diameter	3.38	4.43	0.44	9.67	nr	**14.10**
25mm diameter	3.59	4.69	0.44	9.67	nr	**14.36**
32mm diameter	3.92	5.13	0.44	9.67	nr	**14.80**
40mm diameter	4.09	5.35	0.44	9.67	nr	**15.02**
50mm diameter	4.55	5.95	0.44	9.67	nr	**15.62**
65mm diameter	5.12	6.70	0.44	9.67	nr	**16.37**
80mm diameter	5.59	7.31	0.52	11.43	nr	**18.74**
100mm diameter	6.88	8.99	0.52	11.43	nr	**20.42**
125mm diameter	7.86	10.28	0.52	11.43	nr	**21.71**
150mm diameter	9.17	11.99	0.52	11.43	nr	**23.42**
200mm diameter	12.31	16.10	0.60	13.19	nr	**29.29**
250mm diameter	14.57	19.05	0.60	13.19	nr	**32.24**
300mm diameter	16.03	20.96	0.60	13.19	nr	**34.15**
350mm diameter	17.86	23.35	0.60	13.19	nr	**36.54**
Tees						
15mm diameter	1.92	2.51	0.44	9.67	nr	**12.18**
20mm diameter	2.03	2.65	0.44	9.67	nr	**12.32**
25mm diameter	2.15	2.81	0.44	9.67	nr	**12.48**
32mm diameter	2.35	3.08	0.44	9.67	nr	**12.75**
40mm diameter	2.46	3.22	0.44	9.67	nr	**12.89**
50mm diameter	2.72	3.56	0.44	9.67	nr	**13.23**
65mm diameter	3.07	4.02	0.44	9.67	nr	**13.69**
80mm diameter	3.36	4.39	0.52	11.43	nr	**15.82**
100mm diameter	4.13	5.40	0.52	11.43	nr	**16.83**
125mm diameter	4.72	6.17	0.52	11.43	nr	**17.60**
150mm diameter	5.50	7.19	0.52	11.43	nr	**18.62**
200mm diameter	7.39	9.67	0.60	13.19	nr	**22.86**
250mm diameter	8.74	11.42	0.60	13.19	nr	**24.61**
300mm diameter	9.62	12.59	0.60	13.19	nr	**25.77**
350mm diameter	10.72	14.01	0.60	13.19	nr	**27.20**

T: MECHANICAL/COOLING/HEATING SYSTEMS

Item	Net Price £	Material £	Labour hours	Labour £	Unit	Total rate £
T31 : LOW TEMPERATURE HOT WATER HEATING (cont'd)						
Plantroom pipework (cont'd)						
30mm thick (cont'd)						
Valves						
15mm diameter	10.52	13.76	0.78	17.15	nr	30.91
20mm diameter	11.16	14.59	0.78	17.15	nr	31.74
25mm diameter	11.83	15.47	0.78	17.15	nr	32.62
32mm diameter	12.96	16.95	0.78	17.15	nr	34.09
40mm diameter	13.56	17.73	0.78	17.15	nr	34.88
50mm diameter	15.18	19.85	0.78	17.15	nr	37.00
65mm diameter	17.16	22.44	0.78	17.15	nr	39.59
80mm diameter	18.73	24.50	0.92	20.22	nr	44.72
100mm diameter	23.40	30.60	0.92	20.22	nr	50.82
125mm diameter	26.80	35.04	0.92	20.22	nr	55.26
150mm diameter	31.33	40.97	0.92	20.22	nr	61.20
200mm diameter	42.67	55.80	1.12	24.62	nr	80.42
250mm diameter	50.59	66.16	1.12	24.62	nr	90.78
300mm diameter	55.04	71.98	1.12	24.62	nr	96.60
350mm diameter	61.07	79.86	1.12	24.62	nr	104.48
Pumps						
15mm diameter	19.49	25.48	2.34	51.44	nr	76.92
20mm diameter	20.68	27.04	2.34	51.44	nr	78.48
25mm diameter	21.90	28.64	2.34	51.44	nr	80.08
32mm diameter	24.00	31.38	2.34	51.44	nr	82.82
40mm diameter	25.10	32.83	2.34	51.44	nr	84.27
50mm diameter	28.12	36.77	2.34	51.44	nr	88.20
65mm diameter	31.76	41.54	2.34	51.44	nr	92.97
80mm diameter	34.69	45.37	2.76	60.67	nr	106.04
100mm diameter	43.33	56.66	2.76	60.67	nr	117.33
125mm diameter	49.63	64.90	2.76	60.67	nr	125.57
150mm diameter	58.03	75.89	2.76	60.67	nr	136.56
200mm diameter	79.02	103.33	3.36	73.86	nr	177.19
250mm diameter	93.70	122.52	3.36	73.86	nr	196.38
300mm diameter	101.94	133.30	3.36	73.86	nr	207.16
350mm diameter	113.08	147.87	3.36	73.86	nr	221.73
Expansion bellows						
15mm diameter	15.59	20.38	1.05	23.08	nr	43.47
20mm diameter	16.54	21.62	1.05	23.08	nr	44.70
25mm diameter	17.52	22.91	1.05	23.08	nr	45.99
32mm diameter	19.20	25.11	1.05	23.08	nr	48.19
40mm diameter	20.09	26.27	1.05	23.08	nr	49.35
50mm diameter	22.49	29.41	1.05	23.08	nr	52.49
65mm diameter	25.42	33.24	1.05	23.08	nr	56.32
80mm diameter	27.76	36.30	1.26	27.70	nr	63.99
100mm diameter	34.67	45.33	1.26	27.70	nr	73.03
125mm diameter	39.71	51.92	1.26	27.70	nr	79.62
150mm diameter	46.43	60.71	1.26	27.70	nr	88.41
200mm diameter	63.22	82.67	1.53	33.63	nr	116.30
250mm diameter	74.95	98.01	1.53	33.63	nr	131.64
300mm diameter	81.55	106.64	1.53	33.63	nr	140.28
350mm diameter	90.47	118.30	1.53	33.63	nr	151.93

T: MECHANICAL/COOLING/HEATING SYSTEMS

Item	Net Price £	Material £	Labour hours	Labour £	Unit	Total rate £
40mm thick						
15mm diameter	7.12	9.31	0.44	9.67	m	**18.98**
20mm diameter	7.48	9.78	0.44	9.67	m	**19.45**
25mm diameter	7.96	10.40	0.44	9.67	m	**20.08**
32mm diameter	8.34	10.91	0.44	9.67	m	**20.58**
40mm diameter	8.87	11.60	0.44	9.67	m	**21.27**
50mm diameter	9.82	12.84	0.44	9.67	m	**22.51**
65mm diameter	10.94	14.31	0.44	9.67	m	**23.98**
80mm diameter	11.77	15.39	0.52	11.43	m	**26.82**
100mm diameter	17.24	22.55	0.52	11.43	m	**33.98**
125mm diameter	16.15	21.12	0.52	11.43	m	**32.55**
150mm diameter	18.84	24.64	0.52	11.43	m	**36.07**
200mm diameter	24.90	32.56	0.60	13.19	m	**45.75**
250mm diameter	29.18	38.16	0.60	13.19	m	**51.35**
300mm diameter	32.27	42.20	0.60	13.19	m	**55.39**
350mm diameter	36.06	47.15	0.60	13.19	m	**60.34**
400mm diameter	40.54	53.01	0.60	13.19	m	**66.20**
Extra over for fittings plantroom insulation						
Flange/union						
15mm diameter	8.22	10.75	0.58	12.75	nr	**23.50**
20mm diameter	8.58	11.22	0.58	12.75	nr	**23.97**
25mm diameter	9.18	12.00	0.58	12.75	nr	**24.75**
32mm diameter	9.66	12.63	0.58	12.75	nr	**25.38**
40mm diameter	10.25	13.40	0.58	12.75	nr	**26.15**
50mm diameter	11.42	14.94	0.58	12.75	nr	**27.69**
65mm diameter	12.76	16.68	0.58	12.75	nr	**29.43**
80mm diameter	13.76	18.00	0.67	14.73	nr	**32.73**
100mm diameter	17.24	22.55	0.67	14.73	nr	**37.28**
125mm diameter	19.30	25.23	0.67	14.73	nr	**39.96**
150mm diameter	22.52	29.45	0.67	14.73	nr	**44.18**
200mm diameter	30.22	39.51	0.87	19.12	nr	**58.64**
250mm diameter	35.42	46.32	0.87	19.12	nr	**65.45**
300mm diameter	38.76	50.69	0.87	19.12	nr	**69.81**
350mm diameter	43.20	56.49	0.87	19.12	nr	**75.62**
400mm diameter	48.44	63.35	0.87	19.12	nr	**82.47**
Bends						
15mm diameter	3.91	5.12	0.44	9.67	nr	**14.79**
20mm diameter	4.10	5.37	0.44	9.67	nr	**15.04**
25mm diameter	4.38	5.73	0.44	9.67	nr	**15.40**
32mm diameter	4.58	5.99	0.44	9.67	nr	**15.67**
40mm diameter	4.88	6.39	0.44	9.67	nr	**16.06**
50mm diameter	5.40	7.06	0.44	9.67	nr	**16.73**
65mm diameter	6.02	7.88	0.44	9.67	nr	**17.55**
80mm diameter	6.47	8.46	0.52	11.43	nr	**19.89**
100mm diameter	7.98	10.44	0.52	11.43	nr	**21.87**
125mm diameter	8.88	11.61	0.52	11.43	nr	**23.04**
150mm diameter	10.36	13.54	0.52	11.43	nr	**24.97**
200mm diameter	13.69	17.90	0.60	13.19	nr	**31.09**
250mm diameter	15.94	20.84	0.60	13.19	nr	**34.03**
300mm diameter	17.75	23.21	0.60	13.19	nr	**36.40**
350mm diameter	19.84	25.94	0.60	13.19	nr	**39.13**
400mm diameter	22.30	29.16	0.60	13.19	nr	**42.34**

T: MECHANICAL/COOLING/HEATING SYSTEMS

Item	Net Price £	Material £	Labour hours	Labour £	Unit	Total rate £
T31 : LOW TEMPERATURE HOT WATER HEATING (cont'd)						
Plantroom pipework (cont'd)						
40mm thick (cont'd)						
Extra over for fittings plantroom insulation (cont'd)						
Tees						
15mm diameter	2.35	3.08	0.44	9.67	nr	12.75
20mm diameter	2.46	3.22	0.44	9.67	nr	12.89
25mm diameter	2.63	3.44	0.44	9.67	nr	13.11
32mm diameter	2.75	3.59	0.44	9.67	nr	13.27
40mm diameter	2.93	3.83	0.44	9.67	nr	13.50
50mm diameter	3.24	4.24	0.44	9.67	nr	13.91
65mm diameter	3.61	4.72	0.44	9.67	nr	14.40
80mm diameter	3.89	5.08	0.52	11.43	nr	16.51
100mm diameter	4.79	6.26	0.52	11.43	nr	17.69
125mm diameter	5.33	6.97	0.52	11.43	nr	18.40
150mm diameter	6.22	8.13	0.52	11.43	nr	19.56
200mm diameter	8.22	10.75	0.60	13.19	nr	23.94
250mm diameter	9.64	12.60	0.60	13.19	nr	25.79
300mm diameter	10.64	13.92	0.60	13.19	nr	27.11
350mm diameter	11.90	15.57	0.60	13.19	nr	28.76
400mm diameter	13.38	17.50	0.60	13.19	nr	30.69
Valves						
15mm diameter	13.06	17.07	0.78	17.15	nr	34.22
20mm diameter	13.63	17.83	0.78	17.15	nr	34.97
25mm diameter	14.57	19.05	0.78	17.15	nr	36.20
32mm diameter	15.34	20.05	0.78	17.15	nr	37.20
40mm diameter	16.27	21.28	0.78	17.15	nr	38.42
50mm diameter	18.13	23.71	0.78	17.15	nr	40.86
65mm diameter	20.26	26.49	0.78	17.15	nr	43.63
80mm diameter	21.86	28.59	0.92	20.22	nr	48.81
100mm diameter	27.40	35.82	0.92	20.22	nr	56.05
125mm diameter	30.64	40.06	0.92	20.22	nr	60.29
150mm diameter	35.77	46.78	0.92	20.22	nr	67.00
200mm diameter	47.99	62.75	1.12	24.62	nr	87.37
250mm diameter	56.26	73.56	1.12	24.62	nr	98.18
300mm diameter	61.56	80.50	1.12	24.62	nr	105.12
350mm diameter	68.62	89.73	1.12	24.62	nr	114.35
400mm diameter	76.94	100.62	1.12	24.62	nr	125.24
Pumps						
15mm diameter	24.17	31.60	2.34	51.44	nr	83.04
20mm diameter	25.24	33.00	2.34	51.44	nr	84.44
25mm diameter	26.99	35.29	2.34	51.44	nr	86.73
32mm diameter	28.40	37.14	2.34	51.44	nr	88.58
40mm diameter	30.13	39.40	2.34	51.44	nr	90.84
50mm diameter	33.59	43.92	2.34	51.44	nr	95.36
65mm diameter	37.51	49.05	2.34	51.44	nr	100.49
80mm diameter	40.49	52.94	2.76	60.67	nr	113.61
100mm diameter	50.72	66.33	2.76	60.67	nr	127.00
125mm diameter	56.74	74.19	2.76	60.67	nr	134.86
150mm diameter	66.25	86.64	2.76	60.67	nr	147.31
200mm diameter	88.87	116.22	3.36	73.86	nr	190.07
250mm diameter	104.18	136.24	3.36	73.86	nr	210.10

T: MECHANICAL/COOLING/HEATING SYSTEMS

Item	Net Price £	Material £	Labour hours	Labour £	Unit	Total rate £
300mm diameter	113.99	149.06	3.36	73.86	nr	222.92
350mm diameter	127.06	166.15	3.36	73.86	nr	240.01
400mm diameter	142.49	186.33	3.36	73.86	nr	260.19
Expansion bellows						
15mm diameter	19.33	25.28	1.05	23.08	nr	48.36
20mm diameter	20.20	26.41	1.05	23.08	nr	49.49
25mm diameter	21.59	28.23	1.05	23.08	nr	51.31
32mm diameter	22.72	29.70	1.05	23.08	nr	52.79
40mm diameter	24.10	31.51	1.05	23.08	nr	54.59
50mm diameter	26.87	35.13	1.05	23.08	nr	58.22
65mm diameter	30.01	39.25	1.05	23.08	nr	62.33
80mm diameter	32.39	42.35	1.26	27.70	nr	70.05
100mm diameter	40.58	53.07	1.26	27.70	nr	80.77
125mm diameter	45.38	59.35	1.26	27.70	nr	87.04
150mm diameter	52.99	69.30	1.26	27.70	nr	96.99
200mm diameter	71.09	92.96	1.53	33.63	nr	126.59
250mm diameter	83.35	109.00	1.53	33.63	nr	142.63
300mm diameter	91.19	119.24	1.53	33.63	nr	152.88
350mm diameter	101.64	132.91	1.53	33.63	nr	166.54
400mm diameter	113.99	149.06	1.53	33.63	nr	182.69
50mm thick						
15mm diameter	9.29	12.15	0.44	9.67	m	21.82
20mm diameter	9.79	12.80	0.44	9.67	m	22.48
25mm diameter	10.37	13.56	0.44	9.67	m	23.23
32mm diameter	10.90	14.25	0.44	9.67	m	23.92
40mm diameter	11.40	14.91	0.44	9.67	m	24.58
50mm diameter	12.58	16.45	0.44	9.67	m	26.12
65mm diameter	13.52	17.68	0.44	9.67	m	27.36
80mm diameter	14.52	18.99	0.52	11.43	m	30.42
100mm diameter	17.94	23.46	0.52	11.43	m	34.89
125mm diameter	19.86	25.97	0.52	11.43	m	37.40
150mm diameter	22.81	29.83	0.52	11.43	m	41.26
200mm diameter	29.90	39.10	0.60	13.19	m	52.29
250mm diameter	34.56	45.19	0.60	13.19	m	58.38
300mm diameter	37.62	49.19	0.60	13.19	m	62.38
350mm diameter	41.93	54.83	0.60	13.19	m	68.02
400mm diameter	46.90	61.32	0.60	13.19	m	74.51
Extra over for fittings plantroom insulation						
Flange/union						
15mm diameter	10.94	14.31	0.58	12.75	nr	27.06
20mm diameter	11.54	15.10	0.58	12.75	nr	27.85
25mm diameter	12.23	15.99	0.58	12.75	nr	28.74
32mm diameter	12.84	16.79	0.58	12.75	nr	29.54
40mm diameter	13.45	17.59	0.58	12.75	nr	30.34
50mm diameter	14.93	19.52	0.58	12.75	nr	32.27
65mm diameter	16.13	21.09	0.58	12.75	nr	33.84
80mm diameter	17.29	22.61	0.67	14.73	nr	37.34
100mm diameter	21.62	28.28	0.67	14.73	nr	43.01
125mm diameter	24.05	31.45	0.67	14.73	nr	46.17
150mm diameter	27.67	36.19	0.67	14.73	nr	50.91
200mm diameter	36.78	48.10	0.87	19.12	nr	67.22
250mm diameter	42.50	55.58	0.87	19.12	nr	74.71
300mm diameter	45.86	59.98	0.87	19.12	nr	79.10
350mm diameter	50.99	66.68	0.87	19.12	nr	85.80
400mm diameter	56.89	74.40	0.87	19.12	nr	93.52

Item	Net Price £	Material £	Labour hours	Labour £	Unit	Total rate £
T31 : LOW TEMPERATURE HOT WATER HEATING (cont'd)						
Plantroom pipework (cont'd)						
50mm thick (cont'd)						
Extra over for fittings plantroom insulation (cont'd)						
Bend						
15mm diameter	5.11	6.68	0.44	9.67	nr	**16.36**
20mm diameter	5.39	7.05	0.44	9.67	nr	**16.72**
25mm diameter	5.70	7.45	0.44	9.67	nr	**17.13**
32mm diameter	6.00	7.85	0.44	9.67	nr	**17.52**
40mm diameter	6.26	8.19	0.44	9.67	nr	**17.86**
50mm diameter	6.92	9.05	0.44	9.67	nr	**18.73**
65mm diameter	7.44	9.73	0.44	9.67	nr	**19.40**
80mm diameter	7.98	10.44	0.52	11.43	nr	**21.87**
100mm diameter	9.86	12.90	0.52	11.43	nr	**24.33**
125mm diameter	10.92	14.28	0.52	11.43	nr	**25.71**
150mm diameter	12.55	16.41	0.52	11.43	nr	**27.84**
200mm diameter	16.45	21.51	0.60	13.19	nr	**34.70**
250mm diameter	19.01	24.86	0.60	13.19	nr	**38.05**
300mm diameter	20.69	27.05	0.60	13.19	nr	**40.24**
350mm diameter	23.06	30.16	0.60	13.19	nr	**43.35**
400mm diameter	25.79	33.72	0.60	13.19	nr	**46.91**
Tee						
15mm diameter	3.07	4.02	0.44	9.67	nr	**13.69**
20mm diameter	3.23	4.22	0.44	9.67	nr	**13.89**
25mm diameter	3.42	4.47	0.44	9.67	nr	**14.14**
32mm diameter	3.60	4.71	0.44	9.67	nr	**14.38**
40mm diameter	3.76	4.91	0.44	9.67	nr	**14.58**
50mm diameter	4.15	5.43	0.44	9.67	nr	**15.10**
65mm diameter	4.46	5.84	0.44	9.67	nr	**15.51**
80mm diameter	4.79	6.26	0.52	11.43	nr	**17.69**
100mm diameter	5.92	7.74	0.52	11.43	nr	**19.17**
125mm diameter	6.55	8.57	0.52	11.43	nr	**20.00**
150mm diameter	7.52	9.84	0.52	11.43	nr	**21.27**
200mm diameter	9.86	12.90	0.60	13.19	nr	**26.09**
250mm diameter	11.41	14.92	0.60	13.19	nr	**28.11**
300mm diameter	12.42	16.24	0.60	13.19	nr	**29.43**
350mm diameter	13.84	18.09	0.60	13.19	nr	**31.28**
400mm diameter	15.48	20.24	0.60	13.19	nr	**33.43**
Valves						
15mm diameter	17.39	22.74	0.78	17.15	nr	**39.88**
20mm diameter	18.32	23.96	0.78	17.15	nr	**41.11**
25mm diameter	19.42	25.39	0.78	17.15	nr	**42.54**
32mm diameter	20.39	26.66	0.78	17.15	nr	**43.81**
40mm diameter	21.36	27.93	0.78	17.15	nr	**45.08**
50mm diameter	23.70	30.99	0.78	17.15	nr	**48.14**
65mm diameter	25.61	33.49	0.78	17.15	nr	**50.63**
80mm diameter	27.47	35.92	0.92	20.22	nr	**56.14**
100mm diameter	34.34	44.91	0.92	20.22	nr	**65.13**
125mm diameter	38.18	49.93	0.92	20.22	nr	**70.16**
150mm diameter	43.94	57.46	0.92	20.22	nr	**77.69**
200mm diameter	58.43	76.40	1.12	24.62	nr	**101.02**

T: MECHANICAL/COOLING/HEATING SYSTEMS

Item	Net Price £	Material £	Labour hours	Labour £	Unit	Total rate £
250mm diameter	67.51	88.28	1.12	24.62	nr	**112.90**
300mm diameter	72.85	95.27	1.12	24.62	nr	**119.89**
350mm diameter	80.98	105.89	1.12	24.62	nr	**130.51**
400mm diameter	90.35	118.15	1.12	24.62	nr	**142.77**
Pumps						
15mm diameter	32.20	42.10	2.34	51.44	nr	**93.54**
20mm diameter	33.94	44.38	2.34	51.44	nr	**95.81**
25mm diameter	35.95	47.01	2.34	51.44	nr	**98.45**
32mm diameter	37.76	49.38	2.34	51.44	nr	**100.82**
40mm diameter	39.56	51.74	2.34	51.44	nr	**103.17**
50mm diameter	43.88	57.39	2.34	51.44	nr	**108.82**
65mm diameter	35.42	46.32	2.34	51.44	nr	**97.76**
80mm diameter	50.88	66.53	2.76	60.67	nr	**127.20**
100mm diameter	63.59	83.15	2.76	60.67	nr	**143.82**
125mm diameter	70.72	92.47	2.76	60.67	nr	**153.14**
150mm diameter	81.38	106.42	2.76	60.67	nr	**167.09**
200mm diameter	108.19	141.48	3.36	73.86	nr	**215.34**
250mm diameter	125.02	163.48	3.36	73.86	nr	**237.34**
300mm diameter	134.90	176.41	3.36	73.86	nr	**250.27**
350mm diameter	149.96	196.10	3.36	73.86	nr	**269.96**
400mm diameter	167.32	218.79	3.36	73.86	nr	**292.65**
Expansion bellows						
15mm diameter	25.75	33.68	1.05	23.08	nr	**56.76**
20mm diameter	27.14	35.50	1.05	23.08	nr	**58.58**
25mm diameter	28.76	37.61	1.05	23.08	nr	**60.69**
32mm diameter	30.20	39.50	1.05	23.08	nr	**62.58**
40mm diameter	31.64	41.38	1.05	23.08	nr	**64.46**
50mm diameter	35.11	45.91	1.05	23.08	nr	**69.00**
65mm diameter	37.93	49.60	1.05	23.08	nr	**72.68**
80mm diameter	40.70	53.23	1.26	27.70	nr	**80.92**
100mm diameter	50.87	66.52	1.26	27.70	nr	**94.22**
125mm diameter	56.57	73.97	1.26	27.70	nr	**101.67**
150mm diameter	65.11	85.14	1.26	27.70	nr	**112.84**
200mm diameter	86.56	113.19	1.53	33.63	nr	**146.82**
250mm diameter	100.01	130.78	1.53	33.63	nr	**164.41**
300mm diameter	107.93	141.13	1.53	33.63	nr	**174.77**
350mm diameter	119.96	156.87	1.53	33.63	nr	**190.51**
400mm diameter	133.85	175.03	1.53	33.63	nr	**208.66**
Mineral fibre sectional insulation; bright class O foil faced; bright class O foil taped joints; 0.8mm polyisobutylene sheeting; welded joints						
External pipework						
20mm thick						
15mm diameter	3.96	5.18	0.30	6.59	m	**11.77**
20mm diameter	4.22	5.52	0.30	6.59	m	**12.12**
25mm diameter	4.55	5.95	0.30	6.59	m	**12.54**
32mm diameter	5.00	6.54	0.30	6.59	m	**13.14**
40mm diameter	5.33	6.97	0.30	6.59	m	**13.56**
50mm diameter	6.02	7.88	0.30	6.59	m	**14.47**

T: MECHANICAL/COOLING/HEATING SYSTEMS

Item	Net Price £	Material £	Labour hours	Labour £	Unit	Total rate £
T31 : LOW TEMPERATURE HOT WATER HEATING (cont'd)						
External pipework (cont'd)						
20mm thick (cont'd)						
Extra over for fittings external insulation						
Flange/union						
15mm diameter	6.07	7.94	0.75	16.49	nr	24.43
20mm diameter	6.44	8.43	0.75	16.49	nr	24.91
25mm diameter	6.92	9.05	0.75	16.49	nr	25.54
32mm diameter	7.54	9.85	0.75	16.49	nr	26.34
40mm diameter	7.96	10.40	0.75	16.49	nr	26.89
50mm diameter	8.92	11.66	0.75	16.49	nr	28.15
Bends						
15mm diameter	1.00	1.30	0.30	6.59	nr	7.90
20mm diameter	1.06	1.38	0.30	6.59	nr	7.98
25mm diameter	1.14	1.49	0.30	6.59	nr	8.09
32mm diameter	1.25	1.63	0.30	6.59	nr	8.23
40mm diameter	1.33	1.74	0.30	6.59	nr	8.34
50mm diameter	1.50	1.96	0.30	6.59	nr	8.56
Tees						
15mm diameter	1.00	1.30	0.30	6.59	nr	7.90
20mm diameter	1.06	1.38	0.30	6.59	nr	7.98
25mm diameter	1.14	1.49	0.30	6.59	nr	8.09
32mm diameter	1.25	1.63	0.30	6.59	nr	8.23
40mm diameter	1.33	1.74	0.30	6.59	nr	8.34
50mm diameter	1.50	1.96	0.30	6.59	nr	8.56
Valves						
15mm diameter	9.64	12.60	1.03	22.64	nr	35.24
20mm diameter	10.24	13.39	1.03	22.64	nr	36.03
25mm diameter	10.99	14.37	1.03	22.64	nr	37.02
32mm diameter	11.96	15.64	1.03	22.64	nr	38.29
40mm diameter	12.64	16.52	1.03	22.64	nr	39.17
50mm diameter	14.16	18.52	1.03	22.64	nr	41.16
Expansion bellows						
15mm diameter	14.28	18.67	1.42	31.21	nr	49.89
20mm diameter	15.17	19.83	1.42	31.21	nr	51.05
25mm diameter	16.30	21.31	1.42	31.21	nr	52.52
32mm diameter	17.74	23.19	1.42	31.21	nr	54.41
40mm diameter	18.72	24.48	1.42	31.21	nr	55.69
50mm diameter	20.98	27.43	1.42	31.21	nr	58.64
25mm thick						
15mm diameter	4.37	5.71	0.30	6.59	m	12.31
20mm diameter	4.69	6.14	0.30	6.59	m	12.73
25mm diameter	5.16	6.75	0.30	6.59	m	13.34
32mm diameter	5.59	7.31	0.30	6.59	m	13.91
40mm diameter	5.94	7.77	0.30	6.59	m	14.36
50mm diameter	6.71	8.77	0.30	6.59	m	15.37
65mm diameter	7.61	9.95	0.30	6.59	m	16.54

Material Costs/Prices for Measured Works – Mechanical Installations

T: MECHANICAL/COOLING/HEATING SYSTEMS

Item	Net Price £	Material £	Labour hours	Labour £	Unit	Total rate £
80mm diameter	8.32	10.87	0.40	8.79	m	**19.67**
100mm diameter	10.51	13.75	0.40	8.79	m	**22.54**
125mm diameter	12.10	15.82	0.40	8.79	m	**24.61**
150mm diameter	14.20	18.56	0.40	8.79	m	**27.36**
200mm diameter	19.26	25.19	0.50	10.99	m	**36.18**
250mm diameter	22.99	30.07	0.50	10.99	m	**41.06**
300mm diameter	25.01	32.70	0.50	10.99	m	**43.69**
Extra over for fittings external insulation						
Flange/union						
15mm diameter	6.68	8.74	0.75	16.49	nr	**25.23**
20mm diameter	7.20	9.42	0.75	16.49	nr	**25.90**
25mm diameter	7.90	10.33	0.75	16.49	nr	**26.81**
32mm diameter	8.46	11.06	0.75	16.49	nr	**27.55**
40mm diameter	9.05	11.83	0.75	16.49	nr	**28.32**
50mm diameter	10.14	13.26	0.75	16.49	nr	**29.75**
65mm diameter	11.47	15.00	0.75	16.49	nr	**31.49**
80mm diameter	12.48	16.32	0.89	19.56	nr	**35.88**
100mm diameter	15.50	20.27	0.89	19.56	nr	**39.84**
125mm diameter	17.95	23.48	0.89	19.56	nr	**43.04**
150mm diameter	20.89	27.32	0.89	19.56	nr	**46.88**
200mm diameter	28.27	36.97	1.15	25.28	nr	**62.25**
250mm diameter	33.65	44.00	1.15	25.28	nr	**69.28**
300mm diameter	37.38	48.88	1.15	25.28	nr	**74.16**
Bends						
15mm diameter	1.09	1.43	0.30	6.59	nr	**8.02**
20mm diameter	1.18	1.54	0.30	6.59	nr	**8.13**
25mm diameter	1.28	1.68	0.30	6.59	nr	**8.27**
32mm diameter	1.39	1.82	0.30	6.59	nr	**8.41**
40mm diameter	1.49	1.95	0.30	6.59	nr	**8.54**
50mm diameter	1.68	2.20	0.30	6.59	nr	**8.79**
65mm diameter	1.91	2.50	0.30	6.59	nr	**9.09**
80mm diameter	2.08	2.71	0.40	8.79	nr	**11.51**
100mm diameter	2.63	3.44	0.40	8.79	nr	**12.23**
125mm diameter	3.02	3.95	0.40	8.79	nr	**12.75**
150mm diameter	3.55	4.64	0.40	8.79	nr	**13.44**
200mm diameter	4.81	6.29	0.50	10.99	nr	**17.28**
250mm diameter	5.75	7.52	0.50	10.99	nr	**18.51**
300mm diameter	6.25	8.18	0.50	10.99	nr	**19.17**
Tees						
15mm diameter	1.09	1.43	0.30	6.59	nr	**8.02**
20mm diameter	1.18	1.54	0.30	6.59	nr	**8.13**
25mm diameter	1.28	1.68	0.30	6.59	nr	**8.27**
32mm diameter	1.39	1.82	0.30	6.59	nr	**8.41**
40mm diameter	1.49	1.95	0.30	6.59	nr	**8.54**
50mm diameter	1.68	2.20	0.30	6.59	nr	**8.79**
65mm diameter	1.91	2.50	0.30	6.59	nr	**9.09**
80mm diameter	2.08	2.71	0.40	8.79	nr	**11.51**
100mm diameter	2.63	3.44	0.40	8.79	nr	**12.23**
125mm diameter	3.02	3.95	0.40	8.79	nr	**12.75**
150mm diameter	3.55	4.64	0.40	8.79	nr	**13.44**
200mm diameter	4.81	6.29	0.50	10.99	nr	**17.28**
250mm diameter	5.75	7.52	0.50	10.99	nr	**18.51**
300mm diameter	6.25	8.18	0.50	10.99	nr	**19.17**

T: MECHANICAL/COOLING/HEATING SYSTEMS

Item	Net Price £	Material £	Labour hours	Labour £	Unit	Total rate £
T31 : LOW TEMPERATURE HOT WATER HEATING (cont'd)						
External pipework (cont'd)						
25mm thick (cont'd)						
Extra over for fittings external insulation (cont'd)						
Valves						
15mm diameter	10.61	13.87	1.03	22.64	nr	36.51
20mm diameter	11.44	14.95	1.03	22.64	nr	37.60
25mm diameter	12.54	16.40	1.03	22.64	nr	39.04
32mm diameter	13.44	17.58	1.03	22.64	nr	40.22
40mm diameter	14.36	18.78	1.03	22.64	nr	41.42
50mm diameter	16.12	21.07	1.03	22.64	nr	43.72
65mm diameter	18.23	23.84	1.03	22.64	nr	46.48
80mm diameter	19.81	25.91	1.25	27.48	nr	53.38
100mm diameter	24.64	32.22	1.25	27.48	nr	59.69
125mm diameter	28.50	37.27	1.25	27.48	nr	64.75
150mm diameter	33.19	43.40	1.25	27.48	nr	70.88
200mm diameter	44.90	58.72	1.55	34.07	nr	92.79
250mm diameter	53.44	69.88	1.55	34.07	nr	103.95
300mm diameter	59.36	77.63	1.55	34.07	nr	111.70
Expansion bellows						
15mm diameter	15.72	20.56	1.42	31.21	nr	51.77
20mm diameter	16.94	22.16	1.42	31.21	nr	53.37
25mm diameter	18.58	24.29	1.42	31.21	nr	55.51
32mm diameter	19.92	26.05	1.42	31.21	nr	57.26
40mm diameter	21.29	27.84	1.42	31.21	nr	59.05
50mm diameter	23.87	31.21	1.42	31.21	nr	62.43
65mm diameter	27.00	35.31	1.42	31.21	nr	66.52
80mm diameter	29.35	38.38	1.75	38.47	nr	76.85
100mm diameter	36.49	47.72	1.75	38.47	nr	86.19
125mm diameter	42.23	55.22	1.75	38.47	nr	93.69
150mm diameter	49.16	64.29	1.75	38.47	nr	102.76
200mm diameter	66.52	86.98	2.17	47.70	nr	134.68
250mm diameter	79.16	103.52	2.17	47.70	nr	151.22
300mm diameter	87.94	114.99	3.17	69.68	nr	184.67
30mm thick						
15mm diameter	5.38	7.03	0.30	6.59	m	13.62
20mm diameter	5.72	7.49	0.30	6.59	m	14.08
25mm diameter	6.07	7.94	0.30	6.59	m	14.53
32mm diameter	6.58	8.60	0.30	6.59	m	15.19
40mm diameter	6.94	9.07	0.30	6.59	m	15.66
50mm diameter	7.81	10.22	0.30	6.59	m	16.81
65mm diameter	8.78	11.49	0.30	6.59	m	18.08
80mm diameter	9.55	12.49	0.40	8.79	m	21.28
100mm diameter	11.94	15.61	0.40	8.79	m	24.41
125mm diameter	13.67	17.87	0.40	8.79	m	26.67
150mm diameter	15.88	20.76	0.40	8.79	m	29.55
200mm diameter	21.38	27.96	0.50	10.99	m	38.95
250mm diameter	25.37	33.17	0.50	10.99	m	44.16
300mm diameter	27.36	35.78	0.50	10.99	m	46.77
350mm diameter	29.92	39.12	0.50	10.99	m	50.11

T: MECHANICAL/COOLING/HEATING SYSTEMS

Item	Net Price £	Material £	Labour hours	Labour £	Unit	Total rate £
Extra over for fittings external insulation						
Flange/union						
15mm diameter	8.17	10.69	0.75	16.49	nr	27.17
20mm diameter	8.66	11.33	0.75	16.49	nr	27.82
25mm diameter	9.18	12.00	0.75	16.49	nr	28.49
32mm diameter	10.01	13.09	0.75	16.49	nr	29.57
40mm diameter	10.46	13.68	0.75	16.49	nr	30.17
50mm diameter	11.68	15.27	0.75	16.49	nr	31.75
65mm diameter	13.14	17.18	0.75	16.49	nr	33.67
80mm diameter	14.32	18.72	0.89	19.56	nr	38.28
100mm diameter	17.63	23.05	0.89	19.56	nr	42.62
125mm diameter	20.14	26.33	0.89	19.56	nr	45.90
150mm diameter	23.40	30.60	0.89	19.56	nr	50.16
200mm diameter	31.27	40.89	1.15	25.28	nr	66.17
250mm diameter	37.03	48.43	1.15	25.28	nr	73.70
300mm diameter	40.57	53.05	1.15	25.28	nr	78.33
350mm diameter	44.82	58.61	1.15	25.28	nr	83.89
Bends						
15mm diameter	1.34	1.76	0.30	6.59	nr	8.35
20mm diameter	1.43	1.87	0.30	6.59	nr	8.46
25mm diameter	1.52	1.99	0.30	6.59	nr	8.59
32mm diameter	1.64	2.15	0.30	6.59	nr	8.74
40mm diameter	1.73	2.26	0.30	6.59	nr	8.85
50mm diameter	1.96	2.56	0.30	6.59	nr	9.15
65mm diameter	2.20	2.87	0.30	6.59	nr	9.47
80mm diameter	2.39	3.12	0.40	8.79	nr	11.92
100mm diameter	2.99	3.91	0.40	8.79	nr	12.70
125mm diameter	3.42	4.47	0.40	8.79	nr	13.27
150mm diameter	3.97	5.19	0.40	8.79	nr	13.99
200mm diameter	5.35	7.00	0.50	10.99	nr	17.99
250mm diameter	6.35	8.30	0.50	10.99	nr	19.29
300mm diameter	6.84	8.94	0.50	10.99	nr	19.94
350mm diameter	7.48	9.78	0.50	10.99	nr	20.77
Tees						
15mm diameter	1.34	1.76	0.30	6.59	nr	8.35
20mm diameter	1.43	1.87	0.30	6.59	nr	8.46
25mm diameter	1.52	1.99	0.30	6.59	nr	8.59
32mm diameter	1.64	2.15	0.30	6.59	nr	8.74
40mm diameter	1.73	2.26	0.30	6.59	nr	8.85
50mm diameter	1.96	2.56	0.30	6.59	nr	9.15
65mm diameter	2.20	2.87	0.30	6.59	nr	9.47
80mm diameter	2.39	3.12	0.40	8.79	nr	11.92
100mm diameter	2.99	3.91	0.40	8.79	nr	12.70
125mm diameter	3.42	4.47	0.40	8.79	nr	13.27
150mm diameter	3.97	5.19	0.40	8.79	nr	13.99
200mm diameter	5.35	7.00	0.50	10.99	nr	17.99
250mm diameter	6.35	8.30	0.50	10.99	nr	19.29
300mm diameter	6.84	8.94	0.50	10.99	nr	19.94
350mm diameter	7.48	9.78	0.50	10.99	nr	20.77

T: MECHANICAL/COOLING/HEATING SYSTEMS

Item	Net Price £	Material £	Labour hours	Labour £	Unit	Total rate £
T31 : LOW TEMPERATURE HOT WATER HEATING (cont'd)						
External pipework (cont'd)						
30mm thick (cont'd)						
Extra over for fittings external insulation (cont'd)						
Valves						
15mm diameter	12.98	16.98	1.03	22.64	nr	39.62
20mm diameter	13.75	17.98	1.03	22.64	nr	40.62
25mm diameter	14.58	19.07	1.03	22.64	nr	41.71
32mm diameter	15.89	20.78	1.03	22.64	nr	43.42
40mm diameter	16.63	21.75	1.03	22.64	nr	44.39
50mm diameter	18.54	24.24	1.03	22.64	nr	46.89
65mm diameter	20.87	27.29	1.03	22.64	nr	49.93
80mm diameter	22.74	29.74	1.25	27.48	nr	57.21
100mm diameter	27.98	36.59	1.25	27.48	nr	64.07
125mm diameter	31.98	41.82	1.25	27.48	nr	69.30
150mm diameter	37.15	48.58	1.25	27.48	nr	76.06
200mm diameter	49.67	64.95	1.55	34.07	nr	99.02
250mm diameter	58.81	76.91	1.55	34.07	nr	110.98
300mm diameter	64.44	84.27	1.55	34.07	nr	118.34
350mm diameter	71.18	93.09	1.55	34.07	nr	127.16
Expansion bellows						
15mm diameter	19.22	25.14	1.42	31.21	nr	56.35
20mm diameter	20.38	26.65	1.42	31.21	nr	57.86
25mm diameter	21.60	28.25	1.42	31.21	nr	59.46
32mm diameter	23.54	30.79	1.42	31.21	nr	62.00
40mm diameter	24.64	32.22	1.42	31.21	nr	63.43
50mm diameter	27.46	35.90	1.42	31.21	nr	67.12
65mm diameter	30.92	40.44	1.42	31.21	nr	71.65
80mm diameter	33.70	44.06	1.75	38.47	nr	82.53
100mm diameter	41.46	54.22	1.75	38.47	nr	92.68
125mm diameter	47.38	61.95	1.75	38.47	nr	100.42
150mm diameter	55.04	71.98	1.75	38.47	nr	110.45
200mm diameter	73.58	96.22	2.17	47.70	nr	143.92
250mm diameter	87.13	113.94	2.17	47.70	nr	161.64
300mm diameter	95.46	124.83	2.17	47.70	nr	172.53
350mm diameter	105.46	137.90	2.17	47.70	nr	185.60
40mm thick						
15mm diameter	6.73	8.80	0.30	6.59	m	15.40
20mm diameter	6.96	9.10	0.30	6.59	m	15.70
25mm diameter	7.44	9.73	0.30	6.59	m	16.32
32mm diameter	7.91	10.34	0.30	6.59	m	16.94
40mm diameter	8.29	10.84	0.30	6.59	m	17.44
50mm diameter	9.24	12.08	0.30	6.59	m	18.68
65mm diameter	10.30	13.46	0.30	6.59	m	20.06
80mm diameter	11.16	14.59	0.40	8.79	m	23.39
100mm diameter	13.94	18.23	0.40	8.79	m	27.03
125mm diameter	15.71	20.54	0.40	8.79	m	29.33
150mm diameter	18.13	23.71	0.40	8.79	m	32.50
200mm diameter	24.19	31.64	0.50	10.99	m	42.63
250mm diameter	28.32	37.03	0.50	10.99	m	48.02

T: MECHANICAL/COOLING/HEATING SYSTEMS

Item	Net Price £	Material £	Labour hours	Labour £	Unit	Total rate £
300mm diameter	30.67	40.11	0.50	10.99	m	51.10
350mm diameter	33.70	44.06	0.50	10.99	m	55.05
400mm diameter	37.51	49.05	0.50	10.99	m	60.04
Extra over for fittings external insulation						
Flange/union						
15mm diameter	10.06	13.15	0.75	16.49	nr	29.64
20mm diameter	10.50	13.73	0.75	16.49	nr	30.22
25mm diameter	11.20	14.64	0.75	16.49	nr	31.13
32mm diameter	11.80	15.43	0.75	16.49	nr	31.91
40mm diameter	12.47	16.30	0.75	16.49	nr	32.79
50mm diameter	13.81	18.06	0.75	16.49	nr	34.55
65mm diameter	15.38	20.12	0.75	16.49	nr	36.60
80mm diameter	16.58	21.69	0.89	19.56	nr	41.25
100mm diameter	20.42	26.71	0.89	19.56	nr	46.27
125mm diameter	22.84	29.86	0.89	19.56	nr	49.43
150mm diameter	26.47	34.62	0.89	19.56	nr	54.18
200mm diameter	34.90	45.63	1.15	25.28	nr	70.91
250mm diameter	40.90	53.48	1.15	25.28	nr	78.76
300mm diameter	44.95	58.78	1.15	25.28	nr	84.06
350mm diameter	49.86	65.20	1.15	25.28	nr	90.48
400mm diameter	55.84	73.02	1.15	25.28	nr	98.29
Bends						
15mm diameter	1.68	2.20	0.30	6.59	nr	8.79
20mm diameter	1.74	2.28	0.30	6.59	nr	8.87
25mm diameter	1.86	2.43	0.30	6.59	nr	9.03
32mm diameter	1.98	2.59	0.30	6.59	nr	9.18
40mm diameter	2.08	2.71	0.30	6.59	nr	9.31
50mm diameter	2.32	3.03	0.30	6.59	nr	9.62
65mm diameter	2.58	3.37	0.30	6.59	nr	9.97
80mm diameter	2.80	3.66	0.40	8.79	nr	12.45
100mm diameter	3.49	4.57	0.40	8.79	nr	13.36
125mm diameter	3.92	5.13	0.40	8.79	nr	13.92
150mm diameter	4.54	5.93	0.40	8.79	nr	14.72
200mm diameter	6.05	7.91	0.50	10.99	nr	18.90
250mm diameter	7.08	9.26	0.50	10.99	nr	20.25
300mm diameter	7.67	10.03	0.50	10.99	nr	21.02
350mm diameter	8.42	11.02	0.50	10.99	nr	22.01
400mm diameter	9.38	12.27	0.50	10.99	nr	23.26
Tees						
15mm diameter	1.68	2.20	0.30	6.59	nr	8.79
20mm diameter	1.74	2.28	0.30	6.59	nr	8.87
25mm diameter	1.86	2.43	0.30	6.59	nr	9.03
32mm diameter	1.98	2.59	0.30	6.59	nr	9.18
40mm diameter	2.08	2.71	0.30	6.59	nr	9.31
50mm diameter	2.32	3.03	0.30	6.59	nr	9.62
65mm diameter	2.58	3.37	0.30	6.59	nr	9.97
80mm diameter	2.80	3.66	0.40	8.79	nr	12.45
100mm diameter	3.49	4.57	0.40	8.79	nr	13.36
125mm diameter	3.92	5.13	0.40	8.79	nr	13.92
150mm diameter	4.54	5.93	0.40	8.79	nr	14.72
200mm diameter	6.05	7.91	0.50	10.99	nr	18.90
250mm diameter	7.08	9.26	0.50	10.99	nr	20.25
300mm diameter	7.67	10.03	0.50	10.99	nr	21.02
350mm diameter	8.42	11.02	0.50	10.99	nr	22.01
400mm diameter	9.38	12.27	0.50	10.99	nr	23.26

T: MECHANICAL/COOLING/HEATING SYSTEMS

Item	Net Price £	Material £	Labour hours	Labour £	Unit	Total rate £
T31 : LOW TEMPERATURE HOT WATER HEATING (cont'd)						
External pipework (cont'd)						
40mm thick (cont'd)						
Extra over for fittings external insulation (cont'd)						
Valves						
15mm diameter	15.96	20.87	1.03	22.64	nr	43.51
20mm diameter	16.68	21.81	1.03	22.64	nr	44.45
25mm diameter	17.78	23.26	1.03	22.64	nr	45.90
32mm diameter	18.73	24.50	1.03	22.64	nr	47.14
40mm diameter	19.80	25.89	1.03	22.64	nr	48.53
50mm diameter	21.94	28.68	1.03	22.64	nr	51.33
65mm diameter	24.43	31.95	1.03	22.64	nr	54.59
80mm diameter	26.33	34.43	1.25	27.48	nr	61.91
100mm diameter	32.44	42.42	1.25	27.48	nr	69.89
125mm diameter	36.28	47.44	1.25	27.48	nr	74.91
150mm diameter	42.05	54.98	1.25	27.48	nr	82.46
200mm diameter	55.43	72.48	1.55	34.07	nr	106.55
250mm diameter	64.94	84.93	1.55	34.07	nr	119.00
300mm diameter	71.40	93.37	1.55	34.07	nr	127.44
350mm diameter	79.19	103.55	1.55	34.07	nr	137.62
400mm diameter	88.67	115.95	1.55	34.07	nr	150.02
Expansion bellows						
15mm diameter	23.65	30.93	1.42	31.21	nr	62.14
20mm diameter	24.71	32.31	1.42	31.21	nr	63.52
25mm diameter	26.34	34.44	1.42	31.21	nr	65.66
32mm diameter	27.74	36.28	1.42	31.21	nr	67.49
40mm diameter	29.33	38.35	1.42	31.21	nr	69.57
50mm diameter	32.51	42.51	1.42	31.21	nr	73.72
65mm diameter	36.19	47.33	1.42	31.21	nr	78.54
80mm diameter	39.01	51.01	1.75	38.47	nr	89.48
100mm diameter	48.06	62.85	1.75	38.47	nr	101.31
125mm diameter	53.74	70.27	1.75	38.47	nr	108.74
150mm diameter	62.30	81.47	1.75	38.47	nr	119.94
200mm diameter	82.12	107.38	2.17	47.70	nr	155.08
250mm diameter	96.22	125.82	2.17	47.70	nr	173.52
300mm diameter	105.78	138.33	2.17	47.70	nr	186.03
350mm diameter	117.32	153.42	2.17	47.70	nr	201.12
400mm diameter	131.36	171.78	2.17	47.70	nr	219.48
50mm thick						
15mm diameter	8.87	11.60	0.30	6.59	m	18.19
20mm diameter	9.30	12.16	0.30	6.59	m	18.76
25mm diameter	9.84	12.87	0.30	6.59	m	19.46
32mm diameter	10.62	13.89	0.30	6.59	m	20.48
40mm diameter	10.79	14.11	0.30	6.59	m	20.70
50mm diameter	11.95	15.63	0.30	6.59	m	22.22
65mm diameter	13.03	17.04	0.30	6.59	m	23.64
80mm diameter	13.94	18.23	0.40	8.79	m	27.03
100mm diameter	17.26	22.57	0.40	8.79	m	31.36
125mm diameter	19.31	25.25	0.40	8.79	m	34.04
150mm diameter	22.08	28.87	0.40	8.79	m	37.67
200mm diameter	29.24	38.24	0.50	10.99	m	49.23

T: MECHANICAL/COOLING/HEATING SYSTEMS

Item	Net Price £	Material £	Labour hours	Labour £	Unit	Total rate £
250mm diameter	33.77	44.16	0.50	10.99	m	55.15
300mm diameter	36.22	47.36	0.50	10.99	m	58.35
350mm diameter	39.74	51.97	0.50	10.99	m	62.96
400mm diameter	44.03	57.57	0.50	10.99	m	68.56
Extra over for fittings external insulation						
Flange/union						
15mm diameter	13.07	17.09	0.75	16.49	nr	33.58
20mm diameter	13.75	17.98	0.75	16.49	nr	34.47
25mm diameter	14.53	19.00	0.75	16.49	nr	35.49
32mm diameter	15.26	19.96	0.75	16.49	nr	36.45
40mm diameter	15.96	20.87	0.75	16.49	nr	37.36
50mm diameter	17.60	23.02	0.75	16.49	nr	39.51
65mm diameter	19.03	24.89	0.75	16.49	nr	41.37
80mm diameter	20.40	26.68	0.89	19.56	nr	46.24
100mm diameter	25.08	32.80	0.89	19.56	nr	52.36
125mm diameter	27.88	36.45	0.89	19.56	nr	56.02
150mm diameter	31.92	41.74	0.89	19.56	nr	61.30
200mm diameter	41.76	54.61	1.15	25.28	nr	79.89
250mm diameter	48.26	63.11	1.15	25.28	nr	88.39
300mm diameter	52.36	68.46	1.15	25.28	nr	93.74
350mm diameter	57.94	75.76	1.15	25.28	nr	101.04
400mm diameter	64.56	84.42	1.15	25.28	nr	109.70
Bend						
15mm diameter	2.22	2.90	0.30	6.59	nr	9.50
20mm diameter	2.33	3.04	0.30	6.59	nr	9.64
25mm diameter	2.46	3.22	0.30	6.59	nr	9.81
32mm diameter	2.58	3.37	0.30	6.59	nr	9.97
40mm diameter	2.70	3.53	0.30	6.59	nr	10.13
50mm diameter	2.99	3.91	0.30	6.59	nr	10.50
65mm diameter	3.26	4.27	0.30	6.59	nr	10.86
80mm diameter	3.48	4.55	0.40	8.79	nr	13.34
100mm diameter	4.31	5.63	0.40	8.79	nr	14.43
125mm diameter	4.82	6.31	0.40	8.79	nr	15.10
150mm diameter	5.52	7.22	0.40	8.79	nr	16.01
200mm diameter	7.31	9.56	0.50	10.99	nr	20.55
250mm diameter	8.45	11.05	0.50	10.99	nr	22.04
300mm diameter	9.06	11.85	0.50	10.99	nr	22.84
350mm diameter	9.94	12.99	0.50	10.99	nr	23.98
400mm diameter	11.00	14.39	0.50	10.99	nr	25.38
Tee						
15mm diameter	2.22	2.90	0.30	6.59	nr	9.50
20mm diameter	2.33	3.04	0.30	6.59	nr	9.64
25mm diameter	2.46	3.22	0.30	6.59	nr	9.81
32mm diameter	2.58	3.37	0.30	6.59	nr	9.97
40mm diameter	2.70	3.53	0.30	6.59	nr	10.13
50mm diameter	2.99	3.91	0.30	6.59	nr	10.50
65mm diameter	3.26	4.27	0.30	6.59	nr	10.86
80mm diameter	3.48	4.55	0.40	8.79	nr	13.34
100mm diameter	4.31	5.63	0.40	8.79	nr	14.43
125mm diameter	4.82	6.31	0.40	8.79	nr	15.10
150mm diameter	5.52	7.22	0.40	8.79	nr	16.01
200mm diameter	7.31	9.56	0.50	10.99	nr	20.55
250mm diameter	8.45	11.05	0.50	10.99	nr	22.04
300mm diameter	9.06	11.85	0.50	10.99	nr	22.84
350mm diameter	9.94	12.99	0.50	10.99	nr	23.98
400mm diameter	11.00	14.39	0.50	10.99	nr	25.38

T: MECHANICAL/COOLING/HEATING SYSTEMS

Item	Net Price £	Material £	Labour hours	Labour £	Unit	Total rate £
T31 : LOW TEMPERATURE HOT WATER HEATING (cont'd)						
External pipework (cont'd)						
50mm thick (cont'd)						
Extra over for fittings external insulation (cont'd)						
Valves						
15mm diameter	20.76	27.15	1.03	22.64	nr	49.79
20mm diameter	21.84	28.56	1.03	22.64	nr	51.20
25mm diameter	23.09	30.19	1.03	22.64	nr	52.83
32mm diameter	24.24	31.70	1.03	22.64	nr	54.34
40mm diameter	25.36	33.16	1.03	22.64	nr	55.80
50mm diameter	27.96	36.56	1.03	22.64	nr	59.20
65mm diameter	30.24	39.54	1.03	22.64	nr	62.19
80mm diameter	32.40	42.37	1.25	27.48	nr	69.85
100mm diameter	39.84	52.10	1.25	27.48	nr	79.58
125mm diameter	44.28	57.90	1.25	27.48	nr	85.38
150mm diameter	50.69	66.28	1.25	27.48	nr	93.76
200mm diameter	66.32	86.73	1.55	34.07	nr	120.80
250mm diameter	76.64	100.23	1.55	34.07	nr	134.30
300mm diameter	83.15	108.73	1.55	34.07	nr	142.80
350mm diameter	92.02	120.33	1.55	34.07	nr	154.40
400mm diameter	102.53	134.07	1.55	34.07	nr	168.14
Expansion bellows						
15mm diameter	30.76	40.22	1.42	31.21	nr	71.43
20mm diameter	32.35	42.31	1.42	31.21	nr	73.52
25mm diameter	34.20	44.72	1.42	31.21	nr	75.94
32mm diameter	35.92	46.97	1.42	31.21	nr	78.18
40mm diameter	37.56	49.12	1.42	31.21	nr	80.33
50mm diameter	41.42	54.17	1.42	31.21	nr	85.38
65mm diameter	44.80	58.58	1.42	31.21	nr	89.79
80mm diameter	48.01	62.78	1.75	38.47	nr	101.25
100mm diameter	59.02	77.17	1.75	38.47	nr	115.64
125mm diameter	65.60	85.79	1.75	38.47	nr	124.26
150mm diameter	75.10	98.20	1.75	38.47	nr	136.67
200mm diameter	98.26	128.49	2.17	47.70	nr	176.19
250mm diameter	113.54	148.48	2.17	47.70	nr	196.18
300mm diameter	123.19	161.09	2.17	47.70	nr	208.80
350mm diameter	136.32	178.26	2.17	47.70	nr	225.96
400mm diameter	151.90	198.63	2.17	47.70	nr	246.33

Material Costs/Prices for Measured Works – Mechanical Installations

T: MECHANICAL/COOLING/HEATING SYSTEMS

Item	Net Price £	Material £	Labour hours	Labour £	Unit	Total rate £
T33 : STEAM HEATING						
Y10 - PIPELINES						
For pipework prices refer to Section T31 - Low Temperature Hot Water Heating						
Y11 - PIPELINE ANCILLARIES						
Steam traps and accessories						
Cast iron; inverted bucket type; steam trap pressure range up to 17 bar at 210°C; screwed ends						
1/2" dia.	135.25	173.23	0.85	23.04	nr	**196.26**
3/4" dia.	199.53	255.56	1.13	30.63	nr	**286.18**
1" dia.	312.02	399.63	1.35	36.59	nr	**436.22**
11/2" dia.	575.83	737.52	1.80	48.79	nr	**786.30**
2" dia.	899.20	1151.69	2.18	59.09	nr	**1210.77**
Cast iron; inverted bucket type; steam trap pressure range up to 17 bar at 210°C; flanged ends to BS 4504 PN16; bolted connections						
15mm dia.	326.75	418.50	1.15	31.17	nr	**449.67**
20mm dia.	378.99	485.41	1.25	33.88	nr	**519.29**
25mm dia.	579.85	742.67	1.33	36.05	nr	**778.71**
40mm dia.	897.24	1149.18	1.46	39.57	nr	**1188.75**
50mm dia.	1094.11	1401.33	1.60	43.37	nr	**1444.69**
Steam traps and strainers						
Stainless steel; thermodynamic trap with pressure range up to 42 bar; temperature range to 400°C; screwed ends to steel						
15mm dia.	90.40	115.78	0.84	22.78	nr	**138.56**
20mm dia.	136.59	174.94	1.14	30.91	nr	**205.85**
Stainless steel; thermodynamic trap with pressure range up to 24 bar; temperature range to 288°C; flanged ends to DIN 2456 PN64; bolted connections						
15mm dia.	587.69	752.71	1.24	33.63	nr	**786.33**
20mm dia.	634.78	813.02	1.34	36.33	nr	**849.35**
25mm dia.	700.23	896.85	1.40	37.96	nr	**934.81**
Malleable iron pipeline strainer; max steam working pressure 14 bar and temperature range to 230°C; screwed ends to steel						
1/2" dia.	13.93	17.84	0.84	22.78	nr	**40.62**
3/4" dia.	18.62	23.85	1.14	30.91	nr	**54.75**
1" dia.	27.52	35.25	1.30	35.25	nr	**70.49**
11/2" dia.	45.47	58.24	1.50	40.70	nr	**98.93**
2" dia.	81.02	103.77	1.74	47.22	nr	**150.99**

T: MECHANICAL/COOLING/HEATING SYSTEMS

Item	Net Price £	Material £	Labour hours	Labour £	Unit	Total rate £
T33 : STEAM HEATING (cont'd)						
Steam traps and strainers (cont'd)						
Bronze pipeline strainer; max steam working pressure 25 bar; flanged ends to BS 4504 PN25; bolted connections						
15mm dia.	164.70	210.95	1.24	33.63	nr	244.57
20mm dia.	200.88	257.29	1.34	36.33	nr	293.62
25mm dia.	230.33	295.00	1.40	37.96	nr	332.97
32mm dia.	357.55	457.95	1.46	39.63	nr	497.57
40mm dia.	405.77	519.71	1.54	41.76	nr	561.47
50mm dia.	624.05	799.28	1.64	44.50	nr	843.78
65mm dia.	691.00	885.03	2.50	67.76	nr	952.78
80mm dia.	859.73	1101.13	2.91	78.79	nr	1179.92
100mm dia.	1489.14	1907.28	3.51	95.10	nr	2002.38
Balanced pressure thermostatic steam trap and strainer; max working pressure up to 13 bar; screwed ends to steel						
1/2" dia.	135.23	173.20	1.26	34.18	nr	207.38
3/4" dia.	135.59	173.66	1.71	46.41	nr	220.07
Bimetallic thermostatic steam trap and strainer; max working pressure up to 21 bar; flanged ends						
15mm	175.92	225.32	1.24	33.63	nr	258.94
20mm	193.26	247.53	1.34	36.33	nr	283.86
Sight glasses						
Pressed brass; straight; single window; screwed ends to steel						
15mm dia.	41.51	53.17	0.84	22.78	nr	75.94
20mm dia.	46.07	59.01	1.14	30.91	nr	89.91
25mm dia.	57.57	73.74	1.30	35.25	nr	108.98
Gunmetal; straight; double window; screwed ends to steel						
15mm dia.	66.95	85.75	0.84	22.78	nr	108.52
20mm dia.	73.65	94.33	1.14	30.91	nr	125.24
25mm dia.	91.05	116.62	1.30	35.25	nr	151.86
32mm dia.	149.99	192.11	1.35	36.63	nr	228.73
40mm dia.	149.99	192.11	1.74	47.22	nr	239.33
50mm dia.	182.13	233.27	2.08	56.46	nr	289.74
SG Iron flanged; BS 4504, PN 25						
15mm dia.	136.59	174.94	1.00	27.10	nr	202.05
20mm dia.	160.71	205.84	1.25	33.88	nr	239.72
25mm dia.	204.91	262.45	1.50	40.70	nr	303.14
32mm dia.	226.31	289.86	1.70	46.09	nr	335.95
40mm dia.	295.96	379.06	2.00	54.21	nr	433.27
50mm dia.	357.55	457.95	2.30	62.45	nr	520.40
Check valve and sight glass; gun metal; screwed						
15mm dia.	67.61	86.59	0.84	22.78	nr	109.37
20mm dia.	71.63	91.74	1.14	30.91	nr	122.65
25mm dia.	120.52	154.36	1.30	35.25	nr	189.61

T: MECHANICAL/COOLING/HEATING SYSTEMS

Item	Net Price £	Material £	Labour hours	Labour £	Unit	Total rate £
Pressure reducing valves						
Pressure reducing valve for steam; maximum range of 17 bar and 232°C; screwed ends to steel						
15mm dia.	506.20	648.34	0.87	23.58	nr	**671.92**
20mm dia.	547.71	701.50	0.91	24.66	nr	**726.17**
25mm dia.	590.56	756.38	1.35	36.59	nr	**792.97**
Pressure reducing valve for steam; maximum range of 17 bar and 232°C; flanged ends to BS 4504 PN 25						
25mm dia.	711.09	910.76	1.70	46.08	nr	**956.83**
32mm dia.	807.52	1034.26	1.87	50.68	nr	**1084.95**
40mm dia.	964.20	1234.94	2.12	57.46	nr	**1292.40**
50mm dia.	1112.85	1425.33	2.57	69.66	nr	**1494.98**
Safety and relief valves						
Bronze safety valve; 'pop' type; side outlet; including easing lever; working pressure saturated steam up to 20.7 bar; screwed ends to steel						
15mm dia.	154.00	197.24	0.32	8.67	nr	**205.91**
20mm dia.	174.08	222.96	0.40	10.84	nr	**233.80**
Bronze safety valve; 'pop' type; side outlet; including easing lever; working pressure saturated steam up to 17.2 bar; screwed ends to steel						
25mm dia.	247.73	317.29	0.47	12.74	nr	**330.03**
32mm dia.	295.96	379.06	0.56	15.18	nr	**394.24**
Bronze safety valve; 'pop' type; side outlet; including easing lever; working pressure saturated steam up to 13.8 bar; screwed ends to steel						
40mm dia.	383.00	490.54	0.64	17.35	nr	**507.89**
50mm dia.	532.99	682.65	0.76	20.60	nr	**703.25**
65mm dia.	757.98	970.81	0.94	25.48	nr	**996.29**
80mm dia.	982.97	1258.98	1.10	29.81	nr	**1288.79**

T: MECHANICAL/COOLING/HEATING SYSTEMS

Item	Net Price £	Material £	Labour hours	Labour £	Unit	Total rate £
T33 : STEAM HEATING (cont'd)						
EQUIPMENT						
Y23 - CALORIFIERS						
Non-storage calorifiers; mild steel shell construction with indirect steam heating for secondary LPHW at 82°C flow and 71°C return to BS 853; maximum test on shell 11 bar, tubes 26 bar						
Horizontal/vertical, for steam at 3 bar-5.5 bar						
88 kW capacity	499.04	639.17	8.00	216.83	nr	855.99
176 kW capacity	726.52	930.52	12.05	326.55	nr	1257.07
293 kW capacity	791.91	1014.27	14.08	381.74	nr	1396.01
586 kW capacity	1155.88	1480.44	37.04	1003.83	nr	2484.27
879 kW capacity	1438.81	1842.81	40.00	1084.14	nr	2926.95
1465 kW capacity	1755.86	2248.89	45.45	1231.98	nr	3480.86

T: MECHANICAL/COOLING/HEATING SYSTEMS

Item	Net Price £	Material £	Labour hours	Labour £	Unit	Total rate £
T42 : LOCAL HEATING UNITS						
Warm air unit heater for connection to LTHW or steam supplies; suitable for heights up to 3m; recirculating type; mild steel casing; heating coil; adjustable discharge louvre; axial fan; horizontal or vertical discharge; normal speed; entering air temperature 15°C; complete with enclosures; includes fixing in position; includes connections to primary heating supply; electrical work elsewhere						
Low pressure hot water						
7.5 kW, 265 l/sec	330.96	423.89	6.53	176.99	nr	**600.88**
15.4 kW, 575 l/sec	400.63	513.12	7.54	204.36	nr	**717.48**
26.9 kW, 1040 l/sec	542.87	695.30	8.65	234.45	nr	**929.75**
48.0 kW, 1620 l/sec	715.60	916.53	9.35	253.42	nr	**1169.95**
Steam, 2 Bar						
9.2 kW, 265 l/sec	473.19	606.06	6.53	176.99	nr	**783.04**
18.8 kW, 575 l/sec	512.39	656.26	6.82	184.85	nr	**841.11**
34.8 kW, 1040 l/sec	589.32	754.80	6.82	184.85	nr	**939.64**
51.6 kW, 1625 l/sec	798.36	1022.53	7.10	192.43	nr	**1214.97**

T: MECHANICAL/COOLING/HEATING SYSTEMS

Item	Net Price £	Material £	Labour hours	Labour £	Unit	Total rate £
T60: CENTRAL REFRIGERATION PLANT						
CHILLERS						
Air cooled						
Selection of air cooled chillers based on chilled water flow and return temperatures 6°C and 12°C, and an outdoor temperature of 35°C						
Air cooled liquid chiller; refrigerant 407C; scroll compressors; twin circuit; integral controls; includes placing in position; electrical work elsewhere						
Cooling load						
100 kW	18110.02	23195.13	8.00	216.83	nr	23411.96
150 kW	20429.23	26165.55	8.00	216.83	nr	26382.38
200 kW	25932.60	33214.21	8.00	216.83	nr	33431.04
Air cooled liquid chiller; refrigerant 407C; reciprocating compressors; twin circuit; integral controls; includes placing in position; electrical work elsewhere						
Cooling Load						
250 kW	34083.44	43653.73	8.00	216.83	nr	43870.56
400 kW	47969.58	61438.96	8.00	216.83	nr	61655.79
550 kW	61111.78	78271.36	8.00	216.83	nr	78488.18
700 kW	76713.24	98253.55	9.00	243.93	nr	98497.48
Air cooled liquid chiller; refrigerant R134a; screw compressors; twin circuit; integral controls; includes placing in position; electrical work elsewhere						
Cooling load						
250 kW	38690.50	49554.41	8.00	216.83	nr	49771.23
400 kW	48811.00	62516.64	8.00	216.83	nr	62733.47
600 kW	63406.35	81210.22	8.00	216.83	nr	81427.05
800 kW	94280.99	120754.15	9.00	243.93	nr	120998.08
1000 kW	109660.61	140452.21	9.00	243.93	nr	140696.14
1200 kW	126937.05	162579.70	10.00	271.04	nr	162850.74
Air cooled liquid chiller; ductable for indoor installation; refrigerant 407C; scroll compressors; integral controls; includes placing in position; electrical work elsewhere						
Cooling load						
40 kW	12718.69	16289.97	6.00	162.62	nr	16452.59
80 kW	18470.78	23657.19	6.00	162.62	nr	23819.81

T: MECHANICAL/COOLING/HEATING SYSTEMS

Item	Net Price £	Material £	Labour hours	Labour £	Unit	Total rate £
Higher efficiency air cooled						
Selection of air cooled chillers based on chilled water flow and return temperatures of 6°C and 12°C and an outdoor temperature of 25°C						
These machines have significantly higher part load operating efficiencies than conventional air cooled machines						
Air cooled liquid chiller, refrigerant R410A; scroll compressors; complete with free cooling facility; integral controls; including placing in position; electrical work elsewhere						
Cooling load						
250 kW	35380.80	43157.50	8.00	216.83	nr	**43374.33**
300 kW	37739.52	46034.67	8.00	216.83	nr	**46251.49**
350 kW	40098.24	48911.83	8.00	216.83	nr	**49128.66**
400 kW	43636.32	53227.58	8.00	216.83	nr	**53444.41**
450 kW	47174.40	57543.33	8.00	216.83	nr	**57760.16**
500 kW	50712.48	61859.08	8.00	216.83	nr	**62075.91**
600 kW	54250.56	66174.83	8.00	216.83	nr	**66391.66**
650 kW	58968.00	71929.17	9.00	243.93	nr	**72173.10**
700 kW	70761.60	86315.00	9.00	243.93	nr	**86558.93**
750 kW	75479.04	92069.33	9.00	243.93	nr	**92313.26**
Water cooled						
Selection of water cooled chillers based on chilled water flow and return temperatures of 6°C and 12°C, and condenser entering and leaving temperatures of 27°C and 33°C						
Water cooled liquid chiller; refrigerant 407C; reciprocating compressors; twin circuit; integral controls; includes placing in position; electrical work elsewhere						
Cooling load						
200 kw	19010.82	24348.87	8.00	216.83	nr	**24565.70**
350 kW	31881.88	40833.99	8.00	216.83	nr	**41050.82**
500 kW	40808.04	52266.53	8.00	216.83	nr	**52483.36**
650 kW	52640.49	67421.41	9.00	243.93	nr	**67665.34**
750 kW	57010.02	73017.86	9.00	243.93	nr	**73261.80**
Water cooled condenserless liquid chiller; refrigerant 407C; reciprocating compressors; twin circuit; integral controls; includes placing in position; electrical work elsewhere						
Cooling load						
200 kW	16735.30	21434.40	8.00	216.83	nr	**21651.23**
350 kW	29377.80	37626.79	8.00	216.83	nr	**37843.62**
500 kW	37129.79	47555.46	8.00	216.83	nr	**47772.29**
650 kW	45410.61	58161.46	9.00	243.93	nr	**58405.39**
750 kW	51672.48	66181.60	9.00	243.93	nr	**66425.53**

T: MECHANICAL/COOLING/HEATING SYSTEMS

Item	Net Price £	Material £	Labour hours	Labour £	Unit	Total rate £
T60: CENTRAL REFRIGERATION PLANT (cont'd)						
CHILLERS (cont'd)						
Water cooled (cont'd)						
Water cooled liquid chiller; refrigerant R134a; screw compressors; twin circuit; integral controls; includes placing in position; electrical work elsewhere						
Cooling load						
300 kW	31341.85	40142.33	8.00	216.83	nr	40359.16
500 kW	41321.18	52923.75	8.00	216.83	nr	53140.58
700 kW	60470.91	77450.54	9.00	243.93	nr	77694.47
900 kW	69484.47	88995.01	9.00	243.93	nr	89238.95
1100 kW	83807.56	107339.88	10.00	271.04	nr	107610.92
1300 kW	93374.59	119593.24	10.00	271.04	nr	119864.28
Water cooled liquid chiller; refrigerant R134a; centrifugal compressors; twin circuit; integral controls; includes placing in position; electrical work elsewhere						
Cooling load						
700 kW	59604.85	76341.30	9.00	243.93	nr	76585.23
1000 kW	84029.40	107624.02	10.00	271.04	nr	107895.06
1300 kW	110526.68	141561.47	10.00	271.04	nr	141832.50
1600 kW	132654.41	169902.44	11.00	298.14	nr	170200.58
1900 kW	161560.53	206925.11	11.00	298.14	nr	207223.25
2200 kW	192483.35	246530.75	13.00	352.35	nr	246883.10
2500 kW	198869.58	254710.17	13.00	352.35	nr	255062.51
3000 kW	242004.67	309957.16	15.00	406.55	nr	310363.71
3500 kW	329395.25	421886.14	15.00	406.55	nr	422292.69
4000 kW	349846.88	448080.39	20.00	542.07	nr	448622.46
4500 kW	373249.64	478054.41	20.00	542.07	nr	478596.48
5000 kW	455211.91	583030.86	25.00	677.59	nr	583708.45
Absorption						
Absorption chiller, for operation using low pressure steam; selection based on chilled water flow and return temperatures of 6°C and 12°C, steam at 1 bar gauge and condenser entering and leaving temperatures of 27°C and 33°C; integral controls; includes placing in position; electrical work elsewhere						
Cooling load						
400 kW	57153.44	73201.55	8.00	216.83	nr	73418.38
700 kW	72571.15	92948.40	9.00	243.93	nr	93192.33
1000 kW	82793.61	106041.23	10.00	271.04	nr	106312.26
1300 kW	98805.13	126548.62	12.00	325.24	nr	126873.86
1600 kW	116935.31	149769.58	14.00	379.45	nr	150149.02
2000 kW	138590.25	177505.01	15.00	406.55	nr	177911.56

T: MECHANICAL/COOLING/HEATING SYSTEMS

Item	Net Price £	Material £	Labour hours	Labour £	Unit	Total rate £
Absorption chiller, for operation using low pressure hot water; selection based on chilled water flow and return temperatures of 6°C and 12°C, cooling water temperatures of 27°C and 33°C and hot water at 90°C; integral controls; includes placing in position; electrical work elsewhere						
Cooling load						
700 kW	82793.61	106041.23	9.00	243.93	nr	**106285.16**
1000 kW	103495.09	132555.48	10.00	271.04	nr	**132826.51**
1300 kW	120371.56	154170.69	12.00	325.24	nr	**154495.93**
1600 kW	138590.25	177505.01	14.00	379.45	nr	**177884.46**
HEAT REJECTION						
Dry air liquid coolers						
Dry air liquid cooler; selection based on fluid temperatures 45°C on, 40°C off at 32°C dry bulb ambient temperature; includes 20% ethylene glycol; includes placing in position; electrical work elsewhere						
Flat coil configuration						
500kW	13575.44	17387.29	15.00	406.55	nr	**17793.84**
Extra for inverter panels (factory wired and mounted on units)	7216.17	9242.40	15.00	406.55	nr	**9648.95**
800kW	22966.60	29415.39	15.00	406.55	nr	**29821.94**
Extra for inverter panels (factory wired and mounted on units)	13261.38	16985.04	15.00	406.55	nr	**17391.60**
1100kW	29489.09	37769.33	15.00	406.55	nr	**38175.88**
Extra for inverter panels (factory wired and mounted on units)	14431.10	18483.21	15.00	406.55	nr	**18889.76**
1400kW	37458.29	47976.20	15.00	406.55	nr	**48382.76**
Extra over for inverter panels (factory wired and mounted on units)	19892.68	25478.35	15.00	406.55	nr	**25884.90**
1700kW	44233.03	56653.22	15.00	406.55	nr	**57059.78**
Extra for inverter panels (factory wired and mounted on units)	20225.31	25904.37	15.00	406.55	nr	**26310.93**
2000kW	51743.48	66272.53	15.00	406.55	nr	**66679.08**
Extra for inverter panels (factory wired and mounted on units)	26619.20	34093.61	15.00	406.55	nr	**34500.16**
Note : heat rejection capacities above 500kW require multiple units. Prices are therefore for total number of units						
'Vee' type coil configuration						
500kW	13116.70	16799.74	15.00	406.55	nr	**17206.29**
Extra for inverter panels (factory wired and mounted on units)	5075.80	6501.03	15.00	406.55	nr	**6907.59**
800kW	20120.26	25769.83	15.00	406.55	nr	**26176.38**
Extra for inverter panels (factory wired and mounted on units)	11754.80	15055.43	15.00	406.55	nr	**15461.98**
1100kW	28451.69	36440.64	15.00	406.55	nr	**36847.19**

T: MECHANICAL/COOLING/HEATING SYSTEMS

Item	Net Price £	Material £	Labour hours	Labour £	Unit	Total rate £
T60: CENTRAL REFRIGERATION PLANT (cont'd)						
HEAT REJECTION (cont'd)						
Dry air liquid coolers (cont'd)						
Dry air liquid cooler; selection based on fluid (cont'd)						
'Vee' type coil configuration (cont'd)						
Extra for inverter panels (factory wired and mounted on units)	14213.78	18204.87	15.00	406.55	nr	18611.42
1400kW	36709.90	47017.67	15.00	406.55	nr	47424.23
Extra for inverter panels (factory wired and mounted on units)	17673.80	22636.43	15.00	406.55	nr	23042.98
1700kW	42171.18	54012.43	15.00	406.55	nr	54418.98
Extra for inverter panels (factory wired and mounted on units)	23508.81	30109.85	15.00	406.55	nr	30516.40
2000kW	55064.86	70526.52	15.00	406.55	nr	70933.07
Extra for inverter panels (factory wired and mounted on units)	26510.69	33954.63	15.00	406.55	nr	34361.18
Note : Heat rejection capacities above 1100kW require multiple units. Prices are for total number of units.						
Air cooled condensers						
Air cooled condenser; refrigerant 407C; selection based on condensing temperature of 45°C at 32°C dry bulb ambient; includes placing in position; electrical work elsewhere						
Flat coil configuration						
500kW	14491.06	18560.00	15.00	406.55	nr	18966.56
Extra for inverter panels (factory wired and mounted on units)	6630.69	8492.52	15.00	406.55	nr	8899.07
800kW	23774.03	30449.54	15.00	406.55	nr	30856.09
Extra for inverter panels (factory wired and mounted on units)	10150.98	13001.27	15.00	406.55	nr	13407.83
1100kW	31827.30	40764.09	15.00	406.55	nr	41170.64
Extra for inverter panels (factory wired and mounted on units)	14430.79	18482.81	15.00	406.55	nr	18889.36
1400kW	39735.91	50893.36	15.00	406.55	nr	51299.91
Extra for inverter panels (factory wired and mounted on units)	17746.13	22729.07	15.00	406.55	nr	23135.62
1700kW	51743.48	66272.53	15.00	406.55	nr	66679.08
Extra for inverter panels (factory wired and mounted on units)	26619.20	34093.61	15.00	406.55	nr	34500.16
2000kW	59603.86	76340.03	15.00	406.55	nr	76746.58
Extra for inverter panels (factory wired and mounted on units)	26619.20	34093.61	15.00	406.55	nr	34500.16

T: MECHANICAL/COOLING/HEATING SYSTEMS

Item	Net Price £	Material £	Labour hours	Labour £	Unit	Total rate £
Note : Heat rejection capacities above 500kW require multiple units. Prices are for total number of units.						
'Vee' type coil configuration						
500kW	14225.84	18220.31	15.00	406.55	nr	18626.87
Extra for inverter panels (factory wired and mounted on units)	5075.49	6500.64	15.00	406.55	nr	6907.19
800kW	21399.04	27407.68	15.00	406.55	nr	27814.23
Extra for inverter panels (factory wired and mounted on units)	10392.10	13310.10	15.00	406.55	nr	13716.65
1100kW	31405.36	40223.67	15.00	406.55	nr	40630.22
Extra for inverter panels (factory wired and mounted on units)	17673.80	22636.43	15.00	406.55	nr	23042.98
1400kW	36565.23	46832.38	15.00	406.55	nr	47238.93
Extra for inverter panels (factory wired and mounted on units)	20784.19	26620.18	15.00	406.55	nr	27026.74
1700kW	47108.03	60335.49	15.00	406.55	nr	60742.05
Extra for inverter panels (factory wired and mounted on units)	26522.75	33970.07	15.00	406.55	nr	34376.63
2000kW	54847.86	70248.59	15.00	406.55	nr	70655.14
Extra for inverter panels (factory wired and mounted on units)	31176.29	39930.28	15.00	406.55	nr	40336.83
Note : Heat rejection capacities above 1100kW require multiple units. Prices are for total number of units.						
Cooling towers						
Cooling towers; forced draught, centrifugal fan, counterflow design; based on water temperatures of 35°C on and 29°C off at 21°C wet bulb ambient temperature; includes placing in position; electrical work elsewhere						
Open circuit type						
900kW	9652.61	12362.97	20.00	542.07	nr	12905.04
Extra for stainless steel construction	4327.58	5542.72	-	-	nr	5542.72
Extra for intake and discharge sound attenuation	4657.60	5965.41	-	-	nr	5965.41
Extra for fan dampers for capacity control	1080.67	1384.11	-	-	nr	1384.11
1500kW	14976.29	19181.48	20.00	542.07	nr	19723.55
Extra for stainless steel construction	6932.10	8878.56	-	-	nr	8878.56
Extra for intake and discharge sound attenuation	6979.56	8939.35	-	-	nr	8939.35
Extra for fan dampers for capacity control	1096.87	1404.86	-	-	nr	1404.86
2100kW	21246.18	27211.89	20.00	542.07	nr	27753.96
Extra for stainless steel construction	9831.79	12592.46	-	-	nr	12592.46
Extra for intake and discharge sound attenuation	10363.51	13273.48	-	-	nr	13273.48
Extra for fan dampers for capacity control	1252.49	1604.18	-	-	nr	1604.18
2700kW	25658.51	32863.16	20.00	542.07	nr	33405.23
Extra for stainless steel construction	11993.59	15361.27	-	-	nr	15361.27
Extra for intake and discharge sound attenuation	13961.60	17881.88	-	-	nr	17881.88
Extra for fan dampers for capacity control	1949.69	2497.14	-	-	nr	2497.14
3300kW	31619.64	40498.12	20.00	542.07	nr	41040.19
Extra for stainless steel construction	14638.00	18748.20	-	-	nr	18748.20
Extra for intake and discharge sound attenuation	13167.27	16864.51	-	-	nr	16864.51
Extra for fan dampers for capacity control	1594.87	2042.69	-	-	nr	2042.69

T: MECHANICAL/COOLING/HEATING SYSTEMS

Item	Net Price £	Material £	Labour hours	Labour £	Unit	Total rate £
T60: CENTRAL REFRIGERATION PLANT (cont'd)						
HEAT REJECTION (cont'd)						
Cooling towers (cont'd)						
Open circuit type (cont'd)						
3900kW	35638.54	45645.49	20.00	542.07	nr	46187.56
Extra for stainless steel construction	16499.31	21132.15	-	-	nr	21132.15
Extra for intake and discharge sound attenuation	13420.01	17188.21	-	-	nr	17188.21
Extra for fan dampers for capacity control	1594.87	2042.69	-	-	nr	2042.69
4500kW	40108.15	51370.12	23.00	623.38	nr	51993.50
Extra for stainless steel construction	18703.65	23955.45	-	-	nr	23955.45
Extra for intake and discharge sound attenuation	17798.73	22796.44	-	-	nr	22796.44
Extra for fan dampers for capacity control	2509.95	3214.72	-	-	nr	3214.72
5100kW	46330.73	59339.94	23.00	623.38	nr	59963.32
Extra for stainless steel construction	21346.74	27340.69	-	-	nr	27340.69
Extra for intake and discharge sound attenuation	17751.42	22735.84	-	-	nr	22735.84
Extra for fan dampers for capacity control	2509.95	3214.72	-	-	nr	3214.72
5700kW	48981.36	62734.84	30.00	813.11	nr	63547.94
Extra for stainless steel construction	22871.70	29293.84	-	-	nr	29293.84
Extra for intake and discharge sound attenuation	26287.24	33668.43	-	-	nr	33668.43
Extra for fan dampers for capacity control	2760.20	3535.24	-	-	nr	3535.24
6300kW	59058.51	75641.55	30.00	813.11	nr	76454.65
Extra for stainless steel construction	27341.54	35018.77	-	-	nr	35018.77
Extra for intake and discharge sound attenuation	26666.97	34154.79	-	-	nr	34154.79
Extra for fan dampers for capacity control	2760.20	3535.24	-	-	nr	3535.24
Closed circuit type (includes 20% ethylene glycol)						
900kW	26908.51	34464.15	20.00	542.07	nr	35006.22
Extra for stainless steel construction	21910.45	28062.69	-	-	nr	28062.69
Extra for intake and discharge sound attenuation	8096.34	10369.71	-	-	nr	10369.71
Extra for fan dampers for capacity control	1096.87	1404.86	-	-	nr	1404.86
1500kW	50027.18	64074.31	20.00	542.07	nr	64616.38
Extra for stainless steel construction	39885.53	51084.99	-	-	nr	51084.99
Extra for intake and discharge sound attenuation	12924.50	16553.57	-	-	nr	16553.57
Extra for fan dampers for capacity control	1594.87	2042.69	-	-	nr	2042.69
2100kW	60965.87	78084.48	20.00	542.07	nr	78626.55
Extra for stainless steel construction	54187.16	69402.37	-	-	nr	69402.37
Extra for intake and discharge sound attenuation	15557.70	19926.15	-	-	nr	19926.15
Extra for fan dampers for capacity control	1594.87	2042.69	-	-	nr	2042.69
2700kW	65731.79	84188.62	25.00	677.59	nr	84866.21
Extra for stainless steel construction	67140.65	85993.07	-	-	nr	85993.07
Extra for intake and discharge sound attenuation	23220.78	29740.94	-	-	nr	29740.94
Extra for fan dampers for capacity control	2412.84	3090.34	-	-	nr	3090.34
3300kW	101275.71	129712.92	25.00	677.59	nr	130390.50
Extra for stainless steel construction	83759.62	107278.48	-	-	nr	107278.48
Extra for intake and discharge sound attenuation	30888.81	39562.08	-	-	nr	39562.08
Extra for fan dampers for capacity control	2760.20	3535.24	-	-	nr	3535.24
3900kW	111509.74	142820.56	25.00	677.59	nr	143498.15
Extra for stainless steel construction	95918.06	122850.89	-	-	nr	122850.89
Extra for intake and discharge sound attenuation	31394.29	40209.49	-	-	nr	40209.49
Extra for fan dampers for capacity control	2760.20	3535.24	-	-	nr	3535.24
4500kW	133499.18	170984.41	40.00	1084.14	nr	172068.55
Extra for stainless steel construction	99506.83	127447.35	-	-	nr	127447.35

T: MECHANICAL/COOLING/HEATING SYSTEMS

Item	Net Price £	Material £	Labour hours	Labour £	Unit	Total rate £
Extra for intake and discharge sound attenuation	44871.57	57471.06	-	-	nr	**57471.06**
Extra for fan dampers for capacity control	4825.68	6180.68	-	-	nr	**6180.68**
5100kW	148893.78	190701.66	40.00	1084.14	nr	**191785.80**
Extra for stainless steel construction	128591.41	164698.59	-	-	nr	**164698.59**
Extra for intake and discharge sound attenuation	44871.57	57471.06	-	-	nr	**57471.06**
Extra for fan dampers for capacity control	4825.68	6180.68	-	-	nr	**6180.68**
5700kW	183594.83	235146.42	40.00	1084.14	nr	**236230.56**
Extra for stainless steel construction	132826.01	170122.23	-	-	nr	**170122.23**
Extra for intake and discharge sound attenuation	50162.88	64248.12	-	-	nr	**64248.12**
Extra for fan dampers for capacity control	5520.40	7070.47	-	-	nr	**7070.47**
6300kW	196109.71	251175.36	40.00	1084.14	nr	**252259.50**
Extra for stainless steel construction	151931.39	194592.20	-	-	nr	**194592.20**
Extra for intake and discharge sound attenuation	50162.88	64248.12	-	-	nr	**64248.12**
Extra for fan dampers for capacity control	5520.40	7070.47	-	-	nr	**7070.47**

T: MECHANICAL/COOLING/HEATING SYSTEMS

Item	Net Price £	Material £	Labour hours	Labour £	Unit	Total rate £
T61 : CHILLED WATER						
SCREWED STEEL						
Y10 - PIPELINES						
For pipework prices refer to Section T31 - Low Temperature Hot Water Heating, with the exception of chilled water blocks within brackets as detailed hereafter. For minimum fixing distances, refer to the Tables and Memoranda to the rear of the book						
FIXINGS						
For steel pipes; black malleable iron						
Oversized pipe clip, to contain 30mm insulation block for vapour barrier						
15mm dia	2.38	2.90	0.10	2.71	nr	5.61
20mm dia	2.40	2.93	0.11	2.98	nr	5.91
25mm dia	2.53	3.09	0.12	3.25	nr	6.34
32mm dia	2.60	3.17	0.14	3.79	nr	6.97
40mm dia	2.75	3.35	0.15	4.07	nr	7.42
50mm dia	3.92	4.78	0.16	4.34	nr	9.12
65mm dia	4.23	5.16	0.30	8.13	nr	13.29
80mm dia	4.49	5.48	0.35	9.49	nr	14.96
100mm dia	7.12	8.69	0.40	10.84	nr	19.53
125mm dia	8.26	10.08	0.60	16.27	nr	26.34
150mm dia	12.92	16.55	0.77	20.87	nr	37.42
200mm dia	16.49	21.12	0.90	24.39	nr	45.51
250mm dia	20.36	26.08	1.10	29.81	nr	55.89
300mm dia	24.23	31.03	1.25	33.88	nr	64.91
350mm dia	28.11	36.00	1.50	40.66	nr	76.66
400mm dia	31.48	40.32	1.75	47.43	nr	87.75
Screw on backplate (Male), black malleable iron; plugged and screwed						
M12	0.32	0.41	0.10	2.71	nr	3.12
Screw on backplate (Female), black malleable iron; plugged and screwed						
M12	0.32	0.41	0.10	2.71	nr	3.12
Extra Over channel sections for fabricated hangers and brackets						
Galvanised steel; including inserts, bolts, nuts, washers; fixed to backgrounds						
41 x 21mm	2.50	3.20	0.29	7.86	m	11.06
41 x 41mm	3.90	5.00	0.29	7.86	m	12.86
Threaded rods; metric thread; including nuts, washers etc						
12mm dia x 600mm long	2.70	3.46	0.18	4.88	nr	8.34

T: MECHANICAL/COOLING/HEATING SYSTEMS

Item	Net Price £	Material £	Labour hours	Labour £	Unit	Total rate £
For plastic pipework suitable for chilled water systems, refer to ABS pipework details in Section S10 - Cold Water with the exception of chilled water blocks within brackets as detailed for the aforementioned steel pipe. For minimum fixing distances, refer to the Tables and Memoranda to the rear of the book						
For copper pipework, refer to Section S10 – Cold Water with the exception of chilled water blocks within brackets as detailed hereafter. For minimum fixing distances, refer to the Tables and Memoranda to the rear of the book						
FIXINGS						
For Copper pipework						
Oversized pipe clip, to contain 30mm insulation block for vapour barrier						
15mm dia	2.38	2.90	0.10	2.71	nr	5.61
22mm dia	2.40	2.93	0.11	2.98	nr	5.91
28mm dia	2.53	3.09	0.12	3.25	nr	6.34
35mm dia	2.60	3.17	0.14	3.79	nr	6.97
42mm dia	2.75	3.35	0.15	4.07	nr	7.42
54mm dia	3.92	4.78	0.16	4.34	nr	9.12
67mm dia	4.23	5.16	0.30	8.13	nr	13.29
76mm dia	4.49	5.48	0.35	9.49	nr	14.96
108mm dia	7.12	8.69	0.40	10.84	nr	19.53
133mm dia	8.26	10.08	0.60	16.27	nr	26.34
159mm dia	12.92	16.55	0.77	20.87	nr	37.42
Screw on backplate, female						
All sizes 15mm to 54mm x 10mm	1.38	1.77	0.10	2.71	nr	4.48
Screw on backplate, male						
All sizes 15mm to 54mm x 10mm	1.23	1.58	0.10	2.71	nr	4.29
Extra Over channel sections for fabricated hangers and brackets						
Galvanised steel; including inserts, bolts, nuts, washers; fixed to backgrounds						
41 x 21mm	2.50	3.20	0.29	7.86	m	11.06
41 x 41mm	3.90	5.00	0.29	7.86	m	12.86
Threaded rods; metric thread; including nuts, washers, etc						
10mm dia x 600mm long for ring clips up to 54mm	1.31	1.68	0.18	4.88	nr	6.56
12mm dia x 600mm long for ring clips from 54mm	2.70	3.46	0.18	4.88	nr	8.34

T: MECHANICAL/COOLING/HEATING SYSTEMS

Item	Net Price £	Material £	Labour hours	Labour £	Unit	Total rate £
T61 : CHILLED WATER (cont'd)						
Y11 - PIPELINE ANCILLARIES						
For prices for ancillaries refer to Section T31 – Low Temperature Hot Water Heating						
Y22 - HEAT EXCHANGERS						
Plate heat exchanger; for use in CHW systems; painted carbon steel frame, stainless steel plates, nitrile rubber gaskets, design pressure of 10 bar and operating temperature of 110/135°C						
Primary side; 13°C in, 8°C out; secondary side; 6°C in, 11°C out						
264 kW, 12.60 l/s	4542.30	5817.73	10.00	271.04	nr	6088.77
290 kW, 13.85 l/s	4788.00	6132.42	12.00	325.24	nr	6457.66
316 kW, 15.11 l/s	4984.35	6383.91	12.00	325.24	nr	6709.15
350 kW, 16.69 l/s	5360.25	6865.35	16.00	433.66	nr	7299.01
395 kW, 18.88 l/s	5736.15	7346.80	16.00	433.66	nr	7780.46
454 kW, 21.69 l/s	6230.70	7980.22	10.00	271.04	nr	8251.25
475 kW, 22.66 l/s	6351.45	8134.87	10.00	271.04	nr	8405.91
527 kW, 25.17 l/s	6882.75	8815.36	10.00	271.04	nr	9086.39
554 kW, 26.43 l/s	7065.45	9049.36	10.00	271.04	nr	9320.39
580 kW, 27.68 l/s	7313.25	9366.74	10.00	271.04	nr	9637.77
633 kW, 30.19 l/s	7746.90	9922.15	10.00	271.04	nr	10193.19
661 kW, 31.52 l/s	7933.80	10161.53	10.00	271.04	nr	10432.57
713 kW, 34.04 l/s	8469.30	10847.39	12.00	325.24	nr	11172.64
740 kW, 35.28 l/s	8654.10	11084.08	12.00	325.24	nr	11409.33
804 kW, 38.33 l/s	9214.80	11802.22	12.00	325.24	nr	12127.47
1925 kW, 91.82 l/s	15468.60	19812.03	15.00	406.55	nr	20218.58
2710 kW, 129.26 l/s	20546.40	26315.62	15.00	406.55	nr	26722.18
3100 kW, 147.87 l/s	22890.00	29317.28	15.00	406.55	nr	29723.84
Note : For temperature conditions different to those above, the cost of the units can vary significantly, and therefore the manufacturers advice should be sought.						
Y24 - TRACE HEATING						
Trace heating; for freeze protection or temperature maintenance of pipework; to BS 6351; including fixing to parent structures by plastic pull ties						
Straight laid						
15mm	20.31	26.01	0.27	7.32	m	33.33
25mm	20.31	26.01	0.27	7.32	m	33.33
28mm	20.31	26.01	0.27	7.32	m	33.33
32mm	20.31	26.01	0.30	8.13	m	34.14
35mm	20.31	26.01	0.31	8.40	m	34.41
50mm	20.31	26.01	0.34	9.22	m	35.23
100mm	20.31	26.01	0.40	10.84	m	36.85
150mm	20.31	26.01	0.40	10.84	m	36.85

T: MECHANICAL/COOLING/HEATING SYSTEMS

Item	Net Price £	Material £	Labour hours	Labour £	Unit	Total rate £
Helically wound						
15mm	25.87	33.13	1.00	27.10	m	**60.24**
25mm	25.87	33.13	1.00	27.10	m	**60.24**
28mm	25.87	33.13	1.00	27.10	m	**60.24**
32mm	25.87	33.13	1.00	27.10	m	**60.24**
35mm	25.87	33.13	1.00	27.10	m	**60.24**
50mm	25.87	33.13	1.00	27.10	m	**60.24**
100mm	25.87	33.13	1.00	27.10	m	**60.24**
150mm	25.87	33.13	1.00	27.10	m	**60.24**
Accessories for trace heating; weatherproof; polycarbonate enclosure to IP standards; fully installed						
Connection junction box						
100 x 100 x 75mm	44.46	56.94	1.40	37.94	nr	**94.89**
Single air thermostat						
150 x 150 x 75mm	97.41	124.76	1.42	38.49	nr	**163.25**
Single capillary thermostat						
150 x 150 x 75mm	140.15	179.50	1.46	39.57	nr	**219.07**
Twin capillary thermostat						
150 x 150 x 75mm	251.43	322.03	1.46	39.57	nr	**361.60**
EQUIPMENT						
PRESSURISATION UNITS						
Chilled water packaged pressurisation unit complete with expansion vessel(s), interconnecting pipework and necessary isolating and drain valves; includes placing in position; electrical work elsewhere						
Selection based on a final working pressure of 4 bar, a 3m static head and system operating temperatures of 6°/12°C						
System volume						
1800 litres	1593.65	2041.13	8.00	216.83	nr	**2257.96**
4500 litres	1593.65	2041.13	8.00	216.83	nr	**2257.96**
7200 litres	1670.76	2139.89	10.00	271.04	nr	**2410.93**
9900 litres	1670.76	2139.89	10.00	271.04	nr	**2410.93**
15300 litres	1735.02	2222.20	13.00	352.35	nr	**2574.54**
22500 litres	1839.12	2355.53	20.00	542.07	nr	**2897.60**
27000 litres	1839.12	2355.53	20.00	542.07	nr	**2897.60**

T: MECHANICAL/COOLING/HEATING SYSTEMS

Item	Net Price £	Material £	Labour hours	Labour £	Unit	Total rate £
T61 : CHILLED WATER (cont'd)						
CHILLED BEAMS						
Static (passive) beams; based on water at 14°C flow and 16°C return, 24°C room temperature; 600mm wide coil providing 350-400W/m output						
Static cooled beam for exposed installation with standard casing	117.97	151.09	4.00	108.41	m	259.51
Static cooled beam for installation above open grid or perforated ceiling	80.89	103.60	4.00	108.41	m	212.02
Ventilated (active) beams; based on water at 14°C flow and 16°C return, 24°C room temperature; air supply at 10l/s/linear metre; 300mm wide beam providing 250-350W/m output unless stated otherwise; all exposed beams c/w standard casing; electrical work elsewhere						
Ventilated cooled beam flush mounted within a false ceiling; closed type with integrated secondary air circulation; 600mm wide beam providing 400W/m output	183.13	234.55	4.50	121.97	m	356.52
Ventilated cooled beam flush mounted within a false ceiling; open type	161.78	207.21	4.00	108.41	m	315.62
Ventilated cooled beam for exposed mounting with standard casing	161.78	207.21	4.00	108.41	m	315.62
Ventilated cooled beam flush mounted within a false ceiling; open type; with recessed integrated flush mounted 28W or 35W T5 light fittings	293.23	375.57	4.00	108.41	m	483.98
Ventilated cooled beam for exposed mounting with recessed integrated flush mounted 28W or 35W T5 light fittings	293.23	375.57	4.00	108.41	m	483.98
Ventilated cooled beam for exposed mounting with recessed integrated flush mounted direct and indirect 28W or 35W T5 light fittings	320.20	410.11	4.00	108.41	m	518.52
LEAK DETECTION						
Leak detection system consisting of a central control module connected by a leader cable to water sensing cables						
Control Modules						
Alarm Only	321.77	412.12	4.00	108.37	nr	520.49
Alarm and location	2110.30	2702.85	8.00	216.74	nr	2919.59
Cables						
Sensing - 3m length	88.13	112.88	4.00	108.37	nr	221.24
Sensing - 7.5m length	129.85	166.31	4.00	108.37	nr	274.68
Sensing - 15m length	224.50	287.54	8.00	216.74	nr	504.27
Leader - 3.5m length	39.55	50.66	2.00	54.18	nr	104.84
End terminal						
End terminal	15.36	19.67	0.05	1.35	nr	21.03

T: MECHANICAL/COOLING/HEATING SYSTEMS

Item	Net Price £	Material £	Labour hours	Labour £	Unit	Total rate £
ENERGY METERS						
Ultrasonic						
Energy meter for measuring energy use in chilled water systems; includes ultrasonic flow meter (with sensor and signal converter), energy calculator, pair of temperature sensors with brass pockets, and 3m of interconnecting cable; includes fixing in position; electrical work elsewhere						
Pipe size (flanged connections to PN16); maximum flow rate						
50mm, 36m³/hr	1122.84	1438.12	1.80	48.79	nr	1486.91
65mm, 60m³/hr	1237.71	1585.25	2.32	62.88	nr	1648.13
80mm, 100m³/hr	1387.38	1776.94	2.56	69.39	nr	1846.33
125mm, 250m³/hr	1607.41	2058.75	3.60	97.57	nr	2156.33
150mm, 360m³/hr	1744.93	2234.89	4.80	130.10	nr	2364.99
200mm, 600m³/hr	1949.60	2497.03	6.24	169.13	nr	2666.15
250mm, 1000m³/hr	2249.72	2881.42	9.60	260.19	nr	3141.61
300mm, 1500m³/hr	2644.50	3387.05	10.80	292.72	nr	3679.77
350mm, 2000m³/hr	3186.50	4081.24	13.20	357.77	nr	4439.00
400mm, 2500m³/hr	3647.61	4671.82	15.60	422.81	nr	5094.64
500mm, 3000m³/hr	4139.45	5301.77	24.00	650.48	nr	5952.25
600mm, 3500m³/hr	4650.72	5956.60	28.00	758.90	nr	6715.49
Electromagnetic						
Energy meter for measuring energy use in chilled water systems; includes electro-magnetic flow meter (with sensor and signal converter), energy calculator, pair of temperature sensors with brass pockets, and 3m of interconnecting cable; includes fixing in position; electrical work elsewhere						
Pipe size (flanged connections to PN40); maximum flow rate						
25mm, 17.7m³/hr	845.73	1083.21	1.48	40.11	nr	1123.32
40mm, 45m³/hr	853.82	1093.57	1.55	42.01	nr	1135.58
Pipe size (flanged connections to PN16); maximum flow rate						
50mm, 70m³/hr	863.38	1105.81	1.80	48.79	nr	1154.60
65mm, 120m³/hr	867.06	1110.52	2.32	62.88	nr	1173.40
80mm, 180m³/hr	871.47	1116.17	2.56	69.39	nr	1185.56
125mm, 450m³/hr	948.69	1215.08	3.60	97.57	nr	1312.65
150mm, 625m³/hr	1006.79	1289.49	4.80	130.10	nr	1419.58
200mm, 1100m³/hr	1083.27	1387.45	6.24	169.13	nr	1556.57
250mm, 1750m³/hr	1215.65	1556.99	9.60	260.19	nr	1817.19
300mm, 2550m³/hr	1549.53	1984.62	10.80	292.72	nr	2277.34
350mm, 3450m³/hr	2015.79	2581.80	13.20	357.77	nr	2939.57
400mm, 4500m³/hr	2299.66	2945.38	15.60	422.81	nr	3368.19

T: MECHANICAL/COOLING/HEATING SYSTEMS

Item	Net Price £	Material £	Labour hours	Labour £	Unit	Total rate £
T70 : LOCAL COOLING UNITS						
Split system with ceiling void evaporator unit and external condensing unit						
Ceiling mounted 4 way blow cassette heat pump unit with remote fan speed and load control; refrigerant 470C; includes outdoor unit						
Cooling 3.6kW, heating 4.1kW	1470.24	1883.07	35.00	948.62	nr	2831.69
Cooling 4.9kW, heating 5.5kW	1620.19	2075.12	35.00	948.62	nr	3023.75
Cooling 7.1kW, heating 8.2kW	1961.09	2511.74	35.00	948.62	nr	3460.37
Cooling 10kW, heating 11.2kW	2319.57	2970.88	35.00	948.62	nr	3919.50
Cooling 12.20kW, heating 14.60kW	2537.47	3249.97	35.00	948.62	nr	4198.59
Ceiling mounted 4 way blow cooling only unit with remote fan speed and load control; refrigerant 470C; includes outdoor unit						
Cooling 3.80kW	1330.82	1704.50	35.00	948.62	nr	2653.12
Cooling 5.20kW	1486.63	1904.06	35.00	948.62	nr	2852.68
Cooling 7.10kW	1833.40	2348.20	35.00	948.62	nr	3296.82
Cooling 10kw	2129.79	2727.81	35.00	948.62	nr	3676.44
Cooling 12.2kW	2257.49	2891.37	35.00	948.62	nr	3839.99
In ceiling, ducted heat pump unit with remote fan speed and load control; refrigerant 407C; includes outdoor unit						
Cooling 3.60kW, heating 4.10kW	1077.78	1380.41	35.00	948.62	nr	2329.03
Cooling 4.90kW, heating 5.50kW	1239.45	1587.48	35.00	948.62	nr	2536.10
Cooling 7.10kW, heating 8.20kW	1239.45	1587.48	35.00	948.62	nr	2536.10
Cooling 10kW, heating 11.20kW	1873.23	2399.21	35.00	948.62	nr	3347.84
Cooling 12.20kW, heating 14.50kW	2498.82	3200.46	35.00	948.62	nr	4149.09
In ceiling, ducted cooling only unit with remote fan speed and load control; refrigerant 407C; includes outdoor unit						
Cooling 3.70kW	972.35	1245.38	35.00	948.62	nr	2194.00
Cooling 4.90kW	1156.28	1480.95	35.00	948.62	nr	2429.57
Cooling 7.10kW	1745.54	2235.67	35.00	948.62	nr	3184.29
Cooling 10kW	1990.38	2549.26	35.00	948.62	nr	3497.88
Cooling 12.3kW	2218.82	2841.84	35.00	948.62	nr	3790.47
Room Units						
Ceiling mounted 4 way blow cassette heat pump unit with remote fan speed and load control; refrigerant 407C; excludes outdoor unit						
Cooling 3.6kW, heating 4.1kW	927.83	1188.36	17.00	460.76	nr	1649.11
Cooling 4.9kW, heating 5.5kW	950.09	1216.87	17.00	460.76	nr	1677.63
Cooling 7.1kW, heating 8.2kW	1028.58	1317.39	17.00	460.76	nr	1778.15
Cooling 10kW, heating 11.2kW	1114.10	1426.93	17.00	460.76	nr	1887.69
Cooling 12.20kW, heating 14.60kW	1207.82	1546.96	17.00	460.76	nr	2007.72
Ceiling mounted 4 way blow cooling unit with remote fan speed and load control; refrigerant 407C; excludes outdoor unit						
Cooling 3.80kW	876.28	1122.33	17.00	460.76	nr	1583.09
Cooling 5.20kW	883.31	1131.33	17.00	460.76	nr	1592.09
Cooling 7.10Kw	1028.58	1317.39	17.00	460.76	nr	1778.15
Cooling 10kW	1114.10	1426.93	17.00	460.76	nr	1887.69
Cooling 12.2kW	1207.82	1546.96	17.00	460.76	nr	2007.72

T: MECHANICAL/COOLING/HEATING SYSTEMS

Item	Net Price £	Material £	Labour hours	Labour £	Unit	Total rate £
In ceiling, ducted heat pump unit with remote fan speed and load control; refrigerant 407C; excludes outdoor unit						
Cooling 3.60kW, heating 4.10kW	535.38	685.71	17.00	460.76	nr	**1146.47**
Cooling 4.90kW, heating 5.50kW	569.35	729.22	17.00	460.76	nr	**1189.98**
Cooling 7.10kW, heating 8.20kW	940.72	1204.86	17.00	460.76	nr	**1665.62**
Cooling 10kW, heating 11.20kW	974.69	1248.37	17.00	460.76	nr	**1709.13**
Cooling 12.20kW, heating 14.50kW	1169.16	1497.45	17.00	460.76	nr	**1958.21**
In ceiling, ducted cooling unit only with remote fan speed and load control; refrigerant 407C; excludes outdoor unit						
Cooling 3.70kW	517.80	663.19	17.00	460.76	nr	**1123.95**
Cooling 4.90kW	552.95	708.21	17.00	460.76	nr	**1168.97**
Cooling 7.10kW	940.72	1204.86	17.00	460.76	nr	**1665.62**
Cooling 10kW	974.69	1248.37	17.00	460.76	nr	**1709.13**
Cooling 12.3kW	1169.16	1497.45	17.00	460.76	nr	**1958.21**
External condensing units suitable for connection to multiple indoor units; inverter driven; refrigerant 407C						
Cooling only						
9kW	2025.53	2594.28	17.00	460.76	nr	**3055.04**
Heat pump						
Cooling 5.20kW, heating 6.10kW	1443.29	1848.55	17.00	460.76	nr	**2309.31**
Cooling 6.80kW, heating 2.50kW	1840.43	2357.20	17.00	460.76	nr	**2817.96**
Cooling 8kW, heating 9.60kW	2130.95	2729.30	17.00	460.76	nr	**3190.06**
Cooling 14.50kW, heating 16.50kW	3536.76	4529.85	21.00	569.17	nr	**5099.02**

U: VENTILATION/AIR CONDITIONING SYSTEMS

Item	Net Price £	Material £	Labour hours	Labour £	Unit	Total rate £
U10 : DUCTWORK : CIRCULAR						
Y30 - AIR DUCTLINES						
Galvanised sheet metal DW144 class B spirally wound circular section ductwork; including all necessary stiffeners, joints, couplers in the running length and duct supports						
Straight duct						
80mm dia.	3.75	6.38	0.87	27.51	m	33.89
100mm dia.	3.87	6.58	0.87	27.51	m	34.09
160mm dia.	5.37	9.13	0.87	27.51	m	36.64
200mm dia.	6.86	11.66	0.87	27.51	m	39.17
250mm dia.	8.37	14.23	1.21	38.26	m	52.49
315mm dia.	10.15	17.26	1.21	38.26	m	55.52
355mm dia.	13.81	23.48	1.21	38.26	m	61.74
400mm dia.	15.50	26.36	1.21	38.26	m	64.62
450mm dia.	17.14	29.14	1.21	38.26	m	67.40
500mm dia.	18.65	31.71	1.21	38.26	m	69.97
630mm dia.	33.81	57.49	1.39	43.95	m	101.44
710mm dia.	37.38	63.56	1.39	43.95	m	107.51
800mm dia.	42.83	72.83	1.44	45.53	m	118.36
900mm dia.	53.23	90.51	1.46	46.16	m	136.68
1000mm dia.	65.65	111.63	1.65	52.17	m	163.80
1120mm dia.	78.36	133.24	2.43	76.83	m	210.08
1250mm dia.	85.77	145.84	2.43	76.83	m	222.68
1400mm dia.	96.75	164.51	2.77	87.59	m	252.10
1600mm dia.	110.95	188.66	3.06	96.75	m	285.41
Extra over fittings; circular duct class B						
End cap						
80mm dia.	1.56	2.65	0.15	4.74	nr	7.40
100mm dia.	1.69	2.87	0.15	4.74	nr	7.62
160mm dia.	2.60	4.42	0.15	4.74	nr	9.16
200mm dia.	3.08	5.24	0.20	6.32	nr	11.56
250mm dia.	4.32	7.35	0.29	9.17	nr	16.52
315mm dia.	5.34	9.08	0.29	9.17	nr	18.25
355mm dia.	8.18	13.91	0.44	13.91	nr	27.82
400mm dia.	8.49	14.44	0.44	13.91	nr	28.35
450mm dia.	9.01	15.32	0.44	13.91	nr	29.23
500mm dia.	9.48	16.12	0.44	13.91	nr	30.03
630mm dia.	23.51	39.98	0.58	18.34	nr	58.31
710mm dia.	27.49	46.74	0.69	21.82	nr	68.56
800mm dia.	36.31	61.74	0.81	25.61	nr	87.35
900mm dia.	41.00	69.72	0.92	29.09	nr	98.81
1000mm dia.	56.36	95.83	1.04	32.88	nr	128.72
1120mm dia.	62.38	106.07	1.16	36.68	nr	142.75
1250mm dia.	68.77	116.94	1.16	36.68	nr	153.61
1400mm dia.	87.51	148.80	1.16	36.68	nr	185.48
1600mm dia.	98.89	168.15	1.16	36.68	nr	204.83

U: VENTILATION/AIR CONDITIONING SYSTEMS

Item	Net Price £	Material £	Labour hours	Labour £	Unit	Total rate £
Reducer						
80mm dia.	4.93	8.38	0.29	9.17	nr	17.55
100mm dia.	5.18	8.81	0.29	9.17	nr	17.98
160mm dia.	6.79	11.55	0.29	9.17	nr	20.72
200mm dia.	7.73	13.14	0.44	13.91	nr	27.06
250mm dia.	9.49	16.14	0.58	18.34	nr	34.48
315mm dia.	12.03	20.46	0.58	18.34	nr	38.79
355mm dia.	15.64	26.59	0.87	27.51	nr	54.10
400mm dia.	17.56	29.86	0.87	27.51	nr	57.37
450mm dia.	18.83	32.02	0.87	27.51	nr	59.53
500mm dia.	19.69	33.48	0.87	27.51	nr	60.99
630mm dia.	52.04	88.49	0.87	27.51	nr	116.00
710mm dia.	56.47	96.02	0.96	30.35	nr	126.37
800mm dia.	72.49	123.26	1.06	33.52	nr	156.78
900mm dia.	76.14	129.47	1.16	36.68	nr	166.15
1000mm dia.	98.43	167.37	1.25	39.52	nr	206.89
1120mm dia.	107.57	182.91	3.47	109.72	nr	292.63
1250mm dia.	119.31	202.87	3.47	109.72	nr	312.59
1400mm dia.	148.92	253.22	4.05	128.06	nr	381.28
1600mm dia.	148.44	252.40	4.62	146.08	nr	398.49
90° segmented radius bend						
80mm dia.	2.65	4.51	0.29	9.17	nr	13.68
100mm dia.	3.03	13.79	0.29	9.17	nr	22.96
160mm dia.	5.08	8.64	0.29	9.17	nr	17.81
200mm dia.	6.75	11.48	0.44	13.91	nr	25.39
250mm dia.	9.89	16.82	0.58	18.34	nr	35.16
315mm dia.	11.42	19.42	0.58	18.34	nr	37.76
355mm dia.	11.65	19.81	0.87	27.51	nr	47.32
400mm dia.	13.60	23.13	0.87	27.51	nr	50.63
450mm dia.	15.40	26.19	0.87	27.51	nr	53.69
500mm dia.	15.94	27.10	0.87	27.51	nr	54.61
630mm dia.	35.32	60.06	0.87	27.51	nr	87.57
710mm dia.	49.86	84.78	0.96	30.35	nr	115.14
800mm dia.	54.60	92.84	1.06	33.52	nr	126.36
900mm dia.	57.96	98.55	1.16	36.68	nr	135.23
1000mm dia.	112.09	190.60	1.25	39.52	nr	230.12
1120mm dia.	122.39	208.11	3.47	109.72	nr	317.83
1250mm dia.	158.65	269.77	3.47	109.72	nr	379.48
1400mm dia.	329.67	560.56	4.05	128.06	nr	688.62
1600mm dia.	329.68	560.58	4.62	146.08	nr	706.66
45° radius bend						
80mm dia.	2.78	4.73	0.29	9.17	nr	13.90
100mm dia.	2.93	4.98	0.29	9.17	nr	14.15
160mm dia.	4.45	7.57	0.29	9.17	nr	16.74
200mm dia.	5.73	9.74	0.40	12.65	nr	22.39
250mm dia.	7.92	13.47	0.58	18.34	nr	31.81
315mm dia.	10.56	17.96	0.58	18.34	nr	36.30
355mm dia.	11.20	19.04	0.87	27.51	nr	46.55
400mm dia.	13.34	22.68	0.87	27.51	nr	50.19
450mm dia.	13.43	22.84	0.87	27.51	nr	50.34
500mm dia.	14.31	24.33	0.87	27.51	nr	51.84
630mm dia.	39.20	66.65	0.87	27.51	nr	94.16
710mm dia.	49.36	83.93	0.96	30.35	nr	114.29
800mm dia.	58.16	98.89	1.06	33.52	nr	132.41
900mm dia.	62.84	106.85	1.16	36.68	nr	143.53
1000mm dia.	105.69	179.71	1.25	39.52	nr	219.24

U: VENTILATION/AIR CONDITIONING SYSTEMS

Item	Net Price £	Material £	Labour hours	Labour £	Unit	Total rate £
U10 : DUCTWORK : CIRCULAR (cont'd)						
Galvanised sheet metal DW144 class B (cont'd)						
Extra over fittings; circular duct (cont'd)						
45° radius bend (cont'd)						
1120mm dia.	114.17	194.13	3.47	109.72	nr	303.85
1250mm dia.	126.47	215.05	3.47	109.72	nr	324.77
1400mm dia.	175.34	298.14	4.05	128.06	nr	426.20
1600mm dia.	184.43	313.60	4.62	146.08	nr	459.68
90° equal twin bend						
80mm dia.	7.79	13.25	0.58	18.34	nr	31.59
100mm dia.	8.08	13.74	0.58	18.34	nr	32.08
160mm dia.	14.22	24.18	0.58	18.34	nr	42.52
200mm dia.	19.83	33.72	0.87	27.51	nr	61.23
250mm dia.	30.41	51.71	1.16	36.68	nr	88.39
315mm dia.	42.06	71.52	1.16	36.68	nr	108.20
355mm dia.	46.52	79.10	1.73	54.70	nr	133.80
400mm dia.	51.51	87.59	1.73	54.70	nr	142.29
450mm dia.	55.49	94.35	1.73	54.70	nr	149.06
500mm dia.	59.47	101.12	1.73	54.70	nr	155.82
630mm dia.	112.25	190.87	1.73	54.70	nr	245.57
710mm dia.	149.70	254.55	1.82	57.55	nr	312.09
800mm dia.	182.80	310.83	1.93	61.03	nr	371.85
900mm dia.	210.28	357.56	2.02	63.87	nr	421.43
1000mm dia.	291.04	494.88	2.11	66.72	nr	561.60
1120mm dia.	336.40	572.01	4.62	146.08	nr	718.09
1250mm dia.	402.28	684.03	4.62	146.08	nr	830.11
1400mm dia.	651.32	1107.49	4.62	146.08	nr	1253.57
1600mm dia.	692.80	1178.02	4.62	146.08	nr	1324.10
Conical branch						
80mm dia.	10.22	17.38	0.58	18.34	nr	35.72
100mm dia.	10.47	17.80	0.58	18.34	nr	36.14
160mm dia.	11.37	19.33	0.58	18.34	nr	37.67
200mm dia.	12.11	20.59	0.87	27.51	nr	48.10
250mm dia.	15.83	26.92	1.16	36.68	nr	63.60
315mm dia.	17.67	30.05	1.16	36.68	nr	66.72
355mm dia.	19.19	32.63	1.73	54.70	nr	87.33
400mm dia.	19.83	33.72	1.73	54.70	nr	88.42
450mm dia.	24.48	41.63	1.73	54.70	nr	96.33
500mm dia.	25.23	42.90	1.73	54.70	nr	97.60
630mm dia.	49.36	83.93	1.73	54.70	nr	138.63
710mm dia.	54.94	93.42	1.82	57.55	nr	150.97
800mm dia.	71.27	121.19	1.93	61.03	nr	182.21
900mm dia.	75.48	128.34	2.02	63.87	nr	192.22
1000mm dia.	95.46	162.32	2.11	66.72	nr	229.03
1120mm dia.	133.36	226.76	4.62	146.08	nr	372.84
1250mm dia.	133.36	226.76	5.20	164.42	nr	391.18
1400mm dia .	160.77	273.37	5.20	164.42	nr	437.79
1600mm dia.	189.48	322.19	5.20	164.42	nr	486.61

U: VENTILATION/AIR CONDITIONING SYSTEMS

Item	Net Price £	Material £	Labour hours	Labour £	Unit	**Total rate £**
45° branch						
80mm dia.	8.12	13.81	0.58	18.34	nr	**32.15**
100mm dia.	8.33	14.16	0.58	18.34	nr	**32.50**
160mm dia.	12.68	21.56	0.58	18.34	nr	**39.90**
200mm dia.	13.16	22.38	0.87	27.51	nr	**49.89**
250mm dia.	13.82	23.50	1.16	36.68	nr	**60.18**
315mm dia.	14.73	25.05	1.16	36.68	nr	**61.72**
355mm dia.	18.97	32.26	1.73	54.70	nr	**86.96**
400mm dia.	19.73	33.55	1.73	54.70	nr	**88.25**
450mm dia.	20.58	34.99	1.73	54.70	nr	**89.69**
500mm dia.	21.49	36.54	1.73	54.70	nr	**91.24**
630mm dia.	41.31	70.24	1.73	54.70	nr	**124.94**
710mm dia.	46.09	78.37	1.82	57.55	nr	**135.92**
800mm dia.	65.95	112.14	2.13	67.35	nr	**179.49**
900mm dia.	74.67	126.97	2.31	73.04	nr	**200.01**
1000mm dia.	91.87	156.21	2.31	73.04	nr	**229.25**
1120mm dia.	114.10	194.01	4.62	146.08	nr	**340.09**
1250mm dia.	121.91	207.29	4.62	146.08	nr	**353.37**
1400mm dia.	140.82	239.45	4.62	146.08	nr	**385.53**
1600mm dia.	172.35	293.06	4.62	146.08	nr	**439.14**

For galvanised sheet metal DW144 class C rates, refer to galvanised sheet metal DW144 class B

U: VENTILATION/AIR CONDITIONING SYSTEMS

Item	Net Price £	Material £	Labour hours	Labour £	Unit	Total rate £
U10 : DUCTWORK : FLAT OVAL						
Y30 - AIR DUCTLINES						
Galvanised sheet metal DW144 class B spirally wound flat oval section ductwork; including all necessary stiffeners, joints, couplers in the running length and duct supports						
Straight duct						
345 x 102mm	19.42	33.02	2.71	85.69	m	118.71
427 x 102mm	22.30	37.92	2.99	94.54	m	132.46
508 x 102mm	24.59	41.81	3.14	99.28	m	141.10
559 x 152mm	33.13	56.33	3.43	108.45	m	164.79
531 x 203mm	33.64	57.20	3.43	108.45	m	165.65
851 x 203mm	43.59	74.12	5.72	180.86	m	254.98
582 x 254mm	36.59	62.22	3.62	114.46	m	176.68
823 x 254mm	44.20	75.16	5.80	183.39	m	258.55
1303 x 254mm	105.44	179.29	8.13	257.06	m	436.35
632 x 305mm	40.04	68.08	3.93	124.26	m	192.35
1275 x 305mm	97.33	165.50	8.13	257.06	m	422.56
765 x 356mm	46.39	78.88	5.72	180.86	m	259.74
1247 x 356mm	98.40	167.32	8.13	257.06	m	424.38
1727 x 356mm	132.41	225.15	10.41	329.16	m	554.30
737 x 406mm	47.27	80.38	5.72	180.86	m	261.24
818 x 406mm	49.72	84.54	6.21	196.36	m	280.90
978 x 406mm	54.80	93.18	6.92	218.80	m	311.99
1379 x 406mm	107.12	182.14	8.75	276.67	m	458.81
1699 x 406mm	108.98	185.31	10.41	329.16	m	514.46
709 x 457mm	47.21	80.27	5.72	180.86	m	261.14
1189 x 457mm	90.55	153.97	8.80	278.25	m	432.22
1671 x 457mm	130.43	221.78	10.31	325.99	m	547.77
678 x 508mm	46.72	79.44	5.72	180.86	m	260.30
919 x 508mm	54.80	93.18	7.30	230.82	m	324.00
1321 x 508mm	106.93	181.82	8.75	276.67	m	458.49
Extra over fittings; flat oval duct class B						
End cap						
345 x 102mm	17.83	30.32	0.20	6.32	nr	36.64
427 x 102mm	18.36	31.22	0.20	6.32	nr	37.54
508 x 102mm	18.89	32.12	0.20	6.32	nr	38.44
559 x 152mm	27.24	46.32	0.29	9.17	nr	55.49
531 x 203mm	27.35	46.51	0.29	9.17	nr	55.68
851 x 203mm	30.50	51.86	0.44	13.91	nr	65.77
582 x 254mm	28.13	47.83	0.44	13.91	nr	61.74
823 x 254mm	30.02	51.05	0.44	13.91	nr	64.96
1303 x 254mm	97.28	165.41	0.69	21.82	nr	187.23
632 x 305mm	28.16	47.88	0.69	21.82	nr	69.70
1275 x 305mm	85.82	145.93	0.69	21.82	nr	167.74
765 x 356mm	29.62	50.37	0.69	21.82	nr	72.18
1247 x 356mm	83.34	141.71	0.69	21.82	nr	163.53
1727 x 356mm	123.81	210.52	0.69	21.82	nr	232.34
737 x 406mm	30.46	51.79	1.04	32.88	nr	84.68
818 x 406mm	31.27	53.17	0.69	21.82	nr	74.99
978 x 406mm	32.86	55.87	0.69	21.82	nr	77.69
1379 x 406mm	114.10	194.01	1.04	32.88	nr	226.90
1699 x 406mm	124.55	211.78	1.04	32.88	nr	244.67
709 x 457mm	29.65	50.42	1.04	32.88	nr	83.30

U: VENTILATION/AIR CONDITIONING SYSTEMS

Item	Net Price £	Material £	Labour hours	Labour £	Unit	Total rate £
1189 x 457mm	81.38	138.38	1.04	32.88	nr	**171.26**
1671 x 457mm	121.27	206.21	1.04	32.88	nr	**239.09**
678 x 508mm	39.80	67.68	1.04	32.88	nr	**100.56**
919 x 508mm	43.02	73.15	1.04	32.88	nr	**106.03**
1321 x 508mm	121.24	206.15	1.04	32.88	nr	**239.04**
Reducer						
345 x 102mm	33.40	56.79	0.95	30.04	nr	**86.83**
427 x 102mm	34.03	57.86	1.06	33.52	nr	**91.38**
508 x 102mm	34.80	59.17	1.13	35.73	nr	**94.90**
559 x 152mm	48.39	82.28	1.26	39.84	nr	**122.12**
531 x 203mm	48.72	82.84	1.26	39.84	nr	**122.68**
851 x 203mm	63.13	107.34	1.34	42.37	nr	**149.71**
582 x 254mm	51.49	87.55	1.34	42.37	nr	**129.92**
823 x 254mm	64.85	110.27	1.34	42.37	nr	**152.64**
1303 x 254mm	145.89	248.07	1.34	42.37	nr	**290.44**
632 x 305mm	56.64	96.31	0.70	22.13	nr	**118.44**
1275 x 305mm	148.60	252.68	1.16	36.68	nr	**289.35**
765 x 356mm	60.79	103.37	1.16	36.68	nr	**140.04**
1247 x 356mm	151.15	257.01	1.16	36.68	nr	**293.69**
1727 x 356mm	155.47	264.36	1.25	39.52	nr	**303.88**
737 x 406mm	66.94	113.82	1.16	36.68	nr	**150.50**
818 x 406mm	67.93	115.51	1.27	40.16	nr	**155.66**
978 x 406mm	73.95	125.74	1.44	45.53	nr	**171.27**
1379 x 406mm	153.81	261.54	1.44	45.53	nr	**307.07**
1699 x 406mm	174.37	296.50	1.44	45.53	nr	**342.03**
709 x 457mm	67.63	115.00	1.16	36.68	nr	**151.68**
1189 x 457mm	135.23	229.94	1.34	42.37	nr	**272.31**
1671 x 457mm	152.10	258.63	1.44	45.53	nr	**304.16**
678 x 508mm	71.78	122.05	1.16	36.68	nr	**158.73**
919 x 508mm	76.95	130.84	1.26	39.84	nr	**170.68**
1321 x 508mm	151.80	258.12	1.44	45.53	nr	**303.65**
90 ° radius bend						
345 x 102mm	43.64	74.20	0.29	9.17	nr	**83.37**
427 x 102mm	42.01	71.43	0.58	18.34	nr	**89.77**
508 x 102mm	40.76	69.31	0.58	18.34	nr	**87.65**
559 x 152mm	56.27	95.68	0.58	18.34	nr	**114.02**
531 x 203mm	56.86	96.68	0.87	27.51	nr	**124.19**
851 x 203mm	77.60	131.95	0.87	27.51	nr	**159.46**
582 x 254mm	54.25	92.25	0.87	27.51	nr	**119.75**
823 x 254mm	80.75	137.31	0.87	27.51	nr	**164.81**
1303 x 254mm	181.67	308.91	0.96	30.35	nr	**339.26**
632 x 305mm	51.65	87.82	0.87	27.51	nr	**115.33**
1275 x 305mm	227.64	387.07	0.96	30.35	nr	**417.43**
765 x 356mm	60.49	102.86	0.87	27.51	nr	**130.36**
1247 x 356mm	228.82	389.08	0.96	30.35	nr	**419.44**
1727 x 356mm	282.99	481.19	1.25	39.52	nr	**520.71**
737 x 406mm	62.32	105.97	0.96	30.35	nr	**136.32**
818 x 406mm	88.92	151.20	0.87	27.51	nr	**178.71**
978 x 406mm	80.19	136.35	0.96	30.35	nr	**166.71**
1379 x 406mm	206.64	351.37	1.16	36.68	nr	**388.04**
1699 x 406mm	349.45	594.20	1.25	39.52	nr	**633.72**
709 x 457mm	71.26	121.17	0.87	27.51	nr	**148.68**
1189 x 457mm	170.16	289.34	0.96	30.35	nr	**319.69**
1671 x 457mm	351.27	597.29	1.25	39.52	nr	**636.82**
678 x 508mm	68.87	117.11	0.87	27.51	nr	**144.61**
919 x 508mm	78.50	133.48	0.96	30.35	nr	**163.83**
1321 x 508mm	213.36	362.79	1.16	36.68	nr	**399.47**

U: VENTILATION/AIR CONDITIONING SYSTEMS

Item	Net Price £	Material £	Labour hours	Labour £	Unit	Total rate £
U10 : DUCTWORK : FLAT OVAL (cont'd)						
Galvanised sheet metal DW144 class B (cont'd)						
Extra over fittings; flat oval duct (cont'd)						
45 ° radius bend						
345 x 102mm	30.97	52.66	0.79	24.98	nr	**77.64**
427 x 102mm	30.55	51.95	0.85	26.88	nr	**78.82**
508 x 102mm	30.13	51.23	0.95	30.04	nr	**81.27**
559 x 152mm	41.71	70.92	0.79	24.98	nr	**95.90**
531 x 203mm	41.98	71.38	0.85	26.88	nr	**98.26**
851 x 203mm	58.91	100.17	0.98	30.99	nr	**131.16**
582 x 254mm	41.13	69.94	0.76	24.03	nr	**93.97**
823 x 254mm	61.25	104.15	0.95	30.04	nr	**134.19**
1303 x 254mm	174.89	297.38	1.16	36.68	nr	**334.06**
632 x 305mm	44.77	76.13	0.58	18.34	nr	**94.47**
1275 x 305mm	200.53	340.98	1.16	36.68	nr	**377.66**
765 x 356mm	46.36	78.83	0.87	27.51	nr	**106.34**
1247 x 356mm	201.20	342.12	1.16	36.68	nr	**378.79**
1727 x 356mm	198.56	337.63	1.26	39.84	nr	**377.47**
737 x 406mm	48.79	82.96	0.69	21.82	nr	**104.78**
818 x 406mm	67.17	114.21	0.78	24.66	nr	**138.88**
978 x 406mm	63.62	108.18	0.87	27.51	nr	**135.69**
1379 x 406mm	192.59	327.48	1.16	36.68	nr	**364.15**
1699 x 406mm	232.10	394.66	1.27	40.16	nr	**434.81**
709 x 457mm	54.66	92.94	0.81	25.61	nr	**118.55**
1189 x 457mm	158.69	269.83	0.95	30.04	nr	**299.87**
1671 x 457mm	230.09	391.24	1.26	39.84	nr	**431.08**
678 x 508mm	53.30	90.63	0.92	29.09	nr	**119.72**
919 x 508mm	62.43	106.15	1.10	34.78	nr	**140.94**
1321 x 508mm	195.51	332.44	1.25	39.52	nr	**371.97**
90 ° hard bend with turning vanes						
345 x 102mm	33.74	57.37	0.55	17.39	nr	**74.76**
427 x 102mm	34.44	58.56	1.16	36.68	nr	**95.24**
508 x 102mm	35.12	59.72	1.16	36.68	nr	**96.40**
559 x 152mm	51.85	88.16	1.16	36.68	nr	**124.84**
531 x 203mm	52.74	89.68	1.73	54.70	nr	**144.38**
851 x 203mm	68.62	116.68	1.73	54.70	nr	**171.38**
582 x 254mm	58.97	100.27	1.73	54.70	nr	**154.97**
823 x 254mm	74.53	126.73	1.73	54.70	nr	**181.43**
1303 x 254mm	157.41	267.66	1.82	57.55	nr	**325.20**
632 x 305mm	65.76	111.82	1.73	54.70	nr	**166.52**
1275 x 305mm	158.07	268.78	1.82	57.55	nr	**326.33**
765 x 356mm	74.26	126.27	1.73	54.70	nr	**180.97**
1247 x 356mm	162.61	276.50	1.82	57.55	nr	**334.05**
1727 x 356mm	206.49	351.11	1.82	57.55	nr	**408.66**
737 x 406mm	80.05	136.12	1.73	54.70	nr	**190.82**
818 x 406mm	81.68	138.89	1.73	54.70	nr	**193.59**
978 x 406mm	90.94	154.63	1.73	54.70	nr	**209.33**
1379 x 406mm	174.64	296.95	1.82	57.55	nr	**354.50**
1699 x 406mm	203.47	345.98	2.11	66.72	nr	**412.69**
709 x 457mm	81.75	139.01	1.73	54.70	nr	**193.71**
1189 x 457mm	143.39	243.82	1.82	57.55	nr	**301.36**
1671 x 457mm	203.91	346.72	2.11	66.72	nr	**413.44**
678 x 508mm	87.25	148.36	1.82	57.55	nr	**205.91**
919 x 508mm	96.58	164.22	1.82	57.55	nr	**221.77**
1321 x 508mm	175.37	298.20	2.11	66.72	nr	**364.91**

U: VENTILATION/AIR CONDITIONING SYSTEMS

Item	Net Price £	Material £	Labour hours	Labour £	Unit	Total rate £
90 ° branch						
345 x 102mm	44.71	76.02	0.58	18.34	nr	**94.36**
427 x 102mm	43.85	74.56	0.58	18.34	nr	**92.90**
508 x 102mm	42.99	73.10	1.16	36.68	nr	**109.78**
559 x 152mm	45.78	77.84	1.16	36.68	nr	**114.52**
531 x 203mm	46.17	78.51	1.16	36.68	nr	**115.18**
851 x 203mm	71.68	121.88	1.73	54.70	nr	**176.58**
582 x 254mm	52.22	88.79	1.73	54.70	nr	**143.50**
823 x 254mm	71.25	121.15	1.73	54.70	nr	**175.85**
1303 x 254mm	124.74	212.11	1.82	57.55	nr	**269.65**
632 x 305mm	68.46	116.41	1.73	54.70	nr	**171.11**
1275 x 305mm	126.41	214.94	1.82	57.55	nr	**272.49**
765 x 356mm	67.79	115.27	1.73	54.70	nr	**169.97**
1247 x 356mm	127.27	216.41	1.82	57.55	nr	**273.95**
1727 x 356mm	214.79	365.22	2.11	66.72	nr	**431.94**
737 x 406mm	74.20	126.17	1.73	54.70	nr	**180.87**
818 x 406mm	72.01	122.44	1.73	54.70	nr	**177.15**
978 x 406mm	79.01	134.35	1.82	57.55	nr	**191.89**
1379 x 406mm	167.21	284.32	1.93	61.03	nr	**345.35**
1699 x 406mm	259.60	441.42	2.11	66.72	nr	**508.14**
709 x 457mm	82.60	140.45	1.73	54.70	nr	**195.15**
1189 x 457mm	137.61	233.99	1.82	57.55	nr	**291.54**
1671 x 457mm	218.11	370.87	2.11	66.72	nr	**437.59**
678 x 508mm	79.89	135.84	1.73	54.70	nr	**190.54**
919 x 508mm	88.36	150.25	1.82	57.55	nr	**207.79**
1321 x 508mm	174.58	296.85	2.11	66.72	nr	**363.57**
45 ° branch						
345 x 102mm	35.62	60.57	0.58	18.34	nr	**78.91**
427 x 102mm	37.54	63.83	0.58	18.34	nr	**82.17**
508 x 102mm	39.65	67.42	1.16	36.68	nr	**104.10**
559 x 152mm	56.86	96.68	1.73	54.70	nr	**151.38**
531 x 203mm	56.49	96.05	1.73	54.70	nr	**150.76**
851 x 203mm	83.64	142.22	1.73	54.70	nr	**196.92**
582 x 254mm	63.00	107.12	1.73	54.70	nr	**161.83**
823 x 254mm	84.77	144.14	1.82	57.55	nr	**201.69**
1303 x 254mm	194.67	331.01	1.92	60.71	nr	**391.72**
632 x 305mm	70.01	119.04	1.73	54.70	nr	**173.74**
1275 x 305mm	192.06	326.57	1.82	57.55	nr	**384.12**
765 x 356mm	80.55	136.97	1.73	54.70	nr	**191.67**
1247 x 356mm	192.82	327.87	1.82	57.55	nr	**385.41**
1727 x 356mm	271.22	461.18	1.82	57.55	nr	**518.72**
737 x 406mm	85.41	145.23	1.73	54.70	nr	**199.93**
818 x 406mm	89.63	152.41	1.73	54.70	nr	**207.11**
978 x 406mm	105.13	178.76	1.73	54.70	nr	**233.46**
1379 x 406mm	210.20	357.42	1.93	61.03	nr	**418.44**
1699 x 406mm	262.01	445.52	2.19	69.25	nr	**514.76**
709 x 457mm	84.58	143.82	1.73	54.70	nr	**198.52**
1189 x 457mm	165.65	281.67	1.82	57.55	nr	**339.21**
1671 x 457mm	257.79	438.34	2.11	66.72	nr	**505.06**
678 x 508mm	87.63	149.00	1.73	54.70	nr	**203.71**
919 x 508mm	104.71	178.05	1.82	57.55	nr	**235.59**
1321 x 508mm	202.84	344.91	1.93	61.03	nr	**405.93**

For access door rates refer to ancillaries in U10 : DUCTWORK: RECTANGULAR: CLASS B

U: VENTILATION/AIR CONDITIONING SYSTEMS

Item	Net Price £	Material £	Labour hours	Labour £	Unit	Total rate £
U10 : DUCTWORK : FLEXIBLE						
Y30 : AIR DUCTLINES						
Aluminium foil flexible ductwork, DW 144 class B; multiply aluminium polyester laminate fabric, with high tensile steel wire helix						
Duct						
102mm dia	1.12	1.90	0.33	10.43	m	12.34
152mm dia	1.64	2.79	0.33	10.43	m	13.22
203mm dia	2.16	3.67	0.33	10.43	m	14.10
254mm dia	2.71	4.61	0.33	10.43	m	15.05
304mm dia	3.32	5.65	0.33	10.43	m	16.08
355mm dia	4.53	7.70	0.33	10.43	m	18.14
406mm dia	5.03	8.55	0.33	10.43	m	18.98
Insulated aluminium foil flexible ductwork, DW144 class B; laminate construction of aluminium and polyester multiply inner core with 25mm insulation; outer layer of multiply aluminium polyester laminate, with high tensile steel wire helix						
Duct						
102mm dia	2.51	4.27	0.50	15.81	m	20.08
152mm dia	3.27	5.55	0.50	15.81	m	21.36
203mm dia	3.97	6.74	0.50	15.81	m	22.55
254mm dia	4.89	8.32	0.50	15.81	m	24.13
304mm dia	6.29	10.70	0.50	15.81	m	26.51
355mm dia	7.80	13.26	0.50	15.81	m	29.07
406mm dia	8.64	14.70	0.50	15.81	m	30.51

U: VENTILATION/AIR CONDITIONING SYSTEMS

Item	Net Price £	Material £	Labour hours	Labour £	Unit	Total rate £
U10 : DUCTWORK : PLASTIC						
Y30 - AIR DUCTLINES						
Rigid grey PVC DW 154 circular section ductwork; solvent welded or filler rod welded joints; excludes couplers and supports (these are detailed separately); ductwork to conform to current HSE regulations						
Straight duct (standard length 6m)						
110mm	7.99	13.59	0.17	5.38	m	18.96
160mm	15.41	26.20	0.25	7.90	m	34.11
200mm	19.26	32.75	0.33	10.43	m	43.18
225mm	24.97	42.46	0.42	13.28	m	55.74
250mm	24.22	41.18	0.50	15.81	m	56.99
315mm	30.48	51.83	0.58	18.34	m	70.17
355mm	41.01	69.73	0.67	21.18	m	90.92
400mm	51.39	87.38	0.75	23.71	m	111.10
450mm	64.45	109.59	0.83	26.24	m	135.83
500mm	78.90	134.16	0.92	29.09	m	163.25
600mm	117.99	200.63	1.00	31.62	m	232.25
Extra for supports (BZP finish)						
Horizontal - Maximum 2.4m centres						
Vertical - Maximum 4.0m centres						
Duct Size						
110mm	7.25	12.33	0.17	5.38	m	17.70
160mm	7.98	13.57	0.25	7.90	m	21.47
200mm	8.51	14.47	0.33	10.43	m	24.90
225mm	8.88	15.10	0.42	13.28	m	28.38
250mm	9.59	16.31	0.50	15.81	m	32.12
315mm	10.13	17.22	0.58	18.34	m	35.56
355mm	13.34	22.68	0.67	21.18	m	43.87
400mm	13.71	23.31	0.75	23.71	m	47.03
450mm	15.95	27.12	0.83	26.24	m	53.36
500mm	16.39	27.87	0.92	29.09	m	56.96
600mm	17.93	30.49	1.00	31.62	m	62.11
Note - These are maximum figures and may be reduced subject to local conditions (i.e. a high number of changes of direction)						
Extra over fittings; Rigid grey PVC						
90° Bend						
110mm	18.13	30.83	0.34	10.75	m	41.58
160mm	23.71	40.32	0.50	15.81	m	56.13
200mm	29.29	49.80	0.66	20.87	m	70.67
225mm	34.87	59.29	0.84	26.56	m	85.85
250mm	39.99	68.00	1.00	31.62	m	99.62
315mm	63.24	107.53	1.16	36.68	m	144.21
355mm	85.55	145.47	1.34	42.37	m	187.84
400mm	106.46	181.02	1.50	47.43	m	228.45
450mm	303.61	516.25	1.66	52.49	m	568.74
500mm	358.93	610.32	1.84	58.18	m	668.50
600mm	607.21	1032.49	2.00	63.24	m	1095.73

U: VENTILATION/AIR CONDITIONING SYSTEMS

Item	Net Price £	Material £	Labour hours	Labour £	Unit	Total rate £
U10 : DUCTWORK : PLASTIC (cont'd)						
Rigid grey PVC DW 154 circular section (cont'd)						
Extra over fittings; Rigid grey PVC (cont'd)						
45° Bend						
110mm	13.49	22.94	0.34	10.75	m	**33.69**
160mm	17.67	30.05	0.50	15.81	m	**45.86**
200mm	20.92	35.57	0.66	20.87	m	**56.44**
225mm	24.19	41.13	0.84	26.56	m	**67.69**
250mm	26.96	45.84	1.00	31.62	m	**77.46**
315mm	41.86	71.18	1.16	36.68	m	**107.86**
355mm	54.85	93.27	1.34	42.37	m	**135.64**
400mm	66.51	113.09	1.50	47.43	m	**160.52**
450mm	214.83	365.29	1.66	52.49	m	**417.78**
500mm	241.30	410.30	1.84	58.18	m	**468.48**
600mm	369.17	627.73	2.00	63.24	m	**690.97**
Tee						
110mm	18.13	30.83	0.51	16.13	m	**46.95**
160mm	23.71	40.32	0.75	23.71	m	**64.03**
200mm	29.17	49.60	0.99	31.30	m	**80.90**
225mm	34.87	59.29	1.26	39.84	m	**99.13**
250mm	39.99	68.00	1.50	47.43	m	**115.43**
315mm	63.24	107.53	1.74	55.02	m	**162.55**
355mm	85.55	145.47	2.01	63.55	m	**209.02**
400mm	106.46	181.02	2.25	71.14	m	**252.17**
450mm	303.61	516.25	2.49	78.73	m	**594.98**
500mm	358.93	610.32	2.76	87.27	m	**697.59**
Coupler						
110mm	8.47	14.40	0.34	10.75	m	**25.15**
160mm	11.77	20.01	0.50	15.81	m	**35.82**
200mm	15.88	27.00	0.66	20.87	m	**47.87**
225mm	21.05	35.79	0.84	26.56	m	**62.35**
250mm	24.35	41.40	1.00	31.62	m	**73.02**
315mm	32.69	55.59	1.16	36.68	m	**92.26**
355mm	32.80	55.77	1.34	42.37	m	**98.14**
400mm	35.71	60.72	1.50	47.43	m	**108.15**
450mm	79.44	135.08	1.66	52.49	m	**187.57**
500mm	90.18	153.34	1.84	58.18	m	**211.52**
Damper						
110mm	50.22	85.39	0.34	10.75	m	**96.14**
160mm	57.65	98.03	0.50	15.81	m	**113.84**
200mm	64.16	109.10	0.66	20.87	m	**129.97**
225mm	68.33	116.19	0.84	26.56	m	**142.75**
250mm	70.69	120.20	1.00	31.62	m	**151.82**
315mm	83.69	142.30	1.16	36.68	m	**178.98**
355mm	88.80	150.99	1.34	42.37	m	**193.36**
400mm	96.24	163.64	1.50	47.43	m	**211.07**
Reducer						
160 x 110	18.98	32.27	0.42	13.28	m	**45.55**
200 x 110	28.48	48.43	0.50	15.81	m	**64.24**
200 x 160	25.49	43.34	0.58	18.34	m	**61.68**
225 x 200	34.66	58.94	0.75	23.71	m	**82.65**
250 x 160	34.66	58.94	0.75	23.71	m	**82.65**

U: VENTILATION/AIR CONDITIONING SYSTEMS

Item	Net Price £	Material £	Labour hours	Labour £	Unit	Total rate £
250 x 200	35.71	60.72	0.83	26.24	m	86.96
250 x 225	36.53	62.11	0.92	29.09	m	91.20
315 x 200	38.59	65.62	0.92	29.09	m	94.71
315 x 250	41.89	71.23	1.10	34.78	m	106.01
355 x 200	57.37	97.55	1.10	34.78	m	132.33
355 x 250	46.01	78.23	1.17	36.99	m	115.23
355 x 315	54.52	92.70	1.25	39.52	m	132.23
400 x 225	64.17	109.11	1.17	36.99	m	146.11
400 x 315	69.54	118.24	1.33	42.05	m	160.30
400 x 355	67.68	115.08	1.42	44.90	m	159.98
450 x 315	69.54	118.24	1.45	45.85	m	164.09
Flange						
110mm	12.49	21.24	0.34	10.75	m	31.99
160mm	14.60	24.83	0.50	15.81	m	40.64
200mm	16.71	28.41	0.66	20.87	m	49.28
225mm	16.92	28.77	0.84	26.56	m	55.33
250mm	17.45	29.67	1.00	31.62	m	61.29
315mm	26.22	44.58	1.16	36.68	m	81.26
355mm	28.77	48.92	1.34	42.37	m	91.29
400mm	32.10	54.58	1.50	47.43	m	102.01

Polypropylene (PPS) DW154 circular section ductwork; filler rod welded joints; excludes couplers and supports (these are detailed separately); ductwork to conform to current HSE regulations

Item	Net Price £	Material £	Labour hours	Labour £	Unit	Total rate £
Straight duct (standard length 6m)						
110mm	10.35	17.60	0.21	6.64	m	24.24
160mm	17.03	28.96	0.31	9.80	m	38.76
200mm	21.16	35.98	0.41	12.96	m	48.94
225mm	27.52	46.79	0.52	16.44	m	63.24
250mm	30.66	52.13	0.63	19.92	m	72.05
315mm	54.70	93.01	0.73	23.08	m	116.09
355mm	61.70	104.91	0.84	26.56	m	131.47
400mm	82.28	139.91	0.94	29.72	m	169.63

Extra for supports (BZP finish)
Horizontal - Maximum 2.4m centres
Vertical - Maximum 4.0m centres

Item	Net Price £	Material £	Labour hours	Labour £	Unit	Total rate £
Duct Size						
110mm	7.25	12.33	0.21	6.64	m	18.97
160mm	7.98	13.57	0.31	9.80	m	23.37
200mm	8.51	14.47	0.41	12.96	m	27.43
225mm	8.88	15.10	0.52	16.44	m	31.54
250mm	9.59	16.31	0.63	19.92	m	36.23
315mm	10.13	17.22	0.73	23.08	m	40.31
355mm	13.34	22.68	0.84	26.56	m	49.24
400mm	13.71	23.31	0.94	29.72	m	53.03

Note - These are maximum figures and may be reduced subject to local conditions (i.e. a high number of changes of direction)

U: VENTILATION/AIR CONDITIONING SYSTEMS

Item	Net Price £	Material £	Labour hours	Labour £	Unit	Total rate £
U10 : DUCTWORK : PLASTIC (cont'd)						
Polypropylene (PPS) DW154 circular (cont'd)						
Extra over fittings; Polypropylene (DW 154)						
90° Bend						
110mm	27.17	46.20	0.43	13.60	m	59.80
160mm	39.99	68.00	0.63	19.92	m	87.92
200mm	47.39	80.58	0.83	26.24	m	106.82
225mm	63.24	107.53	1.05	33.20	m	140.73
250mm	67.87	115.40	1.25	39.52	m	154.93
315mm	148.79	253.00	1.45	45.85	m	298.85
355mm	204.58	347.86	1.68	53.12	m	400.98
400mm	218.52	371.57	1.88	59.44	m	431.01
45° Bend						
110mm	20.92	35.57	0.43	13.60	m	49.17
160mm	33.47	56.91	0.63	19.92	m	76.83
200mm	38.13	64.84	0.83	26.24	m	91.08
225mm	47.43	80.65	1.05	33.20	m	113.85
250mm	53.02	90.15	1.25	39.52	m	129.68
315mm	134.84	229.28	1.45	45.85	m	275.13
355mm	153.44	260.91	1.68	53.12	m	314.03
400mm	162.71	276.67	1.88	59.44	m	336.11
Tee						
110mm	94.30	160.35	0.64	20.24	m	180.58
160mm	135.30	230.06	0.94	29.72	m	259.78
200mm	164.03	278.91	1.24	39.21	m	318.12
225mm	182.46	310.25	1.58	49.96	m	360.21
250mm	215.27	366.04	1.88	59.44	m	425.48
315mm	297.28	505.49	2.18	68.93	m	574.42
355mm	369.06	627.54	2.51	79.36	m	706.91
400mm	430.56	732.12	2.81	88.85	m	820.97
Coupler						
110mm	20.92	35.57	0.43	13.60	m	49.17
160mm	25.56	43.46	0.63	19.92	m	63.38
200mm	30.68	52.17	0.83	26.24	m	78.41
225mm	33.95	57.73	1.05	33.20	m	90.93
250mm	35.81	60.89	1.25	39.52	m	100.41
315mm	48.81	83.00	1.45	45.85	m	128.84
355mm	63.24	107.53	1.68	53.12	m	160.65
400mm	68.83	117.04	1.88	59.44	m	176.48
Damper						
110mm	90.41	153.73	0.43	13.60	m	167.33
160mm	106.45	181.01	0.63	19.92	m	200.93
200mm	118.39	201.31	0.83	26.24	m	227.55
225mm	125.99	214.23	1.05	33.20	m	247.43
250mm	133.44	226.90	1.25	39.52	m	266.42
315mm	152.50	259.31	1.45	45.85	m	305.16
355mm	166.93	283.84	1.68	53.12	m	336.96
400mm	182.73	310.71	1.88	59.44	m	370.15

U: VENTILATION/AIR CONDITIONING SYSTEMS

Item	Net Price £	Material £	Labour hours	Labour £	Unit	Total rate £
Reducer						
160 x 110	66.96	113.86	0.53	16.76	m	**130.62**
200 x 160	73.46	124.91	0.73	23.08	m	**147.99**
225 x 200	102.29	173.93	0.94	29.72	m	**203.65**
250 x 200	97.65	166.04	1.04	32.88	m	**198.93**
250 x 225	124.49	211.68	1.15	36.36	m	**248.04**
315 x 200	159.02	270.39	1.15	36.36	m	**306.76**
315 x 250	132.84	225.88	1.38	43.63	m	**269.51**
355 x 200	183.18	311.48	1.38	43.63	m	**355.11**
355 x 250	149.72	254.58	1.46	46.16	m	**300.74**
355 x 315	171.09	290.92	1.56	49.33	m	**340.24**
400 x 315	181.32	308.31	1.66	52.49	m	**360.80**
400 x 355	214.83	365.29	1.78	56.28	m	**421.57**
Flange						
110mm	19.80	33.67	0.43	13.60	m	**47.26**
160mm	24.15	41.06	0.63	19.92	m	**60.98**
200mm	28.26	48.05	0.83	26.24	m	**74.30**
225mm	29.74	50.57	1.05	33.20	m	**83.77**
250mm	33.12	56.32	1.25	39.52	m	**95.84**
315mm	37.38	63.56	1.45	45.85	m	**109.41**
355mm	42.04	71.48	1.68	53.12	m	**124.60**
400mm	46.62	79.27	1.88	59.44	m	**138.72**

U: VENTILATION/AIR CONDITIONING SYSTEMS

Item	Net Price £	Material £	Labour hours	Labour £	Unit	Total rate £
U10 : DUCTWORK : RECTANGULAR – CLASS B						
Y30 - AIR DUCTLINES						
Galvanised sheet metal DW144 class B rectangular section ductwork; including all necessary stiffeners, joints, couplers in the running length and duct supports						
Ductwork up to 400mm longest side						
Sum of two sides 200mm	14.47	24.60	1.19	37.63	m	62.23
Sum of two sides 300mm	15.47	26.30	1.19	37.63	m	63.93
Sum of two sides 400mm	13.45	22.87	1.16	36.68	m	59.55
Sum of two sides 500mm	14.51	24.67	1.16	36.68	m	61.35
Sum of two sides 600mm	15.44	26.25	1.27	40.16	m	66.41
Sum of two sides 700mm	16.36	27.82	1.27	40.16	m	67.97
Sum of two sides 800mm	17.36	29.52	1.27	40.16	m	69.67
Extra Over fittings; Rectangular ductwork class B; up to 400mm longest side						
End Cap						
Sum of two sides 200mm	9.61	16.34	0.38	12.02	nr	28.36
Sum of two sides 300mm	10.70	18.19	0.38	12.02	nr	30.21
Sum of two sides 400mm	11.79	20.05	0.38	12.02	nr	32.06
Sum of two sides 500mm	12.88	21.90	0.38	12.02	nr	33.92
Sum of two sides 600mm	13.97	23.75	0.38	12.02	nr	35.77
Sum of two sides 700mm	15.06	25.61	0.38	12.02	nr	37.62
Sum of two sides 800mm	16.16	27.48	0.38	12.02	nr	39.49
Reducer						
Sum of two sides 200mm	15.72	26.73	1.40	44.27	nr	71.00
Sum of two sides 300mm	17.75	30.18	1.40	44.27	nr	74.45
Sum of two sides 400mm	29.38	49.96	1.42	44.90	nr	94.86
Sum of two sides 500mm	31.89	54.23	1.42	44.90	nr	99.12
Sum of two sides 600mm	34.40	58.49	1.69	53.44	nr	111.93
Sum of two sides 700mm	36.90	62.74	1.69	53.44	nr	116.18
Sum of two sides 800mm	39.39	66.98	1.92	60.71	nr	127.69
Offset						
Sum of two sides 200mm	23.20	39.45	1.63	51.54	nr	90.99
Sum of two sides 300mm	26.23	44.60	1.63	51.54	nr	96.14
Sum of two sides 400mm	39.13	66.54	1.65	52.17	nr	118.71
Sum of two sides 500mm	42.63	72.49	1.65	52.17	nr	124.66
Sum of two sides 600mm	45.53	77.42	1.92	60.71	nr	138.13
Sum of two sides 700mm	48.72	82.84	1.92	60.71	nr	143.55
Sum of two sides 800mm	51.54	87.64	1.92	60.71	nr	148.35
Square to round						
Sum of two sides 200mm	20.41	34.70	1.63	51.54	nr	86.24
Sum of two sides 300mm	22.94	39.01	1.63	51.54	nr	90.55
Sum of two sides 400mm	30.71	52.22	1.65	52.17	nr	104.39
Sum of two sides 500mm	33.46	56.89	1.65	52.17	nr	109.07
Sum of two sides 600mm	36.21	61.57	1.92	60.71	nr	122.28
Sum of two sides 700mm	38.97	66.26	1.92	60.71	nr	126.97
Sum of two sides 800mm	41.70	70.91	1.92	60.71	nr	131.61

U: VENTILATION/AIR CONDITIONING SYSTEMS

Item	Net Price £	Material £	Labour hours	Labour £	Unit	Total rate £
90° radius bend						
Sum of two sides 200mm	15.35	26.10	1.22	38.58	nr	**64.68**
Sum of two sides 300mm	16.56	28.16	1.22	38.58	nr	**66.73**
Sum of two sides 400mm	28.24	48.02	1.25	39.52	nr	**87.54**
Sum of two sides 500mm	30.14	51.25	1.25	39.52	nr	**90.77**
Sum of two sides 600mm	32.63	55.48	1.33	42.05	nr	**97.54**
Sum of two sides 700mm	34.82	59.21	1.33	42.05	nr	**101.26**
Sum of two sides 800mm	37.26	63.36	1.40	44.27	nr	**107.62**
45° radius bend						
Sum of two sides 200mm	16.58	28.19	0.89	28.14	nr	**56.33**
Sum of two sides 300mm	18.19	30.93	1.12	35.41	nr	**66.34**
Sum of two sides 400mm	29.55	50.25	1.10	34.78	nr	**85.03**
Sum of two sides 500mm	31.76	54.00	1.10	34.78	nr	**88.79**
Sum of two sides 600mm	34.27	58.27	1.16	36.68	nr	**94.95**
Sum of two sides 700mm	36.63	62.28	1.16	36.68	nr	**98.96**
Sum of two sides 800mm	39.11	66.50	1.22	38.58	nr	**105.08**
90° mitre bend						
Sum of two sides 200mm	25.57	43.48	1.29	40.79	nr	**84.27**
Sum of two sides 300mm	28.00	47.61	1.29	40.79	nr	**88.40**
Sum of two sides 400mm	41.02	69.75	1.29	40.79	nr	**110.54**
Sum of two sides 500mm	44.31	75.34	1.29	40.79	nr	**116.13**
Sum of two sides 600mm	48.47	82.42	1.39	43.95	nr	**126.37**
Sum of two sides 700mm	52.33	88.98	1.39	43.95	nr	**132.93**
Sum of two sides 800mm	56.53	96.12	1.46	46.16	nr	**142.29**
Branch						
Sum of two sides 200mm	25.40	43.19	0.92	29.09	nr	**72.28**
Sum of two sides 300mm	28.21	47.97	0.92	29.09	nr	**77.06**
Sum of two sides 400mm	35.50	60.36	0.95	30.04	nr	**90.40**
Sum of two sides 500mm	38.68	65.77	0.95	30.04	nr	**95.81**
Sum of two sides 600mm	41.80	71.08	1.03	32.57	nr	**103.64**
Sum of two sides 700mm	44.91	76.36	1.03	32.57	nr	**108.93**
Sum of two sides 800mm	48.02	81.65	1.03	32.57	nr	**114.22**
Grille neck						
Sum of two sides 200mm	29.07	49.43	1.10	34.78	nr	**84.21**
Sum of two sides 300mm	32.51	55.28	1.10	34.78	nr	**90.06**
Sum of two sides 400mm	35.96	61.15	1.16	36.68	nr	**97.82**
Sum of two sides 500mm	39.40	67.00	1.16	36.68	nr	**103.67**
Sum of two sides 600mm	42.84	72.84	1.18	37.31	nr	**110.16**
Sum of two sides 700mm	46.28	78.69	1.18	37.31	nr	**116.00**
Sum of two sides 800mm	49.73	84.56	1.18	37.31	nr	**121.87**
Ductwork 401 to 600mm longest side						
Sum of two sides 600mm	17.36	29.52	1.27	40.16	m	**69.67**
Sum of two sides 700mm	18.67	31.75	1.27	40.16	m	**71.90**
Sum of two sides 800mm	19.94	33.91	1.27	40.16	m	**74.06**
Sum of two sides 900mm	21.12	35.91	1.27	40.16	m	**76.07**
Sum of two sides 1000mm	22.30	37.92	1.37	43.32	m	**81.24**
Sum of two sides 1100mm	23.66	40.23	1.37	43.32	m	**83.55**
Sum of two sides 1200mm	24.85	42.25	1.37	43.32	m	**85.57**

U: VENTILATION/AIR CONDITIONING SYSTEMS

Item	Net Price £	Material £	Labour hours	Labour £	Unit	Total rate £
U10 : DUCTWORK : RECTANGULAR – CLASS B (cont'd)						
Galvanised sheet metal DW144 (cont'd)						
Extra over fittings; Ductwork 401 to 600mm longest side						
End Cap						
Sum of two sides 600mm	14.12	24.01	0.38	12.02	nr	36.02
Sum of two sides 700mm	15.24	25.91	0.38	12.02	nr	37.93
Sum of two sides 800mm	16.35	27.80	0.38	12.02	nr	39.82
Sum of two sides 900mm	17.47	29.71	0.58	18.34	nr	48.04
Sum of two sides 1000mm	18.58	31.59	0.58	18.34	nr	49.93
Sum of two sides 1100mm	19.70	33.50	0.58	18.34	nr	51.84
Sum of two sides 1200mm	20.82	35.40	0.58	18.34	nr	53.74
Reducer						
Sum of two sides 600mm	33.37	56.74	1.69	53.44	nr	110.18
Sum of two sides 700mm	35.87	60.99	1.69	53.44	nr	114.43
Sum of two sides 800mm	38.31	65.14	1.92	60.71	nr	125.85
Sum of two sides 900mm	40.81	69.39	1.92	60.71	nr	130.10
Sum of two sides 1000mm	43.29	73.61	2.18	68.93	nr	142.54
Sum of two sides 1100mm	45.95	78.13	2.18	68.93	nr	147.06
Sum of two sides 1200mm	48.44	82.37	2.18	68.93	nr	151.30
Offset						
Sum of two sides 600mm	46.22	78.59	1.92	60.71	nr	139.30
Sum of two sides 700mm	49.71	84.53	1.92	60.71	nr	145.23
Sum of two sides 800mm	52.52	89.30	1.92	60.71	nr	150.01
Sum of two sides 900mm	55.35	94.12	1.92	60.71	nr	154.82
Sum of two sides 1000mm	58.49	99.46	2.18	68.93	nr	168.39
Sum of two sides 1100mm	61.41	104.42	2.18	68.93	nr	173.35
Sum of two sides 1200mm	64.18	109.13	2.18	68.93	nr	178.06
Square to round						
Sum of two sides 600mm	35.07	59.63	1.33	42.05	nr	101.69
Sum of two sides 700mm	37.81	64.29	1.33	42.05	nr	106.34
Sum of two sides 800mm	40.51	68.88	1.40	44.27	nr	113.15
Sum of two sides 900mm	43.23	73.51	1.40	44.27	nr	117.77
Sum of two sides 1000mm	45.97	78.17	1.82	57.55	nr	135.71
Sum of two sides 1100mm	48.75	82.89	1.82	57.55	nr	140.44
Sum of two sides 1200mm	51.48	87.54	1.82	57.55	nr	145.08
90 ° radius bend						
Sum of two sides 600mm	32.51	55.28	1.16	36.68	nr	91.96
Sum of two sides 700mm	34.40	58.49	1.16	36.68	nr	95.17
Sum of two sides 800mm	37.11	63.10	1.22	38.58	nr	101.68
Sum of two sides 900mm	39.87	67.79	1.22	38.58	nr	106.37
Sum of two sides 1000mm	42.21	71.77	1.40	44.27	nr	116.04
Sum of two sides 1100mm	45.10	76.69	1.40	44.27	nr	120.95
Sum of two sides 1200mm	47.87	81.40	1.40	44.27	nr	125.66

U: VENTILATION/AIR CONDITIONING SYSTEMS

Item	Net Price £	Material £	Labour hours	Labour £	Unit	Total rate £
45 ° bend						
Sum of two sides 600mm	34.37	58.44	1.16	36.68	nr	95.12
Sum of two sides 700mm	36.57	62.18	1.39	43.95	nr	106.13
Sum of two sides 800mm	39.20	66.65	1.46	46.16	nr	112.82
Sum of two sides 900mm	41.87	71.19	1.46	46.16	nr	117.36
Sum of two sides 1000mm	44.32	75.36	1.88	59.44	nr	134.80
Sum of two sides 1100mm	47.17	80.21	1.88	59.44	nr	139.65
Sum of two sides 1200mm	49.84	84.75	1.88	59.44	nr	144.19
90 ° mitre bend						
Sum of two sides 600mm	53.84	91.55	1.39	43.95	nr	135.50
Sum of two sides 700mm	57.05	97.01	2.16	68.30	nr	165.30
Sum of two sides 800mm	61.55	104.66	2.26	71.46	nr	176.12
Sum of two sides 900mm	66.10	112.40	2.26	71.46	nr	183.85
Sum of two sides 1000mm	70.19	119.35	3.01	95.17	nr	214.52
Sum of two sides 1100mm	74.94	127.43	3.01	95.17	nr	222.60
Sum of two sides 1200mm	79.59	135.33	3.01	95.17	nr	230.51
Branch						
Sum of two sides 600mm	42.42	72.13	1.03	32.57	nr	104.70
Sum of two sides 700mm	45.61	77.55	1.03	32.57	nr	110.12
Sum of two sides 800mm	48.81	83.00	1.03	32.57	nr	115.56
Sum of two sides 900mm	52.00	88.42	1.03	32.57	nr	120.99
Sum of two sides 1000mm	55.19	93.84	1.29	40.79	nr	134.63
Sum of two sides 1100mm	58.51	99.49	1.29	40.79	nr	140.28
Sum of two sides 1200mm	61.71	104.93	1.29	40.79	nr	145.72
Grille neck						
Sum of two sides 600mm	43.58	74.10	1.18	37.31	nr	111.41
Sum of two sides 700mm	47.14	80.16	1.18	37.31	nr	117.47
Sum of two sides 800mm	50.71	86.23	1.18	37.31	nr	123.54
Sum of two sides 900mm	54.27	92.28	1.18	37.31	nr	129.59
Sum of two sides 1000mm	57.84	98.35	1.44	45.53	nr	143.88
Sum of two sides 1100mm	61.41	104.42	1.44	45.53	nr	149.95
Sum of two sides 1200mm	64.97	110.47	1.44	45.53	nr	156.01
Ductwork 601 to 800mm longest side						
Sum of two sides 900mm	23.76	40.40	1.27	40.16	m	80.56
Sum of two sides 1000mm	24.94	42.41	1.37	43.32	m	85.73
Sum of two sides 1100mm	26.13	44.43	1.37	43.32	m	87.75
Sum of two sides 1200mm	27.49	46.74	1.37	43.32	m	90.06
Sum of two sides 1300mm	28.68	48.77	1.40	44.27	m	93.03
Sum of two sides 1400mm	29.86	50.77	1.40	44.27	m	95.04
Sum of two sides 1500mm	31.04	52.78	1.48	46.80	m	99.58
Sum of two sides 1600mm	32.23	54.80	1.55	49.01	m	103.81
Extra over fittings: Ductwork 601 to 800mm longest side						
End Cap						
Sum of two sides 900mm	17.47	29.71	0.58	18.34	nr	48.04
Sum of two sides 1000mm	18.58	31.59	0.58	18.34	nr	49.93
Sum of two sides 1100mm	19.70	33.50	0.58	18.34	nr	51.84
Sum of two sides 1200mm	20.82	35.40	0.58	18.34	nr	53.74
Sum of two sides 1300mm	21.93	37.29	0.58	18.34	nr	55.63
Sum of two sides 1400mm	22.88	38.90	0.58	18.34	nr	57.24
Sum of two sides 1500mm	26.61	45.25	0.58	18.34	nr	63.59
Sum of two sides 1600mm	30.34	51.59	0.58	18.34	nr	69.93

U: VENTILATION/AIR CONDITIONING SYSTEMS

Item	Net Price £	Material £	Labour hours	Labour £	Unit	Total rate £
U10 : DUCTWORK : RECTANGULAR – CLASS B (cont'd)						
Galvanised sheet metal DW144 (cont'd)						
Extra over fittings: Ductwork 601 to 800mm longest side (cont'd)						
Reducer						
Sum of two sides 900mm	41.18	70.02	1.92	60.71	nr	130.73
Sum of two sides 1000mm	43.67	74.26	2.18	68.93	nr	143.19
Sum of two sides 1100mm	46.16	78.49	2.18	68.93	nr	147.42
Sum of two sides 1200mm	48.81	83.00	2.18	68.93	nr	151.93
Sum of two sides 1300mm	51.30	87.23	2.30	72.72	nr	159.95
Sum of two sides 1400mm	53.47	90.92	2.30	72.72	nr	163.64
Sum of two sides 1500mm	61.18	104.03	2.47	78.10	nr	182.13
Sum of two sides 1600mm	68.88	117.12	2.47	78.10	nr	195.22
Offset						
Sum of two sides 900mm	56.85	96.67	1.92	60.71	nr	157.38
Sum of two sides 1000mm	59.49	101.16	2.18	68.93	nr	170.09
Sum of two sides 1100mm	62.12	105.63	2.18	68.93	nr	174.56
Sum of two sides 1200mm	64.85	110.27	2.18	68.93	nr	179.20
Sum of two sides 1300mm	67.43	114.66	2.30	72.72	nr	187.38
Sum of two sides 1400mm	69.97	118.98	2.30	72.72	nr	191.70
Sum of two sides 1500mm	80.36	136.64	2.47	78.10	nr	214.74
Sum of two sides 1600mm	90.73	154.28	2.47	78.10	nr	232.37
Square to round						
Sum of two sides 900mm	42.89	72.93	1.40	44.27	nr	117.20
Sum of two sides 1000mm	45.62	77.57	1.82	57.55	nr	135.12
Sum of two sides 1100mm	48.36	82.23	1.82	57.55	nr	139.78
Sum of two sides 1200mm	51.14	86.96	1.82	57.55	nr	144.50
Sum of two sides 1300mm	53.86	91.58	2.15	67.98	nr	159.56
Sum of two sides 1400mm	56.20	95.56	2.15	67.98	nr	163.54
Sum of two sides 1500mm	65.46	111.31	2.38	75.25	nr	186.56
Sum of two sides 1600mm	74.71	127.04	2.38	75.25	nr	202.29
90 ° radius bend						
Sum of two sides 900mm	37.50	63.76	1.22	38.58	nr	102.34
Sum of two sides 1000mm	40.31	68.54	1.40	44.27	nr	112.81
Sum of two sides 1100mm	43.11	73.30	1.40	44.27	nr	117.57
Sum of two sides 1200mm	46.00	78.22	1.40	44.27	nr	122.48
Sum of two sides 1300mm	48.81	83.00	1.91	60.39	nr	143.39
Sum of two sides 1400mm	50.75	86.29	1.91	60.39	nr	146.69
Sum of two sides 1500mm	58.79	99.97	2.11	66.72	nr	166.68
Sum of two sides 1600mm	66.82	113.62	2.11	66.72	nr	180.34
45 ° bend						
Sum of two sides 900mm	41.25	70.14	1.22	38.58	nr	108.72
Sum of two sides 1000mm	43.93	74.70	1.40	44.27	nr	118.96
Sum of two sides 1100mm	46.62	79.27	1.88	59.44	nr	138.72
Sum of two sides 1200mm	49.47	84.12	1.88	59.44	nr	143.56
Sum of two sides 1300mm	52.16	88.69	2.26	71.46	nr	160.15
Sum of two sides 1400mm	54.25	92.25	2.26	71.46	nr	163.70
Sum of two sides 1500mm	62.16	105.70	2.49	78.73	nr	184.43
Sum of two sides 1600mm	70.08	119.16	2.49	78.73	nr	197.89

U: VENTILATION/AIR CONDITIONING SYSTEMS

Item	Net Price £	Material £	Labour hours	Labour £	Unit	Total rate £
90 ° mitre bend						
Sum of two sides 900mm	62.54	106.34	1.22	38.58	nr	144.92
Sum of two sides 1000mm	67.45	114.69	1.40	44.27	nr	158.96
Sum of two sides 1100mm	72.36	123.04	3.01	95.17	nr	218.21
Sum of two sides 1200mm	77.35	131.52	3.01	95.17	nr	226.70
Sum of two sides 1300mm	82.26	139.87	3.67	116.04	nr	255.92
Sum of two sides 1400mm	86.10	146.40	3.67	116.04	nr	262.45
Sum of two sides 1500mm	98.93	168.22	4.07	128.69	nr	296.91
Sum of two sides 1600mm	111.76	190.03	4.07	128.69	nr	318.72
Branch						
Sum of two sides 900mm	53.43	90.85	1.22	38.58	nr	129.43
Sum of two sides 1000mm	56.70	96.41	1.40	44.27	nr	140.68
Sum of two sides 1100mm	59.97	101.97	1.29	40.79	nr	142.76
Sum of two sides 1200mm	63.36	107.74	1.29	40.79	nr	148.52
Sum of two sides 1300mm	66.63	113.30	1.39	43.95	nr	157.25
Sum of two sides 1400mm	69.48	118.14	1.39	43.95	nr	162.09
Sum of two sides 1500mm	79.70	135.52	1.64	51.86	nr	187.38
Sum of two sides 1600mm	89.93	152.92	1.64	51.86	nr	204.77
Grille neck						
Sum of two sides 900mm	54.27	92.28	1.22	38.58	nr	130.86
Sum of two sides 1000mm	57.84	98.35	1.40	44.27	nr	142.62
Sum of two sides 1100mm	61.41	104.42	1.44	45.53	nr	149.95
Sum of two sides 1200mm	64.97	110.47	1.44	45.53	nr	156.01
Sum of two sides 1300mm	68.54	116.54	1.69	53.44	nr	169.98
Sum of two sides 1400mm	71.62	121.78	1.69	53.44	nr	175.22
Sum of two sides 1500mm	83.01	141.15	1.79	56.60	nr	197.75
Sum of two sides 1600mm	94.41	160.53	1.79	56.60	nr	217.13
Ductwork 801 to 1000mm longest side						
Sum of two sides 1100mm	37.95	64.53	1.37	43.32	m	107.85
Sum of two sides 1200mm	40.17	68.30	1.37	43.32	m	111.62
Sum of two sides 1300mm	42.39	72.08	1.40	44.27	m	116.35
Sum of two sides 1400mm	44.78	76.14	1.40	44.27	m	120.41
Sum of two sides 1500mm	47.00	79.92	1.48	46.80	m	126.71
Sum of two sides 1600mm	49.22	83.69	1.55	49.01	m	132.70
Sum of two sides 1700mm	51.44	87.47	1.55	49.01	m	136.48
Sum of two sides 1800mm	53.83	91.53	1.61	50.91	m	142.44
Sum of two sides 1900mm	56.05	95.31	1.61	50.91	m	146.21
Sum of two sides 2000mm	58.27	99.08	1.61	50.91	m	149.99
Extra over fittings; Ductwork 801 to 1000mm longest side						
End Cap						
Sum of two sides 1100mm	19.70	33.50	1.44	45.53	nr	79.03
Sum of two sides 1200mm	20.82	35.40	1.44	45.53	nr	80.93
Sum of two sides 1300mm	21.93	37.29	1.44	45.53	nr	82.82
Sum of two sides 1400mm	22.88	38.90	1.44	45.53	nr	84.44
Sum of two sides 1500mm	26.61	45.25	1.44	45.53	nr	90.78
Sum of two sides 1600mm	30.34	51.59	1.44	45.53	nr	97.12
Sum of two sides 1700mm	34.07	57.93	1.44	45.53	nr	103.46
Sum of two sides 1800mm	37.79	64.26	1.44	45.53	nr	109.79
Sum of two sides 1900mm	41.52	70.60	1.44	45.53	nr	116.13
Sum of two sides 2000mm	45.24	76.93	1.44	45.53	nr	122.46

U: VENTILATION/AIR CONDITIONING SYSTEMS

Item	Net Price £	Material £	Labour hours	Labour £	Unit	Total rate £
U10 : DUCTWORK : RECTANGULAR – CLASS B (cont'd)						
Galvanised sheet metal DW144 (cont'd)						
Extra over fittings; Ductwork 801 to 1000mm longest side (cont'd)						
Reducer						
Sum of two sides 1100mm	37.85	64.36	1.44	45.53	nr	109.89
Sum of two sides 1200mm	39.70	67.51	1.44	45.53	nr	113.04
Sum of two sides 1300mm	41.56	70.67	1.69	53.44	nr	124.10
Sum of two sides 1400mm	43.23	73.51	1.69	53.44	nr	126.94
Sum of two sides 1500mm	50.30	85.53	2.47	78.10	nr	163.63
Sum of two sides 1600mm	57.38	97.57	2.47	78.10	nr	175.67
Sum of two sides 1700mm	64.45	109.59	2.47	78.10	nr	187.69
Sum of two sides 1800mm	71.66	121.85	2.59	81.89	nr	203.74
Sum of two sides 1900mm	78.73	133.87	2.71	85.69	nr	219.56
Sum of two sides 2000mm	85.80	145.89	2.71	85.69	nr	231.58
Offset						
Sum of two sides 1100mm	58.01	98.64	1.44	45.53	nr	144.17
Sum of two sides 1200mm	59.10	100.49	1.44	45.53	nr	146.02
Sum of two sides 1300mm	60.02	102.06	1.69	53.44	nr	155.49
Sum of two sides 1400mm	60.40	102.70	1.69	53.44	nr	156.14
Sum of two sides 1500mm	68.78	116.95	2.47	78.10	nr	195.05
Sum of two sides 1600mm	76.98	130.90	2.47	78.10	nr	208.99
Sum of two sides 1700mm	85.01	144.55	2.59	81.89	nr	226.44
Sum of two sides 1800mm	92.91	157.98	2.61	82.53	nr	240.51
Sum of two sides 1900mm	101.82	173.13	2.71	85.69	nr	258.82
Sum of two sides 2000mm	109.34	185.92	2.71	85.69	nr	271.61
Square to round						
Sum of two sides 1100mm	39.92	67.88	1.44	45.53	nr	113.41
Sum of two sides 1200mm	42.02	71.45	1.44	45.53	nr	116.98
Sum of two sides 1300mm	44.12	75.02	1.69	53.44	nr	128.46
Sum of two sides 1400mm	45.84	77.95	1.69	53.44	nr	131.38
Sum of two sides 1500mm	54.46	92.60	2.38	75.25	nr	167.86
Sum of two sides 1600mm	63.08	107.26	2.38	75.25	nr	182.51
Sum of two sides 1700mm	71.70	121.92	2.55	80.63	nr	202.55
Sum of two sides 1800mm	80.34	136.61	2.55	80.63	nr	217.24
Sum of two sides 1900mm	88.97	151.28	2.83	89.48	nr	240.77
Sum of two sides 2000mm	97.59	165.94	2.83	89.48	nr	255.42
90 ° radius bend						
Sum of two sides 1100mm	30.16	51.28	1.44	45.53	nr	96.82
Sum of two sides 1200mm	32.44	55.16	1.44	45.53	nr	100.69
Sum of two sides 1300mm	34.73	59.05	1.69	53.44	nr	112.49
Sum of two sides 1400mm	36.76	62.51	1.69	53.44	nr	115.94
Sum of two sides 1500mm	44.26	75.26	2.11	66.72	nr	141.98
Sum of two sides 1600mm	51.77	88.03	2.11	66.72	nr	154.75
Sum of two sides 1700mm	59.27	100.78	2.26	71.46	nr	172.24
Sum of two sides 1800mm	66.85	113.67	2.26	71.46	nr	185.13
Sum of two sides 1900mm	72.95	124.04	2.48	78.42	nr	202.46
Sum of two sides 2000mm	80.44	136.78	2.48	78.42	nr	215.19

U: VENTILATION/AIR CONDITIONING SYSTEMS

Item	Net Price £	Material £	Labour hours	Labour £	Unit	Total rate £
45 ° bend						
Sum of two sides 1100mm	39.68	67.47	1.44	45.53	nr	**113.00**
Sum of two sides 1200mm	42.09	71.57	1.44	45.53	nr	**117.10**
Sum of two sides 1300mm	44.49	75.65	1.69	53.44	nr	**129.09**
Sum of two sides 1400mm	46.72	79.44	1.69	53.44	nr	**132.88**
Sum of two sides 1500mm	54.36	92.43	2.49	78.73	nr	**171.16**
Sum of two sides 1600mm	61.98	105.39	2.49	78.73	nr	**184.12**
Sum of two sides 1700mm	69.60	118.35	2.67	84.42	nr	**202.77**
Sum of two sides 1800mm	77.38	131.58	2.67	84.42	nr	**216.00**
Sum of two sides 1900mm	84.28	143.31	3.06	96.75	nr	**240.06**
Sum of two sides 2000mm	91.89	156.25	3.06	96.75	nr	**253.00**
90 ° mitre bend						
Sum of two sides 1100mm	61.49	104.56	1.44	45.53	nr	**150.09**
Sum of two sides 1200mm	65.62	111.58	1.44	45.53	nr	**157.11**
Sum of two sides 1300mm	69.73	118.57	1.69	53.44	nr	**172.00**
Sum of two sides 1400mm	73.42	124.84	1.69	53.44	nr	**178.28**
Sum of two sides 1500mm	85.37	145.16	4.07	128.69	nr	**273.85**
Sum of two sides 1600mm	97.32	165.48	4.07	128.69	nr	**294.17**
Sum of two sides 1700mm	109.27	185.80	2.80	88.53	nr	**274.33**
Sum of two sides 1800mm	121.26	206.19	2.67	84.42	nr	**290.61**
Sum of two sides 1900mm	131.66	223.87	2.95	93.28	nr	**317.15**
Sum of two sides 2000mm	143.65	244.26	2.95	93.28	nr	**337.54**
Branch						
Sum of two sides 1100mm	60.10	102.19	1.44	45.53	nr	**147.72**
Sum of two sides 1200mm	63.36	107.74	1.44	45.53	nr	**153.27**
Sum of two sides 1300mm	66.63	113.30	1.64	51.86	nr	**165.15**
Sum of two sides 1400mm	69.60	118.35	1.64	51.86	nr	**170.20**
Sum of two sides 1500mm	79.83	135.74	1.64	51.86	nr	**187.60**
Sum of two sides 1600mm	90.05	153.12	1.64	51.86	nr	**204.97**
Sum of two sides 1700mm	100.28	170.51	1.69	53.44	nr	**223.95**
Sum of two sides 1800mm	110.64	188.13	1.69	53.44	nr	**241.57**
Sum of two sides 1900mm	120.86	205.51	1.85	58.50	nr	**264.00**
Sum of two sides 2000mm	131.09	222.90	1.85	58.50	nr	**281.40**
Grille neck						
Sum of two sides 1100mm	61.41	104.42	1.44	45.53	nr	**149.95**
Sum of two sides 1200mm	64.97	110.47	1.44	45.53	nr	**156.01**
Sum of two sides 1300mm	68.54	116.54	1.69	53.44	nr	**169.98**
Sum of two sides 1400mm	71.62	121.78	1.69	53.44	nr	**175.22**
Sum of two sides 1500mm	83.01	141.15	1.79	56.60	nr	**197.75**
Sum of two sides 1600mm	94.41	160.53	1.79	56.60	nr	**217.13**
Sum of two sides 1700mm	105.80	179.90	1.86	58.81	nr	**238.71**
Sum of two sides 1800mm	117.19	199.27	2.02	63.87	nr	**263.14**
Sum of two sides 1900mm	128.59	218.65	2.02	63.87	nr	**282.52**
Sum of two sides 2000mm	139.98	238.02	2.02	63.87	nr	**301.89**

U: VENTILATION/AIR CONDITIONING SYSTEMS

Item	Net Price £	Material £	Labour hours	Labour £	Unit	Total rate £
U10 : DUCTWORK : RECTANGULAR – CLASS B (cont'd)						
Galvanised sheet metal DW144 (cont'd)						
Ductwork 1001 to 1250mm longest side						
Sum of two sides 1300mm	50.76	86.31	1.40	44.27	m	130.58
Sum of two sides 1400mm	53.48	90.94	1.40	44.27	m	135.20
Sum of two sides 1500mm	56.02	95.26	1.48	46.80	m	142.05
Sum of two sides 1600mm	58.74	99.88	1.55	49.01	m	148.89
Sum of two sides 1700mm	61.29	104.22	1.55	49.01	m	153.23
Sum of two sides 1800mm	63.83	108.54	1.61	50.91	m	159.44
Sum of two sides 1900mm	66.37	112.85	1.61	50.91	m	163.76
Sum of two sides 2000mm	69.09	117.48	1.61	50.91	m	168.39
Sum of two sides 2100mm	71.64	121.82	2.17	68.61	m	190.43
Sum of two sides 2200mm	74.18	126.13	2.19	69.25	m	195.38
Sum of two sides 2300mm	77.09	131.08	2.19	69.25	m	200.33
Sum of two sides 2400mm	79.63	135.40	2.38	75.25	m	210.65
Sum of two sides 2500mm	82.35	140.03	2.38	75.25	m	215.28
Extra over fittings; Ductwork 1001 to 1250mm longest side						
End Cap						
Sum of two sides 1300mm	22.33	37.97	1.69	53.44	nr	91.41
Sum of two sides 1400mm	23.31	39.64	1.69	53.44	nr	93.07
Sum of two sides 1500mm	27.07	46.03	1.69	53.44	nr	99.47
Sum of two sides 1600mm	30.82	52.41	1.69	53.44	nr	105.84
Sum of two sides 1700mm	34.58	58.80	1.69	53.44	nr	112.24
Sum of two sides 1800mm	38.34	65.19	1.69	53.44	nr	118.63
Sum of two sides 1900mm	42.09	71.57	1.69	53.44	nr	125.01
Sum of two sides 2000mm	45.85	77.96	1.69	53.44	nr	131.40
Sum of two sides 2100mm	49.61	84.36	1.69	53.44	nr	137.79
Sum of two sides 2200mm	53.36	90.73	1.69	53.44	nr	144.17
Sum of two sides 2300mm	57.12	97.13	1.69	53.44	nr	150.56
Sum of two sides 2400mm	60.88	103.52	1.69	53.44	nr	156.96
Sum of two sides 2500mm	64.63	109.90	1.69	53.44	nr	163.33
Reducer						
Sum of two sides 1300mm	40.33	68.58	1.69	53.44	nr	122.01
Sum of two sides 1400mm	41.84	71.14	1.69	53.44	nr	124.58
Sum of two sides 1500mm	48.91	83.17	2.47	78.10	nr	161.26
Sum of two sides 1600mm	56.11	95.41	2.47	78.10	nr	173.51
Sum of two sides 1700mm	63.16	107.40	2.47	78.10	nr	185.50
Sum of two sides 1800mm	70.23	119.42	2.59	81.89	nr	201.31
Sum of two sides 1900mm	77.28	131.41	2.71	85.69	nr	217.09
Sum of two sides 2000mm	84.48	143.65	2.59	81.89	nr	225.54
Sum of two sides 2100mm	91.55	155.67	2.92	92.33	nr	248.00
Sum of two sides 2200mm	98.60	167.66	2.92	92.33	nr	259.99
Sum of two sides 2300mm	105.80	179.90	2.92	92.33	nr	272.23
Sum of two sides 2400mm	112.86	191.90	3.12	98.65	nr	290.56
Sum of two sides 2500mm	120.06	204.15	3.12	98.65	nr	302.80

U: VENTILATION/AIR CONDITIONING SYSTEMS

Item	Net Price £	Material £	Labour hours	Labour £	Unit	Total rate £
Offset						
Sum of two sides 1300mm	80.64	137.12	1.69	53.44	nr	190.55
Sum of two sides 1400mm	83.35	141.73	1.69	53.44	nr	195.16
Sum of two sides 1500mm	93.10	158.31	2.47	78.10	nr	236.40
Sum of two sides 1600mm	102.78	174.77	2.47	78.10	nr	252.86
Sum of two sides 1700mm	112.14	190.68	2.59	81.89	nr	272.57
Sum of two sides 1800mm	121.34	206.32	2.61	82.53	nr	288.85
Sum of two sides 1900mm	130.35	221.64	2.71	85.69	nr	307.33
Sum of two sides 2000mm	139.23	236.74	2.71	85.69	nr	322.43
Sum of two sides 2100mm	147.86	251.42	2.92	92.33	nr	343.75
Sum of two sides 2200mm	157.83	268.37	3.26	103.08	nr	371.45
Sum of two sides 2300mm	169.31	287.89	3.26	103.08	nr	390.97
Sum of two sides 2400mm	180.75	307.34	3.48	110.03	nr	417.38
Sum of two sides 2500mm	192.23	326.86	3.48	110.03	nr	436.90
Square to round						
Sum of two sides 1300mm	42.53	72.32	1.69	53.44	nr	125.75
Sum of two sides 1400mm	44.21	75.17	1.69	53.44	nr	128.61
Sum of two sides 1500mm	52.83	89.83	2.38	75.25	nr	165.08
Sum of two sides 1600mm	61.45	104.49	2.38	75.25	nr	179.74
Sum of two sides 1700mm	70.06	119.13	2.55	80.63	nr	199.76
Sum of two sides 1800mm	78.67	133.77	2.55	80.63	nr	214.40
Sum of two sides 1900mm	87.27	148.39	2.83	89.48	nr	237.87
Sum of two sides 2000mm	95.91	163.08	2.83	89.48	nr	252.57
Sum of two sides 2100mm	104.51	177.71	3.85	121.73	nr	299.44
Sum of two sides 2200mm	113.12	192.35	4.18	132.17	nr	324.52
Sum of two sides 2300mm	121.75	207.02	4.22	133.43	nr	340.45
Sum of two sides 2400mm	130.36	221.66	4.68	147.98	nr	369.64
Sum of two sides 2500mm	138.98	236.32	4.70	148.61	nr	384.93
90 ° radius bend						
Sum of two sides 1300mm	28.31	48.14	1.69	53.44	nr	101.57
Sum of two sides 1400mm	28.82	49.01	1.69	53.44	nr	102.44
Sum of two sides 1500mm	36.53	62.11	2.11	66.72	nr	128.83
Sum of two sides 1600mm	44.28	75.29	2.11	66.72	nr	142.01
Sum of two sides 1700mm	52.00	88.42	2.19	69.25	nr	157.67
Sum of two sides 1800mm	59.71	101.53	2.19	69.25	nr	170.78
Sum of two sides 1900mm	67.42	114.64	2.48	78.42	nr	193.06
Sum of two sides 2000mm	75.17	127.82	2.26	71.46	nr	199.28
Sum of two sides 2100mm	82.89	140.94	2.48	78.42	nr	219.36
Sum of two sides 2200mm	90.60	154.05	2.48	78.42	nr	232.47
Sum of two sides 2300mm	94.78	161.16	2.48	78.42	nr	239.58
Sum of two sides 2400mm	102.48	174.25	3.90	123.31	nr	297.57
Sum of two sides 2500mm	110.21	187.40	3.90	123.31	nr	310.71
45 ° bend						
Sum of two sides 1300mm	41.43	70.45	1.69	53.44	nr	123.88
Sum of two sides 1400mm	42.77	72.73	1.69	53.44	nr	126.16
Sum of two sides 1500mm	50.51	85.89	2.49	78.73	nr	164.62
Sum of two sides 1600mm	58.39	99.29	2.49	78.73	nr	178.02
Sum of two sides 1700mm	66.13	112.45	2.67	84.42	nr	196.87
Sum of two sides 1800mm	73.88	125.62	2.67	84.42	nr	210.05
Sum of two sides 1900mm	81.62	138.78	3.06	96.75	nr	235.54
Sum of two sides 2000mm	89.50	152.18	3.06	96.75	nr	248.94
Sum of two sides 2100mm	97.24	165.34	4.05	128.06	nr	293.40
Sum of two sides 2200mm	104.98	178.51	4.05	128.06	nr	306.56
Sum of two sides 2300mm	111.01	188.76	4.39	138.81	nr	327.57
Sum of two sides 2400mm	118.75	201.92	4.85	153.35	nr	355.27
Sum of two sides 2500mm	126.62	215.30	4.85	153.35	nr	368.66

U: VENTILATION/AIR CONDITIONING SYSTEMS

Item	Net Price £	Material £	Labour hours	Labour £	Unit	Total rate £
U10 : DUCTWORK : RECTANGULAR – CLASS B (cont'd)						
Galvanised sheet metal DW144 (cont'd)						
Extra over fittings; Ductwork 1001 to 1250mm longest side (cont'd)						
90 ° mitre bend						
Sum of two sides 1300mm	86.06	146.33	1.69	53.44	nr	199.77
Sum of two sides 1400mm	89.57	152.30	1.69	53.44	nr	205.74
Sum of two sides 1500mm	103.59	176.14	2.80	88.53	nr	264.68
Sum of two sides 1600mm	117.64	200.03	2.80	88.53	nr	288.57
Sum of two sides 1700mm	131.67	223.89	2.95	93.28	nr	317.17
Sum of two sides 1800mm	145.70	247.75	2.95	93.28	nr	341.02
Sum of two sides 1900mm	159.73	271.60	4.05	128.06	nr	399.66
Sum of two sides 2000mm	173.76	295.46	4.05	128.06	nr	423.52
Sum of two sides 2100mm	187.80	319.33	4.07	128.69	nr	448.02
Sum of two sides 2200mm	201.82	343.17	4.07	128.69	nr	471.86
Sum of two sides 2300mm	211.86	360.24	4.39	138.81	nr	499.05
Sum of two sides 2400mm	225.96	384.22	4.85	153.35	nr	537.57
Sum of two sides 2500mm	240.06	408.19	4.85	153.35	nr	561.55
Branch						
Sum of two sides 1300mm	68.18	115.93	1.44	45.53	nr	161.46
Sum of two sides 1400mm	71.11	120.91	1.44	45.53	nr	166.45
Sum of two sides 1500mm	81.43	138.46	1.64	51.86	nr	190.32
Sum of two sides 1600mm	91.88	156.23	1.64	51.86	nr	208.09
Sum of two sides 1700mm	102.19	173.76	1.64	51.86	nr	225.62
Sum of two sides 1800mm	112.51	191.31	1.64	51.86	nr	243.17
Sum of two sides 1900mm	122.83	208.86	1.69	53.44	nr	262.29
Sum of two sides 2000mm	133.27	226.61	1.69	53.44	nr	280.05
Sum of two sides 2100mm	143.59	244.16	1.85	58.50	nr	302.65
Sum of two sides 2200mm	153.91	261.71	1.85	58.50	nr	320.20
Sum of two sides 2300mm	164.35	279.46	2.61	82.53	nr	361.98
Sum of two sides 2400mm	174.67	297.01	2.61	82.53	nr	379.53
Sum of two sides 2500mm	185.11	314.76	2.61	82.53	nr	397.28
Grille neck						
Sum of two sides 1300mm	70.51	119.89	1.79	56.60	nr	176.49
Sum of two sides 1400mm	73.74	125.39	1.79	56.60	nr	181.98
Sum of two sides 1500mm	85.29	145.03	1.79	56.60	nr	201.62
Sum of two sides 1600mm	96.83	164.65	1.79	56.60	nr	221.25
Sum of two sides 1700mm	108.38	184.29	1.86	58.81	nr	243.10
Sum of two sides 1800mm	119.93	203.93	2.02	63.87	nr	267.80
Sum of two sides 1900mm	131.47	223.55	2.02	63.87	nr	287.42
Sum of two sides 2000mm	143.02	243.19	2.02	63.87	nr	307.06
Sum of two sides 2100mm	154.57	262.83	2.61	82.53	nr	345.35
Sum of two sides 2200mm	166.11	282.45	2.61	82.53	nr	364.98
Sum of two sides 2300mm	177.66	302.09	2.61	82.53	nr	384.62
Sum of two sides 2400mm	189.21	321.73	2.88	91.06	nr	412.79
Sum of two sides 2500mm	200.75	341.35	2.88	91.06	nr	432.41

U: VENTILATION/AIR CONDITIONING SYSTEMS

Item	Net Price £	Material £	Labour hours	Labour £	Unit	Total rate £
Ductwork 1251 to 1600mm longest side						
Sum of two sides 1700mm	77.03	130.98	1.55	49.01	m	179.99
Sum of two sides 1800mm	79.97	135.98	1.61	50.91	m	186.89
Sum of two sides 1900mm	82.97	141.08	1.61	50.91	m	191.99
Sum of two sides 2000mm	85.80	145.89	1.61	50.91	m	196.80
Sum of two sides 2100mm	88.62	150.69	2.17	68.61	m	219.30
Sum of two sides 2200mm	91.44	155.48	2.19	69.25	m	224.73
Sum of two sides 2300mm	94.43	160.57	2.19	69.25	m	229.81
Sum of two sides 2400mm	97.39	165.60	2.38	75.25	m	240.85
Sum of two sides 2500mm	100.21	170.40	2.38	75.25	m	245.65
Sum of two sides 2600mm	103.04	175.21	2.64	83.47	m	258.68
Sum of two sides 2700mm	105.85	179.99	2.66	84.11	m	264.09
Sum of two sides 2800mm	108.83	185.05	2.95	93.28	m	278.33
Sum of two sides 2900mm	121.39	206.41	2.96	93.59	m	300.00
Sum of two sides 3000mm	124.21	211.20	3.15	99.60	m	310.80
Sum of two sides 3100mm	127.18	216.25	3.15	99.60	m	315.85
Sum of two sides 3200mm	130.00	221.05	3.18	100.55	m	321.60
Extra over fittings; Ductwork 1251 to 1600mm longest side						
End Cap						
Sum of two sides 1700mm	34.58	58.80	0.58	18.34	nr	77.14
Sum of two sides 1800mm	38.34	65.19	0.58	18.34	nr	83.53
Sum of two sides 1900mm	42.09	71.57	0.58	18.34	nr	89.91
Sum of two sides 2000mm	45.85	77.96	0.58	18.34	nr	96.30
Sum of two sides 2100mm	49.61	84.36	0.87	27.51	nr	111.86
Sum of two sides 2200mm	53.36	90.73	0.87	27.51	nr	118.24
Sum of two sides 2300mm	57.12	97.13	0.87	27.51	nr	124.63
Sum of two sides 2400mm	60.88	103.52	0.87	27.51	nr	131.03
Sum of two sides 2500mm	64.63	109.90	0.87	27.51	nr	137.40
Sum of two sides 2600mm	68.39	116.29	0.87	27.51	nr	143.80
Sum of two sides 2700mm	72.14	122.67	0.87	27.51	nr	150.17
Sum of two sides 2800mm	75.89	129.04	1.16	36.68	nr	165.72
Sum of two sides 2900mm	79.65	135.44	1.16	36.68	nr	172.11
Sum of two sides 3000mm	127.78	434.55	1.73	54.70	nr	489.25
Sum of two sides 3100mm	86.39	146.90	1.80	56.91	nr	203.81
Sum of two sides 3200mm	89.09	151.49	1.80	56.91	nr	208.40
Reducer						
Sum of two sides 1700mm	44.18	75.12	2.47	78.10	nr	153.22
Sum of two sides 1800mm	50.83	86.43	2.59	81.89	nr	168.32
Sum of two sides 1900mm	57.54	97.84	2.71	85.69	nr	183.53
Sum of two sides 2000mm	64.15	109.08	2.71	85.69	nr	194.77
Sum of two sides 2100mm	70.75	120.30	2.92	92.33	nr	212.63
Sum of two sides 2200mm	77.36	131.54	2.92	92.33	nr	223.87
Sum of two sides 2300mm	84.07	142.95	2.92	92.33	nr	235.28
Sum of two sides 2400mm	90.72	154.26	3.12	98.65	nr	252.91
Sum of two sides 2500mm	97.32	165.48	3.12	98.65	nr	264.13
Sum of two sides 2600mm	103.93	176.72	3.12	98.65	nr	275.37
Sum of two sides 2700mm	110.53	187.94	3.12	98.65	nr	286.59
Sum of two sides 2800mm	117.18	199.25	3.95	124.90	nr	324.15
Sum of two sides 2900mm	116.25	197.67	3.97	125.53	nr	323.20
Sum of two sides 3000mm	122.85	208.89	4.52	142.92	nr	351.81
Sum of two sides 3100mm	127.78	217.27	4.52	142.92	nr	360.19
Sum of two sides 3200mm	132.28	224.93	4.52	142.92	nr	367.85

U: VENTILATION/AIR CONDITIONING SYSTEMS

Item	Net Price £	Material £	Labour hours	Labour £	Unit	Total rate £
U10 : DUCTWORK : RECTANGULAR – CLASS B (cont'd)						
Galvanised sheet metal DW144 (cont'd)						
Extra over fittings; Ductwork 1251 to 1600mm longest side (cont'd)						
Offset						
Sum of two sides 1700mm	110.18	187.35	2.59	81.89	nr	269.24
Sum of two sides 1800mm	122.32	207.99	2.61	82.53	nr	290.52
Sum of two sides 1900mm	130.51	221.92	2.71	85.69	nr	307.60
Sum of two sides 2000mm	138.37	235.28	2.71	85.69	nr	320.97
Sum of two sides 2100mm	146.00	248.26	2.92	92.33	nr	340.58
Sum of two sides 2200mm	153.40	260.84	3.26	103.08	nr	363.92
Sum of two sides 2300mm	160.63	273.13	3.26	103.08	nr	376.21
Sum of two sides 2400mm	172.22	292.84	3.47	109.72	nr	402.56
Sum of two sides 2500mm	179.05	304.45	3.48	110.03	nr	414.49
Sum of two sides 2600mm	190.48	323.89	3.49	110.35	nr	434.24
Sum of two sides 2700mm	201.90	343.31	3.50	110.67	nr	453.97
Sum of two sides 2800mm	213.39	362.84	4.34	137.23	nr	500.07
Sum of two sides 2900mm	213.45	362.95	4.76	150.51	nr	513.45
Sum of two sides 3000mm	224.88	382.38	5.32	168.21	nr	550.60
Sum of two sides 3100mm	233.79	397.53	5.35	169.16	nr	566.69
Sum of two sides 3200mm	242.06	411.59	5.35	169.16	nr	580.76
Square to round						
Sum of two sides 1700mm	51.08	86.86	2.55	80.63	nr	167.48
Sum of two sides 1800mm	59.27	100.78	2.55	80.63	nr	181.41
Sum of two sides 1900mm	67.41	114.62	2.83	89.48	nr	204.10
Sum of two sides 2000mm	75.56	128.48	2.83	89.48	nr	217.96
Sum of two sides 2100mm	83.72	142.36	3.85	121.73	nr	264.09
Sum of two sides 2200mm	91.88	156.23	4.18	132.17	nr	288.40
Sum of two sides 2300mm	100.01	170.06	4.22	133.43	nr	303.49
Sum of two sides 2400mm	108.21	184.00	4.68	147.98	nr	331.98
Sum of two sides 2500mm	116.37	197.87	4.70	148.61	nr	346.48
Sum of two sides 2600mm	124.52	211.73	4.70	148.61	nr	360.34
Sum of two sides 2700mm	132.68	225.61	4.71	148.93	nr	374.53
Sum of two sides 2800mm	140.87	239.53	8.19	258.96	nr	498.49
Sum of two sides 2900mm	141.37	240.38	8.62	272.56	nr	512.94
Sum of two sides 3000mm	149.53	254.26	8.75	276.67	nr	530.93
Sum of two sides 3100mm	155.61	264.60	8.75	276.67	nr	541.26
Sum of two sides 3200mm	161.14	274.00	8.75	276.67	nr	550.67
90 ° radius bend						
Sum of two sides 1700mm	109.57	186.31	2.19	69.25	nr	255.56
Sum of two sides 1800mm	119.74	203.60	2.19	69.25	nr	272.85
Sum of two sides 1900mm	135.06	229.65	2.26	71.46	nr	301.11
Sum of two sides 2000mm	150.12	255.26	2.26	71.46	nr	326.72
Sum of two sides 2100mm	165.19	280.89	2.48	78.42	nr	359.30
Sum of two sides 2200mm	180.25	306.49	2.48	78.42	nr	384.91
Sum of two sides 2300mm	195.57	332.54	2.48	78.42	nr	410.96
Sum of two sides 2400mm	205.34	349.16	3.90	123.31	nr	472.47
Sum of two sides 2500mm	220.38	374.73	3.90	123.31	nr	498.04
Sum of two sides 2600mm	235.41	400.29	4.26	134.70	nr	534.98
Sum of two sides 2700mm	250.44	425.84	4.55	143.87	nr	569.71
Sum of two sides 2800mm	259.90	441.93	4.55	143.87	nr	585.80
Sum of two sides 2900mm	260.66	443.22	6.87	217.22	nr	660.44
Sum of two sides 3000mm	275.67	468.74	7.00	221.33	nr	690.08

U: VENTILATION/AIR CONDITIONING SYSTEMS

Item	Net Price £	Material £	Labour hours	Labour £	Unit	Total rate £
Sum of two sides 3100mm	287.33	488.57	7.00	221.33	nr	709.90
Sum of two sides 3200mm	298.12	506.92	7.00	221.33	nr	728.25
45 ° bend						
Sum of two sides 1700mm	52.65	89.53	2.67	84.42	nr	173.95
Sum of two sides 1800mm	57.66	98.04	2.67	84.42	nr	182.47
Sum of two sides 1900mm	65.32	111.07	3.06	96.75	nr	207.82
Sum of two sides 2000mm	72.84	123.86	3.06	96.75	nr	220.61
Sum of two sides 2100mm	80.38	136.68	4.05	128.06	nr	264.73
Sum of two sides 2200mm	87.91	149.48	4.05	128.06	nr	277.54
Sum of two sides 2300mm	95.55	162.47	4.39	138.81	nr	301.28
Sum of two sides 2400mm	100.37	170.67	4.85	153.35	nr	324.02
Sum of two sides 2500mm	107.89	183.45	4.85	153.35	nr	336.81
Sum of two sides 2600mm	115.40	196.22	4.87	153.99	nr	350.21
Sum of two sides 2700mm	122.91	208.99	4.87	153.99	nr	362.98
Sum of two sides 2800mm	127.57	216.92	8.81	278.57	nr	495.48
Sum of two sides 2900mm	127.46	216.73	8.81	278.57	nr	495.30
Sum of two sides 3000mm	134.96	229.48	9.31	294.37	nr	523.86
Sum of two sides 3100mm	140.78	239.38	9.31	294.37	nr	533.75
Sum of two sides 3200mm	146.18	248.56	9.39	296.90	nr	545.47
90 ° mitre bend						
Sum of two sides 1700mm	115.13	195.76	2.67	84.42	nr	280.19
Sum of two sides 1800mm	122.85	208.89	2.80	88.53	nr	297.43
Sum of two sides 1900mm	137.06	233.05	2.95	93.28	nr	326.33
Sum of two sides 2000mm	151.30	257.27	2.95	93.28	nr	350.54
Sum of two sides 2100mm	165.54	281.48	4.05	128.06	nr	409.54
Sum of two sides 2200mm	179.78	305.69	4.05	128.06	nr	433.75
Sum of two sides 2300mm	193.98	329.84	4.39	138.81	nr	468.65
Sum of two sides 2400mm	201.63	342.85	4.85	153.35	nr	496.20
Sum of two sides 2500mm	215.91	367.13	4.85	153.35	nr	520.48
Sum of two sides 2600mm	230.21	391.44	4.87	153.99	nr	545.43
Sum of two sides 2700mm	244.49	415.73	4.87	153.99	nr	569.71
Sum of two sides 2800mm	252.01	428.51	8.81	278.57	nr	707.08
Sum of two sides 2900mm	253.02	430.23	14.81	468.28	nr	898.51
Sum of two sides 3000mm	267.35	454.60	15.20	480.61	nr	935.21
Sum of two sides 3100mm	279.08	474.54	15.60	493.26	nr	967.80
Sum of two sides 3200mm	290.26	493.55	15.60	493.26	nr	986.81
Branch						
Sum of two sides 1700mm	102.19	173.76	1.69	53.44	nr	227.20
Sum of two sides 1800mm	112.51	191.31	1.69	53.44	nr	244.75
Sum of two sides 1900mm	122.95	209.06	1.85	58.50	nr	267.56
Sum of two sides 2000mm	133.27	226.61	1.85	58.50	nr	285.11
Sum of two sides 2100mm	143.59	244.16	2.61	82.53	nr	326.68
Sum of two sides 2200mm	153.91	261.71	2.61	82.53	nr	344.23
Sum of two sides 2300mm	164.35	279.46	2.61	82.53	nr	361.98
Sum of two sides 2400mm	200.75	341.35	2.88	91.06	nr	432.41
Sum of two sides 2500mm	184.99	314.55	2.88	91.06	nr	405.62
Sum of two sides 2600mm	195.31	332.10	2.88	91.06	nr	423.16
Sum of two sides 2700mm	205.62	349.63	2.88	91.06	nr	440.70
Sum of two sides 2800mm	215.94	367.18	3.94	124.58	nr	491.76
Sum of two sides 2900mm	226.38	384.93	3.94	124.58	nr	509.51
Sum of two sides 3000mm	236.70	402.48	4.83	152.72	nr	555.20
Sum of two sides 3100mm	244.95	416.51	4.83	152.72	nr	569.23
Sum of two sides 3200mm	252.46	429.28	4.83	152.72	nr	582.00

U: VENTILATION/AIR CONDITIONING SYSTEMS

Item	Net Price £	Material £	Labour hours	Labour £	Unit	Total rate £
U10 : DUCTWORK : RECTANGULAR – CLASS B (cont'd)						
Galvanised sheet metal DW144 (cont'd)						
Extra over fittings; Ductwork 1251 to 1600mm longest side (cont'd)						
Grille neck						
Sum of two sides 1700mm	108.38	184.29	1.86	58.81	nr	243.10
Sum of two sides 1800mm	119.93	203.93	2.02	63.87	nr	267.80
Sum of two sides 1900mm	131.47	223.55	2.02	63.87	nr	287.42
Sum of two sides 2000mm	143.02	243.19	2.02	63.87	nr	307.06
Sum of two sides 2100mm	154.57	262.83	2.61	82.53	nr	345.35
Sum of two sides 2200mm	166.11	282.45	2.61	82.53	nr	364.98
Sum of two sides 2300mm	177.66	302.09	2.61	82.53	nr	384.62
Sum of two sides 2400mm	189.21	321.73	2.88	91.06	nr	412.79
Sum of two sides 2500mm	200.75	341.35	2.88	91.06	nr	432.41
Sum of two sides 2600mm	212.30	360.99	2.88	91.06	nr	452.05
Sum of two sides 2700mm	223.85	380.63	2.88	91.06	nr	471.69
Sum of two sides 2800mm	235.39	400.25	3.94	124.58	nr	524.83
Sum of two sides 2900mm	246.94	419.89	4.12	130.27	nr	550.16
Sum of two sides 3000mm	258.49	439.53	5.00	158.10	nr	597.63
Sum of two sides 3100mm	267.70	455.19	5.00	158.10	nr	613.29
Sum of two sides 3200mm	276.08	469.44	5.00	158.10	nr	627.54
Ductwork 1601 to 2000mm longest side						
Sum of two sides 2100mm	99.82	169.73	2.17	68.61	m	238.35
Sum of two sides 2200mm	102.93	175.02	2.17	68.61	m	243.63
Sum of two sides 2300mm	106.09	180.39	2.19	69.25	m	249.64
Sum of two sides 2400mm	109.15	185.60	2.38	75.25	m	260.85
Sum of two sides 2500mm	112.31	190.97	2.38	75.25	m	266.22
Sum of two sides 2600mm	115.29	196.04	2.64	83.47	m	279.51
Sum of two sides 2700mm	118.27	201.10	2.66	84.11	m	285.21
Sum of two sides 2800mm	121.25	206.17	2.95	93.28	m	299.45
Sum of two sides 2900mm	124.41	211.54	2.96	93.59	m	305.14
Sum of two sides 3000mm	127.39	216.61	2.96	93.59	m	310.20
Sum of two sides 3100mm	137.65	234.06	2.96	93.59	m	327.65
Sum of two sides 3200mm	140.63	239.12	3.15	99.60	m	338.72
Sum of two sides 3300mm	153.55	261.09	3.15	99.60	m	360.69
Sum of two sides 3400mm	156.52	266.14	3.15	99.60	m	365.74
Sum of two sides 3500mm	159.50	271.21	3.15	99.60	m	370.81
Sum of two sides 3600mm	162.49	276.29	3.18	100.55	m	376.84
Sum of two sides 3700mm	165.65	281.67	3.18	100.55	m	382.22
Sum of two sides 3800mm	168.62	286.72	3.18	100.55	m	387.27
Sum of two sides 3900mm	171.60	291.79	3.18	100.55	m	392.33
Sum of two sides 4000mm	174.59	296.87	3.18	100.55	m	397.42
Extra over fittings; Ductwork 1601 to 2000mm longest side						
End Cap						
Sum of two sides 2100mm	49.61	84.36	0.87	27.51	nr	111.86
Sum of two sides 2200mm	53.36	90.73	0.87	27.51	nr	118.24
Sum of two sides 2300mm	57.12	97.13	0.87	27.51	nr	124.63
Sum of two sides 2400mm	60.88	103.52	0.87	27.51	nr	131.03
Sum of two sides 2500mm	64.63	109.90	0.87	27.51	nr	137.40
Sum of two sides 2600mm	68.39	116.29	0.87	27.51	nr	143.80

U: VENTILATION/AIR CONDITIONING SYSTEMS

Item	Net Price £	Material £	Labour hours	Labour £	Unit	Total rate £
Sum of two sides 2700mm	72.14	122.67	0.87	27.51	nr	**150.17**
Sum of two sides 2800mm	75.89	129.04	1.16	36.68	nr	**165.72**
Sum of two sides 2900mm	79.65	135.44	1.16	36.68	nr	**172.11**
Sum of two sides 3000mm	83.41	141.83	1.73	54.70	nr	**196.53**
Sum of two sides 3100mm	86.39	146.90	1.80	56.91	nr	**203.81**
Sum of two sides 3200mm	89.09	151.49	1.80	56.91	nr	**208.40**
Sum of two sides 3300mm	91.80	156.09	1.80	56.91	nr	**213.01**
Sum of two sides 3400mm	94.49	160.67	1.80	56.91	nr	**217.58**
Sum of two sides 3500mm	97.20	165.28	1.80	56.91	nr	**222.19**
Sum of two sides 3600mm	99.90	169.87	1.80	56.91	nr	**226.78**
Sum of two sides 3700mm	102.61	174.48	1.80	56.91	nr	**231.39**
Sum of two sides 3800mm	105.31	179.07	1.80	56.91	nr	**235.98**
Sum of two sides 3900mm	108.01	183.66	1.80	56.91	nr	**240.57**
Sum of two sides 4000mm	110.72	188.27	1.80	56.91	nr	**245.18**
Reducer						
Sum of two sides 2100mm	69.98	118.99	2.61	82.53	nr	**201.52**
Sum of two sides 2200mm	76.50	130.08	2.61	82.53	nr	**212.61**
Sum of two sides 2300mm	83.11	141.32	2.61	82.53	nr	**223.84**
Sum of two sides 2400mm	89.53	152.24	2.88	91.06	nr	**243.30**
Sum of two sides 2500mm	96.15	163.49	3.12	98.65	nr	**262.14**
Sum of two sides 2600mm	102.66	174.56	3.12	98.65	nr	**273.21**
Sum of two sides 2700mm	109.17	185.63	3.12	98.65	nr	**284.28**
Sum of two sides 2800mm	115.69	196.72	3.95	124.90	nr	**321.61**
Sum of two sides 2900mm	122.30	207.96	3.97	125.53	nr	**333.48**
Sum of two sides 3000mm	128.82	219.04	4.52	142.92	nr	**361.96**
Sum of two sides 3100mm	127.95	217.56	4.52	142.92	nr	**360.48**
Sum of two sides 3200mm	132.35	225.05	4.52	142.92	nr	**367.96**
Sum of two sides 3300mm	129.07	219.47	4.52	142.92	nr	**362.39**
Sum of two sides 3400mm	133.47	226.95	4.52	142.92	nr	**369.87**
Sum of two sides 3500mm	137.87	234.43	4.52	142.92	nr	**377.35**
Sum of two sides 3600mm	142.28	241.93	4.52	142.92	nr	**384.85**
Sum of two sides 3700mm	153.62	261.21	4.52	142.92	nr	**404.13**
Sum of two sides 3800mm	158.03	268.71	4.52	142.92	nr	**411.63**
Sum of two sides 3900mm	162.43	276.19	4.52	142.92	nr	**419.11**
Sum of two sides 4000mm	166.84	283.69	4.52	142.92	nr	**426.61**
Offset						
Sum of two sides 2100mm	163.56	278.11	2.61	82.53	nr	**360.64**
Sum of two sides 2200mm	175.81	298.94	2.61	82.53	nr	**381.47**
Sum of two sides 2300mm	182.83	310.88	2.61	82.53	nr	**393.41**
Sum of two sides 2400mm	200.30	340.59	2.88	91.06	nr	**431.65**
Sum of two sides 2500mm	207.09	352.13	3.48	110.03	nr	**462.17**
Sum of two sides 2600mm	213.50	363.03	3.49	110.35	nr	**473.38**
Sum of two sides 2700mm	219.66	373.51	3.50	110.67	nr	**484.17**
Sum of two sides 2800mm	225.57	383.55	4.34	137.23	nr	**520.78**
Sum of two sides 2900mm	231.28	393.26	4.76	150.51	nr	**543.77**
Sum of two sides 3000mm	236.68	402.45	5.32	168.21	nr	**570.66**
Sum of two sides 3100mm	237.15	403.25	5.35	169.16	nr	**572.41**
Sum of two sides 3200mm	245.49	417.43	5.35	169.16	nr	**586.59**
Sum of two sides 3300mm	242.23	411.88	5.35	169.16	nr	**581.05**
Sum of two sides 3400mm	250.57	426.06	5.35	169.16	nr	**595.23**
Sum of two sides 3500mm	258.91	440.25	5.35	169.16	nr	**609.41**
Sum of two sides 3600mm	267.24	454.41	5.35	169.16	nr	**623.57**
Sum of two sides 3700mm	282.44	480.26	5.35	169.16	nr	**649.42**
Sum of two sides 3800mm	290.78	494.44	5.35	169.16	nr	**663.60**
Sum of two sides 3900mm	299.12	508.62	5.35	169.16	nr	**677.78**
Sum of two sides 4000mm	307.46	522.80	5.35	169.16	nr	**691.96**

U: VENTILATION/AIR CONDITIONING SYSTEMS

Item	Net Price £	Material £	Labour hours	Labour £	Unit	Total rate £
U10 : DUCTWORK : RECTANGULAR – CLASS B (cont'd)						
Galvanised sheet metal DW144 (cont'd)						
Extra over fittings; Ductwork 1601 to 2000mm longest side (cont'd)						
Square to round						
Sum of two sides 2100mm	102.41	174.14	2.61	82.53	nr	256.66
Sum of two sides 2200mm	112.28	190.92	2.61	82.53	nr	273.44
Sum of two sides 2300mm	122.12	207.65	2.61	82.53	nr	290.18
Sum of two sides 2400mm	132.04	224.52	2.88	91.06	nr	315.58
Sum of two sides 2500mm	141.87	241.23	4.70	148.61	nr	389.84
Sum of two sides 2600mm	151.73	258.00	4.70	148.61	nr	406.61
Sum of two sides 2700mm	161.58	274.75	4.71	148.93	nr	423.67
Sum of two sides 2800mm	171.44	291.51	8.19	258.96	nr	550.47
Sum of two sides 2900mm	181.28	308.24	8.19	258.96	nr	567.21
Sum of two sides 3000mm	191.13	324.99	8.19	258.96	nr	583.95
Sum of two sides 3100mm	192.82	327.87	8.19	258.96	nr	586.83
Sum of two sides 3200mm	199.39	339.04	8.19	258.96	nr	598.00
Sum of two sides 3300mm	198.39	337.34	8.19	258.96	nr	596.30
Sum of two sides 3400mm	205.09	348.73	8.62	272.56	nr	621.29
Sum of two sides 3500mm	211.78	360.11	8.62	272.56	nr	632.66
Sum of two sides 3600mm	218.47	371.48	8.62	272.56	nr	644.04
Sum of two sides 3700mm	228.57	388.66	8.62	272.56	nr	661.21
Sum of two sides 3800mm	235.26	400.03	8.75	276.67	nr	676.70
Sum of two sides 3900mm	241.96	411.42	8.75	276.67	nr	688.09
Sum of two sides 4000mm	248.65	422.80	8.75	276.67	nr	699.47
90 ° radius bend						
Sum of two sides 2100mm	149.71	254.56	2.61	82.53	nr	337.09
Sum of two sides 2200mm	287.48	488.83	2.61	82.53	nr	571.35
Sum of two sides 2300mm	311.05	528.90	2.61	82.53	nr	611.43
Sum of two sides 2400mm	318.77	542.03	2.88	91.06	nr	633.09
Sum of two sides 2500mm	342.27	581.99	3.90	123.31	nr	705.30
Sum of two sides 2600mm	365.34	621.22	4.26	134.70	nr	755.91
Sum of two sides 2700mm	388.42	660.46	4.55	143.87	nr	804.33
Sum of two sides 2800mm	411.50	699.71	4.55	143.87	nr	843.57
Sum of two sides 2900mm	435.00	739.67	6.87	217.22	nr	956.89
Sum of two sides 3000mm	458.07	778.89	6.87	217.22	nr	996.12
Sum of two sides 3100mm	463.16	787.55	6.87	217.22	nr	1004.77
Sum of two sides 3200mm	479.91	816.03	6.87	217.22	nr	1033.25
Sum of two sides 3300mm	479.27	814.94	6.87	217.22	nr	1032.16
Sum of two sides 3400mm	496.03	843.44	7.00	221.33	nr	1064.77
Sum of two sides 3500mm	512.78	871.92	7.00	221.33	nr	1093.26
Sum of two sides 3600mm	529.54	900.42	7.00	221.33	nr	1121.75
Sum of two sides 3700mm	567.20	964.46	7.00	221.33	nr	1185.79
Sum of two sides 3800mm	583.96	992.95	7.00	221.33	nr	1214.29
Sum of two sides 3900mm	600.72	1021.45	7.00	221.33	nr	1242.79
Sum of two sides 4000mm	617.47	1049.93	7.00	221.33	nr	1271.27
45 ° bend						
Sum of two sides 2100mm	72.04	122.50	2.61	82.53	nr	205.02
Sum of two sides 2200mm	205.60	349.60	2.61	82.53	nr	432.12
Sum of two sides 2300mm	221.41	376.48	2.61	82.53	nr	459.01
Sum of two sides 2400mm	228.84	389.12	2.88	91.06	nr	480.18
Sum of two sides 2500mm	244.62	415.95	4.85	153.35	nr	569.30
Sum of two sides 2600mm	260.08	442.23	4.87	153.99	nr	596.22

U: VENTILATION/AIR CONDITIONING SYSTEMS

Item	Net Price £	Material £	Labour hours	Labour £	Unit	Total rate £
Sum of two sides 2700mm	275.53	468.51	4.87	153.99	nr	**622.49**
Sum of two sides 2800mm	290.99	494.79	8.81	278.57	nr	**773.36**
Sum of two sides 2900mm	306.77	521.63	8.81	278.57	nr	**800.19**
Sum of two sides 3000mm	322.22	547.90	9.31	294.37	nr	**842.27**
Sum of two sides 3100mm	327.54	556.94	9.31	294.37	nr	**851.32**
Sum of two sides 3200mm	338.78	576.05	9.31	294.37	nr	**870.43**
Sum of two sides 3300mm	340.95	579.74	9.31	294.37	nr	**874.12**
Sum of two sides 3400mm	352.19	598.86	9.31	294.37	nr	**893.23**
Sum of two sides 3500mm	363.44	617.99	9.39	296.90	nr	**914.89**
Sum of two sides 3600mm	347.68	591.19	9.39	296.90	nr	**888.09**
Sum of two sides 3700mm	399.90	679.98	9.39	296.90	nr	**976.89**
Sum of two sides 3800mm	411.14	699.09	9.39	296.90	nr	**996.00**
Sum of two sides 3900mm	422.39	718.22	9.39	296.90	nr	**1015.13**
Sum of two sides 4000mm	433.63	737.34	9.39	296.90	nr	**1034.24**
90° mitre bend						
Sum of two sides 2100mm	330.23	561.52	2.61	82.53	nr	**644.04**
Sum of two sides 2200mm	348.78	593.06	2.61	82.53	nr	**675.58**
Sum of two sides 2300mm	376.20	639.68	2.61	82.53	nr	**722.21**
Sum of two sides 2400mm	384.97	654.60	2.88	91.06	nr	**745.66**
Sum of two sides 2500mm	412.42	701.27	4.85	153.35	nr	**854.62**
Sum of two sides 2600mm	440.00	748.17	4.87	153.99	nr	**902.15**
Sum of two sides 2700mm	467.60	795.10	4.87	153.99	nr	**949.08**
Sum of two sides 2800mm	495.19	842.01	8.81	278.57	nr	**1120.58**
Sum of two sides 2900mm	522.64	888.69	14.81	468.28	nr	**1356.97**
Sum of two sides 3000mm	550.23	935.60	15.20	480.61	nr	**1416.21**
Sum of two sides 3100mm	557.13	947.33	15.20	480.61	nr	**1427.94**
Sum of two sides 3200mm	578.39	983.48	15.20	480.61	nr	**1464.09**
Sum of two sides 3300mm	583.01	991.34	15.20	480.61	nr	**1471.95**
Sum of two sides 3400mm	604.28	1027.51	15.20	480.61	nr	**1508.12**
Sum of two sides 3500mm	625.56	1063.69	15.60	493.26	nr	**1556.95**
Sum of two sides 3600mm	646.83	1099.86	15.60	493.26	nr	**1593.12**
Sum of two sides 3700mm	674.78	1147.38	15.60	493.26	nr	**1640.64**
Sum of two sides 3800mm	696.06	1183.57	15.60	493.26	nr	**1676.83**
Sum of two sides 3900mm	717.33	1219.73	15.60	493.26	nr	**1712.99**
Sum of two sides 4000mm	738.61	1255.92	15.60	493.26	nr	**1749.18**
Branch						
Sum of two sides 2100mm	147.46	250.74	2.61	82.53	nr	**333.26**
Sum of two sides 2200mm	157.77	268.27	2.61	82.53	nr	**350.80**
Sum of two sides 2300mm	168.23	286.05	2.61	82.53	nr	**368.58**
Sum of two sides 2400mm	178.42	303.38	2.88	91.06	nr	**394.45**
Sum of two sides 2500mm	188.87	321.15	2.88	91.06	nr	**412.21**
Sum of two sides 2600mm	199.21	338.73	2.88	91.06	nr	**429.80**
Sum of two sides 2700mm	209.54	356.30	2.88	91.06	nr	**447.36**
Sum of two sides 2800mm	219.88	373.88	3.94	124.58	nr	**498.46**
Sum of two sides 2900mm	230.34	391.67	3.94	124.58	nr	**516.25**
Sum of two sides 3000mm	240.67	409.23	3.94	124.58	nr	**533.81**
Sum of two sides 3100mm	248.94	423.29	3.94	124.58	nr	**547.87**
Sum of two sides 3200mm	256.46	436.08	3.94	124.58	nr	**560.66**
Sum of two sides 3300mm	264.12	449.10	3.94	124.58	nr	**573.68**
Sum of two sides 3400mm	271.65	461.91	4.83	152.72	nr	**614.63**
Sum of two sides 3500mm	279.17	474.70	4.83	152.72	nr	**627.42**
Sum of two sides 3600mm	286.70	487.50	4.83	152.72	nr	**640.22**
Sum of two sides 3700mm	297.76	506.31	4.83	152.72	nr	**659.03**
Sum of two sides 3800mm	305.29	519.11	4.83	152.72	nr	**671.83**
Sum of two sides 3900mm	312.82	531.91	4.83	152.72	nr	**684.63**
Sum of two sides 4000mm	320.35	544.72	4.83	152.72	nr	**697.44**

U: VENTILATION/AIR CONDITIONING SYSTEMS

Item	Net Price £	Material £	Labour hours	Labour £	Unit	Total rate £
U10 : DUCTWORK : RECTANGULAR – CLASS B (cont'd)						
Galvanised sheet metal DW144 (cont'd)						
Extra over fittings; Ductwork 1601 to 2000mm longest side (cont'd)						
Grille neck						
Sum of two sides 2100mm	154.57	262.83	2.61	82.53	nr	345.35
Sum of two sides 2200mm	166.11	282.45	2.61	82.53	nr	364.98
Sum of two sides 2300mm	177.66	302.09	2.61	82.53	nr	384.62
Sum of two sides 2400mm	189.21	321.73	2.88	91.06	nr	412.79
Sum of two sides 2500mm	200.75	341.35	2.88	91.06	nr	432.41
Sum of two sides 2600mm	212.30	360.99	2.88	91.06	nr	452.05
Sum of two sides 2700mm	223.85	380.63	2.88	91.06	nr	471.69
Sum of two sides 2800mm	235.39	400.25	3.94	124.58	nr	524.83
Sum of two sides 2900mm	246.94	419.89	4.12	130.27	nr	550.16
Sum of two sides 3000mm	258.49	439.53	4.12	130.27	nr	569.80
Sum of two sides 3100mm	267.70	455.19	4.12	130.27	nr	585.46
Sum of two sides 3200mm	276.08	469.44	4.12	130.27	nr	599.71
Sum of two sides 3300mm	284.47	483.71	4.12	130.27	nr	613.98
Sum of two sides 3400mm	292.85	497.96	5.00	158.10	nr	656.05
Sum of two sides 3500mm	301.24	512.22	5.00	158.10	nr	670.32
Sum of two sides 3600mm	309.63	526.49	5.00	158.10	nr	684.58
Sum of two sides 3700mm	318.02	540.75	5.00	158.10	nr	698.85
Sum of two sides 3800mm	326.40	555.00	5.00	158.10	nr	713.10
Sum of two sides 3900mm	334.79	569.27	5.00	158.10	nr	727.37
Sum of two sides 4000mm	343.18	583.54	5.00	158.10	nr	741.63
Ductwork 2001 to 2500mm longest side						
Sum of two sides 2500mm	147.65	251.06	2.38	75.25	m	326.31
Sum of two sides 2600mm	151.45	257.52	2.64	83.47	m	341.00
Sum of two sides 2700mm	154.71	263.07	2.66	84.11	m	347.17
Sum of two sides 2800mm	158.76	269.95	2.95	93.28	m	363.23
Sum of two sides 2900mm	162.02	275.50	2.96	93.59	m	369.09
Sum of two sides 3000mm	165.10	280.73	3.15	99.60	m	380.33
Sum of two sides 3100mm	168.33	286.23	3.15	99.60	m	385.83
Sum of two sides 3200mm	171.42	291.48	3.15	99.60	m	391.08
Sum of two sides 3300mm	174.67	297.01	3.15	99.60	m	396.61
Sum of two sides 3400mm	177.75	302.24	3.15	99.60	m	401.84
Sum of two sides 3500mm	189.65	322.48	2.66	84.11	m	406.58
Sum of two sides 3600mm	192.73	327.71	3.18	100.55	m	428.26
Sum of two sides 3700mm	205.92	350.14	3.18	100.55	m	450.69
Sum of two sides 3800mm	209.01	355.40	3.18	100.55	m	455.95
Sum of two sides 3900mm	212.10	360.65	3.18	100.55	m	461.20
Sum of two sides 4000mm	216.11	367.47	3.18	100.55	m	468.02

U: VENTILATION/AIR CONDITIONING SYSTEMS

Item	Net Price £	Material £	Labour hours	Labour £	Unit	Total rate £
Extra over fittings; Ductwork 2001 to 2500mm longest side						
End Cap						
Sum of two sides 2500mm	64.63	109.90	0.87	27.51	nr	137.40
Sum of two sides 2600mm	68.39	116.29	0.87	27.51	nr	143.80
Sum of two sides 2700mm	72.14	122.67	0.87	27.51	nr	150.17
Sum of two sides 2800mm	75.89	129.04	1.16	36.68	nr	165.72
Sum of two sides 2900mm	79.65	135.44	1.16	36.68	nr	172.11
Sum of two sides 3000mm	83.41	141.83	1.73	54.70	nr	196.53
Sum of two sides 3100mm	86.39	146.90	1.73	54.70	nr	201.60
Sum of two sides 3200mm	89.09	151.49	1.73	54.70	nr	206.19
Sum of two sides 3300mm	91.80	156.09	1.73	54.70	nr	210.80
Sum of two sides 3400mm	94.49	160.67	1.80	56.91	nr	217.58
Sum of two sides 3500mm	97.20	165.28	1.80	56.91	nr	222.19
Sum of two sides 3600mm	99.90	169.87	1.80	56.91	nr	226.78
Sum of two sides 3700mm	102.61	174.48	1.80	56.91	nr	231.39
Sum of two sides 3800mm	105.31	179.07	1.80	56.91	nr	235.98
Sum of two sides 3900mm	108.01	183.66	1.80	56.91	nr	240.57
Sum of two sides 4000mm	110.72	188.27	1.80	56.91	nr	245.18
Reducer						
Sum of two sides 2500mm	76.98	130.90	3.12	98.65	nr	229.55
Sum of two sides 2600mm	83.39	141.79	3.12	98.65	nr	240.45
Sum of two sides 2700mm	89.94	152.93	3.12	98.65	nr	251.58
Sum of two sides 2800mm	96.22	163.61	3.95	124.90	nr	288.51
Sum of two sides 2900mm	102.76	174.73	3.97	125.53	nr	300.26
Sum of two sides 3000mm	109.20	185.68	3.97	125.53	nr	311.21
Sum of two sides 3100mm	113.95	193.76	3.97	125.53	nr	319.29
Sum of two sides 3200mm	118.27	201.10	3.97	125.53	nr	326.63
Sum of two sides 3300mm	122.71	208.65	3.97	125.53	nr	334.18
Sum of two sides 3400mm	127.03	216.00	4.52	142.92	nr	358.92
Sum of two sides 3500mm	125.03	212.60	4.52	142.92	nr	355.52
Sum of two sides 3600mm	129.35	219.94	4.52	142.92	nr	362.86
Sum of two sides 3700mm	125.83	213.96	4.52	142.92	nr	356.88
Sum of two sides 3800mm	130.15	221.30	4.52	142.92	nr	364.22
Sum of two sides 3900mm	134.48	228.67	4.52	142.92	nr	371.59
Sum of two sides 4000mm	138.87	236.13	4.52	142.92	nr	379.05
Offset						
Sum of two sides 2500mm	195.46	332.36	3.48	110.03	nr	442.39
Sum of two sides 2600mm	207.61	353.02	3.49	110.35	nr	463.37
Sum of two sides 2700mm	212.09	360.63	3.50	110.67	nr	471.30
Sum of two sides 2800mm	231.91	394.34	4.34	137.23	nr	531.56
Sum of two sides 2900mm	236.14	401.53	4.76	150.51	nr	552.04
Sum of two sides 3000mm	239.97	408.04	5.32	168.21	nr	576.25
Sum of two sides 3100mm	241.05	409.88	5.35	169.16	nr	579.04
Sum of two sides 3200mm	241.17	410.08	5.35	169.16	nr	579.24
Sum of two sides 3300mm	241.10	409.96	5.32	168.21	nr	578.18
Sum of two sides 3400mm	240.67	409.23	5.32	168.21	nr	577.44
Sum of two sides 3500mm	239.11	406.58	5.32	168.21	nr	574.79
Sum of two sides 3600mm	247.32	420.54	5.35	169.16	nr	589.70
Sum of two sides 3700mm	243.72	414.42	5.35	169.16	nr	583.58
Sum of two sides 3800mm	251.94	428.39	5.35	169.16	nr	597.56
Sum of two sides 3900mm	260.16	442.37	5.35	169.16	nr	611.53
Sum of two sides 4000mm	268.35	456.30	5.35	169.16	nr	625.46

U: VENTILATION/AIR CONDITIONING SYSTEMS

Item	Net Price £	Material £	Labour hours	Labour £	Unit	Total rate £
U10 : DUCTWORK : RECTANGULAR – CLASS B (cont'd)						
Galvanised sheet metal DW144 (cont'd)						
Extra over fittings; Ductwork 2001 to 2500mm longest side (cont'd)						
Square to round						
Sum of two sides 2500mm	122.36	208.06	4.70	148.61	nr	356.67
Sum of two sides 2600mm	132.13	224.67	4.70	148.61	nr	373.28
Sum of two sides 2700mm	141.89	241.27	4.71	148.93	nr	390.19
Sum of two sides 2800mm	151.68	257.91	8.19	258.96	nr	516.87
Sum of two sides 2900mm	161.44	274.51	8.19	258.96	nr	533.47
Sum of two sides 3000mm	171.21	291.12	8.19	258.96	nr	550.08
Sum of two sides 3100mm	178.53	303.57	8.19	258.96	nr	562.53
Sum of two sides 3200mm	185.14	314.81	8.62	272.56	nr	587.37
Sum of two sides 3300mm	191.74	326.03	8.62	272.56	nr	598.59
Sum of two sides 3400mm	198.35	337.27	8.62	272.56	nr	609.83
Sum of two sides 3500mm	198.28	337.15	8.62	272.56	nr	609.71
Sum of two sides 3600mm	204.89	348.39	8.75	276.67	nr	625.06
Sum of two sides 3700mm	203.54	346.10	8.75	276.67	nr	622.76
Sum of two sides 3800mm	210.14	357.32	8.75	276.67	nr	633.99
Sum of two sides 3900mm	216.76	368.57	8.75	276.67	nr	645.24
Sum of two sides 4000mm	223.33	379.75	8.75	276.67	nr	656.41
90 ° radius bend						
Sum of two sides 2500mm	289.21	491.77	3.90	123.31	nr	615.08
Sum of two sides 2600mm	302.03	513.57	4.26	134.70	nr	648.26
Sum of two sides 2700mm	325.27	553.08	4.55	143.87	nr	696.95
Sum of two sides 2800mm	326.87	555.80	4.55	143.87	nr	699.67
Sum of two sides 2900mm	350.03	595.18	6.87	217.22	nr	812.41
Sum of two sides 3000mm	372.80	633.90	6.87	217.22	nr	851.13
Sum of two sides 3100mm	390.58	664.13	6.87	217.22	nr	881.36
Sum of two sides 3200mm	407.03	692.11	6.87	217.22	nr	909.33
Sum of two sides 3300mm	423.88	720.76	6.87	217.22	nr	937.98
Sum of two sides 3400mm	440.33	748.73	6.87	217.22	nr	965.95
Sum of two sides 3500mm	440.09	748.32	7.00	221.33	nr	969.65
Sum of two sides 3600mm	456.54	776.29	7.00	221.33	nr	997.63
Sum of two sides 3700mm	452.08	768.71	7.00	221.33	nr	990.04
Sum of two sides 3800mm	468.53	796.68	7.00	221.33	nr	1018.01
Sum of two sides 3900mm	484.98	824.65	7.00	221.33	nr	1045.98
Sum of two sides 4000mm	489.13	831.71	7.00	221.33	nr	1053.04
45 ° bend						
Sum of two sides 2500mm	217.10	369.15	4.85	153.35	nr	522.51
Sum of two sides 2600mm	227.29	386.48	4.87	153.99	nr	540.46
Sum of two sides 2700mm	242.95	413.11	4.87	153.99	nr	567.09
Sum of two sides 2800mm	247.30	420.50	8.81	278.57	nr	699.07
Sum of two sides 2900mm	262.90	447.03	8.81	278.57	nr	725.60
Sum of two sides 3000mm	278.20	473.05	9.31	294.37	nr	767.42
Sum of two sides 3100mm	290.22	493.48	9.31	294.37	nr	787.86
Sum of two sides 3200mm	301.30	512.32	9.31	294.37	nr	806.70
Sum of two sides 3300mm	312.70	531.71	9.31	294.37	nr	826.08
Sum of two sides 3400mm	323.79	550.57	9.31	294.37	nr	844.94
Sum of two sides 3500mm	326.44	555.07	9.31	294.37	nr	849.45
Sum of two sides 3600mm	337.53	573.93	9.39	296.90	nr	870.83
Sum of two sides 3700mm	337.78	574.35	9.39	296.90	nr	871.26
Sum of two sides 3800mm	348.86	593.19	9.39	296.90	nr	890.10

U: VENTILATION/AIR CONDITIONING SYSTEMS

Item	Net Price £	Material £	Labour hours	Labour £	Unit	Total rate £
Sum of two sides 3900mm	359.94	612.03	9.39	296.90	nr	908.94
Sum of two sides 4000mm	364.87	620.42	9.39	296.90	nr	917.32
90° mitre bend						
Sum of two sides 2500mm	345.96	588.26	4.85	153.35	nr	741.62
Sum of two sides 2600mm	366.75	623.61	4.87	153.99	nr	777.60
Sum of two sides 2700mm	394.04	670.02	4.87	153.99	nr	824.00
Sum of two sides 2800mm	394.75	671.23	8.81	278.57	nr	949.79
Sum of two sides 2900mm	422.05	717.65	14.81	468.28	nr	1185.93
Sum of two sides 3000mm	449.52	764.35	14.81	468.28	nr	1232.64
Sum of two sides 3100mm	471.95	802.49	15.20	480.61	nr	1283.11
Sum of two sides 3200mm	493.11	838.47	15.20	480.61	nr	1319.09
Sum of two sides 3300mm	514.09	874.15	15.20	480.61	nr	1354.76
Sum of two sides 3400mm	535.26	910.15	15.20	480.61	nr	1390.76
Sum of two sides 3500mm	534.24	908.41	15.20	480.61	nr	1389.02
Sum of two sides 3600mm	555.40	944.39	15.60	493.26	nr	1437.65
Sum of two sides 3700mm	556.45	946.18	15.60	493.26	nr	1439.44
Sum of two sides 3800mm	577.61	982.16	15.60	493.26	nr	1475.42
Sum of two sides 3900mm	598.78	1018.15	15.60	493.26	nr	1511.41
Sum of two sides 4000mm	604.17	1027.32	15.60	493.26	nr	1520.58
Branch						
Sum of two sides 2500mm	189.22	321.75	2.88	91.06	nr	412.81
Sum of two sides 2600mm	199.54	339.29	2.88	91.06	nr	430.36
Sum of two sides 2700mm	210.00	357.08	2.88	91.06	nr	448.14
Sum of two sides 2800mm	220.17	374.37	3.94	124.58	nr	498.95
Sum of two sides 2900mm	230.63	392.16	3.94	124.58	nr	516.74
Sum of two sides 3000mm	240.97	409.74	3.94	124.58	nr	534.32
Sum of two sides 3100mm	249.23	423.79	3.94	124.58	nr	548.37
Sum of two sides 3200mm	256.76	436.59	3.94	124.58	nr	561.17
Sum of two sides 3300mm	264.41	449.60	3.94	124.58	nr	574.18
Sum of two sides 3400mm	271.94	462.40	3.94	124.58	nr	586.98
Sum of two sides 3500mm	279.83	475.82	4.83	152.72	nr	628.54
Sum of two sides 3600mm	287.35	488.60	4.83	152.72	nr	641.32
Sum of two sides 3700mm	295.01	501.63	4.83	152.72	nr	654.35
Sum of two sides 3800mm	302.54	514.43	4.83	152.72	nr	667.15
Sum of two sides 3900mm	310.06	527.22	4.83	152.72	nr	679.94
Sum of two sides 4000mm	317.70	540.21	4.83	152.72	nr	692.93
Grille neck						
Sum of two sides 2500mm	200.75	341.35	2.88	91.06	nr	432.41
Sum of two sides 2600mm	212.30	360.99	2.88	91.06	nr	452.05
Sum of two sides 2700mm	223.85	380.63	2.88	91.06	nr	471.69
Sum of two sides 2800mm	235.39	400.25	3.94	124.58	nr	524.83
Sum of two sides 2900mm	246.94	419.89	3.94	124.58	nr	544.47
Sum of two sides 3000mm	258.49	439.53	3.94	124.58	nr	564.11
Sum of two sides 3100mm	267.70	455.19	4.12	130.27	nr	585.46
Sum of two sides 3200mm	276.08	469.44	4.12	130.27	nr	599.71
Sum of two sides 3300mm	284.47	483.71	4.12	130.27	nr	613.98
Sum of two sides 3400mm	292.85	497.96	4.12	130.27	nr	628.23
Sum of two sides 3500mm	301.24	512.22	5.00	158.10	nr	670.32
Sum of two sides 3600mm	309.63	526.49	5.00	158.10	nr	684.58
Sum of two sides 3700mm	318.02	540.75	5.00	158.10	nr	698.85
Sum of two sides 3800mm	326.40	555.00	5.00	158.10	nr	713.10
Sum of two sides 3900mm	334.79	569.27	5.00	158.10	nr	727.37
Sum of two sides 4000mm	343.18	583.54	5.00	158.10	nr	741.63

U: VENTILATION/AIR CONDITIONING SYSTEMS

Item	Net Price £	Material £	Labour hours	Labour £	Unit	Total rate £
U10 : DUCTWORK : RECTANGULAR – CLASS B (cont'd)						
Galvanised sheet metal DW144 (cont'd)						
Ductwork 2501 to 4000mm longest side						
Sum of two sides 3000mm	234.84	399.32	2.38	75.25	m	474.57
Sum of two sides 3100mm	240.48	408.91	2.38	75.25	m	484.16
Sum of two sides 3200mm	245.80	417.95	2.38	75.25	m	493.21
Sum of two sides 3300mm	251.11	426.98	2.38	75.25	m	502.24
Sum of two sides 3400mm	257.71	438.20	2.64	83.47	m	521.68
Sum of two sides 3500mm	263.02	447.23	2.66	84.11	m	531.34
Sum of two sides 3600mm	268.34	456.28	2.95	93.28	m	549.56
Sum of two sides 3700mm	275.11	467.79	2.96	93.59	m	561.38
Sum of two sides 3800mm	280.43	476.84	3.15	99.60	m	576.44
Sum of two sides 3900mm	295.39	502.28	3.15	99.60	m	601.88
Sum of two sides 4000mm	300.70	511.30	3.15	99.60	m	610.90
Sum of two sides 4100mm	306.20	520.66	3.35	105.92	m	626.58
Sum of two sides 4200mm	311.52	529.70	3.35	105.92	m	635.63
Sum of two sides 4300mm	316.83	538.73	3.60	113.83	m	652.56
Sum of two sides 4400mm	324.31	551.45	3.60	113.83	m	665.28
Sum of two sides 4500mm	329.62	560.48	3.60	113.83	m	674.31
Extra over fittings; Ductwork 2501 to 4000mm longest side						
End Cap						
Sum of two sides 3000mm	84.36	143.44	1.73	54.70	nr	198.15
Sum of two sides 3100mm	87.37	148.56	1.73	54.70	nr	203.26
Sum of two sides 3200mm	90.10	153.20	1.73	54.70	nr	207.91
Sum of two sides 3300mm	92.84	157.86	1.73	54.70	nr	212.56
Sum of two sides 3400mm	95.57	162.51	1.73	54.70	nr	217.21
Sum of two sides 3500mm	98.31	167.16	1.73	54.70	nr	221.87
Sum of two sides 3600mm	101.04	171.81	1.73	54.70	nr	226.51
Sum of two sides 3700mm	103.78	176.47	1.73	54.70	nr	231.17
Sum of two sides 3800mm	106.51	181.11	1.73	54.70	nr	235.81
Sum of two sides 3900mm	109.25	185.77	1.80	56.91	nr	242.68
Sum of two sides 4000mm	111.98	190.41	1.80	56.91	nr	247.32
Sum of two sides 4100mm	114.72	195.07	1.80	56.91	nr	251.98
Sum of two sides 4200mm	117.45	199.71	1.80	56.91	nr	256.62
Sum of two sides 4300mm	120.19	204.37	1.88	59.44	nr	263.81
Sum of two sides 4400mm	122.92	209.01	1.88	59.44	nr	268.45
Sum of two sides 4500mm	125.66	213.67	1.88	59.44	nr	273.11
Reducer						
Sum of two sides 3000mm	73.04	124.20	3.12	98.65	nr	222.85
Sum of two sides 3100mm	76.67	130.37	3.12	98.65	nr	229.02
Sum of two sides 3200mm	79.76	135.62	3.12	98.65	nr	234.27
Sum of two sides 3300mm	82.85	140.88	3.12	98.65	nr	239.53
Sum of two sides 3400mm	85.90	146.06	3.12	98.65	nr	244.71
Sum of two sides 3500mm	88.99	151.32	3.12	98.65	nr	249.97
Sum of two sides 3600mm	92.07	156.55	3.95	124.90	nr	281.45
Sum of two sides 3700mm	95.25	161.96	3.97	125.53	nr	287.49
Sum of two sides 3800mm	98.34	167.22	4.52	142.92	nr	310.13
Sum of two sides 3900mm	95.07	161.66	4.52	142.92	nr	304.57
Sum of two sides 4000mm	98.15	166.89	4.52	142.92	nr	309.81
Sum of two sides 4100mm	101.37	172.37	4.52	142.92	nr	315.29
Sum of two sides 4200mm	104.45	177.60	4.52	142.92	nr	320.52

U: VENTILATION/AIR CONDITIONING SYSTEMS

Item	Net Price £	Material £	Labour hours	Labour £	Unit	Total rate £
Sum of two sides 4300mm	107.53	182.84	4.92	155.57	nr	338.41
Sum of two sides 4400mm	110.70	188.23	4.92	155.57	nr	343.80
Sum of two sides 4500mm	113.78	193.47	5.12	161.89	nr	355.36
Offset						
Sum of two sides 3000mm	233.48	397.00	3.48	110.03	nr	507.04
Sum of two sides 3100mm	229.41	390.08	3.48	110.03	nr	500.12
Sum of two sides 3200mm	223.92	380.75	3.48	110.03	nr	490.78
Sum of two sides 3300mm	217.86	370.44	3.48	110.03	nr	480.48
Sum of two sides 3400mm	239.18	406.70	3.49	110.35	nr	517.05
Sum of two sides 3500mm	232.57	395.46	3.50	110.67	nr	506.12
Sum of two sides 3600mm	225.38	383.23	3.50	110.67	nr	493.90
Sum of two sides 3700mm	247.36	420.61	4.76	150.51	nr	571.11
Sum of two sides 3800mm	239.59	407.39	5.32	168.21	nr	575.61
Sum of two sides 3900mm	254.35	432.49	5.35	169.16	nr	601.65
Sum of two sides 4000mm	245.33	417.15	5.35	169.16	nr	586.32
Sum of two sides 4100mm	235.82	400.98	5.85	184.97	nr	585.96
Sum of two sides 4200mm	225.65	383.69	5.85	184.97	nr	568.66
Sum of two sides 4300mm	232.16	394.76	6.15	194.46	nr	589.22
Sum of two sides 4400mm	256.25	435.72	6.15	194.46	nr	630.18
Sum of two sides 4500mm	245.22	416.97	6.30	199.20	nr	616.17
Square to round						
Sum of two sides 3000mm	169.22	287.74	4.70	148.61	nr	436.35
Sum of two sides 3100mm	176.56	300.22	4.70	148.61	nr	448.83
Sum of two sides 3200mm	183.07	311.29	4.70	148.61	nr	459.90
Sum of two sides 3300mm	189.58	322.36	4.70	148.61	nr	470.97
Sum of two sides 3400mm	196.10	333.44	4.71	148.93	nr	482.37
Sum of two sides 3500mm	202.60	344.50	4.71	148.93	nr	493.42
Sum of two sides 3600mm	209.11	355.57	8.19	258.96	nr	614.53
Sum of two sides 3700mm	215.61	366.62	8.62	272.56	nr	639.18
Sum of two sides 3800mm	222.12	377.69	8.62	272.56	nr	650.25
Sum of two sides 3900mm	273.23	464.59	8.62	272.56	nr	737.15
Sum of two sides 4000mm	281.02	477.84	8.75	276.67	nr	754.51
Sum of two sides 4100mm	288.80	491.07	11.23	355.08	nr	846.15
Sum of two sides 4200mm	296.60	504.33	11.23	355.08	nr	859.42
Sum of two sides 4300mm	304.39	517.58	11.25	355.72	nr	873.29
Sum of two sides 4400mm	312.17	530.81	11.25	355.72	nr	886.52
Sum of two sides 4500mm	319.97	544.07	11.26	356.03	nr	900.10
90° radius bend						
Sum of two sides 3000mm	455.20	774.01	3.90	123.31	nr	897.33
Sum of two sides 3100mm	477.31	811.61	3.90	123.31	nr	934.92
Sum of two sides 3200mm	496.90	844.92	4.26	134.70	nr	979.62
Sum of two sides 3300mm	516.49	878.23	4.26	134.70	nr	1012.93
Sum of two sides 3400mm	501.48	852.71	4.26	134.70	nr	987.40
Sum of two sides 3500mm	520.77	885.51	4.55	143.87	nr	1029.37
Sum of two sides 3600mm	540.04	918.27	4.55	143.87	nr	1062.14
Sum of two sides 3700mm	523.41	890.00	6.87	217.22	nr	1107.22
Sum of two sides 3800mm	542.37	922.24	6.87	217.22	nr	1139.46
Sum of two sides 3900mm	501.64	852.98	6.87	217.22	nr	1070.20
Sum of two sides 4000mm	520.30	884.71	7.00	221.33	nr	1106.04
Sum of two sides 4100mm	539.54	917.42	7.20	227.66	nr	1145.08
Sum of two sides 4200mm	558.20	949.15	7.20	227.66	nr	1176.81
Sum of two sides 4300mm	576.86	980.88	7.41	234.30	nr	1215.18
Sum of two sides 4400mm	532.43	905.33	7.41	234.30	nr	1139.63
Sum of two sides 4500mm	550.61	936.25	7.55	238.72	nr	1174.97

U: VENTILATION/AIR CONDITIONING SYSTEMS

Item	Net Price £	Material £	Labour hours	Labour £	Unit	Total rate £
U10 : DUCTWORK : RECTANGULAR – CLASS B (cont'd)						
Galvanised sheet metal DW144 (cont'd)						
Extra over fittings; Ductwork 2501 to 4000mm longest side (cont'd)						
45 ° bend						
Sum of two sides 3000mm	219.94	373.98	4.85	153.35	nr	527.33
Sum of two sides 3100mm	230.93	392.67	4.85	153.35	nr	546.02
Sum of two sides 3200mm	240.67	409.23	4.85	153.35	nr	562.58
Sum of two sides 3300mm	250.41	425.79	4.87	153.99	nr	579.78
Sum of two sides 3400mm	242.61	412.53	4.87	153.99	nr	566.51
Sum of two sides 3500mm	252.19	428.82	4.87	153.99	nr	582.80
Sum of two sides 3600mm	261.78	445.13	8.81	278.57	nr	723.69
Sum of two sides 3700mm	253.15	430.45	8.81	278.57	nr	709.02
Sum of two sides 3800mm	262.58	446.49	9.31	294.37	nr	740.86
Sum of two sides 3900mm	241.47	410.59	9.31	294.37	nr	704.97
Sum of two sides 4000mm	250.75	426.37	9.31	294.37	nr	720.75
Sum of two sides 4100mm	260.30	442.61	10.01	316.51	nr	759.12
Sum of two sides 4200mm	269.58	458.39	10.01	316.51	nr	774.90
Sum of two sides 4300mm	278.86	474.17	9.31	294.37	nr	768.54
Sum of two sides 4400mm	256.20	435.64	10.52	332.63	nr	768.27
Sum of two sides 4500mm	265.24	451.01	10.52	332.63	nr	783.64
90 ° mitre bend						
Sum of two sides 3000mm	604.26	1027.47	4.85	153.35	nr	1180.82
Sum of two sides 3100mm	633.04	1076.41	4.85	153.35	nr	1229.76
Sum of two sides 3200mm	660.01	1122.27	4.87	153.99	nr	1276.25
Sum of two sides 3300mm	686.97	1168.11	4.87	153.99	nr	1322.10
Sum of two sides 3400mm	670.84	1140.68	4.87	153.99	nr	1294.67
Sum of two sides 3500mm	697.46	1185.95	8.81	278.57	nr	1464.51
Sum of two sides 3600mm	724.09	1231.23	8.81	278.57	nr	1509.79
Sum of two sides 3700mm	711.16	1209.24	14.81	468.28	nr	1677.52
Sum of two sides 3800mm	737.49	1254.01	14.81	468.28	nr	1722.29
Sum of two sides 3900mm	686.66	1167.58	14.81	468.28	nr	1635.86
Sum of two sides 4000mm	712.64	1211.76	15.20	480.61	nr	1692.37
Sum of two sides 4100mm	738.28	1255.36	16.30	515.39	nr	1770.75
Sum of two sides 4200mm	764.26	1299.53	16.50	521.72	nr	1821.25
Sum of two sides 4300mm	790.24	1343.71	17.01	537.84	nr	1881.55
Sum of two sides 4400mm	742.32	1262.23	17.01	537.84	nr	1800.07
Sum of two sides 4500mm	767.82	1305.59	17.01	537.84	nr	1843.43
Branch						
Sum of two sides 3000mm	249.25	423.82	2.88	91.06	nr	514.88
Sum of two sides 3100mm	257.04	437.17	2.88	91.06	nr	528.23
Sum of two sides 3200mm	265.65	451.71	2.88	91.06	nr	542.77
Sum of two sides 3300mm	273.42	464.92	2.88	91.06	nr	555.98
Sum of two sides 3400mm	281.16	478.08	2.88	91.06	nr	569.14
Sum of two sides 3500mm	288.93	491.29	3.94	124.58	nr	615.87
Sum of two sides 3600mm	296.70	504.50	3.94	124.58	nr	629.08
Sum of two sides 3700mm	304.57	517.88	3.94	124.58	nr	642.46
Sum of two sides 3800mm	312.33	531.08	4.83	152.72	nr	683.80
Sum of two sides 3900mm	320.44	544.87	4.83	152.72	nr	697.59
Sum of two sides 4000mm	328.21	558.08	4.83	152.72	nr	710.80
Sum of two sides 4100mm	336.11	571.51	5.44	172.01	nr	743.52
Sum of two sides 4200mm	343.88	584.73	5.44	172.01	nr	756.74

U: VENTILATION/AIR CONDITIONING SYSTEMS

Item	Net Price £	Material £	Labour hours	Labour £	Unit	Total rate £
Sum of two sides 4300mm	351.65	597.94	5.85	184.97	nr	782.91
Sum of two sides 4400mm	359.50	611.29	5.85	184.97	nr	796.26
Sum of two sides 4500mm	367.27	624.50	5.85	184.97	nr	809.47
Grille neck						
Sum of two sides 3000mm	263.22	447.57	2.88	91.06	nr	538.64
Sum of two sides 3100mm	272.59	463.51	2.88	91.06	nr	554.57
Sum of two sides 3200mm	281.14	478.04	2.88	91.06	nr	569.11
Sum of two sides 3300mm	289.69	492.58	2.88	91.06	nr	583.65
Sum of two sides 3400mm	298.23	507.10	3.94	124.58	nr	631.68
Sum of two sides 3500mm	306.77	521.63	3.94	124.58	nr	646.21
Sum of two sides 3600mm	315.32	536.16	3.94	124.58	nr	660.74
Sum of two sides 3700mm	323.86	550.69	4.12	130.27	nr	680.96
Sum of two sides 3800mm	332.41	565.22	4.12	130.27	nr	695.49
Sum of two sides 3900mm	340.95	579.74	4.12	130.27	nr	710.02
Sum of two sides 4000mm	349.50	594.28	5.00	158.10	nr	752.38
Sum of two sides 4100mm	258.04	438.77	5.00	158.10	nr	596.86
Sum of two sides 4200mm	366.59	623.34	5.00	158.10	nr	781.44
Sum of two sides 4300mm	375.14	637.88	5.23	165.37	nr	803.25
Sum of two sides 4400mm	383.68	652.40	5.23	165.37	nr	817.77
Sum of two sides 4500mm	392.22	666.92	5.39	170.43	nr	837.35

Y30 - ANCILLARIES

Access doors, hollow steel construction; 25mm mineral wool insulation; removable or hinged; fixed with cams; including sub-frame and integral sealing gaskets

Item	Net Price £	Material £	Labour hours	Labour £	Unit	Total rate £
Rectangular duct						
150 x 150mm	15.05	25.60	1.25	39.52	nr	65.12
200 x 200mm	16.27	27.66	1.25	39.52	nr	67.19
300 x 150mm	16.72	28.43	1.25	39.52	nr	67.95
300 x 300mm	19.15	32.57	1.25	39.52	nr	72.09
400 x 400mm	21.84	37.14	1.35	42.73	nr	79.87
450 x 300mm	21.84	37.14	1.50	47.48	nr	84.61
450 x 450mm	24.08	40.95	1.50	47.48	nr	88.42

Access doors, hollow steel construction; 25mm mineral wool insulation; removable or hinged; fixed with cams; including sub-frame and integral sealing gaskets

Item	Net Price £	Material £	Labour hours	Labour £	Unit	Total rate £
Flat oval duct						
235 x 90mm	30.58	51.99	1.25	39.52	nr	91.51
235 x 140mm	32.59	55.42	1.35	42.73	nr	98.15
335 x 235mm	37.24	63.32	1.50	47.48	nr	110.80
535 x 235mm	41.89	71.23	1.50	47.48	nr	118.70

U: VENTILATION/AIR CONDITIONING SYSTEMS

Item	Net Price £	Material £	Labour hours	Labour £	Unit	Total rate £
U10 : DUCTWORK : RECTANGULAR – CLASS C						
Y30 - AIR DUCTLINES						
Galvanised sheet metal DW144 class C rectangular section ductwork; including all necessary stiffeners, joints, couplers in the running length and duct supports						
Ductwork up to 400mm longest side						
Sum of two sides 200mm	14.97	25.45	1.19	37.63	m	**63.08**
Sum of two sides 300mm	16.19	27.53	1.16	36.68	m	**64.21**
Sum of two sides 400mm	14.50	24.66	1.17	36.99	m	**61.65**
Sum of two sides 500mm	15.82	26.90	1.17	36.99	m	**63.89**
Sum of two sides 600mm	17.00	28.91	1.19	37.63	m	**66.53**
Sum of two sides 700mm	18.19	30.93	1.19	37.63	m	**68.56**
Sum of two sides 800mm	19.45	33.07	1.19	37.63	m	**70.70**
Extra over fittings; Ductwork up to 400mm longest side						
End Cap						
Sum of two sides 200mm	9.66	16.43	0.38	12.02	nr	**28.44**
Sum of two sides 300mm	10.77	18.31	0.38	12.02	nr	**30.33**
Sum of two sides 400mm	11.89	20.22	0.38	12.02	nr	**32.23**
Sum of two sides 500mm	13.00	22.10	0.38	12.02	nr	**34.12**
Sum of two sides 600mm	14.12	24.01	0.38	12.02	nr	**36.02**
Sum of two sides 700mm	15.24	25.91	0.38	12.02	nr	**37.93**
Sum of two sides 800mm	16.35	27.80	0.38	12.02	nr	**39.82**
Reducer						
Sum of two sides 200mm	15.72	26.73	1.40	44.27	nr	**71.00**
Sum of two sides 300mm	17.75	30.18	1.40	44.27	nr	**74.45**
Sum of two sides 400mm	29.36	49.92	1.42	44.90	nr	**94.82**
Sum of two sides 500mm	31.86	54.17	1.42	44.90	nr	**99.07**
Sum of two sides 600mm	34.36	58.43	1.69	53.44	nr	**111.86**
Sum of two sides 700mm	36.87	62.69	1.69	53.44	nr	**116.13**
Sum of two sides 800mm	39.34	66.89	1.92	60.71	nr	**127.60**
Offset						
Sum of two sides 200mm	25.57	43.48	1.63	51.54	nr	**95.02**
Sum of two sides 300mm	28.00	47.61	1.63	51.54	nr	**99.15**
Sum of two sides 400mm	40.99	69.70	1.65	52.17	nr	**121.87**
Sum of two sides 500mm	44.25	75.24	1.65	52.17	nr	**127.41**
Sum of two sides 600mm	48.41	82.32	1.92	60.71	nr	**143.02**
Sum of two sides 700mm	52.25	88.84	1.92	60.71	nr	**149.55**
Sum of two sides 800mm	56.45	95.99	1.92	60.71	nr	**156.70**
Square to round						
Sum of two sides 200mm	29.32	49.86	1.22	38.58	nr	**88.43**
Sum of two sides 300mm	32.88	55.91	1.22	38.58	nr	**94.48**
Sum of two sides 400mm	36.45	61.98	1.25	39.52	nr	**101.50**
Sum of two sides 500mm	40.01	68.03	1.25	39.52	nr	**107.56**
Sum of two sides 600mm	43.58	74.10	1.33	42.05	nr	**116.16**
Sum of two sides 700mm	47.14	80.16	1.33	42.05	nr	**122.21**
Sum of two sides 800mm	50.71	86.23	1.40	44.27	nr	**130.49**

U: VENTILATION/AIR CONDITIONING SYSTEMS

Item	Net Price £	Material £	Labour hours	Labour £	Unit	Total rate £
90 ° radius bend						
Sum of two sides 200mm	20.41	34.70	1.10	34.78	nr	69.49
Sum of two sides 300mm	22.94	39.01	1.10	34.78	nr	73.79
Sum of two sides 400mm	30.69	52.18	1.12	35.41	nr	87.60
Sum of two sides 500mm	33.43	56.84	1.12	35.41	nr	92.26
Sum of two sides 600mm	36.18	61.52	1.16	36.68	nr	98.20
Sum of two sides 700mm	38.93	66.20	1.16	36.68	nr	102.87
Sum of two sides 800mm	41.65	70.82	1.22	38.58	nr	109.40
45 ° radius bend						
Sum of two sides 200mm	23.20	39.45	1.29	40.79	nr	80.24
Sum of two sides 300mm	26.23	44.60	1.29	40.79	nr	85.39
Sum of two sides 400mm	39.09	66.47	1.29	40.79	nr	107.26
Sum of two sides 500mm	42.58	72.40	1.29	40.79	nr	113.19
Sum of two sides 600mm	45.46	77.30	1.39	43.95	nr	121.25
Sum of two sides 700mm	48.64	82.71	1.39	43.95	nr	126.66
Sum of two sides 800mm	51.44	87.47	1.46	46.16	nr	133.63
90 ° mitre bend						
Sum of two sides 200mm	15.35	26.10	2.04	64.50	nr	90.60
Sum of two sides 300mm	16.56	28.16	2.04	64.50	nr	92.66
Sum of two sides 400mm	28.21	47.97	2.09	66.08	nr	114.05
Sum of two sides 500mm	30.10	51.18	2.09	66.08	nr	117.27
Sum of two sides 600mm	32.57	55.38	2.15	67.98	nr	123.36
Sum of two sides 700mm	34.75	59.09	2.15	67.98	nr	127.07
Sum of two sides 800mm	37.18	63.22	2.26	71.46	nr	134.68
Branch						
Sum of two sides 200mm	25.47	43.31	0.92	29.09	nr	72.40
Sum of two sides 300mm	28.32	48.15	0.92	29.09	nr	77.24
Sum of two sides 400mm	35.68	60.67	0.95	30.04	nr	90.71
Sum of two sides 500mm	38.93	66.20	0.95	30.04	nr	96.23
Sum of two sides 600mm	42.09	71.57	1.03	32.57	nr	104.14
Sum of two sides 700mm	45.25	76.94	1.03	32.57	nr	109.51
Sum of two sides 800mm	48.41	82.32	1.03	32.57	nr	114.88
Grille neck						
Sum of two sides 200mm	16.58	28.19	1.10	34.78	nr	62.97
Sum of two sides 300mm	18.19	30.93	1.10	34.78	nr	65.71
Sum of two sides 400mm	29.54	50.23	1.16	36.68	nr	86.91
Sum of two sides 500mm	31.74	53.97	1.16	36.68	nr	90.65
Sum of two sides 600mm	34.24	58.22	1.18	37.31	nr	95.53
Sum of two sides 700mm	36.59	62.22	1.18	37.31	nr	99.53
Sum of two sides 800mm	39.06	66.42	1.18	37.31	nr	103.73
Ductwork 401 to 600mm longest side						
Sum of two sides 600mm	23.56	40.06	1.17	36.99	m	77.06
Sum of two sides 700mm	25.91	44.06	1.17	36.99	m	81.05
Sum of two sides 800mm	28.21	47.97	1.17	36.99	m	84.96
Sum of two sides 900mm	30.43	51.74	1.27	40.16	m	91.90
Sum of two sides 1000mm	32.64	55.50	1.48	46.80	m	102.30
Sum of two sides 1100mm	35.04	59.58	1.49	47.11	m	106.69
Sum of two sides 1200mm	37.26	63.36	1.49	47.11	m	110.47

U: VENTILATION/AIR CONDITIONING SYSTEMS

Item	Net Price £	Material £	Labour hours	Labour £	Unit	Total rate £
U10 : DUCTWORK : RECTANGULAR – CLASS C (cont'd)						
Galvanised sheet metal DW144 (cont'd)						
Extra over fittings: Ductwork 401 to 600mm longest side						
End Cap						
Sum of two sides 600mm	14.12	24.01	0.38	12.02	nr	36.02
Sum of two sides 700mm	15.24	25.91	0.38	12.02	nr	37.93
Sum of two sides 800mm	16.35	27.80	0.38	12.02	nr	39.82
Sum of two sides 900mm	17.47	29.71	0.38	12.02	nr	41.72
Sum of two sides 1000mm	18.58	31.59	0.38	12.02	nr	43.61
Sum of two sides 1100mm	19.70	33.50	0.38	12.02	nr	45.51
Sum of two sides 1200mm	20.82	35.40	0.38	12.02	nr	47.42
Reducer						
Sum of two sides 600mm	30.59	52.01	1.69	53.44	nr	105.45
Sum of two sides 700mm	32.61	55.45	1.69	53.44	nr	108.89
Sum of two sides 800mm	34.60	58.83	1.92	60.71	nr	119.54
Sum of two sides 900mm	36.62	62.27	1.92	60.71	nr	122.98
Sum of two sides 1000mm	38.64	65.70	2.18	68.93	nr	134.63
Sum of two sides 1100mm	40.83	69.43	2.18	68.93	nr	138.36
Sum of two sides 1200mm	42.85	72.86	2.18	68.93	nr	141.79
Offset						
Sum of two sides 600mm	42.97	73.07	1.92	60.71	nr	133.77
Sum of two sides 700mm	45.91	78.06	1.92	60.71	nr	138.77
Sum of two sides 800mm	47.53	80.82	1.92	60.71	nr	141.53
Sum of two sides 900mm	49.00	83.32	1.92	60.71	nr	144.03
Sum of two sides 1000mm	51.04	86.79	2.18	68.93	nr	155.72
Sum of two sides 1100mm	52.31	88.95	2.18	68.93	nr	157.88
Sum of two sides 1200mm	53.28	90.60	2.18	68.93	nr	159.53
Square to round						
Sum of two sides 600mm	32.28	54.89	1.33	42.05	nr	96.94
Sum of two sides 700mm	34.55	58.75	1.33	42.05	nr	100.80
Sum of two sides 800mm	36.78	62.54	1.40	44.27	nr	106.81
Sum of two sides 900mm	39.05	66.40	1.40	44.27	nr	110.67
Sum of two sides 1000mm	41.31	70.24	1.82	57.55	nr	127.79
Sum of two sides 1100mm	43.63	74.19	1.82	57.55	nr	131.73
Sum of two sides 1200mm	45.89	78.03	1.82	57.55	nr	135.58
90 ° radius bend						
Sum of two sides 600mm	28.72	48.83	1.16	36.68	nr	85.51
Sum of two sides 700mm	29.41	50.01	1.16	36.68	nr	86.69
Sum of two sides 800mm	31.41	53.41	1.22	38.58	nr	91.98
Sum of two sides 900mm	33.46	56.89	1.22	38.58	nr	95.47
Sum of two sides 1000mm	34.68	58.97	1.40	44.27	nr	103.24
Sum of two sides 1100mm	36.82	62.61	1.40	44.27	nr	106.87
Sum of two sides 1200mm	38.84	66.04	1.40	44.27	nr	110.31

U: VENTILATION/AIR CONDITIONING SYSTEMS

Item	Net Price £	Material £	Labour hours	Labour £	Unit	Total rate £
45 ° bend						
Sum of two sides 600mm	32.16	54.68	1.39	43.95	nr	**98.63**
Sum of two sides 700mm	33.71	57.32	1.39	43.95	nr	**101.27**
Sum of two sides 800mm	35.95	61.13	1.46	46.16	nr	**107.29**
Sum of two sides 900mm	38.20	64.95	1.46	46.16	nr	**111.12**
Sum of two sides 1000mm	40.04	68.08	1.88	59.44	nr	**127.53**
Sum of two sides 1100mm	42.46	72.20	1.88	59.44	nr	**131.64**
Sum of two sides 1200mm	44.70	76.01	1.88	59.44	nr	**135.45**
90 ° mitre bend						
Sum of two sides 600mm	49.81	84.70	2.15	67.98	nr	**152.68**
Sum of two sides 700mm	51.63	87.79	2.15	67.98	nr	**155.77**
Sum of two sides 800mm	55.34	94.10	2.26	71.46	nr	**165.56**
Sum of two sides 900mm	59.12	100.53	2.26	71.46	nr	**171.99**
Sum of two sides 1000mm	61.92	105.29	3.03	95.81	nr	**201.09**
Sum of two sides 1100mm	65.84	111.95	3.03	95.81	nr	**207.76**
Sum of two sides 1200mm	69.65	118.43	3.03	95.81	nr	**214.24**
Branch						
Sum of two sides 600mm	42.42	72.13	1.03	32.57	nr	**104.70**
Sum of two sides 700mm	45.61	77.55	1.03	32.57	nr	**110.12**
Sum of two sides 800mm	48.81	83.00	1.03	32.57	nr	**115.56**
Sum of two sides 900mm	52.00	88.42	1.03	32.57	nr	**120.99**
Sum of two sides 1000mm	55.19	93.84	1.29	40.79	nr	**134.63**
Sum of two sides 1100mm	58.51	99.49	1.29	40.79	nr	**140.28**
Sum of two sides 1200mm	61.71	104.93	1.29	40.79	nr	**145.72**
Grille neck						
Sum of two sides 600mm	43.58	74.10	1.18	37.31	nr	**111.41**
Sum of two sides 700mm	47.14	80.16	1.18	37.31	nr	**117.47**
Sum of two sides 800mm	50.71	86.23	1.18	37.31	nr	**123.54**
Sum of two sides 900mm	54.27	92.28	1.18	37.31	nr	**129.59**
Sum of two sides 1000mm	57.84	98.35	1.44	45.53	nr	**143.88**
Sum of two sides 1100mm	61.41	104.42	1.44	45.53	nr	**149.95**
Sum of two sides 1200mm	64.97	110.47	1.44	45.53	nr	**156.01**
Ductwork 601 to 800mm longest side						
Sum of two sides 900mm	33.07	56.23	1.27	40.16	m	**96.39**
Sum of two sides 1000mm	35.29	60.01	1.48	46.80	m	**106.80**
Sum of two sides 1100mm	37.51	63.78	1.49	47.11	m	**110.89**
Sum of two sides 1200mm	39.90	67.85	1.49	47.11	m	**114.96**
Sum of two sides 1300mm	42.12	71.62	1.51	47.74	m	**119.37**
Sum of two sides 1400mm	44.34	75.39	1.55	49.01	m	**124.40**
Sum of two sides 1500mm	46.56	79.17	1.61	50.91	m	**130.08**
Sum of two sides 1600mm	48.78	82.94	1.62	51.22	m	**134.17**
Extra over fittings; Ductwork 601 to 800mm longest side						
End Cap						
Sum of two sides 900mm	17.47	29.71	0.38	12.02	nr	**41.72**
Sum of two sides 1000mm	18.58	31.59	0.38	12.02	nr	**43.61**
Sum of two sides 1100mm	19.70	33.50	0.38	12.02	nr	**45.51**
Sum of two sides 1200mm	20.82	35.40	0.38	12.02	nr	**47.42**
Sum of two sides 1300mm	21.93	37.29	0.38	12.02	nr	**49.30**
Sum of two sides 1400mm	22.88	38.90	0.38	12.02	nr	**50.92**
Sum of two sides 1500mm	26.61	45.25	0.38	12.02	nr	**57.26**
Sum of two sides 1600mm	30.34	51.59	0.38	12.02	nr	**63.60**

U: VENTILATION/AIR CONDITIONING SYSTEMS

Item	Net Price £	Material £	Labour hours	Labour £	Unit	Total rate £
U10 : DUCTWORK : RECTANGULAR – CLASS C (cont'd)						
Galvanised sheet metal DW144 (cont'd)						
Extra over fittings; Ductwork 601 to 800mm longest side (cont'd)						
Reducer						
Sum of two sides 900mm	36.99	62.90	1.92	60.71	nr	123.61
Sum of two sides 1000mm	39.01	66.33	2.18	68.93	nr	135.26
Sum of two sides 1100mm	41.04	69.78	2.18	68.93	nr	138.71
Sum of two sides 1200mm	43.23	73.51	2.18	68.93	nr	142.44
Sum of two sides 1300mm	45.25	76.94	2.30	72.72	nr	149.67
Sum of two sides 1400mm	46.95	79.83	2.30	72.72	nr	152.56
Sum of two sides 1500mm	54.19	92.14	2.47	78.10	nr	170.24
Sum of two sides 1600mm	61.43	104.45	2.47	78.10	nr	182.55
Offset						
Sum of two sides 900mm	51.96	88.35	1.92	60.71	nr	149.06
Sum of two sides 1000mm	53.25	90.55	2.18	68.93	nr	159.48
Sum of two sides 1100mm	54.36	92.43	2.18	68.93	nr	161.36
Sum of two sides 1200mm	55.41	94.22	2.18	68.93	nr	163.15
Sum of two sides 1300mm	56.16	95.49	2.47	78.10	nr	173.59
Sum of two sides 1400mm	57.25	97.35	2.47	78.10	nr	175.45
Sum of two sides 1500mm	65.51	111.39	2.47	78.10	nr	189.49
Sum of two sides 1600mm	73.59	125.13	2.47	78.10	nr	203.23
Square to round						
Sum of two sides 900mm	38.70	65.80	1.40	44.27	nr	110.07
Sum of two sides 1000mm	40.97	69.66	1.82	57.55	nr	127.21
Sum of two sides 1100mm	43.24	73.52	1.82	57.55	nr	131.07
Sum of two sides 1200mm	45.55	77.45	1.82	57.55	nr	135.00
Sum of two sides 1300mm	47.81	81.30	2.32	73.36	nr	154.65
Sum of two sides 1400mm	49.68	84.47	2.32	73.36	nr	157.83
Sum of two sides 1500mm	58.47	99.42	2.56	80.95	nr	180.37
Sum of two sides 1600mm	67.26	114.37	2.58	81.58	nr	195.95
90 ° radius bend						
Sum of two sides 900mm	29.63	50.38	1.22	38.58	nr	88.96
Sum of two sides 1000mm	31.55	53.65	1.40	44.27	nr	97.91
Sum of two sides 1100mm	33.48	56.93	1.40	44.27	nr	101.20
Sum of two sides 1200mm	35.51	60.38	1.40	44.27	nr	104.65
Sum of two sides 1300mm	37.43	63.65	1.91	60.39	nr	124.04
Sum of two sides 1400mm	37.93	64.50	1.91	60.39	nr	124.89
Sum of two sides 1500mm	45.06	76.62	2.11	66.72	nr	143.34
Sum of two sides 1600mm	52.17	88.71	2.11	66.72	nr	155.43
45 ° bend						
Sum of two sides 900mm	36.84	62.64	1.46	46.16	nr	108.81
Sum of two sides 1000mm	39.04	66.38	1.88	59.44	nr	125.83
Sum of two sides 1100mm	41.24	70.12	1.88	59.44	nr	129.57
Sum of two sides 1200mm	43.60	74.14	1.88	59.44	nr	133.58
Sum of two sides 1300mm	45.81	77.89	2.26	71.46	nr	149.35
Sum of two sides 1400mm	47.11	80.10	2.44	77.15	nr	157.26
Sum of two sides 1500mm	54.52	92.70	2.44	77.15	nr	169.86
Sum of two sides 1600mm	61.93	105.30	2.68	84.74	nr	190.04

Material Costs/Prices for Measured Works – Mechanical Installations

U: VENTILATION/AIR CONDITIONING SYSTEMS

Item	Net Price £	Material £	Labour hours	Labour £	Unit	Total rate £
90 ° mitre bend						
Sum of two sides 900mm	53.69	91.29	2.26	71.46	nr	**162.75**
Sum of two sides 1000mm	57.63	97.99	3.03	95.81	nr	**193.80**
Sum of two sides 1100mm	61.55	104.66	3.03	95.81	nr	**200.46**
Sum of two sides 1200mm	65.57	111.49	3.03	95.81	nr	**207.30**
Sum of two sides 1300mm	69.49	118.16	3.85	121.73	nr	**239.89**
Sum of two sides 1400mm	71.62	121.78	3.85	121.73	nr	**243.52**
Sum of two sides 1500mm	83.42	141.85	4.25	134.38	nr	**276.23**
Sum of two sides 1600mm	95.21	161.89	4.26	134.70	nr	**296.59**
Branch						
Sum of two sides 900mm	53.43	90.85	1.03	32.57	nr	**123.42**
Sum of two sides 1000mm	56.70	96.41	1.29	40.79	nr	**137.20**
Sum of two sides 1100mm	59.97	101.97	1.29	40.79	nr	**142.76**
Sum of two sides 1200mm	63.36	107.74	1.29	40.79	nr	**148.52**
Sum of two sides 1300mm	66.63	113.30	1.39	43.95	nr	**157.25**
Sum of two sides 1400mm	69.48	118.14	1.39	43.95	nr	**162.09**
Sum of two sides 1500mm	79.70	135.52	1.64	51.86	nr	**187.38**
Sum of two sides 1600mm	89.93	152.92	1.64	51.86	nr	**204.77**
Grille neck						
Sum of two sides 900mm	54.27	92.28	1.18	37.31	nr	**129.59**
Sum of two sides 1000mm	57.84	98.35	1.44	45.53	nr	**143.88**
Sum of two sides 1100mm	61.41	104.42	1.44	45.53	nr	**149.95**
Sum of two sides 1200mm	64.97	110.47	1.44	45.53	nr	**156.01**
Sum of two sides 1300mm	68.54	116.54	1.69	53.44	nr	**169.98**
Sum of two sides 1400mm	71.62	121.78	1.69	53.44	nr	**175.22**
Sum of two sides 1500mm	83.01	141.15	1.79	56.60	nr	**197.75**
Sum of two sides 1600mm	94.41	160.53	1.79	56.60	nr	**217.13**
Ductwork 801 to 1000mm longest side						
Sum of two sides 1100mm	37.95	64.53	1.49	47.11	m	**111.64**
Sum of two sides 1200mm	40.17	68.30	1.49	47.11	m	**115.42**
Sum of two sides 1300mm	42.39	72.08	1.51	47.74	m	**119.82**
Sum of two sides 1400mm	44.78	76.14	1.55	49.01	m	**125.15**
Sum of two sides 1500mm	47.00	79.92	1.61	50.91	m	**130.82**
Sum of two sides 1600mm	49.22	83.69	1.62	51.22	m	**134.92**
Sum of two sides 1700mm	51.44	87.47	1.74	55.02	m	**142.48**
Sum of two sides 1800mm	53.83	91.53	1.76	55.65	m	**147.18**
Sum of two sides 1900mm	56.05	95.31	1.81	57.23	m	**152.54**
Sum of two sides 2000mm	58.27	99.08	1.82	57.55	m	**156.63**
Extra over fittings; Ductwork 801 to 1000mm longest side						
End Cap						
Sum of two sides 1100mm	19.70	33.50	0.38	12.02	nr	**45.51**
Sum of two sides 1200mm	20.82	35.40	0.38	12.02	nr	**47.42**
Sum of two sides 1300mm	21.93	37.29	0.38	12.02	nr	**49.30**
Sum of two sides 1400mm	22.88	38.90	0.38	12.02	nr	**50.92**
Sum of two sides 1500mm	26.61	45.25	0.38	12.02	nr	**57.26**
Sum of two sides 1600mm	30.34	51.59	0.38	12.02	nr	**63.60**
Sum of two sides 1700mm	34.07	57.93	0.58	18.34	nr	**76.27**
Sum of two sides 1800mm	37.79	64.26	0.58	18.34	nr	**82.60**
Sum of two sides 1900mm	41.52	70.60	0.58	18.34	nr	**88.94**
Sum of two sides 2000mm	45.24	76.93	0.58	18.34	nr	**95.26**

U: VENTILATION/AIR CONDITIONING SYSTEMS

Item	Net Price £	Material £	Labour hours	Labour £	Unit	Total rate £
U10 : DUCTWORK : RECTANGULAR – CLASS C (cont'd)						
Galvanised sheet metal DW144 (cont'd)						
Extra over fittings; Ductwork 801 to 1000mm longest side (cont'd)						
Reducer						
Sum of two sides 1100mm	37.85	64.36	2.18	68.93	nr	133.29
Sum of two sides 1200mm	39.70	67.51	2.18	68.93	nr	136.44
Sum of two sides 1300mm	41.56	70.67	2.30	72.72	nr	143.39
Sum of two sides 1400mm	43.23	73.51	2.30	72.72	nr	146.23
Sum of two sides 1500mm	50.30	85.53	2.47	78.10	nr	163.63
Sum of two sides 1600mm	57.38	97.57	2.47	78.10	nr	175.67
Sum of two sides 1700mm	64.45	109.59	2.59	81.89	nr	191.48
Sum of two sides 1800mm	71.66	121.85	2.59	81.89	nr	203.74
Sum of two sides 1900mm	78.73	133.87	2.71	85.69	nr	219.56
Sum of two sides 2000mm	85.80	145.89	2.71	85.69	nr	231.58
Offset						
Sum of two sides 1100mm	58.01	98.64	2.18	68.93	nr	167.57
Sum of two sides 1200mm	59.10	100.49	2.18	68.93	nr	169.42
Sum of two sides 1300mm	60.02	102.06	2.47	78.10	nr	180.16
Sum of two sides 1400mm	60.40	102.70	2.47	78.10	nr	180.80
Sum of two sides 1500mm	68.78	116.95	2.47	78.10	nr	195.05
Sum of two sides 1600mm	76.98	130.90	2.47	78.10	nr	208.99
Sum of two sides 1700mm	85.01	144.55	2.61	82.53	nr	227.08
Sum of two sides 1800mm	92.91	157.98	2.61	82.53	nr	240.51
Sum of two sides 1900mm	101.82	173.13	2.71	85.69	nr	258.82
Sum of two sides 2000mm	109.34	185.92	2.71	85.69	nr	271.61
Square to round						
Sum of two sides 1100mm	39.92	67.88	1.82	57.55	nr	125.43
Sum of two sides 1200mm	42.02	71.45	1.82	57.55	nr	129.00
Sum of two sides 1300mm	44.12	75.02	2.32	73.36	nr	148.38
Sum of two sides 1400mm	45.84	77.95	2.32	73.36	nr	151.30
Sum of two sides 1500mm	54.46	92.60	2.56	80.95	nr	173.55
Sum of two sides 1600mm	63.08	107.26	2.58	81.58	nr	188.84
Sum of two sides 1700mm	71.70	121.92	2.84	89.80	nr	211.72
Sum of two sides 1800mm	80.34	136.61	2.84	89.80	nr	226.41
Sum of two sides 1900mm	88.97	151.28	3.13	98.97	nr	250.25
Sum of two sides 2000mm	97.59	165.94	3.13	98.97	nr	264.91
90 ° radius bend						
Sum of two sides 1100mm	30.16	51.28	1.40	44.27	nr	95.55
Sum of two sides 1200mm	32.44	55.16	1.40	44.27	nr	99.43
Sum of two sides 1300mm	34.73	59.05	1.91	60.39	nr	119.45
Sum of two sides 1400mm	36.76	62.51	1.91	60.39	nr	122.90
Sum of two sides 1500mm	44.26	75.26	2.11	66.72	nr	141.98
Sum of two sides 1600mm	51.77	88.03	2.11	66.72	nr	154.75
Sum of two sides 1700mm	59.27	100.78	2.55	80.63	nr	181.41
Sum of two sides 1800mm	66.85	113.67	2.55	80.63	nr	194.30
Sum of two sides 1900mm	72.95	124.04	2.80	88.53	nr	212.58
Sum of two sides 2000mm	80.44	136.78	2.80	88.53	nr	225.31

Material Costs/Prices for Measured Works – Mechanical Installations

U: VENTILATION/AIR CONDITIONING SYSTEMS

Item	Net Price £	Material £	Labour hours	Labour £	Unit	Total rate £
45 ° bend						
Sum of two sides 1100mm	39.68	67.47	1.88	59.44	nr	126.92
Sum of two sides 1200mm	42.09	71.57	1.88	59.44	nr	131.01
Sum of two sides 1300mm	44.49	75.65	2.26	71.46	nr	147.11
Sum of two sides 1400mm	46.72	79.44	2.44	77.15	nr	156.59
Sum of two sides 1500mm	54.36	92.43	2.68	84.74	nr	177.17
Sum of two sides 1600mm	61.98	105.39	2.69	85.06	nr	190.45
Sum of two sides 1700mm	69.60	118.35	2.96	93.59	nr	211.94
Sum of two sides 1800mm	77.38	131.58	2.96	93.59	nr	225.17
Sum of two sides 1900mm	84.28	143.31	3.26	103.08	nr	246.39
Sum of two sides 2000mm	91.89	156.25	3.26	103.08	nr	259.33
90 ° mitre bend						
Sum of two sides 1100mm	61.49	104.56	3.03	95.81	nr	200.36
Sum of two sides 1200mm	65.62	111.58	3.03	95.81	nr	207.39
Sum of two sides 1300mm	69.73	118.57	3.85	121.73	nr	240.30
Sum of two sides 1400mm	73.42	124.84	3.85	121.73	nr	246.58
Sum of two sides 1500mm	85.37	145.16	4.25	134.38	nr	279.54
Sum of two sides 1600mm	97.32	165.48	4.26	134.70	nr	300.18
Sum of two sides 1700mm	109.27	185.80	4.68	147.98	nr	333.78
Sum of two sides 1800mm	121.26	206.19	4.68	147.98	nr	354.17
Sum of two sides 1900mm	131.66	223.87	4.87	153.99	nr	377.86
Sum of two sides 2000mm	143.65	244.26	4.87	153.99	nr	398.25
Branch						
Sum of two sides 1100mm	60.10	102.19	1.29	40.79	nr	142.98
Sum of two sides 1200mm	63.36	107.74	1.29	40.79	nr	148.52
Sum of two sides 1300mm	66.63	113.30	1.39	43.95	nr	157.25
Sum of two sides 1400mm	69.60	118.35	1.39	43.95	nr	162.30
Sum of two sides 1500mm	79.83	135.74	1.64	51.86	nr	187.60
Sum of two sides 1600mm	90.05	153.12	1.64	51.86	nr	204.97
Sum of two sides 1700mm	100.28	170.51	1.69	53.44	nr	223.95
Sum of two sides 1800mm	110.64	188.13	1.69	53.44	nr	241.57
Sum of two sides 1900mm	120.86	205.51	1.85	58.50	nr	264.00
Sum of two sides 2000mm	131.09	222.90	1.85	58.50	nr	281.40
Grille neck						
Sum of two sides 1100mm	61.41	104.42	1.44	45.53	nr	149.95
Sum of two sides 1200mm	64.97	110.47	1.44	45.53	nr	156.01
Sum of two sides 1300mm	68.54	116.54	1.69	53.44	nr	169.98
Sum of two sides 1400mm	71.62	121.78	1.69	53.44	nr	175.22
Sum of two sides 1500mm	83.01	141.15	1.79	56.60	nr	197.75
Sum of two sides 1600mm	94.41	160.53	1.79	56.60	nr	217.13
Sum of two sides 1700mm	105.80	179.90	1.86	58.81	nr	238.71
Sum of two sides 1800mm	117.19	199.27	1.86	58.81	nr	258.08
Sum of two sides 1900mm	128.59	218.65	2.02	63.87	nr	282.52
Sum of two sides 2000mm	139.98	238.02	2.02	63.87	nr	301.89

U: VENTILATION/AIR CONDITIONING SYSTEMS

Item	Net Price £	Material £	Labour hours	Labour £	Unit	Total rate £
U10 : DUCTWORK : RECTANGULAR – CLASS C (cont'd)						
Galvanised sheet metal DW144 (cont'd)						
Ductwork 1001 to 1250mm longest side						
Sum of two sides 1300mm	62.63	106.49	1.51	47.74	m	154.24
Sum of two sides 1400mm	65.60	111.54	1.55	49.01	m	160.55
Sum of two sides 1500mm	68.42	116.34	1.61	50.91	m	167.25
Sum of two sides 1600mm	71.42	121.44	1.62	51.22	m	172.66
Sum of two sides 1700mm	74.24	126.24	1.74	55.02	m	181.25
Sum of two sides 1800mm	77.06	131.03	1.76	55.65	m	186.68
Sum of two sides 1900mm	79.89	135.84	1.81	57.23	m	193.07
Sum of two sides 2000mm	82.89	140.94	1.82	57.55	m	198.49
Sum of two sides 2100mm	85.70	145.72	2.53	80.00	m	225.72
Sum of two sides 2200mm	97.87	166.42	2.55	80.63	m	247.05
Sum of two sides 2300mm	101.05	171.82	2.56	80.95	m	252.77
Sum of two sides 2400mm	103.87	176.62	2.76	87.27	m	263.89
Sum of two sides 2500mm	106.87	181.72	2.77	87.59	m	269.30
Extra over fittings; Ductwork 1001 to 1250mm longest side						
End Cap						
Sum of two sides 1300mm	22.33	37.97	0.38	12.02	nr	49.98
Sum of two sides 1400mm	23.31	39.64	0.38	12.02	nr	51.65
Sum of two sides 1500mm	27.07	46.03	0.38	12.02	nr	58.04
Sum of two sides 1600mm	30.82	52.41	0.38	12.02	nr	64.42
Sum of two sides 1700mm	34.58	58.80	0.58	18.34	nr	77.14
Sum of two sides 1800mm	38.34	65.19	0.58	18.34	nr	83.53
Sum of two sides 1900mm	42.09	71.57	0.58	18.34	nr	89.91
Sum of two sides 2000mm	45.85	77.96	0.58	18.34	nr	96.30
Sum of two sides 2100mm	49.61	84.36	0.87	27.51	nr	111.86
Sum of two sides 2200mm	53.36	90.73	0.87	27.51	nr	118.24
Sum of two sides 2300mm	57.12	97.13	0.87	27.51	nr	124.63
Sum of two sides 2400mm	60.88	103.52	0.87	27.51	nr	131.03
Sum of two sides 2500mm	64.63	109.90	0.87	27.51	nr	137.40
Reducer						
Sum of two sides 1300mm	33.21	56.47	2.30	72.72	nr	129.19
Sum of two sides 1400mm	34.57	58.78	2.30	72.72	nr	131.51
Sum of two sides 1500mm	41.46	70.50	2.47	78.10	nr	148.60
Sum of two sides 1600mm	48.50	82.47	2.47	78.10	nr	160.57
Sum of two sides 1700mm	55.39	94.18	2.59	81.89	nr	176.08
Sum of two sides 1800mm	62.28	105.90	2.59	81.89	nr	187.79
Sum of two sides 1900mm	69.81	118.70	2.71	85.69	nr	204.39
Sum of two sides 2000mm	76.21	129.59	2.71	85.69	nr	215.27
Sum of two sides 2100mm	83.11	141.32	2.92	92.33	nr	233.65
Sum of two sides 2200mm	84.39	143.50	2.92	92.33	nr	235.82
Sum of two sides 2300mm	91.42	155.45	2.92	92.33	nr	247.78
Sum of two sides 2400mm	98.32	167.18	3.12	98.65	nr	265.83
Sum of two sides 2500mm	105.35	179.13	3.12	98.65	nr	277.79

U: VENTILATION/AIR CONDITIONING SYSTEMS

Item	Net Price £	Material £	Labour hours	Labour £	Unit	Total rate £
Offset						
Sum of two sides 1300mm	78.07	132.75	2.47	78.10	nr	210.85
Sum of two sides 1400mm	80.48	136.85	2.47	78.10	nr	214.95
Sum of two sides 1500mm	89.41	152.03	2.47	78.10	nr	230.13
Sum of two sides 1600mm	98.22	167.01	2.47	78.10	nr	245.11
Sum of two sides 1700mm	106.68	181.40	2.61	82.53	nr	263.92
Sum of two sides 1800mm	114.92	195.41	2.61	82.53	nr	277.93
Sum of two sides 1900mm	122.93	209.03	2.71	85.69	nr	294.72
Sum of two sides 2000mm	130.77	222.36	2.71	85.69	nr	308.05
Sum of two sides 2100mm	138.32	235.20	2.92	92.33	nr	327.52
Sum of two sides 2200mm	136.58	232.24	3.26	103.08	nr	335.32
Sum of two sides 2300mm	148.04	251.72	3.26	103.08	nr	354.80
Sum of two sides 2400mm	159.47	271.16	3.47	109.72	nr	380.88
Sum of two sides 2500mm	170.93	290.65	3.48	110.03	nr	400.68
Square to round						
Sum of two sides 1300mm	35.41	60.21	2.32	73.36	nr	133.57
Sum of two sides 1400mm	36.94	62.81	2.32	73.36	nr	136.17
Sum of two sides 1500mm	45.38	77.16	2.56	80.95	nr	158.11
Sum of two sides 1600mm	53.84	91.55	2.58	81.58	nr	173.13
Sum of two sides 1700mm	62.29	105.92	2.84	89.80	nr	195.72
Sum of two sides 1800mm	70.73	120.27	2.84	89.80	nr	210.07
Sum of two sides 1900mm	87.27	148.39	3.13	98.97	nr	247.36
Sum of two sides 2000mm	87.63	149.00	3.13	98.97	nr	247.97
Sum of two sides 2100mm	96.07	163.36	4.26	134.70	nr	298.05
Sum of two sides 2200mm	98.91	168.18	4.27	135.01	nr	303.20
Sum of two sides 2300mm	107.37	182.57	4.30	135.96	nr	318.53
Sum of two sides 2400mm	115.81	196.92	4.77	150.82	nr	347.74
Sum of two sides 2500mm	124.27	211.31	4.79	151.46	nr	362.76
90° radius bend						
Sum of two sides 1300mm	15.00	25.51	1.91	60.39	nr	85.90
Sum of two sides 1400mm	14.75	25.08	1.91	60.39	nr	85.47
Sum of two sides 1500mm	22.14	37.65	2.11	66.72	nr	104.36
Sum of two sides 1600mm	29.57	50.28	2.11	66.72	nr	117.00
Sum of two sides 1700mm	36.96	62.85	2.55	80.63	nr	143.47
Sum of two sides 1800mm	44.35	75.41	2.55	80.63	nr	156.04
Sum of two sides 1900mm	51.75	87.99	2.80	88.53	nr	176.53
Sum of two sides 2000mm	59.18	100.63	2.80	88.53	nr	189.16
Sum of two sides 2100mm	66.57	113.19	2.80	88.53	nr	201.73
Sum of two sides 2200mm	63.11	107.31	2.80	88.53	nr	195.84
Sum of two sides 2300mm	65.09	110.68	4.35	137.54	nr	248.22
Sum of two sides 2400mm	72.45	123.19	4.35	137.54	nr	260.74
Sum of two sides 2500mm	79.84	135.76	4.35	137.54	nr	273.30
45° bend						
Sum of two sides 1300mm	34.18	58.12	2.26	71.46	nr	129.58
Sum of two sides 1400mm	35.12	59.72	2.44	77.15	nr	136.87
Sum of two sides 1500mm	42.70	72.61	2.68	84.74	nr	157.35
Sum of two sides 1600mm	50.40	85.70	2.69	85.06	nr	170.75
Sum of two sides 1700mm	57.96	98.55	2.96	93.59	nr	192.15
Sum of two sides 1800mm	65.54	111.44	2.96	93.59	nr	205.04
Sum of two sides 1900mm	73.11	124.31	3.26	103.08	nr	227.39
Sum of two sides 2000mm	80.81	137.41	3.26	103.08	nr	240.49
Sum of two sides 2100mm	88.38	150.28	7.50	237.14	nr	387.42
Sum of two sides 2200mm	90.06	153.14	7.50	237.14	nr	390.28
Sum of two sides 2300mm	94.96	161.47	7.55	238.72	nr	400.19
Sum of two sides 2400mm	102.52	174.32	8.13	257.06	nr	431.39
Sum of two sides 2500mm	110.20	187.38	8.30	262.44	nr	449.82

U: VENTILATION/AIR CONDITIONING SYSTEMS

Item	Net Price £	Material £	Labour hours	Labour £	Unit	Total rate £
U10 : DUCTWORK : RECTANGULAR – CLASS C (cont'd)						
Galvanised sheet metal DW144 (cont'd)						
Extra over fittings; Ductwork 1001 to 1250mm longest side (cont'd)						
90 ° mitre bend						
Sum of two sides 1300mm	76.21	129.59	3.85	121.73	nr	251.32
Sum of two sides 1400mm	79.21	134.69	3.85	121.73	nr	256.42
Sum of two sides 1500mm	93.30	158.65	4.25	134.38	nr	293.03
Sum of two sides 1600mm	107.39	182.60	4.26	134.70	nr	317.30
Sum of two sides 1700mm	131.67	223.89	4.68	147.98	nr	371.87
Sum of two sides 1800mm	145.70	247.75	4.68	147.98	nr	395.72
Sum of two sides 1900mm	159.73	271.60	4.87	153.99	nr	425.59
Sum of two sides 2000mm	163.74	278.42	4.87	153.99	nr	432.41
Sum of two sides 2100mm	177.82	302.36	7.50	237.14	nr	539.51
Sum of two sides 2200mm	179.29	304.86	7.50	237.14	nr	542.01
Sum of two sides 2300mm	186.98	317.94	7.55	238.72	nr	556.66
Sum of two sides 2400mm	201.12	341.98	8.13	257.06	nr	599.04
Sum of two sides 2500mm	215.24	365.99	8.30	262.44	nr	628.43
Branch						
Sum of two sides 1300mm	68.18	115.93	1.39	43.95	nr	159.88
Sum of two sides 1400mm	71.11	120.91	1.39	43.95	nr	164.86
Sum of two sides 1500mm	81.43	138.46	1.64	51.86	nr	190.32
Sum of two sides 1600mm	91.88	156.23	1.64	51.86	nr	208.09
Sum of two sides 1700mm	102.19	173.76	1.69	53.44	nr	227.20
Sum of two sides 1800mm	112.51	191.31	1.69	53.44	nr	244.75
Sum of two sides 1900mm	122.83	208.86	1.85	58.50	nr	267.35
Sum of two sides 2000mm	133.27	226.61	1.85	58.50	nr	285.11
Sum of two sides 2100mm	143.59	244.16	2.61	82.53	nr	326.68
Sum of two sides 2200mm	153.91	261.71	2.61	82.53	nr	344.23
Sum of two sides 2300mm	164.35	279.46	2.61	82.53	nr	361.98
Sum of two sides 2400mm	174.67	297.01	2.88	91.06	nr	388.07
Sum of two sides 2500mm	185.11	314.76	2.88	91.06	nr	405.82
Grille neck						
Sum of two sides 1300mm	70.51	119.89	1.69	53.44	nr	173.33
Sum of two sides 1400mm	73.74	125.39	1.69	53.44	nr	178.82
Sum of two sides 1500mm	85.29	145.03	1.79	56.60	nr	201.62
Sum of two sides 1600mm	96.83	164.65	1.79	56.60	nr	221.25
Sum of two sides 1700mm	108.38	184.29	1.86	58.81	nr	243.10
Sum of two sides 1800mm	119.93	203.93	1.86	58.81	nr	262.74
Sum of two sides 1900mm	131.47	223.55	2.02	63.87	nr	287.42
Sum of two sides 2000mm	143.02	243.19	2.02	63.87	nr	307.06
Sum of two sides 2100mm	154.57	262.83	2.61	82.53	nr	345.35
Sum of two sides 2200mm	166.11	282.45	2.80	88.53	nr	370.98
Sum of two sides 2300mm	177.66	302.09	2.80	88.53	nr	390.62
Sum of two sides 2400mm	189.21	321.73	3.06	96.75	nr	418.48
Sum of two sides 2500mm	200.75	341.35	3.06	96.75	nr	438.11

U: VENTILATION/AIR CONDITIONING SYSTEMS

Item	Net Price £	Material £	Labour hours	Labour £	Unit	Total rate £
Ductwork 1251 to 1600mm longest side						
Sum of two sides 1700mm	79.73	135.57	1.74	55.02	m	190.59
Sum of two sides 1800mm	82.84	140.86	1.76	55.65	m	196.51
Sum of two sides 1900mm	86.00	146.23	1.81	57.23	m	203.46
Sum of two sides 2000mm	88.98	151.30	1.82	57.55	m	208.85
Sum of two sides 2100mm	91.96	156.37	2.53	80.00	m	236.36
Sum of two sides 2200mm	94.94	161.43	2.55	80.63	m	242.06
Sum of two sides 2300mm	98.10	166.81	2.56	80.95	m	247.75
Sum of two sides 2400mm	101.21	172.10	2.76	87.27	m	259.36
Sum of two sides 2500mm	104.20	177.18	2.77	87.59	m	264.76
Sum of two sides 2600mm	116.69	198.42	2.97	93.91	m	292.33
Sum of two sides 2700mm	119.66	203.47	2.99	94.54	m	298.01
Sum of two sides 2800mm	122.85	208.89	3.30	104.34	m	313.24
Sum of two sides 2900mm	126.01	214.26	3.31	104.66	m	318.92
Sum of two sides 3000mm	128.99	219.33	3.53	111.62	m	330.95
Sum of two sides 3100mm	132.12	224.65	3.55	112.25	m	336.90
Sum of two sides 3200mm	135.10	229.72	3.56	112.56	m	342.29
Extra over fittings; Ductwork 1251 to 1600mm longest side						
End Cap						
Sum of two sides 1700mm	34.58	58.80	0.58	18.34	nr	77.14
Sum of two sides 1800mm	38.34	65.19	0.58	18.34	nr	83.53
Sum of two sides 1900mm	42.09	71.57	0.58	18.34	nr	89.91
Sum of two sides 2000mm	45.85	77.96	0.58	18.34	nr	96.30
Sum of two sides 2100mm	49.61	84.36	0.87	27.51	nr	111.86
Sum of two sides 2200mm	53.36	90.73	0.87	27.51	nr	118.24
Sum of two sides 2300mm	57.12	97.13	0.87	27.51	nr	124.63
Sum of two sides 2400mm	60.88	103.52	0.87	27.51	nr	131.03
Sum of two sides 2500mm	64.63	109.90	0.87	27.51	nr	137.40
Sum of two sides 2600mm	68.39	116.29	0.87	27.51	nr	143.80
Sum of two sides 2700mm	72.14	122.67	0.87	27.51	nr	150.17
Sum of two sides 2800mm	75.89	129.04	1.16	36.68	nr	165.72
Sum of two sides 2900mm	79.65	135.44	1.16	36.68	nr	172.11
Sum of two sides 3000mm	77.76	132.22	1.73	54.70	nr	186.92
Sum of two sides 3100mm	86.39	146.90	1.73	54.70	nr	201.60
Sum of two sides 3200mm	89.09	151.49	1.73	54.70	nr	206.19
Reducer						
Sum of two sides 1700mm	48.31	82.15	2.59	81.89	nr	164.04
Sum of two sides 1800mm	54.82	93.21	2.59	81.89	nr	175.11
Sum of two sides 1900mm	61.44	104.47	2.71	85.69	nr	190.16
Sum of two sides 2000mm	45.85	155.92	2.71	85.69	nr	241.61
Sum of two sides 2100mm	74.46	126.61	2.92	92.33	nr	218.94
Sum of two sides 2200mm	80.97	137.68	2.92	92.33	nr	230.01
Sum of two sides 2300mm	87.60	148.95	2.92	92.33	nr	241.28
Sum of two sides 2400mm	94.11	160.02	3.12	98.65	nr	258.67
Sum of two sides 2500mm	100.62	171.09	3.12	98.65	nr	269.74
Sum of two sides 2600mm	99.53	169.24	3.16	99.92	nr	269.16
Sum of two sides 2700mm	106.04	180.31	3.16	99.92	nr	280.23
Sum of two sides 2800mm	112.52	191.33	4.00	126.48	nr	317.80
Sum of two sides 2900mm	125.83	213.96	4.01	126.79	nr	340.75
Sum of two sides 3000mm	132.35	264.57	4.56	144.18	nr	408.76
Sum of two sides 3100mm	137.18	233.26	4.56	144.18	nr	377.44
Sum of two sides 3200mm	141.59	240.76	4.56	144.18	nr	384.94

U: VENTILATION/AIR CONDITIONING SYSTEMS

Item	Net Price £	Material £	Labour hours	Labour £	Unit	Total rate £
U10 : DUCTWORK : RECTANGULAR – CLASS C (cont'd)						
Galvanised sheet metal DW144 (cont'd)						
Extra over fittings; Ductwork 1251 to 1600mm longest side (cont'd)						
Offset						
Sum of two sides 1700mm	118.36	201.26	2.61	82.53	nr	283.78
Sum of two sides 1800mm	130.61	222.09	2.61	82.53	nr	304.61
Sum of two sides 1900mm	138.71	235.86	2.71	85.69	nr	321.55
Sum of two sides 2000mm	146.45	249.02	2.71	85.69	nr	334.71
Sum of two sides 2100mm	153.94	261.76	2.92	92.33	nr	354.08
Sum of two sides 2200mm	161.17	274.05	3.26	103.08	nr	377.13
Sum of two sides 2300mm	168.21	286.02	3.26	103.08	nr	389.10
Sum of two sides 2400mm	179.85	305.81	3.47	109.72	nr	415.53
Sum of two sides 2500mm	186.45	317.04	3.48	110.03	nr	427.07
Sum of two sides 2600mm	186.61	317.31	3.49	110.35	nr	427.66
Sum of two sides 2700mm	198.10	336.85	3.50	110.67	nr	447.51
Sum of two sides 2800mm	209.56	356.33	4.33	136.91	nr	493.24
Sum of two sides 2900mm	227.79	387.33	4.74	149.88	nr	537.20
Sum of two sides 3000mm	239.28	406.87	5.31	167.90	nr	574.76
Sum of two sides 3100mm	248.27	422.15	5.34	168.85	nr	591.00
Sum of two sides 3200mm	256.60	436.32	5.35	169.16	nr	605.48
Square to round						
Sum of two sides 1700mm	52.05	88.50	2.84	89.80	nr	178.30
Sum of two sides 1800mm	60.12	102.23	2.84	89.80	nr	192.03
Sum of two sides 1900mm	68.15	115.88	3.13	98.97	nr	214.85
Sum of two sides 2000mm	76.20	129.57	3.13	98.97	nr	228.54
Sum of two sides 2100mm	84.24	143.24	4.26	134.70	nr	277.94
Sum of two sides 2200mm	92.29	156.93	4.27	135.01	nr	291.94
Sum of two sides 2300mm	100.32	170.58	4.30	135.96	nr	306.54
Sum of two sides 2400mm	108.38	184.29	4.77	150.82	nr	335.11
Sum of two sides 2500mm	116.43	197.98	4.79	151.46	nr	349.43
Sum of two sides 2600mm	116.86	198.71	4.95	156.51	nr	355.22
Sum of two sides 2700mm	124.91	212.39	4.95	156.51	nr	368.91
Sum of two sides 2800mm	132.94	226.05	8.49	268.45	nr	494.50
Sum of two sides 2900mm	144.31	245.38	8.88	280.78	nr	526.16
Sum of two sides 3000mm	152.36	259.07	9.02	285.21	nr	544.28
Sum of two sides 3100mm	158.34	269.24	9.02	285.21	nr	554.44
Sum of two sides 3200mm	163.75	278.44	9.09	287.42	nr	565.86
90 ° radius bend						
Sum of two sides 1700mm	118.69	201.82	2.55	80.63	nr	282.45
Sum of two sides 1800mm	128.42	218.36	2.55	80.63	nr	298.99
Sum of two sides 1900mm	143.59	244.16	2.80	88.53	nr	332.69
Sum of two sides 2000mm	158.51	269.53	2.80	88.53	nr	358.06
Sum of two sides 2100mm	173.42	294.88	2.61	82.53	nr	377.41
Sum of two sides 2200mm	188.34	320.25	2.62	82.84	nr	403.09
Sum of two sides 2300mm	203.51	346.04	2.63	83.16	nr	429.20
Sum of two sides 2400mm	212.76	361.77	4.34	137.23	nr	499.00
Sum of two sides 2500mm	227.63	387.06	4.35	137.54	nr	524.60
Sum of two sides 2600mm	228.85	389.13	4.53	143.24	nr	532.37
Sum of two sides 2700mm	243.73	414.43	4.53	143.24	nr	557.67
Sum of two sides 2800mm	251.79	428.14	7.13	225.44	nr	653.58
Sum of two sides 2900mm	280.23	476.50	7.17	226.71	nr	703.21

U: VENTILATION/AIR CONDITIONING SYSTEMS

Item	Net Price £	Material £	Labour hours	Labour £	Unit	Total rate £
Sum of two sides 3000mm	295.06	501.71	7.26	229.56	nr	731.27
Sum of two sides 3100mm	306.55	521.25	7.26	229.56	nr	750.81
Sum of two sides 3200mm	317.16	539.29	7.31	231.14	nr	770.43
45 ° bend						
Sum of two sides 1700mm	57.08	97.06	2.96	93.59	nr	190.65
Sum of two sides 1800mm	61.86	105.19	2.96	93.59	nr	198.78
Sum of two sides 1900mm	69.43	118.06	3.26	103.08	nr	221.14
Sum of two sides 2000mm	76.88	130.73	3.26	103.08	nr	233.80
Sum of two sides 2100mm	84.33	143.39	7.50	237.14	nr	380.54
Sum of two sides 2200mm	91.77	156.04	7.50	237.14	nr	393.19
Sum of two sides 2300mm	99.35	168.93	7.55	238.72	nr	407.66
Sum of two sides 2400mm	103.89	176.65	8.13	257.06	nr	433.72
Sum of two sides 2500mm	111.31	189.27	8.30	262.44	nr	451.71
Sum of two sides 2600mm	111.44	189.49	8.56	270.66	nr	460.15
Sum of two sides 2700mm	118.87	202.12	8.62	272.56	nr	474.68
Sum of two sides 2800mm	122.82	208.84	9.09	287.42	nr	496.26
Sum of two sides 2900mm	137.02	232.99	9.09	287.42	nr	520.40
Sum of two sides 3000mm	144.42	245.57	9.62	304.18	nr	549.75
Sum of two sides 3100mm	150.15	255.31	9.62	304.18	nr	559.49
Sum of two sides 3200mm	155.45	264.32	9.62	304.18	nr	568.50
90 ° mitre bend						
Sum of two sides 1700mm	122.55	208.38	4.68	147.98	nr	356.36
Sum of two sides 1800mm	130.06	221.15	4.68	147.98	nr	369.13
Sum of two sides 1900mm	144.34	245.43	4.87	153.99	nr	399.42
Sum of two sides 2000mm	158.66	269.78	4.87	153.99	nr	423.77
Sum of two sides 2100mm	172.99	294.15	7.50	237.14	nr	531.29
Sum of two sides 2200mm	187.31	318.50	7.50	237.14	nr	555.64
Sum of two sides 2300mm	201.59	342.78	7.55	238.72	nr	581.50
Sum of two sides 2400mm	208.92	355.24	8.13	257.06	nr	612.31
Sum of two sides 2500mm	223.27	379.64	8.30	262.44	nr	642.08
Sum of two sides 2600mm	221.46	376.57	8.56	270.66	nr	647.23
Sum of two sides 2700mm	235.81	400.97	8.62	272.56	nr	673.52
Sum of two sides 2800mm	241.89	411.30	15.20	480.61	nr	891.92
Sum of two sides 2900mm	266.84	453.73	15.20	480.61	nr	934.34
Sum of two sides 3000mm	281.23	478.20	15.60	493.26	nr	971.46
Sum of two sides 3100mm	293.02	498.25	16.04	507.17	nr	1005.42
Sum of two sides 3200mm	304.24	517.32	16.04	507.17	nr	1024.50
Branch						
Sum of two sides 1700mm	105.34	179.12	1.69	53.44	nr	232.55
Sum of two sides 1800mm	115.65	196.65	1.69	53.44	nr	250.09
Sum of two sides 1900mm	126.11	214.43	1.85	58.50	nr	272.93
Sum of two sides 2000mm	136.45	232.02	1.85	58.50	nr	290.51
Sum of two sides 2100mm	146.78	249.58	2.61	82.53	nr	332.11
Sum of two sides 2200mm	157.12	267.16	2.61	82.53	nr	349.69
Sum of two sides 2300mm	167.58	284.95	2.61	82.53	nr	367.48
Sum of two sides 2400mm	177.90	302.50	2.88	91.06	nr	393.56
Sum of two sides 2500mm	188.23	320.06	2.88	91.06	nr	411.13
Sum of two sides 2600mm	198.57	337.64	2.88	91.06	nr	428.71
Sum of two sides 2700mm	208.90	355.21	2.88	91.06	nr	446.27
Sum of two sides 2800mm	219.22	372.76	3.94	124.58	nr	497.34
Sum of two sides 2900mm	233.03	396.24	3.94	124.58	nr	520.82
Sum of two sides 3000mm	243.36	413.80	4.83	152.72	nr	566.53
Sum of two sides 3100mm	251.62	427.85	4.83	152.72	nr	580.57
Sum of two sides 3200mm	259.15	440.65	4.83	152.72	nr	593.37

U: VENTILATION/AIR CONDITIONING SYSTEMS

Item	Net Price £	Material £	Labour hours	Labour £	Unit	Total rate £
U10 : DUCTWORK : RECTANGULAR – CLASS C (cont'd)						
Galvanised sheet metal DW144 (cont'd)						
Extra over fittings; Ductwork 1251 to 1600mm longest side (cont'd)						
Grille neck						
Sum of two sides 1700mm	108.38	184.29	1.86	58.81	nr	243.10
Sum of two sides 1800mm	119.93	203.93	1.86	58.81	nr	262.74
Sum of two sides 1900mm	131.47	223.55	2.02	63.87	nr	287.42
Sum of two sides 2000mm	143.02	243.19	2.02	63.87	nr	307.06
Sum of two sides 2100mm	154.57	262.83	2.80	88.53	nr	351.36
Sum of two sides 2200mm	166.11	282.45	2.80	88.53	nr	370.98
Sum of two sides 2300mm	177.66	302.09	2.80	88.53	nr	390.62
Sum of two sides 2400mm	189.21	321.73	3.06	96.75	nr	418.48
Sum of two sides 2500mm	200.75	341.35	3.06	96.75	nr	438.11
Sum of two sides 2600mm	212.30	360.99	3.08	97.39	nr	458.38
Sum of two sides 2700mm	223.85	380.63	3.08	97.39	nr	478.02
Sum of two sides 2800mm	235.39	400.25	4.13	130.59	nr	530.84
Sum of two sides 2900mm	246.94	419.89	4.13	130.59	nr	550.48
Sum of two sides 3000mm	258.49	439.53	5.02	158.73	nr	598.26
Sum of two sides 3100mm	267.70	455.19	5.02	158.73	nr	613.92
Sum of two sides 3200mm	276.08	469.44	5.02	158.73	nr	628.17
Ductwork 1601 to 2000mm longest side						
Sum of two sides 2100mm	106.93	181.82	2.53	80.00	m	261.82
Sum of two sides 2200mm	110.38	187.69	2.55	80.63	m	268.32
Sum of two sides 2300mm	113.87	193.62	2.55	80.63	m	274.25
Sum of two sides 2400mm	117.28	199.42	2.56	80.95	m	280.37
Sum of two sides 2500mm	120.77	205.35	2.76	87.27	m	292.62
Sum of two sides 2600mm	124.09	211.00	2.77	87.59	m	298.59
Sum of two sides 2700mm	127.41	216.65	2.97	93.91	m	310.55
Sum of two sides 2800mm	130.73	222.29	2.99	94.54	m	316.83
Sum of two sides 2900mm	134.23	228.24	3.30	104.34	m	332.59
Sum of two sides 3000mm	137.54	233.87	3.31	104.66	m	338.53
Sum of two sides 3100mm	157.90	268.49	3.53	111.62	m	380.11
Sum of two sides 3200mm	161.22	274.14	3.53	111.62	m	385.75
Sum of two sides 3300mm	164.72	280.09	3.53	111.62	m	391.70
Sum of two sides 3400mm	168.04	285.73	3.55	112.25	m	397.98
Sum of two sides 3500mm	171.35	291.36	3.55	112.25	m	403.61
Sum of two sides 3600mm	174.67	297.01	3.55	112.25	m	409.25
Sum of two sides 3700mm	178.17	302.96	3.56	112.56	m	415.52
Sum of two sides 3800mm	181.49	308.60	3.56	112.56	m	421.17
Sum of two sides 3900mm	184.80	314.23	3.56	112.56	m	426.79
Sum of two sides 4000mm	188.13	319.89	3.56	112.56	m	432.46

U: VENTILATION/AIR CONDITIONING SYSTEMS

Item	Net Price £	Material £	Labour hours	Labour £	Unit	Total rate £
Extra over fittings; Ductwork 1601 to 2000mm longest side						
End Cap						
Sum of two sides 2100mm	50.27	85.48	0.87	27.51	nr	**112.99**
Sum of two sides 2200mm	54.05	91.91	0.87	27.51	nr	**119.41**
Sum of two sides 2300mm	57.85	98.37	0.87	27.51	nr	**125.88**
Sum of two sides 2400mm	61.63	104.79	0.87	27.51	nr	**132.30**
Sum of two sides 2500mm	65.42	111.24	0.87	27.51	nr	**138.75**
Sum of two sides 2600mm	69.21	117.68	0.87	27.51	nr	**145.19**
Sum of two sides 2700mm	73.00	124.13	0.87	27.51	nr	**151.64**
Sum of two sides 2800mm	76.78	130.56	1.16	36.68	nr	**167.23**
Sum of two sides 2900mm	80.57	137.00	1.16	36.68	nr	**173.68**
Sum of two sides 3000mm	84.36	143.44	1.73	54.70	nr	**198.15**
Sum of two sides 3100mm	87.37	148.56	1.73	54.70	nr	**203.26**
Sum of two sides 3200mm	90.10	153.20	1.73	54.70	nr	**207.91**
Sum of two sides 3300mm	92.84	157.86	1.73	54.70	nr	**212.56**
Sum of two sides 3400mm	95.57	162.51	1.73	54.70	nr	**217.21**
Sum of two sides 3500mm	98.31	167.16	1.73	54.70	nr	**221.87**
Sum of two sides 3600mm	101.04	171.81	1.73	54.70	nr	**226.51**
Sum of two sides 3700mm	103.78	176.47	1.73	54.70	nr	**231.17**
Sum of two sides 3800mm	106.51	181.11	1.73	54.70	nr	**235.81**
Sum of two sides 3900mm	109.25	185.77	1.73	54.70	nr	**240.47**
Sum of two sides 4000mm	111.98	190.41	1.73	54.70	nr	**245.11**
Reducer						
Sum of two sides 2100mm	69.60	118.35	2.92	92.33	nr	**210.67**
Sum of two sides 2200mm	76.10	129.40	2.92	92.33	nr	**221.73**
Sum of two sides 2300mm	82.70	140.62	2.92	92.33	nr	**232.95**
Sum of two sides 2400mm	89.10	151.50	3.12	98.65	nr	**250.16**
Sum of two sides 2500mm	95.70	162.73	3.12	98.65	nr	**261.38**
Sum of two sides 2600mm	102.19	173.76	3.16	99.92	nr	**273.68**
Sum of two sides 2700mm	108.69	184.81	3.16	99.92	nr	**284.73**
Sum of two sides 2800mm	115.18	195.85	4.00	126.48	nr	**322.33**
Sum of two sides 2900mm	121.78	207.07	4.01	126.79	nr	**333.87**
Sum of two sides 3000mm	128.28	218.12	4.01	126.79	nr	**344.92**
Sum of two sides 3100mm	119.58	203.33	4.01	126.79	nr	**330.12**
Sum of two sides 3200mm	123.97	210.80	4.56	144.18	nr	**354.98**
Sum of two sides 3300mm	135.29	230.04	4.56	144.18	nr	**374.23**
Sum of two sides 3400mm	139.68	237.51	4.56	144.18	nr	**381.69**
Sum of two sides 3500mm	144.07	244.97	4.56	144.18	nr	**389.16**
Sum of two sides 3600mm	148.46	252.44	4.56	144.18	nr	**396.62**
Sum of two sides 3700mm	152.95	260.07	4.56	144.18	nr	**404.26**
Sum of two sides 3800mm	157.34	267.54	4.56	144.18	nr	**411.72**
Sum of two sides 3900mm	161.73	275.00	4.56	144.18	nr	**419.19**
Sum of two sides 4000mm	166.12	282.47	4.56	144.18	nr	**426.65**
Offset						
Sum of two sides 2100mm	163.22	277.54	2.92	92.33	nr	**369.86**
Sum of two sides 2200mm	175.45	298.33	3.26	103.08	nr	**401.41**
Sum of two sides 2300mm	182.42	310.18	3.26	103.08	nr	**413.26**
Sum of two sides 2400mm	199.91	339.92	3.47	109.72	nr	**449.64**
Sum of two sides 2500mm	206.63	351.35	3.48	110.03	nr	**461.38**
Sum of two sides 2600mm	212.97	362.13	3.49	110.35	nr	**472.48**
Sum of two sides 2700mm	219.07	372.50	3.50	110.67	nr	**483.17**
Sum of two sides 2800mm	224.91	382.43	4.33	136.91	nr	**519.34**
Sum of two sides 2900mm	230.54	392.01	4.74	149.88	nr	**541.88**
Sum of two sides 3000mm	235.87	401.07	5.31	167.90	nr	**568.97**
Sum of two sides 3100mm	224.69	382.06	5.34	168.85	nr	**550.90**

U: VENTILATION/AIR CONDITIONING SYSTEMS

Item	Net Price £	Material £	Labour hours	Labour £	Unit	Total rate £
U10 : DUCTWORK : RECTANGULAR – CLASS C (cont'd)						
Galvanised sheet metal DW144 (cont'd)						
Extra over fittings; Ductwork 1601 to 2000mm longest side (cont'd)						
Offset (cont'd)						
Sum of two sides 3200mm	232.99	396.17	5.35	169.16	nr	565.33
Sum of two sides 3300mm	248.17	421.98	5.35	169.16	nr	591.15
Sum of two sides 3400mm	256.48	436.11	5.35	169.16	nr	605.28
Sum of two sides 3500mm	264.79	450.24	5.35	169.16	nr	619.41
Sum of two sides 3600mm	273.10	464.37	5.35	169.16	nr	633.54
Sum of two sides 3700mm	281.45	478.57	5.35	169.16	nr	647.73
Sum of two sides 3800mm	289.76	492.70	5.35	169.16	nr	661.86
Sum of two sides 3900mm	298.07	506.83	5.35	169.16	nr	676.00
Sum of two sides 4000mm	306.38	520.96	5.35	169.16	nr	690.13
Square to round						
Sum of two sides 2100mm	102.03	173.49	4.26	134.70	nr	308.19
Sum of two sides 2200mm	111.89	190.26	4.27	135.01	nr	325.27
Sum of two sides 2300mm	121.71	206.95	4.30	135.96	nr	342.92
Sum of two sides 2400mm	207.89	353.49	4.77	150.82	nr	504.32
Sum of two sides 2500mm	141.42	240.47	4.79	151.46	nr	391.92
Sum of two sides 2600mm	151.26	257.20	4.95	156.51	nr	413.71
Sum of two sides 2700mm	161.10	273.93	4.95	156.51	nr	430.45
Sum of two sides 2800mm	170.93	290.65	8.49	268.45	nr	559.09
Sum of two sides 2900mm	180.75	307.34	8.88	280.78	nr	588.12
Sum of two sides 3000mm	190.59	324.08	9.02	285.21	nr	609.28
Sum of two sides 3100mm	184.46	313.65	9.02	285.21	nr	598.86
Sum of two sides 3200mm	191.13	324.99	9.09	287.42	nr	612.41
Sum of two sides 3300mm	201.21	342.13	9.09	287.42	nr	629.55
Sum of two sides 3400mm	207.89	353.49	9.09	287.42	nr	640.91
Sum of two sides 3500mm	214.56	364.83	9.09	287.42	nr	652.25
Sum of two sides 3600mm	221.24	376.19	9.09	287.42	nr	663.61
Sum of two sides 3700mm	227.90	387.52	9.09	287.42	nr	674.94
Sum of two sides 3800mm	234.57	398.86	9.09	287.42	nr	686.28
Sum of two sides 3900mm	241.25	410.22	9.09	287.42	nr	697.64
Sum of two sides 4000mm	247.93	421.58	9.09	287.42	nr	708.99
90 ° radius bend						
Sum of two sides 2100mm	151.06	256.86	2.61	82.53	nr	339.39
Sum of two sides 2200mm	288.85	491.15	2.62	82.84	nr	574.00
Sum of two sides 2300mm	312.93	532.10	2.63	83.16	nr	615.26
Sum of two sides 2400mm	320.17	544.41	4.34	137.23	nr	681.64
Sum of two sides 2500mm	344.22	585.30	4.35	137.54	nr	722.85
Sum of two sides 2600mm	367.86	625.50	4.53	143.24	nr	768.74
Sum of two sides 2700mm	391.49	665.68	4.53	143.24	nr	808.92
Sum of two sides 2800mm	415.13	705.88	7.13	225.44	nr	931.32
Sum of two sides 2900mm	439.18	746.77	7.17	226.71	nr	973.48
Sum of two sides 3000mm	462.81	786.95	7.26	229.56	nr	1016.51
Sum of two sides 3100mm	450.63	766.24	7.26	229.56	nr	995.80
Sum of two sides 3200mm	467.94	795.68	7.31	231.14	nr	1026.81
Sum of two sides 3300mm	506.16	860.66	7.31	231.14	nr	1091.80
Sum of two sides 3400mm	523.47	890.10	7.31	231.14	nr	1121.23
Sum of two sides 3500mm	540.78	919.53	7.31	231.14	nr	1150.67
Sum of two sides 3600mm	558.10	948.98	7.31	231.14	nr	1180.12

U: VENTILATION/AIR CONDITIONING SYSTEMS

Item	Net Price £	Material £	Labour hours	Labour £	Unit	Total rate £
Sum of two sides 3700mm	575.83	979.13	7.31	231.14	nr	1210.27
Sum of two sides 3800mm	593.15	1008.58	7.31	231.14	nr	1239.72
Sum of two sides 3900mm	610.46	1038.01	7.31	231.14	nr	1269.15
Sum of two sides 4000mm	627.78	1067.46	7.31	231.14	nr	1298.60
45° bend						
Sum of two sides 2100mm	72.79	123.77	7.50	237.14	nr	360.91
Sum of two sides 2200mm	206.35	350.87	7.50	237.14	nr	588.02
Sum of two sides 2300mm	222.45	378.25	7.55	238.72	nr	616.97
Sum of two sides 2400mm	229.61	390.42	8.13	257.06	nr	647.49
Sum of two sides 2500mm	245.70	417.78	8.30	262.44	nr	680.22
Sum of two sides 2600mm	261.45	444.56	8.56	270.66	nr	715.22
Sum of two sides 2700mm	277.20	471.35	8.62	272.56	nr	743.90
Sum of two sides 2800mm	292.96	498.14	9.09	287.42	nr	785.56
Sum of two sides 2900mm	309.04	525.49	9.09	287.42	nr	812.90
Sum of two sides 3000mm	324.80	552.28	9.62	304.18	nr	856.46
Sum of two sides 3100mm	321.02	545.86	9.62	304.18	nr	850.03
Sum of two sides 3200mm	332.56	565.48	9.62	304.18	nr	869.66
Sum of two sides 3300mm	358.08	608.87	9.62	304.18	nr	913.05
Sum of two sides 3400mm	369.62	628.49	9.62	304.18	nr	932.67
Sum of two sides 3500mm	381.16	648.12	9.62	304.18	nr	952.29
Sum of two sides 3600mm	392.71	667.76	9.62	304.18	nr	971.93
Sum of two sides 3700mm	404.58	687.94	9.62	304.18	nr	992.12
Sum of two sides 3800mm	416.13	707.58	9.62	304.18	nr	1011.76
Sum of two sides 3900mm	427.67	727.20	9.62	304.18	nr	1031.38
Sum of two sides 4000mm	439.21	746.82	9.62	304.18	nr	1051.00
90° mitre bend						
Sum of two sides 2100mm	329.33	559.99	7.50	237.14	nr	797.13
Sum of two sides 2200mm	347.79	591.38	7.50	237.14	nr	828.52
Sum of two sides 2300mm	375.17	637.93	7.55	238.72	nr	876.66
Sum of two sides 2400mm	383.77	652.55	8.13	257.06	nr	909.62
Sum of two sides 2500mm	411.17	699.15	8.30	262.44	nr	961.58
Sum of two sides 2600mm	438.72	745.99	8.56	270.66	nr	1016.65
Sum of two sides 2700mm	466.26	792.82	8.62	272.56	nr	1065.38
Sum of two sides 2800mm	493.80	839.65	15.20	480.61	nr	1320.26
Sum of two sides 2900mm	521.20	886.24	15.20	480.61	nr	1366.85
Sum of two sides 3000mm	548.74	933.07	15.60	493.26	nr	1426.33
Sum of two sides 3100mm	534.13	908.22	16.04	507.17	nr	1415.40
Sum of two sides 3200mm	555.35	944.31	16.04	507.17	nr	1451.48
Sum of two sides 3300mm	588.20	1000.16	16.04	507.17	nr	1507.34
Sum of two sides 3400mm	609.42	1036.25	16.04	507.17	nr	1543.42
Sum of two sides 3500mm	630.65	1072.34	16.04	507.17	nr	1579.52
Sum of two sides 3600mm	651.87	1108.43	16.04	507.17	nr	1615.60
Sum of two sides 3700mm	672.95	1144.27	16.04	507.17	nr	1651.44
Sum of two sides 3800mm	694.17	1180.35	16.04	507.17	nr	1687.52
Sum of two sides 3900mm	715.40	1216.45	16.04	507.17	nr	1723.62
Sum of two sides 4000mm	736.62	1252.53	16.04	507.17	nr	1759.71
Branch						
Sum of two sides 2100mm	149.45	254.12	2.61	82.53	nr	336.65
Sum of two sides 2200mm	159.86	271.82	2.61	82.53	nr	354.35
Sum of two sides 2300mm	170.42	289.78	2.61	82.53	nr	372.30
Sum of two sides 2400mm	180.69	307.24	2.88	91.06	nr	398.30
Sum of two sides 2500mm	191.24	325.18	2.88	91.06	nr	416.24
Sum of two sides 2600mm	201.68	342.93	2.88	91.06	nr	434.00
Sum of two sides 2700mm	212.10	360.65	2.88	91.06	nr	451.71
Sum of two sides 2800mm	222.53	378.39	3.94	124.58	nr	502.97
Sum of two sides 2900mm	233.09	396.34	3.94	124.58	nr	520.92

U: VENTILATION/AIR CONDITIONING SYSTEMS

Item	Net Price £	Material £	Labour hours	Labour £	Unit	Total rate £
U10 : DUCTWORK : RECTANGULAR – CLASS C (cont'd)						
Galvanised sheet metal DW144 (cont'd)						
Extra over fittings; Ductwork 1601 to 2000mm longest side (cont'd)						
Branch (cont'd)						
Sum of two sides 3000mm	243.52	414.08	4.83	152.72	nr	566.80
Sum of two sides 3100mm	251.87	428.27	4.83	152.72	nr	581.00
Sum of two sides 3200mm	259.49	441.23	4.83	152.72	nr	593.95
Sum of two sides 3300mm	270.65	460.21	4.83	152.72	nr	612.93
Sum of two sides 3400mm	278.28	473.18	4.83	152.72	nr	625.90
Sum of two sides 3500mm	285.90	486.14	4.83	152.72	nr	638.86
Sum of two sides 3600mm	293.52	499.10	4.83	152.72	nr	651.82
Sum of two sides 3700mm	301.27	512.27	4.83	152.72	nr	664.99
Sum of two sides 3800mm	308.89	525.23	4.83	152.72	nr	677.95
Sum of two sides 3900mm	316.52	538.20	4.83	152.72	nr	690.92
Sum of two sides 4000mm	324.14	551.16	4.83	152.72	nr	703.88
Grille neck						
Sum of two sides 2100mm	157.88	268.46	2.80	88.53	nr	356.99
Sum of two sides 2200mm	169.59	288.37	2.80	88.53	nr	376.90
Sum of two sides 2300mm	181.29	308.26	2.80	88.53	nr	396.80
Sum of two sides 2400mm	192.99	328.16	3.06	96.75	nr	424.91
Sum of two sides 2500mm	204.70	348.07	3.06	96.75	nr	444.82
Sum of two sides 2600mm	216.41	367.98	3.08	97.39	nr	465.37
Sum of two sides 2700mm	212.10	360.65	3.08	97.39	nr	458.04
Sum of two sides 2800mm	222.53	378.39	4.13	130.59	nr	508.97
Sum of two sides 2900mm	233.09	396.34	4.13	130.59	nr	526.93
Sum of two sides 3000mm	263.22	447.57	5.02	158.73	nr	606.30
Sum of two sides 3100mm	272.59	463.51	5.02	158.73	nr	622.24
Sum of two sides 3200mm	281.14	478.04	5.02	158.73	nr	636.77
Sum of two sides 3300mm	289.69	492.58	5.02	158.73	nr	651.31
Sum of two sides 3400mm	298.23	507.10	5.02	158.73	nr	665.83
Sum of two sides 3500mm	306.77	521.63	5.02	158.73	nr	680.35
Sum of two sides 3600mm	315.32	536.16	5.02	158.73	nr	694.89
Sum of two sides 3700mm	323.86	550.69	5.02	158.73	nr	709.41
Sum of two sides 3800mm	332.41	565.22	5.02	158.73	nr	723.95
Sum of two sides 3900mm	340.95	579.74	5.02	158.73	nr	738.47
Sum of two sides 4000mm	349.50	594.28	5.02	158.73	nr	753.01
Ductwork 2001 to 2500mm longest side						
Sum of two sides 2500mm	150.55	247.79	2.77	87.59	m	335.38
Sum of two sides 2600mm	154.99	255.10	2.97	93.91	m	349.01
Sum of two sides 2700mm	159.03	261.75	2.99	94.54	m	356.29
Sum of two sides 2800mm	163.64	269.34	3.30	104.34	m	373.68
Sum of two sides 2900mm	167.67	275.97	3.31	104.66	m	380.63
Sum of two sides 3000mm	171.58	282.41	3.53	111.62	m	394.02
Sum of two sides 3100mm	175.60	289.03	3.55	112.25	m	401.27
Sum of two sides 3200mm	179.51	295.46	3.56	112.56	m	408.03
Sum of two sides 3300mm	183.55	302.11	3.56	112.56	m	414.67
Sum of two sides 3400mm	187.45	308.53	3.56	112.56	m	421.09
Sum of two sides 3500mm	205.15	337.66	3.56	112.56	m	450.23
Sum of two sides 3600mm	209.05	344.08	3.56	112.56	m	456.65
Sum of two sides 3700mm	213.09	350.73	3.56	112.56	m	463.30
Sum of two sides 3800mm	217.00	357.17	3.56	112.56	m	469.73

U: VENTILATION/AIR CONDITIONING SYSTEMS

Item	Net Price £	Material £	Labour hours	Labour £	Unit	Total rate £
Sum of two sides 3900mm	220.91	363.60	3.56	112.56	m	476.17
Sum of two sides 4000mm	225.50	383.44	3.56	112.56	m	496.00
Extra over fittings; Ductwork 2001 to 2500mm longest side						
End Cap						
Sum of two sides 2500mm	48.09	81.77	0.87	27.51	nr	109.28
Sum of two sides 2600mm	50.87	86.50	0.87	27.51	nr	114.01
Sum of two sides 2700mm	53.66	91.24	0.87	27.51	nr	118.75
Sum of two sides 2800mm	56.45	95.99	1.16	36.68	nr	132.66
Sum of two sides 2900mm	59.23	100.71	1.16	36.68	nr	137.39
Sum of two sides 3000mm	62.01	105.44	1.73	54.70	nr	160.14
Sum of two sides 3100mm	64.23	109.22	1.73	54.70	nr	163.92
Sum of two sides 3200mm	66.24	112.63	1.73	54.70	nr	167.33
Sum of two sides 3300mm	68.25	116.05	1.73	54.70	nr	170.75
Sum of two sides 3400mm	70.26	119.47	1.73	54.70	nr	174.17
Sum of two sides 3500mm	72.27	122.89	1.73	54.70	nr	177.59
Sum of two sides 3600mm	74.28	126.30	1.73	54.70	nr	181.01
Sum of two sides 3700mm	76.29	129.72	1.73	54.70	nr	184.42
Sum of two sides 3800mm	78.30	133.14	1.73	54.70	nr	187.84
Sum of two sides 3900mm	80.31	136.56	1.73	54.70	nr	191.26
Sum of two sides 4000mm	82.32	139.98	1.73	54.70	nr	194.68
Reducer						
Sum of two sides 2500mm	34.58	58.80	3.12	98.65	nr	157.45
Sum of two sides 2600mm	38.38	65.26	3.16	99.92	nr	165.18
Sum of two sides 2700mm	42.29	71.91	3.16	99.92	nr	171.83
Sum of two sides 2800mm	45.99	78.20	4.00	126.48	nr	204.68
Sum of two sides 2900mm	49.90	84.85	4.01	126.79	nr	211.64
Sum of two sides 3000mm	53.71	91.33	4.56	144.18	nr	235.51
Sum of two sides 3100mm	56.30	95.73	4.56	144.18	nr	239.92
Sum of two sides 3200mm	58.56	99.57	4.56	144.18	nr	243.76
Sum of two sides 3300mm	60.93	103.60	4.56	144.18	nr	247.79
Sum of two sides 3400mm	63.20	107.46	4.56	144.18	nr	251.65
Sum of two sides 3500mm	54.97	93.47	4.56	144.18	nr	237.65
Sum of two sides 3600mm	57.24	97.33	4.56	144.18	nr	241.51
Sum of two sides 3700mm	64.71	110.03	4.56	144.18	nr	254.22
Sum of two sides 3800mm	66.98	113.89	4.56	144.18	nr	258.08
Sum of two sides 3900mm	69.25	117.75	4.56	144.18	nr	261.93
Sum of two sides 4000mm	71.59	121.73	4.56	144.18	nr	265.91
Offset						
Sum of two sides 2500mm	129.80	220.71	3.48	110.03	nr	330.74
Sum of two sides 2600mm	138.15	234.91	3.49	110.35	nr	345.26
Sum of two sides 2700mm	137.80	234.31	3.50	110.67	nr	344.98
Sum of two sides 2800mm	154.84	263.29	4.33	136.91	nr	400.20
Sum of two sides 2900mm	154.08	261.99	4.74	149.88	nr	411.87
Sum of two sides 3000mm	152.80	259.82	5.31	167.90	nr	427.72
Sum of two sides 3100mm	149.28	253.83	5.34	168.85	nr	422.68
Sum of two sides 3200mm	144.84	246.28	5.35	169.16	nr	415.45
Sum of two sides 3300mm	140.02	238.09	5.35	169.16	nr	407.25
Sum of two sides 3400mm	134.74	229.11	5.35	169.16	nr	398.27
Sum of two sides 3500mm	123.61	210.18	5.35	169.16	nr	379.35
Sum of two sides 3600mm	128.40	218.33	5.35	169.16	nr	387.49
Sum of two sides 3700mm	138.35	235.25	5.35	169.16	nr	404.41
Sum of two sides 3800mm	143.13	243.38	5.35	169.16	nr	412.54
Sum of two sides 3900mm	147.92	251.52	5.35	169.16	nr	420.68
Sum of two sides 4000mm	152.73	259.70	5.35	169.16	nr	428.86

U: VENTILATION/AIR CONDITIONING SYSTEMS

Item	Net Price £	Material £	Labour hours	Labour £	Unit	Total rate £
U10 : DUCTWORK : RECTANGULAR – CLASS C (cont'd)						
Galvanised sheet metal DW144 (cont'd)						
Extra over fittings; Ductwork 2001 to 2500mm longest side (cont'd)						
Square to round						
Sum of two sides 2500mm	64.46	109.61	4.79	151.46	nr	**261.06**
Sum of two sides 2600mm	70.63	120.10	4.95	156.51	nr	**276.61**
Sum of two sides 2700mm	76.79	130.57	4.95	156.51	nr	**287.09**
Sum of two sides 2800mm	82.96	141.06	8.49	268.45	nr	**409.51**
Sum of two sides 2900mm	89.12	151.54	8.88	280.78	nr	**432.32**
Sum of two sides 3000mm	95.29	162.03	9.02	285.21	nr	**447.23**
Sum of two sides 3100mm	99.64	169.43	9.02	285.21	nr	**454.63**
Sum of two sides 3200mm	103.48	175.96	9.09	287.42	nr	**463.37**
Sum of two sides 3300mm	107.32	182.48	9.09	287.42	nr	**469.90**
Sum of two sides 3400mm	111.16	189.01	9.09	287.42	nr	**476.43**
Sum of two sides 3500mm	104.25	177.26	9.09	287.42	nr	**464.68**
Sum of two sides 3600mm	108.09	183.79	9.09	287.42	nr	**471.21**
Sum of two sides 3700mm	114.48	194.66	9.09	287.42	nr	**482.08**
Sum of two sides 3800mm	118.32	201.19	9.09	287.42	nr	**488.61**
Sum of two sides 3900mm	122.17	207.74	9.09	287.42	nr	**495.15**
Sum of two sides 4000mm	125.99	214.23	9.09	287.42	nr	**501.65**
90 ° radius bend						
Sum of two sides 2500mm	166.25	282.69	4.35	137.54	nr	**420.23**
Sum of two sides 2600mm	170.75	290.34	4.53	143.24	nr	**433.57**
Sum of two sides 2700mm	186.19	316.59	4.53	143.24	nr	**459.83**
Sum of two sides 2800mm	178.50	303.52	7.13	225.44	nr	**528.96**
Sum of two sides 2900mm	193.70	329.36	7.17	226.71	nr	**556.07**
Sum of two sides 3000mm	208.58	354.67	7.26	229.56	nr	**584.22**
Sum of two sides 3100mm	219.80	373.74	7.26	229.56	nr	**603.30**
Sum of two sides 3200mm	230.03	391.14	7.31	231.14	nr	**622.27**
Sum of two sides 3300mm	240.59	409.09	7.31	231.14	nr	**640.23**
Sum of two sides 3400mm	250.82	426.49	7.31	231.14	nr	**657.63**
Sum of two sides 3500mm	233.12	396.39	7.31	231.14	nr	**627.53**
Sum of two sides 3600mm	243.37	413.82	7.31	231.14	nr	**644.96**
Sum of two sides 3700mm	269.27	457.86	7.31	231.14	nr	**689.00**
Sum of two sides 3800mm	279.50	475.26	7.31	231.14	nr	**706.39**
Sum of two sides 3900mm	289.74	492.67	7.31	231.14	nr	**723.80**
Sum of two sides 4000mm	286.28	486.78	7.31	231.14	nr	**717.92**
45 ° bend						
Sum of two sides 2500mm	138.14	234.89	8.30	262.44	nr	**497.33**
Sum of two sides 2600mm	143.22	243.53	8.56	270.66	nr	**514.19**
Sum of two sides 2700mm	153.96	261.79	8.62	272.56	nr	**534.35**
Sum of two sides 2800mm	152.75	259.73	9.09	287.42	nr	**547.15**
Sum of two sides 2900mm	163.37	277.79	9.09	287.42	nr	**565.21**
Sum of two sides 3000mm	173.74	295.42	9.62	304.18	nr	**599.60**
Sum of two sides 3100mm	181.70	308.96	9.62	304.18	nr	**613.14**
Sum of two sides 3200mm	188.98	321.34	9.62	304.18	nr	**625.51**
Sum of two sides 3300mm	196.50	334.12	9.62	304.18	nr	**638.30**
Sum of two sides 3400mm	203.77	346.49	9.62	304.18	nr	**650.66**
Sum of two sides 3500mm	196.66	334.40	9.62	304.18	nr	**638.57**
Sum of two sides 3600mm	203.94	346.78	9.62	304.18	nr	**650.95**
Sum of two sides 3700mm	221.69	376.96	9.62	304.18	nr	**681.13**
Sum of two sides 3800mm	228.97	389.34	9.62	304.18	nr	**693.51**

U: VENTILATION/AIR CONDITIONING SYSTEMS

Item	Net Price £	Material £	Labour hours	Labour £	Unit	Total rate £
Sum of two sides 3900mm	236.25	401.71	9.62	304.18	nr	**705.89**
Sum of two sides 4000mm	236.67	402.43	9.62	304.18	nr	**706.61**
90 ° mitre bend						
Sum of two sides 2500mm	190.95	324.69	8.30	262.44	nr	**587.13**
Sum of two sides 2600mm	199.97	340.02	8.56	270.66	nr	**610.69**
Sum of two sides 2700mm	217.38	369.63	8.62	272.56	nr	**642.19**
Sum of two sides 2800mm	207.18	352.28	15.20	480.61	nr	**832.90**
Sum of two sides 2900mm	224.32	381.43	15.20	480.61	nr	**862.04**
Sum of two sides 3000mm	241.59	410.79	15.20	480.61	nr	**891.41**
Sum of two sides 3100mm	255.15	433.85	16.04	507.17	nr	**941.02**
Sum of two sides 3200mm	267.79	455.34	16.04	507.17	nr	**962.52**
Sum of two sides 3300mm	280.27	476.57	16.04	507.17	nr	**983.74**
Sum of two sides 3400mm	292.91	498.06	16.04	507.17	nr	**1005.23**
Sum of two sides 3500mm	270.23	459.49	16.04	507.17	nr	**966.67**
Sum of two sides 3600mm	282.86	480.97	16.04	507.17	nr	**988.14**
Sum of two sides 3700mm	304.82	518.31	16.04	507.17	nr	**1025.48**
Sum of two sides 3800mm	317.46	539.80	16.04	507.17	nr	**1046.97**
Sum of two sides 3900mm	330.09	561.28	16.04	507.17	nr	**1068.45**
Sum of two sides 4000mm	325.41	553.32	16.04	507.17	nr	**1060.49**
Branch						
Sum of two sides 2500mm	144.31	245.38	2.88	91.06	nr	**336.45**
Sum of two sides 2600mm	152.08	258.59	2.88	91.06	nr	**349.66**
Sum of two sides 2700mm	159.95	271.98	2.88	91.06	nr	**363.04**
Sum of two sides 2800mm	167.60	284.98	3.94	124.58	nr	**409.56**
Sum of two sides 2900mm	175.48	298.38	3.94	124.58	nr	**422.96**
Sum of two sides 3000mm	183.26	311.61	4.83	152.72	nr	**464.33**
Sum of two sides 3100mm	189.50	322.22	4.83	152.72	nr	**474.94**
Sum of two sides 3200mm	195.21	331.93	4.83	152.72	nr	**484.65**
Sum of two sides 3300mm	201.02	341.81	4.83	152.72	nr	**494.53**
Sum of two sides 3400mm	206.74	351.54	4.83	152.72	nr	**504.26**
Sum of two sides 3500mm	212.71	361.69	4.83	152.72	nr	**514.41**
Sum of two sides 3600mm	218.43	371.41	4.83	152.72	nr	**524.13**
Sum of two sides 3700mm	226.80	385.65	4.83	152.72	nr	**538.37**
Sum of two sides 3800mm	232.51	395.36	4.83	152.72	nr	**548.08**
Sum of two sides 3900mm	238.22	405.06	4.83	152.72	nr	**557.79**
Sum of two sides 4000mm	244.03	414.94	4.83	152.72	nr	**567.66**
Grille neck						
Sum of two sides 2500mm	150.48	255.87	3.06	96.75	nr	**352.63**
Sum of two sides 2600mm	159.09	270.51	3.08	97.39	nr	**367.90**
Sum of two sides 2700mm	167.69	285.14	3.08	97.39	nr	**382.52**
Sum of two sides 2800mm	176.29	299.76	4.13	130.59	nr	**430.35**
Sum of two sides 2900mm	184.90	314.40	4.13	130.59	nr	**444.99**
Sum of two sides 3000mm	193.50	329.02	5.02	158.73	nr	**487.75**
Sum of two sides 3100mm	200.39	340.74	5.02	158.73	nr	**499.47**
Sum of two sides 3200mm	206.67	351.42	5.02	158.73	nr	**510.15**
Sum of two sides 3300mm	212.96	362.11	5.02	158.73	nr	**520.84**
Sum of two sides 3400mm	219.24	372.79	5.02	158.73	nr	**531.52**
Sum of two sides 3500mm	225.52	383.47	5.02	158.73	nr	**542.20**
Sum of two sides 3600mm	231.80	394.15	5.02	158.73	nr	**552.88**
Sum of two sides 3700mm	238.08	404.83	5.02	158.73	nr	**563.55**
Sum of two sides 3800mm	244.37	415.52	5.02	158.73	nr	**574.25**
Sum of two sides 3900mm	250.64	426.18	5.02	158.73	nr	**584.91**
Sum of two sides 4000mm	256.92	436.86	5.02	158.73	nr	**595.59**

U: VENTILATION/AIR CONDITIONING SYSTEMS

Item	Net Price £	Material £	Labour hours	Labour £	Unit	Total rate £
U10 : DUCTWORK : RECTANGULAR – CLASS C (cont'd)						
Y30 - DUCTWORK ANCILLARIES: ACCESS DOORS						
Refer to ancillaries in U10: DUCTWORK: RECTANGULAR: CLASS B for details of access doors						

U: VENTILATION/AIR CONDITIONING SYSTEMS

Item	Net Price £	Material £	Labour hours	Labour £	Unit	Total rate £
U10 : DUCTWORK : VOLUME/FIRE DAMPERS						
Y30 - DUCTWORK ANCILLARIES: VOLUME CONTROL AND FIRE						
DAMPERS						
Volume control damper; opposed blade; galvanised steel casing; aluminium aerofoil blades; manually operated						
Rectangular						
Sum of two sides 200mm	14.02	23.84	1.60	50.59	nr	**74.43**
Sum of two sides 300mm	15.00	25.51	1.60	50.59	nr	**76.10**
Sum of two sides 400mm	16.40	27.89	1.60	50.59	nr	**78.48**
Sum of two sides 500mm	17.95	30.52	1.60	50.59	nr	**81.11**
Sum of two sides 600mm	19.84	33.74	1.70	53.77	nr	**87.51**
Sum of two sides 700mm	21.81	37.09	2.10	66.43	nr	**103.51**
Sum of two sides 800mm	23.84	40.54	2.15	67.98	nr	**108.52**
Sum of two sides 900mm	26.15	44.46	2.30	72.72	nr	**117.19**
Sum of two sides 1000mm	28.39	48.27	2.40	75.89	nr	**124.16**
Sum of two sides 1100mm	30.92	52.58	2.60	82.21	nr	**134.79**
Sum of two sides 1200mm	34.97	59.46	2.80	88.53	nr	**148.00**
Sum of two sides 1300mm	37.72	64.14	3.10	98.02	nr	**162.16**
Sum of two sides 1400mm	40.66	69.14	3.25	102.76	nr	**171.90**
Sum of two sides 1500mm	44.03	74.87	3.40	107.51	nr	**182.37**
Sum of two sides 1600mm	47.32	80.46	3.45	109.09	nr	**189.55**
Sum of two sides 1700mm	50.48	85.84	3.60	113.83	nr	**199.66**
Sum of two sides 1800mm	54.26	92.26	3.90	123.31	nr	**215.58**
Sum of two sides 1900mm	57.76	98.21	4.20	132.85	nr	**231.07**
Sum of two sides 2000mm	61.97	105.37	4.33	136.91	nr	**242.28**
Circular						
100mm dia.	18.72	31.83	0.80	25.30	nr	**57.13**
160mm dia.	22.30	37.92	0.90	28.46	nr	**66.38**
200mm dia.	24.18	41.12	1.05	33.21	nr	**74.33**
250mm dia.	26.92	45.77	1.20	37.96	nr	**83.73**
315mm dia.	31.12	52.92	1.35	42.73	nr	**95.65**
350mm dia.	32.74	55.67	1.65	52.18	nr	**107.85**
400mm dia.	35.75	60.79	1.90	60.11	nr	**120.90**
450mm dia.	38.63	65.69	2.10	66.43	nr	**132.11**
500mm dia.	42.07	71.53	2.95	93.27	nr	**164.81**
650mm dia.	53.14	90.36	4.55	143.87	nr	**234.23**
700mm dia.	57.41	97.62	5.20	164.42	nr	**262.04**
800mm dia.	66.67	113.36	5.80	183.39	nr	**296.76**
900mm dia.	76.69	130.40	6.40	202.36	nr	**332.76**
1000mm dia.	87.28	148.41	7.00	221.33	nr	**369.74**
Flat oval						
345 x 102mm	32.67	55.55	1.20	37.94	nr	**93.49**
508 x 102mm	35.62	60.57	1.60	50.59	nr	**111.16**
559 x 152mm	40.24	68.42	1.90	60.11	nr	**128.53**
531 x 203mm	43.95	74.73	1.90	60.11	nr	**134.84**
851 x 203mm	50.55	85.95	4.55	143.87	nr	**229.82**
582 x 254mm	48.79	82.96	2.10	66.43	nr	**149.39**
823 x 254mm	54.54	92.74	4.10	129.64	nr	**222.38**
632 x 305mm	53.77	91.43	2.95	93.27	nr	**184.70**
765 x 356mm	57.62	97.98	4.55	143.87	nr	**241.84**
737 x 406mm	60.15	102.28	4.55	143.87	nr	**246.15**

U: VENTILATION/AIR CONDITIONING SYSTEMS

Item	Net Price £	Material £	Labour hours	Labour £	Unit	Total rate £
U10 : DUCTWORK : VOLUME/FIRE DAMPERS (cont'd)						
DAMPERS (cont'd)						
Volume control damper; opposed (cont'd)						
Flat oval (cont'd)						
818 x 406mm	64.07	108.94	5.20	164.42	nr	273.36
978 x 406mm	66.73	113.47	5.50	173.91	nr	287.37
709 x 457mm	66.67	113.36	4.50	142.43	nr	255.79
678 x 508mm	70.53	119.93	4.55	143.87	nr	263.80
919 x 508mm	76.28	129.71	6.00	189.72	nr	319.42
Fire damper; galvanised steel casing; stainless steel folding shutter; fusible link with manual reset; BS 476 4 hour fire rated						
Rectangular						
Sum of two sides 200mm	32.68	55.57	1.60	50.59	nr	106.16
Sum of two sides 300mm	32.68	55.57	1.60	50.59	nr	106.16
Sum of two sides 400mm	32.68	55.57	1.60	50.59	nr	106.16
Sum of two sides 500mm	40.53	68.92	1.60	50.59	nr	119.51
Sum of two sides 600mm	44.38	75.46	1.70	53.77	nr	129.24
Sum of two sides 700mm	48.16	81.89	2.10	66.43	nr	148.32
Sum of two sides 800mm	52.36	89.03	2.15	68.00	nr	157.03
Sum of two sides 900mm	56.42	95.94	2.30	72.85	nr	168.79
Sum of two sides 1000mm	60.49	102.86	2.40	76.01	nr	178.86
Sum of two sides 1100mm	64.26	109.27	2.60	82.34	nr	191.61
Sum of two sides 1200mm	68.61	116.66	2.80	88.57	nr	205.23
Sum of two sides 1300mm	73.02	124.16	3.10	98.02	nr	222.18
Sum of two sides 1400mm	77.43	131.66	3.25	102.76	nr	234.42
Sum of two sides 1500mm	81.70	138.92	3.40	107.55	nr	246.47
Sum of two sides 1600mm	86.18	146.54	3.45	109.09	nr	255.62
Sum of two sides 1700mm	90.73	154.28	3.60	113.83	nr	268.10
Sum of two sides 1800mm	95.49	162.37	3.90	123.31	nr	285.68
Sum of two sides 1900mm	99.97	169.99	4.20	132.85	nr	302.84
Sum of two sides 2000mm	104.66	177.96	4.33	136.91	nr	314.87
Sum of two sides 2100mm	111.02	188.78	4.43	140.07	nr	328.85
Sum of two sides 2200mm	117.30	199.45	4.55	143.87	nr	343.32
Circular						
100mm dia.	36.60	62.23	0.80	25.30	nr	87.53
160mm dia.	38.50	65.46	0.90	28.46	nr	93.92
200mm dia.	40.18	68.32	1.05	33.20	nr	101.52
250mm dia.	44.60	75.84	1.20	37.96	nr	113.80
315mm dia.	51.60	87.74	1.35	42.73	nr	130.47
355mm dia.	54.41	92.52	1.65	52.18	nr	144.70
400mm dia.	59.87	101.80	1.90	60.11	nr	161.91
450mm dia.	65.26	110.97	2.10	66.43	nr	177.39
500mm dia.	71.14	120.97	2.95	93.27	nr	214.24
630mm dia.	89.84	152.76	4.55	143.87	nr	296.63
710mm dia.	103.64	176.23	5.20	164.42	nr	340.65
800mm dia.	110.85	188.49	5.80	183.39	nr	371.88
900mm dia.	126.20	214.59	6.40	202.36	nr	416.95
1000mm dia.	142.52	242.34	7.00	221.33	nr	463.67

U: VENTILATION/AIR CONDITIONING SYSTEMS

Item	Net Price £	Material £	Labour hours	Labour £	Unit	Total rate £
Flat oval						
345 x 102mm	49.17	83.61	1.20	37.96	nr	121.57
427 x 102mm	53.28	90.60	1.35	42.73	nr	133.33
508 x 102mm	55.06	93.62	1.60	50.59	nr	144.21
559 x 152mm	59.72	101.55	1.90	60.11	nr	161.66
531 x 203mm	62.52	106.31	1.90	60.11	nr	166.42
851 x 203mm	78.09	132.78	4.55	143.87	nr	276.65
582 x 254mm	81.44	138.48	2.10	66.43	nr	204.90
632 x 305mm	88.09	149.79	2.95	93.28	nr	243.06
765 x 356mm	97.80	166.30	4.55	143.87	nr	310.16
737 x 406mm	101.89	173.25	4.55	143.87	nr	317.12
818 x 406mm	108.05	183.73	5.20	164.42	nr	348.15
978 x 406mm	117.32	199.49	5.50	173.91	nr	373.39
709 x 457mm	102.61	174.48	4.50	142.43	nr	316.90
678 x 508mm	106.64	181.33	4.55	143.87	nr	325.20
Smoke/fire damper; galvanised steel casing; stainless steel folding shutter; fusible link and 24V d.c. electro-magnetic shutter release mechanism; spring operated; BS 476 4 hour fire rating						
Rectangular						
Sum of two sides 200mm	205.12	348.78	1.60	50.59	nr	399.37
Sum of two sides 300mm	205.69	349.75	1.60	50.59	nr	400.34
Sum of two sides 400mm	206.25	350.70	1.60	50.59	nr	401.29
Sum of two sides 500mm	211.29	359.27	1.60	50.59	nr	409.86
Sum of two sides 600mm	216.66	368.40	1.70	53.77	nr	422.18
Sum of two sides 700mm	222.02	377.52	2.10	66.43	nr	443.94
Sum of two sides 800mm	227.61	387.02	2.15	68.00	nr	455.02
Sum of two sides 900mm	233.43	396.92	2.30	72.85	nr	469.77
Sum of two sides 1000mm	239.45	407.16	2.40	76.01	nr	483.16
Sum of two sides 1100mm	245.62	417.65	2.60	82.34	nr	499.99
Sum of two sides 1200mm	251.99	428.48	2.80	88.57	nr	517.05
Sum of two sides 1300mm	258.57	439.67	3.10	98.02	nr	537.69
Sum of two sides 1400mm	265.30	451.11	3.25	102.76	nr	553.87
Sum of two sides 1500mm	272.23	462.89	3.40	107.55	nr	570.44
Sum of two sides 1600mm	279.38	475.05	3.45	109.09	nr	584.14
Sum of two sides 1700mm	286.66	487.43	3.60	113.83	nr	601.26
Sum of two sides 1800mm	294.16	500.18	3.90	123.31	nr	623.50
Sum of two sides 1900mm	301.79	513.16	4.20	132.85	nr	646.01
Sum of two sides 2000mm	309.64	526.51	4.33	136.91	nr	663.42
Circular						
100mm dia.	74.25	126.25	0.80	25.30	nr	151.55
160mm dia.	74.25	126.25	0.90	28.46	nr	154.71
200mm dia.	74.25	126.25	1.05	33.20	nr	159.45
250mm dia.	78.66	133.75	1.20	37.96	nr	171.71
315mm dia.	83.22	141.51	1.35	42.73	nr	184.24
355mm dia.	94.23	160.23	1.65	52.18	nr	212.40
400mm dia.	103.40	175.82	1.90	60.11	nr	235.93
450mm dia.	108.16	183.91	2.10	66.43	nr	250.34
500mm dia.	117.55	199.88	2.95	93.27	nr	293.15
630mm dia.	143.97	244.80	4.55	143.87	nr	388.67
710mm dia.	176.33	299.83	5.20	164.42	nr	464.25
800mm dia.	155.40	264.24	5.80	183.39	nr	447.63
900mm dia.	195.81	332.95	6.40	202.36	nr	535.31
1000mm dia.	217.80	370.34	7.00	221.33	nr	591.68

U: VENTILATION/AIR CONDITIONING SYSTEMS

Item	Net Price £	Material £	Labour hours	Labour £	Unit	Total rate £
U10 : DUCTWORK : VOLUME/FIRE DAMPERS (cont'd)						
DAMPERS (cont'd)						
Smoke/fire damper; galvanised steel (cont'd)						
Flat oval						
531 x 203mm	125.96	214.18	1.90	60.11	nr	**274.29**
851 x 203mm	138.74	235.91	4.55	143.87	nr	**379.78**
582 x 254mm	134.79	229.19	2.10	66.43	nr	**295.62**
632 x 305mm	143.61	244.19	2.95	93.28	nr	**337.47**
765 x 356mm	154.92	263.42	4.55	143.87	nr	**407.29**
737 x 406mm	161.77	275.07	4.55	143.87	nr	**418.94**
818 x 406mm	166.52	283.15	5.20	164.42	nr	**447.57**
978 x 406mm	173.60	295.19	5.50	173.91	nr	**469.09**
709 x 457mm	166.20	282.60	4.50	142.43	nr	**425.03**
678 x 508mm	172.90	294.00	4.55	143.87	nr	**437.86**

U: VENTILATION/AIR CONDITIONING SYSTEMS

Item	Net Price £	Material £	Labour hours	Labour £	Unit	Total rate £
U10 : PLANT/EQUIPMENT						
Y41 - FANS						
Axial flow fan; including ancillaries, anti vibration mountings, mounting feet, matching flanges, flexible connectors and clips; 415V, 3 phase, 50Hz motor; includes fixing in position; electrical work elsewhere						
Aerofoil blade fan unit; short duct case						
315mm dia.; 0.47 m3/s duty; 147 Pa	445.20	570.21	4.50	121.97	nr	692.17
500mm dia.; 1.89 m3/s duty; 500 Pa	601.55	770.46	1.00	27.10	nr	797.56
560mm dia.; 2.36 m3/s duty; 147 Pa	539.54	691.04	5.50	149.07	nr	840.11
710mm dia.; 5.67 m3/s duty; 245 Pa	837.40	1072.53	6.00	162.62	nr	1235.15
Aerofoil blade fan unit; long duct case						
315mm dia.; 0.47 m3/s duty; 147 Pa	286.20	366.56	4.50	121.97	nr	488.53
500mm dia.; 1.89 m3/s duty; 500 Pa	707.56	906.24	5.00	135.52	nr	1041.75
560mm dia.; 2.36 m3/s duty; 147 Pa	443.08	567.49	5.50	149.07	nr	716.56
710mm dia.; 5.67 m3/s duty; 245 Pa	710.20	909.62	6.00	162.62	nr	1072.24
Aerofoil blade fan unit; two stage parallel fan arrangement; long duct case						
315mm; 0.47m3/s @ 500 Pa	1046.22	1339.99	4.50	121.97	nr	1461.95
355mm; 0.83m3/s @ 147 Pa	1144.80	1466.25	4.75	128.74	nr	1594.99
710mm; 3.77m3/s @ 431 Pa	2579.17	3303.38	6.00	162.62	nr	3466.00
710mm; 6.61m3/s @ 500 Pa	2723.22	3487.87	6.00	162.62	nr	3650.49
Axial flow fan; suitable for operation at 300°C for 90 minutes; including ancillaries, anti vibration mountings, mounting feet, matching flanges, flexible connectors and clips; 415V, 3 phase, 50Hz motor; includes fixing in position; electrical work elsewhere						
450mm; 2.0m³/s @ 300Pa	1399.20	1792.08	5.00	135.52	nr	1927.60
630mm; 4.6m³/s @ 200Pa	1587.88	2033.74	5.50	149.07	nr	2182.81
800mm; 9.0m³/s @ 300Pa	2923.26	3744.08	6.50	176.17	nr	3920.25
1000mm; 15.0m³/s @ 400Pa	3927.73	5030.60	7.50	203.28	nr	5233.87
Bifurcated fan; suitable for temperature up to 200°C with motor protection to IP55; including ancillaries, anti vibration mountings, mounting feet, matching flanges, flexible connectors and clips; 415V, 3 phase, 50Hz motor; includes fixing in position; electrical work elsewhere						
300mm; 0.50m³/s @ 100Pa	961.42	1231.38	4.50	121.97	nr	1353.34
400mm; 1.97m³/s @ 200Pa	1399.20	1792.08	5.00	135.52	nr	1927.60
630mm; 3.86m³/s @ 200Pa	1632.41	2090.77	5.50	149.07	nr	2239.84
800mm; 6.10m³/s @ 400Pa	2445.42	3132.07	6.50	176.17	nr	3308.24

U: VENTILATION/AIR CONDITIONING SYSTEMS

Item	Net Price £	Material £	Labour hours	Labour £	Unit	Total rate £
U10 : PLANT/EQUIPMENT (cont'd)						
Duct mounted in line fan with backward curved centrifugal impellor; including ancillaries, matching flanges, flexible connectors and clips; 415V, 3phase, 50Hz motor; includes fixing in position; electrical work elsewhere						
0.5m³/s @ 200Pa	1071.82	1372.78	4.50	121.97	nr	1494.74
1.0m³/s @ 300Pa	1321.34	1692.36	5.00	135.52	nr	1827.88
3.0m³/s @ 500Pa	2183.34	2796.40	5.50	149.07	nr	2945.47
5.0m³/s @ 750Pa	2892.21	3704.31	6.50	176.17	nr	3880.49
7.0m³/s @ 1000Pa	3516.02	4503.28	7.00	189.72	nr	4693.01
Twin fan extract unit; belt driven; located internally; complete with anti-vibration mounts and non return shutter; including ancillaries, matching flanges, flexible connectors and clips; 3 phase, 50Hz motor; includes fixing in position; electrical work elsewhere						
0.25m³/s @ 150Pa	1478.99	1894.28	4.50	121.97	nr	2016.24
Extra for external unit	198.22	253.88	-	-	nr	253.88
0.50m³/s @ 200Pa	1720.59	2203.71	5.00	135.52	nr	2339.23
Extra for external unit	198.22	253.88	-	-	nr	253.88
1.00m³/s @ 200Pa	1962.05	2512.97	5.00	135.52	nr	2648.49
Extra for external unit	198.22	253.88	-	-	nr	253.88
1.50m³/s @ 250Pa	5988.37	7669.84	5.50	149.07	nr	7818.91
Extra for external unit	198.22	253.88	-	-	nr	253.88
2.00m³/s @ 250Pa	2544.00	3258.33	6.50	176.17	nr	3434.50
Extra for external unit	198.22	253.88	-	-	nr	253.88
Extra for auto changeover panel	186.56	238.94	2.50	67.76	nr	306.70
Roof mounted extract fan; including ancillaries, fibreglass cowling, fitted shutters and bird guard; 415V, 3 phase, 50Hz motor; includes fixing in position; electrical work elsewhere						
Flat roof installation, fixed to curb						
315mm; 900rpm	2664.76	3413.00	4.50	121.97	nr	3534.96
315mm; 1380rpm	493.02	631.46	4.50	121.97	nr	753.42
400mm; 900rpm	693.63	888.39	5.50	149.07	nr	1037.46
400mm; 1360rpm	598.16	766.12	5.50	149.07	nr	915.19
800mm; 530rpm	1993.86	2553.72	7.00	189.72	nr	2743.44
800mm; 700rpm	1873.02	2398.95	7.00	189.72	nr	2588.67
800mm; 920rpm	1661.55	2128.10	7.00	189.72	nr	2317.82
1000mm; 470rpm	2658.48	3404.95	8.00	216.83	nr	3621.78
1000mm; 570rpm	2537.64	3250.18	8.00	216.83	nr	3467.01
1000mm; 710rpm	2468.77	3161.98	8.00	216.83	nr	3378.80

U: VENTILATION/AIR CONDITIONING SYSTEMS

Item	Net Price £	Material £	Labour hours	Labour £	Unit	Total rate £
Pitched roof installation; including purlin mounting box						
315mm; 900rpm	548.08	701.98	4.50	121.97	nr	**823.94**
315mm; 1380rpm	548.08	701.98	4.50	121.97	nr	**823.94**
400mm; 900rpm	760.15	973.59	5.50	149.07	nr	**1122.66**
400mm; 1360rpm	664.68	851.32	5.50	149.07	nr	**1000.38**
800mm; 530rpm	2126.38	2723.45	7.00	189.72	nr	**2913.17**
800mm; 700rpm	2005.54	2568.68	7.00	189.72	nr	**2758.40**
800mm; 920rpm	1794.07	2297.83	7.00	189.72	nr	**2487.55**
1000mm; 470rpm	2854.48	3655.99	8.00	216.83	nr	**3872.82**
1000mm; 570rpm	2733.64	3501.22	8.00	216.83	nr	**3718.05**
1000mm; 710rpm	2664.76	3413.00	8.00	216.83	nr	**3629.83**
Centrifugal fan; single speed for internal domestic kitchens/utility rooms; fitted with standard overload protection; complete with housing; includes placing in position; electrical work elsewhere						
Window mounted						
245m3/hr	81.44	104.31	0.50	13.55	nr	**117.86**
500m3/hr	260.75	667.93	0.50	13.55	nr	**681.48**
Wall mounted						
245m3/hr	192.57	367.84	0.83	22.59	nr	**390.42**
500m3/hr	420.80	1223.77	0.83	22.59	nr	**1246.36**
Centrifugal fan; various speeds, simultaneous ventilation from separate areas fitted with standard overload protection; complete with housing; includes placing in position; ducting and electrical work elsewhere						
Fan unit						
147-300m3/hr	120.23	153.99	1.00	27.10	nr	**181.09**
175-411m3/hr	198.49	254.22	1.00	27.10	nr	**281.33**
Toilet extract units; centrifugal fan; various speeds for internal domestic bathrooms/ W.Cs, with built in filter; complete with housing; includes placing in position; electrical work elsewhere						
Fan unit; fixed to wall; including shutter						
Single speed 85m3/hr	80.51	103.12	0.75	20.33	nr	**123.44**
Two speed 60-85m3/hr	100.11	128.22	0.83	22.59	nr	**150.81**
Humidity controlled; autospeed; fixed to wall; including shutter						
30-60-85m3/hr	162.51	208.14	1.00	27.10	nr	**235.24**

U: VENTILATION/AIR CONDITIONING SYSTEMS

Item	Net Price £	Material £	Labour hours	Labour £	Unit	Total rate £
U10 : PLANT/EQUIPMENT (cont'd)						
Y42 - AIR FILTRATION						
High efficiency duct mounted filters; 99.997% H13 (EU13); tested to BS 3928						
Standard; 1700m³/ hr air volume; continuous rating up to 80°C; sealed wood case, aluminium spacers, neoprene gaskets; water repellent filter media; includes placing in position						
610 x 610 x 292mm	176.40	299.94	1.00	31.62	nr	331.56
Side withdrawal frame	84.60	143.85	2.50	79.05	nr	222.90
High capacity; 3400m³/hr air volume; continuous rating up to 80°C; anti-corrosion coated mild steel frame, polyurethane sealant and neoprene gaskets; water repellent filter media; includes placing in position						
610 x 610 x 292mm	320.17	544.40	1.00	31.62	nr	576.02
Side withdrawal frame	84.60	143.85	2.50	79.05	nr	222.90
Bag Filters; 40/60% F5 (EU5); tested to BSEN 779						
Duct mounted bag filter; continuous rating up to 60°C; rigid filter assembly; sealed into one piece coated mild steel header with sealed pocket separators; includes placing in position						
6 pocket, 592 x 592 x 25mm header; pockets 380mm long; 1690m³/hr	57.98	98.59	1.00	31.62	nr	130.21
Side withdrawal frame	58.66	99.75	2.00	63.24	nr	162.99
6 pocket, 592 x 592 x 25mm header; pockets 500mm long; 2550m³/hr	61.91	105.27	1.50	47.43	nr	152.70
Side withdrawal frame	58.66	99.75	2.50	79.05	nr	178.80
6 pocket, 592 x 592 x 25mm header; pockets 635mm long; 3380m³/hr	65.44	111.27	1.50	47.43	nr	158.70
Side withdrawal frame	58.66	99.75	3.00	94.86	nr	194.61
Bag Filters; 80/90% F7, (EU7); tested to BSEN 779						
Duct mounted bag filter; continuous rating up to 60°C; rigid filter assembly; sealed into one piece coated mild steel header with sealed pocket separators; includes placing in position						
6 pocket, 592 x 592 x 25mm header; pockets 500mm long; 1688m³/hr	80.72	137.25	1.00	31.62	nr	168.87
Side withdrawal frame	58.66	99.75	2.00	63.24	nr	162.99
6 pocket, 592 x 592 x 25mm header; pockets 635mm long; 2047m³/hr	86.60	147.26	1.50	47.43	nr	194.69
Side withdrawal frame	58.66	99.75	2.50	79.05	nr	178.80
6 pocket, 592 x 592 x 25mm header; pockets 762mm long; 2729m³/hr	101.40	172.42	1.50	47.43	nr	219.84
Side withdrawal frame	58.66	99.75	3.00	94.86	nr	194.61

U: VENTILATION/AIR CONDITIONING SYSTEMS

Item	Net Price £	Material £	Labour hours	Labour £	Unit	Total rate £
Grease filters, washable; minimum 65%						
Double sided extract unit; lightweight stainless steel construction; demountable composite filter media of woven metal mat and expanded metal mesh supports; for mounting on hood and extract systems (hood not included); includes placing in position						
500 x 686 x 565mm, 4080m³/hr	359.30	610.95	2.00	63.24	nr	674.19
1000 x 686 x 565mm, 8160m³/hr;	547.93	931.69	3.00	94.86	nr	1026.55
1500 x 686 x 565mm, 12240m³/hr;	751.56	1277.94	3.50	110.67	nr	1388.61
Panel filters; 82% G3 (EU3); tested to BS EN779						
Modular duct mounted filter panels; continuous rating up to 100°C; graduated density media; rigid cardboard frame; includes placing in position						
596 x 596 x 47mm, 2360m³/hr	5.04	8.57	1.00	31.62	nr	40.19
Side withdrawal frame	51.48	87.54	2.50	79.05	nr	166.58
596 x 287 x 47mm, 1140m³/hr	3.59	6.10	1.00	31.62	nr	37.72
Side withdrawal frame	51.48	87.54	2.50	79.05	nr	166.58
Panel filters; 90% G4 (EU4); tested to BS EN779						
Modular duct mounted filter panels; continuous rating up to 100°C; pleated media with wire support; rigid cardboard frame; includes placing in position						
596 x 596 x 47mm, 2560m³/hr	10.29	17.49	1.00	31.62	nr	49.11
side withdrawal frame	64.93	110.41	3.00	94.86	nr	205.27
596 x 287 x 47mm, 1230m3/hr	7.94	13.50	1.00	31.62	nr	45.12
Side withdrawal frame	51.48	87.54	3.00	94.86	nr	182.39
Carbon filters; standard duty disposable carbon filters; steel frame with bonded carbon panels; for fixing to ductwork; including placing in position						
12 panels						
597 x 597 x 298mm, 1460m³/hr	306.22	520.69	0.33	10.43	nr	531.12
597 x 597 x 451mm, 2200m³/hr	345.21	586.99	0.33	10.43	nr	597.43
597 x 597 x 597mm, 2930m³/hr	384.97	654.59	0.33	10.43	nr	665.02
8 panels						
451 x 451 x 298mm, 740m³/hr	230.86	392.54	0.29	9.17	nr	401.71
451 x 451 x 451mm, 1105m³/hr	255.51	434.46	0.29	9.17	nr	443.63
451 x 451 x 597mm, 1460m³/hr	279.04	474.47	0.29	9.17	nr	483.64
6 panels						
298 x 298 x 298mm, 365m³/hr	163.58	278.15	0.25	7.90	nr	286.05
298 x 298 x 451mm, 550m³/hr	175.14	297.81	0.25	7.90	nr	305.71
298 x 298 x 597mm, 780m³/hr	186.20	316.61	0.25	7.90	nr	324.51

U: VENTILATION/AIR CONDITIONING SYSTEMS

Item	Net Price £	Material £	Labour hours	Labour £	Unit	Total rate £
U10 : SILENCERS/ACOUSTIC TREATMENT						
Y45 - SILENCERS/ACOUSTIC TREATMENT						
Attenuators; DW144 galvanised construction c/w splitters; self securing; fitted to ductwork						
To suit rectangular ducts; unit length 600mm						
100 x 100mm	83.23	141.52	0.75	23.71	nr	**165.24**
150 x 150mm	86.89	147.75	0.75	23.71	nr	**171.46**
200 x 200mm	90.53	153.94	0.75	23.71	nr	**177.65**
300 x 300mm	101.28	172.21	0.75	23.71	nr	**195.93**
400 x 400mm	113.70	193.33	1.00	31.62	nr	**224.95**
500 x 500mm	132.70	225.64	1.25	39.52	nr	**265.16**
600 x 300mm	127.83	217.36	1.25	39.52	nr	**256.88**
600 x 600mm	178.91	304.21	1.25	39.52	nr	**343.74**
700 x 300mm	135.53	230.45	1.50	47.43	nr	**277.88**
700 x 700mm	201.81	343.15	1.50	47.43	nr	**390.58**
800 x 300mm	140.39	238.72	2.00	63.24	nr	**301.95**
800 x 800mm	232.82	395.88	2.00	63.24	nr	**459.12**
1000 x 1000mm	298.03	506.76	3.00	94.86	nr	**601.62**
To suit rectangular ducts; unit length 1200mm						
200 x 200mm	99.83	169.75	1.00	31.62	nr	**201.37**
300 x 300mm	123.42	209.86	1.00	31.62	nr	**241.48**
400 x 400mm	148.40	252.34	1.33	42.05	nr	**294.39**
500 x 500mm	179.11	304.56	1.66	52.49	nr	**357.04**
600 x 300mm	170.79	290.41	1.66	52.49	nr	**342.90**
600 x 600mm	249.87	424.87	1.66	52.49	nr	**477.36**
700 x 300mm	183.07	311.29	2.00	63.24	nr	**374.53**
700 x 700mm	285.95	486.22	2.00	63.24	nr	**549.46**
800 x 300mm	191.40	325.45	2.66	84.11	nr	**409.56**
800 x 800mm	329.95	561.04	2.66	84.11	nr	**645.15**
1000 x 1000mm	446.64	759.46	4.00	126.48	nr	**885.93**
1300 x 1300mm	766.70	1303.68	8.00	252.95	nr	**1556.63**
1500 x 1500mm	877.56	1492.19	8.00	252.95	nr	**1745.14**
1800 x 1800mm	1200.72	2041.68	10.66	337.06	nr	**2378.74**
2000 x 2000mm	1347.35	2291.01	13.33	421.48	nr	**2712.49**
To suit rectangular ducts; unit length 1800mm						
200 x 200mm	125.99	214.23	1.00	31.62	nr	**245.85**
300 x 300mm	167.42	284.68	1.00	31.62	nr	**316.30**
400 x 400mm	200.13	340.30	1.33	42.05	nr	**382.35**
500 x 500mm	233.23	396.58	1.66	52.49	nr	**449.07**
600 x 300mm	243.93	414.77	1.66	52.49	nr	**467.26**
600 x 600mm	340.85	579.57	1.66	52.49	nr	**632.06**
700 x 300mm	258.59	439.70	2.00	63.24	nr	**502.94**
700 x 700mm	407.25	692.48	2.00	63.24	nr	**755.72**
800 x 300mm	273.46	464.99	2.66	84.11	nr	**549.09**
800 x 800mm	473.85	805.73	2.66	84.11	nr	**889.83**
1000 x 1000mm	636.92	1083.01	4.00	126.48	nr	**1209.48**
1300 x 1300mm	1064.02	1809.24	8.00	252.95	nr	**2062.19**
1500 x 1500mm	1205.80	2050.32	8.00	252.95	nr	**2303.27**
1800 x 1800mm	1895.61	3223.26	10.66	337.06	nr	**3560.32**
2000 x 2000mm	1674.04	2846.50	13.33	421.48	nr	**3267.99**
2300 x 2300mm	2369.48	4029.02	16.00	505.91	nr	**4534.92**
2500 x 2500mm	3305.12	5619.96	18.66	590.01	nr	**6209.97**

U: VENTILATION/AIR CONDITIONING SYSTEMS

Item	Net Price £	Material £	Labour hours	Labour £	Unit	Total rate £
To suit rectangular ducts; unit length 2400mm						
500 x 500mm	301.60	512.83	2.08	65.86	nr	578.70
600 x 300mm	287.94	489.61	2.08	65.86	nr	555.47
600 x 600mm	428.86	729.23	2.08	65.86	nr	795.09
700 x 300mm	312.11	530.71	2.50	79.05	nr	609.75
700 x 700mm	506.55	861.33	2.50	79.05	nr	940.38
800 x 300mm	329.95	561.04	3.33	105.29	nr	666.33
800 x 800mm	586.23	996.81	3.33	105.29	nr	1102.11
1000 x 1000mm	790.92	1344.86	5.00	158.10	nr	1502.96
1300 x 1300mm	1281.05	2178.27	10.00	316.19	nr	2494.46
1500 x 1500mm	1462.08	2486.09	10.00	316.19	nr	2802.28
1800 x 1800mm	2086.12	3547.20	13.33	421.42	nr	3968.62
2000 x 2000mm	2340.98	3980.56	16.66	526.78	nr	4507.33
2300 x 2300mm	2896.31	4924.83	20.00	632.38	nr	5557.21
2500 x 2500mm	4066.84	6915.17	23.32	737.49	nr	7652.66
To suit circular ducts; unit length 600mm						
100mm dia.	72.29	122.92	0.75	23.71	nr	146.63
200mm dia.	85.06	144.63	0.75	23.71	nr	168.35
250mm dia.	96.42	163.95	0.75	23.71	nr	187.67
315mm dia.	112.23	190.83	1.00	31.62	nr	222.45
355mm dia.	129.86	220.81	1.00	31.62	nr	252.43
400mm dia.	139.59	237.36	1.00	31.62	nr	268.98
450mm dia.	168.57	286.63	1.00	31.62	nr	318.25
500mm dia.	185.60	315.59	1.26	39.78	nr	355.37
630mm dia.	218.83	372.09	1.50	47.43	nr	419.52
710mm dia.	238.29	405.18	1.50	47.43	nr	452.61
800mm dia.	265.05	450.69	2.00	63.24	nr	513.92
1000mm dia.	355.80	605.00	3.00	94.86	nr	699.85
To suit circular ducts; unit length 1200mm						
100mm dia.	106.34	180.82	1.00	31.62	nr	212.44
200mm dia.	116.68	198.40	1.00	31.62	nr	230.02
250mm dia.	131.08	222.89	1.00	31.62	nr	254.50
315mm dia.	149.93	254.94	1.33	42.05	nr	296.99
355mm dia.	179.72	305.59	1.33	42.05	nr	347.65
400mm dia.	190.67	324.21	1.33	42.05	nr	366.26
450mm dia.	220.87	375.56	1.33	42.05	nr	417.62
500mm dia.	226.34	384.86	1.66	52.49	nr	437.35
630mm dia.	299.72	509.64	2.00	63.24	nr	572.88
710mm dia.	331.33	563.39	2.00	63.24	nr	626.63
800mm dia.	330.72	562.35	2.66	84.11	nr	646.46
1000mm dia.	443.97	754.92	4.00	126.48	nr	881.39
1250mm dia.	836.34	1422.10	8.00	252.95	nr	1675.05
1400mm dia.	909.31	1546.17	8.00	252.95	nr	1799.13
1600mm dia.	1008.09	1714.14	10.66	337.06	nr	2051.20
To suit circular ducts; unit length 1800mm						
200mm dia.	163.50	278.01	1.25	39.52	nr	317.54
250mm dia.	182.96	311.10	1.25	39.52	nr	350.63
315mm dia.	217.42	369.70	1.66	52.49	nr	422.18
355mm dia.	256.34	435.88	1.66	52.49	nr	488.36
400mm dia.	270.94	460.70	1.66	52.49	nr	513.19
450mm dia.	265.25	451.03	1.66	52.49	nr	503.51
500mm dia.	628.89	1069.35	2.08	65.77	nr	1135.12
630mm dia.	329.50	560.28	2.50	79.05	nr	639.32
710mm dia.	385.45	655.41	2.50	79.05	nr	734.46
800mm dia.	398.83	678.16	3.33	105.29	nr	783.45
1000mm dia.	1169.23	1988.14	5.00	158.10	nr	2146.23

U: VENTILATION/AIR CONDITIONING SYSTEMS

Item	Net Price £	Material £	Labour hours	Labour £	Unit	Total rate £
U10 : SILENCERS/ACOUSTIC TREATMENT (cont'd)						
TREATMENT (cont'd)						
Attenuators; DW144 galvanised (cont'd)						
To suit circular ducts; unit length 1800mm (cont'd)						
1250mm dia.	1058.29	1799.50	6.66	210.58	nr	**2010.08**
1400mm dia.	1041.27	1770.55	6.66	210.58	nr	**1981.14**
1600mm dia.	645.24	1097.15	10.00	316.19	nr	**1413.35**

Material Costs/Prices for Measured Works – Mechanical Installations 435

U: VENTILATION/AIR CONDITIONING SYSTEMS

Item	Net Price £	Material £	Labour hours	Labour £	Unit	Total rate £
U10 : GRILLES/DIFFUSERS/LOUVRES						
Y46 - GRILLES/DIFFUSERS/LOUVRES						
Supply grilles; single deflection; extruded aluminium alloy frame and adjustable horizontal vanes; silver grey polyester powder coated; screw fixed						
Rectangular; for duct, ceiling and sidewall applications						
100 x 100mm	15.61	26.54	0.60	18.97	nr	**45.51**
150 x 150mm	16.43	27.94	0.60	18.97	nr	**46.91**
200 x 150mm	17.21	29.26	0.65	20.55	nr	**49.82**
200 x 200mm	17.99	30.59	0.72	22.77	nr	**53.36**
300 x 100mm	17.21	29.26	0.72	22.77	nr	**52.03**
300 x 150mm	17.99	30.59	0.80	25.30	nr	**55.89**
300 x 200mm	21.12	35.91	0.88	27.82	nr	**63.74**
300 x 300mm	22.68	38.56	1.04	32.88	nr	**71.45**
400 x 100mm	18.77	31.92	0.88	27.82	nr	**59.74**
400 x 150mm	20.34	34.59	0.94	29.72	nr	**64.31**
400 x 200mm	25.03	42.56	1.04	32.88	nr	**75.44**
400 x 300mm	28.16	47.88	1.12	35.41	nr	**83.30**
600 x 200mm	31.29	53.20	1.26	39.84	nr	**93.05**
600 x 300mm	39.89	67.83	1.40	44.27	nr	**112.10**
600 x 400mm	49.28	83.79	1.61	50.91	nr	**134.70**
600 x 500mm	57.10	97.09	1.76	55.65	nr	**152.74**
600 x 600mm	69.61	118.36	2.17	68.61	nr	**186.98**
800 x 300mm	48.49	82.45	1.76	55.65	nr	**138.10**
800 x 400mm	58.66	99.74	2.17	68.61	nr	**168.36**
800 x 600mm	80.56	136.98	3.00	94.86	nr	**231.84**
1000 x 300mm	53.97	91.77	2.60	82.21	nr	**173.98**
1000 x 400mm	67.27	114.38	3.00	94.86	nr	**209.24**
1000 x 600mm	93.86	159.60	3.80	120.15	nr	**279.75**
1000 x 800mm	108.14	183.88	3.80	120.15	nr	**304.03**
1200 x 600mm	109.50	186.19	4.61	145.76	nr	**331.96**
1200 x 800mm	137.97	234.60	4.61	145.76	nr	**380.37**
1200 x 1000mm	164.26	279.30	4.61	145.76	nr	**425.07**
Rectangular; for duct, ceiling and sidewall applications; including opposed blade damper volume regulator						
100 x 100mm	25.03	42.56	0.72	22.78	nr	**65.34**
150 x 150mm	25.81	43.89	0.72	22.78	nr	**66.67**
200 x 150mm	28.16	47.88	0.83	26.26	nr	**74.15**
200 x 200mm	29.72	50.54	0.90	28.46	nr	**79.00**
300 x 100mm	29.72	50.54	0.90	28.46	nr	**79.00**
300 x 150mm	31.29	53.20	0.98	30.99	nr	**84.19**
300 x 200mm	34.42	58.53	1.06	33.53	nr	**92.06**
300 x 300mm	38.33	65.18	1.20	37.96	nr	**103.13**
400 x 100mm	32.07	54.53	1.06	33.53	nr	**88.06**
400 x 150mm	35.20	59.85	1.13	35.73	nr	**95.58**
400 x 200mm	38.33	65.18	1.20	37.96	nr	**103.13**
400 x 300mm	43.80	74.48	1.34	42.39	nr	**116.86**
600 x 200mm	51.62	87.77	1.50	47.48	nr	**135.25**
600 x 300mm	57.10	97.09	1.66	52.52	nr	**149.61**
600 x 400mm	71.96	122.36	1.80	56.97	nr	**179.33**
600 x 500mm	82.91	140.98	2.00	63.24	nr	**204.22**

U: VENTILATION/AIR CONDITIONING SYSTEMS

Item	Net Price £	Material £	Labour hours	Labour £	Unit	Total rate £
U10 : GRILLES/DIFFUSERS/LOUVRES (cont'd)						
Supply grilles; single deflection (cont'd)						
Rectangular; for duct, ceiling and sidewall (cont'd)						
600 x 600mm	89.17	151.62	2.60	82.34	nr	233.97
800 x 300mm	84.47	143.63	2.00	63.24	nr	206.87
800 x 400mm	96.99	164.92	2.60	82.34	nr	247.26
800 x 600mm	125.93	214.13	3.61	114.15	nr	328.28
1000 x 300mm	97.77	166.25	3.00	94.95	nr	261.20
1000 x 400mm	118.11	200.83	3.61	114.15	nr	314.98
1000 x 600mm	155.65	264.66	4.61	145.71	nr	410.37
1000 x 800mm	192.99	328.16	4.61	145.71	nr	473.87
1200 x 600mm	196.33	333.84	5.62	177.64	nr	511.47
1200 x 800mm	227.73	387.23	6.10	192.88	nr	580.10
1200 x 1000mm	261.11	443.99	6.50	205.52	nr	649.51
Supply grilles; double deflection; extruded aluminium alloy frame and adjustable horizontal and vertical vanes; white polyester powder coated; screw fixed						
Rectangular; for duct, ceiling and sidewall applications						
100 x 100mm	14.15	24.06	0.88	27.82	nr	51.89
150 x 150mm	14.86	25.27	0.88	27.82	nr	53.09
200 x 150mm	17.21	29.26	1.08	34.15	nr	63.41
200 x 200mm	20.34	34.59	1.25	39.52	nr	74.11
300 x 100mm	20.34	34.59	1.25	39.52	nr	74.11
300 x 150mm	20.34	34.59	1.50	47.43	nr	82.01
300 x 200mm	22.68	38.56	1.75	55.33	nr	93.90
300 x 300mm	28.16	47.88	2.15	67.98	nr	115.86
400 x 100mm	25.03	42.56	1.75	55.33	nr	97.89
400 x 150mm	25.03	42.56	1.95	61.66	nr	104.22
400 x 200mm	25.03	42.56	2.15	67.98	nr	110.54
400 x 300mm	33.63	57.18	2.55	80.63	nr	137.81
600 x 200mm	38.33	65.18	3.01	95.17	nr	160.35
600 x 300mm	47.71	81.13	3.36	106.24	nr	187.37
600 x 400mm	57.88	98.42	3.80	120.15	nr	218.57
600 x 500mm	69.61	118.36	4.20	132.80	nr	251.16
600 x 600mm	82.13	139.65	4.51	142.60	nr	282.25
800 x 300mm	66.48	113.04	4.20	132.80	nr	245.84
800 x 400mm	82.13	139.65	4.51	142.60	nr	282.25
800 x 600mm	118.11	200.83	5.10	161.26	nr	362.09
1000 x 300mm	87.60	148.95	4.80	151.77	nr	300.73
1000 x 400mm	109.50	186.19	5.10	161.26	nr	347.45
1000 x 600mm	155.65	264.66	5.72	180.86	nr	445.53
1000 x 800mm	192.99	328.16	5.72	180.86	nr	509.02
1200 x 600mm	183.81	312.55	6.33	200.15	nr	512.70
1200 x 800mm	240.79	409.43	6.33	200.15	nr	609.58
1200 x 1000mm	273.86	465.67	6.33	200.15	nr	665.82

U: VENTILATION/AIR CONDITIONING SYSTEMS

Item	Net Price £	Material £	Labour hours	Labour £	Unit	Total rate £
Rectangular; for duct, ceiling and sidewall applications; including opposed blade damper volume regulator						
100 x 100mm	26.30	44.72	1.00	31.62	nr	76.34
150 x 150mm	28.16	47.88	1.00	31.62	nr	79.50
200 x 150mm	30.50	51.86	1.26	39.84	nr	91.70
200 x 200mm	34.42	58.53	1.43	45.22	nr	103.74
300 x 100mm	32.07	54.53	1.43	45.22	nr	99.75
300 x 150mm	35.98	61.18	1.68	53.12	nr	114.30
300 x 200mm	39.11	66.50	1.93	61.03	nr	127.53
300 x 300mm	43.80	74.48	2.31	73.04	nr	147.52
400 x 100mm	38.33	65.18	1.93	61.03	nr	126.20
400 x 150mm	42.24	71.82	2.14	67.67	nr	139.49
400 x 200mm	45.37	77.15	2.31	73.04	nr	150.19
400 x 300mm	53.19	90.44	2.77	87.59	nr	178.03
600 x 200mm	58.66	99.74	3.25	102.76	nr	202.51
600 x 300mm	69.61	118.36	3.62	114.46	nr	232.82
600 x 400mm	80.56	136.98	3.99	126.16	nr	263.14
600 x 500mm	87.60	148.95	4.44	140.39	nr	289.34
600 x 600mm	106.38	180.89	4.94	156.20	nr	337.09
800 x 300mm	90.73	154.28	4.44	140.39	nr	294.66
800 x 400mm	108.72	184.87	4.94	156.20	nr	341.06
800 x 600mm	166.60	283.28	5.71	180.55	nr	463.83
1000 x 300mm	115.76	196.84	5.20	164.42	nr	361.26
1000 x 400mm	140.79	239.40	5.71	180.55	nr	419.94
1000 x 600mm	214.32	364.43	6.53	206.47	nr	570.90
1000 x 800mm	224.88	382.38	6.53	206.47	nr	588.85
1200 x 600mm	259.68	441.55	7.34	232.08	nr	673.64
1200 x 800mm	263.37	447.83	8.80	278.25	nr	726.08
1200 x 1000mm	296.58	504.30	8.80	278.25	nr	782.55
Floor grille suitable for mounting in raised access floors; heavy duty; extruded aluminium; standard mill finish; complete with opposed blade volume control damper						
Diffuser						
600mm x 600mm	220.57	375.05	0.70	22.13	nr	397.19
Extra for nylon coated black finish	30.50	51.86	-	-	nr	51.86
Exhaust grilles; aluminium						
0 ° fixed blade core						
150 x 150mm	12.51	21.27	0.60	19.04	nr	40.32
200 x 200mm	16.90	28.74	0.72	22.78	nr	51.52
250 x 250mm	18.30	31.12	0.80	25.30	nr	56.41
300 x 300mm	21.90	37.24	1.00	31.62	nr	68.86
350 x 350mm	29.45	50.08	1.20	37.94	nr	88.02
0 ° fixed blade core; including opposed blade damper volume regulator						
150 x 150mm	20.25	34.43	0.62	19.61	nr	54.05
200 x 200mm	25.03	42.56	0.72	22.78	nr	65.34
250 x 250mm	30.11	51.20	0.80	25.30	nr	76.49
300 x 300mm	35.67	60.65	1.00	31.62	nr	92.27
350 x 350mm	47.29	80.41	1.20	37.94	nr	118.35

U: VENTILATION/AIR CONDITIONING SYSTEMS

Item	Net Price £	Material £	Labour hours	Labour £	Unit	Total rate £
U10 : GRILLES/DIFFUSERS/LOUVRES (cont'd)						
Exhaust grilles; aluminium (cont'd)						
45 ° fixed blade core						
150 x 150mm	4.87	8.28	0.62	19.61	nr	27.89
200 x 200mm	5.41	9.20	0.72	22.78	nr	31.98
250 x 250mm	6.22	10.58	0.80	25.30	nr	35.87
300 x 300mm	8.66	14.73	1.00	31.62	nr	46.34
350 x 350mm	10.28	17.48	1.20	37.94	nr	55.42
45 ° fixed blade core; including opposed blade damper volume regulator						
150 x 150mm	8.12	13.81	0.62	19.61	nr	33.42
200 x 200mm	9.47	16.10	0.72	22.78	nr	38.88
250 x 250mm	10.28	17.48	0.80	25.30	nr	42.78
300 x 300mm	13.53	23.01	1.00	31.62	nr	54.63
350 x 350mm	14.61	24.84	1.20	37.94	nr	62.79
Eggcrate core						
150 x 150mm	5.41	9.20	0.62	19.61	nr	28.81
200 x 200mm	6.22	10.58	1.00	31.62	nr	42.20
250 x 250mm	7.57	12.87	0.80	25.30	nr	38.17
300 x 300mm	9.47	16.10	1.00	31.62	nr	47.72
350 x 350mm	11.09	18.86	1.20	37.94	nr	56.80
Eggcrate core; including opposed blade damper volume regulator						
150 x 150mm	6.22	10.58	0.62	19.61	nr	30.19
200 x 200mm	9.20	15.64	0.72	22.78	nr	38.43
250 x 250mm	11.09	18.86	0.80	25.30	nr	44.15
300 x 300mm	12.71	21.61	1.00	31.62	nr	53.23
350 x 350mm	15.69	26.68	1.20	37.94	nr	64.62
Mesh/perforated plate core						
150 x 150mm	8.12	13.81	0.62	19.60	nr	33.41
200 x 200mm	8.39	14.27	0.72	22.77	nr	37.03
250 x 250mm	10.01	17.02	0.80	25.30	nr	42.32
300 x 300mm	11.63	19.78	1.00	31.62	nr	51.39
350 x 350mm	13.26	22.55	1.20	37.94	nr	60.49
Mesh/perforated plate core; including opposed blade damper volume regulator						
150 x 150mm	13.91	23.65	0.62	19.60	nr	43.26
200 x 200mm	17.58	29.89	0.72	22.77	nr	52.66
250 x 250mm	27.03	45.96	0.80	25.30	nr	71.26
300 x 300mm	33.06	56.21	0.80	25.30	nr	81.51
350 x 350mm	46.23	78.61	1.20	37.94	nr	116.55
Plastic air diffusion system						
Eggcrate grilles						
150 x 150mm	3.75	4.80	0.62	16.81	nr	21.62
200 x 200mm	5.91	7.57	0.72	19.53	nr	27.10
250 x 250mm	6.57	8.41	0.80	21.68	nr	30.10
300 x 300mm	7.27	9.31	1.00	27.10	nr	36.41

U: VENTILATION/AIR CONDITIONING SYSTEMS

Item	Net Price £	Material £	Labour hours	Labour £	Unit	Total rate £
Single deflection grilles						
150 x 150mm	5.91	7.57	0.62	16.81	nr	**24.38**
200 x 200mm	5.91	7.57	0.72	19.53	nr	**27.10**
250 x 250mm	6.57	8.41	0.80	21.68	nr	**30.10**
300 x 300mm	7.27	9.31	1.00	27.10	nr	**36.41**
Double deflection grilles						
150 x 150mm	5.91	7.57	0.62	16.81	nr	**24.38**
200 x 200mm	9.66	12.37	0.72	19.53	nr	**31.90**
250 x 250mm	10.91	13.97	0.80	21.68	nr	**35.66**
300 x 300mm	14.97	19.17	1.00	27.10	nr	**46.28**
Door transfer grilles						
150 x 150mm	5.50	7.04	0.62	16.81	nr	**23.86**
200 x 200mm	6.50	8.33	0.72	19.53	nr	**27.85**
250 x 250mm	9.94	12.73	0.80	21.68	nr	**34.41**
300 x 300mm	12.02	15.40	1.00	27.10	nr	**42.50**
Opposed blade dampers						
150 x 150mm	2.76	3.54	0.62	16.81	nr	**20.35**
200 x 200mm	3.61	4.62	0.72	19.53	nr	**24.15**
250 x 250mm	3.61	4.62	0.80	21.68	nr	**26.31**
300 x 300mm	4.63	5.93	1.00	27.10	nr	**33.03**
Ceiling mounted diffusers; circular aluminium multi-core diffuser						
Circular; for ceiling mounting						
141mm dia. neck	12.76	21.70	0.80	25.30	nr	**46.99**
197mm dia. neck	15.68	26.66	1.10	34.78	nr	**61.45**
309mm dia. neck	24.07	40.93	1.40	44.29	nr	**85.21**
365mm dia. neck	30.26	51.45	1.50	47.48	nr	**98.93**
457mm dia. neck	49.22	83.69	2.00	63.24	nr	**146.93**
Circular; for ceiling mounting; including louvre damper volume control						
141mm dia. neck	14.22	24.18	1.00	31.62	nr	**55.80**
197mm dia. neck	17.14	29.14	1.20	37.96	nr	**67.10**
309mm dia. neck	25.89	44.02	1.60	50.59	nr	**94.61**
365mm dia. neck	33.18	56.42	1.90	60.11	nr	**116.53**
457mm dia. neck	55.79	94.86	2.40	76.01	nr	**170.87**
Ceiling mounted diffusers; rectangular aluminium multi-cone diffuser; four way flow						
Rectangular; for ceiling mounting						
150 x 150 mm neck	9.98	16.97	1.80	56.97	nr	**73.94**
300 x 150 mm neck	34.95	59.43	2.30	72.72	nr	**132.15**
300 x 300 mm neck	18.26	31.05	2.80	88.53	nr	**119.58**
450 x 150 mm neck	44.18	75.12	2.80	88.53	nr	**163.66**
450 x 300 mm neck	49.63	84.39	3.20	101.18	nr	**185.57**
450 x 450 mm neck	28.69	48.78	3.40	107.55	nr	**156.33**
600 x 150 mm neck	53.39	90.78	3.20	101.18	nr	**191.96**
600 x 300 mm neck	61.00	103.72	3.50	110.67	nr	**214.39**
600 x 600 mm neck	73.26	124.57	4.00	126.48	nr	**251.05**

U: VENTILATION/AIR CONDITIONING SYSTEMS

Item	Net Price £	Material £	Labour hours	Labour £	Unit	Total rate £
U10 : GRILLES/DIFFUSERS/LOUVRES (cont'd)						
Ceiling mounted diffusers (cont'd)						
Rectangular; for ceiling mounting; including opposed blade damper volume regulator						
150 x 150 mm neck	9.98	16.97	1.80	56.97	nr	73.94
300 x 150 mm neck	34.95	59.43	2.30	72.72	nr	132.15
300 x 300 mm neck	30.89	52.52	2.80	88.57	nr	141.09
450 x 150 mm neck	56.90	96.75	2.80	88.53	nr	185.29
450 x 300 mm neck	65.89	112.04	3.30	104.34	nr	216.38
450 x 450 mm neck	48.71	82.83	3.51	110.95	nr	193.77
600 x 150 mm neck	67.25	114.35	3.30	104.34	nr	218.69
600 x 300 mm neck	79.19	134.65	4.00	126.48	nr	261.13
600 x 600 mm neck	100.12	170.24	5.62	177.64	nr	347.88
Slot diffusers; continuous aluminium slot diffuser with flanged frame (1500mm sections)						
Diffuser						
1 slot	11.36	19.32	3.76	118.87	m	138.19
2 slot	20.02	34.04	3.76	118.84	m	152.88
3 slot	26.78	45.54	3.76	118.84	m	164.37
4 slot	33.27	56.57	4.50	142.43	m	199.00
6 slot	47.34	80.50	4.50	142.43	m	222.92
8 slot	58.00	98.62	5.20	164.42	m	263.04
Diffuser; including equalizing deflector						
1 slot	11.94	20.30	5.26	166.42	m	186.72
2 slot	14.54	24.72	5.26	166.42	m	191.14
3 slot	17.89	30.42	5.26	166.42	m	196.84
4 slot	20.14	34.25	6.33	200.12	m	234.37
6 slot	26.54	45.13	6.33	200.12	m	245.25
8 slot	32.51	55.28	7.20	227.66	m	282.94
Extra over for ends						
1 slot	6.16	10.47	1.00	31.62	nr	42.09
2 slot	5.93	10.08	1.00	31.62	nr	41.70
3 slot	6.78	11.53	1.00	31.62	nr	43.15
4 slot	7.63	12.97	1.30	41.10	nr	54.08
6 slot	9.32	15.85	1.40	44.27	nr	60.11
8 slot	2.92	4.97	1.40	44.27	nr	49.23
Plenum boxes; 1.0m long; circular spigot; including cord operated flap damper						
1 slot	14.11	23.99	2.75	87.10	nr	111.10
2 slot	14.11	23.99	2.75	87.10	nr	111.10
3 slot	15.06	25.61	2.75	87.10	nr	112.71
4 slot	15.29	26.00	3.51	110.95	nr	136.94
6 slot	16.23	27.60	3.51	110.95	nr	138.54
8 slot	16.97	28.86	4.20	132.80	nr	161.66
Plenum boxes; 2.0m long; circular spigot; including cord operated flap damper						
1 slot	18.11	54.79	3.26	102.99	nr	157.78
2 slot	18.11	30.79	3.26	102.99	nr	133.79
3 slot	19.05	32.39	3.26	102.99	nr	135.39
4 slot	19.76	33.60	3.76	118.87	nr	152.47
6 slot	20.23	34.40	3.76	118.87	nr	153.27
8 slot	21.32	36.25	4.10	129.64	nr	165.89

U: VENTILATION/AIR CONDITIONING SYSTEMS

Item	Net Price £	Material £	Labour hours	Labour £	Unit	Total rate £
Perforated diffusers; rectangular face aluminium perforated diffuser; quick release face plate; for integration with rectangular ceiling tiles						
Circular spigot; rectangular diffuser						
150mm dia. spigot; 300 x 300 diffuser	35.57	60.48	1.00	31.62	nr	92.10
300mm dia. spigot; 600 x 600 diffuser	53.82	91.51	1.40	44.29	nr	135.80
Circular spigot; rectangular diffuser; including louvre damper volume regulator						
150mm dia. spigot; 300 x 300 diffuser	43.45	73.88	1.00	31.62	nr	105.50
300mm dia. spigot; 600 x 600 diffuser	62.44	106.17	1.60	50.59	nr	156.76
Rectangular spigot; rectangular diffuser						
150 x 150mm dia. spigot; 300 x 300mm diffuser	35.15	59.77	1.00	27.10	nr	86.87
300 x 150mm dia. spigot; 600 x 300mm diffuser	46.49	79.05	1.20	32.52	nr	111.57
300 x 300mm dia. spigot; 600 x 600mm diffuser	50.66	86.14	1.40	37.94	nr	124.09
600 x 300mm dia. spigot; 1200 x 600mm diffuser	53.82	91.51	1.60	43.37	nr	134.88
Rectangular spigot; rectangular diffuser; including opposed blade damper volume regulator						
150 x 150mm dia. spigot; 300 x 300mm diffuser	40.10	68.19	1.20	32.52	nr	100.71
300 x 150mm dia. spigot; 600 x 300mm diffuser	55.49	94.35	1.40	37.96	nr	132.32
300 x 300mm dia. spigot; 600 x 600mm diffuser	60.58	103.01	1.60	43.37	nr	146.37
600 x 300mm dia. spigot; 1200 x 600mm diffuser	68.05	115.71	1.80	48.79	nr	164.50
Floor swirl diffuser; manual adjustment of air discharge direction; complete with damper and dirt trap						
Plastic Diffuser						
150 dia	11.97	20.35	0.50	15.81	nr	36.16
200 dia	15.53	26.41	0.50	15.81	nr	42.22
Aluminium Diffuser						
150 dia	14.56	24.76	0.50	15.81	nr	40.57
200 dia	18.44	31.36	0.50	15.81	nr	47.16
Plastic air diffusion system						
Cellular diffusers						
300 x 300mm	6.12	7.84	2.80	75.92	nr	83.76
600 x 600mm	12.29	15.74	4.00	108.41	nr	124.15
Multi-cone diffusers						
300 x 300mm	4.97	6.37	2.80	75.92	nr	82.29
450 x 450mm	9.76	12.50	3.40	92.19	nr	104.69
500 x 500mm	10.45	13.38	3.80	103.06	nr	116.44
600 x 600mm	13.26	16.98	4.00	108.41	nr	125.40
625 x 625mm	23.08	29.56	4.26	115.33	nr	144.89
Opposed blade dampers						
300 x 300mm	5.98	7.66	1.20	32.54	nr	40.20
450 x 450mm	8.57	10.98	1.50	40.70	nr	51.67
600 x 600mm	16.77	21.48	2.60	70.58	nr	92.06

U: VENTILATION/AIR CONDITIONING SYSTEMS

Item	Net Price £	Material £	Labour hours	Labour £	Unit	Total rate £
U10 : GRILLES/DIFFUSERS/LOUVRES (cont'd)						
Plastic air diffusion system (cont'd)						
Plenum boxes						
300mm	3.56	4.56	2.80	75.92	nr	80.48
450mm	5.12	6.56	3.40	92.19	nr	98.75
600mm	8.53	10.93	4.00	108.41	nr	119.34
Plenum spigot reducer						
600mm	4.71	6.03	1.00	27.10	nr	33.14
Blanking kits for cellular diffusers						
300mm	1.34	1.72	0.88	23.86	nr	25.58
600mm	4.83	6.19	1.10	29.82	nr	36.00
Blanking kits for multi-cone diffusers						
300mm	1.34	1.72	0.88	23.86	nr	25.58
450mm	2.87	3.68	0.90	24.40	nr	28.07
600mm	4.83	6.19	1.10	29.82	nr	36.00
Acoustic louvres; opening mounted; 300mm deep steel louvres with blades packed with acoustic infill; 12mm galvanised mesh birdscreen; screw fixing in opening						
Louvre units; self finished galvanised steel						
900 high x 600 wide	114.54	146.70	3.00	81.31	nr	228.01
900 high x 900 wide	142.12	182.03	3.00	81.31	nr	263.34
900 high x 1200 wide	168.25	215.49	3.34	90.53	nr	306.02
900 high x 1500 wide	219.59	281.25	3.34	90.53	nr	371.77
900 high x 1800 wide	246.21	315.34	3.34	90.53	nr	405.87
900 high x 2100 wide	272.82	349.43	3.34	90.53	nr	439.95
900 high x 2400 wide	298.96	382.90	3.68	99.74	nr	482.65
900 high x 2700 wide	337.94	432.83	3.68	99.74	nr	532.57
900 high x 3000 wide	361.70	463.26	3.68	99.74	nr	563.00
1200 high x 600 wide	150.19	192.36	3.00	81.31	nr	273.67
1200 high x 900 wide	184.89	236.81	3.34	90.53	nr	327.33
1200 high x 1200 wide	219.10	280.62	3.34	90.53	nr	371.15
1200 high x 1500 wide	290.41	371.95	3.34	90.53	nr	462.48
1200 high x 1800 wide	324.63	415.78	3.68	99.74	nr	515.52
1200 high x 2100 wide	359.80	460.83	3.68	99.74	nr	560.57
1200 high x 2400 wide	394.02	504.66	3.68	99.74	nr	604.40
1500 high x 600 wide	186.31	238.62	3.00	81.31	nr	319.93
1500 high x 900 wide	228.62	292.81	3.34	90.53	nr	383.34
1500 high x 1200 wide	270.92	346.99	3.34	90.53	nr	437.52
1500 high x 1500 wide	361.23	462.66	3.68	99.74	nr	562.40
1500 high x 1800 wide	403.53	516.84	3.68	99.74	nr	616.58
1500 high x 2100 wide	446.31	571.63	4.00	108.41	nr	680.04
1800 high x 600 wide	221.48	283.67	3.34	90.53	nr	374.20
1800 high x 900 wide	271.87	348.21	3.34	90.53	nr	438.73
1800 high x 1200 wide	322.72	413.34	3.68	99.74	nr	513.08
1800 high x 1500 wide	432.52	553.97	3.68	99.74	nr	653.71

U: VENTILATION/AIR CONDITIONING SYSTEMS

Item	Net Price £	Material £	Labour hours	Labour £	Unit	Total rate £
Louvre units; polyester powder coated steel						
900 high x 600 wide	166.36	213.07	3.00	81.31	nr	294.38
900 high x 900 wide	219.10	280.62	3.00	81.31	nr	361.93
900 high x 1200 wide	270.92	346.99	3.34	90.53	nr	437.52
900 high x 1500 wide	347.92	445.61	3.34	90.53	nr	536.14
900 high x 1800 wide	400.19	512.56	3.34	90.53	nr	603.09
900 high x 2100 wide	452.96	580.15	3.34	90.53	nr	670.67
900 high x 2400 wide	504.76	646.49	3.68	99.74	nr	746.23
900 high x 2700 wide	568.93	728.68	3.68	99.74	nr	828.42
900 high x 3000 wide	618.36	791.99	3.68	99.74	nr	891.73
1200 high x 600 wide	218.63	280.02	3.00	81.31	nr	361.33
1200 high x 900 wide	287.55	368.29	3.34	90.53	nr	458.82
1200 high x 1200 wide	356.47	456.56	3.34	90.53	nr	547.09
1200 high x 1500 wide	461.51	591.10	3.34	90.53	nr	681.62
1200 high x 1800 wide	529.95	678.75	3.68	99.74	nr	778.50
1200 high x 2100 wide	599.34	767.63	3.68	99.74	nr	867.37
1200 high x 2400 wide	667.79	855.30	3.68	99.74	nr	955.04
1500 high x 600 wide	271.87	348.21	3.00	81.31	nr	429.52
1500 high x 900 wide	356.94	457.17	3.34	90.53	nr	547.69
1500 high x 1200 wide	442.02	566.13	3.34	90.53	nr	656.66
1500 high x 1500 wide	575.11	736.60	3.68	99.74	nr	836.34
1500 high x 1800 wide	660.66	846.17	3.68	99.74	nr	945.91
1500 high x 2100 wide	745.74	955.14	4.00	108.41	nr	1063.55
1800 high x 600 wide	324.14	415.16	3.34	90.53	nr	505.68
1800 high x 900 wide	425.86	545.44	3.34	90.53	nr	635.96
1800 high x 1200 wide	528.05	676.32	3.68	99.74	nr	776.06
1800 high x 1500 wide	689.18	882.69	3.68	99.74	nr	982.44
Weather louvres; opening mounted; 300mm deep galvanised steel louvres; screw fixing in position						
Louvre units; including 12mm galvanised mesh birdscreen						
900 x 600mm	97.23	124.53	2.25	61.05	nr	185.58
900 x 900mm	140.48	179.93	2.25	61.05	nr	240.97
900 x 1200mm	169.03	216.49	2.50	67.76	nr	284.25
900 x 1500mm	205.71	263.47	2.50	67.76	nr	331.23
900 x 1800mm	242.40	310.46	2.50	67.76	nr	378.22
900 x 2100mm	303.83	389.14	2.50	67.76	nr	456.90
900 x 2400mm	326.40	418.05	2.76	74.87	nr	492.92
900 x 2700mm	368.48	471.95	2.76	74.87	nr	546.82
900 x 3000mm	402.42	515.42	2.76	74.87	nr	590.29
1200 x 600mm	131.68	168.65	2.25	61.05	nr	229.70
1200 x 900mm	183.99	235.65	2.50	67.76	nr	303.41
1200 x 1200mm	236.72	303.19	2.50	67.76	nr	370.95
1200 x 1500mm	276.32	353.91	2.50	67.76	nr	421.67
1200 x 1800mm	361.37	462.84	2.76	74.87	nr	537.71
1200 x 2100mm	410.97	526.37	2.76	74.87	nr	601.24
1200 x 2400mm	429.19	549.70	2.76	74.87	nr	624.57
1500 x 600mm	155.63	199.33	2.25	61.05	nr	260.37
1500 x 900mm	214.47	274.69	2.50	67.76	nr	342.45
1500 x 1200mm	282.50	361.82	2.50	67.76	nr	429.58
1500 x 1500mm	331.21	424.21	2.76	74.87	nr	499.08
1500 x 1800mm	390.54	500.20	2.76	74.87	nr	575.07
1500 x 2100mm	488.45	625.60	3.00	81.39	nr	706.99
1800 x 600mm	174.10	222.99	2.50	67.76	nr	290.74
1800 x 900mm	240.67	308.25	2.50	67.76	nr	376.01
1800 x 1200mm	315.58	404.19	2.76	74.87	nr	479.06
1800 x 1500mm	373.30	478.12	3.00	81.39	nr	559.51

U: VENTILATION/AIR CONDITIONING SYSTEMS

Item	Net Price £	Material £	Labour hours	Labour £	Unit	Total rate £
U10 : THERMAL INSULATION						
Y50 - THERMAL INSULATION						
Concealed Ductwork						
Flexible wrap; 20kg-45kg Bright Class O aluminium foil faced; Bright Class O foil taped joints; 62mm metal pins and washers; aluminium bands						
40mm thick insulation	8.93	11.68	0.40	8.79	m²	**20.47**
Semi-rigid slab; 45kg Bright Class O aluminium foil faced mineral fibre; Bright Class O foil taped joints; 62mm metal pins and washers; aluminium bands						
40mm thick insulation	11.69	15.29	0.65	14.29	m²	**29.57**
Plantroom Ductwork						
Semi-rigid slab; 45kg Bright Class O aluminium foil faced mineral fibre; Bright Class O foil taped joints; 62mm metal pins and washers; 22 swg plain/embossed aluminium cladding; pop riveted						
50mm thick insulation	29.49	38.56	1.50	32.97	m²	**71.54**
External Ductwork						
Semi-rigid slab; 45kg Bright Class O aluminium foil faced mineral fibre; Bright Class O foil taped joints; 62mm metal pins and washers; 0.8mm polyisobutylene sheeting; welded joints						
50mm thick insulation	21.51	28.13	1.25	27.48	m²	**55.61**

U: VENTILATION/AIR CONDITIONING SYSTEMS

Item	Net Price £	Material £	Labour hours	Labour £	Unit	Total rate £
U14 : DUCTWORK : FIRE RATED						
Y30 - DUCTLINES						
The relevant BS requires that the fire rating of ductwork meets 3 criteria; stability (hours), integrity (hours) and insulation (hours). The least of the 3 periods defines the fire rating. The BS does however allow stability and integrity to be considered in isolation. Rates are therefore provided for both types of system.						
Care should to be taken when using the rates within this section to ensure that the requirements for stability, integrity and insulation are known and the appropriate rates are used						
High density single layer mineral wool fire rated ductwork slab, in accordance with BS476, Part 24 (ISO 6944: 1985), ducts "Type A" and "Type B"; 165kg class O foil faced mineral fibre; 100mm wide bright class O foil taped joints; welded pins; includes protection to all supports.						
1/2 hour stability, integrity and insulation						
25mm thick, vertical and horizontal ductwork	32.09	41.96	1.25	27.48	m²	**69.44**
1 hour stability, integrity and insulation						
30mm thick, vertical ductwork	37.69	49.29	1.50	32.97	m²	**82.26**
40mm thick, horizontal ductwork	44.32	57.96	1.50	32.97	m²	**90.93**
1½ hour stability, integrity and insulation						
50mm thick, vertical ductwork	52.93	69.22	1.75	38.47	m²	**107.68**
70mm thick, horizontal ductwork	66.22	86.59	1.75	38.47	m²	**125.06**
2 hour stability, integrity and insulation						
70mm, vertical ductwork	68.23	89.22	2.00	43.96	m²	**133.19**
90mm horizontal ductwork	81.50	106.58	2.00	43.96	m²	**150.54**
Kitchen extract, 1 hour stability, integrity and insulation						
90mm, vertical and horizontal	81.50	106.58	2.00	43.96	m²	**150.54**

U: VENTILATION/AIR CONDITIONING SYSTEMS

Item	Net Price £	Material £	Labour hours	Labour £	Unit	Total rate £
U14 : DUCTWORK : FIRE RATED (cont'd)						
Galvanised sheet metal rectangular section ductwork to BS476 Part 24 (ISO 6944:1985), ducts "Type A" and "Type B"; provides 2 hours stability and 2 hours integrity at 1100°C (no rating for insulation); including all necessary stiffeners, joints and supports in the running length						
Ductwork up to 600mm longest side						
Sum of two sides 200mm	59.72	101.55	2.91	92.01	m	193.57
Sum of two sides 300mm	64.84	110.25	2.99	94.54	m	204.79
Sum of two sides 400mm	69.94	118.93	3.17	100.23	m	219.16
Sum of two sides 500mm	75.06	127.63	3.37	106.56	m	234.19
Sum of two sides 600mm	80.16	136.31	3.54	111.93	m	248.24
Sum of two sides 700mm	85.28	145.01	3.72	117.62	m	262.63
Sum of two sides 800mm	90.40	153.71	3.90	123.31	m	277.02
Sum of two sides 900mm	95.50	162.38	5.04	159.36	m	321.75
Sum of two sides 1000mm	100.62	171.08	5.58	176.44	m	347.52
Sum of two sides 1100mm	105.72	179.76	5.84	184.66	m	364.42
Sum of two sides 1200mm	110.83	188.46	6.11	193.19	m	381.65
Extra over fittings; Ductwork up to 600mm longest side						
End Cap						
Sum of two sides 200mm	16.96	28.83	0.81	25.61	nr	54.44
Sum of two sides 300mm	17.98	30.58	0.84	26.56	nr	57.14
Sum of two sides 400mm	19.02	32.34	0.87	27.51	nr	59.85
Sum of two sides 500mm	20.05	34.09	0.90	28.46	nr	62.55
Sum of two sides 600mm	21.07	35.83	0.93	29.41	nr	65.24
Sum of two sides 700mm	22.11	37.60	0.96	30.35	nr	67.95
Sum of two sides 800mm	23.17	39.39	0.98	30.99	nr	70.38
Sum of two sides 900mm	24.19	41.14	1.17	36.99	nr	78.13
Sum of two sides 1000mm	25.22	42.88	1.22	38.58	nr	81.46
Sum of two sides 1100mm	26.26	44.65	1.25	39.52	nr	84.17
Sum of two sides 1200mm	27.28	46.39	1.28	40.47	nr	86.86
Reducer						
Sum of two sides 200mm	67.26	114.36	2.23	70.51	nr	184.87
Sum of two sides 300mm	70.35	119.62	2.37	74.94	nr	194.56
Sum of two sides 400mm	73.43	124.85	2.51	79.36	nr	204.22
Sum of two sides 500mm	76.53	130.13	2.65	83.79	nr	213.92
Sum of two sides 600mm	79.61	135.37	2.79	88.22	nr	223.58
Sum of two sides 700mm	82.70	140.62	2.87	90.75	nr	231.37
Sum of two sides 800mm	85.79	145.88	3.01	95.17	nr	241.05
Sum of two sides 900mm	88.88	151.14	3.06	96.75	nr	247.89
Sum of two sides 1000mm	91.96	156.37	3.17	100.23	nr	256.60
Sum of two sides 1100mm	95.04	161.60	3.22	101.81	nr	263.42
Sum of two sides 1200mm	98.14	166.88	3.28	103.71	nr	270.59

U: VENTILATION/AIR CONDITIONING SYSTEMS

Item	Net Price £	Material £	Labour hours	Labour £	Unit	Total rate £
Offset						
Sum of two sides 200mm	162.32	276.01	2.95	93.28	nr	369.29
Sum of two sides 300mm	166.35	282.85	3.11	98.34	nr	381.19
Sum of two sides 400mm	170.37	289.69	3.26	103.08	nr	392.77
Sum of two sides 500mm	174.41	296.56	3.42	108.14	nr	404.69
Sum of two sides 600mm	178.43	303.40	3.57	112.88	nr	416.28
Sum of two sides 700mm	182.47	310.26	3.73	117.94	nr	428.20
Sum of two sides 800mm	186.49	317.10	3.23	102.13	nr	419.23
Sum of two sides 900mm	190.53	323.97	3.45	109.09	nr	433.05
Sum of two sides 1000mm	194.55	330.81	3.67	116.04	nr	446.85
Sum of two sides 1100mm	198.57	337.65	3.78	119.52	nr	457.17
Sum of two sides 1200mm	202.62	344.53	3.89	123.00	nr	467.53
90° radius bend						
Sum of two sides 200mm	65.81	111.91	2.06	65.14	nr	177.04
Sum of two sides 300mm	70.81	120.40	2.21	69.88	nr	190.28
Sum of two sides 400mm	75.78	128.85	2.36	74.62	nr	203.47
Sum of two sides 500mm	80.78	137.36	2.51	79.36	nr	216.73
Sum of two sides 600mm	85.77	145.83	2.66	84.11	nr	229.94
Sum of two sides 700mm	90.73	154.28	2.81	88.85	nr	243.13
Sum of two sides 800mm	95.73	162.77	2.97	93.91	nr	256.68
Sum of two sides 900mm	100.70	171.22	2.99	94.54	nr	265.76
Sum of two sides 1000mm	105.70	179.74	3.04	96.12	nr	275.86
Sum of two sides 1100mm	110.67	188.19	3.07	97.07	nr	285.26
Sum of two sides 1200mm	115.65	196.66	3.09	97.70	nr	294.36
45° radius bend						
Sum of two sides 200mm	81.15	137.98	1.52	48.06	nr	186.04
Sum of two sides 300mm	83.17	141.43	1.59	50.27	nr	191.70
Sum of two sides 400mm	85.20	144.87	1.66	52.49	nr	197.36
Sum of two sides 500mm	87.21	148.29	1.73	54.70	nr	202.99
Sum of two sides 600mm	89.22	151.71	1.80	56.91	nr	208.63
Sum of two sides 700mm	91.25	155.15	1.87	59.13	nr	214.28
Sum of two sides 800mm	93.24	158.55	1.93	61.03	nr	219.58
Sum of two sides 900mm	95.27	161.99	1.99	62.92	nr	224.92
Sum of two sides 1000mm	97.28	165.41	2.05	64.82	nr	230.23
Sum of two sides 1100mm	99.29	168.84	2.11	66.72	nr	235.55
Sum of two sides 1200mm	101.30	172.26	2.17	68.61	nr	240.87
90° mitre bend						
Sum of two sides 200mm	68.23	116.02	2.47	78.10	nr	194.11
Sum of two sides 300mm	79.08	134.47	2.65	83.79	nr	218.26
Sum of two sides 400mm	89.92	152.90	2.84	89.80	nr	242.70
Sum of two sides 500mm	100.79	171.38	3.02	95.49	nr	266.87
Sum of two sides 600mm	111.63	189.82	3.20	101.18	nr	291.00
Sum of two sides 700mm	122.49	208.27	3.38	106.87	nr	315.14
Sum of two sides 800mm	133.34	226.73	3.56	112.56	nr	339.29
Sum of two sides 900mm	144.19	245.18	3.59	113.51	nr	358.70
Sum of two sides 1000mm	155.06	263.66	3.66	115.73	nr	379.39
Sum of two sides 1100mm	165.89	282.07	3.69	116.67	nr	398.75
Sum of two sides 1200mm	176.77	300.57	3.72	117.62	nr	418.20

U: VENTILATION/AIR CONDITIONING SYSTEMS

Item	Net Price £	Material £	Labour hours	Labour £	Unit	Total rate £
U14 : DUCTWORK : FIRE RATED (cont'd)						
Galvanised sheet metal rectangular (cont'd)						
Extra over fittings; Ductwork up to 600mm longest side (cont'd)						
Branch (Side-on Shoe)						
Sum of two sides 200mm	25.33	43.06	0.98	30.99	nr	74.05
Sum of two sides 300mm	25.47	43.32	1.06	33.52	nr	76.83
Sum of two sides 400mm	25.61	43.55	1.14	36.05	nr	79.59
Sum of two sides 500mm	25.76	43.80	1.22	38.58	nr	82.37
Sum of two sides 600mm	25.88	44.01	1.30	41.10	nr	85.11
Sum of two sides 700mm	26.03	44.26	1.38	43.63	nr	87.89
Sum of two sides 800mm	26.15	44.46	1.46	46.16	nr	90.63
Sum of two sides 900mm	26.35	44.81	1.37	43.32	nr	88.13
Sum of two sides 1000mm	26.42	44.92	1.46	46.16	nr	91.09
Sum of two sides 1100mm	26.57	45.18	1.50	47.43	nr	92.60
Sum of two sides 1200mm	26.70	45.41	1.55	49.01	nr	94.41
Ductwork 601 to 800mm longest side						
Sum of two sides 900mm	95.49	162.36	5.04	159.36	m	321.72
Sum of two sides 1000mm	100.60	171.06	5.58	176.44	m	347.50
Sum of two sides 1100mm	105.72	179.76	5.84	184.66	m	364.42
Sum of two sides 1200mm	110.83	188.46	6.11	193.19	m	381.65
Sum of two sides 1300mm	115.97	197.18	6.38	201.73	m	398.92
Sum of two sides 1400mm	121.08	205.88	6.65	210.27	m	416.15
Sum of two sides 1500mm	126.21	214.61	6.91	218.49	m	433.10
Sum of two sides 1600mm	131.33	223.31	7.18	227.03	m	450.33
Extra over fittings; Ductwork 601 to 800mm longest side						
End Cap						
Sum of two sides 900mm	22.48	38.22	1.17	36.99	nr	75.21
Sum of two sides 1000mm	24.08	40.95	1.22	38.58	nr	79.53
Sum of two sides 1100mm	25.70	43.71	1.25	39.52	nr	83.23
Sum of two sides 1200mm	27.30	46.42	1.28	40.47	nr	86.89
Sum of two sides 1300mm	28.90	49.15	1.31	41.42	nr	90.57
Sum of two sides 1400mm	30.51	51.88	1.34	42.37	nr	94.25
Sum of two sides 1500mm	32.12	54.61	1.36	43.00	nr	97.61
Sum of two sides 1600mm	33.72	57.34	1.39	43.95	nr	101.29
Reducer						
Sum of two sides 900mm	84.12	143.03	3.06	96.75	nr	239.79
Sum of two sides 1000mm	88.80	151.00	3.17	100.23	nr	251.23
Sum of two sides 1100mm	93.47	158.94	3.22	101.81	nr	260.76
Sum of two sides 1200mm	98.16	166.91	3.28	103.71	nr	270.62
Sum of two sides 1300mm	102.84	174.87	3.33	105.29	nr	280.16
Sum of two sides 1400mm	107.53	182.84	3.38	106.87	nr	289.71
Sum of two sides 1500mm	112.20	190.78	3.44	108.77	nr	299.55
Sum of two sides 1600mm	116.88	198.75	3.49	110.35	nr	309.10

U: VENTILATION/AIR CONDITIONING SYSTEMS

Item	Net Price £	Material £	Labour hours	Labour £	Unit	Total rate £
Offset						
Sum of two sides 900mm	178.90	304.20	3.45	109.09	nr	413.29
Sum of two sides 1000mm	187.42	318.69	3.67	116.04	nr	434.73
Sum of two sides 1100mm	195.94	333.17	3.78	119.52	nr	452.69
Sum of two sides 1200mm	204.46	347.66	3.89	123.00	nr	470.65
Sum of two sides 1300mm	212.96	362.12	4.00	126.48	nr	488.59
Sum of two sides 1400mm	221.49	376.62	4.11	129.95	nr	506.58
Sum of two sides 1500mm	230.00	391.09	4.23	133.75	nr	524.84
Sum of two sides 1600mm	238.79	406.03	4.34	137.23	nr	543.26
90° radius bend						
Sum of two sides 900mm	89.45	152.10	2.99	94.54	nr	246.64
Sum of two sides 1000mm	98.24	167.04	3.04	96.12	nr	263.17
Sum of two sides 1100mm	107.00	181.94	3.07	97.07	nr	279.01
Sum of two sides 1200mm	115.79	196.89	3.09	97.70	nr	294.59
Sum of two sides 1300mm	124.56	211.81	3.12	98.65	nr	310.46
Sum of two sides 1400mm	133.34	226.73	3.15	99.60	nr	326.33
Sum of two sides 1500mm	142.13	241.67	3.17	100.23	nr	341.90
Sum of two sides 1600mm	150.92	256.62	3.20	101.18	nr	357.80
45° bend						
Sum of two sides 900mm	88.82	151.02	1.74	55.02	nr	206.04
Sum of two sides 1000mm	92.99	158.11	1.85	58.50	nr	216.61
Sum of two sides 1100mm	97.16	165.21	1.91	60.39	nr	225.60
Sum of two sides 1200mm	101.32	172.28	1.96	61.97	nr	234.25
Sum of two sides 1300mm	105.48	179.35	2.02	63.87	nr	243.22
Sum of two sides 1400mm	109.65	186.44	2.07	65.45	nr	251.89
Sum of two sides 1500mm	113.61	193.51	2.13	67.35	nr	260.86
Sum of two sides 1600mm	117.98	200.60	2.18	68.93	nr	269.53
90° mitre bend						
Sum of two sides 900mm	133.92	227.71	3.59	113.51	nr	341.23
Sum of two sides 1000mm	148.20	252.00	3.66	115.73	nr	367.73
Sum of two sides 1100mm	162.47	276.26	3.69	116.67	nr	392.94
Sum of two sides 1200mm	176.77	300.57	3.72	117.62	nr	418.20
Sum of two sides 1300mm	191.03	324.82	3.75	118.57	nr	443.39
Sum of two sides 1400mm	205.31	349.10	3.78	119.52	nr	468.62
Sum of two sides 1500mm	219.58	373.37	3.81	120.47	nr	493.83
Sum of two sides 1600mm	233.86	397.65	3.84	121.42	nr	519.07
Branch (Side-on Shoe)						
Sum of two sides 900mm	24.85	42.26	1.37	43.32	nr	85.58
Sum of two sides 1000mm	25.47	43.32	1.46	46.16	nr	89.48
Sum of two sides 1100mm	26.10	44.37	1.50	47.43	nr	91.80
Sum of two sides 1200mm	26.72	45.43	1.55	49.01	nr	94.44
Sum of two sides 1300mm	27.34	46.48	1.59	50.27	nr	96.76
Sum of two sides 1400mm	27.97	47.56	1.63	51.54	nr	99.10
Sum of two sides 1500mm	26.86	45.68	1.68	53.12	nr	98.80
Sum of two sides 1600mm	29.21	49.67	1.72	54.38	nr	104.06

U: VENTILATION/AIR CONDITIONING SYSTEMS

Item	Net Price £	Material £	Labour hours	Labour £	Unit	Total rate £
U14 : DUCTWORK : FIRE RATED (cont'd)						
Galvanised sheet metal rectangular (cont'd)						
Ductwork 801 to 1000mm longest side						
Sum of two sides 1100mm	105.72	179.76	5.84	184.66	m	364.42
Sum of two sides 1200mm	110.83	188.46	6.11	193.19	m	381.65
Sum of two sides 1300mm	115.95	197.16	6.38	201.73	m	398.89
Sum of two sides 1400mm	121.07	205.86	6.65	210.27	m	416.13
Sum of two sides 1500mm	126.18	214.56	6.91	218.49	m	433.05
Sum of two sides 1600mm	131.30	223.26	7.18	227.03	m	450.29
Sum of two sides 1700mm	136.43	231.98	7.45	235.56	m	467.55
Sum of two sides 1800mm	141.55	240.68	7.71	243.78	m	484.47
Sum of two sides 1900mm	146.66	249.38	7.98	252.32	m	501.71
Sum of two sides 2000mm	151.78	258.08	8.25	260.86	m	518.94
Extra over fittings; Ductwork 801 to 1000mm longest side						
End Cap						
Sum of two sides 1100mm	25.70	43.71	1.25	39.52	nr	83.23
Sum of two sides 1200mm	27.30	46.42	1.28	40.47	nr	86.89
Sum of two sides 1300mm	28.90	49.15	1.31	41.42	nr	90.57
Sum of two sides 1400mm	30.51	51.88	1.34	42.37	nr	94.25
Sum of two sides 1500mm	32.12	54.61	1.36	43.00	nr	97.61
Sum of two sides 1600mm	33.71	57.32	1.39	43.95	nr	101.27
Sum of two sides 1700mm	35.32	60.05	1.42	44.90	nr	104.95
Sum of two sides 1800mm	36.92	62.78	1.45	45.85	nr	108.63
Sum of two sides 1900mm	38.52	65.49	1.48	46.80	nr	112.29
Sum of two sides 2000mm	40.12	68.22	1.51	47.74	nr	115.97
Reducer						
Sum of two sides 1100mm	93.47	158.94	3.22	101.81	nr	260.76
Sum of two sides 1200mm	98.17	166.93	3.28	103.71	nr	270.64
Sum of two sides 1300mm	102.88	174.94	3.33	105.29	nr	280.23
Sum of two sides 1400mm	107.59	182.95	3.38	106.87	nr	289.83
Sum of two sides 1500mm	112.28	190.92	3.44	108.77	nr	299.69
Sum of two sides 1600mm	117.02	198.98	3.49	110.35	nr	309.33
Sum of two sides 1700mm	121.70	206.94	3.54	111.93	nr	318.87
Sum of two sides 1800mm	126.41	214.95	3.60	113.83	nr	328.78
Sum of two sides 1900mm	131.13	222.96	3.65	115.41	nr	338.37
Sum of two sides 2000mm	138.17	234.95	3.70	116.99	nr	351.94
Offset						
Sum of two sides 1100mm	194.40	330.55	3.78	119.52	nr	450.07
Sum of two sides 1200mm	202.62	344.53	3.89	123.00	nr	467.53
Sum of two sides 1300mm	210.83	358.49	4.00	126.48	nr	484.97
Sum of two sides 1400mm	219.05	372.47	4.11	129.95	nr	502.42
Sum of two sides 1500mm	227.26	386.43	4.23	133.75	nr	520.18
Sum of two sides 1600mm	235.49	400.43	4.34	137.23	nr	537.66
Sum of two sides 1700mm	243.70	414.39	4.45	140.71	nr	555.09
Sum of two sides 1800mm	251.92	428.37	4.56	144.18	nr	572.55
Sum of two sides 1900mm	260.13	442.32	4.67	147.66	nr	589.98
Sum of two sides 2000mm	272.46	463.28	5.53	174.85	nr	638.13

U: VENTILATION/AIR CONDITIONING SYSTEMS

Item	Net Price £	Material £	Labour hours	Labour £	Unit	Total rate £
90 ° radius bend						
Sum of two sides 1100mm	106.84	181.67	3.07	97.07	nr	278.74
Sum of two sides 1200mm	115.64	196.63	3.09	97.70	nr	294.34
Sum of two sides 1300mm	124.44	211.60	3.12	98.65	nr	310.25
Sum of two sides 1400mm	133.25	226.57	3.15	99.60	nr	326.17
Sum of two sides 1500mm	142.05	241.53	3.17	100.23	nr	341.77
Sum of two sides 1600mm	150.84	256.48	3.20	101.18	nr	357.66
Sum of two sides 1700mm	159.64	271.44	3.22	101.81	nr	373.26
Sum of two sides 1800mm	168.44	286.41	3.25	102.76	nr	389.17
Sum of two sides 1900mm	177.24	301.38	3.28	103.71	nr	405.09
Sum of two sides 2000mm	185.77	315.89	3.30	104.34	nr	420.23
45 ° bend						
Sum of two sides 1100mm	97.79	166.29	1.91	60.39	nr	226.68
Sum of two sides 1200mm	101.88	173.24	1.96	61.97	nr	235.22
Sum of two sides 1300mm	105.97	180.20	2.02	63.87	nr	244.07
Sum of two sides 1400mm	110.07	187.15	2.07	65.45	nr	252.60
Sum of two sides 1500mm	114.16	194.11	2.13	67.35	nr	261.46
Sum of two sides 1600mm	118.25	201.06	2.18	68.93	nr	269.99
Sum of two sides 1700mm	122.34	208.02	2.24	70.83	nr	278.85
Sum of two sides 1800mm	126.43	214.97	2.30	72.72	nr	287.70
Sum of two sides 1900mm	130.50	221.91	2.35	74.31	nr	296.21
Sum of two sides 2000mm	134.61	228.89	2.76	87.27	nr	316.15
90 ° mitre bend						
Sum of two sides 1100mm	162.47	276.26	3.69	116.67	nr	392.94
Sum of two sides 1200mm	176.77	300.57	3.72	117.62	nr	418.20
Sum of two sides 1300mm	191.03	324.82	3.75	118.57	nr	443.39
Sum of two sides 1400mm	205.31	349.10	3.78	119.52	nr	468.62
Sum of two sides 1500mm	219.60	373.41	3.81	120.47	nr	493.88
Sum of two sides 1600mm	233.86	397.65	3.84	121.42	nr	519.07
Sum of two sides 1700mm	248.14	421.94	3.87	122.37	nr	544.30
Sum of two sides 1800mm	262.43	446.22	3.90	123.31	nr	569.54
Sum of two sides 1900mm	276.70	470.49	3.93	124.26	nr	594.75
Sum of two sides 2000mm	290.98	494.77	3.96	125.21	nr	619.99
Branch (Side-on Shoe)						
Sum of two sides 1100mm	26.07	44.33	1.50	47.43	nr	91.76
Sum of two sides 1200mm	26.68	45.36	1.55	49.01	nr	94.37
Sum of two sides 1300mm	27.32	46.46	1.59	50.27	nr	96.74
Sum of two sides 1400mm	27.96	47.54	1.63	51.54	nr	99.08
Sum of two sides 1500mm	28.59	48.62	1.68	53.12	nr	101.74
Sum of two sides 1600mm	29.21	49.67	1.72	54.38	nr	104.06
Sum of two sides 1700mm	29.85	50.75	1.77	55.97	nr	106.72
Sum of two sides 1800mm	30.48	51.83	1.81	57.23	nr	109.06
Sum of two sides 1900mm	31.12	52.91	1.86	58.81	nr	111.72
Sum of two sides 2000mm	32.05	54.50	1.90	60.08	nr	114.57
Ductwork 1001 to 1250mm longest side						
Sum of two sides 1300mm	131.42	223.47	6.38	201.73	m	425.20
Sum of two sides 1400mm	137.09	233.11	6.65	210.27	m	443.38
Sum of two sides 1500mm	142.76	242.75	6.91	218.49	m	461.24
Sum of two sides 1600mm	147.01	249.98	6.91	218.49	m	468.47
Sum of two sides 1700mm	151.25	257.19	7.45	235.56	m	492.75
Sum of two sides 1800mm	156.90	266.78	7.71	243.78	m	510.57
Sum of two sides 1900mm	162.55	276.40	7.98	252.32	m	528.72
Sum of two sides 2000mm	168.22	286.04	8.25	260.86	m	546.90

U: VENTILATION/AIR CONDITIONING SYSTEMS

Item	Net Price £	Material £	Labour hours	Labour £	Unit	Total rate £
U14 : DUCTWORK : FIRE RATED (cont'd)						
Galvanised sheet metal rectangular (cont'd)						
Ductwork 1001 to 1250mm longest side (cont'd)						
Sum of two sides 2100mm	173.88	295.66	9.62	304.18	m	599.84
Sum of two sides 2200mm	178.12	302.87	9.62	304.18	m	607.05
Sum of two sides 2300mm	182.37	310.10	10.07	318.41	m	628.51
Sum of two sides 2400mm	188.03	319.72	10.47	331.05	m	650.77
Sum of two sides 2500mm	193.68	329.34	10.87	343.70	m	673.04
Extra over fittings; Ductwork 1001 to 1250mm longest side						
End Cap						
Sum of two sides 1300mm	28.92	49.17	1.31	41.42	nr	90.59
Sum of two sides 1400mm	30.51	51.88	1.34	42.37	nr	94.25
Sum of two sides 1500mm	32.09	54.56	1.36	43.00	nr	97.57
Sum of two sides 1600mm	33.71	57.32	1.39	43.95	nr	101.27
Sum of two sides 1700mm	35.32	60.05	1.42	44.90	nr	104.95
Sum of two sides 1800mm	36.90	62.74	1.45	45.85	nr	108.58
Sum of two sides 1900mm	38.52	65.49	1.48	46.80	nr	112.29
Sum of two sides 2000mm	40.12	68.22	1.51	47.74	nr	115.97
Sum of two sides 2100mm	41.73	70.95	2.66	84.11	nr	155.06
Sum of two sides 2200mm	43.34	73.69	2.80	88.53	nr	162.22
Sum of two sides 2300mm	44.94	76.42	2.95	93.28	nr	169.69
Sum of two sides 2400mm	46.53	79.13	3.10	98.02	nr	177.15
Sum of two sides 2500mm	48.14	81.86	3.24	102.45	nr	184.30
Reducer						
Sum of two sides 1300mm	102.84	174.87	3.33	105.29	nr	280.16
Sum of two sides 1400mm	107.47	182.75	3.38	106.87	nr	289.62
Sum of two sides 1500mm	112.10	190.62	3.44	108.77	nr	299.39
Sum of two sides 1600mm	116.76	198.54	3.49	110.35	nr	308.89
Sum of two sides 1700mm	121.42	206.46	3.54	111.93	nr	318.39
Sum of two sides 1800mm	126.08	214.38	3.60	113.83	nr	328.21
Sum of two sides 1900mm	130.72	222.27	3.65	115.41	nr	337.68
Sum of two sides 2000mm	135.38	230.19	3.70	116.99	nr	347.19
Sum of two sides 2100mm	140.01	238.07	3.75	118.57	nr	356.64
Sum of two sides 2200mm	144.65	245.96	3.80	120.15	nr	366.12
Sum of two sides 2300mm	149.31	253.88	3.85	121.73	nr	375.62
Sum of two sides 2400mm	153.97	261.80	3.90	123.31	nr	385.12
Sum of two sides 2500mm	158.63	269.72	3.95	124.90	nr	394.62
Offset						
Sum of two sides 1300mm	210.13	357.30	4.00	126.48	nr	483.77
Sum of two sides 1400mm	218.47	371.48	4.11	129.95	nr	501.44
Sum of two sides 1500mm	226.80	385.65	4.23	133.75	nr	519.40
Sum of two sides 1600mm	235.13	399.81	4.34	137.23	nr	537.04
Sum of two sides 1700mm	243.46	413.97	4.45	140.71	nr	554.68
Sum of two sides 1800mm	251.79	428.14	4.56	144.18	nr	572.32
Sum of two sides 1900mm	260.12	442.30	4.67	147.66	nr	589.96
Sum of two sides 2000mm	268.45	456.46	5.53	174.85	nr	631.32
Sum of two sides 2100mm	276.78	470.63	5.76	182.13	nr	652.75
Sum of two sides 2200mm	284.97	484.56	5.99	189.40	nr	673.96
Sum of two sides 2300mm	293.44	498.95	6.22	196.67	nr	695.62
Sum of two sides 2400mm	301.75	513.09	6.45	203.94	nr	717.04
Sum of two sides 2500mm	310.10	527.28	6.68	211.22	nr	738.50

U: VENTILATION/AIR CONDITIONING SYSTEMS

Item	Net Price £	Material £	Labour hours	Labour £	Unit	Total rate £
90 ° radius bend						
Sum of two sides 1300mm	124.28	211.32	3.12	98.65	nr	309.98
Sum of two sides 1400mm	138.29	235.15	3.15	99.60	nr	334.75
Sum of two sides 1500mm	141.94	241.35	3.17	100.23	nr	341.58
Sum of two sides 1600mm	150.75	256.34	3.20	101.18	nr	357.52
Sum of two sides 1700mm	159.58	271.35	3.22	101.81	nr	373.17
Sum of two sides 1800mm	168.40	286.34	3.25	102.76	nr	389.10
Sum of two sides 1900mm	177.23	301.35	3.28	103.71	nr	405.07
Sum of two sides 2000mm	186.06	316.37	3.30	104.34	nr	420.71
Sum of two sides 2100mm	194.87	331.36	3.32	104.98	nr	436.33
Sum of two sides 2200mm	203.70	346.37	3.34	105.61	nr	451.98
Sum of two sides 2300mm	212.53	361.38	3.36	106.24	nr	467.62
Sum of two sides 2400mm	221.36	376.40	3.38	106.87	nr	483.27
Sum of two sides 2500mm	230.18	391.38	3.40	107.51	nr	498.89
45 ° bend						
Sum of two sides 1300mm	100.67	171.18	2.02	63.87	nr	235.05
Sum of two sides 1400mm	105.19	178.87	2.07	65.45	nr	244.32
Sum of two sides 1500mm	109.73	186.58	2.13	67.35	nr	253.93
Sum of two sides 1600mm	114.26	194.29	2.18	68.93	nr	263.22
Sum of two sides 1700mm	118.79	201.98	2.24	70.83	nr	272.81
Sum of two sides 1800mm	123.32	209.70	2.30	72.72	nr	282.42
Sum of two sides 1900mm	127.84	217.39	2.35	74.31	nr	291.69
Sum of two sides 2000mm	132.39	225.12	2.76	87.27	nr	312.39
Sum of two sides 2100mm	136.93	232.83	2.89	91.38	nr	324.21
Sum of two sides 2200mm	141.47	240.55	3.01	95.17	nr	335.72
Sum of two sides 2300mm	145.99	248.24	3.13	98.97	nr	347.20
Sum of two sides 2400mm	150.53	255.95	3.26	103.08	nr	359.03
Sum of two sides 2500mm	155.06	263.66	3.37	106.56	nr	370.22
90 ° mitre bend						
Sum of two sides 1300mm	190.59	324.08	3.75	118.57	nr	442.65
Sum of two sides 1400mm	204.74	348.14	3.78	119.52	nr	467.66
Sum of two sides 1500mm	219.44	373.14	3.81	120.47	nr	493.60
Sum of two sides 1600mm	233.85	397.63	3.84	121.42	nr	519.05
Sum of two sides 1700mm	248.25	422.12	3.87	122.37	nr	544.49
Sum of two sides 1800mm	262.64	446.59	3.90	123.31	nr	569.91
Sum of two sides 1900mm	277.06	471.11	3.93	124.26	nr	595.37
Sum of two sides 2000mm	291.48	495.62	3.96	125.21	nr	620.84
Sum of two sides 2100mm	305.88	520.12	3.99	126.16	nr	646.28
Sum of two sides 2200mm	320.30	544.63	4.02	127.11	nr	671.74
Sum of two sides 2300mm	334.71	569.13	4.05	128.06	nr	697.18
Sum of two sides 2400mm	349.10	593.60	4.08	129.01	nr	722.60
Sum of two sides 2500mm	363.50	618.09	4.11	129.95	nr	748.04
Branch (Side-on Shoe)						
Sum of two sides 1300mm	27.38	46.55	1.59	50.27	nr	96.83
Sum of two sides 1400mm	28.00	47.61	1.63	51.54	nr	99.15
Sum of two sides 1500mm	28.62	48.66	1.68	53.12	nr	101.79
Sum of two sides 1600mm	29.24	49.72	1.72	54.38	nr	104.11
Sum of two sides 1700mm	29.88	50.80	1.77	55.97	nr	106.77
Sum of two sides 1800mm	30.51	51.88	1.81	57.23	nr	109.11
Sum of two sides 1900mm	31.13	52.93	1.86	58.81	nr	111.75
Sum of two sides 2000mm	31.77	54.01	1.90	60.08	nr	114.09
Sum of two sides 2100mm	32.39	55.07	2.58	81.58	nr	136.65
Sum of two sides 2200mm	33.01	56.13	2.61	82.53	nr	138.65
Sum of two sides 2300mm	33.64	57.20	2.64	83.47	nr	140.68
Sum of two sides 2400mm	34.28	58.28	2.88	91.06	nr	149.35
Sum of two sides 2500mm	34.90	59.34	2.91	92.01	nr	151.35

U: VENTILATION/AIR CONDITIONING SYSTEMS

Item	Net Price £	Material £	Labour hours	Labour £	Unit	Total rate £
U14 : DUCTWORK : FIRE RATED (cont'd)						
Galvanised sheet metal rectangular (cont'd)						
Ductwork 1251 to 2000mm longest side						
Sum of two sides 1700mm	144.11	245.05	7.45	235.56	m	480.61
Sum of two sides 1800mm	153.36	260.77	7.71	243.78	m	504.55
Sum of two sides 1900mm	162.59	276.47	7.98	252.32	m	528.79
Sum of two sides 2000mm	171.83	292.17	8.25	260.86	m	553.03
Sum of two sides 2100mm	181.06	307.87	9.62	304.18	m	612.05
Sum of two sides 2200mm	190.30	323.58	9.66	305.44	m	629.02
Sum of two sides 2300mm	199.52	339.25	10.07	318.41	m	657.66
Sum of two sides 2400mm	208.75	354.96	10.47	331.05	m	686.01
Sum of two sides 2500mm	217.98	370.66	10.87	343.70	m	714.36
Sum of two sides 2600mm	227.23	386.38	11.27	356.35	m	742.73
Sum of two sides 2700mm	236.45	402.06	11.67	369.00	m	771.06
Sum of two sides 2800mm	245.70	417.78	12.08	381.96	m	799.74
Sum of two sides 2900mm	254.92	433.46	12.48	394.61	m	828.07
Sum of two sides 3000mm	264.15	449.16	12.88	407.26	m	856.42
Sum of two sides 3100mm	273.39	464.86	13.26	419.27	m	884.13
Sum of two sides 3200mm	282.62	480.57	13.69	432.87	m	913.43
Sum of two sides 3300mm	291.87	496.29	14.09	445.51	m	941.80
Sum of two sides 3400mm	301.09	511.97	14.49	458.16	m	970.13
Sum of two sides 3500mm	310.32	527.67	14.89	470.81	m	998.48
Sum of two sides 3600mm	319.55	543.35	15.29	483.46	m	1026.81
Sum of two sides 3700mm	328.79	559.07	15.69	496.11	m	1055.18
Sum of two sides 3800mm	338.01	574.75	16.09	508.75	m	1083.50
Sum of two sides 3900mm	347.26	590.47	16.49	521.40	m	1111.88
Sum of two sides 4000mm	356.48	606.15	16.89	534.05	m	1140.20
Extra over fittings; Ductwork 1251 to 2000mm longest side						
End Cap						
Sum of two sides 1700mm	52.46	89.20	1.42	44.90	nr	134.10
Sum of two sides 1800mm	55.84	94.94	1.45	45.85	nr	140.79
Sum of two sides 1900mm	59.21	100.68	1.48	46.80	nr	147.48
Sum of two sides 2000mm	62.59	106.42	1.51	47.74	nr	154.16
Sum of two sides 2100mm	65.96	112.16	2.66	84.11	nr	196.27
Sum of two sides 2200mm	69.34	117.90	2.80	88.53	nr	206.43
Sum of two sides 2300mm	72.70	123.61	2.95	93.28	nr	216.89
Sum of two sides 2400mm	76.07	129.35	3.10	98.02	nr	227.37
Sum of two sides 2500mm	79.45	135.09	3.24	102.45	nr	237.54
Sum of two sides 2600mm	82.82	140.83	3.39	107.19	nr	248.02
Sum of two sides 2700mm	86.18	146.55	3.54	111.93	nr	258.48
Sum of two sides 2800mm	89.56	152.28	3.68	116.36	nr	268.64
Sum of two sides 2900mm	92.96	158.07	3.83	121.10	nr	279.17
Sum of two sides 3000mm	96.32	163.78	3.98	125.84	nr	289.63
Sum of two sides 3100mm	99.70	169.52	4.12	130.27	nr	299.79
Sum of two sides 3200mm	103.07	175.26	4.27	135.01	nr	310.28
Sum of two sides 3300mm	106.43	180.98	4.42	139.76	nr	320.74
Sum of two sides 3400mm	109.81	186.72	4.57	144.50	nr	331.22
Sum of two sides 3500mm	112.59	191.45	4.72	149.24	nr	340.69
Sum of two sides 3600mm	114.86	195.30	4.87	153.99	nr	349.29
Sum of two sides 3700mm	118.23	201.04	5.02	158.73	nr	359.77
Sum of two sides 3800mm	113.71	193.35	5.17	163.47	nr	356.82
Sum of two sides 3900mm	126.67	215.39	5.32	168.21	nr	383.60
Sum of two sides 4000mm	130.05	221.13	5.47	172.96	nr	394.08

U: VENTILATION/AIR CONDITIONING SYSTEMS

Item	Net Price £	Material £	Labour hours	Labour £	Unit	Total rate £
Reducer						
Sum of two sides 1700mm	113.29	192.64	3.54	111.93	nr	**304.57**
Sum of two sides 1800mm	127.09	216.10	3.60	113.83	nr	**329.93**
Sum of two sides 1900mm	140.87	239.54	3.65	115.41	nr	**354.95**
Sum of two sides 2000mm	154.68	263.02	3.70	116.99	nr	**380.01**
Sum of two sides 2100mm	168.47	286.46	2.92	92.33	nr	**378.79**
Sum of two sides 2200mm	182.26	309.92	3.10	98.02	nr	**407.94**
Sum of two sides 2300mm	196.06	333.38	3.29	104.03	nr	**437.40**
Sum of two sides 2400mm	209.86	356.84	3.48	110.03	nr	**466.87**
Sum of two sides 2500mm	223.64	380.27	3.66	115.73	nr	**496.00**
Sum of two sides 2600mm	237.45	403.76	3.85	121.73	nr	**525.49**
Sum of two sides 2700mm	251.24	427.19	4.03	127.43	nr	**554.62**
Sum of two sides 2800mm	265.03	450.66	4.22	133.43	nr	**584.09**
Sum of two sides 2900mm	278.83	474.12	4.40	139.12	nr	**613.24**
Sum of two sides 3000mm	299.50	509.26	4.59	145.13	nr	**654.39**
Sum of two sides 3100mm	299.50	509.26	4.78	151.14	nr	**660.40**
Sum of two sides 3200mm	313.31	532.74	4.96	156.83	nr	**689.57**
Sum of two sides 3300mm	333.99	567.91	5.15	162.84	nr	**730.75**
Sum of two sides 3400mm	347.79	591.37	5.34	168.85	nr	**760.22**
Sum of two sides 3500mm	361.58	614.83	5.53	174.85	nr	**789.68**
Sum of two sides 3600mm	375.37	638.27	5.72	180.86	nr	**819.13**
Sum of two sides 3700mm	389.18	661.75	5.91	186.87	nr	**848.62**
Sum of two sides 3800mm	402.96	685.19	6.10	192.88	nr	**878.06**
Sum of two sides 3900mm	416.76	708.65	6.29	198.88	nr	**907.53**
Sum of two sides 4000mm	430.56	732.11	6.48	204.89	nr	**937.00**
Offset						
Sum of two sides 1700mm	113.29	192.64	4.45	140.71	nr	**333.34**
Sum of two sides 1800mm	122.63	208.52	4.56	144.18	nr	**352.71**
Sum of two sides 1900mm	131.98	224.41	4.67	147.66	nr	**372.07**
Sum of two sides 2000mm	141.28	240.23	5.53	174.85	nr	**415.08**
Sum of two sides 2100mm	150.62	256.11	5.76	182.13	nr	**438.24**
Sum of two sides 2200mm	159.96	272.00	5.99	189.40	nr	**461.39**
Sum of two sides 2300mm	169.30	287.88	6.22	196.67	nr	**484.55**
Sum of two sides 2400mm	178.62	303.72	6.45	203.94	nr	**507.66**
Sum of two sides 2500mm	187.96	319.60	6.68	211.22	nr	**530.82**
Sum of two sides 2600mm	197.30	335.49	6.91	218.49	nr	**553.98**
Sum of two sides 2700mm	206.64	351.37	7.14	225.76	nr	**577.14**
Sum of two sides 2800mm	215.99	367.26	7.37	233.03	nr	**600.29**
Sum of two sides 2900mm	225.30	383.10	7.60	240.31	nr	**623.40**
Sum of two sides 3000mm	234.64	398.98	7.83	247.58	nr	**646.56**
Sum of two sides 3100mm	243.99	414.87	8.06	254.85	nr	**669.72**
Sum of two sides 3200mm	253.33	430.75	8.29	262.12	nr	**692.88**
Sum of two sides 3300mm	262.64	446.59	8.52	269.40	nr	**715.99**
Sum of two sides 3400mm	271.97	462.45	8.75	276.67	nr	**739.12**
Sum of two sides 3500mm	281.31	478.34	8.98	283.94	nr	**762.28**
Sum of two sides 3600mm	290.65	494.22	9.21	291.21	nr	**785.44**
Sum of two sides 3700mm	299.98	510.09	9.44	298.49	nr	**808.57**
Sum of two sides 3800mm	309.31	525.95	9.67	305.76	nr	**831.71**
Sum of two sides 3900mm	318.65	541.83	9.90	313.03	nr	**854.86**
Sum of two sides 4000mm	328.00	557.72	10.13	320.30	nr	**878.02**

U: VENTILATION/AIR CONDITIONING SYSTEMS

Item	Net Price £	Material £	Labour hours	Labour £	Unit	Total rate £
U14 : DUCTWORK : FIRE RATED (cont'd)						
Galvanised sheet metal rectangular (cont'd)						
Extra over fittings; Ductwork 1251 to 2000mm longest side (cont'd)						
90 ° radius bend						
Sum of two sides 1700mm	157.67	268.09	3.22	101.81	nr	369.91
Sum of two sides 1800mm	171.91	292.31	3.25	102.76	nr	395.07
Sum of two sides 1900mm	186.16	316.55	3.28	103.71	nr	420.26
Sum of two sides 2000mm	200.42	340.79	3.30	104.34	nr	445.14
Sum of two sides 2100mm	214.66	365.01	3.32	104.98	nr	469.99
Sum of two sides 2200mm	228.91	389.23	3.34	105.61	nr	494.84
Sum of two sides 2300mm	243.16	413.47	3.36	106.24	nr	519.71
Sum of two sides 2400mm	257.39	437.66	3.38	106.87	nr	544.54
Sum of two sides 2500mm	271.66	461.93	3.40	107.51	nr	569.43
Sum of two sides 2600mm	285.92	486.17	3.42	108.14	nr	594.30
Sum of two sides 2700mm	300.17	510.41	3.44	108.77	nr	619.18
Sum of two sides 2800mm	314.42	534.63	3.46	109.40	nr	644.03
Sum of two sides 2900mm	328.64	558.82	3.48	110.03	nr	668.85
Sum of two sides 3000mm	342.90	583.06	3.50	110.67	nr	693.73
Sum of two sides 3100mm	357.14	607.28	3.52	111.30	nr	718.58
Sum of two sides 3200mm	371.40	631.52	3.54	111.93	nr	743.45
Sum of two sides 3300mm	385.64	655.74	3.56	112.56	nr	768.30
Sum of two sides 3400mm	400.06	680.25	3.58	113.20	nr	793.45
Sum of two sides 3500mm	414.15	704.22	3.60	113.83	nr	818.05
Sum of two sides 3600mm	428.40	728.44	3.62	114.46	nr	842.90
Sum of two sides 3700mm	442.64	752.65	3.64	115.09	nr	867.75
Sum of two sides 3800mm	456.89	776.89	3.66	115.73	nr	892.62
Sum of two sides 3900mm	471.15	801.13	3.68	116.36	nr	917.49
Sum of two sides 4000mm	485.39	825.35	3.70	116.99	nr	942.34
45 ° bend						
Sum of two sides 1700mm	116.88	198.75	2.24	70.83	nr	269.57
Sum of two sides 1800mm	129.80	220.71	2.30	72.72	nr	293.44
Sum of two sides 1900mm	142.69	242.64	2.35	74.31	nr	316.94
Sum of two sides 2000mm	155.63	264.63	2.76	87.27	nr	351.90
Sum of two sides 2100mm	214.66	365.01	2.89	91.38	nr	456.39
Sum of two sides 2200mm	181.44	308.52	3.01	95.17	nr	403.69
Sum of two sides 2300mm	194.37	330.51	3.13	98.97	nr	429.48
Sum of two sides 2400mm	207.27	352.43	3.26	103.08	nr	455.51
Sum of two sides 2500mm	220.19	374.40	3.37	106.56	nr	480.95
Sum of two sides 2600mm	233.08	396.32	3.49	110.35	nr	506.67
Sum of two sides 2700mm	246.01	418.31	3.62	114.46	nr	532.77
Sum of two sides 2800mm	258.92	440.26	3.74	118.26	nr	558.51
Sum of two sides 2900mm	271.82	462.20	3.86	122.05	nr	584.25
Sum of two sides 3000mm	284.73	484.15	3.98	125.84	nr	609.99
Sum of two sides 3100mm	297.66	506.14	4.10	129.64	nr	635.78
Sum of two sides 3200mm	310.55	528.06	4.22	133.43	nr	661.49
Sum of two sides 3300mm	323.47	550.03	4.34	137.23	nr	687.26
Sum of two sides 3400mm	336.38	571.97	4.46	141.02	nr	712.99
Sum of two sides 3500mm	349.30	593.94	4.58	144.82	nr	738.76
Sum of two sides 3600mm	362.22	615.91	4.70	148.61	nr	764.52
Sum of two sides 3700mm	375.11	637.83	4.82	152.40	nr	790.24
Sum of two sides 3800mm	388.03	659.80	4.94	156.20	nr	816.00
Sum of two sides 3900mm	400.92	681.72	5.06	159.99	nr	841.71
Sum of two sides 4000mm	413.86	703.71	5.18	163.79	nr	867.50

U: VENTILATION/AIR CONDITIONING SYSTEMS

Item	Net Price £	Material £	Labour hours	Labour £	Unit	Total rate £
90° mitre bend						
Sum of two sides 1700mm	236.14	401.53	3.87	122.37	nr	**523.90**
Sum of two sides 1800mm	276.95	470.92	3.90	123.31	nr	**594.24**
Sum of two sides 1900mm	317.78	540.34	3.93	124.26	nr	**664.60**
Sum of two sides 2000mm	358.56	609.69	3.96	125.21	nr	**734.90**
Sum of two sides 2100mm	399.38	679.10	3.82	120.79	nr	**799.89**
Sum of two sides 2200mm	440.19	748.50	3.83	121.10	nr	**869.60**
Sum of two sides 2300mm	481.00	817.89	3.83	121.10	nr	**938.99**
Sum of two sides 2400mm	521.82	887.28	3.84	121.42	nr	**1008.70**
Sum of two sides 2500mm	562.63	956.68	3.85	121.73	nr	**1078.41**
Sum of two sides 2600mm	603.41	1026.03	3.85	121.73	nr	**1147.76**
Sum of two sides 2700mm	644.23	1095.44	3.86	122.05	nr	**1217.49**
Sum of two sides 2800mm	685.04	1164.84	3.86	122.05	nr	**1286.89**
Sum of two sides 2900mm	725.85	1234.23	3.87	122.37	nr	**1356.59**
Sum of two sides 3000mm	766.66	1303.62	3.87	122.37	nr	**1425.99**
Sum of two sides 3100mm	807.48	1373.02	3.88	122.68	nr	**1495.70**
Sum of two sides 3200mm	848.27	1442.39	3.88	122.68	nr	**1565.07**
Sum of two sides 3300mm	889.10	1511.80	3.89	123.00	nr	**1634.80**
Sum of two sides 3400mm	929.91	1581.20	3.90	123.31	nr	**1704.51**
Sum of two sides 3500mm	970.72	1650.59	3.91	123.63	nr	**1774.22**
Sum of two sides 3600mm	1011.53	1719.98	3.92	123.95	nr	**1843.93**
Sum of two sides 3700mm	1052.34	1789.38	3.93	124.26	nr	**1913.64**
Sum of two sides 3800mm	1093.14	1858.75	3.94	124.58	nr	**1983.33**
Sum of two sides 3900mm	1133.96	1928.16	3.95	124.90	nr	**2053.06**
Sum of two sides 4000mm	1174.77	1997.56	3.96	125.21	nr	**2122.77**
Branch (Side-on Shoe)						
Sum of two sides 1700mm	26.77	45.52	1.77	55.97	nr	**101.49**
Sum of two sides 1800mm	29.25	49.74	1.81	57.23	nr	**106.97**
Sum of two sides 1900mm	31.75	53.99	1.86	58.81	nr	**112.80**
Sum of two sides 2000mm	34.25	58.24	1.90	60.08	nr	**118.31**
Sum of two sides 2100mm	36.73	62.46	2.58	81.58	nr	**144.04**
Sum of two sides 2200mm	39.22	66.68	2.61	82.53	nr	**149.21**
Sum of two sides 2300mm	41.72	70.93	2.65	83.79	nr	**154.72**
Sum of two sides 2400mm	44.20	75.16	2.68	84.74	nr	**159.89**
Sum of two sides 2500mm	46.70	79.40	2.71	85.69	nr	**165.09**
Sum of two sides 2600mm	49.19	83.65	2.75	86.95	nr	**170.60**
Sum of two sides 2700mm	51.66	87.85	2.78	87.90	nr	**175.75**
Sum of two sides 2800mm	54.16	92.10	2.81	88.85	nr	**180.95**
Sum of two sides 2900mm	56.66	96.34	2.84	89.80	nr	**186.14**
Sum of two sides 3000mm	59.14	100.57	2.87	90.75	nr	**191.31**
Sum of two sides 3100mm	61.63	104.79	2.90	91.70	nr	**196.49**
Sum of two sides 3200mm	64.13	109.04	2.93	92.64	nr	**201.68**
Sum of two sides 3300mm	66.61	113.26	2.93	92.64	nr	**205.90**
Sum of two sides 3400mm	69.11	117.51	3.00	94.86	nr	**212.36**
Sum of two sides 3500mm	71.60	121.75	3.03	95.81	nr	**217.56**
Sum of two sides 3600mm	74.07	125.95	3.06	96.75	nr	**222.71**
Sum of two sides 3700mm	76.57	130.20	3.09	97.70	nr	**227.90**
Sum of two sides 3800mm	79.07	134.45	3.12	98.65	nr	**233.10**
Sum of two sides 3900mm	81.57	138.69	3.15	99.60	nr	**238.30**
Sum of two sides 4000mm	84.04	142.90	3.18	100.55	nr	**243.44**

U: VENTILATION/AIR CONDITIONING SYSTEMS

Item	Net Price £	Material £	Labour hours	Labour £	Unit	Total rate £
U14 : DUCTWORK : FIRE RATED (cont'd)						
Rectangular section ductwork to BS476 Part 24 (ISO 6944:1985) (Duraduct LT), ducts "Type A" and "Type B"; manufactured from 6mm thick laminate fire board consisting of steel circular hole punched facings pressed to a fibre cement core; provides up to 4 hours stability, 4 hours integrity and 32 minutes insulation; including all necessary stiffeners, joints and supports in the running length						
Ductwork up to 600mm longest side						
Sum of two sides 200mm	198.47	337.47	4.00	126.48	m	463.95
Sum of two sides 400mm	198.47	337.47	4.00	126.48	m	463.95
Sum of two sides 600mm	198.47	337.47	4.00	126.48	m	463.95
Sum of two sides 800mm	280.36	476.72	5.50	173.91	m	650.62
Sum of two sides 1000mm	280.36	476.72	5.50	173.91	m	650.62
Sum of two sides 1200mm	390.40	663.83	6.00	189.72	m	853.54
Extra over fittings; Ductwork up to 600mm longest side						
End Cap						
Sum of two sides 200 mm	47.10	80.09	0.81	25.61	m	105.70
Sum of two sides 400mm	47.10	80.09	0.87	27.51	m	107.60
Sum of two sides 600mm	47.10	80.09	0.93	29.41	m	109.49
Sum of two sides 800mm	76.08	129.36	0.98	30.99	m	160.35
Sum of two sides 1000mm	76.08	129.36	1.22	38.58	m	167.94
Sum of two sides 1200mm	115.92	197.11	1.28	40.47	m	237.58
Reducer						
Sum of two sides 200 mm	41.65	70.82	2.23	70.51	m	141.33
Sum of two sides 400mm	41.65	70.82	2.51	79.36	m	150.19
Sum of two sides 600mm	41.65	70.82	2.79	88.22	m	159.04
Sum of two sides 800mm	81.51	138.60	3.01	95.17	m	233.77
Sum of two sides 1000mm	81.51	138.60	3.17	100.23	m	238.83
Sum of two sides 1200mm	103.25	175.56	3.28	103.71	m	279.28
Offset						
Sum of two sides 200 mm	41.65	70.82	2.95	93.28	m	164.10
Sum of two sides 400mm	41.65	70.82	3.26	103.08	m	173.90
Sum of two sides 600mm	41.65	70.82	3.57	112.88	m	183.70
Sum of two sides 800mm	81.51	138.60	3.23	102.13	m	240.73
Sum of two sides 1000mm	81.51	138.60	3.67	116.04	m	254.64
Sum of two sides 1200mm	103.25	175.56	3.89	123.00	m	298.56
90 ° radius bend						
Sum of two sides 200 mm	139.47	237.15	2.06	65.14	m	302.29
Sum of two sides 400mm	139.47	237.15	2.36	74.62	m	311.77
Sum of two sides 600mm	139.47	237.15	2.66	84.11	m	321.26
Sum of two sides 800mm	163.02	277.20	2.97	93.91	m	371.10
Sum of two sides 1000mm	163.02	277.20	3.04	96.12	m	373.32
Sum of two sides 1200mm	184.76	314.16	3.09	97.70	m	411.87

Material Costs/Prices for Measured Works – Mechanical Installations

U: VENTILATION/AIR CONDITIONING SYSTEMS

Item	Net Price £	Material £	Labour hours	Labour £	Unit	Total rate £
45 ° radius bend						
Sum of two sides 200 mm	139.47	237.15	1.52	48.06	m	285.21
Sum of two sides 400mm	139.47	237.15	1.66	52.49	m	289.64
Sum of two sides 600mm	139.47	237.15	1.80	56.91	m	294.07
Sum of two sides 800mm	163.02	277.20	1.93	61.03	m	338.22
Sum of two sides 1000mm	163.02	277.20	2.05	64.82	m	342.02
Sum of two sides 1200mm	184.76	314.16	2.17	68.61	m	382.78
90 ° mitre bend						
Sum of two sides 200 mm	139.47	237.15	2.47	78.10	m	315.25
Sum of two sides 400mm	139.47	237.15	2.84	89.80	m	326.95
Sum of two sides 600mm	139.47	237.15	3.20	101.18	m	338.33
Sum of two sides 800mm	163.02	277.20	3.56	112.56	m	389.76
Sum of two sides 1000mm	163.02	277.20	3.66	115.73	m	392.92
Sum of two sides 1200mm	184.76	314.16	3.72	117.62	m	431.79
Branch						
Sum of two sides 200 mm	50.72	86.24	0.98	30.99	m	117.23
Sum of two sides 400mm	50.72	86.24	1.14	36.05	m	122.29
Sum of two sides 600mm	50.72	86.24	1.30	41.10	m	127.35
Sum of two sides 800mm	63.39	107.79	1.46	46.16	m	153.95
Sum of two sides 1000mm	63.39	107.79	1.46	46.16	m	153.95
Sum of two sides 1200mm	85.12	144.74	1.55	49.01	m	193.75
Ductwork 601 to 1000mm longest side						
Sum of two sides 1000mm	280.36	476.72	6.00	189.72	m	666.43
Sum of two sides 1100mm	390.40	663.83	6.00	189.72	m	853.54
Sum of two sides 1300mm	390.40	663.83	6.00	189.72	m	853.54
Sum of two sides 1500mm	390.40	663.83	6.00	189.72	m	853.54
Sum of two sides 1700mm	490.89	834.70	6.00	189.72	m	1024.41
Sum of two sides 1900mm	490.89	834.70	6.00	189.72	m	1024.41
Extra over fittings; Ductwork 601 to 1000mm longest side						
End Cap						
Sum of two sides 1000mm	76.08	129.36	1.25	39.52	m	168.89
Sum of two sides 1100mm	115.92	197.11	1.25	39.52	m	236.63
Sum of two sides 1300mm	115.92	197.11	1.31	41.42	m	238.53
Sum of two sides 1500mm	115.92	197.11	1.36	43.00	m	240.11
Sum of two sides 1700mm	204.68	348.03	1.42	44.90	m	392.93
Sum of two sides 1900mm	204.68	348.03	1.48	46.80	m	394.83
Reducer						
Sum of two sides 1000mm	81.51	138.60	3.22	101.81	m	240.41
Sum of two sides 1100mm	103.25	175.56	3.22	101.81	m	277.38
Sum of two sides 1300mm	103.25	175.56	3.33	105.29	m	280.86
Sum of two sides 1500mm	103.25	175.56	3.44	108.77	m	284.33
Sum of two sides 1700mm	126.80	215.61	3.54	111.93	m	327.54
Sum of two sides 1900mm	126.80	215.61	3.65	115.41	m	331.02
Offset						
Sum of two sides 1000mm	81.51	138.60	3.78	119.52	m	258.12
Sum of two sides 1100mm	103.25	175.56	3.78	119.52	m	295.08
Sum of two sides 1300mm	103.25	175.56	4.00	126.48	m	302.04
Sum of two sides 1500mm	103.25	175.56	4.23	133.75	m	309.31
Sum of two sides 1700mm	126.80	215.61	4.45	140.71	m	356.31
Sum of two sides 1900mm	126.80	215.61	4.67	147.66	m	363.27

U: VENTILATION/AIR CONDITIONING SYSTEMS

Item	Net Price £	Material £	Labour hours	Labour £	Unit	Total rate £
U14 : DUCTWORK : FIRE RATED (cont'd)						
Rectangular section ductwork to BS476 (cont'd)						
Extra over fittings; Ductwork 601 to 1000mm longest side (cont'd)						
90 ° radius bend						
Sum of two sides 1000mm	163.02	277.20	3.78	119.52	m	396.72
Sum of two sides 1100mm	184.76	314.16	3.07	97.07	m	411.23
Sum of two sides 1300mm	184.76	314.16	3.12	98.65	m	412.81
Sum of two sides 1500mm	184.76	314.16	3.17	100.23	m	414.40
Sum of two sides 1700mm	242.72	412.72	3.22	101.81	m	514.53
Sum of two sides 1900mm	242.72	412.72	3.28	103.71	m	516.43
45 ° bend						
Sum of two sides 1000mm	163.02	277.20	3.78	119.52	m	396.72
Sum of two sides 1100mm	184.76	314.16	1.91	60.39	m	374.55
Sum of two sides 1300mm	184.76	314.16	2.02	63.87	m	378.03
Sum of two sides 1500mm	184.76	314.16	2.13	67.35	m	381.51
Sum of two sides 1700mm	242.72	412.72	2.24	70.83	m	483.54
Sum of two sides 1900mm	242.72	412.72	2.35	74.31	m	487.02
90 ° mitre bend						
Sum of two sides 1000mm	163.02	277.20	3.78	119.52	m	396.72
Sum of two sides 1100mm	184.76	314.16	3.69	116.67	m	430.84
Sum of two sides 1300mm	184.76	314.16	3.75	118.57	m	432.73
Sum of two sides 1500mm	184.76	314.16	3.81	120.47	m	434.63
Sum of two sides 1700mm	242.72	412.72	3.87	122.37	m	535.08
Sum of two sides 1900mm	242.72	412.72	3.93	124.26	m	536.98
Branch						
Sum of two sides 1000mm	63.39	107.79	3.78	119.52	m	227.31
Sum of two sides 1100mm	85.12	144.74	1.50	47.43	m	192.17
Sum of two sides 1300mm	85.12	144.74	1.59	50.27	m	195.01
Sum of two sides 1500mm	85.12	144.74	1.68	53.12	m	197.86
Sum of two sides 1700mm	108.68	184.80	1.77	55.97	m	240.76
Sum of two sides 1900mm	108.68	184.80	1.86	58.81	m	243.61
Ductwork 1001 to 1250mm longest side						
Sum of two sides 1300mm	390.40	663.83	6.00	189.72	m	853.54
Sum of two sides 1500mm	390.40	663.83	6.00	189.72	m	853.54
Sum of two sides 1700mm	490.89	834.70	6.00	189.72	m	1024.41
Sum of two sides 1900mm	490.89	834.70	6.00	189.72	m	1024.41
Sum of two sides 2100mm	654.00	1112.05	6.50	205.52	m	1317.57
Sum of two sides 2300mm	654.00	1112.05	6.50	205.52	m	1317.57
Sum of two sides 2500mm	654.00	1112.05	6.50	205.52	m	1317.57

U: VENTILATION/AIR CONDITIONING SYSTEMS

Item	Net Price £	Material £	Labour hours	Labour £	Unit	Total rate £
Extra over fittings; Ductwork 1001 to 1250mm longest side						
End Cap						
Sum of two sides 1300mm	115.92	197.11	1.31	41.42	m	**238.53**
Sum of two sides 1500mm	115.92	197.11	1.36	43.00	m	**240.11**
Sum of two sides 1700mm	204.68	348.03	1.42	44.90	m	**392.93**
Sum of two sides 1900mm	204.68	348.03	1.48	46.80	m	**394.83**
Sum of two sides 2100mm	278.94	474.30	2.66	84.11	m	**558.41**
Sum of two sides 2300mm	278.94	474.30	2.95	93.28	m	**567.58**
Sum of two sides 2500mm	278.94	474.30	3.24	102.45	m	**576.75**
Reducer						
Sum of two sides 1300mm	103.25	175.56	3.33	105.29	m	**280.86**
Sum of two sides 1500mm	103.25	175.56	3.44	108.77	m	**284.33**
Sum of two sides 1700mm	126.80	215.61	3.54	111.93	m	**327.54**
Sum of two sides 1900mm	126.80	215.61	3.65	115.41	m	**331.02**
Sum of two sides 2100mm	150.33	255.62	3.75	118.57	m	**374.19**
Sum of two sides 2300mm	150.33	255.62	3.85	121.73	m	**377.35**
Sum of two sides 2500mm	150.33	255.62	3.95	124.90	m	**380.51**
Offset						
Sum of two sides 1300mm	103.25	175.56	4.00	126.48	m	**302.04**
Sum of two sides 1500mm	103.25	175.56	4.23	133.75	m	**309.31**
Sum of two sides 1700mm	126.80	215.61	4.45	140.71	m	**356.31**
Sum of two sides 1900mm	126.80	215.61	4.67	147.66	m	**363.27**
Sum of two sides 2100mm	150.33	255.62	5.76	182.13	m	**437.74**
Sum of two sides 2300mm	150.33	255.62	6.22	196.67	m	**452.29**
Sum of two sides 2500mm	150.33	255.62	6.68	211.22	m	**466.83**
90° radius bend						
Sum of two sides 1300mm	184.76	314.16	3.12	98.65	m	**412.81**
Sum of two sides 1500mm	184.76	314.16	3.17	100.23	m	**414.40**
Sum of two sides 1700mm	242.72	412.72	3.22	101.81	m	**514.53**
Sum of two sides 1900mm	242.72	412.72	3.28	103.71	m	**516.43**
Sum of two sides 2100mm	248.15	421.95	3.32	104.98	m	**526.92**
Sum of two sides 2300mm	248.15	421.95	3.36	106.24	m	**528.19**
Sum of two sides 2500mm	248.15	421.95	3.40	107.51	m	**529.45**
45° bend						
Sum of two sides 1300mm	184.76	314.16	2.02	63.87	m	**378.03**
Sum of two sides 1500mm	184.76	314.16	2.13	67.35	m	**381.51**
Sum of two sides 1700mm	242.72	412.72	2.24	70.83	m	**483.54**
Sum of two sides 1900mm	242.72	412.72	2.35	74.31	m	**487.02**
Sum of two sides 2100mm	248.15	421.95	2.89	91.38	m	**513.33**
Sum of two sides 2300mm	248.15	421.95	3.13	98.97	m	**520.92**
Sum of two sides 2500mm	248.15	421.95	3.37	106.56	m	**528.51**
90° mitre bend						
Sum of two sides 1300mm	184.76	314.16	3.75	118.57	m	**432.73**
Sum of two sides 1500mm	184.76	314.16	3.81	120.47	m	**434.63**
Sum of two sides 1700mm	242.72	412.72	3.87	122.37	m	**535.08**
Sum of two sides 1900mm	242.72	412.72	3.93	124.26	m	**536.98**
Sum of two sides 2100mm	248.15	421.95	3.99	126.16	m	**548.11**
Sum of two sides 2300mm	248.15	421.95	4.05	128.06	m	**550.01**
Sum of two sides 2500mm	248.15	421.95	4.11	129.95	m	**551.90**

U: VENTILATION/AIR CONDITIONING SYSTEMS

Item	Net Price £	Material £	Labour hours	Labour £	Unit	Total rate £
U14 : DUCTWORK : FIRE RATED (cont'd)						
Rectangular section ductwork to BS476 (cont'd)						
Extra over fittings; Ductwork 1001 to 1250mm longest side (cont'd)						
Branch						
Sum of two sides 1300mm	85.12	144.74	1.59	50.27	m	195.01
Sum of two sides 1500mm	85.12	144.74	1.68	53.12	m	197.86
Sum of two sides 1700mm	108.68	184.80	1.77	55.97	m	240.76
Sum of two sides 1900mm	108.68	184.80	1.86	58.81	m	243.61
Sum of two sides 2100mm	114.10	194.01	2.58	81.58	m	275.59
Sum of two sides 2300mm	114.10	194.01	2.64	83.47	m	277.49
Sum of two sides 2500mm	114.10	194.01	2.91	92.01	m	286.03
Ductwork 1251 to 2000mm longest side						
Sum of two sides 1800mm	490.89	834.70	6.00	189.72	m	1024.41
Sum of two sides 2000mm	490.89	834.70	6.00	189.72	m	1024.41
Sum of two sides 2200mm	654.00	1112.05	6.50	205.52	m	1317.57
Sum of two sides 2400mm	654.00	1112.05	6.50	205.52	m	1317.57
Sum of two sides 2600mm	774.82	1317.49	6.66	210.58	m	1528.07
Sum of two sides 2800mm	774.82	1317.49	6.66	210.58	m	1528.07
Sum of two sides 3000mm	774.82	1317.49	6.66	210.58	m	1528.07
Sum of two sides 3200mm	931.43	1583.78	9.00	284.57	m	1868.36
Sum of two sides 3400mm	931.43	1583.78	9.00	284.57	m	1868.36
Sum of two sides 3600mm	1043.07	1773.62	11.70	369.94	m	2143.56
Sum of two sides 3800mm	1043.07	1773.62	11.70	369.94	m	2143.56
Sum of two sides 4000mm	1043.07	1773.62	11.70	369.94	m	2143.56
Extra over fittings; Ductwork 1251 to 2000mm longest sides						
End Cap						
Sum of two sides 1800mm	204.68	348.03	1.45	45.85	m	393.88
Sum of two sides 2000mm	204.68	348.03	1.51	47.74	m	395.78
Sum of two sides 2200mm	278.94	474.30	2.80	88.53	m	562.84
Sum of two sides 2400mm	278.94	474.30	3.10	98.02	m	572.32
Sum of two sides 2600mm	382.17	649.83	3.39	107.19	m	757.02
Sum of two sides 2800mm	382.17	649.83	3.68	116.36	m	766.19
Sum of two sides 3000mm	382.17	649.83	3.98	125.84	m	775.68
Sum of two sides 3200mm	516.21	877.75	4.27	135.01	m	1012.77
Sum of two sides 3400mm	516.21	877.75	4.57	144.50	m	1022.25
Sum of two sides 3600mm	635.74	1081.00	4.87	153.99	m	1234.99
Sum of two sides 3800mm	635.74	1081.00	5.17	163.47	m	1244.47
Sum of two sides 4000mm	635.74	1081.00	5.47	172.96	m	1253.96
Reducer						
Sum of two sides 1800mm	126.80	215.61	3.60	113.83	m	329.44
Sum of two sides 2000mm	126.80	215.61	3.70	116.99	m	332.60
Sum of two sides 2200mm	150.33	255.62	3.10	98.02	m	353.64
Sum of two sides 2400mm	150.33	255.62	3.48	110.03	m	365.65
Sum of two sides 2600mm	208.29	354.17	3.85	121.73	m	475.91
Sum of two sides 2800mm	208.29	354.17	4.22	133.43	m	487.61
Sum of two sides 3000mm	208.29	354.17	4.59	145.13	m	499.30
Sum of two sides 3200mm	231.84	394.22	4.96	156.83	m	551.05

U: VENTILATION/AIR CONDITIONING SYSTEMS

Item	Net Price £	Material £	Labour hours	Labour £	Unit	Total rate £
Sum of two sides 3400mm	231.84	394.22	5.34	168.85	m	563.06
Sum of two sides 3600mm	253.58	431.18	5.72	180.86	m	612.04
Sum of two sides 3800mm	253.58	431.18	6.10	192.88	m	624.06
Sum of two sides 4000mm	253.58	431.18	6.48	204.89	m	636.07
Offset						
Sum of two sides 1800mm	126.80	215.61	4.56	144.18	m	359.79
Sum of two sides 2000mm	126.80	215.61	5.53	174.85	m	390.46
Sum of two sides 2200mm	150.33	255.62	5.99	189.40	m	445.02
Sum of two sides 2400mm	150.33	255.62	6.45	203.94	m	459.56
Sum of two sides 2600mm	208.29	354.17	6.91	218.49	m	572.66
Sum of two sides 2800mm	208.29	354.17	7.37	233.03	m	587.21
Sum of two sides 3000mm	208.29	354.17	7.83	247.58	m	601.75
Sum of two sides 3200mm	231.84	394.22	8.29	262.12	m	656.34
Sum of two sides 3400mm	231.84	394.22	8.75	276.67	m	670.88
Sum of two sides 3600mm	253.58	431.18	9.21	291.21	m	722.40
Sum of two sides 3800mm	253.58	431.18	9.67	305.76	m	736.94
Sum of two sides 4000mm	253.58	431.18	10.13	320.30	m	751.48
90 ° radius bend						
Sum of two sides 1800mm	242.72	412.72	3.25	102.76	m	515.48
Sum of two sides 2000mm	242.72	412.72	3.30	104.34	m	517.06
Sum of two sides 2200mm	248.15	421.95	3.34	105.61	m	527.56
Sum of two sides 2400mm	248.15	421.95	3.38	106.87	m	528.82
Sum of two sides 2600mm	380.37	646.77	3.42	108.14	m	754.91
Sum of two sides 2800mm	380.37	646.77	3.46	109.40	m	756.18
Sum of two sides 3000mm	380.37	646.77	3.50	110.67	m	757.44
Sum of two sides 3200mm	456.44	776.12	3.54	111.93	m	888.05
Sum of two sides 3400mm	456.44	776.12	3.58	113.20	m	889.32
Sum of two sides 3600mm	534.31	908.53	3.62	114.46	m	1022.99
Sum of two sides 3800mm	534.31	908.53	3.66	115.73	m	1024.26
Sum of two sides 4000mm	534.31	908.53	3.70	116.99	m	1025.52
45 ° bend						
Sum of two sides 1800mm	242.72	412.72	2.30	72.72	m	485.44
Sum of two sides 2000mm	242.72	412.72	2.76	87.27	m	499.99
Sum of two sides 2200mm	248.15	421.95	3.01	95.17	m	517.12
Sum of two sides 2400mm	248.15	421.95	3.26	103.08	m	525.03
Sum of two sides 2600mm	380.37	646.77	3.49	110.35	m	757.12
Sum of two sides 2800mm	380.37	646.77	3.74	118.26	m	765.03
Sum of two sides 3000mm	380.37	646.77	3.98	125.84	m	772.62
Sum of two sides 3200mm	456.44	776.12	4.22	133.43	m	909.55
Sum of two sides 3400mm	456.44	776.12	4.46	141.02	m	917.14
Sum of two sides 3600mm	534.31	908.53	4.70	148.61	m	1057.14
Sum of two sides 3800mm	534.31	908.53	4.94	156.20	m	1064.73
Sum of two sides 4000mm	534.31	908.53	5.18	163.79	m	1072.32
90 ° mitre band						
Sum of two sides 1800mm	242.72	412.72	3.90	123.31	m	536.03
Sum of two sides 2000mm	242.72	412.72	3.96	125.21	m	537.93
Sum of two sides 2200mm	248.15	421.95	3.83	121.10	m	543.05
Sum of two sides 2400mm	248.15	421.95	3.84	121.42	m	543.37
Sum of two sides 2600mm	380.37	646.77	3.85	121.73	m	768.51
Sum of two sides 2800mm	380.37	646.77	3.86	122.05	m	768.82
Sum of two sides 3000mm	380.37	646.77	3.87	122.37	m	769.14
Sum of two sides 3200mm	456.44	776.12	3.88	122.68	m	898.80
Sum of two sides 3400mm	456.44	776.12	3.90	123.31	m	899.44
Sum of two sides 3600mm	534.31	908.53	3.92	123.95	m	1032.48
Sum of two sides 3800mm	534.31	908.53	3.94	124.58	m	1033.11
Sum of two sides 4000mm	534.31	908.53	3.96	125.21	m	1033.74

U: VENTILATION/AIR CONDITIONING SYSTEMS

Item	Net Price £	Material £	Labour hours	Labour £	Unit	Total rate £
U14 : DUCTWORK : FIRE RATED (cont'd)						
Rectangular section ductwork to BS476 (cont'd)						
Extra over fittings; Ductwork 1251 to 2000mm longest sides (cont'd)						
Branch						
Sum of two sides 1800mm	108.68	184.80	1.81	57.23	m	**242.03**
Sum of two sides 2000mm	108.68	184.80	1.90	60.08	m	**244.87**
Sum of two sides 2200mm	114.10	194.01	2.61	82.53	m	**276.54**
Sum of two sides 2400mm	114.10	194.01	2.68	84.74	m	**278.75**
Sum of two sides 2600mm	135.84	230.98	2.75	86.95	m	**317.93**
Sum of two sides 2800mm	135.84	230.98	2.81	88.85	m	**319.83**
Sum of two sides 3000mm	135.84	230.98	2.87	90.75	m	**321.73**
Sum of two sides 3200mm	141.29	240.25	2.93	92.64	m	**332.89**
Sum of two sides 3400mm	141.29	240.25	3.00	94.86	m	**335.10**
Sum of two sides 3600mm	181.13	307.99	3.06	96.75	m	**404.74**
Sum of two sides 3800mm	181.13	307.99	3.12	98.65	m	**406.64**
Sum of two sides 4000mm	181.13	307.99	3.18	100.55	m	**408.54**

U: VENTILATION/AIR CONDITIONING SYSTEMS

Item	Net Price £	Material £	Labour hours	Labour £	Unit	Total rate £
U30 : LOW VELOCITY AIR CONDITIONING						
Y40 - AIR HANDLING UNITS						
Supply air handling unit; inlet with motorised damper, LTHW frost coil (at -5°C to +5°C), panel filter (EU4), bag filter (EU6), cooling coil (at 28°C db/20°C wb to 12°C db/11.5°C wb), LTHW heating coil (at 5°C to 21°C), supply fan, outlet plenum; includes access sections; all units located internally; Includes placing in position and fitting of sections together; electrical work elsewhere.						
Volume, external pressure						
2 m³/s at 350 Pa	4392.72	5626.15	40.00	1084.14	nr	**6710.29**
2 m³/s at 700 Pa	4692.99	6010.73	40.00	1084.14	nr	**7094.87**
5 m³/s at 350 Pa	7407.02	9486.84	65.00	1761.73	nr	**11248.56**
5 m³/s at 700 Pa	7666.13	9818.70	65.00	1761.73	nr	**11580.43**
8 m³/s at 350 Pa	11094.13	14209.25	77.00	2086.97	nr	**16296.22**
8 m/³s at 700 Pa	11267.06	14430.74	77.00	2086.97	nr	**16517.71**
10 m/³ at 350 Pa	12216.77	15647.12	100.00	2710.35	nr	**18357.47**
10 m³/s at 700 Pa	12467.56	15968.33	100.00	2710.35	nr	**18678.68**
13 m³/s at 350 Pa	15266.66	19553.39	108.00	2927.18	nr	**22480.56**
13 m³/s at 700 Pa	15738.18	20157.30	108.00	2927.18	nr	**23084.48**
15 m³/s at 350 Pa	17113.28	21918.52	120.00	3252.42	nr	**25170.94**
15 m³/s at 700 Pa	17424.66	22317.33	120.00	3252.42	nr	**25569.75**
18 m³/s at 350 Pa	19419.18	24871.89	133.00	3604.77	nr	**28476.66**
18 m³/s at 700 Pa	19776.16	25329.11	133.00	3604.77	nr	**28933.87**
20 m³/s at 350 Pa	22503.53	28822.30	142.00	3848.70	nr	**32670.99**
20 m³/s at 700 Pa	22827.15	29236.79	142.00	3848.70	nr	**33085.48**
Extra for inlet and discharge attenuators at 900mm long						
2 m³/s at 350 Pa	1351.18	1730.58	5.00	135.52	nr	**1866.10**
5 m³/s at 350 Pa	2835.81	3632.08	10.00	271.04	nr	**3903.11**
10 m³/s at 700 Pa	3964.58	5077.79	13.00	352.35	nr	**5430.14**
15 m³/s at 700 Pa	5909.05	7568.25	16.00	433.66	nr	**8001.91**
20 m³/s at 700 Pa	7561.05	9684.12	20.00	542.07	nr	**10226.19**
Extra for locating units externally						
2 m³/s at 350 Pa	1001.15	1282.26	-	-	nr	-
5 m³/s at 350 Pa	1270.27	1626.95	-	-	nr	-
10 m³/s at 700 Pa	1996.85	2557.55	-	-	nr	-
15 m³/s at 700 Pa	2836.42	3632.86	-	-	nr	-
20 m³/s at 700 Pa	4874.76	6243.54	-	-	nr	-

U: VENTILATION/AIR CONDITIONING SYSTEMS

Item	Net Price £	Material £	Labour hours	Labour £	Unit	Total rate £
U30 : LOW VELOCITY AIR CONDITIONING (cont'd)						
Modular air handling unit with supply and extract sections. Supply side; inlet with motorised damper, LTHW frost coil (at -5°C to 5°C), panel filter (EU4), bag filter (EU6), cooling coil at 28°Cdb/20°Cwb to 12°Cdb/ 11.5°C wb), LTHW heating coil (at 5°C to 21°C), supply fan, outlet plenum. Extract side; inlet with motorised damper, extract fan; includes access sections; placing in position and fitting of sections together; electrical work elsewhere.						
2 m³/s at 350 Pa	6136.47	7859.53	50.00	1355.17	nr	9214.70
2 m³/s at 700 Pa	6433.40	8239.83	50.00	1355.17	nr	9595.01
5 m³/s at 350 Pa	10575.90	13545.51	86.00	2330.90	nr	15876.41
5 m³/s at 700 Pa	10774.41	13799.76	86.00	2330.90	nr	16130.66
8 m³/s at 350 Pa	14686.15	18809.87	105.00	2845.87	nr	21655.74
8 m³/s at 700 Pa	15038.69	19261.40	105.00	2845.87	nr	22107.27
10 m³/s at 350 Pa	16958.13	21719.80	120.00	3252.42	nr	24972.22
10 m³/s at 700 Pa	17323.45	22187.70	120.00	3252.42	nr	25440.12
13 m³/s at 350 Pa	21255.78	27224.19	130.00	3523.45	nr	30747.65
13 m³/s at 700 Pa	21922.48	28078.09	130.00	3523.45	nr	31601.55
15 m³/s at 350 Pa	23751.85	30421.13	145.00	3930.01	nr	34351.14
15 m³/s at 700 Pa	24348.48	31185.29	145.00	3930.01	nr	35115.30
18 m³/s at 350 Pa	26575.98	34038.25	160.00	4336.56	nr	38374.81
18 m³/s at 700 Pa	27198.19	34835.17	160.00	4336.56	nr	39171.73
20 m³/s at 350 Pa	29046.47	37202.43	175.00	4743.11	nr	41945.54
20 m³/s at 700 Pa	30056.25	38495.74	175.00	4743.11	nr	43238.86
Extra for inlet and discharge attenuators at 900mm long						
2 m³/s at 350 Pa	2463.27	3154.93	8.00	216.83	nr	3371.76
5 m³/s at 350 Pa	4864.25	6230.08	10.00	271.04	nr	6501.12
10 m³/s at 700 Pa	6839.30	8759.71	13.00	352.35	nr	9112.05
15 m³/s at 700 Pa	11498.93	14727.71	16.00	433.66	nr	15161.37
20 m³/s at 700 Pa	13309.40	17046.55	20.00	542.07	nr	17588.62
Extra for locating units externally						
2 m³/s at 350 Pa	1971.74	2525.38	-	-	nr	-
5 m³/s at 350 Pa	2799.85	3586.02	-	-	nr	-
10 m³/s at 700 Pa	3927.10	5029.79	-	-	nr	-
15 m³/s at 700 Pa	7329.06	9386.99	-	-	nr	-
20 m³/s at 700 Pa	11171.54	14308.40	-	-	nr	-
Extra for humidifier, self generating type						
2 m³/s at 350 Pa (10kg/hr)	1618.00	2072.32	5.00	135.52	nr	2207.84
5 m³/s at 350 Pa (18kg/hr)	2002.31	2564.54	5.00	135.52	nr	2700.06
10 m³/s at 700 Pa (30kg/hr)	2278.53	2918.32	6.00	162.62	nr	3080.94
15 m³/s at 700 Pa (60kg/hr)	4219.69	5404.54	8.00	216.83	nr	5621.36
20 m³/s at 700 Pa (90kg/hr)	6344.28	8125.69	10.00	271.04	nr	8396.73
Extra for mixing box						
2 m³/s at 350 Pa	810.71	1038.35	4.00	108.41	nr	1146.76
5 m³/s at 350 Pa	1123.20	1438.58	4.00	108.41	nr	1547.00
10 m³/s at 700 Pa	1560.26	1998.37	5.00	135.52	nr	2133.88
15 m³/s at 700 Pa	2082.38	2667.09	6.00	162.62	nr	2829.71
20 m³/s at 700 Pa	3121.62	3998.14	6.00	162.62	nr	4160.76

U: VENTILATION/AIR CONDITIONING SYSTEMS

Item	Net Price £	Material £	Labour hours	Labour £	Unit	Total rate £
Extra for runaround coil, including pump and associated pipework; typical outputs in brackets. (based on minimal distance between the supply and extract units.)						
2 m³/s at 350 Pa (26kW)	2427.01	3108.49	30.00	813.11	nr	**3921.60**
5 m³/s at 350 Pa (37kW)	4229.59	5417.22	30.00	813.11	nr	**6230.32**
10 m³/s at 700 Pa (85kW)	7260.82	9299.59	40.00	1084.14	nr	**10383.73**
15 m³/s at 700 Pa (151kW)	10216.24	13084.86	50.00	1355.17	nr	**14440.03**
20 m³/s at 700 Pa (158kW)	13694.08	17539.24	60.00	1626.21	nr	**19165.45**
Extra for thermal wheel (typical outputs in brackets)						
2 m³/s at 350 Pa (37kW)	5248.13	6721.75	12.00	325.24	nr	**7046.99**
5 m³/s at 350 Pa (65kW)	7537.58	9654.06	12.00	325.24	nr	**9979.30**
10 m³/s at 700 Pa (127kW)	10051.93	12874.41	15.00	406.55	nr	**13280.96**
15 m³/s at 700 Pa (160kW)	18833.03	24121.16	17.00	460.76	nr	**24581.92**
20 m³/s at 700 Pa (262kW)	21999.17	28176.32	19.00	514.97	nr	**28691.28**
Extra for plate heat exchanger, including additional filtration in extract leg (typical outputs in brackets)						
2 m³/s at 350 Pa (25kW)	2927.04	3748.92	12.00	325.24	nr	**4074.17**
5 m³/s at 350 Pa (51kW)	5721.97	7328.64	12.00	325.24	nr	**7653.88**
10 m³/s at 700 Pa (98kW)	8395.71	10753.14	15.00	406.55	nr	**11159.69**
15 m³/s at 700 Pa (160kW)	14498.71	18569.80	17.00	460.76	nr	**19030.56**
20 m³/s at 700 Pa (190kW)	18723.86	23981.33	19.00	514.97	nr	**24496.30**
Extra for electric heating in lieu of LTHW						
2 m³/s at 350 Pa	1102.69	1412.31	-	-	nr	-
5 m³/s at 350 Pa	2030.69	2600.89	-	-	nr	-
10 m³/s at 700 Pa	2850.62	3651.05	-	-	nr	-
15 m³/s at 700 Pa	2718.51	3481.84	-	-	nr	-
20 m³/s at 700 Pa	3308.06	4236.93	-	-	nr	-

U: VENTILATION/AIR CONDITIONING SYSTEMS

Item	Net Price £	Material £	Labour hours	Labour £	Unit	Total rate £
U31 : VAV AIR CONDITIONING						
VAV TERMINAL BOXES						
VAV terminal box; integral acoustic silencer; factory installed and pre wired control components (excluding electronic controller); selected at 200Pa at entry to unit; includes fixing in position; electrical work elsewhere						
80 l/s - 110 l/s	219.36	373.00	2.00	63.24	nr	436.23
Extra for secondary silencer	53.52	91.00	0.50	15.81	nr	106.81
Extra for 2 row LTHW heating coil	21.62	36.76	-	-	nr	36.76
150 l/s - 190 l/s	230.83	392.50	2.00	63.24	nr	455.74
Extra for secondary silencer	58.52	99.51	0.50	15.81	nr	115.32
Extra for 2 row LTHW heating coil	25.94	44.11	-	-	nr	44.11
250 l/s - 310 l/s	257.59	438.00	2.00	63.24	nr	501.24
Extra for secondary silencer	77.63	132.00	0.50	15.81	nr	147.81
Extra for 2 row LTHW heating coil	14.49	24.64	-	-	nr	24.64
420 l/s - 520 l/s	279.06	474.51	2.00	63.24	nr	537.75
Extra for secondary silencer	88.22	150.01	0.50	15.81	nr	165.82
Extra for 2 row LTHW heating coil	35.32	60.06	-	-	nr	60.06
650 l/s - 790 l/s	337.87	574.51	2.00	63.24	nr	637.75
Extra for secondary silencer	112.92	192.01	0.50	15.81	nr	207.82
Extra for 2 row LTHW heating coil	46.12	78.42	-	-	nr	78.42
1130 l/s - 1370 l/s	392.27	667.01	2.00	63.24	nr	730.25
Extra for secondary silencer	157.61	268.00	0.50	15.81	nr	283.81
Extra for 2 row LTHW heating coil	58.74	99.88	-	-	nr	99.88
Extra for electric heater & thyristor controls, 3kw/1ph (per box)	344.04	585.00	-	-	nr	585.00
Extra for fitting free issue box controller	-	-	2.13	67.35	nr	67.35
Fan assisted VAV terminal box; factory installed and pre wired control components (excluding electronic controller); selected at 40 Pa external static pressure; includes fixing in position, electrical work elsewhere						
100 l/s - 175 l/s	485.48	825.50	3.00	94.86	nr	920.36
Extra for secondary silencer	57.93	98.50	0.50	15.81	nr	114.31
Extra for 1 row LTHW heating coil	30.62	52.07	-	-	nr	52.07
170 l/s - 360 l/s	529.59	900.50	3.00	94.86	nr	995.36
Extra for secondary silencer	73.22	124.50	0.50	15.81	nr	140.31
Extra for 1 row LTHW heating coil	34.93	59.39	-	-	nr	59.39
300 l/s - 640 l/s	607.22	1032.50	3.00	94.86	nr	1127.36
Extra for secondary silencer	112.33	191.00	0.50	15.81	nr	206.81
Extra for 1 row LTHW heating coil	38.91	66.16	-	-	nr	66.16
620 l/s - 850 l/s	607.22	1032.50	3.00	94.86	nr	1127.36
Extra for secondary silencer	112.33	191.00	0.50	15.81	nr	206.81
Extra for 1 row LTHW heating coil	38.91	66.16	-	-	nr	66.16
Extra for electric heater plus thyristor controls' 3kw/lph (per box)	344.04	585.00	-	-	nr	585.00
Extra for fitting free issue controller	-	-	2.13	67.35	nr	67.35

U: VENTILATION/AIR CONDITIONING SYSTEMS

Item	Net Price £	Material £	Labour hours	Labour £	Unit	Total rate £
U41 : FAN COIL AIR CONDITIONING						
FAN COIL UNITS						
All selections based on summer return air condition of 23°C @ 50% RH, CHW @ 6°/12°C, LTHW @ 82°/71°C (where applicable), medium speed, external resistance of 30Pa.						
All selections are based on heating and cooling units. For waterside control units there is no significant reduction in cost between 4 pipe heating and cooling and 2 pipe cooling only units (excluding controls). For airside control units, there is a marginal reduction (less than 5%) between 4 pipe heating and cooling units and 2 pipe cooling only units (excluding controls).						
Ceiling void mounted horizontal waterside control fan coil unit; cooling coil ; LTHW heating coil; multi tapped speed transformer; fine wire mesh filter; includes fixing in position; electrical work elsewhere						
Total cooling load, heating load						
2800 W, 1000 W – PWS Size 2	320.30	410.24	4.00	108.41	nr	**518.65**
4000 W, 1700 W – PWS Size 3	372.40	476.97	4.00	108.41	nr	**585.38**
4500 W, 1900 W – PWS Size 4	411.60	527.17	4.00	108.41	nr	**635.59**
6000 W, 2600 W – PWS Size 5	482.20	617.60	4.00	108.41	nr	**726.01**
Ceiling void mounted horizontal waterside control fan coil unit; cooling coil ; electric heating coil; multi tapped speed transformer; fine wire mesh filter; includes fixing in position; electrical work elsewhere + thyristor and 2 No. HTCO's						
Total cooling load, heating load						
2800W, 1500W – PWS Size 2	470.30	602.36	4.00	108.41	nr	**710.77**
4000W, 2000W – PWS Size 3	522.40	669.08	4.00	108.41	nr	**777.50**
4500W, 2000W – PWS Size 4	561.60	719.29	4.00	108.41	nr	**827.71**
6000W, 3000W – PWS Size 5	632.20	809.72	4.00	108.41	nr	**918.13**
Ceiling void mounted horizontal airside control fan coil unit; cooling coil ; LHTW heating coil; multi tapped speed transformer; fine wire mesh filter, damper actuator & fixing kit; includes fixing in position; electrical work elsewhere						
Total cooling load, heating load						
2600W, 2200W – PAS Size 2	653.70	837.25	4.00	108.41	nr	**945.67**
3600W, 3200W – PAS Size 3	700.50	897.19	4.00	108.41	nr	**1005.61**
4000W, 3600W – PAS Size 4	771.70	988.39	4.00	108.41	nr	**1096.80**
5400W, 5000W – PAS Size 5	844.80	1082.01	4.00	108.41	nr	**1190.43**

U: VENTILATION/AIR CONDITIONING SYSTEMS

Item	Net Price £	Material £	Labour hours	Labour £	Unit	Total rate £
U41 : FAN COIL AIR CONDITIONING (cont'd)						
FAN COIL UNITS (cont'd)						
Ceiling void mounted horizontal airside control fan coil unit; cooling coil ; electric heating coil; multi tapped speed transformer; fine wire mesh filter, damper actuator & fixing kit; includes fixing in position; electrical work elsewhere + thyristor and 2 No. HTCO's						
Total cooling load, heating load						
2600W, 1500W – PAS Size 2	773.70	990.95	4.00	108.41	nr	1099.36
3600W, 2000W – PAS Size 3	820.50	1050.89	4.00	108.41	nr	1159.30
4000W, 2000W – PAS Size 4	891.70	1142.08	4.00	108.41	nr	1250.49
5400W, 3000W – PAS Size 5	964.80	1235.71	4.00	108.41	nr	1344.12
Ceiling void mounted slimline horizontal waterside control fan coil unit, 170mm deep; cooling coil; LTHW heating coil; multi tapped speed transformer; fine wire mesh filter; includes fixing in position; electrical work elsewhere						
Total cooling load, heating load						
1100W, 1500W.	520.72	666.93	3.50	94.86	nr	761.80
3200W, 3700W	637.58	816.61	4.00	108.41	nr	925.02
Ceiling void mounted slimline horizontal waterside control fan coil unit, 170mm deep; cooling coil; electric heating coil; multi tapped speed transformer; fine wire mesh filter; includes fixing in position; electrical work elsewhere						
Total cooling load, heating load - NOT WITHIN OUR RANGE OF MANUFACTURE						
1100W, 1000W.	639.07	818.51	3.50	94.86	nr	913.38
3400W, 2000W	755.93	968.19	4.00	108.41	nr	1076.60
Ceiling void mounted slimline horizontal airside control fan coil unit, 170mm deep; cooling coil; LTHW heating coil; multi tapped speed transformer; fine wire mesh filter, damper actuator & fixing kit; includes fixing in position; electrical work elsewhere						
Total cooling load, heating load - NOT WITHIN OUR RANGE OF MANUFACTURE						
1000W, 1600W.	563.48	721.70	3.50	94.86	nr	816.56
3000W, 3300W	877.93	1124.44	4.00	108.41	nr	1232.86
4500W, 4500W	1125.40	1441.40	4.00	108.41	nr	1549.82
Ceiling void mounted slimline horizontal airside control fan coil unit, 170mm deep; cooling coil; electric heating coil; multi tapped speed transformer; fine wire mesh filter, damper actuator & fixing kit; includes fixing in position; electrical work elsewhere						
Total cooling load, heating load - NOT WITHIN OUR RANGE OF MANUFACTURE						
1000W, 1000W.	830.05	1063.12	3.50	94.86	nr	1157.98
3000W, 2000W	1093.40	1400.42	4.00	108.41	nr	1508.83
4500W, 3000W	1388.70	1778.63	4.00	108.41	nr	1887.05

U: VENTILATION/AIR CONDITIONING SYSTEMS

Item	Net Price £	Material £	Labour hours	Labour £	Unit	Total rate £
Low level perimeter waterside control fan coil unit; cooling coil; LTHW heating coil; multi tapped speed transformer; fine wire mesh filter; includes fixing in position; electrical work elsewhere - NOT WITHIN OUR RANGE OF MANUFACTURE						
Total cooling load, heating load						
1700W, 1400W	399.06	511.11	3.50	94.86	nr	605.97
Extra over for standard cabinet	212.30	271.91	1.00	27.10	nr	299.02
2200W, 1900W	462.91	592.89	3.50	94.86	nr	687.75
Extra over for standard cabinet	258.60	331.21	1.00	27.10	nr	358.32
2600W, 2200W	510.80	654.23	3.50	94.86	nr	749.09
Extra over for standard cabinet	264.98	339.38	1.00	27.10	nr	366.49
3900W, 3200W	670.43	858.68	3.50	94.86	nr	953.54
Extra over for standard cabinet	288.92	370.05	1.00	27.10	nr	397.15
4600W, 3900W	766.19	981.33	3.50	94.86	nr	1076.19
Extra over for standard cabinet	379.91	486.58	1.00	27.10	nr	513.69
Low level perimeter waterside control fan coil unit; cooling coil; electric heating coil; multi tapped speed transformer; fine wire mesh filter; includes fixing in position; electrical work elsewhere - NOT WITHIN OUR RANGE OF MANUFACTURE						
Total cooling load, heating load						
1700W, 1500W	518.79	664.46	3.50	94.86	nr	759.32
Extra over for standard cabinet	212.30	271.91	1.00	27.10	nr	299.02
2200W, 2000W	542.73	695.12	3.50	94.86	nr	789.99
Extra over for standard cabinet	258.60	331.21	1.00	27.10	nr	358.32
2600W, 2000W	678.41	868.90	3.50	94.86	nr	963.76
Extra over for standard cabinet	264.98	339.38	1.00	27.10	nr	366.49
3800W, 3000W	806.11	1032.46	3.50	94.86	nr	1127.32
Extra over for standard cabinet	288.92	370.05	1.00	27.10	nr	397.15
4600W, 4000W	981.70	1257.35	3.50	94.86	nr	1352.21
Extra over for standard cabinet	379.91	486.58	1.00	27.10	nr	513.69
Low level perimeter airside control fan coil unit; cooling coil; LTHW heating coil; multi tapped speed transformer; fine wire mesh filter; damper actuator & fixing kit; includes fixing in position; electrical work elsewhere - NOT WITHIN OUR RANGE OF MANUFACTURE						
Total cooling load, heating load						
1200W, 1400W	515.59	660.36	3.50	94.86	nr	755.22
Extra over for standard cabinet	212.30	271.91	1.00	27.10	nr	299.02
1800W, 2000W	576.25	738.06	3.50	94.86	nr	832.92
Extra over for standard cabinet	258.60	331.21	1.00	27.10	nr	358.32
2200W, 2400W	620.95	795.31	3.50	94.86	nr	890.17
Extra over for standard cabinet	264.98	339.38	-	-	nr	339.38
3200W, 3600W	806.11	1032.46	3.50	94.86	nr	1127.32
Extra over for standard cabinet	288.92	370.05	1.00	27.10	nr	397.15

U: VENTILATION/AIR CONDITIONING SYSTEMS

Item	Net Price £	Material £	Labour hours	Labour £	Unit	Total rate £
U41 : FAN COIL AIR CONDITIONING (cont'd)						
FAN COIL UNITS (cont'd)						
Low level perimeter airside control fan coil unit; cooling coil; electric heating coil; multi tapped speed transformer; fine wire mesh filter; damper actuator & fixing kit; includes fixing in position; electrical work elsewhere - NOT WITHIN OUR RANGE OF MANUFACTURE						
Total cooling load, heating load						
1250W, 1500W	702.35	899.56	3.50	94.86	nr	994.43
Extra over for standard cabinet	212.30	271.91	1.00	27.10	nr	299.02
1900W, 2000W	758.23	971.13	3.50	94.86	nr	1066.00
Extra over for standard cabinet	258.60	331.21	1.00	27.10	nr	358.32
2300W, 2000W	866.77	1110.15	3.50	94.86	nr	1205.01
Extra over for standard cabinet	264.98	339.38	1.00	27.10	nr	366.49
3300W, 3000W	989.67	1267.56	3.50	94.86	nr	1362.42
Extra over for standard cabinet	288.92	370.05	1.00	27.10	nr	397.15
Typical DDC control pack for airside control units; controller, return air sensor (damper and damper actuator included in above rates)						
Heating and cooling, per unit	170.00	217.73	2.00	54.21	nr	271.94
Cooling only, per unit	200.00	256.16	2.00	54.21	nr	310.37
Typical DDC control pack for waterside control units; controller, return air sensor, four port valves and actuators						
Heating and cooling, per unit	150.00	192.12	2.00	54.21	nr	246.33
Cooling only, per unit	130.00	166.50	2.00	54.21	nr	220.71
Note - Care needs to be taken when using these controls prices, as they can vary significantly, depending on the equipment manufacturer specified and the ° of control required						

U: VENTILATION/AIR CONDITIONING SYSTEMS

Item	Net Price £	Material £	Labour hours	Labour £	Unit	Total rate £
U70 - AIR CURTAINS						
The selection of air curtains requires consideration of the particular conditions involved; climatic conditions, wind influence, construction and position all influence selection; consultation with a specialist manufacturer is therefore advisable.						
Commercial grade air curtains; recessed or exposed units with rigid sheet steel casing; aluminium grilles; high quality motor/ centrifugal fan assembly; includes fixing in position; electrical work elsewhere						
Ambient temperature; 240V single phase supply; mounting height 2.40m						
1000 x 590 x 270mm	2317.00	2967.59	12.05	326.55	nr	**3294.14**
1500 x 590 x 270mm	2966.00	3798.82	12.05	326.55	nr	**4125.37**
2000 x 590 x 270mm	3589.00	4596.76	12.05	326.60	nr	**4923.35**
2500 x 590 x 270mm	3989.00	5109.07	13.00	352.35	nr	**5461.42**
Ambient temperature; 240V single phase supply; mounting height 2.80m						
1000 x 590 x 270mm	2678.00	3429.96	16.13	437.15	nr	**3867.11**
1500 x 590 x 270mm	3418.00	4377.74	16.13	437.15	nr	**4814.89**
2000 x 590 x 270mm	4261.00	5457.45	16.13	437.15	nr	**5894.60**
2500 x 590 x 270mm	4928.00	6311.73	17.10	463.47	nr	**6775.20**
Ambient temperature 240V single phase supply; mounting height 3.30m						
1000 x 774 x 370mm	3498.00	4480.20	17.24	467.30	nr	**4947.51**
1500 x 774 x 370mm	4731.00	6059.42	17.24	467.30	nr	**6526.72**
2000 x 774 x 370mm	5851.00	7493.90	17.24	467.30	nr	**7961.20**
2500 x 774 x 370mm	6926.00	8870.75	18.30	495.99	nr	**9366.75**
Ambient temperature; 240V single phase supply; mounting height 4.00m						
1000 x 774 x 370mm	3864.00	4948.97	19.10	517.68	nr	**5466.65**
1500 x 774 x 370mm	5069.00	6492.32	19.10	517.68	nr	**7010.00**
2000 x 774 x 370mm	6310.00	8081.78	19.10	517.68	nr	**8599.46**
2500 x 774 x 370mm	7395.00	9471.44	19.90	539.36	nr	**10010.80**
Water heated; 240V single phase supply; mounting height 2.40m						
1000 x 590 x 270mm; 2.30 - 9.40kW output	2727.00	3492.71	12.05	326.55	nr	**3819.26**
1500 x 590 x 270mm; 3.50 - 14.20kW output	3490.00	4469.96	12.05	326.55	nr	**4796.51**
2000 x 590 x 270mm; 4.70 - 19.00kW output	4223.00	5408.78	12.05	326.55	nr	**5735.32**
2500 x 590 x 270mm; 5.90 - 23.70kW output	4694.00	6012.03	13.00	352.35	nr	**6364.37**
Water heated; 240V single phase supply; mounting height 2.80m						
1000 x 590 x 270mm; 3.30 - 11.90kW output	3151.00	4035.77	16.13	437.15	nr	**4472.92**
1500 x 590 x 270mm; 5.00 - 17.90kW output	4023.00	5152.62	16.13	437.15	nr	**5589.77**
2000 x 590 x 270mm; 6.70 - 23.90kW output	5014.00	6421.88	16.13	437.15	nr	**6859.03**
2500 x 590 x 270mm; 8.30 - 29.80kW output	5798.00	7426.02	17.10	463.47	nr	**7889.49**

U: VENTILATION/AIR CONDITIONING SYSTEMS

Item	Net Price £	Material £	Labour hours	Labour £	Unit	Total rate £
U70 - AIR CURTAINS (cont'd)						
Commercial grade air curtains (cont'd)						
Water heated; 240V single phase supply; mounting height 3.30m						
1000 x 774 x 370mm; 6.10 - 21.80kW output	4116.00	5271.73	17.24	467.30	nr	5739.03
1500 x 774 x 370mm; 9.20 - 32.80kW output	5566.00	7128.88	17.24	467.30	nr	7596.18
2000 x 774 x 370mm; 12.30 - 43.70kW output	6884.00	8816.96	17.24	467.30	nr	9284.26
2500 x 774 x 370mm; 15.30 - 54.60kW output	8149.00	10437.16	18.30	495.99	nr	10933.15
Water heated; 240V single phase supply; mounting height 4.00m						
1000 x 774 x 370mm; 7.20 - 24.20kW output	4546.00	5822.47	19.10	517.68	nr	6340.15
1500 x 774 x 370mm; 10.90 - 36.30kW output	5964.00	7638.63	19.10	517.68	nr	8156.31
2000 x 774 x 370mm; 14.50 - 48.40kW output	7424.00	9508.58	19.10	517.68	nr	10026.26
2500 x 774 x 370mm; 18.10 - 60.60kW output	8701.00	11144.15	19.90	539.36	nr	11683.51
Electrically heated; 415V three phase supply; mounting height 2.40m						
1000 x 590 x 270mm; 2.30 - 9.40kW output	3350.00	4290.65	12.05	326.55	nr	4617.19
1500 x 590 x 270mm; 3.50 - 14.20kW output	4167.00	5337.05	12.05	326.55	nr	5663.60
2000 x 590 x 270mm; 4.70 - 19.00kW output	4967.00	6361.68	12.05	326.41	nr	6688.09
2500 x 590 x 270mm; 5.90 - 23.70kW output	5647.00	7232.62	13.00	352.20	nr	7584.82
Electrically heated; 415V three phase supply; mounting height 2.80m						
1000 x 590 x 270mm; 3.30 - 11.90kW output	3944.00	5051.44	16.13	436.97	nr	5488.40
1500 x 590 x 270mm; 5.00 - 17.90kW output	4949.00	6338.63	16.13	437.15	nr	6775.78
2000 x 590 x 270mm; 6.70 - 23.90kW output	6123.00	7842.28	16.13	437.15	nr	8279.43
2500 x 590 x 270mm; 8.30 - 29.80kW output	6844.00	8765.73	17.10	463.47	nr	9229.20
Electrically heated; 415V three phase supply; mounting height 3.30m						
1000 x 774 x 370mm; 6.10 - 21.80kW output	5992.00	7674.49	17.24	467.07	nr	8141.56
1500 x 774 x 370mm; 9.20 - 32.80kW output	8029.00	10283.46	17.24	467.26	nr	10750.73
2000 x 774 x 370mm; 12.30 - 43.70kW output	9690.00	12410.86	17.24	467.26	nr	12878.12
2500 x 774 x 370mm; 15.30 - 54.60kW output	11471.00	14691.94	18.30	495.99	nr	15187.94
Electrically heated; 415V three phase supply; mounting height 4.00m						
1000 x 774 x 370mm; 7.20 - 24.20kW output	6613.00	8469.86	19.10	517.46	nr	8987.32
1500 x 774 x 370mm; 10.90 - 36.30kW output	8580.00	10989.18	19.10	517.68	nr	11506.86
2000 x 774 x 370mm; 14.50 - 48.40kW output	10377.00	13290.76	19.10	517.68	nr	13808.43
2500 x 774 x 370mm; 18.10 - 60.60kW output	12292.00	15743.47	19.90	539.36	nr	16282.83
Industrial grade air curtains; recessed or exposed units with rigid sheet steel casing; aluminium grilles; high quality motor/ centrifugal fan assembly; includes fixing in position; electrical work elsewhere						
Ambient temperature; 415V three phase supply; including wiring between multiple units; horizontally or vertically mounted; opening maximum 6.00m						
1500 x 585 x 853mm; 1.6A supply	2415.00	3093.11	17.24	467.30	nr	3560.41
2000 x 585 x 853mm; 2.1A supply	2945.00	3771.93	17.24	467.30	nr	4239.23

U: VENTILATION/AIR CONDITIONING SYSTEMS

Item	Net Price £	Material £	Labour hours	Labour £	Unit	Total rate £
Water heated; 415V three phase supply; including wiring between multiple units; horizontally or vertically mounted; opening maximum 6.00m						
1500 x 585 x 956mm; 1.6A supply; 34.80kW output	2795.00	3579.81	17.24	467.30	nr	**4047.11**
2000 x 585 x 956mm; 2.1A supply; 50.70kW output	3375.00	4322.67	17.24	467.30	nr	**4789.97**
Water heated; 415V three phase supply; including wiring between multiple units; vertically mounted in single bank for openings maximum 6.00m wide or opposing twin banks for openings maximum 10.00m wide						
1500 x 585mm; 1.6A supply; 41.1kW output	2795.00	3579.81	17.24	467.30	nr	**4047.11**
2000 x 585mm; 2.1A supply; 57.7kW output	3375.00	4322.67	17.24	467.30	nr	**4789.97**
Remote mounted electronic controller unit; 415V three phase supply; excluding wiring to units						
Five speed, 7A	425.00	544.34	5.00	135.52	nr	**679.85**

ESSENTIAL READING FROM TAYLOR AND FRANCIS

Heat and Mass Transfer in Buildings

Second Edition

Keith J. Moss

The second edition of this reliable text provides readers with a thorough understanding of the design procedures that are essential in designing new buildings and building refurbishment.

Covering the fundamentals of heat and mass transfer as essential underpinning knowledge, this edition has been thoroughly updated and reflects the need for new building design and building refurbishment to feature low energy consumption and sustainable characteristics.

New additions include:

- extended and updated worked examples

- two new appendices covering renewable energy systems and sustainable building engineering – with startling conclusions.

This book is an invaluable guide for HND and degree level students of building services engineering, as well as building, built environment, building engineering and architecture courses.

2007: 234x156 mm: 314 pages
Hb: 978-0-415-40907-0: **£90.00**
Pb: 978-0-415-40908-7: **£29.99**

To Order: Tel: +44 (0) 1235 400524 **Fax:** +44 (0) 1235 400525
or Post: Taylor and Francis Customer Services,
Bookpoint Ltd, Unit T1, 200 Milton Park, Abingdon, Oxon, OX14 4TA UK
Email: book.orders@tandf.co.uk

**For a complete listing of all our titles visit:
www.tandf.co.uk**

Electrical Installations
Material Costs/Measured Work Prices

DIRECTIONS

The following explanations are given for each of the column headings and letter codes.

Unit	Prices for each unit are given as singular (i.e. 1 metre, 1 nr) unless stated otherwise
Net price	Industry tender prices, plus nominal allowance for fixings, waste and applicable trade discounts.
Material cost	Net price plus percentage allowance for overheads, profit and preliminaries.
Labour norms	In man-hours for each operation.
Labour cost	Labour constant multiplied by the appropriate all-in man-hour cost based on gang rate. (See also relevant Rates of Wages Section) plus percentage allowance for overheads, profit and preliminaries
Measured work Price (total rate)	Material cost plus Labour cost.

MATERIAL COSTS

The Material Costs given are based at Second Quarter 2008 but exclude any charges in respect of VAT. At the time of going to press, the raw material cost of copper and steel was fluctuating but had not been reflected within factory exit prices for finished goods i.e. cable etc. Users of the book are advised to register on the SPON's website www.pricebooks.co.uk/updates to receive the free quarterly updates - alerts will then be provided by e-mail as changes arise.

MEASURED WORK PRICES

These prices are intended to apply to new work in the London area. The prices are for reasonable quantities of work and the user should make suitable adjustments if the quantities are especially small or especially large. Adjustments may also be required for locality (e.g. outside London – refer to cost indices in approximate estimating section for details of adjustment factors) and for the market conditions (e.g. volume of work secured or being tendered) at the time of use.

ELECTRICAL INSTALLATIONS

The labour rate has been based on average gang rates per man hour effective from 7 January 2008 including allowances for all other emoluments and expenses. To this rate has been added 8% and 6% to cover site and head office overheads and preliminary items together with a further 3% for profit, resulting in an inclusive rate of £27.09 per man hour. The rate has been calculated on a working year of 2025 hours; a detailed build-up of the rate is given at the end of these directions.

In calculating the 'Measured Work Prices' the following assumptions have been made:
(a) That the work is carried out as a sub-contract under the Standard Form of Building Contract.
(b) That, unless otherwise stated, the work is being carried out in open areas at a height which would not require more than simple scaffolding.
(c) That the building in which the work is being carried out is no more than six storey's high.

Where these assumptions are not valid, as for example where work is carried out in ducts and similar confined spaces or in multi-storey structures when additional time is needed to get to and from upper floors, then an appropriate adjustment must be made to the prices. Such adjustment will normally be to the labour element only.

DIRECTIONS

LABOUR RATE - ELECTRICAL

The annual cost of a notional eleven man gang

		TECHNICIAN 1 NR	APPROVED ELECTRICIANS 4 NR	ELECTRICIANS 4 NR	LABOURERS 2 NR	SUB-TOTALS
Hourly Rate from 7 January 2008		16.50	14.65	13.51	10.83	
Working hours per annum per man		1,687.50	1,687.50	1,687.50	1,687.50	
x Hourly rate x nr of men = £ per annum		27,843.75	98,887.50	91,192.50	36,551.25	254,475.00
Overtime Rate		24.75	21.98	20.27	16.25	
Overtime hours per annum per man		337.50	337.50	337.50	337.50	
x Hourly rate x nr of men = £ per annum		8,353.13	29,666.25	27,357.75	10,965.38	76,342.50
Total		36,196.88	128,553.75	118,550.25	47,516.63	330,817.50
Incentive schemes (insert percentage)	5.00%	1,809.84	6,427.69	5,927.51	2,375.83	16,540.88
Daily Travel Time Allowance (15-20 miles each way) effective from 07/01/08		4.58	4.58	4.58	4.58	
Days per annum per man		225.00	225.00	225.00	225.00	
x nr of men = £ per annum		1,030.50	4,122.00	4,122.00	2,061.00	11,335.50
Daily Travel Allowance (15-20 miles each way) effective from 07/01/08		3.15	3.15	3.15	3.15	
Days per annum per man		225.00	225.00	225.00	225.00	
x nr of men = £ per annum		708.75	2,835.00	2,835.00	1,417.50	7,796.25
JIB Pension Scheme @ 2.5%		1,113.09	3,972.16	3,676.18	1,490.25	10,251.68
JIB combined benefits scheme (nr of weeks per man)		52.00	52.00	52.00	52.00	
Benefit Credit effective from 1 October 2007		52.00	52.00	52.00	52.00	
x nr of men = £ per annum		2,973.88	10,811.84	10,129.60	4,294.16	28,209.48
Holiday Top-up Funding		48.32	43.13	39.99	32.22	
x nr of men @ 7.5 hrs per day = £ per annum		2,512.64	8,971.04	8,317.92	3,362.32	23,163.92
National Insurance Contributions:						
Annual gross pay (subject to NI) each		39,745.97	141,938.44	131,434.76	55,370.96	
% of NI Contributions		12.80	12.80	12.80	12.80	
£ Contributions/annum		4,380.38	15,339.71	13,995.24	5,417.28	39,132.62

DIRECTIONS

LABOUR RATE - ELECTRICAL

The annual cost of a notional eleven man gang

			SUB-TOTALS
	SUB-TOTAL		467,247.83
	TRAINING (INCLUDING ANY TRADE REGISTRATIONS) - SAY	1.00%	4,672.48
	SEVERANCE PAY AND SUNDRY COSTS - SAY	1.50%	7,078.80
	EMPLOYER'S LIABILITY AND THIRD PARTY INSURANCE - SAY	2.00%	9,579.98
	ANNUAL COST OF NOTIONAL GANG		488,579.09
MEN ACTUALLY WORKING = 10.5	THEREFORE ANNUAL COST PER PRODUCTIVE MAN		46,531.34
AVERAGE NR OF HOURS WORKED PER MAN = 2025	THEREFORE ALL IN MAN HOUR		22.98
	PRELIMINARY ITEMS - SAY	8.00%	1.84
	SITE AND HEAD OFFICE OVERHEADS - SAY	6.00%	1.49
	PROFIT - SAY	3.00%	0.79
	THEREFORE INCLUSIVE MAN HOUR		27.09

Notes:

(1) Hourly wage rates are those effective from 7 January 2008.
(2) The following assumptions have been made in the above calculations:-
    (a) Hourly rates are based on London rate and job reporting own transport.
    (b) The working week of 37.5 hours is made up of 7.5 hours Monday to Friday.
    (c) Five days in the year are lost through sickness or similar reason.
    (d) A working year of 2025 hours.
(3) The incentive scheme addition of 5% is intended to reflect bonus schemes typically in use.
(4) National insurance contributions are those effective from 6 April 2008, paid for 48 weeks. Calculation is based on employer making regular payment into the holiday pay scheme, allowing savings on NI.
(5) Weekly JIB Combined Benefit Credit Scheme are those effective from 1 October 2008.
(6) Paid Holidays with effect from 7 January 2008, for all 30 days (22 Annual and 8 Public) are to be paid at normal earnings level.
(7) Overtime is paid after 37.5 hours.

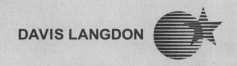

Constructing the best and most valued relationships in the industry

www.davislangdon.com

Offices in Europe & the Middle East, Africa, Asia Pacific, Australasia and the USA

Cost Management | Project Management
Banking Tax and Finance | Building Surveying | Engineering Services | Legal Support Group
Management Consultancy | Specifications and Design Management | VPR

ESSENTIAL READING FROM TAYLOR AND FRANCIS

Building Acoustics

Tor Erik Vigran

Covering all aspects of sound and vibration in buildings, this book explores room acoustics, sound insulation, and noise and vibration problems connected to service equipment and external sources. Measuring techniques are also discussed. It is designed for advanced level engineering studies and as a guide for practitioners.

Selected Contents

1. Oscillatory Motion. Description and Analysis
2. Excitation and Response Relationships
3. Sound Waves in Fluids and Solid Media
4. Room Acoustics
5. Sound Absorbents
6. Sound Transmission. Properties of Simple Walls and Floors
7. Statistical Energy Analysis (SEA)
8. Sound Transmission Through Complex Systems
9. Sound Transmission in Buildings 10. Flanking Transmission

July 2008: 246x174: 384pp
Hb: 978-0-415-42853-8 **£80.00**

To Order: Tel: +44 (0) 1235 400524 **Fax:** +44 (0) 1235 400525
or Post: Taylor and Francis Customer Services,
Bookpoint Ltd, Unit T1, 200 Milton Park, Abingdon, Oxon, OX14 4TA UK
Email: book.orders@tandf.co.uk

For a complete listing of all our titles visit:
www.tandf.co.uk

V: ELECTRICAL SUPPLY/POWER/LIGHTING SYSTEMS

Item	Net Price £	Material £	Labour hours	Labour £	Unit	Total rate £
V10 : ELECTRICAL GENERATION PLANT						
STANDBY GENERATORS						
Standby diesel generating sets; supply and Installation; fixing to base; all supports and fixings; all necessary connections to equipment	-	-	-	-	-	-
Three phase, 400 Volt, four wire 50 Hz packaged standby diesel generating set, complete with radio and television suppressors, daily service fuel tank and associated piping, 4 metres of exhaust pipe and primary exhaust silencer, control panel, mains Failure relay, starting battery with charger, all internal wiring, interconnections, earthing and labels. Rated for standby duty; including UK delivery, installation and commissioning						
60 kVA	14807.42	17524.14	100.00	2710.15	nr	20234.29
100 kVA	20758.49	24567.05	100.00	2710.15	nr	27277.20
150 kVA	23740.69	28096.39	100.00	2710.15	nr	30806.54
315 kVA	42803.08	50656.16	120.00	3252.18	nr	53908.34
500 kVA	50451.63	59708.00	120.00	3252.18	nr	62960.18
750 kVA	85855.83	101607.79	140.00	3794.21	nr	105402.00
1000 kVA	114109.75	135045.47	140.00	3794.21	nr	138839.68
1500kVA	134296.14	158935.45	170.00	4607.26	nr	163542.71
2000kVA	206866.27	244820.02	170.00	4607.26	nr	249427.28
2500kVA	283227.88	335191.70	210.00	5691.31	nr	340883.01
Extra for residential silencer; performance 75dBA at 1m; including connection to exhaust pipe						
60 kVA	795.01	940.87	10.00	271.01	nr	1211.89
100 kVA	949.97	1124.26	10.00	271.01	nr	1395.27
150 kVA	1089.41	1289.28	10.00	271.01	nr	1560.30
315 kVA	2119.24	2508.06	15.00	406.52	nr	2914.58
500 kVA	2710.43	3207.71	15.00	406.52	nr	3614.23
750 kVA	4038.22	4779.12	20.00	542.03	nr	5321.15
1000kVA	5438.74	6436.58	20.00	542.03	nr	6978.61
1500kVA	8664.07	10253.66	20.00	542.03	nr	10795.69
2000kVA	9384.00	11105.68	30.00	813.04	nr	11918.72
2500kVA	10729.67	12698.24	30.00	813.04	nr	13511.29
Synchronization panel for paralleling generators - not generators to mains; including interconnecting cables; commissioning and testing; fixing to backgrounds						
2 x 60 kVA	6622.30	7837.30	80.00	2168.12	nr	10005.42
2 x 100 kVA	7119.35	8425.54	80.00	2168.12	nr	10593.66
2 x 150 kVA	8956.09	10599.27	80.00	2168.12	nr	12767.39
2 x 315 kVA	15225.60	18019.04	80.00	2168.12	nr	20187.16
2 x 500 kVA	21981.41	26014.33	80.00	2168.12	nr	28182.45
2 x 750kVA	24805.07	29356.06	80.00	2168.12	nr	31524.18
2 x 1000kVA	26733.60	31638.41	120.00	3252.18	nr	34890.59
2 x 1500kVA	31647.89	37454.33	120.00	3252.18	nr	40706.51
2 x 2000kVA	34290.38	40581.63	120.00	3252.18	nr	43833.81
2 x 2500kVA	39977.04	47311.63	120.00	3252.18	nr	50563.81

V: ELECTRICAL SUPPLY/POWER/LIGHTING SYSTEMS

Item	Net Price £	Material £	Labour hours	Labour £	Unit	Total rate £
V10 : ELECTRICAL GENERATION PLANT (cont'd)						
STANDBY GENERATORS (cont'd)						
Prefabricated drop-over acoustic housing; performance 85dBA at 1m over the range from 60kVA to 315kVA, 75dBA from 500kVA to 2500kVA						
60kVA	2860.61	3385.45	4.00	108.41	nr	3493.86
100kVA	2959.53	3502.52	7.00	189.71	nr	3692.23
150kVA	4805.83	5687.56	15.00	406.52	nr	6094.08
315kVA	10254.09	12135.41	25.00	677.54	nr	12812.94
500kVA	16936.00	20043.25	40.00	1084.06	nr	21127.31
750kVA	26689.50	31586.23	40.00	1084.06	nr	32670.29
1000kVA	33765.93	39960.96	40.00	1084.06	nr	41045.02
1500kVA	66977.62	79266.01	40.00	1084.06	nr	80350.07
2000kVA	80505.92	95276.34	60.00	1626.09	nr	96902.43
2500kVA	111969.07	132512.03	70.00	1897.11	nr	134409.14
COMBINED HEAT AND POWER (CHP) UNITS						
Gas fired engine; acoustic enclosure complete with exhaust fan and attenuators; exhaust gas attenuation; includes 6m long pipe connections; dry air cooler for secondary water circuit to reject excess heat; controls and panel; commissioning						
Electrical output; Heat output						
82 kW; 132 kW	87947.58	104083.32	-	-	nr	104083.32
100kW, 148kW	93678.50	110865.69	-	-	nr	110865.69
118kW, 181kW	98307.32	116343.76	-	-	nr	116343.76
130kW, 201kW	107087.74	126735.13	-	-	nr	126735.13
140kW, 207kW	113957.14	134864.86	-	-	nr	134864.86
150kW, 208kW	120018.69	142038.52	-	-	nr	142038.52
160kW, 216kW	114618.40	135647.44	-	-	nr	135647.44
198kW, 233kW	149941.24	177450.96	-	-	nr	177450.96
210kW, 319kW	141730.06	167733.27	-	-	nr	167733.27
237kW, 359kW	155836.94	184428.34	-	-	nr	184428.34
307kW, 435kW	211328.21	250100.60	-	-	nr	250100.60
380kW, 500kW	244666.20	289555.11	-	-	nr	289555.11
490kW, 679kW	336801.76	398594.78	-	-	nr	398594.78
501kW, 518kW	310792.20	367813.24	-	-	nr	367813.24
600kW, 873kW	370030.61	437920.13	-	-	nr	437920.13
725kW, 1019kW	500353.40	592153.24	-	-	nr	592153.24
975kW, 1293kW	621309.41	735301.05	-	-	nr	735301.05
1160kW, 1442kW	703415.86	832471.57	-	-	nr	832471.57
1379kW, 1475kW	839745.63	993813.76	-	-	nr	993813.76
1566kW, 1647kW	918600.35	1087135.96	-	-	nr	1087135.96
1600kW, 1625kW	916396.15	1084527.35	-	-	nr	1084527.35
1760kW, 1821kW	1006327.51	1190958.42	-	-	nr	1190958.42
2430kW	1307000.00	1546795.29	-	-	nr	1546795.29
2931kW	1580000.00	1869882.60	-	-	nr	1869882.60
3500kW	1700000.00	2011899.00	-	-	nr	2011899.00
3900kW	2000000.00	2366940.00	-	-	nr	2366940.00
Note: The costs detailed are based on a specialist sub-contract package, as part of the M&E contract works, and include installation						

V: ELECTRICAL SUPPLY/POWER/LIGHTING SYSTEMS

Item	Net Price £	Material £	Labour hours	Labour £	Unit	Total rate £
V11 : HV SUPPLY						
Y61 - HV CABLES						
Cable; 6350/11000 volts, 3 core, XLPE; stranded copper conductors; steel wire armoured; LSOH to BS 7835						
Laid in trench/duct including marker tape (cable tiles measured elsewhere)						
95mm^2	32.95	38.99	0.23	6.23	m	45.23
120mm^2	36.26	42.91	0.23	6.23	m	49.14
150mm^2	43.39	51.35	0.25	6.78	m	58.12
185mm^2	47.10	55.74	0.25	6.78	m	62.52
240mm^2	54.85	64.92	0.27	7.32	m	72.23
300mm^2	62.62	74.11	0.29	7.86	m	81.97
Cable tiles; single width; laid in trench above cables on prepared sand bed (cost of excavation excluded); reinforced concrete covers; concave/convex ends						
914 x 152 x 63/38mm	10.15	12.02	0.11	2.98	m	15.00
914 x 229 x 63/38mm	11.95	14.14	0.11	2.98	m	17.12
914 x 305 x 63/38mm	13.30	15.74	0.11	2.98	m	18.73
Clipped direct to backgrounds including cleats						
95mm^2	41.80	49.47	0.47	12.74	m	62.21
120mm^2	45.11	53.38	0.50	13.55	m	66.93
150mm^2	51.98	61.51	0.53	14.36	m	75.87
185mm^2	56.96	67.41	0.55	14.91	m	82.32
240mm^2	65.86	77.94	0.60	16.26	m	94.20
300mm^2	74.86	88.60	0.68	18.43	m	107.03
Terminations for above cables, including heat-shrink kit and glanding off						
95mm^2	435.86	515.82	4.75	128.73	m	644.55
120mm^2	433.49	513.03	5.30	143.64	m	656.66
150mm^2	466.28	551.83	6.00	162.61	m	714.44
185mm^2	467.41	553.16	6.90	187.00	m	740.16
240mm^2	488.37	577.97	7.43	201.36	m	779.33
300mm^2	526.64	623.26	8.75	237.14	m	860.40
Cable; 6350/11000volts, 3 core, paper insulated; lead sheathed; steel wire armoured; stranded copper conductors; to BS 6480						
Laid in trench/duct including marker tape (cable tiles measured elsewhere)						
95mm^2	35.89	42.47	0.22	5.96	m	48.44
120mm^2	38.73	45.84	0.22	5.96	m	51.80
150mm^2	47.39	56.08	0.24	6.50	m	62.58
185mm^2	49.45	58.53	0.26	7.05	m	65.57
240mm^2	57.38	67.91	0.34	9.21	m	77.13

V: ELECTRICAL SUPPLY/POWER/LIGHTING SYSTEMS

Item	Net Price £	Material £	Labour hours	Labour £	Unit	Total rate £
V11 : HV SUPPLY (cont'd)						
Y61 - HV CABLES (cont'd)						
Cable tiles; single width; laid in trench above cables on prepared sand bed (cost of excavation excluded); reinforced concrete covers; concave/ convex ends						
914 x 152 x 63/38mm	10.15	12.02	0.11	2.98	m	15.00
914 x 229 x 63/38mm	11.95	14.14	0.11	2.98	m	17.12
914 x 305 x 63/38mm	13.30	15.74	0.11	2.98	m	18.73
Clipped direct to backgrounds including cleats						
95mm^2	45.87	54.29	0.55	14.91	m	69.20
120mm^2	48.70	57.63	0.63	17.07	m	74.71
150mm^2	57.06	67.53	0.66	17.89	m	85.41
185mm^2	60.43	71.51	0.72	19.51	m	91.03
240mm^2	69.65	82.42	0.82	22.22	m	104.65
Terminations for above cables, including compound joint and glanding off						
95mm^2	486.60	575.88	4.75	128.73	m	704.61
120mm^2	495.39	586.28	5.30	143.64	m	729.92
150mm^2	517.46	612.40	6.10	165.32	m	777.72
185mm^2	518.56	613.70	6.90	187.00	m	800.70
240mm^2	545.36	645.42	7.43	201.36	m	846.78
300mm^2	578.34	684.45	8.75	237.14	m	921.59

V: ELECTRICAL SUPPLY/POWER/LIGHTING SYSTEMS

Item	Net Price £	Material £	Labour hours	Labour £	Unit	Total rate £
Y70 - HV SWITCHGEAR AND TRANSFORMERS						
H.V. Circuit Breakers; installed on prepared foundations including all supports, fixings and inter panel connections where relevant. Excludes main and multi core cabling and heat shrink cable termination kits.						
Three phase 11kV, 630 Amp, Air or SF6 insulated, with fixed pattern vacuum or SF6 circuit breaker panels; hand charged spring closing operation; prospective fault level up to 25 kA for 3 seconds. Feeders include ammeter with selector switch, 3 pole IDMT, overcurrent and earth fault relays with necessary current relays with necessary current transformers; incomers include 3 phase VT, voltmeter and phase selector switch; Includes IDMT overcurrent and earth fault relays/CTs.						
Single panel with cable chamber	19126.99	22636.21	31.70	859.12	nr	23495.33
Three panel with one incomer and two feeders; with cable chambers	57310.19	67824.89	67.83	1838.29	nr	69663.19
Five panel with two incoming, two feeders and a bus section; with cable chambers	95500.29	113021.73	99.17	2687.66	nr	115709.39
Ring Main Unit (RMU)						
Three phase 11kV, 630 Amp, SF6 insulated RMU with vacuum or SF6 200 Amp circuit breaker tee-off, Includes IDMT overcurrent and earth fault relays/CTs for tee-off and cable boxes						
3 way Ring Main Unit	11531.89	13647.64	67.83	1838.29	nr	15485.94
Extra for,						
Remote actuator to ring switches (per switch)	1741.31	2060.79	-	-	nr	2060.79
Remote tripping of circuit breaker	870.66	1030.40	-	-	nr	1030.40
3 - Phase Neon indicators (per circuit)	102.39	121.17	-	-	nr	121.17
Pressure gauge with alarm contacts (for SF6 only)	224.64	265.85	-	-	nr	265.85
Tripping Batteries						
Battery chargers; switchgear tripping and closing; double wound transformer and earth screen; including fixing to background, commissioning and testing						
Valve regulated lead acid battery						
30 volt; 19 Ah; 3A	2077.49	2458.65	6.50	176.16	nr	2634.81
30 volt; 29 Ah; 3A	3328.30	3938.95	8.50	230.36	nr	4169.31
110 volt; 19 Ah; 3A	3395.18	4018.10	6.50	176.16	nr	4194.26
110 volt; 29 Ah; 3A	3704.55	4384.22	8.50	230.36	nr	4614.58
110 volt; 38 Ah; 3A	3958.65	4684.94	10.00	271.01	nr	4955.95

V: ELECTRICAL SUPPLY/POWER/LIGHTING SYSTEMS

Item	Net Price £	Material £	Labour hours	Labour £	Unit	Total rate £
V11 : HV SUPPLY (cont'd)						
Y70 - HV SWITCHGEAR AND TRANSFORMERS (cont'd)						
Step down transformers; 11 / 0.415kV, Dyn 11, 50Hz. Complete with lifting lugs, mounting skids, provisions for wheels, undrilled gland plates to air-filled cable boxes, off load tapping facility, including UK delivery						
Oil-filled in free breathing ventilated steel tank						
500kVA	8241.76	9753.88	30.00	813.04	nr	10566.92
800kVA	9386.06	11108.12	30.00	813.04	nr	11921.16
1000kVA	10637.88	12589.61	30.00	813.04	nr	13402.66
1250kVA	12926.45	15298.06	35.00	948.55	nr	16246.62
1500kVA	15317.95	18128.34	35.00	948.55	nr	19076.89
2000kVA	19895.08	23545.23	40.00	1084.06	nr	24629.29
MIDEL - filled in gasket-sealed steel tank						
500kVA	10916.33	12919.15	30.00	813.04	nr	13732.19
800kVA	12442.05	14724.79	30.00	813.04	nr	15537.83
1000kVA	14166.08	16765.13	30.00	813.04	nr	17578.18
1250kVA	17122.13	20263.53	35.00	948.55	nr	21212.08
1500kVA	20371.88	24109.51	35.00	948.55	nr	25058.06
2000kVA	26379.36	31219.18	40.00	1084.06	nr	32303.24
Extra for,						
Fluid temperature indicator with 2 N/O contacts	301.88	357.26	2.00	54.20	nr	411.46
Winding temperature indicator with 2 N/O contacts	739.60	875.29	2.00	54.20	nr	929.50
Dehydrating Breather	75.47	89.32	2.00	54.20	nr	143.52
Plain Rollers	226.41	267.95	2.00	54.20	nr	322.15
Pressure relief device with 1 N/O contact	452.81	535.89	2.00	54.20	nr	590.09
Step down transformers; 11 / 0.415kV, Dyn 11, 50Hz. Complete with lifting lugs, mounting skids, provisions for wheels, undrilled gland plates to air-filled cable boxes, off load tapping facility, including delivery						
Cast Resin type in ventilated steel encloure, AN - Air Natural including winding temperture indicator with 2 N/O contacts						
500kVA	11679.18	13821.96	40.00	1084.06	nr	14906.02
800kVA	13586.32	16079.00	40.00	1084.06	nr	17163.06
1000kVA	16435.57	19451.00	40.00	1084.06	nr	20535.06
1250kVA	17789.64	21053.50	45.00	1219.57	nr	22273.07
1600kVA	20181.17	23883.80	45.00	1219.57	nr	25103.37
2000kVA	22851.17	27043.67	50.00	1355.08	nr	28398.75

V: ELECTRICAL SUPPLY/POWER/LIGHTING SYSTEMS

Item	Net Price £	Material £	Labour hours	Labour £	Unit	Total rate £
Cast Resin type in ventilated steel enclosure with temperature controlled fans to achieve 40% increase to AN/AF rating. Includes winding temperature indicator with 2 N/O contacts						
500/700kVA	12823.45	15176.17	42.00	1138.26	nr	**16314.44**
800/1120kVA	14921.32	17658.93	42.00	1138.26	nr	**18797.20**
1000/1400kVA	18461.70	21848.87	42.00	1138.26	nr	**22987.13**
12501750kVA	19224.55	22751.68	47.00	1273.77	nr	**24025.45**
1600/2240kVA	21719.05	25703.84	47.00	1273.77	nr	**26977.61**
2000/2800kVA	24770.46	29315.10	52.00	1409.28	nr	**30724.37**

V: ELECTRICAL SUPPLY/POWER/LIGHTING SYSTEMS

Item	Net Price £	Material £	Labour hours	Labour £	Unit	Total rate £
V20 : LV DISTRIBUTION						
Y60 : CONDUIT AND CABLE TRUNKING						
Heavy gauge, screwed drawn steel; surface fixed on saddles to backgrounds, with standard pattern boxes and fittings including all fixings and supports (forming holes, conduit entry, draw wires etc. and components for earth continuity are included)						
Black enamelled						
20 mm dia.	1.91	2.26	0.49	13.28	m	15.54
25 mm dia.	2.77	3.28	0.56	15.18	m	18.46
32 mm dia.	5.66	6.70	0.64	17.34	m	24.05
38 mm dia.	6.75	7.99	0.73	19.78	m	27.77
50 mm dia.	10.71	12.68	1.04	28.19	m	40.87
Galvanised						
20 mm dia.	2.44	2.89	0.49	13.28	m	16.17
25 mm dia.	3.20	3.78	0.56	15.18	m	18.96
32 mm dia.	5.94	7.02	0.64	17.34	m	24.37
38 mm dia.	8.09	9.58	0.73	19.78	m	29.36
50 mm dia.	12.18	14.41	1.04	28.19	m	42.60
High impact PVC; surface fixed on saddles to backgrounds; with standard pattern boxes and fittings; including all fixings and supports						
Light gauge						
16 mm dia.	0.64	0.76	0.27	7.32	m	8.08
20 mm dia.	0.87	1.03	0.28	7.59	m	8.61
25 mm dia.	1.42	1.68	0.33	8.94	m	10.63
32 mm dia.	1.87	2.22	0.38	10.30	m	12.51
38 mm dia.	2.41	2.85	0.44	11.92	m	14.77
50 mm dia.	2.87	3.39	0.48	13.01	m	16.40
Heavy gauge						
16 mm dia.	1.05	1.24	0.27	7.32	m	8.56
20 mm dia.	1.20	1.42	0.28	7.59	m	9.01
25 mm dia.	1.64	1.94	0.33	8.94	m	10.88
32 mm dia.	2.61	3.09	0.38	10.30	m	13.39
38 mm dia.	3.45	4.08	0.44	11.92	m	16.00
50 mm dia.	5.69	6.74	0.48	13.01	m	19.75
Flexible conduits; including adaptors and lock nuts (for connections to equipment)						
Metallic, PVC covered conduit; not exceeding 1m long; including zinc plated mild steel adaptors, lock nuts and earth conductor						
16 mm dia.	7.60	8.99	0.46	12.47	nr	21.46
20 mm dia.	8.10	9.59	0.42	11.38	nr	20.97
25 mm dia.	11.99	14.20	0.43	11.65	nr	25.85
32 mm dia.	18.53	21.93	0.51	13.82	nr	35.75
38 mm dia.	23.34	27.62	0.56	15.18	nr	42.80
50 mm dia.	79.80	94.44	0.82	22.22	nr	116.66

V: ELECTRICAL SUPPLY/POWER/LIGHTING SYSTEMS

Item	Net Price £	Material £	Labour hours	Labour £	Unit	Total rate £
PVC conduit; not exceeding 1m long; including nylon adaptors, lock nuts						
16 mm dia.	4.57	5.41	0.46	12.47	nr	17.87
20 mm dia.	4.57	5.41	0.48	13.01	nr	18.42
25 mm dia.	5.74	6.79	0.50	13.55	nr	20.34
32 mm.dia.	8.31	9.84	0.58	15.72	nr	25.56
PVC adaptable boxes; fixed to backgrounds; including all supports and fixings (cutting and connecting conduit to boxes is included)						
Square pattern						
75 x 75 x 53 mm	3.15	3.72	0.69	18.70	nr	22.42
100 x 100 x 75 mm	4.07	4.81	0.71	19.24	nr	24.05
150 x 150 x 75 mm	7.29	8.62	0.80	21.68	nr	30.30
Terminal strips to be fixed in metal or polythene adaptable boxes)						
20 Amp high density polythene						
2 way	1.01	1.19	0.23	6.23	nr	7.42
3 way	1.12	1.33	0.23	6.23	nr	7.56
4 way	1.20	1.42	0.23	6.23	nr	7.65
5 way	1.28	1.52	0.23	6.23	nr	7.75
6 way	1.38	1.63	0.25	6.78	nr	8.41
7 way	1.48	1.75	0.25	6.78	nr	8.52
8 way	1.59	1.89	0.29	7.86	nr	9.75
9 way	1.72	2.04	0.30	8.13	nr	10.17
10 way	1.97	2.33	0.34	9.21	nr	11.54
11 way	2.05	2.43	0.34	9.21	nr	11.65
12 way	2.23	2.63	0.34	9.21	nr	11.85
13 way	2.47	2.93	0.37	10.03	nr	12.95
14 way	2.78	3.29	0.37	10.03	nr	13.32
15 way	2.86	3.38	0.39	10.57	nr	13.95
16 way	2.92	3.46	0.45	12.20	nr	15.65
18 way	3.16	3.74	0.45	12.20	nr	15.93
TRUNKING						
Galvanised steel trunking; fixed to backgrounds; jointed with standard connectors (including plates for air gap between trunking and background); earth continuity straps included						
Single compartment						
50 x 50 mm	6.05	7.16	0.39	10.57	m	17.73
75 x 50 mm	8.18	9.68	0.44	11.92	m	21.60
75 x 75 mm	8.28	9.80	0.47	12.74	m	22.54
100 x 50 mm	9.69	11.47	0.50	13.55	m	25.02
100 x 75 mm	11.07	13.10	0.57	15.45	m	28.55
100 x 100 mm	9.91	11.73	0.62	16.80	m	28.53
150 x 50 mm	12.78	15.12	0.78	21.14	m	36.26
150 x 100 mm	9.91	11.73	0.78	21.14	m	32.87
150 x 150 mm	10.63	12.58	0.86	23.31	m	35.89
225 x 75 mm	10.88	12.88	0.88	23.85	m	36.73
225 x 150 mm	10.99	13.00	0.84	22.77	m	35.77
225 x 225 mm	12.38	14.65	0.99	26.83	m	41.48

V: ELECTRICAL SUPPLY/POWER/LIGHTING SYSTEMS

Item	Net Price £	Material £	Labour hours	Labour £	Unit	Total rate £
V20 : LV DISTRIBUTION (cont'd)						
Y60 : CONDUIT AND CABLE TRUNKING (cont'd)						
TRUNKING (cont'd)						
Galvanized steel trunking (cont'd)						
Single compartment (cont'd)						
300 x 75 mm	12.40	14.68	0.96	26.02	m	**40.69**
300 x 100 mm	14.22	16.83	0.99	26.83	m	**43.66**
300 x 150 mm	17.63	20.86	0.99	26.83	m	**47.69**
300 x 225 mm	23.22	27.48	1.09	29.54	m	**57.02**
300 x 300 mm	24.19	28.63	1.16	31.44	m	**60.07**
Double compartment						
50 x 50 mm	9.23	10.92	0.41	11.11	m	**22.03**
75 x 50 mm	8.40	9.94	0.47	12.74	m	**22.68**
75 x 75 mm	9.01	10.66	0.50	13.55	m	**24.21**
100 x 50 mm	9.22	10.91	0.54	14.63	m	**25.55**
100 x 75 mm	9.31	11.02	0.62	16.80	m	**27.83**
100 x 100 mm	10.49	12.41	0.66	17.89	m	**30.30**
150 x 50 mm	10.51	12.44	0.70	18.97	m	**31.41**
150 x 100 mm	12.05	14.27	0.83	22.49	m	**36.76**
150 x 150 mm	14.94	17.68	0.92	24.93	m	**42.62**
Triple compartment						
75 x 50 mm	10.16	12.02	0.54	14.63	m	**26.66**
75 x 75 mm	11.51	13.62	0.58	15.72	m	**29.33**
100 x 50 mm	11.42	13.52	0.61	16.53	m	**30.05**
100 x 75 mm	19.89	23.54	0.70	18.97	m	**42.52**
100 x 100 mm	13.03	15.42	0.74	20.06	m	**35.47**
150 x 50 mm	13.18	15.60	0.79	21.41	m	**37.01**
150 x 100 mm	13.68	16.19	0.78	21.14	m	**37.32**
150 x 150 mm	14.97	17.72	1.01	27.37	m	**45.09**
Four compartment						
100 x 50 mm	12.48	14.77	0.64	17.34	m	**32.12**
100 x 75 mm	13.82	16.35	0.72	19.51	m	**35.87**
100 x 100 mm	15.42	18.25	0.77	20.87	m	**39.12**
150 x 50 mm	14.21	16.81	0.82	22.22	m	**39.04**
150 x 100 mm	15.97	18.89	0.95	25.75	m	**44.64**
150 x 150 mm	22.93	27.13	1.04	28.19	m	**55.32**
Galvanised steel trunking fittings; cutting and jointing trunking to fittings is included						
Stop end						
50 x 50 mm	0.77	0.91	0.19	5.15	nr	**6.06**
75 x 50 mm	1.03	1.22	0.20	5.42	nr	**6.64**
75 x 75 mm	1.03	1.22	0.21	5.69	nr	**6.91**
100 x 50 mm	1.06	1.26	0.31	8.40	nr	**9.66**
100 x 75 mm	1.13	1.34	0.27	7.32	nr	**8.66**
100 x 100 mm	1.13	1.34	0.27	7.32	nr	**8.66**
150 x 50 mm	1.25	1.48	0.28	7.59	nr	**9.07**
150 x 100 mm	1.37	1.62	0.30	8.13	nr	**9.75**
150 x 150 mm	1.40	1.66	0.32	8.67	nr	**10.33**

V: ELECTRICAL SUPPLY/POWER/LIGHTING SYSTEMS

Item	Net Price £	Material £	Labour hours	Labour £	Unit	Total rate £
225 x 75 mm	1.45	1.72	0.35	9.49	nr	11.20
225 x 150 mm	2.09	2.47	0.37	10.03	nr	12.50
225 x 225 mm	2.51	2.97	0.38	10.30	nr	13.27
300 x 75 mm	2.09	2.47	0.42	11.38	nr	13.85
300 x 100 mm	2.51	2.97	0.42	11.38	nr	14.36
300 x 150 mm	2.62	3.10	0.43	11.65	nr	14.75
300 x 225 mm	2.89	3.42	0.45	12.20	nr	15.62
300 x 300 mm	3.15	3.73	0.48	13.01	nr	16.74
Flanged connector						
50 x 50 mm	1.03	1.22	0.19	5.15	nr	6.36
75 x 50 mm	1.56	1.84	0.20	5.42	nr	7.26
75 x 75 mm	1.52	1.80	0.21	5.69	nr	7.49
100 x 50 mm	1.64	1.94	0.26	7.05	nr	8.99
100 x 75 mm	1.68	1.98	0.27	7.32	nr	9.30
100 x 100 mm	1.64	1.94	0.27	7.32	nr	9.26
150 x 50 mm	1.64	1.94	0.28	7.59	nr	9.53
150 x 100 mm	1.75	2.07	0.30	8.13	nr	10.20
150 x 150 mm	1.79	2.12	0.32	8.67	nr	10.80
225 x 75 mm	1.45	1.72	0.35	9.49	nr	11.20
225 x 150 mm	2.09	2.47	0.37	10.03	nr	12.50
225 x 225 mm	2.51	2.97	0.38	10.30	nr	13.27
300 x 75 mm	2.09	2.47	0.42	11.38	nr	13.85
300 x 100 mm	2.51	2.97	0.42	11.38	nr	14.36
300 x 150 mm	2.63	3.11	0.43	11.65	nr	14.77
300 x 225 mm	2.89	3.42	0.45	12.20	nr	15.62
300 x 300 mm	3.15	3.73	0.48	13.01	nr	16.74
Bends 90°; single compartment						
50 x 50 mm	4.31	5.10	0.42	11.38	nr	16.48
75 x 50 mm	5.40	6.40	0.45	12.20	nr	18.59
75 x 75 mm	5.22	6.17	0.48	13.01	nr	19.18
100 x 50 mm	6.20	7.33	0.53	14.36	nr	21.70
100 x 75 mm	5.89	6.97	0.56	15.18	nr	22.15
100 x 100 mm	5.72	6.77	0.58	15.72	nr	22.49
150 x 50 mm	7.01	8.30	0.64	17.34	nr	25.64
150 x 100 mm	9.55	11.30	0.91	24.66	nr	35.96
150 x 150 mm	8.45	10.00	0.89	24.12	nr	34.12
225 x 75 mm	12.78	15.12	0.76	20.60	nr	35.72
225 x 150 mm	16.39	19.40	0.82	22.22	nr	41.62
225 x 225 mm	19.05	22.54	0.83	22.49	nr	45.03
300 x 75 mm	17.59	20.82	0.85	23.04	nr	43.86
300 x 100 mm	17.98	21.28	0.90	24.39	nr	45.67
300 x 150 mm	19.01	22.50	0.96	26.02	nr	48.51
300 x 225 mm	20.48	24.24	0.98	26.56	nr	50.80
300 x 300 mm	21.97	26.00	1.06	28.73	nr	54.73
Bends 90°; double compartment						
50 x 50 mm	4.91	5.81	0.42	11.38	nr	17.19
75 x 50 mm	5.89	6.97	0.45	12.20	nr	19.16
75 x 75 mm	6.71	7.95	0.49	13.28	nr	21.23
100 x 50 mm	5.89	6.97	0.53	14.36	nr	21.33
100 x 75 mm	7.78	9.20	0.56	15.18	nr	24.38
100 x 100 mm	7.87	9.31	0.58	15.72	nr	25.03
150 x 50 mm	9.19	10.88	0.65	17.62	nr	28.49
150 x 100 mm	9.26	10.96	0.69	18.70	nr	29.66
150 x 150 mm	14.63	17.32	0.73	19.78	nr	37.10

V: ELECTRICAL SUPPLY/POWER/LIGHTING SYSTEMS

Item	Net Price £	Material £	Labour hours	Labour £	Unit	Total rate £
V20 : LV DISTRIBUTION (cont'd)						
Y60 : CONDUIT AND CABLE TRUNKING (cont'd)						
TRUNKING (cont'd)						
Galvanized steel trunking fittings (cont'd)						
Bends 90°; triple compartment						
75 x 50 mm	5.25	6.21	0.47	12.74	nr	18.95
75 x 75 mm	6.39	7.56	0.51	13.82	nr	21.38
100 x 50 mm	5.71	6.76	0.56	15.18	nr	21.94
100 x 75 mm	7.65	9.05	0.59	15.99	nr	25.04
100 x 100 mm	8.54	10.10	0.61	16.53	nr	26.64
150 x 50 mm	9.18	10.86	0.68	18.43	nr	29.29
150 x 100 mm	10.18	12.04	0.73	19.78	nr	31.83
150 x 150 mm	14.69	17.38	0.77	20.87	nr	38.25
Bends 90°; four compartment						
100 x 50 mm	7.84	9.27	0.56	15.18	nr	24.45
100 x 75 mm	10.70	12.67	0.59	15.99	nr	28.66
100 x 100 mm	17.95	21.24	0.61	16.53	nr	37.77
150 x 50 mm	18.59	21.99	0.69	18.70	nr	40.69
150 x 100 mm	19.49	23.07	0.73	19.78	nr	42.85
150 x 150 mm	20.93	24.77	0.77	20.87	nr	45.64
Tees; single compartment						
50 x 50 mm	5.04	5.96	0.56	15.18	nr	21.14
75 x 50 mm	5.98	7.08	0.57	15.45	nr	22.53
75 x 75 mm	5.81	6.87	0.60	16.26	nr	23.13
100 x 50 mm	7.02	8.31	0.65	17.62	nr	25.93
100 x 75 mm	7.06	8.35	0.71	19.24	nr	27.59
100 x 100 mm	6.83	8.09	0.72	19.51	nr	27.60
150 x 50 mm	9.95	11.77	0.82	22.22	nr	34.00
150 x 100 mm	10.05	11.90	0.84	22.77	nr	34.66
150 x 150 mm	9.20	10.89	0.91	24.66	nr	35.56
225 x 75 mm	17.22	20.37	0.94	25.48	nr	45.85
225 x 150 mm	23.26	27.52	1.01	27.37	nr	54.90
225 x 225 mm	26.87	31.80	1.02	27.64	nr	59.44
300 x 75 mm	24.85	29.41	1.07	29.00	nr	58.41
300 x 100 mm	26.00	30.76	1.07	29.00	nr	59.76
300 x 150 mm	27.06	32.02	1.14	30.90	nr	62.92
300 x 225 mm	29.41	34.80	1.19	32.25	nr	67.05
300 x 300mm	31.77	37.59	1.26	34.15	nr	71.74
Tees; double compartment						
50 x 50 mm	6.66	7.88	0.56	15.18	nr	23.05
75 x 50 mm	7.93	9.38	0.57	15.45	nr	24.83
75 x 75 mm	9.04	10.70	0.60	16.26	nr	26.96
100 x 50 mm	9.84	11.65	0.65	17.62	nr	29.26
100 x 75 mm	10.44	12.36	0.71	19.24	nr	31.60
100 x 100 mm	11.43	13.53	0.72	19.51	nr	33.05
150 x 50 mm	11.88	14.06	0.82	22.22	nr	36.29
150 x 100 mm	12.44	14.72	0.85	23.04	nr	37.76
150 x 150 mm	13.95	16.51	0.91	24.66	nr	41.17

Material Costs/Prices for Measured Works – Electrical Installations

V: ELECTRICAL SUPPLY/POWER/LIGHTING SYSTEMS

Item	Net Price £	Material £	Labour hours	Labour £	Unit	Total rate £
Tees; triple compartment						
75 x 50 mm	8.48	10.04	0.60	16.26	nr	26.30
75 x 75 mm	10.31	12.21	0.63	17.07	nr	29.28
100 x 50 mm	10.31	12.21	0.68	18.43	nr	30.63
100 x 75 mm	12.33	14.59	0.74	20.06	nr	34.65
100 x 100 mm	13.76	16.28	0.75	20.33	nr	36.61
150 x 50 mm	15.08	17.85	0.87	23.58	nr	41.43
150 x 100 mm	16.39	19.40	0.89	24.12	nr	43.52
150 x 150 mm	23.61	27.94	0.96	26.02	nr	53.96
Tees; four compartment						
100 x 50 mm	14.51	17.18	0.66	17.89	nr	35.06
100 x 75 mm	14.35	16.98	0.72	19.51	nr	36.49
100 x 100 mm	16.27	19.26	0.72	19.51	nr	38.77
150 x 50 mm	11.73	13.88	0.83	22.49	nr	36.38
150 x 100 mm	21.91	25.93	0.85	23.04	nr	48.97
150 x 150 mm	28.00	33.14	0.92	24.93	nr	58.07
Crossovers; single compartment						
50 x 50 mm	6.60	7.81	0.65	17.62	nr	25.42
75 x 50 mm	9.48	11.21	0.66	17.89	nr	29.10
75 x 75 mm	9.13	10.81	0.69	18.70	nr	29.51
100 x 50 mm	11.19	13.24	0.74	20.06	nr	33.29
100 x 75 mm	11.26	13.32	0.80	21.68	nr	35.00
100 x 100 mm	10.84	12.83	0.81	21.95	nr	34.79
150 x 50 mm	13.07	15.47	0.91	24.66	nr	40.14
150 x 100 mm	13.55	16.03	0.94	25.48	nr	41.51
150 x 150 mm	13.07	15.47	0.99	26.83	nr	42.30
225 x 75 mm	23.54	27.86	1.01	27.37	nr	55.23
225 x 150 mm	30.88	36.55	1.08	29.27	nr	65.82
225 x 225 mm	35.28	41.76	1.09	29.54	nr	71.30
300 x 75 mm	33.23	39.33	1.14	30.90	nr	70.22
300 x 100 mm	34.22	40.50	1.16	31.44	nr	71.94
300 x 150 mm	34.22	40.50	1.19	32.25	nr	72.75
300 x 225 mm	38.24	45.26	1.21	32.79	nr	78.05
300 x 300mm	41.29	48.86	1.29	34.96	nr	83.82
Crossovers; double compartment						
50 x 50 mm	7.61	9.01	0.66	17.89	nr	26.89
75 x 50 mm	9.05	10.71	0.66	17.89	nr	28.60
75 x 75 mm	10.29	12.18	0.70	18.97	nr	31.15
100 x 50 mm	11.99	14.19	0.74	20.06	nr	34.24
100 x 75 mm	11.87	14.05	0.80	21.68	nr	35.73
100 x 100 mm	12.99	15.38	0.81	21.95	nr	37.33
150 x 50 mm	14.08	16.66	0.86	23.31	nr	39.97
150 x 100 mm	14.11	16.70	0.94	25.48	nr	42.18
150 x 150 mm	22.10	26.16	1.00	27.10	nr	53.26
Crossovers; triple compartment						
75 x 50 mm	9.05	10.71	0.70	18.97	nr	29.68
75 x 75 mm	10.29	12.18	0.73	19.78	nr	31.96
100 x 50 mm	11.52	13.63	0.79	21.41	nr	35.04
100 x 75 mm	13.74	16.26	0.85	23.04	nr	39.29
100 x 100 mm	15.28	18.08	0.85	23.04	nr	41.12
150 x 50 mm	14.51	17.18	0.97	26.29	nr	43.47
150 x 100 mm	15.98	18.91	0.99	26.83	nr	45.74
150 x 150 mm	26.15	30.95	1.06	28.73	nr	59.67

V: ELECTRICAL SUPPLY/POWER/LIGHTING SYSTEMS

Item	Net Price £	Material £	Labour hours	Labour £	Unit	Total rate £
V20 : LV DISTRIBUTION (cont'd)						
Y60 : CONDUIT AND CABLE TRUNKING (cont'd)						
TRUNKING (cont'd)						
Galvanized steel trunking fittings (cont'd)						
Crossovers; four compartment						
100 x 50 mm	14.47	17.12	0.79	21.41	nr	**38.53**
100 x 75 mm	16.26	19.24	0.85	23.04	nr	**42.28**
100 x 100 mm	18.35	21.72	0.86	23.31	nr	**45.02**
150 x 50 mm	18.53	21.93	0.97	26.29	nr	**48.21**
150 x 100 mm	20.36	24.09	1.00	27.10	nr	**51.19**
150 x 150 mm	31.54	37.33	1.06	28.73	nr	**66.06**
Galvanised steel flush floor trunking; fixed to backgrounds; supports and fixings; standard coupling joints; earth continuity straps included						
Triple compartment						
350 x 60mm	44.34	52.48	1.32	35.77	m	**88.25**
Four compartment						
350 x 60mm	46.17	54.64	1.32	35.77	m	**90.42**
Galvanized steel flush floor trunking; fittings (cutting and jointing trunking to fittings is included)						
Stop end; triple compartment						
350 x 60mm	4.65	5.50	0.53	14.36	nr	**19.87**
Stop end; four compartment						
350 x 60mm	4.65	5.50	0.53	14.36	nr	**19.87**
Rising bend; standard; triple compartment						
350 x 60mm	38.36	45.40	1.30	35.23	nr	**80.63**
Rising bend; standard; four compartment						
350 x 60mm	40.18	47.55	1.30	35.23	nr	**82.78**
Rising bend; skirting; triple compartment						
350 x 60mm	75.24	89.04	1.33	36.05	nr	**125.09**
Rising bend; skirting; four compartment						
350 x 60mm	85.40	101.06	1.33	36.05	nr	**137.11**
Junction box; triple compartment						
350 x 60mm	50.01	59.18	1.16	31.44	nr	**90.62**
Junction box; four compartment						
350 x 60mm	52.01	61.56	1.16	31.44	nr	**93.00**
Body coupler (pair)						
3 and 4 Compartment	2.58	3.06	0.16	4.34	nr	**7.39**

V: ELECTRICAL SUPPLY/POWER/LIGHTING SYSTEMS

Item	Net Price £	Material £	Labour hours	Labour £	Unit	Total rate £
Service outlet module comprising flat lid with flanged carpet trim; twin 13 A outlet and drilled plate for mounting 2 telephone outlets; one blank plate; triple compartment						
3 Compartment	62.03	73.41	0.47	12.74	nr	86.15
Service outlet module comprising flat lid with flanged carpet trim; twin 13 A outlet and drilled plate for mounting 2 telephone outlets; two blank plates; four compartment						
4 Compartment	69.09	81.76	0.47	12.74	nr	94.50
Single compartment PVC trunking; grey finish; clip on lid; fixed to backgrounds; including supports and fixings (standard coupling joints)						
50 x 50mm	9.94	11.77	0.27	7.32	m	19.09
75 x 50mm	10.78	12.76	0.28	7.59	m	20.35
75 x 75mm	12.22	14.46	0.29	7.86	m	22.32
100 x 50mm	13.83	16.37	0.34	9.21	m	25.58
100 x 75mm	15.17	17.96	0.37	10.03	m	27.98
100 x 100mm	16.03	18.98	0.37	10.03	m	29.00
150 x 50mm	13.90	16.45	0.41	11.11	m	27.56
150 x 75mm	24.78	29.33	0.44	11.92	m	41.25
150 x 100mm	29.83	35.30	0.44	11.92	m	47.23
150 x 150mm	30.52	36.12	0.48	13.01	m	49.13
Single compartment PVC trunking; fittings (cutting and jointing trunking to fittings is included)						
Crossover						
50 x 50mm	16.29	19.27	0.29	7.86	nr	27.13
75 x 50mm	18.07	21.39	0.30	8.13	nr	29.52
75 x 75mm	19.61	23.21	0.31	8.40	nr	31.61
100 x 50mm	26.14	30.94	0.35	9.49	nr	40.43
100 x 75mm	32.96	39.01	0.36	9.76	nr	48.76
100 x 100mm	29.15	34.50	0.40	10.84	nr	45.34
150 x 75mm	37.35	44.20	0.45	12.20	nr	56.40
150 x 100mm	44.81	53.04	0.46	12.47	nr	65.50
150 x 150mm	69.56	82.33	0.47	12.74	nr	95.06
Stop end						
50 x 50mm	0.68	0.81	0.12	3.25	nr	4.06
75 x 50mm	0.97	1.14	0.12	3.25	nr	4.40
75 x 75mm	1.27	1.50	0.13	3.52	nr	5.03
100 x 50mm	1.74	2.06	0.16	4.34	nr	6.40
100 x 75mm	2.72	3.22	0.16	4.34	nr	7.55
100 x 100mm	2.73	3.23	0.18	4.88	nr	8.11
150 x 75mm	7.09	8.39	0.20	5.42	nr	13.81
150 x 100mm	8.78	10.39	0.21	5.69	nr	16.08
150 x 150mm	8.95	10.59	0.22	5.96	nr	16.55

V: ELECTRICAL SUPPLY/POWER/LIGHTING SYSTEMS

Item	Net Price £	Material £	Labour hours	Labour £	Unit	Total rate £
V20 : LV DISTRIBUTION (cont'd)						
Y60 : CONDUIT AND CABLE TRUNKING (cont'd)						
TRUNKING (cont'd)						
Single compartment PVC trunking (cont'd)						
Flanged coupling						
50 x 50mm	4.09	4.84	0.32	8.67	nr	13.51
75 x 50mm	4.67	5.52	0.33	8.94	nr	14.46
75 x 75mm	5.60	6.62	0.34	9.21	nr	15.84
100 x 50mm	6.31	7.47	0.44	11.92	nr	19.40
100 x 75mm	7.15	8.46	0.45	12.20	nr	20.65
100 x 100mm	7.62	9.02	0.46	12.47	nr	21.48
150 x 75mm	8.06	9.54	0.57	15.45	nr	24.98
150 x 100mm	8.49	10.04	0.57	15.45	nr	25.49
150 x 150mm	8.90	10.54	0.59	15.99	nr	26.53
Internal coupling						
50 x 50mm	1.44	1.71	0.07	1.90	nr	3.61
75 x 50mm	1.71	2.03	0.07	1.90	nr	3.92
75 x 75mm	1.70	2.01	0.07	1.90	nr	3.91
100 x 50mm	2.29	2.71	0.08	2.17	nr	4.88
100 x 75mm	2.58	3.05	0.08	2.17	nr	5.22
100 x 100mm	2.84	3.36	0.08	2.17	nr	5.52
External coupling						
50 x 50mm	1.58	1.87	0.09	2.44	nr	4.31
75 x 50mm	1.88	2.23	0.09	2.44	nr	4.67
75 x 75mm	1.87	2.22	0.09	2.44	nr	4.66
100 x 50mm	2.51	2.98	0.10	2.71	nr	5.69
100 x 75mm	2.86	3.38	0.10	2.71	nr	6.09
100 x 100mm	3.12	3.70	0.10	2.71	nr	6.41
150 x 75mm	3.57	4.23	0.11	2.98	nr	7.21
150 x 100mm	3.72	4.41	0.11	2.98	nr	7.39
150 x 150mm	3.85	4.56	0.11	2.98	nr	7.54
Angle; flat cover						
50 x 50mm	5.78	6.84	0.18	4.88	nr	11.72
75 x 50mm	7.71	9.13	0.19	5.15	nr	14.28
75 x 75mm	8.87	10.50	0.20	5.42	nr	15.92
100 x 50mm	13.24	15.66	0.23	6.23	nr	21.90
100 x 75mm	20.16	23.86	0.26	7.05	nr	30.90
100 x 100mm	18.38	21.76	0.26	7.05	nr	28.80
150 x 75mm	25.07	29.67	0.30	8.13	nr	37.80
150 x 100mm	29.81	35.28	0.33	8.94	nr	44.22
150 x 150mm	45.18	53.46	0.34	9.21	nr	62.68
Angle; internal or external cover						
50 x 50mm	7.03	8.32	0.18	4.88	nr	13.20
75 x 50mm	9.70	11.49	0.19	5.15	nr	16.63
75 x 75mm	12.45	14.74	0.20	5.42	nr	20.16
100 x 50mm	13.41	15.87	0.23	6.23	nr	22.10
100 x 75mm	21.58	25.54	0.26	7.05	nr	32.59
100 x 100mm	21.67	25.64	0.26	7.05	nr	32.69
150 x 75mm	26.49	31.35	0.30	8.13	nr	39.48
150 x 100mm	31.23	36.96	0.33	8.94	nr	45.91
150 x 150mm	43.86	51.91	0.34	9.21	nr	61.12

V: ELECTRICAL SUPPLY/POWER/LIGHTING SYSTEMS

Item	Net Price £	Material £	Labour hours	Labour £	Unit	Total rate £
Tee; flat cover						
50 x 50mm	4.84	5.72	0.24	6.50	nr	12.23
75 x 50mm	7.38	8.74	0.25	6.78	nr	15.51
75 x 75mm	7.60	8.99	0.26	7.05	nr	16.04
100 x 50mm	16.39	19.40	0.32	8.67	nr	28.07
100 x 75mm	17.42	20.62	0.33	8.94	nr	29.56
100 x 100mm	22.73	26.90	0.34	9.21	nr	36.11
150 x 75mm	30.06	35.57	0.41	11.11	nr	46.68
150 x 100mm	38.57	45.65	0.42	11.38	nr	57.03
150 x 150mm	52.45	62.07	0.44	11.92	nr	74.00
Tee; internal or external cover						
50 x 50mm	13.72	16.23	0.24	6.50	nr	22.74
75 x 50mm	15.08	17.84	0.25	6.78	nr	24.62
75 x 75mm	16.75	19.82	0.26	7.05	nr	26.86
100 x 50mm	21.51	25.45	0.32	8.67	nr	34.13
100 x 75mm	24.58	29.09	0.33	8.94	nr	38.03
100 x 100mm	27.58	32.65	0.34	9.21	nr	41.86
150 x 75mm	35.73	42.28	0.41	11.11	nr	53.39
150 x 100mm	43.05	50.94	0.42	11.38	nr	62.33
150 x 150mm	57.47	68.01	0.44	11.92	nr	79.94
Division Strip (1.8m long)						
50mm	7.44	8.80	0.07	1.90	nr	10.70
75mm	9.52	11.27	0.07	1.90	nr	13.17
100mm	12.12	14.35	0.08	2.17	nr	16.52
PVC miniature trunking; white finish; fixed to backgrounds; including supports and fixing; standard coupling joints						
Single compartment						
16 x 16mm	1.34	1.58	0.20	5.42	m	7.00
25 x 16mm	1.64	1.94	0.21	5.69	m	7.63
38 x 16mm	2.05	2.43	0.24	6.50	m	8.94
38 x 25mm	2.44	2.89	0.25	6.78	m	9.66
Compartmented						
38 x 16mm	2.39	2.82	0.24	6.50	m	9.33
38 x 25mm	2.86	3.38	0.25	6.78	m	10.16
PVC miniature trunking fittings; single compartment; white finish; cutting and jointing trunking to fittings is included						
Coupling						
16 x 16mm	0.41	0.48	0.10	2.71	nr	3.19
25 x 16mm	0.41	0.48	0.10	2.71	nr	3.19
38 x 16mm	0.41	0.48	0.12	3.25	nr	3.73
38 x 25mm	0.98	1.17	0.14	3.79	nr	4.96
Stop end						
16 x 16mm	0.41	0.48	0.12	3.25	nr	3.73
25 x 16mm	0.41	0.48	0.12	3.25	nr	3.73
38 x 16mm	0.41	0.48	0.15	4.07	nr	4.55
38 x 25mm	0.98	1.17	0.17	4.61	nr	5.77

V: ELECTRICAL SUPPLY/POWER/LIGHTING SYSTEMS

Item	Net Price £	Material £	Labour hours	Labour £	Unit	Total rate £
V20 : LV DISTRIBUTION (cont'd)						
Y60 : CONDUIT AND CABLE TRUNKING (cont'd)						
TRUNKING (cont'd)						
PVC miniature trunking fittings (cont'd)						
Bend; flat, internal or external						
16 x 16mm	0.41	0.48	0.18	4.88	nr	5.36
25 x 16mm	0.41	0.48	0.18	4.88	nr	5.36
38 x 16mm	0.41	0.48	0.21	5.69	nr	6.17
38 x 25mm	0.98	1.17	0.23	6.23	nr	7.40
Tee						
16 x 16mm	0.70	0.82	0.23	6.23	nr	7.06
25 x 16mm	0.70	0.82	0.19	5.15	nr	5.97
38 x 16mm	0.70	0.82	0.26	7.05	nr	7.87
38 x 25mm	0.97	1.15	0.29	7.86	nr	9.01
PVC bench trunking; white or grey finish; fixed to backgrounds; including supports and fixings; standard coupling joints						
Trunking						
90 x 90mm	22.53	26.67	0.33	8.94	m	35.61
PVC bench trunking fittings; white or grey finish; cutting and jointing trunking to fittings is included						
Stop end						
90 x 90mm	4.77	5.65	0.09	2.44	nr	8.09
Coupling						
90 x 90mm	2.95	3.50	0.09	2.44	nr	5.93
Internal or external bend						
90 x 90mm	16.20	19.17	0.28	7.59	nr	26.76
Socket plate						
90 x 90mm - 1 gang	0.91	1.08	0.10	2.71	nr	3.79
90 x 90mm - 2 gang	1.10	1.30	0.10	2.71	nr	4.01
PVC underfloor trunking; single compartment; fitted in floor screed; standard coupling joints						
Trunking						
60 x 25mm	11.06	13.09	0.22	5.96	m	19.06
90 x 35mm	15.91	18.83	0.27	7.32	m	26.15
PVC underfloor trunking fittings; single compartment; fitted in floor screed; (cutting and jointing trunking to fittings is included)						
Jointing sleeve						
60 x 25mm	0.74	0.87	0.08	2.17	nr	3.04
90 x 35mm	1.22	1.44	0.10	2.71	nr	4.15

V: ELECTRICAL SUPPLY/POWER/LIGHTING SYSTEMS

Item	Net Price £	Material £	Labour hours	Labour £	Unit	Total rate £
Duct connector 90 x 35mm	0.57	0.67	0.17	4.61	nr	**5.28**
Socket reducer 90 x 35mm	1.02	1.20	0.12	3.25	nr	**4.46**
Vertical access box; 2 compartment Shallow	57.16	67.65	0.37	10.03	nr	**77.67**
Duct bend; vertical 60 x 25mm 90 x 35mm	11.15 12.59	13.20 14.90	0.27 0.35	7.32 9.49	nr nr	**20.51** **24.39**
Duct bend; horizontal 60 x 25mm 90 x 35mm	13.15 13.34	15.56 15.79	0.30 0.37	8.13 10.03	nr nr	**23.69** **25.82**
Zinc coated steel underfloor ducting; fixed to backgrounds; standard coupling joints; earth continuity straps; (Including supports and fixing, packing shims where required)						
Double compartment 150 x 25mm	9.65	11.42	0.57	15.45	m	**26.87**
Triple compartment 225 x 25mm	17.12	20.26	0.93	25.20	m	**45.47**
Zinc coated steel underfloor ducting fittings; (cutting and jointing to fittings is included)						
Stop end; double compartment 150 x 25mm	3.04	3.60	0.31	8.40	nr	**12.00**
Stop end; triple compartment 225 x 25mm	3.47	4.10	0.37	10.03	nr	**14.13**
Rising bend; double compartment; standard trunking 150 x 25mm	19.55	23.14	0.71	19.24	nr	**42.38**
Rising bend; triple compartment; standard trunking 225 x 25mm	35.14	41.59	0.85	23.04	nr	**64.62**
Rising bend; double compartment; to skirting 150 x 25	43.53	51.51	0.90	24.39	nr	**75.91**
Rising bend; triple compartment; to skirting 225 x 25	51.33	60.74	0.95	25.75	nr	**86.49**
Horizontal bend; double compartment 150 x 25mm	29.72	35.18	0.64	17.34	nr	**52.52**
Horizontal bend; triple compartment 225 x 25mm	34.28	40.57	0.77	20.87	nr	**61.44**
Junction or service outlet boxes; terminal; double compartment 150mm	30.75	36.39	0.91	24.66	nr	**61.06**

V: ELECTRICAL SUPPLY/POWER/LIGHTING SYSTEMS

Item	Net Price £	Material £	Labour hours	Labour £	Unit	Total rate £
V20 : LV DISTRIBUTION (cont'd)						
Y60 : CONDUIT AND CABLE TRUNKING (cont'd)						
TRUNKING (cont'd)						
Zinc coated steel underfloor ducting (cont'd)						
Junction or service outlet boxes; terminal; triple compartment						
225mm	35.16	41.61	1.11	30.08	nr	**71.69**
Junction or service outlet boxes; through or angle; double compartment						
150mm	40.85	48.35	0.97	26.29	nr	**74.64**
Junction or service outlet boxes; through or angle; triple compartment						
225mm	45.37	53.69	1.17	31.71	nr	**85.40**
Junction or service outlet boxes; tee; double compartment						
150mm	40.85	48.35	1.02	27.64	nr	**75.99**
Junction or service outlet boxes; tee; triple compartment						
225mm	45.37	53.69	1.22	33.06	nr	**86.76**
Junction or service outlet boxes; cross; double compartment						
up to 150mm	40.85	48.35	1.03	27.91	nr	**76.26**
Junction or service outlet boxes; cross; triple compartment						
225mm	45.37	53.69	1.23	33.33	nr	**87.03**
Plates for junction/inspection boxes; double and triple compartment						
Blank plate	8.93	10.57	0.92	24.93	nr	**35.51**
Conduit entry plate	11.19	13.25	0.86	23.31	nr	**36.55**
Trunking entry plate	11.19	13.25	0.86	23.31	nr	**36.55**
Service outlet box comprising flat lid with flanged carpet trim; twin 13A outlet and drilled plate for mounting 2 telephone outlets and terminal blocks; terminal outlet box; double compartment						
150 x 25mm trunking	63.70	75.38	1.68	45.53	nr	**120.91**
Service outlet box comprising flat lid with flanged carpet trim; twin 13A outlet and drilled plate for mounting 2 telephone outlets and terminal blocks; terminal outlet box; triple compartment						
225 x 25mm trunking	70.91	83.92	1.93	52.31	nr	**136.22**

V: ELECTRICAL SUPPLY/POWER/LIGHTING SYSTEMS

Item	Net Price £	Material £	Labour hours	Labour £	Unit	Total rate £
PVC skirting/dado modular trunking; white (cutting and jointing trunking to fittings and backplates for fixing to walls is included)						
Main carrier/backplate						
50 x 170mm	17.66	20.89	2.02	54.74	m	**75.64**
Extension carrier/backplate						
50 x 42mm	10.74	12.71	0.58	15.72	m	**28.43**
Carrier/backplate						
Including cover seal	6.26	7.41	0.53	14.36	m	**21.77**
Chamfered covers for fixing to backplates						
50 x 42mm	3.63	4.29	0.33	8.94	m	**13.24**
Square covers for fixing to backplates						
50 x 42 mm	7.30	8.64	0.33	8.94	m	**17.58**
Plain covers for fixing to backplates						
85 mm	3.63	4.29	0.34	9.21	m	**13.51**
Retainers-clip to backplates to hold cables						
For chamfered covers	1.01	1.19	0.07	1.90	m	**3.09**
For square-recessed covers	0.87	1.03	0.07	1.90	m	**2.92**
For plain covers	3.36	3.98	0.07	1.90	m	**5.87**
Prepackaged corner assemblies						
Internal ; for 170 x 50 Assy	7.05	8.35	0.51	13.82	nr	**22.17**
Internal; for 215 x 50 Assy	8.77	10.38	0.53	14.36	nr	**24.75**
Internal; for 254 x 50 Assy	10.44	12.36	0.53	14.36	nr	**26.72**
External; for 170 x 50 Assy	7.05	8.35	0.56	15.18	nr	**23.52**
External ; for 215 x 50 Assy	8.77	10.38	0.58	15.72	nr	**26.10**
External ; for 254 x 50 Assy	10.44	12.36	0.58	15.72	nr	**28.08**
Clip on end caps						
170 x 50 Assy	4.23	5.00	0.11	2.98	nr	**7.98**
215 x 50 Assy	5.00	5.91	0.11	2.98	nr	**8.89**
254 x 50 Assy	5.87	6.95	0.11	2.98	nr	**9.93**
Outlet box						
1 Gang; in horizontal trunking; clip in	3.77	4.46	0.34	9.21	nr	**13.67**
2 Gang; in horizontal trunking; clip in	4.71	5.57	0.34	9.21	nr	**14.79**
1 Gang; in vertical trunking; clip in	3.77	4.46	0.34	9.21	nr	**13.67**
Sheet steel adaptable boxes; with plain or knockout sides; fixed to backgrounds; including supports and fixings (cutting and connecting conduit to boxes is included)						
Square pattern - black						
75 x 75 x 37 mm	2.29	2.71	0.69	18.70	nr	**21.41**
75 x 75 x 50 mm	2.18	2.58	0.69	18.70	nr	**21.28**
75 x 75 x 75 mm	2.49	2.95	0.69	18.70	nr	**21.65**
100 x 100 x 50 mm	2.29	2.71	0.71	19.24	nr	**21.95**
150 x 150 x 50 mm	3.48	4.12	0.79	21.41	nr	**25.53**
150 x 150 x 75 mm	4.09	4.85	0.80	21.68	nr	**26.53**
150 x 150 x 100 mm	5.45	6.45	0.80	21.68	nr	**28.13**

V: ELECTRICAL SUPPLY/POWER/LIGHTING SYSTEMS

Item	Net Price £	Material £	Labour hours	Labour £	Unit	Total rate £
V20 : LV DISTRIBUTION (cont'd)						
Y60 : CONDUIT AND CABLE TRUNKING (cont'd)						
TRUNKING (cont'd)						
Sheet steel adaptable boxes (cont'd)						
Square pattern – black (cont'd)						
225 x 225 x 50 mm	6.97	8.25	0.93	25.20	nr	33.46
225 x 225 x 100 mm	9.24	10.93	0.94	25.48	nr	36.41
300 x 300 x 100 mm	9.92	11.74	0.99	26.83	nr	38.58
Square pattern - galvanised						
75 x 75 x 37 mm	3.27	3.87	0.69	18.70	nr	22.57
75 x 75 x 50 mm	3.41	4.04	0.69	18.70	nr	22.74
75 x 75 x 75 mm	2.94	3.48	0.70	18.97	nr	22.45
100 x 100 x 50 mm	3.61	4.27	0.71	19.24	nr	23.52
150 x 150 x 50 mm	4.20	4.97	0.84	22.77	nr	27.74
150 x 150 x 75 mm	5.01	5.94	0.80	21.68	nr	27.62
150 x 150 x 100 mm	6.03	7.14	0.80	21.68	nr	28.82
225 x 225 x 50 mm	7.74	9.16	0.93	25.20	nr	34.37
225 x 225 x 100 mm	9.86	11.67	0.94	25.48	nr	37.15
300 x 300 x 100 mm	15.85	18.75	0.99	26.83	nr	45.59
Rectangular pattern - black						
100 x 75 x 50 mm	3.67	4.34	0.69	18.70	nr	23.04
150 x 75 x 50 mm	3.83	4.54	0.70	18.97	nr	23.51
150 x 75 x 75 mm	4.17	4.93	0.71	19.24	nr	24.17
150 x 100 x 75 mm	9.27	10.98	0.71	19.24	nr	30.22
225 x 75 x 50 mm	8.01	9.48	0.78	21.14	nr	30.62
225 x 150 x 75 mm	12.80	15.15	0.81	21.95	nr	37.10
225 x 150 x 100 mm	24.02	28.43	0.81	21.95	nr	50.38
300 x 150 x 50 mm	24.02	28.43	0.93	25.20	nr	53.64
300 x 150 x 75 mm	24.02	28.43	0.94	25.48	nr	53.91
300 x 150 x 100 mm	24.02	28.43	0.96	26.02	nr	54.45
Rectangular pattern - galvanised						
100 x 75 x 50 mm	5.52	6.54	0.69	18.70	nr	25.24
150 x 75 x 50 mm	8.18	9.68	0.70	18.97	nr	28.65
150 x 75 x 75 mm	10.03	11.87	0.71	19.24	nr	31.11
150 x 100 x 75 mm	16.71	19.77	0.71	19.24	nr	39.02
225 x 75 x 50 mm	15.07	17.83	0.89	24.12	nr	41.95
225 x 150 x 75 mm	21.16	25.04	0.81	21.95	nr	46.99
225 x 150 x 100 mm	39.93	47.26	0.81	21.95	nr	69.21
300 x 150 x 50 mm	39.93	47.26	0.93	25.20	nr	72.46
300 x 150 x 75 mm	39.93	47.26	0.94	25.48	nr	72.73
300 x 150 x 100 mm	39.93	47.26	0.96	26.02	nr	73.27

V: ELECTRICAL SUPPLY/POWER/LIGHTING SYSTEMS

Item	Net Price £	Material £	Labour hours	Labour £	Unit	Total rate £
Y61 - LV CABLES AND WIRING						
ARMOURED CABLE						
Cable; XLPE insulated; PVC sheathed; copper stranded conductors to BS 5467; laid in trench/duct including marker tape (Cable tiles measured elsewhere)						
600/1000 Volt grade; single core (aluminium wire armour)						
25 mm²	2.02	2.39	0.15	4.07	m	6.46
35 mm²	2.25	2.66	0.15	4.07	m	6.72
50 mm²	3.03	3.58	0.17	4.61	m	8.19
70 mm²	4.25	5.03	0.18	4.88	m	9.91
95 mm²	5.71	6.76	0.20	5.42	m	12.18
120 mm²	6.79	8.04	0.22	5.96	m	14.00
150 mm²	8.40	9.94	0.24	6.50	m	16.45
185 mm²	10.08	11.93	0.26	7.05	m	18.98
240 mm²	13.01	15.40	0.30	8.13	m	23.53
300 mm²	15.85	18.75	0.31	8.40	m	27.16
400 mm²	20.60	24.38	0.38	10.30	m	34.68
500 mm²	26.44	31.29	0.44	11.92	m	43.22
630 mm²	33.13	39.20	0.52	14.09	m	53.30
800 mm²	42.69	50.53	0.62	16.80	m	67.33
1000 mm²	49.43	58.50	0.65	17.62	m	76.12
600/1000 Volt grade; two core (galvanised steel wire armour)						
1.5 mm²	0.72	0.85	0.06	1.63	m	2.47
2.5 mm²	0.82	0.98	0.06	1.63	m	2.60
4 mm²	1.07	1.27	0.08	2.17	m	3.43
6 mm²	1.31	1.54	0.08	2.17	m	3.71
10 mm²	1.75	2.08	0.10	2.71	m	4.79
16 mm²	2.63	3.12	0.10	2.71	m	5.83
25 mm²	3.38	4.00	0.15	4.07	m	8.07
35 mm²	5.33	6.31	0.15	4.07	m	10.37
50 mm²	6.02	7.13	0.17	4.61	m	11.74
70 mm²	8.26	9.78	0.18	4.88	m	14.65
95 mm²	11.11	13.14	0.20	5.42	m	18.56
120 mm²	13.60	16.09	0.22	5.96	m	22.06
150 mm²	16.87	19.97	0.24	6.50	m	26.47
185 mm²	21.49	25.43	0.26	7.05	m	32.47
240 mm²	27.52	32.57	0.30	8.13	m	40.70
300 mm²	36.97	43.75	0.31	8.40	m	52.15
400 mm²	48.89	57.86	0.35	9.49	m	67.34
600/1000 Volt grade; three core (galvanised steel wire armour)						
1.5 mm²	0.79	0.94	0.07	1.90	m	2.83
2.5 mm²	1.00	1.18	0.07	1.90	m	3.07
4 mm²	1.23	1.46	0.09	2.44	m	3.90
6 mm²	1.59	1.89	0.10	2.71	m	4.60
10 mm²	2.43	2.87	0.11	2.98	m	5.86
16 mm²	3.40	4.03	0.11	2.98	m	7.01
25 mm²	4.60	5.45	0.16	4.34	m	9.78
35 mm²	6.40	7.57	0.16	4.34	m	11.91
50 mm²	8.35	9.88	0.19	5.15	m	15.03

V: ELECTRICAL SUPPLY/POWER/LIGHTING SYSTEMS

Item	Net Price £	Material £	Labour hours	Labour £	Unit	Total rate £
V20 : LV DISTRIBUTION (cont'd)						
Y61 - LV CABLES AND WIRING (cont'd)						
ARMOURED CABLE (cont'd)						
Cable; XLPE insulated; PVC sheathed (cont'd)						
600/1000 Volt grade; three core (cont'd)						
70 mm²	11.43	13.52	0.21	5.69	m	19.22
95 mm²	15.74	18.63	0.23	6.23	m	24.86
120 mm²	19.58	23.17	0.24	6.50	m	29.68
150 mm²	24.33	28.80	0.27	7.32	m	36.11
185 mm²	29.98	35.48	0.30	8.13	m	43.61
240 mm²	38.35	45.38	0.33	8.94	m	54.33
300 mm²	48.60	57.52	0.35	9.49	m	67.00
400 mm²	57.72	68.30	0.41	11.11	m	79.42
600/1000 Volt grade; four core (galvanised steel wire armour)						
1.5 mm²	0.93	1.10	0.08	2.17	m	3.27
2.5 mm²	1.16	1.37	0.09	2.44	m	3.81
4 mm²	1.44	1.71	0.10	2.71	m	4.42
6 mm²	2.12	2.51	0.10	2.71	m	5.22
10 mm²	2.96	3.51	0.12	3.25	m	6.76
16 mm²	4.28	5.07	0.12	3.25	m	8.32
25 mm²	6.18	7.32	0.18	4.88	m	12.20
35 mm²	8.05	9.52	0.19	5.15	m	14.67
50 mm²	10.67	12.63	0.21	5.69	m	18.32
70 mm²	15.18	17.97	0.23	6.23	m	24.20
95 mm²	20.33	24.06	0.26	7.05	m	31.11
120 mm²	25.53	30.21	0.28	7.59	m	37.80
150 mm²	30.90	36.57	0.32	8.67	m	45.24
185 mm²	37.93	44.89	0.35	9.49	m	54.38
240 mm²	49.04	58.04	0.36	9.76	m	67.79
300 mm²	61.52	72.81	0.40	10.84	m	83.65
400 mm²	117.36	138.89	0.45	12.20	m	151.08
600/1000 Volt grade; seven core (galvanised steel wire armour)						
1.5 mm²	1.33	1.57	0.10	2.71	m	4.28
2.5 mm²	1.85	2.19	0.10	2.71	m	4.90
4 mm²	3.37	3.99	0.11	2.98	m	6.97
600/1000 Volt grade; twelve core (galvanised steel wire armour)						
1.5 mm²	2.12	2.51	0.11	2.98	m	5.49
2.5 mm²	3.17	3.75	0.11	2.98	m	6.73
600/1000 Volt grade; nineteen core (galvanised steel wire armour)						
1.5 mm²	3.19	3.77	0.13	3.52	m	7.30
2.5 mm²	4.79	5.67	0.14	3.79	m	9.47
600/1000 Volt grade; twenty seven core (galvanised steel wire armour)						
1.5 mm²	4.39	5.19	0.14	3.79	m	8.99
2.5 mm²	6.12	7.24	0.16	4.34	m	11.58

V: ELECTRICAL SUPPLY/POWER/LIGHTING SYSTEMS

Item	Net Price £	Material £	Labour hours	Labour £	Unit	Total rate £
600/1000 Volt grade; thirty seven core (galvanised steel wire armour)						
1.5 mm²	5.47	6.47	0.15	4.07	m	10.54
2.5 mm²	7.74	9.16	0.17	4.61	m	13.76
Cable; XLPE insulated; PVC sheathed copper starnded conductors to BS 5467; clipped direct to backgrounds including cleat						
600/1000 Volt grade; single core (aluminium wire armour)						
25 mm²	3.05	3.61	0.35	9.49	m	13.09
35 mm²	3.32	3.93	0.36	9.76	m	13.68
50 mm²	3.36	3.98	0.37	10.03	m	14.00
70 mm²	4.43	5.24	0.39	10.57	m	15.81
95 mm²	5.94	7.03	0.42	11.38	m	18.41
120 mm²	7.02	8.31	0.47	12.74	m	21.04
150 mm²	8.63	10.22	0.51	13.82	m	24.04
185 mm²	10.29	12.18	0.59	15.99	m	28.17
240 mm²	13.48	15.96	0.68	18.43	m	34.38
300 mm²	16.31	19.30	0.74	20.06	m	39.35
400 mm²	23.80	28.16	0.88	23.85	m	52.01
500 mm²	29.99	35.49	0.88	23.85	m	59.34
630 mm²	39.00	46.16	1.05	28.46	m	74.61
800 mm²	48.49	57.39	1.33	36.05	m	93.43
1000 mm²	54.22	64.16	1.40	37.94	m	102.11
600/1000 Volt grade; two core (galvanised steel wire armour)						
1.5 mm²	0.90	1.06	0.20	5.42	m	6.48
2.5 mm²	1.03	1.22	0.20	5.42	m	6.64
4.0 mm²	1.31	1.54	0.21	5.69	m	7.24
6.0 mm²	1.53	1.81	0.22	5.96	m	7.77
10.0 mm²	1.98	2.34	0.24	6.50	m	8.85
16.0 mm²	2.86	3.38	0.25	6.78	m	10.16
25 mm²	2.48	2.94	0.35	9.49	m	12.42
35 mm²	5.55	6.57	0.36	9.76	m	16.33
50 mm²	6.26	7.41	0.37	10.03	m	17.44
70 mm²	8.56	10.13	0.39	10.57	m	20.70
95 mm²	11.18	13.23	0.42	11.38	m	24.62
120 mm²	13.73	16.25	0.47	12.74	m	28.98
150 mm²	17.00	20.12	0.51	13.82	m	33.94
185 mm²	21.66	25.63	0.59	15.99	m	41.62
240 mm²	26.94	31.89	0.68	18.43	m	50.31
300 mm²	37.73	44.65	0.74	20.06	m	64.71
400 mm²	46.97	55.59	0.88	23.85	m	79.44
600/1000 Volt grade; three core (galvanised steel wire armour)						
1.5 mm²	0.97	1.15	0.20	5.42	m	6.57
2.5 mm²	1.14	1.35	0.21	5.69	m	7.05
4.0 mm²	1.46	1.72	0.22	5.96	m	7.68
6.0 mm²	1.82	2.15	0.22	5.96	m	8.12
10.0 mm²	2.65	3.14	0.25	6.78	m	9.92
16.0 mm²	3.63	4.29	0.26	7.05	m	11.34
25 mm²	5.10	6.04	0.37	10.03	m	16.07
35 mm²	6.62	19.92	0.39	10.57	m	30.49
50 mm²	8.65	10.23	0.40	10.84	m	21.07

V: ELECTRICAL SUPPLY/POWER/LIGHTING SYSTEMS

Item	Net Price £	Material £	Labour hours	Labour £	Unit	Total rate £
V20 : LV DISTRIBUTION (cont'd)						
Y61 - LV CABLES AND WIRING (cont'd)						
ARMOURED CABLE (cont'd)						
Cable; XLPE insulated; PVC sheathed (cont'd)						
600/1000 Volt grade; three core (cont'd)						
70 mm^2	11.55	13.66	0.42	11.38	m	25.05
95 mm^2	15.87	18.78	0.45	12.20	m	30.98
120 mm^2	19.70	42.09	0.52	14.09	m	56.19
150 mm^2	24.49	28.99	0.55	14.91	m	43.89
185 mm^2	30.15	35.68	0.63	17.07	m	52.76
240 mm^2	40.05	47.40	0.71	19.24	m	66.64
300 mm^2	50.41	59.66	0.78	21.14	m	80.80
400 mm^2	59.85	70.82	0.87	23.58	m	94.40
600/1000 Volt grade; four core (galvanised steel wire armour)						
1.5 mm^2	1.10	1.30	0.21	5.69	m	7.00
2.5 mm^2	1.38	1.63	0.22	5.96	m	7.60
4.0 mm^2	1.68	1.99	0.22	5.96	m	7.95
6.0 mm^2	2.34	2.77	0.23	6.23	m	9.01
10.0 mm^2	3.20	3.79	0.26	7.05	m	10.83
16.0 mm^2	4.50	5.33	0.26	7.05	m	12.38
25 mm^2	6.41	7.59	0.39	10.57	m	18.15
35 mm^2	8.35	9.88	0.40	10.84	m	20.72
50 mm^2	10.96	12.97	0.41	11.11	m	24.08
70 mm^2	15.31	18.12	0.45	12.20	m	30.32
95 mm^2	20.50	24.26	0.50	13.55	m	37.81
120 mm^2	25.69	30.40	0.54	14.63	m	45.04
150 mm^2	32.43	38.38	0.60	16.26	m	54.64
185 mm^2	39.63	46.90	0.67	18.16	m	65.06
240 mm^2	51.32	60.73	0.75	20.33	m	81.06
300 mm^2	64.39	76.21	0.83	22.49	m	98.70
400 mm^2	120.74	142.89	0.91	24.66	m	167.55
600/1000 Volt grade; seven core (galvanised steel wire armour)						
1.5 mm^2	1.55	1.84	0.20	5.42	m	7.26
2.5 mm^2	2.13	2.52	0.20	5.42	m	7.94
4.0 mm^2	3.84	4.55	0.23	6.23	m	10.78
600/1000 Volt grade; twelve core (galvanised steel wire armour)						
1.5 mm^2	2.26	6.71	0.23	6.23	m	12.94
2.5 mm^2	3.39	4.01	0.24	6.50	m	10.52
600/1000 Volt grade; nineteen core (galvanised steel wire armour)						
1.5 mm^2	3.41	4.04	0.26	7.05	m	11.09
2.5 mm^2	5.02	5.94	0.28	7.59	m	13.53
600/1000 Volt grade; twenty seven core (galvanised steel wire armour)						
1.5 mm^2	4.61	5.46	0.29	7.86	m	13.32
2.5 mm^2	6.41	7.59	0.30	8.13	m	15.72

V: ELECTRICAL SUPPLY/POWER/LIGHTING SYSTEMS

Item	Net Price £	Material £	Labour hours	Labour £	Unit	Total rate £
600/1000 Volt grade; thirty seven core (galvanised steel wire armour)						
1.5 mm²	5.76	6.81	0.32	8.67	m	15.49
2.5 mm²	8.04	9.51	0.33	8.94	m	18.45
Cable termination; brass weatherproof gland with inner and outer seal, shroud, brass lock nut and earth ring (including drilling and cutting mild steel gland plate)						
600/1000 Volt grade; single core (aluminium wire armour)						
25 mm²	4.83	5.72	1.70	46.07	nr	51.79
35 mm²	4.83	5.72	1.79	48.51	nr	54.23
50 mm²	4.83	5.72	2.06	55.83	nr	61.55
70 mm²	5.82	6.89	2.12	57.46	nr	64.34
95 mm²	5.91	7.00	2.39	64.77	nr	71.77
120 mm²	6.14	7.27	2.47	66.94	nr	74.21
150 mm²	8.88	10.51	2.73	73.99	nr	84.49
185 mm²	5.42	6.41	3.05	82.66	nr	89.07
240 mm²	13.80	16.33	3.45	93.50	nr	109.83
300 mm²	13.95	16.51	3.84	104.07	nr	120.58
400 mm²	20.09	23.78	4.21	114.10	nr	137.87
500 mm²	22.30	26.39	5.70	154.48	m	180.87
630 mm²	25.13	29.74	6.20	168.03	m	197.77
800 mm²	42.86	50.72	7.50	203.26	m	253.99
1000 mm²	51.72	61.20	10.00	271.01	m	332.22
600/1000 Volt grade; two core (galvanised steel wire armour)						
1.5 mm²	4.71	5.58	0.58	15.72	nr	21.30
2.5 mm²	4.71	5.58	0.58	15.72	nr	21.30
4 mm²	4.74	5.61	0.58	15.72	nr	21.33
6 mm²	4.83	5.72	0.67	18.16	nr	23.87
10 mm²	5.15	6.10	1.00	27.10	nr	33.20
16 mm²	5.20	6.15	1.11	30.08	nr	36.23
25 mm²	5.91	7.00	1.70	46.07	nr	53.07
35 mm²	5.83	6.90	1.79	48.51	nr	55.41
50 mm²	8.65	10.23	2.06	55.83	nr	66.06
70 mm²	8.86	10.48	2.12	57.46	nr	67.93
95 mm²	13.17	15.58	2.39	64.77	nr	80.36
120 mm²	13.60	16.10	2.47	66.94	nr	83.04
150 mm²	19.26	22.80	2.73	73.99	nr	96.78
185 mm²	19.61	23.20	3.05	82.66	nr	105.86
240 mm²	20.88	24.72	3.45	93.50	nr	118.22
300 mm²	28.27	33.45	3.84	104.07	nr	137.52
400 mm²	44.78	53.00	4.21	114.10	nr	167.09
600/1000 Volt grade; three core (galvanised steel wire armour)						
1.5 mm²	4.88	5.77	0.62	16.80	nr	22.57
2.5 mm²	4.88	5.77	0.62	16.80	nr	22.57
4 mm²	4.74	5.61	0.62	16.80	nr	22.41
6 mm²	5.36	6.34	0.71	19.24	nr	25.58
10 mm²	6.00	7.10	1.06	28.73	nr	35.83
16 mm²	6.16	7.29	1.19	32.25	nr	39.55
25 mm²	8.77	10.38	1.81	49.05	nr	59.44
35 mm²	8.89	10.52	1.99	53.93	nr	64.45
50 mm²	8.97	10.62	2.23	60.44	nr	71.05

V: ELECTRICAL SUPPLY/POWER/LIGHTING SYSTEMS

Item	Net Price £	Material £	Labour hours	Labour £	Unit	Total rate £
V20 : LV DISTRIBUTION (cont'd)						
Y61 - LV CABLES AND WIRING (cont'd)						
ARMOURED CABLE (cont'd)						
Cable termination; brass weatherproof (cont'd)						
600/1000 Volt grade; three core (cont'd)						
70 mm²	13.39	15.84	2.40	65.04	nr	80.89
95 mm²	13.69	16.20	2.63	71.28	nr	87.47
120 mm²	14.34	16.97	2.83	76.70	nr	93.67
150 mm²	20.14	23.83	3.22	87.27	nr	111.10
185 mm²	20.64	24.43	3.44	93.23	nr	117.66
240 mm²	29.60	35.03	3.83	103.80	nr	138.83
300 mm²	30.10	35.62	4.28	115.99	nr	151.61
400 mm²	40.41	47.83	5.00	135.51	nr	183.33
600/1000 Volt grade; four core (galvanised steel wire armour)						
1.5 mm²	5.03	5.95	0.67	18.16	nr	24.11
2.5 mm²	5.03	5.95	0.67	18.16	nr	24.11
4 mm²	5.08	6.02	0.71	19.24	nr	25.26
6 mm²	6.21	7.35	0.76	20.60	nr	27.95
10 mm²	6.21	7.35	1.14	30.90	nr	38.25
16 mm²	6.33	7.49	1.29	34.96	nr	42.45
25 mm²	9.03	10.68	1.99	53.93	nr	64.62
35 mm²	9.19	10.87	2.16	58.54	nr	69.41
50 mm²	13.40	15.86	2.49	67.48	nr	83.34
70 mm²	13.81	16.35	2.65	71.82	nr	88.16
95 mm²	19.63	23.23	2.98	80.76	nr	103.99
120 mm²	20.49	24.25	3.15	85.37	nr	109.62
150 mm²	21.02	24.88	3.50	94.86	nr	119.73
185 mm²	28.72	33.98	3.72	100.82	nr	134.80
240 mm²	31.28	37.02	4.33	117.35	nr	154.37
300 mm²	41.64	49.28	4.86	131.71	nr	180.99
400 mm²	44.52	52.68	5.46	147.97	nr	200.66
600/1000 Volt grade; seven core (galvanised steel wire armour)						
1.5 mm²	5.49	6.49	0.81	21.95	nr	28.44
2.5 mm²	5.49	6.49	0.85	23.04	nr	29.53
4 mm²	2.53	2.99	0.93	25.20	nr	28.20
600/1000 Volt grade; twelve core (galvanised steel wire armour)						
1.5 mm²	6.59	7.80	1.14	30.90	nr	38.69
2.5 mm²	7.25	8.57	1.13	30.62	nr	39.20
600/1000 Volt grade; nineteen core (galvanised steel wire armour)						
1.5 mm²	8.33	9.85	1.54	41.74	nr	51.59
2.5 mm²	8.33	9.85	1.54	41.74	nr	51.59
600/1000 Volt grade; twenty seven core (galvanised steel wire armour)						
1.5 mm²	9.54	11.30	1.94	52.58	nr	63.87
2.5 mm²	12.11	14.33	2.31	62.60	nr	76.94

V: ELECTRICAL SUPPLY/POWER/LIGHTING SYSTEMS

Item	Net Price £	Material £	Labour hours	Labour £	Unit	Total rate £
600/1000 Volt grade; thirty seven core (galvanised steel wire armour)						
1.5 mm²	13.69	16.20	2.53	68.57	nr	84.76
2.5 mm²	13.69	16.20	2.87	77.78	nr	93.98
Cable; XLPE insulated; LSOH sheathed (LSF); copper stranded conductors to BS 6724; laid in trench/duct including marker tape (cable tiles measured elsewhere)						
600/1000 Volt grade; single core (aluminium wire armour)						
50 mm²	3.21	3.80	0.17	4.61	m	8.41
70 mm²	4.01	4.75	0.18	4.88	m	9.63
95 mm²	5.45	6.45	0.20	5.42	m	11.87
120 mm²	6.47	7.66	0.22	5.96	m	13.62
150 mm²	8.00	9.47	0.24	6.50	m	15.98
185 mm²	9.61	11.37	0.26	7.05	m	18.42
240 mm²	12.40	14.68	0.30	8.13	m	22.81
300 mm²	15.10	17.87	0.31	8.40	m	26.27
400 mm²	19.62	23.22	0.35	9.49	m	32.71
500 mm²	25.19	29.81	0.44	11.92	m	41.73
630 mm²	31.55	37.34	0.52	14.09	m	51.44
800 mm²	41.37	48.96	0.62	16.80	m	65.76
1000 mm²	52.79	62.48	0.65	17.62	m	80.10
600/1000 Volt grade; two core (galvanised steel wire armour)						
1.5 mm²	0.68	0.81	0.06	1.63	m	2.44
2.5 mm²	0.81	0.96	0.06	1.63	m	2.59
4 mm²	1.03	1.22	0.08	2.17	m	3.38
6 mm²	1.25	1.48	0.08	2.17	m	3.65
10 mm²	1.68	1.99	0.10	2.71	m	4.70
16 mm²	2.51	2.98	0.10	2.71	m	5.69
25 mm²	3.22	3.81	0.15	4.07	m	7.88
35 mm²	5.08	6.01	0.15	4.07	m	10.08
50 mm²	3.61	4.27	0.17	4.61	m	8.87
70 mm²	7.88	9.32	0.18	4.88	m	14.20
95 mm²	10.58	12.52	0.20	5.42	m	17.94
120 mm²	12.96	15.34	0.22	5.96	m	21.30
150 mm²	16.07	19.02	0.24	6.50	m	25.52
185 mm²	20.47	24.22	0.26	7.05	m	31.27
240 mm²	26.21	31.02	0.30	8.13	m	39.16
300 mm²	35.21	41.67	0.31	8.40	m	50.08
400 mm²	46.57	55.11	0.35	9.49	m	64.60
600/1000 Volt grade; three core (galvanised steel wire armour)						
1.5 mm²	0.76	0.90	0.07	1.90	m	2.80
2.5 mm²	0.95	1.13	0.07	1.90	m	3.02
4 mm²	1.18	1.39	0.09	2.44	m	3.83
6 mm²	1.52	1.80	0.10	2.71	m	4.51
10 mm²	2.32	2.75	0.11	2.98	m	5.73
16 mm²	3.24	3.84	0.11	2.98	m	6.82
25 mm²	4.39	5.19	0.16	4.34	m	9.53
35 mm²	6.10	7.22	0.16	4.34	m	11.55
50 mm²	7.95	9.41	0.19	5.15	m	14.56
70 mm²	10.89	12.89	0.21	5.69	m	18.58

V: ELECTRICAL SUPPLY/POWER/LIGHTING SYSTEMS

Item	Net Price £	Material £	Labour hours	Labour £	Unit	Total rate £
V20 : LV DISTRIBUTION (cont'd)						
Y61 - LV CABLES AND WIRING (cont'd)						
ARMOURED CABLE (cont'd)						
Cable; XLPE insulated; LSOH sheathed (cont'd)						
600/1000 Volt grade; three core (cont'd)						
95 mm²	14.99	17.74	0.23	6.23	m	23.97
120 mm²	18.65	22.07	0.24	6.50	m	28.58
150 mm²	23.18	27.43	0.27	7.32	m	34.75
185 mm²	28.56	33.80	0.30	8.13	m	41.93
240 mm²	36.53	43.23	0.33	8.94	m	52.18
300 mm²	46.29	54.78	0.35	9.49	m	64.27
400 mm²	54.98	65.06	0.41	11.11	m	76.17
600/1000 Volt grade; four core (galvanised steel wire armour)						
1.5 mm²	0.89	1.05	0.08	2.17	m	3.22
2.5 mm²	1.10	1.30	0.09	2.44	m	3.74
4 mm²	1.38	1.63	0.10	2.71	m	4.34
6 mm²	2.02	2.39	0.10	2.71	m	5.10
10 mm²	2.82	3.34	0.12	3.25	m	6.60
16 mm²	4.08	4.82	0.12	3.25	m	8.08
25 mm²	5.90	6.98	0.18	4.88	m	11.86
35 mm²	7.67	9.08	0.19	5.15	m	14.23
50 mm²	10.16	12.03	0.21	5.69	m	17.72
70 mm²	14.47	17.12	0.23	6.23	m	23.35
95 mm²	19.37	22.92	0.26	7.05	m	29.97
120 mm²	24.32	28.78	0.28	7.59	m	36.37
150 mm²	29.44	34.84	0.32	8.67	m	43.51
185 mm²	36.13	42.76	0.35	9.49	m	52.25
240 mm²	46.71	55.27	0.36	9.76	m	65.03
300 mm²	58.60	69.36	0.40	10.84	m	80.20
400 mm²	111.77	132.28	0.45	12.20	m	144.47
600/1000 Volt grade; seven core (galvanised steel wire armour)						
1.5 mm²	1.27	1.51	0.10	2.71	m	4.22
2.5 mm²	1.77	2.09	0.10	2.71	m	4.80
4 mm²	3.21	3.80	0.11	2.98	m	6.78
600/1000 Volt grade; twelve core (galvanised steel wire armour)						
1.5 mm²	2.02	2.39	0.11	2.98	m	5.37
2.5 mm²	3.02	3.57	0.11	2.98	m	6.55
600/1000 Volt grade; nineteen core (galvanised steel wire armour)						
1.5 mm²	3.04	3.60	0.13	3.52	m	7.12
2.5 mm²	4.57	5.41	0.14	3.79	m	9.20
600/1000 Volt grade; twenty seven core (galvanised steel wire armour)						
1.5 mm²	4.18	4.95	0.14	3.79	m	8.75
2.5 mm²	5.83	6.90	0.16	4.34	m	11.24

V: ELECTRICAL SUPPLY/POWER/LIGHTING SYSTEMS

Item	Net Price £	Material £	Labour hours	Labour £	Unit	Total rate £
600/1000 Volt grade; thirty seven core (galvanised steel wire armour)						
1.5 mm²	5.21	6.17	0.15	4.07	m	10.23
2.5 mm²	7.37	8.72	0.17	4.61	m	13.33
Cable; XLPE insulated; LSOH sheathed (LSF) copper stranded conductors to BS 6724; clipped direct to backgrounds including cleat						
600/1000 Volt grade; single core (aluminium wire armour)						
50 mm²	3.50	4.14	0.37	10.03	m	14.17
70 mm²	4.23	5.00	0.39	10.57	m	15.57
95 mm²	5.66	6.70	0.42	11.38	m	18.08
120 mm²	6.69	7.91	0.47	12.74	m	20.65
150 mm²	8.23	9.74	0.51	13.82	m	23.56
185 mm²	9.81	11.61	0.59	15.99	m	27.60
240 mm²	12.84	15.20	0.68	18.43	m	33.62
300 mm²	15.54	18.39	0.74	20.06	m	38.44
400 mm²	22.67	26.83	0.81	21.95	m	48.79
500 mm²	28.57	33.81	0.88	23.85	m	57.66
630 mm²	37.15	43.97	1.05	28.46	m	72.42
800 mm²	46.19	54.67	1.33	36.05	m	90.71
1000 mm²	51.64	61.11	1.40	37.94	m	99.05
600/1000 Volt grade; two core (galvanised steel wire armour)						
1.5 mm²	0.86	1.01	0.20	5.42	m	6.43
2.5 mm²	0.98	1.17	0.20	5.42	m	6.59
4.0 mm²	1.25	1.48	0.21	5.69	m	7.17
6.0 mm²	1.47	1.73	0.22	5.96	m	7.70
10.0 mm²	1.89	2.24	0.24	6.50	m	8.75
16.0 mm²	2.73	3.23	0.25	6.78	m	10.00
25 mm²	3.43	4.06	0.35	9.49	m	13.55
35 mm²	5.30	6.27	0.36	9.76	m	16.02
50 mm²	5.97	7.07	0.37	10.03	m	17.09
70 mm²	8.15	9.65	0.39	10.57	m	20.22
95 mm²	10.66	12.61	0.42	11.38	m	24.00
120 mm²	13.08	15.47	0.47	12.74	m	28.21
150 mm²	16.20	19.17	0.51	13.82	m	32.99
185 mm²	20.63	24.41	0.59	15.99	m	40.40
240 mm²	25.67	30.38	0.68	18.43	m	48.81
300 mm²	35.94	42.54	0.74	20.06	m	62.59
400 mm²	44.74	52.94	0.81	21.95	m	74.90
600/1000 Volt grade; three core (galvanised steel wire armour)						
1.5 mm²	0.93	1.10	0.20	5.42	m	6.52
2.5 mm²	1.09	1.29	0.21	5.69	m	6.98
4.0 mm²	1.39	1.65	0.22	5.96	m	7.61
6.0 mm²	1.73	2.05	0.22	5.96	m	8.01
10.0 mm²	2.54	3.00	0.25	6.78	m	9.78
16.0 mm²	3.46	4.09	0.26	7.05	m	11.14
25 mm²	4.87	5.76	0.37	10.03	m	15.79
35 mm²	6.31	7.47	0.39	10.57	m	18.04
50 mm²	8.24	9.75	0.40	10.84	m	20.59
70 mm²	10.97	12.98	0.42	11.38	m	24.36
95 mm²	15.12	17.89	0.45	12.20	m	30.09

V: ELECTRICAL SUPPLY/POWER/LIGHTING SYSTEMS

Item	Net Price £	Material £	Labour hours	Labour £	Unit	Total rate £
V20 : LV DISTRIBUTION (cont'd)						
Y61 - LV CABLES AND WIRING (cont'd)						
ARMOURED CABLE (cont'd)						
Cable; XLPE insulated; LSOH sheathed (cont'd)						
600/1000 Volt grade; three core (cont'd)						
120 mm²	18.77	22.21	0.52	14.09	m	36.30
150 mm²	23.33	27.61	0.55	14.91	m	42.51
185 mm²	28.72	33.99	0.63	17.07	m	51.06
240 mm²	38.15	45.14	0.71	19.24	m	64.39
300 mm²	48.01	56.82	0.78	21.14	m	77.96
400 mm²	57.00	67.46	0.87	23.58	m	91.03
600/1000 Volt grade; four core (galvanised steel wire armour)						
1.5 mm²	1.06	1.25	0.21	5.69	m	6.94
2.5 mm²	1.32	1.56	0.22	5.96	m	7.52
4.0 mm²	1.60	1.90	0.22	5.96	m	7.86
6.0 mm²	2.24	2.65	0.23	6.23	m	8.88
10.0 mm²	3.05	3.61	0.26	7.05	m	10.66
16.0 mm²	4.29	5.08	0.26	7.05	m	12.12
25 mm²	6.11	7.23	0.39	10.57	m	17.80
35 mm²	7.95	9.41	0.40	10.84	m	20.25
50 mm²	10.44	12.36	0.41	11.11	m	23.47
70 mm²	14.58	17.26	0.45	12.20	m	29.46
95 mm²	19.53	23.11	0.50	13.55	m	36.66
120 mm²	24.47	28.96	0.54	14.63	m	43.60
150 mm²	30.89	36.56	0.60	16.26	m	52.82
185 mm²	37.75	44.68	0.67	18.16	m	62.83
240 mm²	48.88	57.85	0.75	20.33	m	78.17
300 mm²	61.33	72.59	0.83	22.49	m	95.08
400 mm²	114.99	136.09	0.91	24.66	m	160.75
600/1000 Volt grade; seven core (galvanised steel wire armour)						
1.5 mm²	1.49	1.76	0.20	5.42	m	7.18
2.5 mm²	2.03	2.41	0.20	5.42	m	7.83
4.0 mm²	3.66	4.33	0.23	6.23	m	10.56
600/1000 Volt grade; twelve core (galvanised steel wire armour)						
1.5 mm²	2.15	2.55	0.23	6.23	m	8.78
2.5 mm²	3.23	3.82	0.24	6.50	m	10.33
600/1000 Volt grade; nineteen core (galvanised steel wire armour)						
1.5 mm²	3.25	3.85	0.26	7.05	m	10.90
2.5 mm²	4.78	5.66	0.28	7.59	m	13.25
600/1000 Volt grade; twenty seven core (galvanised steel wire armour)						
1.5 mm²	4.40	5.20	0.29	7.86	m	13.06
2.5 mm²	6.11	7.23	0.30	8.13	m	15.36

V: ELECTRICAL SUPPLY/POWER/LIGHTING SYSTEMS

Item	Net Price £	Material £	Labour hours	Labour £	Unit	Total rate £
600/1000 Volt grade; thirty seven core (galvanised steel wire armour)						
1.5 mm²	5.49	6.50	0.32	8.67	m	15.17
2.5 mm²	7.66	9.07	0.33	8.94	m	18.01
Cable termination; brass weatherproof gland with inner and outer seal, shroud, brass lock nut and earth ring (including drilling and cutting mild steel gland plate)						
600/1000 Volt grade; single core (aluminium wire armour)						
25 mm²	4.66	5.51	1.70	46.07	nr	51.58
35 mm²	4.66	5.51	1.79	48.51	nr	54.02
50 mm²	4.66	5.51	2.06	55.83	nr	61.34
70 mm²	5.54	6.56	2.12	57.46	nr	64.02
95 mm²	5.63	6.67	2.39	64.77	nr	71.44
120 mm²	5.85	6.93	2.47	66.94	nr	73.87
150 mm²	8.46	10.02	2.73	73.99	nr	84.00
185 mm²	13.14	15.56	3.05	82.66	nr	98.22
240 mm²	13.29	15.73	3.45	93.50	nr	109.23
300 mm²	19.14	22.65	3.84	104.07	nr	126.72
400 mm²	21.24	25.14	4.21	114.10	nr	139.23
500 mm²	23.93	28.32	5.70	154.48	m	182.80
630 mm²	40.83	48.32	6.20	168.03	m	216.34
800 mm²	49.25	58.29	7.50	203.26	m	261.55
1000 mm²	64.72	76.60	10.00	271.01	m	347.61
600/1000 Volt grade; two core (galvanised steel wire armour)						
1.5 mm²	4.50	5.32	0.58	15.72	nr	21.04
2.5 mm²	4.50	5.32	0.58	15.72	nr	21.04
4 mm²	4.52	5.35	0.58	15.72	nr	21.07
6 mm²	4.60	5.44	0.67	18.16	nr	23.60
10 mm²	4.91	5.81	1.00	27.10	nr	32.91
16 mm²	4.96	5.87	1.11	30.08	nr	35.95
25 mm²	5.63	6.67	1.70	46.07	nr	52.74
35 mm²	5.55	6.57	1.79	48.51	nr	55.09
50 mm²	8.25	9.76	2.06	55.83	nr	65.59
70 mm²	8.44	9.99	2.12	57.46	nr	67.44
95 mm²	12.55	14.85	2.39	64.77	nr	79.62
120 mm²	12.96	15.34	2.47	66.94	nr	82.28
150 mm²	18.35	21.72	2.73	73.99	nr	95.71
185 mm²	18.68	22.10	3.05	82.66	nr	104.76
240 mm²	19.89	23.55	3.45	93.50	nr	117.05
300 mm²	26.92	31.86	3.84	104.07	nr	135.93
400 mm²	42.65	50.48	4.21	114.10	nr	164.58
600/1000 Volt grade; three core (galvanised steel wire armour)						
1.5 mm²	4.65	5.50	0.62	16.80	nr	22.30
2.5 mm²	4.65	5.50	0.62	16.80	nr	22.30
4 mm²	4.52	5.35	0.62	16.80	nr	22.15
6 mm²	5.11	6.04	0.71	19.24	nr	25.28
10 mm²	5.73	6.78	1.06	28.73	nr	35.51
16 mm²	5.81	6.87	1.19	32.25	nr	39.12
25 mm²	8.36	9.89	1.81	49.05	nr	58.95
35 mm²	8.48	10.03	1.99	53.93	nr	63.96
50 mm²	8.54	10.11	2.23	60.44	nr	70.55

V: ELECTRICAL SUPPLY/POWER/LIGHTING SYSTEMS

Item	Net Price £	Material £	Labour hours	Labour £	Unit	Total rate £
V20 : LV DISTRIBUTION (cont'd)						
Y61 - LV CABLES AND WIRING (cont'd)						
ARMOURED CABLE (cont'd)						
Cable termination; brass weatherproof (cont'd)						
600/1000 Volt grade; three core (cont'd)						
70 mm²	12.75	15.09	2.40	65.04	nr	80.14
95 mm²	13.04	15.43	2.63	71.28	nr	86.71
120 mm²	13.66	16.17	2.83	76.70	nr	92.87
150 mm²	19.18	22.70	3.22	87.27	nr	109.97
185 mm²	19.66	23.27	3.44	93.23	nr	116.50
240 mm²	28.20	33.37	3.83	103.80	nr	137.17
300 mm²	28.67	33.93	4.28	115.99	nr	149.92
400 mm²	38.49	45.55	5.00	135.51	nr	181.06
600/1000 Volt grade; four core (galvanised steel wire armour)						
1.5 mm²	4.80	5.68	0.67	18.16	nr	23.83
2.5 mm²	4.80	5.68	0.67	18.16	nr	23.83
4 mm²	4.84	5.73	0.71	19.24	nr	24.97
6 mm²	5.92	7.01	0.76	20.60	nr	27.61
10 mm²	5.92	7.01	1.14	30.90	nr	37.90
16 mm²	6.03	7.13	1.29	34.96	nr	42.09
25 mm²	8.60	10.18	1.99	53.93	nr	64.11
35 mm²	8.75	10.36	2.16	58.54	nr	68.90
50 mm²	12.77	15.11	2.49	67.48	nr	82.59
70 mm²	13.16	15.57	2.65	71.82	nr	87.39
95 mm²	18.70	22.13	2.98	80.76	nr	102.89
120 mm²	19.53	23.11	3.15	85.37	nr	108.48
150 mm²	20.02	23.69	3.50	94.86	nr	118.55
185 mm²	27.36	32.38	3.72	100.82	nr	133.20
240 mm²	29.57	34.99	4.33	117.35	nr	152.34
300 mm²	39.66	46.94	4.86	131.71	nr	178.65
400 mm²	42.40	50.18	5.46	147.97	nr	198.15
600/1000 Volt grade; seven core (galvanised steel wire armour)						
1.5 mm²	5.23	6.19	0.81	21.95	nr	28.14
2.5 mm²	5.23	6.19	0.85	23.04	nr	29.23
4 mm²	4.15	4.91	0.93	25.20	nr	30.12
600/1000 Volt grade; twelve core (galvanised steel wire armour)						
1.5 mm²	6.27	7.42	1.14	30.90	nr	38.31
2.5 mm²	6.90	8.17	1.13	30.62	nr	38.79
600/1000 Volt grade; nineteen core (galvanised steel wire armour)						
1.5 mm²	7.92	9.38	1.54	41.74	nr	51.11
2.5 mm²	7.92	9.38	1.54	41.74	nr	51.11
600/1000 Volt grade; twenty seven core (galvanised steel wire armour)						
1.5 mm²	9.10	10.77	1.94	52.58	nr	63.34
2.5 mm²	11.60	13.73	2.31	62.60	nr	76.34

V: ELECTRICAL SUPPLY/POWER/LIGHTING SYSTEMS

Item	Net Price £	Material £	Labour hours	Labour £	Unit	Total rate £
600/1000 Volt grade; thirty seven core (galvanised steel wire armour)						
1.5 mm²	13.02	15.41	2.53	68.57	nr	83.97
2.5 mm²	13.02	15.41	2.87	77.78	nr	93.19
UN-ARMOURED CABLE						
Cable: XLPE insulated; PVC sheathed 90c copper to CMA Code 6181e; for internal wiring; clipped to backgrounds; (Supports and fixings included)						
300/500 Volt grade; single core						
6.0 mm²	0.50	0.59	0.09	2.44	m	3.03
10 mm²	0.75	0.89	0.10	2.71	m	3.60
16 mm²	1.04	1.23	0.12	3.25	m	4.48
Cable; LSF insulated to CMA Code 6491B; non-sheathed copper; laid/drawn in trunking/ conduit						
450/750 Volt grade; single core						
1.5 mm²	0.10	0.12	0.03	0.81	m	0.93
2.5 mm²	0.19	0.22	0.03	0.81	m	1.03
4.0 mm²	0.29	0.34	0.03	0.81	m	1.15
6.0 mm²	0.40	0.47	0.04	1.08	m	1.56
10.0 mm²	0.64	0.75	0.04	1.08	m	1.84
16.0 mm²	0.95	1.12	0.05	1.36	m	2.48
25.0 mm²	2.23	2.63	0.06	1.63	m	4.26
35.0 mm²	2.01	2.38	0.06	1.63	m	4.01
50.0 mm²	2.70	3.20	0.07	1.90	m	5.09
70.0 mm²	3.96	4.69	0.08	2.17	m	6.86
95.0 mm²	5.46	6.46	0.08	2.17	m	8.63
120.0 mm²	6.78	8.02	0.10	2.71	m	10.73
150.0 mm²	8.43	9.97	0.13	3.52	m	13.49
Cable; twin & earth to CMA code 6242Y; clipped to backgrounds						
300/500 Volt grade; PVC/PVC						
1.5 mm² 2C+E	0.46	0.55	0.01	0.27	m	0.82
1.5 mm² 3C+E	0.80	0.95	0.02	0.54	m	1.49
2.5mm² 2C+E	0.63	0.74	0.02	0.54	m	1.28
4.0mm² 2C+E	0.89	1.05	0.02	0.54	m	1.59
6.0mm² 2C+E	1.16	1.38	0.02	0.54	m	1.92
10.0mm² 2C+E	2.55	3.02	0.03	0.81	m	3.83
16.0mm² 2C+E	3.66	4.33	0.03	0.81	m	5.15
300/500 Volt grade; LSF/LSF						
1.5 mm² 2C+E	0.54	0.64	0.01	0.27	m	0.91
1.5 mm² 3C+E	0.93	1.09	0.02	0.54	m	1.64
2.5mm² 2C+E	0.71	0.84	0.02	0.54	m	1.39
4.0mm² 2C+E	1.10	1.30	0.02	0.54	m	1.84
6.0mm² 2C+E	1.44	1.70	0.02	0.54	m	2.24
10.0mm² 2C+E	3.15	3.73	0.03	0.81	m	4.54
16.0mm² 2C+E	4.58	5.41	0.03	0.81	m	6.23

V: ELECTRICAL SUPPLY/POWER/LIGHTING SYSTEMS

Item	Net Price £	Material £	Labour hours	Labour £	Unit	Total rate £
V20 : LV DISTRIBUTION (cont'd)						
Y61 - LV CABLES AND WIRING (cont'd)						
EARTH CABLE						
Cable; LSF insulated to CMA Code 6491B; non-sheathed copper; laid/drawn in trunking/ conduit						
450/750 Volt grade; single core						
1.5 mm^2	0.10	0.12	0.03	0.81	m	0.93
2.5 mm^2	0.19	0.22	0.03	0.81	m	1.03
4.0 mm^2	0.29	0.34	0.03	0.81	m	1.15
6.0 mm^2	0.40	0.47	0.04	1.08	m	1.56
10.0 mm^2	0.64	0.75	0.04	1.08	m	1.84
16.0 mm^2	0.95	1.12	0.05	1.36	m	2.48
25.0 mm^2	2.23	2.63	0.06	1.63	m	4.26
35.0 mm^2	2.01	2.38	0.06	1.63	m	4.01
50.0 mm^2	2.70	3.20	0.07	1.90	m	5.09
70.0 mm^2	3.96	4.69	0.08	2.17	m	6.86
95.0 mm^2	5.46	6.46	0.08	2.17	m	8.63
120.0 mm^2	6.78	8.02	0.10	2.71	m	10.73
150.0 mm^2	8.43	9.97	0.13	3.52	m	13.49
185.0 mm^2	10.25	12.13	0.16	4.34	m	16.47
240.0 mm^2	13.38	15.83	0.20	5.42	m	21.25
FLEXIBLE CABLE						
Flexible cord; PVC insulated; PVC sheathed; copper stranded to CMA Code 218*Y (laid loose)						
300 Volt grade; two core						
0.50 mm^2	0.13	0.16	0.07	1.90	m	2.05
0.75 mm^2	0.17	0.20	0.07	1.90	m	2.09
300 Volt grade; three core						
0.50 mm^2	0.15	0.18	0.07	1.90	m	2.08
0.75 mm^2	0.18	0.21	0.07	1.90	m	2.11
1.0 mm^2	0.19	0.22	0.07	1.90	m	2.12
1.5 mm^2	0.26	0.31	0.07	1.90	m	2.21
2.5 mm^2	0.44	0.52	0.08	2.17	m	2.69
Flexible cord; PVC insulated; PVC sheathed; copper stranded to CMA Code 318Y (laid loose)						
300/500 Volt grade; two core						
0.75 mm^2	0.15	0.18	0.07	1.90	m	2.08
1.0 mm^2	0.19	0.22	0.07	1.90	m	2.12
1.5 mm^2	0.26	0.31	0.07	1.90	m	2.21
2.5 mm^2	0.44	0.52	0.07	1.90	m	2.42
300/500 Volt grade; three core						
0.75 mm^2	0.19	0.22	0.07	1.90	m	2.12
1.0 mm^2	0.24	0.29	0.07	1.90	m	2.18
1.5 mm^2	0.34	0.40	0.07	1.90	m	2.30
2.5 mm^2	0.56	0.66	0.08	2.17	m	2.83

V: ELECTRICAL SUPPLY/POWER/LIGHTING SYSTEMS

Item	Net Price £	Material £	Labour hours	Labour £	Unit	Total rate £
300/500 Volt grade; four core						
0.75 mm^2	0.26	0.31	0.08	2.17	m	**2.48**
1.0 mm^2	0.35	0.42	0.08	2.17	m	**2.58**
1.5 mm^2	0.52	0.61	0.08	2.17	m	**2.78**
2.5 mm^2	0.80	0.95	0.09	2.44	m	**3.39**
Flexible cord; PVC insulated; PVC sheathed for use in high temperature zones; copper stranded to CMA Code 309Y (laid loose)						
300/500 Volt grade; two core						
0.50 mm^2	0.19	0.22	0.07	1.90	m	**2.12**
0.75 mm^2	0.24	0.29	0.07	1.90	m	**2.18**
1.0 mm^2	0.28	0.33	0.07	1.90	m	**2.22**
1.5 mm^2	0.39	0.46	0.07	1.90	m	**2.35**
2.5 mm^2	0.57	0.68	0.07	1.90	m	**2.57**
300/500 Volt grade; three core						
0.50 mm^2	0.26	0.31	0.07	1.90	m	**2.21**
0.75 mm^2	0.31	0.36	0.07	1.90	m	**2.26**
1.0 mm^2	0.35	0.42	0.07	1.90	m	**2.31**
1.5 mm^2	0.46	0.55	0.07	1.90	m	**2.44**
2.5 mm^2	0.70	0.83	0.07	1.90	m	**2.73**
Flexible cord; rubber insulated; rubber sheathed; copper stranded to CMA code 318 (laid loose)						
300/500 Volt grade; two core						
0.50 mm^2	0.23	0.27	0.07	1.90	m	**2.16**
0.75 mm^2	0.26	0.31	0.07	1.90	m	**2.21**
1.0 mm^2	0.29	0.34	0.07	1.90	m	**2.24**
1.5 mm^2	0.44	0.52	0.07	1.90	m	**2.42**
2.5 mm^2	0.62	0.73	0.07	1.90	m	**2.63**
300/500 Volt grade; three core						
0.50 mm^2	0.36	0.43	0.07	1.90	m	**2.33**
0.75 mm^2	0.40	0.47	0.07	1.90	m	**2.37**
1.0 mm^2	0.52	0.61	0.07	1.90	m	**2.51**
1.5 mm^2	0.70	0.83	0.07	1.90	m	**2.73**
2.5 mm^2	1.08	1.28	0.07	1.90	m	**3.17**
300/500 Volt grade; four core						
0.50 mm^2	0.31	0.36	0.08	2.17	m	**2.53**
0.75 mm^2	0.37	0.44	0.08	2.17	m	**2.61**
1.0 mm^2	0.47	0.56	0.08	2.17	m	**2.73**
1.5 mm^2	0.59	0.70	0.08	2.17	m	**2.87**
2.5 mm^2	0.88	1.04	0.08	2.17	m	**3.21**
Flexible cord; rubber insulated; rubber sheathed; for 90C operation; copper stranded to CMA Code 318 (laid loose)						
450/750 Volt grade; two core						
0.50 mm^2	0.21	0.25	0.07	1.90	m	**2.14**
0.75 mm^2	0.32	0.38	0.07	1.90	m	**2.27**
1.0 mm^2	0.35	0.42	0.07	1.90	m	**2.31**
1.5 mm^2	0.45	0.53	0.07	1.90	m	**2.43**
2.5 mm^2	0.69	0.82	0.07	1.90	m	**2.72**

V: ELECTRICAL SUPPLY/POWER/LIGHTING SYSTEMS

Item	Net Price £	Material £	Labour hours	Labour £	Unit	Total rate £
V20 : LV DISTRIBUTION (cont'd)						
Y61 - LV CABLES AND WIRING (cont'd)						
FLEXIBLE CABLE (cont'd)						
Flexible cord; rubber insulated (cont'd)						
450/750 Volt grade; three core						
0.50 mm²	0.34	0.40	0.07	1.90	m	2.30
0.75 mm²	0.39	0.46	0.07	1.90	m	2.35
1.0 mm²	0.43	0.51	0.07	1.90	m	2.40
1.5 mm²	0.51	0.60	0.07	1.90	m	2.50
2.5 mm²	0.68	0.81	0.07	1.90	m	2.70
450/750 Volt grade; four core						
0.75 mm²	0.51	0.60	0.08	2.17	m	2.77
1.0 mm²	0.65	0.77	0.08	2.17	m	2.94
1.5 mm²	0.94	1.11	0.08	2.17	m	3.27
2.5 mm²	1.47	1.74	0.08	2.17	m	3.91
Heavy flexible cable; rubber insulated; rubber sheathed; copper stranded to CMA Code 638P (laid loose)						
450/750 Volt grade; two core						
1.0 mm²	0.36	0.43	0.08	2.17	m	2.60
1.5 mm²	0.43	0.51	0.08	2.17	m	2.68
2.5 mm²	0.59	0.70	0.08	2.17	m	2.87
450/750 Volt grade; three core						
1.0 mm²	0.43	0.51	0.08	2.17	m	2.68
1.5 mm²	0.46	0.54	0.08	2.17	m	2.71
2.5 mm²	0.62	0.73	0.08	2.17	m	2.90
450/750 Volt grade; four core						
1.0 mm²	0.59	0.70	0.08	2.17	m	2.87
1.5 mm²	0.64	0.76	0.08	2.17	m	2.92
2.5 mm²	0.89	1.05	0.08	2.17	m	3.22
FIRE RATED CABLE						
Cable, mineral insulated; copper sheathed with copper conductors; fixed with clips to backgrounds. BASEC approval to BS 6207 Part 1 1995; complies with BS 6387 Category CWZ						
Light duty 500 Volt grade; bare						
2L 1.0	3.77	4.46	0.23	6.23	m	10.69
2L 1.5	4.30	5.08	0.23	6.23	m	11.32
2L 2.5	5.18	6.14	0.25	6.78	m	12.91
2L 4.0	7.30	8.63	0.25	6.78	m	15.41
3L 1.0	4.44	5.25	0.24	6.50	m	11.76
3L 1.5	5.28	6.25	0.25	6.78	m	13.02
3L 2.5	7.55	8.93	0.25	6.78	m	15.71
4L 1.0	5.08	6.01	0.25	6.78	m	12.78
4L 1.5	5.69	6.73	0.25	6.78	m	13.51
4L 2.5	8.94	10.58	0.26	7.05	m	17.63
7L 1.5	9.00	10.65	0.28	7.59	m	18.24
7L 2.5	11.44	13.53	0.27	7.32	m	20.85

V: ELECTRICAL SUPPLY/POWER/LIGHTING SYSTEMS

Item	Net Price £	Material £	Labour hours	Labour £	Unit	Total rate £
Light duty 500 Volt grade; LSF sheathed						
2L 1.0	4.04	4.79	0.23	6.23	m	**11.02**
2L 1.5	4.50	5.33	0.23	6.23	m	**11.56**
2L 2.5	5.39	6.38	0.25	6.78	m	**13.15**
2L 4.0	7.58	8.98	0.25	6.78	m	**15.75**
3L 1.0	4.80	5.68	0.24	6.50	m	**12.19**
3L 1.5	5.63	6.66	0.25	6.78	m	**13.44**
3L 2.5	7.87	9.32	0.25	6.78	m	**16.09**
4L 1.0	5.45	6.45	0.25	6.78	m	**13.22**
4L 1.5	6.10	7.21	0.25	6.78	m	**13.99**
4L 2.5	9.22	10.91	0.26	7.05	m	**17.95**
7L 1.5	9.90	11.72	0.28	7.59	m	**19.30**
7L 2.5	12.41	14.68	0.27	7.32	m	**22.00**
Heavy duty 750 Volt grade; bare						
1H 10	7.57	8.96	0.25	6.78	m	**15.74**
1H 16	10.25	12.13	0.26	7.05	m	**19.17**
1H 25	14.23	16.84	0.27	7.32	m	**24.16**
1H 35	20.54	24.31	0.32	8.67	m	**32.99**
1H 50	22.49	26.61	0.35	9.49	m	**36.10**
1H 70	30.01	35.52	0.38	10.30	m	**45.82**
1H 95	44.50	52.66	0.41	11.11	m	**63.77**
1H 120	46.81	55.40	0.46	12.47	m	**67.87**
1H 150	57.42	67.95	0.50	13.55	m	**81.51**
1H 185	70.02	82.87	0.56	15.18	m	**98.04**
1H 240	91.10	107.82	0.69	18.70	m	**126.52**
2H 1.5	6.98	8.27	0.25	6.78	m	**15.04**
2H 2.5	8.42	9.97	0.26	7.05	m	**17.02**
2H 4.0	10.32	12.21	0.26	7.05	m	**19.26**
2H 6.0	13.26	15.69	0.29	7.86	m	**23.55**
2H 10.0	16.96	20.07	0.34	9.21	m	**29.28**
2H 16.0	24.24	28.69	0.40	10.84	m	**39.53**
2H 25.0	33.43	39.57	0.44	11.92	m	**51.49**
3H 1.5	7.75	9.17	0.25	6.78	m	**15.95**
3H 2.5	9.61	11.38	0.25	6.78	m	**18.15**
3H 4.0	11.70	13.85	0.27	7.32	m	**21.16**
3H 6.0	14.46	17.11	0.30	8.13	m	**25.24**
3H 10.0	21.43	25.36	0.35	9.49	m	**34.85**
3H 16.0	28.87	34.17	0.41	11.11	m	**45.28**
3H 25.0	42.34	50.10	0.47	12.74	m	**62.84**
4H 1.5	9.42	11.15	0.24	6.50	m	**17.65**
4H 2.5	11.42	13.52	0.26	7.05	m	**20.57**
4H 4.0	14.10	16.69	0.29	7.86	m	**24.55**
4H 6.0	17.83	21.10	0.31	8.40	m	**29.51**
4H 10.0	25.33	29.98	0.37	10.03	m	**40.01**
4H 16.0	36.30	42.96	0.44	11.92	m	**54.88**
4H 25.0	51.82	61.32	0.52	14.09	m	**75.42**
7H 1.5	12.70	15.03	0.30	8.13	m	**23.16**
7H 2.5	16.92	20.02	0.32	8.67	m	**28.70**
12H 2.5	28.87	34.17	0.39	10.57	m	**44.74**
19H 1.5	42.32	50.09	0.42	11.38	m	**61.47**

V: ELECTRICAL SUPPLY/POWER/LIGHTING SYSTEMS

Item	Net Price £	Material £	Labour hours	Labour £	Unit	Total rate £
V20 : LV DISTRIBUTION (cont'd)						
Y61 - LV CABLES AND WIRING (cont'd)						
FIRE RATED CABLE (cont'd)						
Cable, mineral insulated (cont'd)						
Heavy duty 750 Volt grade; LSF sheathed						
1H 10	8.02	9.49	0.25	6.78	m	16.26
1H 16	10.61	12.55	0.26	7.05	m	19.60
1H 25	14.60	17.28	0.27	7.32	m	24.60
1H 35	21.52	25.46	0.32	8.67	m	34.14
1H 50	23.66	28.01	0.35	9.49	m	37.49
1H 70	30.25	35.80	0.38	10.30	m	46.10
1H 95	45.30	53.61	0.41	11.11	m	64.72
1H 120	48.14	56.98	0.46	12.47	m	69.44
1H 150	58.69	69.46	0.50	13.55	m	83.01
1H 185	71.40	84.50	0.56	15.18	m	99.68
1H 240	93.20	110.30	0.68	18.43	m	128.73
2H 1.5	7.22	8.55	0.25	6.78	m	15.32
2H 2.5	8.66	10.25	0.26	7.05	m	17.30
2H 4.0	10.68	12.64	0.26	7.05	m	19.69
2H 6.0	13.74	16.26	0.29	7.86	m	24.12
2H 10.0	17.66	20.90	0.34	9.21	m	30.12
2H 16.0	25.08	29.68	0.40	10.84	m	40.52
2H 25.0	34.39	40.70	0.44	11.92	m	52.63
3H 1.5	7.87	9.32	0.25	6.78	m	16.09
3H 2.5	9.84	11.65	0.25	6.78	m	18.42
3H 4.0	12.18	14.41	0.27	7.32	m	21.73
3H 6.0	15.30	18.11	0.30	8.13	m	26.24
3H 10.0	22.39	26.50	0.35	9.49	m	35.99
3H 16.0	29.90	35.39	0.41	11.11	m	46.50
3H 25.0	43.43	51.40	0.47	12.74	m	64.13
4H 1.5	9.68	11.46	0.24	6.50	m	17.97
4H 2.5	11.86	14.03	0.26	7.05	m	21.08
4H 4.0	14.71	17.41	0.29	7.86	m	25.27
4H 6.0	18.67	22.10	0.31	8.40	m	30.50
4H 10.0	26.48	31.34	0.37	10.03	m	41.37
4H 16.0	37.38	44.24	0.44	11.92	m	56.16
4H 25.0	53.27	63.04	0.52	14.09	m	77.13
7H 1.5	13.12	15.52	0.30	8.13	m	23.65
7H 2.5	17.44	20.64	0.32	8.67	m	29.31
12H 2.5	29.58	35.01	0.39	10.57	m	45.58
19H 1.5	43.37	51.32	0.42	11.38	m	62.71
Cable terminations for M.I. Cable; Polymeric one piece moulding; containing grey sealing compound; testing; phase marking and connection						
Light Duty 500 Volt grade; Brass gland; polymeric one moulding containing grey sealing compound; coloured conductor sleeving; Earth tag; plastic gland shroud						
2L 1.5	8.45	10.00	0.27	7.32	m	17.32
2L 2.5	8.48	11.34	0.27	7.32	m	18.66
3L 1.5	8.78	11.26	0.27	7.32	m	18.58
4L 1.5	8.86	11.35	0.27	7.32	m	18.67

V: ELECTRICAL SUPPLY/POWER/LIGHTING SYSTEMS

Item	Net Price £	Material £	Labour hours	Labour £	Unit	Total rate £
Cable Terminations; for MI copper sheathed cable. Certified for installation in potentially explosive atmospheres; testing; phase marking and connection; BS 6207 Part 2 1995						
Light duty 500 Volt grade; brass gland; brass pot with earth tail; pot closure; sealing compound; conductor sleving; plastic gland shroud; identification markers						
2L 1.0	9.50	11.25	0.39	10.57	nr	21.82
2L 1.5	9.50	11.25	0.41	11.11	nr	22.36
2L 2.5	9.50	11.25	0.41	11.11	nr	22.36
2L 4.0	9.50	11.25	0.46	12.47	nr	23.71
3L 1.0	9.53	11.28	0.43	11.65	nr	22.93
3L 1.5	9.53	11.28	0.43	11.65	nr	22.93
3L 2.5	9.53	11.28	0.44	11.92	nr	23.20
4L 1.0	9.53	11.28	0.47	12.74	nr	24.01
4L 1.5	9.53	11.28	0.47	12.74	nr	24.01
4L 2.5	9.53	11.28	0.50	13.55	nr	24.83
7L 1.0	23.11	27.35	0.69	18.70	nr	46.05
7L 1.5	23.11	27.35	0.70	18.97	nr	46.32
7L 2.5	23.11	27.35	0.74	20.06	nr	47.41
Heavy duty 750 Volt grade; brass gland; brass pot with earth tail; pot closure; sealing compound; conductor sleeving; plastic gland shroud; identification markers						
1H 10	9.56	11.32	0.37	10.03	nr	21.35
1H 16	9.56	11.32	0.39	10.57	nr	21.89
1H 25	9.56	11.32	0.56	15.18	nr	26.50
1H 35	9.56	11.32	0.57	15.45	nr	26.77
1H 50	23.18	27.44	0.60	16.26	nr	43.70
1H 70	23.18	27.44	0.67	18.16	nr	45.60
1H 95	23.18	27.44	0.75	20.33	nr	47.76
1H 120	37.55	44.44	0.94	25.48	nr	69.91
1H 150	37.55	44.44	0.99	26.83	nr	71.27
1H 185	37.55	44.44	1.26	34.15	nr	78.58
1H 240	59.24	70.11	1.37	37.13	nr	107.24
2H 1.5	10.13	11.99	0.42	11.38	nr	23.37
2H 2.5	10.13	11.99	0.42	11.38	nr	23.37
2H 4	10.13	11.99	0.47	12.74	nr	24.72
2H 6	10.13	11.99	0.54	14.63	nr	26.62
2H 10	24.56	29.07	0.58	15.72	nr	44.79
2H 16	24.56	29.07	0.69	18.70	nr	47.77
2H 25	39.79	47.09	0.77	20.87	nr	67.96
3H 1.5	10.14	12.00	0.44	11.92	nr	23.93
3H 2.5	10.14	12.00	0.44	11.92	nr	23.93
3H 4	10.14	12.00	0.57	15.45	nr	27.45
3H 6	24.65	29.17	0.61	16.53	nr	45.70
3H 10	24.65	29.17	0.65	17.62	nr	46.79
3H 16	24.65	29.17	0.78	21.14	nr	50.31
3H 25	62.80	74.32	0.85	23.04	nr	97.35
4H 1.5	10.14	12.00	0.52	14.09	nr	26.09
4H 2.5	10.14	12.00	0.53	14.36	nr	26.36
4H 4	24.65	29.17	0.60	16.26	nr	45.43
4H 6	24.65	29.17	0.65	17.62	nr	46.79
4H 10	24.65	29.17	0.69	18.70	nr	47.87
4H 16	39.79	47.09	0.88	23.85	nr	70.94
4H 25	62.80	74.32	0.93	25.20	nr	99.52

V: ELECTRICAL SUPPLY/POWER/LIGHTING SYSTEMS

Item	Net Price £	Material £	Labour hours	Labour £	Unit	Total rate £
V20 : LV DISTRIBUTION (cont'd)						
Y61 - LV CABLES AND WIRING (cont'd)						
FIRE RATED CABLE (cont'd)						
Cable Terminations; for MI copper (cont'd)						
Heavy duty 750 Volt grade; brass gland (cont'd)						
7H 1.5	24.56	29.07	0.71	19.24	nr	48.31
7H 2.5	24.56	29.07	0.74	20.06	nr	49.13
12H 1.5	39.79	47.09	0.85	23.04	nr	70.13
12H 2.5	39.79	47.09	1.00	27.10	nr	74.19
19H 2.5	62.80	74.32	1.11	30.08	nr	104.40
Cable; FP100; LOSH insulated; non sheathed fire resistant to LPCB Approved to BS 6387 Catergory CWZ; in conduit or trunking including terminations						
450/750 volt grade; single core						
1.0 mm²	0.70	0.82	0.13	3.52	m	4.35
1.5 mm²	0.73	0.87	0.13	3.52	m	4.39
2.5 mm²	0.88	1.04	0.13	3.52	m	4.56
4.0 mm²	0.88	1.04	0.13	3.52	m	4.56
6.0 mm²	1.16	1.38	0.13	3.52	m	4.90
10 mm²	1.92	2.27	0.16	4.34	m	6.61
16 mm²	2.96	3.51	0.16	4.34	m	7.84
Cable; FP200; Insudite insulated; LSOH sheathed screened fire resistant BASEC Approved to BS 7629; fixed with clips to backgrounds						
300/500 volt grade; two core						
1.5 mm²	2.06	2.44	0.13	3.52	m	5.97
2.5 mm²	2.50	2.95	0.13	3.52	m	6.48
4.0 mm²	3.24	3.83	0.13	3.52	m	7.36
300/500 volt grade; three core						
1.5 mm²	2.52	2.98	0.13	3.52	m	6.51
2.5 mm²	2.86	3.38	0.13	3.52	m	6.90
4.0 mm²	3.97	4.70	0.13	3.52	m	8.22
300/500 volt grade; four core						
1.5 mm²	2.81	3.32	0.13	3.52	m	6.85
2.5 mm²	3.48	4.12	0.13	3.52	m	7.64
4.0 mm	4.76	5.64	0.13	3.52	m	9.16
Terminations; including glanding-off, connection to equipment						
Two core						
1.5 mm²	0.88	1.04	0.35	9.49	nr	10.52
2.5 mm²	1.26	1.49	0.35	9.49	nr	10.98
4.0 mm²	1.12	1.32	0.35	9.49	nr	10.81

V: ELECTRICAL SUPPLY/POWER/LIGHTING SYSTEMS

Item	Net Price £	Material £	Labour hours	Labour £	Unit	Total rate £
Three core						
1.5 mm²	0.90	1.07	0.35	9.49	nr	10.55
2.5 mm²	0.90	1.07	0.35	9.49	nr	10.55
4.0 mm²	1.15	1.36	0.35	9.49	nr	10.85
Four core						
1.5 mm²	1.18	1.39	0.35	9.49	nr	10.88
2.5 mm²	1.18	1.39	0.35	9.49	nr	10.88
4.0 mm²	1.18	1.39	0.35	9.49	nr	10.88
Cable; FP400; polymeric insulated; LSOH sheathed fire resistant; armoured; with copper stranded copper conductors; BASEC Approved to BS 7846; fixed with clips to backgrounds						
600/1000 volt grade; two core						
1.5 mm²	2.95	3.49	0.16	4.34	m	7.83
2.5 mm²	3.42	4.05	0.16	4.34	m	8.38
4.0 mm²	3.80	4.50	0.16	4.34	m	8.84
6.0 mm²	4.67	5.52	0.16	4.34	m	9.86
10 mm²	5.05	5.98	0.20	5.42	m	11.40
16 mm²	7.67	9.07	0.20	5.42	m	14.50
25 mm²	10.62	12.57	0.20	5.42	m	17.99
600/1000 volt grade; three core						
1.5 mm²	3.28	3.88	0.16	4.34	m	8.21
2.5 mm²	3.83	4.53	0.16	4.34	m	8.87
4.0 mm²	4.52	5.35	0.16	4.34	m	9.69
6.0 mm²	4.79	5.67	0.16	4.34	m	10.00
10 mm²	5.81	6.87	0.20	5.42	m	12.29
16 mm²	9.52	11.26	0.20	5.42	m	16.68
25 mm²	11.87	14.05	0.20	5.42	m	19.47
600/1000 volt grade; four core						
1.5 mm²	3.70	4.37	0.16	4.34	m	8.71
2.5 mm²	4.48	5.30	0.16	4.34	m	9.63
4.0 mm²	4.94	5.85	0.16	4.34	m	10.19
6.0 mm²	6.22	7.36	0.16	4.34	m	11.69
10 mm²	7.38	8.73	0.20	5.42	m	14.15
16 mm²	10.98	12.99	0.20	5.42	m	18.41
25 mm²	15.94	18.86	0.20	5.42	m	24.28
Terminations; including glanding-off, connection to equipment,						
Two core						
1.5 mm²	4.81	5.69	0.58	15.72	nr	21.41
2.5 mm²	4.81	5.69	0.58	15.72	nr	21.41
4.0 mm²	4.81	5.69	0.58	15.72	nr	21.41
6.0 mm²	6.05	7.16	0.67	18.16	nr	25.32
10 mm²	6.05	7.16	1.00	27.10	nr	34.26
16 mm²	8.52	10.08	1.11	30.08	nr	40.17
25 mm²	8.52	10.08	1.70	46.07	nr	56.16

V: ELECTRICAL SUPPLY/POWER/LIGHTING SYSTEMS

Item	Net Price £	Material £	Labour hours	Labour £	Unit	Total rate £
V20 : LV DISTRIBUTION (cont'd)						
Y61 - LV CABLES AND WIRING (cont'd)						
FIRE RATED CABLE (cont'd)						
Terminations; including glanding-off (cont'd)						
Three core						
1.5 mm²	4.81	5.69	0.62	16.80	nr	22.50
2.5 mm²	4.81	5.69	0.62	16.80	nr	22.50
4.0 mm²	4.81	5.69	0.62	16.80	nr	22.50
6.0 mm²	6.05	7.16	0.71	19.24	nr	26.40
10 mm²	6.05	7.16	1.06	28.73	nr	35.89
16 mm²	8.52	10.08	1.19	32.25	nr	42.33
25 mm²	8.52	10.08	1.81	49.05	nr	59.14
Four core						
1.5 mm²	4.81	5.69	0.67	18.16	nr	23.85
2.5 mm²	4.81	5.69	0.67	18.16	nr	23.85
4.0 mm²	6.05	7.16	0.71	19.24	nr	26.40
6.0 mm²	6.05	7.16	0.76	20.60	nr	27.75
10 mm²	8.52	10.08	1.14	30.90	nr	40.98
16 mm²	8.52	10.08	1.29	34.96	nr	45.04
25 mm²	14.06	16.64	1.99	53.93	nr	70.58
Cable; Firetuff fire resistant to BS 6387; fixed with clips to backgrounds						
Two core						
1.5 mm²	2.10	2.49	0.16	4.34	m	6.82
2.5 mm²	2.53	3.00	0.16	4.34	m	7.33
4.0 mm²	3.28	3.88	0.16	4.34	m	8.21
Three core						
1.5 mm²	2.56	3.02	0.16	4.34	m	7.36
2.5 mm²	2.89	3.42	0.16	4.34	m	7.76
4.0 mm²	4.18	4.94	0.16	4.34	m	9.28
Four core						
1.5 mm²	2.86	3.38	0.16	4.34	m	7.72
2.5 mm²	3.50	4.15	0.16	4.34	m	8.48
4.0 mm²	4.78	5.65	0.16	4.34	m	9.99
MODULAR WIRING						
Modular wiring systems; including commissioning						
Master distribution box; steel; fixed to backgrounds; 6 Port						
4.0mm 18 core armoured home run cable	144.00	170.42	0.90	24.39	nr	194.81
4.0mm 24 core armoured cable home run cable	144.00	170.42	0.95	25.75	nr	196.17
4.0mm 18 core armoured home run cable & data cable	152.99	181.06	0.95	25.75	nr	206.80
6.0mm 18 core armoured home run cable	144.00	170.42	1.00	27.10	nr	197.52
6.0mm 24 core armoured home run cable	144.00	170.42	1.10	29.81	nr	200.23
6.0mm 18 core armoured home run cable & data cable	152.99	181.06	1.10	29.81	nr	210.87

V: ELECTRICAL SUPPLY/POWER/LIGHTING SYSTEMS

Item	Net Price £	Material £	Labour hours	Labour £	Unit	Total rate £
Master distribution box; steel; fixed to backgrounds; 9 Port						
4.0mm 27 core armoured home run cable	179.98	213.00	1.30	35.23	nr	**248.24**
4.0mm 27 core armoured home run cable & data cable	179.98	213.00	1.45	39.30	nr	**252.30**
6.0mm 27 core armoured home run cable	188.99	223.67	1.45	39.30	nr	**262.96**
6.0mm 27 core armoured home run cable & data cable	188.99	223.67	1.55	42.01	nr	**265.67**
Metal clad cable; BSEN 60439 Part 2 1993; BASEC approved						
4.0mm 18 core	12.57	14.88	0.30	8.13	m	**23.01**
4.0mm 24 core	19.34	22.89	0.32	8.67	m	**31.56**
4.0mm 27 core	19.34	22.89	0.32	8.67	m	**31.56**
6.0mm 18 core	16.75	19.83	0.32	8.67	m	**28.50**
6.0mm 27 core	26.03	30.80	0.35	9.49	m	**40.29**
Metal clad data cable						
Single twisted pair	2.44	2.89	0.18	4.88	m	**7.77**
Twin twisted pair	3.94	4.66	0.18	4.88	m	**9.54**
Distribution cables; armoured; BSEN 60439 Part 2 1993; BASEC approved						
3 wire; 6.1 metre long	38.05	45.03	0.92	24.93	nr	**69.96**
4 wire; 6.1 metre long	45.70	54.09	0.96	26.02	nr	**80.11**
Extender cables; armoured; BSEN 60439 Part 2 1993; BASEC approved						
3 Wire						
0.9 metre long	16.63	19.68	0.13	3.52	nr	**23.21**
1.5 metre long	19.50	23.08	0.23	6.23	nr	**29.31**
2.1 metre long	22.37	26.48	0.31	8.40	nr	**34.88**
2.7 metre long	25.21	29.84	0.40	10.84	nr	**40.68**
3.4 metre long	28.08	33.24	0.51	13.82	nr	**47.06**
4.6 metre long	33.84	40.04	0.69	18.70	nr	**58.74**
6.1 metre long	40.98	48.49	0.92	24.93	nr	**73.43**
7.6 metre long	48.15	56.98	1.14	30.90	nr	**87.88**
9.1 metre long	55.33	65.48	1.37	37.13	nr	**102.61**
10.7 metre long	75.43	89.27	1.61	43.63	nr	**132.90**
4 Wire						
0.9 metre long	18.11	21.43	0.14	3.79	nr	**25.22**
1.5 metre long	21.68	25.66	0.24	6.50	nr	**32.16**
2.1 metre long	25.28	29.92	0.32	8.67	nr	**38.59**
2.7 metre long	28.84	34.13	0.43	11.65	nr	**45.79**
3.4 metre long	32.42	38.36	0.51	13.82	nr	**52.19**
4.6 metre long	39.60	46.87	0.67	18.16	nr	**65.02**
6.1 metre long	48.56	57.48	0.92	24.93	nr	**82.41**
7.6 metre long	57.50	68.05	1.22	33.06	nr	**101.11**
9.1 metre long	66.46	78.66	1.46	39.57	nr	**118.22**
10.7 metre long	75.43	89.27	1.71	46.34	nr	**135.61**

V: ELECTRICAL SUPPLY/POWER/LIGHTING SYSTEMS

Item	Net Price £	Material £	Labour hours	Labour £	Unit	Total rate £
V20 : LV DISTRIBUTION (cont'd)						
Y61 - LV CABLES AND WIRING (cont'd)						
MODULAR WIRING (cont'd)						
3 Wire; including twisted pair						
0.9 metre long	19.80	23.43	0.13	3.52	nr	26.95
1.5 metre long	24.35	28.82	0.23	6.23	nr	35.05
2.1 metre long	28.89	34.19	0.31	8.40	nr	42.59
2.7 metre long	33.43	39.56	0.40	10.84	nr	50.40
3.4 metre long	37.96	44.93	0.51	13.82	nr	58.75
4.6 metre long	47.05	55.68	0.69	18.70	nr	74.38
6.1 metre long	58.40	69.11	0.92	24.93	nr	94.05
7.6 metre long	69.73	82.52	1.14	30.90	nr	113.42
9.1 metre long	81.09	95.97	1.37	37.13	nr	133.10
10.7 metre long	92.44	109.40	1.61	43.63	nr	153.04
Extender whip ended cables; armoured; BSEN 60439 Part 2 1993; BASEC approved						
3 wire; 3.0 metre long	23.09	27.33	0.30	8.13	nr	35.46
4 wire; 3.0 metre long	27.02	31.97	0.30	8.13	nr	40.10
T Connectors						
3 wire						
Snap fix	13.89	16.44	0.10	2.71	nr	19.15
0.3 metre flexible cable	15.76	18.66	0.10	2.71	nr	21.37
0.3 metre armoured cable	15.74	18.63	0.15	4.07	nr	22.69
0.3 metre armoured cable with twisted pair	17.56	20.78	0.15	4.07	nr	24.84
4 Wire						
Snap fix	14.50	17.16	0.10	2.71	nr	19.87
0.3 metre flexible cable	16.32	19.32	0.10	2.71	nr	22.03
0.3 metre armoured cable	16.32	19.32	0.18	4.88	nr	24.20
Splitters						
5 wire	22.24	26.32	0.20	5.42	nr	31.74
5 wire converter	17.24	20.40	0.20	5.42	nr	25.82
Switch modules						
3 wire; 6.1 metre long armoured cable	47.20	55.86	0.75	20.33	nr	76.19
4 wire; 6.1 metre long armoured cable	53.11	62.85	0.80	21.68	nr	84.53
Distribution cables; unarmoured; IEC 998 DIN/VDE 0628						
3 wire; 6.1 metre long	16.23	19.20	0.70	18.97	nr	38.17
4 wire; 6.1 metre long	19.40	22.96	0.75	20.33	nr	43.29
Extender cables; unarmoured; IEC 998 DIN/VDE 0628						
3 Wire						
0.9 metre long	11.28	13.34	0.07	1.90	nr	15.24
1.5 metre long	12.17	14.40	0.12	3.25	nr	17.65
2.1 metre long	13.09	15.49	0.17	4.61	nr	20.10
2.7 metre long	16.34	19.33	0.22	5.96	nr	25.29
3.4 metre long	14.92	17.65	0.27	7.32	nr	24.97
4.6 metre long	16.87	19.97	0.37	10.03	nr	30.00

Material Costs/Prices for Measured Works – Electrical Installations

V: ELECTRICAL SUPPLY/POWER/LIGHTING SYSTEMS

Item	Net Price £	Material £	Labour hours	Labour £	Unit	Total rate £
6.1 metre long	19.16	22.68	0.49	13.28	nr	35.96
7.6 metre long	21.44	25.37	0.61	16.53	nr	41.90
9.1 metre long	23.73	28.08	0.73	19.78	nr	47.86
10.7 metre long	26.15	30.94	0.86	23.31	nr	54.25
4 Wire						
0.9 metre long	12.89	15.26	0.08	2.17	nr	17.43
1.5 metre long	14.04	16.61	0.14	3.79	nr	20.41
2.1 metre long	15.18	17.97	0.19	5.15	nr	23.11
2.7 metre long	16.34	19.33	0.24	6.50	nr	25.84
3.4 metre long	17.47	20.67	0.31	8.40	nr	29.07
4.6 metre long	19.95	23.61	0.41	11.11	nr	34.73
6.1 metre long	22.82	27.01	0.55	14.91	nr	41.92
7.6 metre long	25.70	30.41	0.68	18.43	nr	48.84
9.1 metre long	33.64	39.81	0.82	22.22	nr	62.03
10.7 metre long	36.69	43.42	0.96	26.02	nr	69.43
5 Wire						
0.9 metre long	15.20	17.99	0.09	2.44	nr	20.43
1.5 metre long	17.05	20.18	0.15	4.07	nr	24.24
2.1 metre long	18.90	22.37	0.21	5.69	nr	28.06
2.7 metre long	20.73	24.54	0.27	7.32	nr	31.86
3.4 metre long	22.59	26.74	0.34	9.21	nr	35.95
4.6 metre long	25.08	29.68	0.46	12.47	nr	42.15
6.1 metre long	31.20	36.92	0.61	16.53	nr	53.45
7.6 metre long	35.79	42.36	0.76	20.60	nr	62.96
9.1 metre long	40.40	47.82	0.91	24.66	nr	72.48
10.7 metre long	45.34	53.66	1.07	29.00	nr	82.66
Extender whip ended cables; armoured; IEC 998 DIN/VDE 0628						
3 wire; 2.5mm; 3.0 metre long	11.49	13.60	0.30	8.13	nr	21.73
4 wire; 2.5mm; 3.0 metre long	13.50	15.97	0.30	8.13	nr	24.10
T Connectors						
3 wire						
5 pin; direct fix	11.62	13.75	0.10	2.71	nr	16.46
5 pin; 1.5mm flexible cable; 0.3 metre long	14.86	17.59	0.15	4.07	nr	21.65
4 Wire						
5 pin; direct fix	13.51	15.99	0.20	5.42	nr	21.41
5 pin; 1.5mm flexible cable; 0.3 metre long	16.83	19.92	0.20	5.42	nr	25.34
5 Wire						
5 pin; direct fix	15.40	18.23	0.20	5.42	nr	23.65
Splitters						
3 way; 5 pin	8.87	10.49	0.25	6.78	nr	17.27
Switch Modules						
3 wire	25.28	29.92	0.20	5.42	nr	35.34
4 wire	26.34	31.18	0.22	5.96	nr	37.14

V: ELECTRICAL SUPPLY/POWER/LIGHTING SYSTEMS

Item	Net Price £	Material £	Labour hours	Labour £	Unit	Total rate £
V20 : LV DISTRIBUTION (cont'd)						
Y62 - BUSBAR TRUNKING						
MAINS BUSBAR						
Low impedance busbar trunking; fixed to backgrounds including supports, fixings and connections/jointing to equipment						
Straight copper busbar						
1000 Amp TP&N	430.35	509.31	3.41	92.42	m	601.73
1350 Amp TP&N	528.97	626.01	3.58	97.02	m	723.04
2000 Amp TP&N	663.46	785.18	5.00	135.51	m	920.69
2500 Amp TP&N	1002.37	1186.28	5.90	159.90	m	1346.18
Extra for fittings mains bus bar						
IP54 protection						
1000 Amp TP&N	29.35	34.73	2.16	58.54	m	93.27
1350 Amp TP&N	31.54	37.33	2.61	70.73	m	108.07
2000 Amp TP&N	43.75	51.77	3.51	95.13	m	146.90
2500 Amp TP&N	51.95	61.48	3.96	107.32	m	168.80
End cover						
1000 Amp TP&N	33.75	39.95	0.56	15.18	nr	55.12
1350 Amp TP&N	35.21	41.67	0.56	15.18	nr	56.85
2000 Amp TP&N	56.49	66.85	0.66	17.89	nr	84.74
2500 Amp TP&N	57.21	67.71	0.66	17.89	nr	85.60
Edge elbow						
1000 Amp TP&N	483.45	572.15	2.01	54.47	nr	626.62
1350 Amp TP&N	544.08	643.90	2.01	54.47	nr	698.37
2000 Amp TP&N	843.90	998.74	2.40	65.04	nr	1063.78
2500 Amp TP&N	1108.38	1311.74	2.40	65.04	nr	1376.78
Flat elbow						
1000 Amp TP&N	419.44	496.39	2.01	54.47	nr	550.87
1350 Amp TP&N	454.80	538.24	2.01	54.47	nr	592.72
2000 Amp TP&N	645.14	763.50	2.40	65.04	nr	828.55
2500 Amp TP&N	803.48	950.90	2.40	65.04	nr	1015.94
Offset						
1000 Amp TP&N	842.24	996.76	3.00	81.30	nr	1078.07
1350 Amp TP&N	1035.94	1226.01	3.00	81.30	nr	1307.31
2000 Amp TP&N	1632.25	1931.72	3.50	94.86	nr	2026.58
2500 Amp TP&N	1857.95	2198.83	3.50	94.86	nr	2293.69
Edge Z unit						
1000 Amp TP&N	1261.66	1493.14	3.00	81.30	nr	1574.45
1350 Amp TP&N	1615.40	1911.78	3.00	81.30	nr	1993.09
2000 Amp TP&N	2420.58	2864.68	3.50	94.86	nr	2959.54
2500 Amp TP&N	2764.21	3271.36	3.50	94.86	nr	3366.21
Flat Z unit						
1000 Amp TP&N	1083.12	1281.84	3.00	81.30	nr	1363.14
1350 Amp TP&N	1364.42	1614.75	3.00	81.30	nr	1696.05
2000 Amp TP&N	2031.46	2404.18	3.50	94.86	nr	2499.03
2500 Amp TP&N	2442.47	2890.59	3.50	94.86	nr	2985.45

V: ELECTRICAL SUPPLY/POWER/LIGHTING SYSTEMS

Item	Net Price £	Material £	Labour hours	Labour £	Unit	Total rate £
Edge tee						
1000 Amp TP&N	1261.66	1493.14	2.20	59.62	nr	1552.77
1350 Amp TP&N	1615.40	1911.78	2.20	59.62	nr	1971.41
2000 Amp TP&N	2420.58	2864.68	2.60	70.46	nr	2935.14
2500 Amp TP&N	2765.89	3273.35	2.60	70.46	nr	3343.81
Tap off; TP&N integral contactor/breaker						
18 Amp	205.31	242.98	0.82	22.22	nr	265.20
Tap off; TP&N fusable with on-load switch; excludes fuses						
32 Amp	581.16	687.79	0.82	22.22	nr	710.01
63 Amp	592.81	701.58	0.88	23.85	nr	725.42
100 Amp	725.37	858.46	1.18	31.98	nr	890.44
160 Amp	825.04	976.42	1.41	38.21	nr	1014.63
250 Amp	1065.09	1260.50	1.76	47.70	nr	1308.20
315 Amp	1257.38	1488.07	2.06	55.83	nr	1543.90
Tap off; TP&N MCCB						
63 Amp	734.37	869.11	0.88	23.85	nr	892.96
125 Amp	880.17	1041.66	1.18	31.98	nr	1073.64
160 Amp	961.04	1137.36	1.41	38.21	nr	1175.58
250 Amp	1236.01	1462.78	1.76	47.70	nr	1510.48
400 Amp	1575.63	1864.71	2.06	55.83	nr	1920.54
RISING MAINS BUSBAR						
Rising mains busbar; insulated supports, earth continuity bar; including couplers; fixed to backgrounds						
Straight aluminium bar						
200 Amp TP&N	153.66	181.86	2.13	57.73	m	239.58
315 Amp TP&N	171.93	203.47	2.15	58.27	m	261.74
400 Amp TP&N	199.31	235.87	2.15	58.27	m	294.14
630 Amp TP&N	246.48	291.70	2.47	66.94	m	358.64
800 Amp TP&N	369.71	437.54	2.88	78.05	m	515.60
Extra for fittings rising busbar						
End feed unit						
200 Amp TP&N	308.10	364.62	2.57	69.65	nr	434.27
315 Amp TP&N	308.10	364.62	2.76	74.80	nr	439.42
400 Amp TP&N	344.77	408.02	2.76	74.80	nr	482.83
630 Amp TP&N	344.77	408.02	3.64	98.65	nr	506.67
800 Amp TP&N	383.04	453.32	4.54	123.04	nr	576.36
Top feeder unit						
200 Amp TP&N	308.10	364.62	2.57	69.65	nr	434.27
315 Amp TP&N	308.10	364.62	2.76	74.80	nr	439.42
400 Amp TP&N	344.77	408.02	2.76	74.80	nr	482.83
630 Amp TP&N	344.77	408.02	3.64	98.65	nr	506.67
800 Amp TP&N	383.04	453.32	4.54	123.04	nr	576.36
End cap						
200 Amp TP&N	26.40	31.25	0.18	4.88	nr	36.13
315 Amp TP&N	26.40	31.25	0.27	7.32	nr	38.57
400 Amp TP&N	29.35	34.73	0.27	7.32	nr	42.05
630 Amp TP&N	29.35	34.73	0.41	11.11	nr	45.84
800 Amp TP&N	85.09	100.70	0.41	11.11	nr	111.81

V: ELECTRICAL SUPPLY/POWER/LIGHTING SYSTEMS

Item	Net Price £	Material £	Labour hours	Labour £	Unit	Total rate £
V20 : LV DISTRIBUTION (cont'd)						
Y62 – BUSBAR TRUNKING (cont'd)						
RISING MAINS BUSBAR (cont'd)						
Edge elbow						
200 Amp TP&N	36.69	43.42	0.55	14.91	nr	58.32
315 Amp TP&N	36.69	43.42	0.94	25.48	nr	68.89
400 Amp TP&N	275.82	326.42	0.94	25.48	nr	351.90
630 Amp TP&N	275.82	326.42	1.45	39.30	nr	365.72
800 Amp TP&N	261.14	309.05	1.45	39.30	nr	348.35
Flat elbow						
200 Amp TP&N	118.84	140.64	0.55	14.91	nr	155.55
315 Amp TP&N	118.84	140.64	0.94	25.48	nr	166.12
400 Amp TP&N	161.38	190.99	0.94	25.48	nr	216.46
630 Amp TP&N	161.38	190.99	1.45	39.30	nr	230.28
800 Amp TP&N	223.00	263.91	1.45	39.30	nr	303.21
Edge tee						
200 Amp TP&N	167.26	197.94	0.61	16.53	nr	214.47
315 Amp TP&N	167.26	197.94	1.02	27.64	nr	225.59
400 Amp TP&N	234.74	277.81	1.02	27.64	nr	305.45
630 Amp TP&N	234.74	277.81	1.57	42.55	nr	320.35
800 Amp TP&N	334.50	395.87	1.57	42.55	nr	438.42
Flat tee						
200 Amp TP&N	214.20	253.50	0.61	16.53	nr	270.03
315 Amp TP&N	167.26	197.94	1.02	27.64	nr	225.59
400 Amp TP&N	337.43	399.34	1.02	27.64	nr	426.99
630 Amp TP&N	337.43	399.34	1.57	42.55	nr	441.89
800 Amp TP&N	470.94	557.34	1.57	42.55	nr	599.89
Tap off units						
TP&N fusable with on-load switch; excludes fuses						
32 Amp	156.98	185.78	0.82	22.22	nr	208.00
63 Amp	207.48	245.55	0.88	23.85	nr	269.40
100 Amp	278.47	329.56	1.18	31.98	nr	361.54
250 Amp	417.71	494.34	1.41	38.21	nr	532.56
400 Amp	608.81	720.51	2.06	55.83	nr	776.34
TP&N MCCB						
32 Amp	159.72	189.02	0.82	22.22	nr	211.25
63 Amp	221.14	261.71	0.88	23.85	nr	285.56
100 Amp	354.91	420.02	1.18	31.98	nr	452.00
250 Amp	600.62	710.82	1.41	38.21	nr	749.03
400 Amp	1053.82	1247.17	2.06	55.83	nr	1303.00
LIGHTING BUSBAR						
Pre-wired busbar, plug-in trunking for lighting; galvanised sheet steel housing (PE); tin-plated copper conductors with tap-off units at 1m intervals						
Straight lengths - 25 Amp						
2 Pole & PE	25.85	30.59	0.16	4.34	m	34.93
4 Pole & PE	27.96	33.09	0.16	4.34	m	37.42

V: ELECTRICAL SUPPLY/POWER/LIGHTING SYSTEMS

Item	Net Price £	Material £	Labour hours	Labour £	Unit	Total rate £
Straight lengths - 40 Amp						
2 Pole & PE	25.56	30.25	0.16	4.34	m	34.59
4 Pole & PE	34.33	40.62	0.16	4.34	m	44.96
Components for pre-wired busbars, plug-in trunking for lighting						
Plug-in tap off units						
10 Amp with phase selection, 2P & PE; 2m of cable	16.35	19.35	0.10	2.71	nr	22.06
10 Amp 4 Pole & PE; 3m of cable	21.85	25.86	0.10	2.71	nr	28.57
16 Amp 4 Pole & PE; 3m of cable	21.18	25.06	0.10	2.71	nr	27.77
16 Amp with phase selection, 2P & PE; no cable	18.75	22.19	0.10	2.71	nr	24.90
Trunking components						
End feed unit & cover; 4P & PE	27.62	32.68	0.23	6.23	nr	38.92
Centre feed unit	136.89	162.00	0.29	7.86	nr	169.86
Right hand, intermediate terminal box feed unit	28.83	34.11	0.23	6.23	nr	40.35
End cover (for R/hand feed)	7.98	9.45	0.06	1.63	nr	11.07
Flexible elbow unit	65.54	77.56	0.12	3.25	nr	80.81
Fixing bracket - universal	4.78	5.66	0.10	2.71	nr	8.37
Suspension Bracket - Flat	4.21	4.98	0.10	2.71	nr	7.69
UNDERFLOOR BUSBAR						
Pre-wired busbar, plug-in trunking for underfloor power distribution; galvanised sheet steel housing (PE); copper conductors with tap-off units at 300mm intervals						
Straight lengths - 63 Amp						
2 pole & PE	17.37	20.55	0.28	7.59	m	28.14
3 pole & PE; Clean Earth System	21.97	26.00	0.28	7.59	m	33.59
Components for pre-wired busbars, plug-in trunking for underfloor power distribution						
Plug-in tap off units						
32 Amp 2P & PE; 3m metal flexible pre-wired conduit	28.78	34.06	0.25	6.78	nr	40.84
32 Amp 3P & PE; clean earth; 3m metal flexible pre-wired conduit	34.65	41.00	0.28	7.59	nr	48.59
Trunking components						
End feed unit & cover; 2P & PE	31.01	36.70	0.35	9.49	nr	46.18
End feed unit & cover; 3P & PE; clean earth	33.96	40.19	0.38	10.30	nr	50.49
End cover; 2P & PE	9.65	11.42	0.11	2.98	nr	14.40
End cover; 3P & PE	10.38	12.28	0.11	2.98	nr	15.26
Flexible interlink/corner; 2P&PE; 1m long	53.43	63.23	0.34	9.21	nr	72.44
Flexible interlink/corner; 3P&PE; 1m long	60.33	71.39	0.35	9.49	nr	80.88
Flexible interlink/corner; 2P&PE; 2m long	65.20	77.16	0.37	10.03	nr	87.18
Flexible interlink/corner; 3P&PE; 2m long	71.53	84.65	0.37	10.03	nr	94.68

V: ELECTRICAL SUPPLY/POWER/LIGHTING SYSTEMS

Item	Net Price £	Material £	Labour hours	Labour £	Unit	Total rate £
V20 : LV DISTRIBUTION (cont'd)						
Y63 - CABLE SUPPORTS						
LADDER RACK						
Light duty Galvanised Steel Ladder Rack; fixed to backgrounds; including supports, fixings and brackets; earth continuity straps						
Straight lengths						
150 mm wide ladder	19.89	23.54	0.69	18.70	m	42.24
300 mm wide ladder	20.71	24.51	0.88	23.85	m	48.36
450 mm wide ladder	22.11	26.17	1.26	34.15	m	60.32
600 mm wide ladder	23.55	27.87	1.51	40.92	m	68.80
Extra over; (cutting and jointing racking to fittings is included)						
Inside riser bend						
150 mm wide ladder	50.09	59.28	0.33	8.94	nr	68.22
300 mm wide ladder	51.86	61.38	0.56	15.18	nr	76.55
450 mm wide ladder	52.64	62.30	0.85	23.04	nr	85.33
600 mm wide ladder	54.55	64.56	0.99	26.83	nr	91.39
Outside riser bend						
300 mm wide ladder	50.09	59.28	0.43	11.65	nr	70.93
450 mm wide ladder	52.64	62.30	0.73	19.78	nr	82.08
600 mm wide ladder	54.55	64.56	0.86	23.31	nr	87.87
Equal tee						
300 mm wide ladder	60.53	71.64	0.62	16.80	nr	88.44
450 mm wide ladder	67.53	79.92	1.09	29.54	nr	109.46
600 mm wide ladder	71.01	84.04	1.12	30.35	nr	114.39
Unequal tee						
300 mm wide ladder	63.07	74.64	0.57	15.45	nr	90.09
450 mm wide ladder	64.78	76.67	1.17	31.71	nr	108.38
600 mm wide ladder	66.55	78.76	1.17	31.71	nr	110.47
4 way cross overs						
300 mm wide ladder	98.21	116.23	0.72	19.51	nr	135.74
450 mm wide ladder	106.21	125.70	1.13	30.62	nr	156.32
600 mm wide ladder	125.94	149.05	1.29	34.96	nr	184.01
Heavy duty galvanised steel ladder rack; fixed to backgrounds; including supports, fixings and brackets; earth continuity straps						
Straight lengths						
150 mm wide ladder	26.23	31.04	0.68	18.43	m	49.47
300 mm wide ladder	27.65	32.72	0.79	21.41	m	54.13
450 mm wide ladder	29.00	34.33	1.07	29.00	m	63.32
600 mm wide ladder	31.28	37.02	1.24	33.61	m	70.63
750 mm wide ladder	34.67	41.03	1.49	40.38	m	81.41
900 mm wide ladder	35.57	42.09	1.67	45.26	m	87.35

V: ELECTRICAL SUPPLY/POWER/LIGHTING SYSTEMS

Item	Net Price £	Material £	Labour hours	Labour £	Unit	Total rate £
Extra over; (cutting and jointing racking to fittings is included)						
Flat bend						
150 mm wide ladder	42.94	50.82	0.34	9.21	nr	**60.03**
300 mm wide ladder	45.51	53.86	0.39	10.57	nr	**64.43**
450 mm wide ladder	51.01	60.37	0.43	11.65	nr	**72.02**
600 mm wide ladder	57.23	67.73	0.61	16.53	nr	**84.26**
750 mm wide ladder	64.38	76.19	0.82	22.22	nr	**98.42**
900 mm wide ladder	68.55	81.12	0.97	26.29	nr	**107.41**
Inside riser bend						
150 mm wide ladder	59.01	69.84	0.27	7.32	nr	**77.16**
300 mm wide ladder	59.87	70.86	0.45	12.20	nr	**83.05**
450 mm wide ladder	64.71	76.58	0.65	17.62	nr	**94.20**
600 mm wide ladder	68.94	81.58	0.81	21.95	nr	**103.54**
750 mm wide ladder	71.19	84.25	0.92	24.93	nr	**109.18**
900 mm wide ladder	77.02	91.15	1.06	28.73	nr	**119.88**
Outside riser bend						
150 mm wide ladder	59.01	69.84	0.27	7.32	nr	**77.16**
300 mm wide ladder	59.87	70.86	0.33	8.94	nr	**79.80**
450 mm wide ladder	64.71	76.58	0.61	16.53	nr	**93.12**
600 mm wide ladder	68.94	81.58	0.76	20.60	nr	**102.18**
750 mm wide ladder	71.19	84.25	0.94	25.48	nr	**109.73**
900 mm wide ladder	77.49	91.71	1.05	28.46	nr	**120.16**
Equal tee						
150mm wide ladder	65.18	77.14	0.37	10.03	nr	**87.17**
300 mm wide ladder	75.91	89.84	0.57	15.45	nr	**105.28**
450 mm wide ladder	82.26	97.35	0.83	22.49	nr	**119.84**
600 mm wide ladder	91.27	108.02	0.92	24.93	nr	**132.95**
750 mm wide ladder	115.50	136.69	1.13	30.62	nr	**167.31**
900 mm wide ladder	116.06	137.36	1.20	32.52	nr	**169.88**
Unequal tee						
300 mm wide ladder	72.35	85.62	0.57	15.45	nr	**101.07**
450 mm wide ladder	70.75	83.73	1.17	31.71	nr	**115.44**
600 mm wide ladder	75.91	89.84	1.17	31.71	nr	**121.55**
750 mm wide ladder	80.49	95.25	1.25	33.88	nr	**129.13**
900 mm wide ladder	101.13	119.68	1.33	36.05	nr	**155.72**
4 way cross overs						
150 mm wide ladder	102.67	121.51	0.50	13.55	nr	**135.06**
300 mm wide ladder	108.11	127.95	0.67	18.16	nr	**146.10**
450 mm wide ladder	135.11	159.90	0.92	24.93	nr	**184.83**
600 mm wide ladder	144.18	170.64	1.07	29.00	nr	**199.64**
750 mm wide ladder	151.26	179.02	1.25	33.88	nr	**212.89**
900 mm wide ladder	179.62	212.57	1.36	36.86	nr	**249.43**

V: ELECTRICAL SUPPLY/POWER/LIGHTING SYSTEMS

Item	Net Price £	Material £	Labour hours	Labour £	Unit	Total rate £
V20 : LV DISTRIBUTION (cont'd)						
Y63 - CABLE SUPPORTS (cont'd)						
LADDER RACK (cont'd)						
Extra heavy duty galvanised steel ladder rack; fixed to backgrounds; including supports, fixings and brackets; earth continuity straps						
Straight lengths						
150 mm wide ladder	28.50	33.73	0.63	17.07	m	50.80
300 mm wide ladder	29.90	35.39	0.70	18.97	m	54.36
450 mm wide ladder	31.38	37.13	0.83	22.49	m	59.63
600 mm wide ladder	31.81	37.65	0.89	24.12	m	61.77
750 mm wide ladder	31.90	37.75	1.22	33.06	m	70.81
900 mm wide ladder	33.29	39.40	1.44	39.03	m	78.42
Extra over; (cutting and jointing racking to fittings is included)						
Flat bend						
150 mm wide ladder	54.50	64.50	0.36	9.76	nr	74.26
300 mm wide ladder	56.72	67.13	0.39	10.57	nr	77.70
450 mm wide ladder	61.24	72.48	0.43	11.65	nr	84.13
600 mm wide ladder	66.14	78.27	0.61	16.53	nr	94.81
750 mm wide ladder	71.70	84.85	0.82	22.22	nr	107.07
900 mm wide ladder	77.28	91.46	0.97	26.29	nr	117.74
Inside riser bend						
150 mm wide ladder	65.31	77.30	0.36	9.76	nr	87.05
300 mm wide ladder	66.19	78.33	0.39	10.57	nr	88.90
450 mm wide ladder	69.96	82.80	0.43	11.65	nr	94.45
600 mm wide ladder	73.66	87.17	0.61	16.53	nr	103.70
750 mm wide ladder	75.58	89.45	0.82	22.22	nr	111.67
900 mm wide ladder	81.49	96.44	0.97	26.29	nr	122.73
Outside riser bend						
150 mm wide ladder	65.31	77.30	0.36	9.76	nr	87.05
300 mm wide ladder	66.19	78.33	0.39	10.57	nr	88.90
450 mm wide ladder	69.96	82.80	0.41	11.11	nr	93.91
600 mm wide ladder	73.66	87.17	0.57	15.45	nr	102.62
750 mm wide ladder	75.58	89.45	0.82	22.22	nr	111.67
900 mm wide ladder	81.49	96.44	0.93	25.20	nr	121.65
Equal tee						
150 mm wide ladder	78.72	93.16	0.37	10.03	nr	103.19
300 mm wide ladder	86.91	102.85	0.57	15.45	nr	118.30
450 mm wide ladder	92.06	108.95	0.83	22.49	nr	131.45
600 mm wide ladder	99.04	117.21	0.92	24.93	nr	142.14
750 mm wide ladder	117.35	138.88	1.13	30.62	nr	169.51
900 mm wide ladder	119.91	141.91	1.20	32.52	nr	174.43
Unequal tee						
150 mm wide ladder	81.08	95.95	0.37	10.03	nr	105.98
300 mm wide ladder	89.27	105.64	0.57	15.45	nr	121.09
450 mm wide ladder	94.42	111.75	1.17	31.71	nr	143.46
600 mm wide ladder	101.40	120.00	1.17	31.71	nr	151.71
750 mm wide ladder	119.71	397.47	1.25	33.88	nr	431.35
900 mm wide ladder	124.63	147.50	1.33	36.05	nr	183.54

V: ELECTRICAL SUPPLY/POWER/LIGHTING SYSTEMS

Item	Net Price £	Material £	Labour hours	Labour £	Unit	Total rate £
4 way cross overs						
150 mm wide ladder	109.47	129.55	0.50	13.55	nr	143.10
300 mm wide ladder	113.76	134.64	0.67	18.16	nr	152.79
450 mm wide ladder	133.09	157.51	0.92	24.93	nr	182.44
600 mm wide ladder	141.23	167.15	1.07	29.00	nr	196.15
750 mm wide ladder	146.80	173.74	1.25	33.88	nr	207.61
900 mm wide ladder	153.40	181.54	1.36	36.86	nr	218.40
CABLE TRAY						
Galvanised steel cable tray to BS 729; including standard coupling joints, fixings and earth continuity straps (supports and hangers are excluded)						
Light duty tray						
Straight lengths						
50 mm wide	2.15	2.54	0.19	5.15	m	7.69
75 mm wide	2.47	2.92	0.23	6.23	m	9.15
100 mm wide	2.74	3.24	0.31	8.40	m	11.64
150 mm wide	3.48	4.12	0.33	8.94	m	13.06
225 mm wide	5.35	6.33	0.39	10.57	m	16.90
300 mm wide	8.22	9.73	0.49	13.28	m	23.01
450 mm wide	13.05	15.45	0.60	16.26	m	31.71
600 mm wide	17.77	21.03	0.79	21.41	m	42.44
750 mm wide	22.56	26.70	1.04	28.19	m	54.89
900 mm wide	28.12	33.28	1.26	34.15	m	67.43
Extra over; (cutting and jointing tray to fittings is included)						
Straight reducer						
75 mm wide	5.88	6.95	0.22	5.96	nr	12.92
100 mm wide	6.47	7.65	0.25	6.78	nr	14.43
150 mm wide	8.59	10.17	0.27	7.32	nr	17.48
225 mm wide	10.96	12.97	0.34	9.21	nr	22.19
300 mm wide	14.12	16.72	0.39	10.57	nr	27.29
450 mm wide	21.17	25.05	0.49	13.28	nr	38.33
600 mm wide	25.66	30.37	0.54	14.63	nr	45.01
750 mm wide	32.73	38.74	0.61	16.53	nr	55.27
900 mm wide	38.02	45.00	0.69	18.70	nr	63.70
Flat bend; 90°						
50mm wide	3.85	4.55	0.19	5.15	nr	9.70
75 mm wide	3.94	4.66	0.24	6.50	nr	11.17
100 mm wide	4.61	5.46	0.28	7.59	nr	13.05
150 mm wide	4.78	5.66	0.30	8.13	nr	13.79
225 mm wide	6.62	7.83	0.36	9.76	nr	17.59
300 mm wide	9.02	10.67	0.44	11.92	nr	22.59
450 mm wide	14.58	17.26	0.57	15.45	nr	32.71
600 mm wide	22.76	26.94	0.69	18.70	nr	45.64
750 mm wide	32.24	38.15	0.81	21.95	nr	60.10
900 mm wide	47.41	56.11	0.94	25.48	nr	81.59

V: ELECTRICAL SUPPLY/POWER/LIGHTING SYSTEMS

Item	Net Price £	Material £	Labour hours	Labour £	Unit	Total rate £
V20 : LV DISTRIBUTION (cont'd)						
Y63 - CABLE SUPPORTS (cont'd)						
CABLE TRAY (cont'd)						
Adjustable riser						
50 mm wide	6.96	8.24	0.26	7.05	nr	15.29
75 mm wide	7.61	9.01	0.29	7.86	nr	16.87
100 mm wide	8.38	9.92	0.32	8.67	nr	18.59
150 mm wide	10.68	12.64	0.36	9.76	nr	22.39
225 mm wide	13.13	15.54	0.44	11.92	nr	27.47
300 mm wide	16.64	19.69	0.52	14.09	nr	33.78
450 mm wide	21.32	25.23	0.66	17.89	nr	43.12
600 mm wide	26.97	31.92	0.79	21.41	nr	53.33
750 mm wide	34.69	41.06	1.03	27.91	nr	68.97
900 mm wide	40.84	48.33	1.10	29.81	nr	78.14
Inside riser; 90°						
50 mm wide	5.58	6.61	0.28	7.59	nr	14.19
75 mm wide	5.88	6.95	0.31	8.40	nr	15.36
100 mm wide	6.35	13.30	0.33	8.94	nr	22.25
150 mm wide	8.38	9.92	0.37	10.03	nr	19.94
225 mm wide	10.60	12.54	0.44	11.92	nr	24.47
300 mm wide	14.38	17.02	0.53	14.36	nr	31.39
450 mm wide	20.17	23.87	0.67	18.16	nr	42.02
600 mm wide	26.90	31.84	0.79	21.41	nr	53.25
750 mm wide	33.18	39.27	0.95	25.75	nr	65.02
900 mm wide	39.42	46.66	1.11	30.08	nr	76.74
Outside riser; 90°						
50 mm wide	5.58	6.61	0.28	7.59	nr	14.19
75 mm wide	5.88	6.95	0.31	8.40	nr	15.36
100 mm wide	6.35	7.51	0.33	8.94	nr	16.46
150 mm wide	8.38	9.92	0.37	10.03	nr	19.94
225 mm wide	10.60	12.54	0.44	11.92	nr	24.47
300 mm wide	14.38	17.02	0.53	14.36	nr	31.39
450 mm wide	20.17	23.87	0.67	18.16	nr	42.02
600 mm wide	26.90	31.84	0.79	21.41	nr	53.25
750 mm wide	33.18	39.27	0.95	25.75	nr	65.02
900 mm wide	63.02	74.59	1.11	30.08	nr	104.67
Equal tee						
50 mm wide	5.55	6.56	0.30	8.13	nr	14.69
75 mm wide	5.76	6.81	0.31	8.40	nr	15.22
100 mm wide	6.25	7.40	0.35	9.49	nr	16.89
150 mm wide	6.96	8.24	0.36	9.76	nr	18.00
225 mm wide	9.63	11.40	0.74	20.06	nr	31.45
300 mm wide	12.05	14.26	0.54	14.63	nr	28.89
450 mm wide	21.85	25.86	0.71	19.24	nr	45.11
600 mm wide	30.03	35.54	0.92	24.93	nr	60.47
750 mm wide	45.01	53.26	1.19	32.25	nr	85.51
900 mm wide	64.02	75.76	1.44	39.03	nr	114.79

V: ELECTRICAL SUPPLY/POWER/LIGHTING SYSTEMS

Item	Net Price £	Material £	Labour hours	Labour £	Unit	Total rate £
Unequal tee						
75 mm wide	5.76	6.81	0.38	10.30	nr	17.11
100 mm wide	6.25	7.40	0.39	10.57	nr	17.97
150 mm wide	6.96	8.24	0.43	11.65	nr	19.89
225 mm wide	9.63	11.40	0.50	13.55	nr	24.95
300 mm wide	12.05	14.26	0.63	17.07	nr	31.33
450 mm wide	21.85	25.86	0.80	21.68	nr	47.54
600 mm wide	30.03	35.54	1.02	27.64	nr	63.18
750 mm wide	45.01	53.26	1.12	30.35	nr	83.62
900 mm wide	64.02	75.76	1.35	36.59	nr	112.35
4 way crossovers						
50 mm wide	7.72	9.13	0.38	10.30	nr	19.43
75 mm wide	7.87	9.31	0.40	10.84	nr	20.16
100 mm wide	8.63	10.21	0.40	10.84	nr	21.05
150 mm wide	9.68	11.45	0.44	11.92	nr	23.38
225 mm wide	12.90	15.26	0.53	14.36	nr	29.63
300 mm wide	18.34	21.70	0.64	17.34	nr	39.05
450 mm wide	29.95	35.44	0.84	22.77	nr	58.21
600 mm wide	40.04	47.38	1.03	27.91	nr	75.30
750 mm wide	60.27	71.33	1.13	30.62	nr	101.96
900 mm wide	87.83	103.94	1.36	36.86	nr	140.80
Medium duty tray with return flange						
Straight lengths						
75 mm wide	3.67	4.34	0.33	8.94	m	13.29
100 mm wide	4.04	4.78	0.35	9.49	m	14.26
150 mm wide	4.90	5.80	0.39	10.57	m	16.37
225 mm wide	5.83	6.90	0.45	12.20	m	19.09
300 mm wide	8.35	9.89	0.57	15.45	m	25.34
450 mm wide	12.40	14.68	0.69	18.70	m	33.38
600 mm wide	17.35	20.53	0.91	24.66	m	45.19
Extra over; (cutting and jointing tray to fittings is included.)						
Straight reducer						
100 mm wide	11.10	13.14	0.25	6.78	nr	19.92
150 mm wide	12.13	14.36	0.27	7.32	nr	21.67
225 mm wide	13.97	16.53	0.34	9.21	nr	25.75
300 mm wide	16.18	19.15	0.39	10.57	nr	29.72
450 mm wide	20.90	24.73	0.49	13.28	nr	38.01
600 mm wide	25.90	30.65	0.54	14.63	nr	45.29
Flat bend; 90°						
75 mm wide	14.80	17.51	0.24	6.50	nr	24.02
100 mm wide	16.48	19.51	0.28	7.59	nr	27.10
150 mm wide	17.49	20.70	0.30	8.13	nr	28.83
225 mm wide	20.18	23.88	0.36	9.76	nr	33.64
300 mm wide	24.89	29.45	0.44	11.92	nr	41.38
450 mm wide	37.39	44.25	0.57	15.45	nr	59.70
600 mm wide	45.21	53.50	0.69	18.70	nr	72.20
Adjustable bend						
75 mm wide	16.92	20.03	0.29	7.86	nr	27.89
100 mm wide	18.42	21.80	0.32	8.67	nr	30.47
150 mm wide	21.06	24.93	0.36	9.76	nr	34.68
225 mm wide	23.61	27.94	0.44	11.92	nr	39.87
300 mm wide	26.67	31.56	0.52	14.09	nr	45.65

V: ELECTRICAL SUPPLY/POWER/LIGHTING SYSTEMS

Item	Net Price £	Material £	Labour hours	Labour £	Unit	Total rate £
V20 : LV DISTRIBUTION (cont'd)						
Y63 - CABLE SUPPORTS (cont'd)						
CABLE TRAY (cont'd)						
Extra over; (cutting and jointing) (cont'd)						
Adjustable riser						
75 mm wide	15.10	17.88	0.29	7.86	nr	25.73
100 mm wide	15.35	18.17	0.32	8.67	nr	26.84
150 mm wide	16.91	20.01	0.36	9.76	nr	29.77
225 mm wide	17.94	21.23	0.44	11.92	nr	33.15
300 mm wide	19.13	22.64	0.52	14.09	nr	36.73
450 mm wide	25.09	29.69	0.66	17.89	nr	47.58
600 mm wide	30.77	36.42	0.79	21.41	nr	57.83
Inside riser; 90°						
75 mm wide	9.48	11.21	0.31	8.40	nr	19.62
100 mm wide	9.57	11.33	0.33	8.94	nr	20.27
150 mm wide	10.90	12.90	0.37	10.03	nr	22.93
225 mm wide	13.32	15.77	0.44	11.92	nr	27.69
300 mm wide	16.18	19.15	0.53	14.36	nr	33.51
450 mm wide	23.36	27.65	0.67	18.16	nr	45.81
600 mm wide	37.48	44.35	0.79	21.41	nr	65.76
Outside riser; 90°						
75 mm wide	9.48	11.21	0.31	8.40	nr	19.62
100 mm wide	9.57	11.33	0.33	8.94	nr	20.27
150 mm wide	10.90	12.90	0.37	10.03	nr	22.93
225 mm wide	13.32	15.77	0.44	11.92	nr	27.69
300 mm wide	16.18	19.15	0.53	14.36	nr	33.51
450 mm wide	23.36	27.65	0.67	18.16	nr	45.81
600 mm wide	37.48	44.35	0.79	21.41	nr	65.76
Equal tee						
75 mm wide	21.07	24.94	0.31	8.40	nr	33.34
100 mm wide	22.47	26.59	0.35	9.49	nr	36.07
150 mm wide	24.19	28.63	0.36	9.76	nr	38.38
225 mm wide	26.29	31.11	0.74	20.06	nr	51.17
300 mm wide	31.90	37.75	0.54	14.63	nr	52.38
450 mm wide	42.03	49.74	0.71	19.24	nr	68.99
600 mm wide	60.38	71.46	0.92	24.93	nr	96.39
Unequal tee						
100 mm wide	22.47	26.59	0.39	10.57	nr	37.16
150 mm wide	22.47	26.59	0.43	11.65	nr	38.24
225 mm wide	26.29	31.11	0.50	13.55	nr	44.66
300 mm wide	31.90	37.75	0.63	17.07	nr	54.82
450 mm wide	42.03	49.74	0.80	21.68	nr	71.42
600 mm wide	60.38	71.46	1.02	27.64	nr	99.10
4 way crossovers						
75 mm wide	29.88	35.36	0.40	10.84	nr	46.20
100 mm wide	32.21	38.12	0.40	10.84	nr	48.96
150 mm wide	27.99	33.12	0.44	11.92	nr	45.05
225 mm wide	40.47	47.90	0.53	14.36	nr	62.26
300 mm wide	47.06	55.69	0.64	17.34	nr	73.04
450 mm wide	59.94	70.94	0.84	22.77	nr	93.71
600 mm wide	87.37	103.40	1.03	27.91	nr	131.31

V: ELECTRICAL SUPPLY/POWER/LIGHTING SYSTEMS

Item	Net Price £	Material £	Labour hours	Labour £	Unit	Total rate £
Heavy duty tray with return flange						
Straight lengths						
75 mm	6.88	8.14	0.34	9.21	m	**17.36**
100 mm	7.34	8.69	0.36	9.76	m	**18.44**
150 mm	8.48	10.04	0.40	10.84	m	**20.88**
225 mm	9.51	11.26	0.46	12.47	m	**23.72**
300 mm	11.59	13.71	0.58	15.72	m	**29.43**
450 mm	17.52	20.74	0.70	18.97	m	**39.71**
600 mm	21.24	25.14	0.92	24.93	m	**50.07**
750 mm	26.79	31.70	1.01	27.37	m	**59.07**
900 mm	29.63	35.07	1.14	30.90	m	**65.96**
Extra over; (cutting and jointing tray to fittings is included)						
Straight reducer						
100 mm wide	18.20	21.53	0.25	6.78	nr	**28.31**
150 mm wide	18.90	22.37	0.27	7.32	nr	**29.69**
225 mm wide	21.46	25.40	0.34	9.21	nr	**34.62**
300 mm wide	23.93	28.32	0.39	10.57	nr	**38.89**
450 mm wide	34.22	40.50	0.49	13.28	nr	**53.78**
600 mm wide	38.05	45.04	0.54	14.63	nr	**59.67**
750 mm wide	48.25	57.10	0.60	16.26	nr	**73.36**
900 mm wide	52.93	62.65	0.66	17.89	nr	**80.53**
Flat bend; 90°						
75 mm wide	20.36	24.09	0.24	6.50	nr	**30.59**
100 mm wide	23.15	27.40	0.28	7.59	nr	**34.99**
150 mm wide	24.67	29.20	0.30	8.13	nr	**37.33**
225 mm wide	27.90	33.01	0.36	9.76	nr	**42.77**
300 mm wide	31.79	37.62	0.44	11.92	nr	**49.55**
450 mm wide	46.61	55.16	0.57	15.45	nr	**70.61**
600 mm wide	62.30	73.73	0.69	18.70	nr	**92.43**
750 mm wide	83.48	98.80	0.83	22.49	nr	**121.30**
900 mm wide	94.67	112.04	1.01	27.37	nr	**139.41**
Adjustable bend						
75 mm wide	21.17	25.05	0.29	7.86	nr	**32.91**
100 mm wide	23.42	27.72	0.32	8.67	nr	**36.39**
150 mm wide	24.74	29.28	0.36	9.76	nr	**39.04**
225 mm wide	27.23	32.23	0.44	11.92	nr	**44.16**
300 mm wide	30.48	36.07	0.52	14.09	nr	**50.16**
Adjustable riser						
75 mm wide	18.66	22.08	0.29	7.86	nr	**29.94**
100 mm wide	19.40	22.96	0.32	8.67	nr	**31.63**
150 mm wide	21.48	25.42	0.36	9.76	nr	**35.17**
225 mm wide	23.01	27.23	0.44	11.92	nr	**39.16**
300 mm wide	24.92	29.49	0.52	14.09	nr	**43.59**
450 mm wide	30.33	35.89	0.66	17.89	nr	**53.78**
600 mm wide	36.14	42.77	0.79	21.41	nr	**64.18**
750 mm wide	43.75	51.78	1.03	27.91	nr	**79.70**
900 mm wide	50.80	60.12	1.10	29.81	nr	**89.93**

V: ELECTRICAL SUPPLY/POWER/LIGHTING SYSTEMS

Item	Net Price £	Material £	Labour hours	Labour £	Unit	Total rate £
V20 : LV DISTRIBUTION (cont'd)						
Y63 - CABLE SUPPORTS (cont'd)						
CABLE TRAY (cont'd)						
Extra over; (cutting and jointing) (cont'd)						
Inside riser; 90°						
75 mm wide	15.61	18.48	0.31	8.40	nr	26.88
100 mm wide	15.81	18.71	0.33	8.94	nr	27.66
150 mm wide	17.13	20.28	0.37	10.03	nr	30.30
225 mm wide	17.91	21.20	0.44	11.92	nr	33.12
300 mm wide	18.47	21.86	0.53	14.36	nr	36.22
450 mm wide	31.68	37.50	0.67	18.16	nr	55.65
600 mm wide	38.87	46.00	0.79	21.41	nr	67.41
750 mm wide	48.11	56.94	0.95	25.75	nr	82.68
900 mm wide	56.72	67.13	1.11	30.08	nr	97.21
Outside riser; 90°						
75 mm wide	15.61	18.48	0.31	8.40	nr	26.88
100 mm wide	15.81	18.71	0.33	8.94	nr	27.66
150 mm wide	17.13	20.28	0.37	10.03	nr	30.30
225 mm wide	17.91	21.20	0.44	11.92	nr	33.12
300 mm wide	18.47	21.86	0.53	14.36	nr	36.22
450 mm wide	31.68	37.50	0.67	18.16	nr	55.65
600 mm wide	38.87	46.00	0.79	21.41	nr	67.41
750 mm wide	48.11	56.94	0.95	25.75	nr	82.68
900 mm wide	56.72	67.13	1.11	30.08	nr	97.21
Equal tee						
75 mm wide	27.19	32.18	0.31	8.40	nr	40.58
100 mm wide	30.39	35.96	0.35	9.49	nr	45.45
150 mm wide	33.28	39.38	0.36	9.76	nr	49.14
225 mm wide	38.30	45.33	0.74	20.06	nr	65.39
300 mm wide	41.41	49.00	0.54	14.63	nr	63.64
450 mm wide	60.13	71.17	0.71	19.24	nr	90.41
600 mm wide	79.39	93.96	0.92	24.93	nr	118.89
750 mm wide	104.70	123.91	1.19	32.25	nr	156.16
900 mm wide	119.56	141.49	1.45	39.30	nr	180.79
Unequal tee						
75 mm wide	27.19	32.18	0.38	10.30	nr	42.47
100 mm wide	30.36	35.93	0.39	10.57	nr	46.50
150 mm wide	33.28	39.38	0.43	11.65	nr	51.03
225 mm wide	38.30	45.33	0.50	13.55	nr	58.88
300 mm wide	41.41	49.00	0.63	17.07	nr	66.08
450 mm wide	60.13	71.17	0.80	21.68	nr	92.85
600 mm wide	79.39	93.96	1.02	27.64	nr	121.60
750 mm wide	104.70	123.91	1.12	30.35	nr	154.26
900 mm wide	126.64	149.87	1.35	36.59	nr	186.46
4 way crossovers						
75 mm wide	27.59	32.65	0.40	10.84	nr	43.49
100 mm wide	39.54	46.80	0.40	10.84	nr	57.64
150 mm wide	39.54	46.80	0.44	11.92	nr	58.72
225 mm wide	54.98	65.06	0.53	14.36	nr	79.43
300 mm wide	60.63	71.75	0.64	17.34	nr	89.10

V: ELECTRICAL SUPPLY/POWER/LIGHTING SYSTEMS

Item	Net Price £	Material £	Labour hours	Labour £	Unit	Total rate £
450 mm wide	87.21	103.21	0.84	22.77	nr	125.98
600 mm wide	118.53	140.28	1.03	27.91	nr	168.19
750 mm wide	144.56	171.08	1.13	30.62	nr	201.71
900 mm wide	174.38	206.37	1.36	36.86	nr	243.23
GRP cable tray including standard coupling joints and fixings; (supports and hangers excluded)						
Tray						
100 mm wide	22.09	26.14	0.34	9.21	m	35.36
200 mm wide	28.27	33.46	0.39	10.57	m	44.03
400 mm wide	44.34	52.48	0.53	14.36	m	66.84
Cover						
100 mm wide	12.44	14.72	0.10	2.71	m	17.43
200 mm wide	16.48	19.51	0.11	2.98	m	22.49
400 mm wide	28.15	33.32	0.14	3.79	m	37.11
Extra for; (cutting and jointing to fittings included)						
Reducer						
200 mm wide	58.69	69.46	0.23	6.23	nr	75.69
400 mm wide	76,68	90.74	0.30	8.13	nr	98.87
Reducer cover						
200 mm wide	36.73	43.47	0.25	6.78	nr	50.25
400 mm wide	53.69	63.54	0.28	7.59	nr	71.13
Bend						
100 mm wide	49.08	58.08	0.34	9.21	nr	67.29
200 mm wide	56.97	67.42	0.40	10.84	nr	78.26
400 mm wide	73.21	86.64	0.32	8.67	nr	95.31
Bend cover						
100 mm wide	24.38	28.85	0.10	2.71	nr	31.56
200 mm wide	32.57	38.54	0.10	2.71	nr	41.25
400 mm wide	45.01	53.26	0.13	3.52	nr	56.79
Tee						
100 mm wide	62.47	73.93	0.37	10.03	nr	83.96
200 mm wide	68.86	81.50	0.43	11.65	nr	93.15
400 mm wide	85.43	101.11	0.56	15.18	nr	116.28
Tee cover						
100 mm wide	31.46	37.23	0.27	7.32	nr	44.55
200 mm wide	37.88	44.83	0.31	8.40	nr	53.23
400 mm wide	53.43	63.23	0.37	10.03	nr	73.26

V: ELECTRICAL SUPPLY/POWER/LIGHTING SYSTEMS

Item	Net Price £	Material £	Labour hours	Labour £	Unit	Total rate £
V20 : LV DISTRIBUTION (cont'd)						
Y63 - CABLE SUPPORTS (cont'd)						
BASKET TRAY						
Mild Steel Cable Basket; Zinc Plated Including Standard Coupling Joints, Fixings and Earth Continuity Straps (supports and hangers are excluded)						
Basket 54mm deep						
100 mm wide	3.21	3.80	0.22	5.96	m	9.76
150 mm wide	3.58	4.23	0.25	6.78	m	11.01
200 mm wide	3.92	4.64	0.28	7.59	m	12.22
300 mm wide	4.59	5.43	0.34	9.21	m	14.65
450 mm wide	5.57	6.59	0.44	11.92	m	18.52
600 mm wide	6.81	8.06	0.70	18.97	m	27.03
Extra for; (cutting and jointing to fittings is included)						
Reducer						
150 mm wide	11.38	13.46	0.25	6.78	nr	20.24
200 mm wide	13.35	15.79	0.28	7.59	nr	23.38
300 mm wide	13.57	16.06	0.38	10.30	nr	26.36
450 mm wide	14.90	17.64	0.48	13.01	nr	30.65
600 mm wide	18.56	21.97	0.48	13.01	nr	34.98
Bend						
100 mm wide	10.87	12.86	0.23	6.23	nr	19.09
150 mm wide	12.74	15.08	0.26	7.05	nr	22.13
200 mm wide	12.96	15.33	0.30	8.13	nr	23.46
300 mm wide	14.18	16.79	0.35	9.49	nr	26.27
450 mm wide	17.66	20.91	0.50	13.55	nr	34.46
600 mm wide	20.79	24.61	0.58	15.72	nr	40.33
Tee						
100 mm wide	13.66	16.17	0.28	7.59	nr	23.76
150 mm wide	14.49	17.15	0.30	8.13	nr	25.28
200 mm wide	14.79	17.50	0.33	8.94	nr	26.44
300 mm wide	18.60	22.01	0.39	10.57	nr	32.58
450 mm wide	24.37	28.84	0.56	15.18	nr	44.01
600 mm wide	25.20	29.83	0.65	17.62	nr	47.45
Cross over						
100 mm wide	18.63	22.05	0.40	10.84	nr	32.89
150 mm wide	18.96	22.44	0.42	11.38	nr	33.82
200 mm wide	20.59	24.37	0.46	12.47	nr	36.84
300 mm wide	23.39	27.68	0.51	13.82	nr	41.50
450 mm wide	27.41	32.44	0.74	20.06	nr	52.50
600 mm wide	28.30	33.49	0.82	22.22	nr	55.71

V: ELECTRICAL SUPPLY/POWER/LIGHTING SYSTEMS

Item	Net Price £	Material £	Labour hours	Labour £	Unit	Total rate £
Mild steel cable basket; expoxy coated including standard coupling joints, fixings and earth continuity straps (supports and hangers are excluded)						
Basket 54mm deep						
100 mm wide	7.79	9.22	0.22	5.96	m	**15.18**
150 mm wide	8.81	10.43	0.25	6.78	m	**17.21**
200 mm wide	9.81	11.60	0.28	7.59	m	**19.19**
300 mm wide	11.14	13.18	0.34	9.21	m	**22.40**
450 mm wide	13.75	16.27	0.44	11.92	m	**28.19**
600 mm wide	15.65	18.52	0.70	18.97	m	**37.49**
Extra for; (cutting and jointing to fittings is included)						
Reducer						
150 mm wide	16.28	19.27	0.28	7.59	nr	**26.86**
200 mm wide	18.25	21.60	0.28	7.59	nr	**29.19**
300 mm wide	19.46	23.03	0.38	10.30	nr	**33.33**
450 mm wide	22.76	26.94	0.48	13.01	nr	**39.95**
600 mm wide	28.37	33.57	0.48	13.01	nr	**46.58**
Bend						
100 mm wide	15.79	18.69	0.23	6.23	nr	**24.92**
150 mm wide	17.65	20.89	0.26	7.05	nr	**27.94**
200 mm wide	17.87	21.14	0.30	8.13	nr	**29.27**
300 mm wide	20.10	23.78	0.35	9.49	nr	**33.27**
450 mm wide	25.51	30.19	0.50	13.55	nr	**43.74**
600 mm wide	30.62	36.24	0.58	15.72	nr	**51.96**
Tee						
100 mm wide	18.57	21.98	0.28	7.59	nr	**29.57**
150 mm wide	19.41	22.97	0.30	8.13	nr	**31.10**
200 mm wide	19.71	23.32	0.33	8.94	nr	**32.27**
300 mm wide	24.50	28.99	0.39	10.57	nr	**39.56**
450 mm wide	31.22	36.95	0.56	15.18	nr	**52.13**
600 mm wide	35.01	41.43	0.65	17.62	nr	**59.05**
Cross over						
100 mm wide	23.55	27.87	0.40	10.84	nr	**38.71**
150 mm wide	23.81	28.18	0.42	11.38	nr	**39.56**
200 mm wide	25.51	30.19	0.46	12.47	nr	**42.66**
300 mm wide	29.28	34.65	0.51	13.82	nr	**48.47**
450 mm wide	35.27	41.74	0.74	20.06	nr	**61.80**
600 mm wide	38.11	45.11	0.82	22.22	nr	**67.33**

V: ELECTRICAL SUPPLY/POWER/LIGHTING SYSTEMS

Item	Net Price £	Material £	Labour hours	Labour £	Unit	Total rate £
V20 : LV DISTRIBUTION (cont'd)						
Y71 - LV SWITCHGEAR & DISTRIBUTION BOARDS						
LV switchboard components, factory-assembled modular construction to IP41; form 4, type 5; 2400mm high, with front and rear access; top cable entry/exit; includes delivery, offloading, positioning and commissioning (hence separate labour costs are not detailed below); excludes cabling and cable terminations						
Air circuit breakers (ACBs) to BSEN 60947-2, withdrawable type, fitted with adjustable instantaneous and overload protection. Includes enclosure and copper links, assembled into LV switchboard.						
ACB-100 kA fault rated						
4 pole, 6300 A (1600mm wide)	31108.12	36815.53	-	-	nr	36815.53
4 pole, 5000 A (1600mm wide)	23506.59	27819.35	-	-	nr	27819.35
4 pole, 4000 A (1600mm wide)	19720.24	23338.31	-	-	nr	23338.31
4 pole, 3200 A (1600mm wide)	13313.15	15755.71	-	-	nr	15755.71
4 pole, 2500 A (1600mm wide)	10347.01	12245.37	-	-	nr	12245.37
4 pole, 2000 A (1600mm wide)	8066.37	9546.30	-	-	nr	9546.30
4 pole, 1600 A (1600mm wide)	6604.20	7815.87	-	-	nr	7815.87
4 pole, 1250 A (1600mm wide)	5845.18	6917.60	-	-	nr	6917.60
4 pole, 1000 A (1600mm wide)	5774.56	6834.02	-	-	nr	6834.02
4 pole, 800 A (1600mm wide)	5676.88	6718.42	-	-	nr	6718.42
3 pole, 6300 A (1600mm wide)	28186.72	33358.14	-	-	nr	33358.14
3 pole, 5000 A (1600mm wide)	19767.32	23394.03	-	-	nr	23394.03
3 pole, 4000 A (1600mm wide)	16186.89	19156.69	-	-	nr	19156.69
3 pole, 3200 A (1600mm wide)	11115.48	13154.83	-	-	nr	13154.83
3 pole, 2500 A (1600mm wide)	8899.55	10532.35	-	-	nr	10532.35
3 pole, 2000 A (1600mm wide)	6701.29	7930.77	-	-	nr	7930.77
3 pole, 1600 A (1600mm wide)	5447.99	6447.54	-	-	nr	6447.54
3 pole, 1250 A (1600mm wide)	5065.53	5994.91	-	-	nr	5994.91
3 pole, 1000 A (1600mm wide)	4956.68	5866.08	-	-	nr	5866.08
3 pole, 800 A (1600mm wide)	4861.94	5753.97	-	-	nr	5753.97
ACB-65 kA fault rated						
4 pole, 4000 A (1600mm wide)	17304.86	20479.78	-	-	nr	20479.78
4 pole, 3200 A (1600mm wide)	11706.82	13854.67	-	-	nr	13854.67
4 pole, 2500 A (1600mm wide)	9202.58	10890.98	-	-	nr	10890.98
4 pole, 2000 A (1600mm wide)	7319.09	8661.93	-	-	nr	8661.93
4 pole, 1600 A (1600mm wide)	6039.33	7147.37	-	-	nr	7147.37
4 pole, 1250 A (1600mm wide)	5392.09	6381.38	-	-	nr	6381.38
4 pole, 1000 A (1600mm wide)	5289.13	6259.53	-	-	nr	6259.53
4 pole, 800 A (1600mm wide)	5194.40	6147.41	-	-	nr	6147.41
3 pole, 4000 A (1600mm wide)	14659.99	17349.65	-	-	nr	17349.65
3 pole, 3200 A (1600mm wide)	9941.61	11765.60	-	-	nr	11765.60
3 pole, 2500 A (1600mm wide)	7731.58	9150.09	-	-	nr	9150.09
3 pole, 2000 A (1600mm wide)	6145.25	7272.72	-	-	nr	7272.72
3 pole, 1600 A (1600mm wide)	5027.29	5949.64	-	-	nr	5949.64
3 pole, 1250 A (1600mm wide)	4680.14	5538.80	-	-	nr	5538.80
3 pole, 1000 A (1600mm wide)	4574.23	5413.46	-	-	nr	5413.46
3 pole, 800 A (1600mm wide)	4479.49	5301.34	-	-	nr	5301.34

V: ELECTRICAL SUPPLY/POWER/LIGHTING SYSTEMS

Item	Net Price £	Material £	Labour hours	Labour £	Unit	Total rate £
Extra for						
Cable box (one per ACB for form 4, types 6 & 7)	290.63	343.95	-	-	nr	343.95
Opening Coil	86.71	102.62	-	-	nr	102.62
Closing Coil	86.71	102.62	-	-	nr	102.62
Undervoltage Release	170.32	201.57	-	-	nr	201.57
Motor Operator	526.46	623.05	-	-	nr	623.05
Mechnical Interlock (per ACB)	495.50	586.40	-	-	nr	586.40
ACB Fortress/Castell Adaptor Kit (one per ACB)	123.87	146.60	-	-	nr	146.60
Fortress/Castell ACB Lock (one per ACB)	247.74	293.20	-	-	nr	293.20
Fortress/Castell Key	61.94	73.31	-	-	nr	73.31
Moulded case circuit breakers (MCCBs) to BS EN 60947-2; plug-in type, fitted with electronic trip unit. Includes metalwork section and copper links, assembled into LV switchboard						
MCCB-150 kA fault rated						
4 Pole, 630 A (800mm wide, 600mm high)	2933.05	3471.18	-	-	nr	3471.18
4 Pole, 400 A (800mm wide, 400mm high)	2098.38	2483.37	-	-	nr	2483.37
4 Pole, 250 A (800mm wide, 400mm high)	1876.58	2220.87	-	-	nr	2220.87
4 Pole, 160 A (800mm wide, 300mm high)	1275.36	1509.35	-	-	nr	1509.35
4 Pole, 100 A (800mm wide, 200mm high)	1062.33	1257.23	-	-	nr	1257.23
3 Pole, 630 A (800mm wide, 600mm high)	2617.87	3098.17	-	-	nr	3098.17
3 Pole, 400 A (800mm wide, 400mm high)	1999.14	2365.92	-	-	nr	2365.92
3 Pole, 250 A (800mm wide, 400mm high)	1610.99	1906.56	-	-	nr	1906.56
3 Pole, 160 A (800mm wide, 300mm high)	1179.06	1395.38	-	-	nr	1395.38
3 Pole, 100 A (800mm wide, 200mm high)	942.66	1115.61	-	-	nr	1115.61
MCCB-70kA fault rated						
4 Pole, 630 A (800mm wide, 600mm high)	2512.80	2973.82	-	-	nr	2973.82
4 Pole, 400 A (800mm wide, 400mm high)	1727.73	2044.71	-	-	nr	2044.71
4 Pole, 250 A (800mm wide, 400mm high)	1485.49	1758.04	-	-	nr	1758.04
4 Pole, 160 A (800mm wide, 300mm high)	1114.84	1319.38	-	-	nr	1319.38
4 Pole, 100 A (800mm wide, 200mm high)	840.52	994.73	-	-	nr	994.73
3 Pole, 630 A (800mm wide, 600mm high)	2200.53	2604.26	-	-	nr	2604.26
3 Pole, 400 A (800mm wide, 400mm high)	1491.34	1764.95	-	-	nr	1764.95
3 Pole, 250 A (800mm wide, 400mm high)	1316.23	1557.72	-	-	nr	1557.72
3 Pole, 160 A (800mm wide, 300mm high)	971.84	1150.14	-	-	nr	1150.14
3 Pole, 100 A (800mm wide, 200mm high)	715.02	846.21	-	-	nr	846.21
MCCB-45kA fault rated						
4 Pole, 630 A (800mm wide, 600mm high)	2436.92	2884.02	-	-	nr	2884.02
4 Pole, 400 A (800mm wide, 400mm high)	1675.21	1982.56	-	-	nr	1982.56
3 Pole, 630 A (800mm wide, 600mm high)	2107.14	2493.73	-	-	nr	2493.73
3 Pole, 400 A (800mm wide, 400mm high)	1435.88	1699.33	-	-	nr	1699.33
MCCB-36kA fault rated						
4 Pole, 250 A (800mm wide, 400mm high)	1421.30	1682.06	-	-	nr	1682.06
4 Pole, 160 A (800mm wide, 300mm high)	1076.91	1274.49	-	-	nr	1274.49
3 Pole, 250 A (800mm wide, 400mm high)	807.28	955.39	-	-	nr	955.39
3 Pole, 160 A (800mm wide, 300mm high)	710.90	841.33	-	-	nr	841.33

V: ELECTRICAL SUPPLY/POWER/LIGHTING SYSTEMS

Item	Net Price £	Material £	Labour hours	Labour £	Unit	Total rate £
V20 : LV DISTRIBUTION (cont'd)						
Y71 - LV SWITCHGEAR & DISTRIBUTION BOARDS (cont'd)						
Extra for						
Cable box (one per MCCB for form 4, types 6 & 7)	122.88	145.43	-	-	nr	145.43
Shunt trip (for ratings 100A to 630A)	43.01	50.90	-	-	nr	50.90
Undervoltage release (for ratings 100A to 630A)	61.45	72.72	-	-	nr	72.72
Motor operator for 630A MCCB	583.69	690.78	-	-	nr	690.78
Motor operator for 400A MCCB	583.69	690.78	-	-	nr	690.78
Motor operator for 250A MCCB	485.39	574.44	-	-	nr	574.44
Motor operator for 160A/100A MCCB	310.28	367.20	-	-	nr	367.20
Door handle for 630/400A MCCB	76.81	90.91	-	-	nr	90.91
Door handle for 250/160/100A MCCB	61.45	72.72	-	-	nr	72.72
MCCB earth fault protection	460.81	545.36	-	-	nr	545.36
LV Switchboard busbar						
Copper busbar assembled into LV switchboard, ASTA type tested to appropriate fault level. Busbar Length may be estimated by adding the widths of the ACB sections to the width of the MCCB sections. ACB's up to 2000A rating may be stacked two high; larger ratings are one per section. To determine the number of MCCB sections, add together all the MCCB heights and divide by 1800mm, rounding up as necessary						
6000 A (6 x 10mm x 100mm)	2381.66	2818.63	-	-	nr	2818.63
5000 A (4 x 10mm x 100mm)	1958.26	2317.55	-	-	nr	2317.55
4000 A (4 x 10mm x 100mm)	1958.26	2317.55	-	-	nr	2317.55
3200 A (3 x 10mm x 100mm)	1359.53	1608.97	-	-	nr	1608.97
2500 A (2 x 10mm x 100mm)	1147.84	1358.43	-	-	nr	1358.43
2000 A (2 x 10mm x 80mm)	843.50	998.26	-	-	nr	998.26
1600 A (2 x 10mm x 50mm)	611.95	724.23	-	-	nr	724.23
1250 A (2 x 10mm x 40mm)	479.63	567.63	-	-	nr	567.63
1000 A (2 x 10mm x 30mm)	479.63	567.63	-	-	nr	567.63
800 A (2 x 10mm x 20mm)	373.78	442.36	-	-	nr	442.36
630 A (2 x 10mm x 20mm)	373.78	442.36	-	-	nr	442.36
400 A (2 x 10mm x 10mm)	320.61	379.43	-	-	nr	379.43
Automatic power factor correction (PFC); floor standing steel enclosure to IP 42, complete with microprocessor based relay and status indication; includes delivery, offloading, positioning and commissioning; excludes cabling and cable terminations						
Standard PFC (no de-tuning)						
100 kVAr	4669.36	5526.04	-	-	nr	5526.04
200 kVAr	6530.58	7728.74	-	-	nr	7728.74
400 kVAr	11148.07	13193.41	-	-	nr	13193.41
600 kVAr	15270.63	18072.33	-	-	nr	18072.33

V: ELECTRICAL SUPPLY/POWER/LIGHTING SYSTEMS

Item	Net Price £	Material £	Labour hours	Labour £	Unit	Total rate £
PFC with de-tuning reactors						
100 kVAr	7597.52	8991.44	-	-	nr	8991.44
200 kVAr	10617.55	12565.55	-	-	nr	12565.55
400 kVAr	18298.06	21655.21	-	-	nr	21655.21
600 kVAr	26909.19	31846.22	-	-	nr	31846.22
AUTOMATIC TRANSFER SWITCHES						
Automatic transfer switches; steel enclosure; solenoid operating; programmable controller, keypad and LCD display; fixed to backgrounds; including commissioning and testing						
Panel Mounting type 3 pole or 4 pole; overlapping neutral						
100 amp	2884.62	3413.86	2.60	70.46	nr	3484.33
250amp	3853.33	4560.30	3.30	89.44	nr	4649.74
400amp	5092.64	6026.99	4.30	116.54	nr	6143.53
630amp	6543.85	7744.45	5.30	143.64	nr	7888.09
800amp	9437.59	11169.11	5.50	149.06	nr	11318.16
1000amp	15014.34	17769.03	5.83	158.00	nr	17927.03
1600amp	16086.80	19038.24	6.20	168.03	nr	19206.27
2000amp	19763.78	23389.84	6.90	187.00	nr	23576.84
Enclosed type 3 pole or 4 pole; over lapping neutral						
63 amp	2787.75	3299.22	2.60	70.46	nr	3369.68
100 amp	2884.62	3413.86	2.60	70.46	nr	3484.33
125 amp	2955.82	3498.12	2.90	78.59	nr	3576.72
160 amp	3096.58	3664.71	2.90	78.59	nr	3743.30
250amp	3853.33	4560.30	3.30	89.44	nr	4649.74
400amp	6059.44	7171.17	4.30	116.54	nr	7287.70
630amp	8157.07	9653.65	4.84	131.17	nr	9784.82
800amp	11260.28	13326.20	5.12	138.76	nr	13464.96
1000amp	16890.42	19989.31	5.50	149.06	nr	20138.36
1250amp	18297.95	21655.07	6.00	162.61	nr	21817.68
1600amp	19705.48	23320.84	6.20	168.03	nr	23488.87
2000amp	23443.83	27745.07	6.90	187.00	nr	27932.07
3000amp	28193.26	33365.88	6.90	187.00	nr	33552.88
4000amp	34510.00	40841.55	6.90	187.00	nr	41028.55
Enclosed type 3 pole or 4 pole; over lapping Neutral, with single By-Pass						
63 amp	5034.13	5957.74	2.60	70.46	nr	6028.21
100 amp	5180.04	6130.42	2.60	70.46	nr	6200.89
125 amp	5325.96	6303.11	2.90	78.59	nr	6381.71
160 amp	7660.63	9066.13	2.90	78.59	nr	9144.72
250 amp	9951.52	11777.33	3.30	89.44	nr	11866.76
400 amp	11162.63	13210.64	4.30	116.54	nr	13327.17
630 amp	23200.75	27457.39	4.84	131.17	nr	27588.56
800 amp	24514.00	29011.58	5.12	138.76	nr	29150.34
1000 amp	45926.54	54352.68	5.50	149.06	nr	54501.74
1250 amp	46705.74	55274.84	6.00	162.61	nr	55437.45
1600 amp	19705.48	23320.84	6.20	168.03	nr	23488.87

V: ELECTRICAL SUPPLY/POWER/LIGHTING SYSTEMS

Item	Net Price £	Material £	Labour hours	Labour £	Unit	Total rate £
V20 : LV DISTRIBUTION (cont'd)						
Y71 - LV SWITCHGEAR & DISTRIBUTION BOARDS (cont'd)						
AUTOMATIC TRANSFER SWITCHES (cont'd)						
Enclosed type 3 pole or 4 pole; over lapping Neutral, with Dual By-Pass						
63 amp	5922.50	7009.10	2.60	70.46	nr	7079.56
100 amp	6094.17	7212.27	2.60	70.46	nr	7282.73
125 amp	6265.83	7415.42	2.90	78.59	nr	7494.02
160 amp	9012.50	10666.02	2.90	78.59	nr	10744.62
250 amp	11707.67	13855.68	3.30	89.44	nr	13945.11
400 amp	13132.50	15541.92	4.30	116.54	nr	15658.46
630 amp	27295.00	32302.81	4.84	131.17	nr	32433.99
800 amp	28840.00	34131.27	5.12	138.76	nr	34270.03
1000 amp	54031.23	63944.34	5.50	149.06	nr	64093.40
1250 amp	54947.93	65029.23	6.00	162.61	nr	65191.84
1600 amp	58059.81	68712.04	6.20	168.03	nr	68880.07
MCCB panelboards; IP4X construction, 50kA busbars and fully-rated neutral; fitted with doorlock, removable glandplate; form 3b Type2; BSEN 60439-1; including fixing to backgrounds						
Panelboards cubicle with MCCB incomer						
Up to 250A						
4 Way TPN	863.37	1021.77	1.00	27.10	nr	1048.87
Extra over for integral incomer metering	884.27	1046.50	1.50	40.65	nr	1087.16
Up to 630A						
6 way TPN	1553.09	1838.04	2.00	54.20	nr	1892.24
12 way TPN	1754.06	2075.88	2.50	67.75	nr	2143.63
18 Way TPN	2041.85	2416.47	3.00	81.30	nr	2497.78
Extra over for integral incomer metering	1045.04	1236.78	1.50	40.65	nr	1277.43
Up to 800A						
6 way TPN	2657.63	3145.23	2.00	54.20	nr	3199.43
12 way TPN	3188.20	3773.13	2.50	67.75	nr	3840.89
18 Way TPN	3393.98	4016.68	3.00	81.30	nr	4097.98
Extra over for integral incomer metering	1045.04	1236.78	1.50	40.65	nr	1277.43
Up to 1200A						
20 Way TPN	7111.13	8415.80	3.50	94.86	nr	8510.66
Up to 1600A						
28 Way TPN	9209.25	10898.88	3.50	94.86	nr	10993.73
Up to 2000A						
28Way TPN	10024.39	11863.56	4.00	108.41	nr	11971.97

V: ELECTRICAL SUPPLY/POWER/LIGHTING SYSTEMS

Item	Net Price £	Material £	Labour hours	Labour £	Unit	Total rate £
Feeder MCCBs						
Single Pole						
32A	70.74	83.72	0.75	20.33	nr	104.05
63A	72.35	85.62	0.75	20.33	nr	105.95
100A	73.95	87.52	0.75	20.33	nr	107.85
160A	78.78	93.24	1.00	27.10	nr	120.34
Double pole						
32A	106.11	125.57	0.75	20.33	nr	145.90
63A	107.71	127.47	0.75	20.33	nr	147.80
100A	157.56	186.47	0.75	20.33	nr	206.80
160A	196.14	232.13	1.00	27.10	nr	259.23
Triple pole						
32A	141.49	167.45	0.75	20.33	nr	187.78
63A	144.69	171.24	0.75	20.33	nr	191.57
100A	188.10	222.61	0.75	20.33	nr	242.94
160A	242.77	287.31	1.00	27.10	nr	314.41
250A	364.96	431.92	1.00	27.10	nr	459.02
400A	493.58	584.14	1.25	33.88	nr	618.02
630A	810.32	958.98	1.50	40.65	nr	999.64
MCB distribution boards; IP3X external protection enclosure; removable earth and neutral bars and DIN rail; 125/250amp incomers; including fixing to backgrounds						
SP & N						
6 way	59.55	70.48	2.00	54.20	nr	124.68
8 way	70.63	83.59	2.50	67.75	nr	151.34
12 way	81.05	95.92	3.00	81.30	nr	177.22
16 way	96.35	114.03	4.00	108.41	nr	222.43
24 way	202.96	240.20	5.00	135.51	nr	375.71
TP & N						
4 way	431.42	510.57	3.00	81.30	nr	591.88
6 way	446.58	528.51	3.50	94.86	nr	623.37
8 way	467.83	553.66	4.00	108.41	nr	662.07
12 way	498.56	590.04	4.00	108.41	nr	698.44
16 way	570.49	675.16	5.00	135.51	nr	810.67
24 way	718.45	850.27	6.40	173.45	nr	1023.72
Miniature circuit breakers for distribution boards; BS EN 60 898; DIN rail mounting; including connecting to circuit						
SP&N; including connecting of wiring						
6 Amp	10.23	12.11	0.10	2.71	nr	14.82
10 - 40 Amp	10.64	12.59	0.10	2.71	nr	15.30
50 - 63 Amp	11.14	13.19	0.14	3.79	nr	16.98
TP&N; including connecting of wiring						
6 Amp	43.35	51.30	0.30	8.13	nr	59.44
10 - 40 Amp	45.06	53.32	0.45	12.20	nr	65.52
50 - 63 Amp	47.20	55.86	0.45	12.20	nr	68.06

V: ELECTRICAL SUPPLY/POWER/LIGHTING SYSTEMS

Item	Net Price £	Material £	Labour hours	Labour £	Unit	Total rate £
V20 : LV DISTRIBUTION (cont'd)						
Y71 - LV SWITCHGEAR & DISTRIBUTION BOARDS (cont'd)						
AUTOMATIC TRANSFER SWITCHES (cont'd)						
Residual current circuit breakers for distribution boards; DIN rail mounting; including connecting to circuit						
SP&N						
10mA						
6 Amp	67.13	79.45	0.21	5.69	nr	85.14
10 - 32 Amp	65.86	77.94	0.26	7.05	nr	84.99
45 Amp	65.86	77.94	0.26	7.05	nr	84.99
30mA						
6 Amp	67.13	79.45	0.21	5.69	nr	85.14
10 - 40 Amp	65.86	77.94	0.21	5.69	nr	83.63
50 -63 Amp	65.86	77.94	0.26	7.05	nr	84.99
100mA						
6 Amp	124.12	146.90	0.21	5.69	nr	152.59
10 - 40 Amp	124.12	146.90	0.21	5.69	nr	152.59
50 -63 Amp	124.12	146.90	0.26	7.05	nr	153.94
HRC fused distribution boards; IP4X external protection enclosure; including earth and neutral bars; fixing to backgrounds						
SP&N						
20 Amp incomer						
4 way	136.18	161.16	1.00	27.10	nr	188.27
6 way	164.40	194.56	1.20	32.52	nr	227.08
8 way	192.72	228.08	1.40	37.94	nr	266.02
12 way	249.39	295.15	1.80	48.78	nr	343.93
32 Amp incomer						
4 way	163.96	194.04	1.00	27.10	nr	221.14
6 way	215.60	255.16	1.20	32.52	nr	287.68
8 way	253.66	300.20	1.40	37.94	nr	338.14
12 way	326.59	386.51	1.80	48.78	nr	435.29
TP&N						
20 Amp incomer						
4 way	244.08	288.86	1.50	40.65	nr	329.51
6 way	308.70	365.34	2.10	56.91	nr	422.26
8 way	364.99	431.96	2.70	73.17	nr	505.13
12 way	512.07	606.02	3.90	105.70	nr	711.72
32 Amp incomer						
4 way	292.00	345.57	1.50	40.65	nr	386.22
6 way	392.02	463.94	2.10	56.91	nr	520.85
8 way	478.88	566.75	2.70	73.17	nr	639.92
12 way	664.19	786.05	3.90	105.70	nr	891.75

V: ELECTRICAL SUPPLY/POWER/LIGHTING SYSTEMS

Item	Net Price £	Material £	Labour hours	Labour £	Unit	Total rate £
63 Amp incomer						
4 way	620.69	734.56	2.17	58.81	nr	**793.37**
6 way	796.03	942.07	2.83	76.70	nr	**1018.77**
8 way	958.61	1134.48	2.57	69.65	nr	**1204.13**
100 Amp incomer						
4 way	981.62	1161.72	2.40	65.04	nr	**1226.76**
6 way	1283.06	1518.47	2.73	73.99	nr	**1592.45**
8 way	1568.86	1856.70	3.87	104.88	nr	**1961.59**
200 Amp incomer						
4 way	2431.37	2877.46	5.36	145.26	nr	**3022.72**
6 way	3213.57	3803.17	6.17	167.22	nr	**3970.38**
HRC fuse; includes fixing to fuse holder						
2-30 Amp	2.78	3.29	0.10	2.71	nr	**6.00**
35 - 63 Amp	6.02	7.12	0.12	3.25	nr	**10.37**
80 Amp	8.88	10.51	0.15	4.07	nr	**14.57**
100 Amp	8.88	10.51	0.15	4.07	nr	**14.57**
125 Amp	16.15	19.11	0.15	4.07	nr	**23.18**
160 Amp	16.94	20.05	0.15	4.07	nr	**24.11**
200 Amp	17.55	20.76	0.15	4.07	nr	**24.83**
Consumer units; fixed to backgrounds; including supports, fixings, connections/ jointing to equipment						
Switched and insulated; moulded plastic case, 63 Amp 230 Volt SP&N; earth and neutral bars; 30mA RCCB protection; fitted MCB's						
2 way	112.45	133.08	1.67	45.26	nr	**178.34**
4 way	126.18	149.33	1.59	43.09	nr	**192.42**
6 way	137.16	162.32	2.50	67.75	nr	**230.08**
8 way	148.00	175.16	3.00	81.30	nr	**256.46**
12 way	171.93	203.47	4.00	108.41	nr	**311.88**
16 way	207.75	245.86	5.50	149.06	nr	**394.92**
Switched and insulated; moulded plastic case, 100 Amp 230 Volt SP&N; earth and neutral bars; 30mA RCCB protection; fitted MCB's						
2 way	112.45	133.08	1.67	45.26	nr	**178.34**
4 way	126.18	149.33	1.59	43.09	nr	**192.42**
6 way	137.16	162.32	2.50	67.75	nr	**230.08**
8 way	148.00	175.16	3.00	81.30	nr	**256.46**
12 way	171.93	203.47	4.00	108.41	nr	**311.88**
16 way	207.75	245.86	5.50	149.06	nr	**394.92**
Extra for Residual current device; double pole; 230 volt/30mA tripping current						
16 Amp	62.46	73.92	0.22	5.96	nr	**79.88**
30 Amp	61.48	72.76	0.22	5.96	nr	**78.72**
40 Amp	63.43	75.06	0.22	5.96	nr	**81.03**
63 Amp	78.53	89.81	0.22	5.96	nr	**95.77**
80 Amp	87.35	103.38	0.22	5.96	nr	**109.34**
100 Amp	107.51	127.24	0.25	6.78	nr	**134.01**

V: ELECTRICAL SUPPLY/POWER/LIGHTING SYSTEMS

Item	Net Price £	Material £	Labour hours	Labour £	Unit	Total rate £
V20 : LV DISTRIBUTION (cont'd)						
Y71 - LV SWITCHGEAR & DISTRIBUTION BOARDS (cont'd)						
AUTOMATIC TRANSFER SWITCHES (cont'd)						
Residual current device; double pole; 230 volt/100mA tripping current						
63 Amp	71.79	84.96	0.22	5.96	nr	90.92
80 Amp	83.05	98.29	0.22	5.96	nr	104.25
100 Amp	107.53	127.25	0.25	6.78	nr	134.03
Heavy duty fuse switches; with HRC fuses BS 5419; short circuit rating 65kA, 500 volt; including retractable operating switches						
SP&N						
63 Amp	242.69	287.22	1.30	35.23	nr	322.45
100 Amp	354.76	419.85	1.95	52.85	nr	472.70
TP&N						
63 Amp	305.75	361.84	1.83	49.60	nr	411.44
100 Amp	429.64	508.46	2.48	67.21	nr	575.68
200 Amp	662.75	784.34	3.13	84.83	nr	869.17
300 Amp	1151.38	1362.62	4.45	120.60	nr	1483.23
400 Amp	1264.00	1495.90	4.45	120.60	nr	1616.51
600 Amp	1907.94	2257.99	5.72	155.02	nr	2413.01
800 Amp	2980.42	3527.23	7.88	213.56	nr	3740.79
Switch disconnectors to BSEN 60947-3; in sheet steel case; IP41 with door interlock fixed to backgrounds						
Double pole						
20 Amp	44.54	52.71	1.02	27.64	nr	80.35
32 Amp	53.68	63.53	1.02	27.64	nr	91.17
63 Amp	195.89	231.83	1.21	32.79	nr	264.62
100 Amp	179.75	212.73	1.86	50.41	nr	263.14
TP&N						
20 Amp	55.91	66.17	1.29	34.96	nr	101.13
32 Amp	65.05	76.99	1.83	49.60	nr	126.59
63 Amp	222.59	263.42	2.48	67.21	nr	330.63
100 Amp	220.77	261.27	2.48	67.21	nr	328.49
125 Amp	232.17	274.76	2.48	67.21	nr	341.97
160 Amp	534.19	632.20	2.48	67.21	nr	699.41
Enclosed switch disconnector to BSEN 60947-3; enclosure minimum IP55 rating; complete with earth connection bar; fixed to backgrounds.						
TP						
20 Amp	44.54	52.71	1.02	27.64	nr	80.35
32 Amp	53.68	63.53	1.02	27.64	nr	91.17
63 Amp	195.89	231.83	1.21	32.79	nr	264.62

Material Costs/Prices for Measured Works – Electrical Installations

V: ELECTRICAL SUPPLY/POWER/LIGHTING SYSTEMS

Item	Net Price £	Material £	Labour hours	Labour £	Unit	Total rate £
TP&N						
20 Amp	55.91	66.17	1.29	34.96	nr	101.13
32 Amp	65.05	76.99	1.83	49.60	nr	126.59
63 Amp	222.59	263.42	2.48	67.21	nr	330.63
Busbar chambers; fixed to background including all supports, fixings, connections/ jointing to equipment						
Sheet steel case enclosing 4 pole 550 Volt copper bars, detachable metal end plates						
600mm long						
200 Amp	455.41	538.97	2.62	71.01	nr	609.97
300 Amp	585.20	692.57	3.03	82.12	nr	774.68
500 Amp	1002.73	1186.70	4.48	121.41	nr	1308.11
900mm long						
200 Amp	655.97	776.33	3.04	82.39	nr	858.71
300 Amp	773.09	914.93	3.59	97.29	nr	1012.22
500 Amp	1143.54	1353.34	4.42	119.79	nr	1473.13
1350mm long						
200 Amp	896.00	1060.39	3.38	91.60	nr	1152.00
300 Amp	1054.58	1248.07	3.94	106.78	nr	1354.84
500 Amp	1688.08	1997.79	4.82	130.63	nr	2128.42
Contactor relays; pressed steel enclosure; fixed to backgrounds including supports, fixings, connections/jointing to equipment						
Relays						
6 Amp, 415/240 Volt, 4 pole N/O	52.74	62.42	0.52	14.09	nr	76.51
6 Amp, 415/240 Volt, 8 pole N/O	64.50	76.34	0.85	23.04	nr	99.37
Push button stations; heavy gauge pressed steel enclosure; polycarbonate cover; IP65; fixed to backgrounds including supports, fixings,connections/joining to equipment						
Standard units						
One button (start or stop)	66.74	78.98	0.39	10.57	nr	89.55
Two button (start or stop)	70.85	83.85	0.47	12.74	nr	96.59
Three button (forward-reverse-stop)	100.41	118.83	0.57	15.45	nr	134.28

V: ELECTRICAL SUPPLY/POWER/LIGHTING SYSTEMS

Item	Net Price £	Material £	Labour hours	Labour £	Unit	Total rate £
V20 : LV DISTRIBUTION (cont'd)						
Y71 - LV SWITCHGEAR & DISTRIBUTION BOARDS (cont'd)						
AUTOMATIC TRANSFER SWITCHES (cont'd)						
Weatherproof junction boxes; enclosures with rail mounted terminal blocks; side hung door to receive padlock; fixed to backgrounds, including all supports and fixings (Suitable for cable up to 2.5mm²; including glandplates and gaskets)						
Sheet steel with zinc spray finish enclosure						
Overall Size 229 x 152; suitable to receive 3 x 20(A) glands per gland plate	69.74	82.54	1.43	38.76	nr	**121.29**
Overall Size 306 x 306; suitable to receive 14 x 20(A) glands per gland plate	93.56	110.72	2.17	58.81	nr	**169.53**
Overall Size 458 x 382; suitable to receive 18 x 20(A) glands per gland plate	136.22	161.22	3.51	95.13	nr	**256.34**
Overall Size 762 x508; suitable to receive 26 x 20(A) glands per gland plate	144.07	170.50	4.85	131.44	nr	**301.94**
Overall Size 914 x 610; suitable to receive 45 x 20(A) glands per gland plate	159.24	188.45	7.01	189.98	nr	**378.43**
Weatherproof junction boxes; enclosures with rail mounted terminal blocks; screw fixed lid; fixed to backgrounds, including all supports and fixings (suitable for cable up to 2.5mm²; including glandplates and gaskets)						
Glassfibre reinforced polycarbonate enclosure						
Overall Size 190 x 190 x 130	95.24	112.71	1.43	38.76	nr	**151.47**
Overall Size 190 x 190 x 180	139.44	165.02	1.53	41.47	nr	**206.48**
Overall Size 280 x 190 x 130	157.49	186.38	2.17	58.81	nr	**245.19**
Overall Size 280 x 190 x 180	176.44	208.81	2.37	64.23	nr	**273.04**
Overall Size 380 x 190 x 130	196.86	232.97	3.33	90.25	nr	**323.22**
Overall Size 380 x 190 x 180	211.43	250.22	3.30	89.44	nr	**339.66**
Overall Size 380 x 280 x 130	226.02	267.48	4.66	126.29	nr	**393.78**
Overall Size 380 x 280 x 180	243.52	288.20	5.36	145.26	nr	**433.46**
Overall Size 560 x 280 x 130	293.10	346.87	7.01	189.98	nr	**536.85**
Overall Size 560 x 380 x 180	301.85	357.23	7.67	207.87	nr	**565.10**

V: ELECTRICAL SUPPLY/POWER/LIGHTING SYSTEMS

Item	Net Price £	Material £	Labour hours	Labour £	Unit	Total rate £
V21 : GENERAL LIGHTING						
Y73 - LUMINAIRES (GENERAL)						
LUMINAIRES						
Fluorescent Luminaires; surface fixed to backgrounds						
Batten type; surface mounted						
600 mm Single - 18 W	8.47	10.03	0.58	15.72	nr	**25.75**
600 mm Twin - 18 W	15.01	17.76	0.59	15.99	nr	**33.75**
1200 mm Single - 36 W	11.21	13.27	0.76	20.60	nr	**33.87**
1200 mm Twin - 36 W	21.48	25.42	0.77	20.87	nr	**46.29**
1500 mm Single - 58 W	12.69	15.01	0.84	22.77	nr	**37.78**
1500 mm Twin - 58 W	25.32	29.97	0.85	23.04	nr	**53.00**
1800 mm Single - 70 W	15.30	18.11	1.05	28.46	nr	**46.57**
1800 mm Twin - 70 W	27.99	33.12	1.06	28.73	nr	**61.85**
2400 mm Single - 100 W	20.95	24.79	1.25	33.88	nr	**58.67**
2400 mm Twin - 100 W	36.66	43.39	1.27	34.42	nr	**77.81**
Surface mounted, opal diffuser						
600 mm Twin - 18 W	23.28	27.55	0.62	16.80	nr	**44.35**
1200 mm Single - 36 W	19.88	23.52	0.79	21.41	nr	**44.93**
1200 mm Twin - 36 W	30.96	36.64	0.80	21.68	nr	**58.32**
1500 mm Single - 58 W	22.86	27.05	0.88	23.85	nr	**50.90**
1500 mm Twin - 58 W	36.20	42.84	0.90	24.39	nr	**67.23**
1800 mm Single - 70 W	28.51	33.74	1.09	29.54	nr	**63.28**
1800 mm Twin - 70 W	39.16	46.34	1.10	29.81	nr	**76.15**
2400 mm Single - 100 W	37.47	44.35	1.30	35.23	nr	**79.58**
2400 mm Twin - 100 W	54.09	64.01	1.31	35.50	nr	**99.51**
Surface mounted linear fluorescent; T8 lamp; high frequency control gear; low brightness; 65° cut-off; including wedge style louvre						
1200mm, 1 x 36 watt	53.63	63.47	1.09	29.54	nr	**93.01**
1200mm 2 x 36 watt	57.35	67.87	1.09	29.54	nr	**97.41**
Extra for emergency pack	48.38	57.26	0.25	6.78	nr	**64.03**
1500mm, 1 x 58 watt	62.51	73.98	0.90	24.39	nr	**98.37**
1500mm 2 x 58 watt	66.82	79.07	0.90	24.39	nr	**103.47**
Extra for emergency pack	48.28	57.13	0.25	6.78	nr	**63.91**
1800mm, 1 x 70 watt	94.14	111.41	0.90	24.39	nr	**135.81**
1800mm, 2 x 70 watt	108.06	127.88	0.90	24.39	nr	**152.27**
Extra for emergency pack	79.93	94.59	0.25	6.78	nr	**101.37**
Modular recessed linear fluorescent; high frequency control gear; low brightness; 65° cut off; including wedge style louvre; fitted to exposed T grid ceiling						
600 x 600 mm, 3 x 18 watt T8	42.18	49.92	0.84	22.77	nr	**72.68**
600 x 600 mm, 4 x 18 watt T8	43.59	51.59	0.87	23.58	nr	**75.17**
Extra for emergency pack	48.78	57.73	0.25	6.78	nr	**64.51**
300 x 1200 mm, 2 x 36 watt T8	93.16	110.26	0.87	23.58	nr	**133.83**
Extra for emergency pack	55.81	66.04	0.25	6.78	nr	**72.82**
600 x 1200 mm, 3 x 36 watt T8	94.56	111.91	0.89	24.12	nr	**136.03**
600 x 1200 mm, 4 x 36 watt T8	96.76	114.51	0.91	24.66	nr	**139.17**

V: ELECTRICAL SUPPLY/POWER/LIGHTING SYSTEMS

Item	Net Price £	Material £	Labour hours	Labour £	Unit	Total rate £
V21 : GENERAL LIGHTING (cont'd)						
Y73 - LUMINAIRES (GENERAL) (cont'd)						
LUMINAIRES (cont'd)						
Modular recessed linear fluorescent (cont'd)						
Extra for emergency pack	58.86	69.66	0.25	6.78	nr	76.44
600 x 600, 3 x 14 watt T5	61.71	73.03	0.84	22.77	nr	95.79
600 x 600, 4 x 14 watt T5	64.23	76.02	0.87	23.58	nr	99.59
Extra for emergency pack	73.88	87.44	0.25	6.78	nr	94.21
Modular recessed; T8 lamp; high frequency control gear; cross-blade louvre; fitted to exposed T grid ceiling						
600 x 600 mm, 3 x 18 watt	46.06	54.51	0.84	22.77	nr	77.28
600 x 600 mm, 4 x 18 watt	62.33	73.76	0.87	23.58	nr	97.34
Extra for emergency pack	55.44	65.62	0.25	6.78	nr	72.39
Modular recessed compact fluorescent; TCL lamp; high frequency control gear; low brightness; 65° cut-off; including wedge style louvre; fitted to exposed T grid ceiling						
300 x 300 mm, 2 x 18 watt	93.28	110.40	0.75	20.33	nr	130.73
Extra for emergency pack	73.48	86.96	0.25	6.78	nr	93.73
500 x 500 mm, 2 x 36 watt	86.14	101.94	0.82	22.22	nr	124.16
600 x 600 mm, 2 x 36 watt	87.23	103.24	0.82	22.22	nr	125.46
600 x 600 mm, 2 x 40 watt	91.27	108.01	0.82	22.22	nr	130.24
Extra for emergency pack	52.64	62.30	0.25	6.78	nr	69.08
Ceiling recessed asymetric compact fluorescent downlighter; high frequency control gear; TCD lamp in 200 mm diameter luminaire; for wall-washing application						
1 x 18 watt	147.73	174.83	0.75	20.33	nr	195.16
1 x 26 watt	147.73	174.83	0.75	20.33	nr	195.16
2 x 18 watt	164.57	194.76	0.75	20.33	nr	215.09
2 x 26 watt	164.57	194.76	0.75	20.33	nr	215.09
Ceiling recessed asymetric compact fluorescesnt downlights; high frequency control gear; linear 200mm x 600mm luminaire with low glare louvre; for wall washing applications						
1 x 55 watt TCL	59.67	70.61	0.75	20.33	nr	90.94
Wall mounted compact fluorescent uplighter; high frequency control gear; TCL lamp in 300mm x 600mm luminaire						
2 x 36 watt	231.58	274.07	0.84	22.77	nr	296.84
2 x 40 watt	249.05	294.74	0.84	22.77	nr	317.51
2 x 55 watt	249.05	294.74	0.84	22.77	nr	317.51
Suspended linear fluorescent; T5 lamp; high frequency control gear; low brightness; 65° cut-off; 30% uplight, 70% downlight; including wedge style louvre						
1 x 49 watt	158.81	187.95	0.75	20.33	nr	208.27
Extra for emergency pack	87.73	103.83	0.25	6.78	nr	110.60

V: ELECTRICAL SUPPLY/POWER/LIGHTING SYSTEMS

Item	Net Price £	Material £	Labour hours	Labour £	Unit	Total rate £
Semi-recessed 'architectural' linear fluorescent; T5 lamp; high frequency control gear; low brightness, delivers direct, ceiling and graduated wall washing illumination						
600 x 600 mm, 2 x 24 watt	130.10	153.97	0.87	23.58	nr	177.55
600 x 600 mm, 4 x 14 watt	140.01	165.70	0.87	23.58	nr	189.28
500 x 500 mm, 2 x 24 watt	127.92	151.39	0.87	23.58	nr	174.96
Extra for emergency pack	62.30	73.73	0.25	6.78	nr	80.51
Downlighter, recessed; low voltage; mirror reflector with white/chrome bezel; dimmable transformer; for dichroic lamps						
85mm dia x 20/50 watt	15.07	17.83	0.66	17.89	nr	35.72
118mm dia x 50 watt	19.91	23.56	0.66	17.89	nr	41.45
165mm dia x 100 watt	90.96	107.65	0.66	17.89	nr	125.54
High/Low Bay luminaires						
Compact discharge; aluminium reflector						
150 watt	55.59	65.79	1.50	40.65	nr	106.44
250 watt	55.59	65.79	1.50	40.65	nr	106.44
400 watt	59.01	69.84	1.50	40.65	nr	110.49
Sealed discharge; aluminium reflector						
150 watt	171,79	203.31	1.50	40.65	nr	243.96
250 watt	185.42	219.44	1.50	40.65	nr	260.09
400 watt	242.43	286.91	1.50	40.65	nr	327.56
Corrosion resistant GRP body; gasket sealed; acrylic diffuser						
600 mm Single - 18 W	28.74	34.01	0.49	13.28	nr	47.29
600 mm Twin - 18 W	37.42	44.29	0.49	13.28	nr	57.57
1200 mm Single - 36 W	32.35	38.28	0.64	17.34	nr	55.63
1200 mm Twin - 36 W	41.48	49.09	0.64	17.34	nr	66.43
1500 mm Single - 58 W	36.05	42.67	0.72	19.51	nr	62.18
1500 mm Twin - 58 W	44.83	53.06	0.72	19.51	nr	72.57
1800 mm Single - 70 W	53.36	63.15	0.94	25.48	nr	88.63
1800 mm Twin - 70 W	66.23	78.38	0.94	25.48	nr	103.86
Flameproof to IIA/IIB, I.P. 64; Aluminium Body; BS 229 and 899						
600 mm Single - 18 W	322.21	381.33	1.04	28.19	nr	409.52
600 mm Twin - 18 W	401.40	475.05	1.04	28.19	nr	503.24
1200 mm Single - 36 W	352.95	417.70	1.31	35.50	nr	453.21
1200 mm Twin - 36 W	436.37	516.43	1.18	31.98	nr	548.41
1500 mm Single - 58 W	377.86	447.19	1.64	44.45	nr	491.64
1500 mm Twin - 58 W	456.27	539.98	1.64	44.45	nr	584.42
1800 mm Single - 70 W	414.23	490.22	1.97	53.39	nr	543.61
1800 mm Twin - 70 W	477.88	565.55	1.97	53.39	nr	618.94
External Lighting						
Ground mounted 50 watt	354.89	420.00	2.25	60.98	nr	480.98
Ceiling mounted 50 watt	159.45	188.70	2.25	60.98	nr	249.68
Bulkhead; aluminium body and polycarbonate bowl; vandal resistant; IP65						
60 watt	31.23	36.96	0.75	20.33	nr	57.29

V: ELECTRICAL SUPPLY/POWER/LIGHTING SYSTEMS

Item	Net Price £	Material £	Labour hours	Labour £	Unit	Total rate £
V21 : GENERAL LIGHTING (cont'd)						
Y73 - LUMINAIRES (GENERAL) (cont'd)						
LUMINAIRES (cont'd)						
Extra for						
Emergency version	83.94	99.34	0.25	6.78	nr	106.11
2D 2 pin 16 watt	26.38	31.22	0.66	17.89	nr	49.10
2D 2 pin 28 watt	52.68	62.35	0.66	17.89	nr	80.24
Extra for						
Emergency version	62.08	73.47	0.25	6.78	nr	80.24
Photocell	18.51	21.91	0.75	20.33	nr	42.24
1500 mm high circular bollard; polycarbonate visor; vandal resistant; IP54						
50 watt	195.53	231.41	1.75	47.43	nr	278.83
70 watt	198.66	235.10	1.75	47.43	nr	282.53
80 watt	237.76	281.39	1.75	47.43	nr	328.81
Floodlight; enclosed high performance dischargelight; integral control gear; reflector; toughened glass; IP65						
70 watt	86.62	102.51	1.25	33.88	nr	136.38
100 watt	120.69	142.83	1.25	33.88	nr	176.71
150 watt	96.12	113.76	1.25	33.88	nr	147.63
250 watt	168.24	199.10	1.25	33.88	nr	232.98
400 watt	174.28	206.26	1.25	33.88	nr	240.14
Extra for						
Photocell	18.50	21.89	0.75	20.33	nr	42.22
Lighting Track						
Single circuit; 25A 2 P&E steel trunking; low voltage with copper conductors; including couplers and supports; fixed to backgrounds						
Straight track	12.43	14.71	0.50	13.55	m	28.27
Live end feed unit complete with end stop	21.40	25.33	0.33	8.94	nr	34.27
Flexible couplers 0.5m	46.82	55.41	0.33	8.94	nr	64.36
Tap off complete with 0.8m of cable	7.36	8.71	0.25	6.78	nr	15.49
Three circuit; 25A 2 P&E steel trunking; low voltage with copper conductors; including couplers and supports incorporating integral twisted pair comms bus bracket; fixed to backgrounds						
Straight track	20.24	23.96	0.75	20.33	m	44.28
Live end feed unit complete with end stop	26.75	31.66	0.50	13.55	nr	45.21
Flexible couplers 0.5m	53.48	63.29	0.45	12.20	nr	75.49
Tap off complete with 0.8m of cable	9.66	11.43	0.30	8.13	nr	19.57

V: ELECTRICAL SUPPLY/POWER/LIGHTING SYSTEMS

Item	Net Price £	Material £	Labour hours	Labour £	Unit	Total rate £
Y74 - LIGHTING ACCESSORIES						
SWITCHES						
6 Amp metal clad surface mounted switch, gridswitch; one way						
1 Gang	4.68	5.53	0.43	11.65	nr	**17.19**
2 Gang	6.47	7.66	0.55	14.91	nr	**22.57**
3 Gang	10.54	12.47	0.77	20.87	nr	**33.34**
4 Gang	12.50	14.79	0.88	23.85	nr	**38.64**
6 Gang	21.13	25.01	1.10	29.81	nr	**54.82**
8 Gang	24.73	29.26	1.28	34.69	nr	**63.95**
12 Gang	38.17	45.17	1.67	45.26	nr	**90.43**
Extra for						
10 Amp - Two way switch	1.73	2.05	0.03	0.81	nr	**2.86**
20 Amp - Two way switch	2.32	2.75	0.04	1.08	nr	**3.83**
20 Amp - Intermediate	4.38	5.18	0.08	2.17	nr	**7.35**
20 Amp - One way SP switch	1.79	2.11	0.08	2.17	nr	**4.28**
Steel blank plate; 1 Gang	1.11	1.32	0.07	1.90	nr	**3.21**
Steel blank plate; 2 Gang	1.99	2.36	0.08	2.17	nr	**4.52**
6 Amp modular type switch; galvanised steel box, bronze or satin chrome coverplate; metalclad switches; flush mounting; one way						
1 Gang	13.00	15.39	0.43	11.65	nr	**27.04**
2 Gang	17.95	21.25	0.55	14.91	nr	**36.15**
3 Gang	25.92	30.67	0.77	20.87	nr	**51.54**
4 Gang	30.86	36.52	0.88	23.85	nr	**60.37**
6 Gang	52.03	61.58	1.18	31.98	nr	**93.56**
8 Gang	62.08	73.47	1.63	44.18	nr	**117.65**
9 Gang	77.86	92.15	1.83	49.60	nr	**141.75**
12 Gang	92.73	109.74	2.29	62.06	nr	**171.80**
6 Amp modular type swtich; galvanised steel box; bronze or satin chrome coverplate; flush mounting; two way						
1 Gang	13.47	15.94	0.43	11.65	nr	**27.60**
2 Gang	18.92	22.39	0.55	14.91	nr	**37.29**
3 Gang	27.37	32.39	0.77	20.87	nr	**53.26**
4 Gang	32.81	38.83	0.88	23.85	nr	**62.67**
6 Gang	55.07	65.18	1.18	31.98	nr	**97.16**
8 Gang	65.95	78.06	1.63	44.18	nr	**122.23**
9 Gang	82.21	97.29	1.83	49.60	nr	**146.89**
12 Gang	98.51	116.59	2.22	60.17	nr	**176.75**
Plate switches; 10 Amp flush mounted, white plastic fronted; 16mm metal box; fitted brass earth terminal						
1 Gang 1 Way, Single Pole	1.94	2.29	0.28	7.59	nr	**9.88**
1 Gang 2 Way, Single Pole	2.19	2.60	0.33	8.94	nr	**11.54**
2 Gang 2 Way, Single Pole	3.17	15.27	0.44	11.92	nr	**27.20**
3 Gang 2 Way, Single Pole	6.29	7.45	0.56	15.18	nr	**22.62**
1 Gang Intermediate	7.29	8.62	0.43	11.65	nr	**20.28**
1 Gang 1 Way, Double Pole	6.49	7.69	0.33	8.94	nr	**16.63**
1 Gang Single Pole with bell symbol	5.25	6.22	0.23	6.23	nr	**12.45**
1 Gang Single Pole marked "PRESS"	4.32	5.12	0.23	6.23	nr	**11.35**
Time delay switch, suppressed	40.55	47.99	0.49	13.28	nr	**61.27**

V: ELECTRICAL SUPPLY/POWER/LIGHTING SYSTEMS

Item	Net Price £	Material £	Labour hours	Labour £	Unit	Total rate £
V21 : GENERAL LIGHTING (cont'd)						
Y74 - LIGHTING ACCESSORIES (cont'd)						
SWITCHES (cont'd)						
Plate switches; 6 Amp flush mounted white plastic fronted; 25mm metal box; fitted brass earth terminal						
4 Gang 2 Way, Single Pole	14.26	16.88	0.42	11.38	nr	**28.26**
6 Gang 2 Way, Single Way	22.79	26.97	0.47	12.74	nr	**39.71**
Architrave plate switches; 6 Amp flush mounted, white plastic fronted; 27mm metal box; brass earth terminal						
1 Gang 2 Way, Single Pole	2.58	3.05	0.30	8.13	nr	**11.18**
2 Gang 2 Way, Single Pole	5.11	12.11	0.36	9.76	nr	**21.86**
Ceiling switches, white moulded plastic, pull cord; standard unit						
6 Amp, 1 Way, Single Pole	3.70	4.38	0.32	8.67	nr	**13.05**
6 Amp, 2 Way, Single Pole	4.45	5.27	0.34	9.21	nr	**14.48**
16 Amp, 1 Way, Double Pole	6.66	7.88	0.37	10.03	nr	**17.90**
45 Amp, 1 Way, Double Pole with neon indicator	10.34	12.23	0.47	12.74	nr	**24.97**
10 Amp splash proof moulded switch with plain, threaded or PVC entry						
1 Gang, 2 Way Single Pole	15.89	18.80	0.34	9.21	nr	**28.02**
2 Gang, 1 Way Single Pole	17.95	21.25	0.36	9.76	nr	**31.01**
2 Gang, 2 Way Single Pole	22.74	26.91	0.40	10.84	nr	**37.75**
6 Amp watertight switch; metalclad; BS 3676; ingress protected to IP65 surface mounted						
1 Gang, 2 Way; terminal entry	15.33	18.15	0.41	11.11	nr	**29.26**
1 Gang, 2 Way; through entry	15.33	18.15	0.42	11.38	nr	**29.53**
2 Gang, 2 Way; terminal entry	45.19	53.48	0.54	14.63	nr	**68.11**
2 Gang, 2 Way; through entry	45.19	53.48	0.53	14.36	nr	**67.84**
2 Way replacement switch	12.18	14.41	0.10	2.71	nr	**17.12**
15 Amp watertight switch; metalclad; BS 3676; ingress protected to IP65; surface mounted						
1 Gang 2 Way, terminal entry	20.52	24.29	0.42	11.38	nr	**35.67**
1 Gang 2 Way, through entry	20.52	24.29	0.43	11.65	nr	**35.94**
2 Gang 2 Way, terminal entry	48.18	57.02	0.55	14.91	nr	**71.93**
2 Gang 2 Way, through entry	48.18	57.02	0.54	14.63	nr	**71.66**
Intermediate interior only	12.18	14.41	0.11	2.98	nr	**17.39**
2 way interior only	12.18	14.41	0.11	2.98	nr	**17.39**
Double pole interior only	12.18	14.41	0.11	2.98	nr	**17.39**
Electrical accessories; fixed to backgrounds (Including fixings)						
Dimmer switches; rotary action; for individual lights; moulded plastic case; metal backbox; flush mounted						
1 Gang, 1 Way; 250 Watt	12.78	15.12	0.28	7.59	nr	**22.71**
1 Gang, 1 Way; 400 Watt	16.82	19.91	0.28	7.59	nr	**27.49**

V: ELECTRICAL SUPPLY/POWER/LIGHTING SYSTEMS

Material Costs/Prices for Measured Works – Electrical Installations

Item	Net Price £	Material £	Labour hours	Labour £	Unit	Total rate £
Dimmer switches; push on/off action; for individual lights; moulded plastic case; metal backbox; flush mounted						
1 Gang, 2 Way; 250 Watt	21.56	25.52	0.34	9.21	nr	**34.73**
3 Gang, 2 Way; 250 Watt	114.25	135.22	0.48	13.01	nr	**148.23**
4 Gang, 2 Way; 250 Watt	140.46	166.23	0.57	15.45	nr	**181.68**
Dimmer switches; rotary action; metal cald; metal backbox; BS 5518 and BS 800; flush mounted						
1 Gang, 1 Way; 400 Watt	35.12	41.56	0.33	8.94	nr	**50.50**
Ceiling Roses						
Ceiling rose: white moulded plastic; flush fixed to conduit box						
Plug in type; ceiling socket with 2 terminals, loop-in and ceiling plug with 3 terminals and cover	7.41	8.76	0.34	9.21	nr	**17.98**
BC lampholder; white moulded plastic; heat resistent PVC insulated and sheathed cable; flush fixed						
2 Core; 0.75mm^2	1.78	2.11	0.33	8.94	nr	**11.05**
Batten holder: white moulded plastic; 3 terminals; BS 5042; fixed to conduit						
Straight pattern; 2 terminals with loop-in and Earth	5.29	6.26	0.29	7.86	nr	**14.12**
Angled pattern; looped in terminal	5.29	6.26	0.29	7.86	nr	**14.12**
LIGHTING CONTROLS						
Lighting control system; including software, commissioning and testing. **Typical component parts indicated.** **System requirements dependanton final lighting design.**						

V: ELECTRICAL SUPPLY/POWER/LIGHTING SYSTEMS

Item	Net Price £	Material £	Labour hours	Labour £	Unit	Total rate £
V21 : GENERAL LIGHTING (cont'd)						
Y61 – CABLES						
Cable; Twin twisted bus; LSF sheathed; aluminium conductors	1.16	1.37	0.08	2.17	m	3.54
Cable; ELV 4 core 7/0.2; LSF sheathed; alumimium screened; copper conductor	1.52	1.80	0.15	4.07	m	5.86
EQUIPMENT						
Central supervisor controller including software	4992.00	5907.88	12.00	325.22	nr	6233.10
Area control unit	900.00	1065.12	4.00	108.41	nr	1173.53
Lighting control module; plug in; 9 output, 9 channel switching						
Base and lid assembly	150.00	177.52	2.05	55.56	nr	233.08
Lighting control module; plug in; 9 output 9 channel dimming (DSI)						
Base and lid assembly	170.00	201.19	2.05	55.56	nr	256.75
Lighting control module; plug in; 9 output 9 channel dimming (DALI)						
Base and lid assembly	180.00	213.02	2.05	55.56	nr	268.58
Lighting control module; hard wired; 4 circuit switching						
Base and lid assembly	160.00	189.36	1.85	50.14	nr	239.49
Compact lighting control module; 3 output 18 ballast drive; dimmable (DALI)	-	-	-	-	nr	-
Base and lid assembly	190.00	224.86	1.85	50.14	nr	275.00
Presence detectors						
Flush mounted	52.00	61.54	0.60	16.26	nr	77.80
Universal presence detectors with photo cell; flush mounted	57.20	67.69	0.60	16.26	nr	83.96
Scene switch plate; anodised aluminium finish						
4 way	62.40	73.85	1.20	32.52	nr	106.37

V: ELECTRICAL SUPPLY/POWER/LIGHTING SYSTEMS

Item	Net Price £	Material £	Labour hours	Labour £	Unit	Total rate £
V22 : GENERAL LV POWER						
Y74 - ACCESSORIES						
OUTLETS						
Socket outlet: unswitched; 13 Amp metal clad; BS 1363; galvanised steel box and coverplate with white plastic inserts; fixed surface mounted						
1 Gang	5.67	6.71	0.41	11.11	nr	**17.82**
2 Gang	10.44	12.36	0.41	11.11	nr	**23.47**
Socket outlet: switched; 13 Amp metal clad; BS 1363; galvanised steel box and coverplate with white plastic inserts; fixed surface mounted						
1 Gang	5.81	6.88	0.43	11.65	nr	**18.53**
2 Gang	10.40	12.31	0.45	12.20	nr	**24.50**
Socket outlet: switched with neon indicator; 13 Amp metal clad; BS 1363; galvanised steel box and coverplate with white plastic inserts; fixed surface mounted						
1 Gang	12.33	14.59	0.43	11.65	nr	**26.24**
2 Gang	22.42	26.53	0.45	12.20	nr	**38.73**
Socket outlet: unswitched; 13 Amp; BS 1363; white moulded plastic box and coverplate; fixed surface mounted						
1 Gang	4.26	5.04	0.41	11.11	nr	**16.15**
2 Gang	8.41	9.95	0.41	11.11	nr	**21.06**
Socket outlet; switched; 13 Amp; BS 1363; white moulded plastic box and coverplate; fixed surface mounted						
1 Gang	5.14	6.08	0.43	11.65	nr	**17.73**
2 Gang	8.67	10.26	0.45	12.20	nr	**22.45**
Socket outlet: switched with neon indicator; 13 Amp; BS 1363; white moulded plastic box and coverplate; fixed surface mounted						
1 Gang	10.52	12.45	0.43	11.65	nr	**24.10**
2 Gang	14.19	16.79	0.45	12.20	nr	**28.99**
Socket outlet: switched; 13 Amp; BS 1363; galvanised steel box, white moulded coverplate; flush fitted						
1 Gang	5.14	6.08	0.43	11.65	nr	**17.73**
2 Gang	9.95	11.78	0.45	12.20	nr	**23.97**
Socket outlet: switched with neon indicator; 13 Amp; BS 1363; galvanised steel box, white moulded coverplate; flush fixed						
1 Gang	10.52	24.90	0.43	11.65	nr	**36.55**
2 Gang	18.20	43.08	0.45	12.20	nr	**55.28**
Socket outlet: switched; 13 Amp; BS 1363; galvanised steel box, satin chrome coverplate; BS 4662; flush fixed						
1 Gang	17.17	20.32	0.43	11.65	nr	**31.98**
2 Gang	24.36	28.83	0.45	12.20	nr	**41.03**

V: ELECTRICAL SUPPLY/POWER/LIGHTING SYSTEMS

Item	Net Price £	Material £	Labour hours	Labour £	Unit	Total rate £
V22 : GENERAL LV POWER (cont'd)						
Y74 – ACCESSORIES (cont'd)						
OUTLETS (cont'd)						
Socket outlet: switched with neon indicator; 13 Amp; BS 1363; steel backbox, satin chrome coverplate; BS 4662; flush fixed						
1 Gang	13.50	15.98	0.43	11.65	nr	**27.63**
2 Gang	24.36	28.83	0.45	12.20	nr	**41.03**
RCD protected socket outlets, 13 amp, to BS 1363; galvanised steel box, white moulded cover plate; flush fitted						
2 gang, 10 mA tripping (active control)	62.76	74.27	0.45	12.20	nr	**86.47**
2 gang, 30 mA tripping (active control)	55.90	66.15	0.45	12.20	nr	**78.35**
2 gang, 30 mA tripping (passive control)	55.90	66.15	0.45	12.20	nr	**78.35**
Filtered socket outlets, 13 amp, to BS 1363, with separate 'clean earth' terminal; galvanised steel box, white moulded cover plate; flush fitted						
2 gang (spike protected)	47.26	55.93	0.50	13.55	nr	**69.48**
2 gang (spike and RFI protected)	59.82	70.80	0.55	14.91	nr	**85.71**
Replacement filter cassette	16.57	19.62	0.15	4.07	nr	**23.68**
Non-standard socket outlets, 13 amp, to BS 1363, with separate 'clean earth' terminal; for plugs with T-shaped earth pin; galvanised steel box, white moulded cover plate; flush fitted						
1 gang	8.77	10.38	0.43	11.65	nr	**22.04**
2 gang	15.78	18.68	0.43	11.65	nr	**30.33**
2 gang coloured RED	21.92	25.95	0.43	11.65	nr	**37.60**
Weatherproof socket outlet: 40 Amp; switched; single gang; RCD protected; water and dust protected to I.P.66; surface mounted						
40A 30mA tripping current protecting 1 socket	71.41	84.51	0.52	14.09	nr	**98.61**
40A 30mA tripping current protecting 2 sockets	76.16	90.14	0.64	17.34	nr	**107.48**
Plug for weatherproof socket outlet: protected to I.P.66						
13Amp plug	3.12	3.70	0.21	5.69	nr	**9.39**
Floor service outlet box; comprising flat lid with flanged carpet trim; twin 13A switched socket outlets; punched plate for mounting 2 telephone outlets; one blank plate; triple compartment						
3 compartment	32.14	38.04	0.88	23.85	nr	**61.89**
Floor service outlet box; comprising flat lid with flanged carpet trim; twin 13A switched socket outlets; punched plate for mounting 2 telephone outlets; two blank plates; four compartment						
4 compartment	43.84	51.88	0.88	23.85	nr	**75.73**
Floor service outlet box; comprising flat lid with flanged carpet trim; single 13A unswitched socket outlet; single compartment; circular						
1 compartment	46.76	55.34	0.79	21.41	nr	**76.75**

V: ELECTRICAL SUPPLY/POWER/LIGHTING SYSTEMS

Item	Net Price £	Material £	Labour hours	Labour £	Unit	Total rate £
Floor service grommet, comprising flat lid with flanged carpet trim; circular						
Floor Grommet	23.38	27.67	0.49	13.28	nr	**40.95**
POWER POSTS/POLES/PILLARS						
Power Post						
Power post; aluminium painted body; PVC-U cover; 5 nr outlets	265.11	313.75	4.00	108.41	nr	**422.16**
Power Pole						
Power pole; 3.6 metres high; aluminium painted body; PVC-U cover; 6 nr outlets	339.94	402.31	4.00	108.41	nr	**510.71**
Extra for						
Power pole extension bar; 900mm long	37.15	43.97	1.50	40.65	nr	**84.62**
Vertical multi compartment pillar; PVC-U; BS 4678 Part4 EN60529; excludes accessories						
Single						
630mm long	118.41	140.13	2.00	54.20	nr	**194.33**
3000mm long	341.22	403.83	2.00	54.20	nr	**458.03**
Double						
630mm long	118.41	140.13	3.00	81.30	nr	**221.43**
3000mm long	362.15	428.60	3.00	81.30	nr	**509.90**
CONNECTION UNITS						
Connection units: moulded pattern; BS 5733; moulded plastic box; white coverplate; knockout for flex outlet; surface mounted - standard fused						
DP Switched	6.56	7.76	0.49	13.28	nr	**21.04**
Unswitched	6.02	7.13	0.49	13.28	nr	**20.41**
DP Switched with neon indicator	8.32	9.85	0.49	13.28	nr	**23.13**
Connection units: moulded pattern; BS 5733; galvanised steel box; white coverplate; knockout for flex outlet; surface mounted						
DP Switched	8.27	17.55	0.49	13.28	nr	**30.83**
DP Unswitched	7.73	9.14	0.49	13.28	nr	**22.42**
DP Switched with neon indicator	10.04	11.88	0.49	13.28	nr	**25.16**
Connection units: galvanised pressed steel pattern; galvanised steel box; satin chrome or satin brass finish; white moulded plastic inserts; flush mounted - standard fused						
DP Switched	12.10	14.32	0.49	13.28	nr	**27.60**
Unswitched	11.38	13.47	0.49	13.28	nr	**26.75**
DP Switched with neon indicator	16.33	19.32	0.49	13.28	nr	**32.60**
Connection units: galvanised steel box; satin chrome or satin brass finish; white moulded plastic inserts; flex outlet; flush mounted - standard fused						
Switched	11.61	13.74	0.49	13.28	nr	**27.02**
Unswitched	11.02	13.04	0.49	13.28	nr	**26.32**
Switched with neon indicator	14.94	17.68	0.49	13.28	nr	**30.96**

V: ELECTRICAL SUPPLY/POWER/LIGHTING SYSTEMS

Item	Net Price £	Material £	Labour hours	Labour £	Unit	Total rate £
V22 : GENERAL LV POWER (cont'd)						
Y74 – ACCESSORIES (cont'd)						
SHAVER SOCKETS						
Shaver unit: self setting overload device; 200/250 voltage supply; white moulded plastic faceplate; unswitched						
Surface type with moulded plastic box	21.83	25.83	0.55	14.91	nr	40.74
Flush type with galvanised steel box	22.84	27.04	0.57	15.45	nr	42.48
Shaver unit: dual voltage supply unit; white moulded plastic faceplate; unswitched						
Surface type with moulded plastic box	26.31	31.14	0.62	16.80	nr	47.94
Flush type with galvanised steel box	27.36	32.38	0.64	17.34	nr	49.72
COOKER CONTROL UNITS						
Cooker control unit: BS 4177; 45 amp D.P. main switch; 13 Amp switched socket outlet; metal coverplate; plastic inserts; neon indicators						
Surface mounted with mounting box	30.90	36.57	0.61	16.53	nr	53.10
Flush mounted with galvanised steel box	29.53	34.95	0.61	16.53	nr	51.48
Cooker control unit: BS 4177; 45 Amp D.P. main switch; 13 Amp switched socket outlet; moulded plastic box and coverplate; surface mounted						
Standard	21.44	25.38	0.61	16.53	nr	41.91
With neon indicators	25.16	29.77	0.61	16.53	nr	46.30
CONTROL COMPONENTS						
Connector unit : moulded white plastic cover and block; galvanised steel back box; to immersion heaters						
3Kw up to 915mm long; fitted to thermostat	22.98	27.20	0.75	20.33	nr	47.53
Water heater switch : 20 Amp; switched with neon indicator						
DP Switched with neon indicator	11.41	28.81	0.45	12.20	nr	41.00
SWITCH DISCONNECTORS						
Switch disconnectors; moulded plastic enclosure; fixed to backgrounds						
3 pole; IP54; Grey						
16 Amp	19.84	23.48	0.80	21.68	nr	45.16
25 Amp	23.47	27.78	0.80	21.68	nr	49.46
40 Amp	38.23	45.24	0.80	21.68	nr	66.92
63 Amp	59.51	70.43	1.00	27.10	nr	97.53
80 Amp	103.17	122.09	1.25	33.88	nr	155.97
6 pole; IP54; Grey						
25 Amp	32.98	39.03	1.00	27.10	nr	66.13
63 Amp	55.71	65.93	1.25	33.88	nr	99.80
80 Amp	105.59	124.97	1.80	48.78	nr	173.75

V: ELECTRICAL SUPPLY/POWER/LIGHTING SYSTEMS

Item	Net Price £	Material £	Labour hours	Labour £	Unit	Total rate £
3 pole; IP54; Yellow						
16 Amp	21.80	25.80	0.80	21.68	nr	47.49
25 Amp	25.77	30.50	0.80	21.68	nr	52.18
40 Amp	41.81	49.49	0.80	21.68	nr	71.17
63 Amp	65.04	76.98	1.00	27.10	nr	104.08
6 pole; IP54; Yellow						
25 Amp	32.98	39.03	1.00	27.10	nr	66.13
INDUSTRIAL SOCKETS/PLUGS						
Plugs; Splashproof; 100-130 volts, 50-60 Hz; IP 44 (Yellow)						
2 pole and earth						
16 Amp	1.92	2.27	0.55	14.91	nr	17.17
32 Amp	6.90	8.17	0.60	16.26	nr	24.43
3 pole and earth						
16 Amp	7.65	9.05	0.65	17.62	nr	26.67
32 Amp	10.29	12.18	0.72	19.51	nr	31.70
3 pole; neutral and earth						
16 Amp	8.16	9.66	0.72	19.51	nr	29.18
32 Amp	12.38	14.65	0.78	21.14	nr	35.79
Connectors; Splashproof; 100-130 volts, 50-60 Hz; IP 44 (Yellow)						
2 pole and earth						
16 Amp	4.95	5.86	0.42	11.38	nr	17.25
32 Amp	8.57	10.14	0.50	13.55	nr	23.69
3 pole and earth						
16 Amp	9.30	11.00	0.48	13.01	nr	24.01
32 Amp	13.88	16.42	0.58	15.72	nr	32.14
3 pole; neutral and earth						
16 Amp	13.03	15.42	0.52	14.09	nr	29.52
32 Amp	18.28	21.63	0.73	19.78	nr	41.41
Angled sockets; surface mounted; Splashproof; 100-130 volts, 50-60 Hz; IP 44 (Yellow)						
2 pole and earth						
16 Amp	4.18	4.95	0.55	14.91	nr	19.86
32 Amp	10.46	12.38	0.60	16.26	nr	28.65
3 pole and earth						
16 Amp	9.97	11.80	0.65	17.62	nr	29.42
32 Amp	18.73	22.16	0.72	19.51	nr	41.67
3 pole; neutral and earth						
16 Amp	13.84	16.37	0.72	19.51	nr	35.89
32 Amp	18.12	21.44	0.78	21.14	nr	42.58

V: ELECTRICAL SUPPLY/POWER/LIGHTING SYSTEMS

Item	Net Price £	Material £	Labour hours	Labour £	Unit	Total rate £
V22 : GENERAL LV POWER (cont'd)						
Y74 – ACCESSORIES (cont'd)						
INDUSTRIAL SOCKETS/PLUGS (cont'd)						
Plugs; Watertight; 100-130 volts, 50-60 Hz; IP67 (Yellow)						
2 pole and earth						
16 Amp	8.00	9.47	0.55	14.91	nr	24.38
32 Amp	13.93	16.49	0.60	16.26	nr	32.75
63 Amp	38.22	45.23	0.75	20.33	nr	65.56
Connectors; Watertight; 100-130 volts, 50-60 Hz; IP 67 (Yellow)						
2 pole and earth						
16 Amp	16.33	19.32	0.42	11.38	nr	30.71
32 Amp	27.69	32.77	0.50	13.55	nr	46.32
63 Amp	67.01	79.31	0.67	18.16	nr	97.47
Angled sockets; surface mounted; Watertight; 100-130 volts, 50-60 Hz; IP 67 (Yellow)						
2 pole and earth						
16 Amp	13.50	15.98	0.55	14.91	nr	30.89
32 Amp	26.40	31.24	0.60	16.26	nr	47.50
Plugs; Splashproof; 200-250 volts, 50-60 Hz; IP 44 (Blue)						
2 pole and earth						
16 Amp	1.92	2.27	0.55	14.91	nr	17.17
32 Amp	6.91	8.18	0.60	16.26	nr	24.44
63 Amp	33.32	39.43	0.75	20.33	nr	59.76
3 pole and earth						
16 Amp	7.66	9.07	0.65	17.62	nr	26.68
32 Amp	10.29	12.18	0.72	19.51	nr	31.70
63 Amp	33.45	39.58	0.83	22.49	nr	62.08
3 pole; neutral and earth						
16 Amp	8.16	9.66	0.72	19.51	nr	29.18
32 Amp	12.39	14.66	0.78	21.14	nr	35.80
Connectors; Splashproof; 200-250 volts, 50-60 Hz; IP 44 (Blue)						
2 pole and earth						
16 Amp	4.76	5.64	0.42	11.38	nr	17.02
32 Amp	11.83	14.01	0.50	13.55	nr	27.56
63 Amp	41.76	49.42	0.67	18.16	nr	67.58
3 pole and earth						
16 Amp	12.60	14.92	0.48	13.01	nr	27.93
32 Amp	16.66	19.72	0.58	15.72	nr	35.44
63 Amp	34.36	40.66	0.75	20.33	nr	60.99

V: ELECTRICAL SUPPLY/POWER/LIGHTING SYSTEMS

Item	Net Price £	Material £	Labour hours	Labour £	Unit	Total rate £
3 pole; neutral and earth						
16 Amp	14.65	17.34	0.52	14.09	nr	31.43
32 Amp	50.17	59.38	0.73	19.78	nr	79.16
Angled sockets; surface mounted; Splashproof; 200-250 volts, 50-60 Hz; IP 44 (Blue)						
2 pole and earth						
16 Amp	6.13	7.26	0.55	14.91	nr	22.16
32 Amp	9.03	10.69	0.60	16.26	nr	26.95
63 Amp	46.68	55.25	0.75	20.33	nr	75.58
3 pole and earth						
16 Amp	12.16	14.39	0.65	17.62	nr	32.00
32 Amp	21.06	24.92	0.72	19.51	nr	44.43
63 Amp	46.38	54.89	0.83	22.49	nr	77.39
3 pole; neutral and earth						
16 Amp	10.59	12.54	0.72	19.51	nr	32.05
32 Amp	18.53	21.93	0.78	21.14	nr	43.07
Plugs; Watertight; 200-250 volts, 50-60 Hz; IP67 (Blue)						
2 pole and earth						
16 Amp	8.00	9.47	0.41	11.11	nr	20.58
32 Amp	13.92	16.47	0.50	13.55	nr	30.03
63 Amp	38.22	45.23	0.66	17.89	nr	63.12
125 Amp	97.97	115.94	0.86	23.31	nr	139.25
Connectors; Watertight; 200-250 volts, 50-60 Hz; IP 67 (Blue)						
2 pole and earth						
16 Amp	16.11	19.07	0.42	11.38	nr	30.45
32 Amp	25.68	30.39	0.50	13.55	nr	43.94
63 Amp	53.95	63.85	0.67	18.16	nr	82.01
125 Amp	160.39	189.82	0.87	23.58	nr	213.40
Angled sockets; surface mounted; Watertight; 200-250 volts, 50-60 Hz; IP 67 (Blue)						
2 pole and earth						
16 Amp	13.50	15.98	0.55	14.91	nr	30.89
32 Amp	26.39	31.23	0.60	16.26	nr	47.49
125 Amp	138.62	164.05	1.00	27.10	nr	191.15

V: ELECTRICAL SUPPLY/POWER/LIGHTING SYSTEMS

Item	Net Price £	Material £	Labour hours	Labour £	Unit	Total rate £
V32 : UNINTERRUPTIBLE POWER SUPPLY						
Uninterruptible power supply; sheet steel enclosure; self contained battery pack; including installation, testing and commissioning.						
Single phase input and output; 5 year battery life; standard 13A socket outlet connection						
1.0kVA (10 minute supply)	982.67	1162.96	0.30	8.13	nr	1171.09
1.0kVA (30 minute supply)	1719.95	2035.51	0.50	13.55	nr	2049.06
2.0kVA (10 minute supply)	1855.86	2196.35	0.50	13.55	nr	2209.91
2.0kVA (60 minute supply)	3115.73	3687.37	0.50	13.55	nr	3700.92
3.0kVA (10 minute supply)	2424.12	2868.87	0.50	13.55	nr	2882.42
3.0kVA (40 minute supply)	3683.99	4359.89	1.00	27.10	nr	4386.99
5.0kVA (30 minute supply)	5853.07	6926.94	1.00	27.10	nr	6954.04
8.0kVA (10 minute supply)	7471.92	8842.79	2.00	54.20	nr	8897.00
8.0kVA (30 minute supply)	8675.52	10267.22	2.00	54.20	nr	10321.42
Uninterruptible power supply; including final connections and testing and commissioning						
Medium size static; single phase input and output; 10 year battery life; in cubicle						
10.0 kVA (10 minutes supply)	7980.54	9444.73	10.00	271.01	nr	9715.74
10.0 kVA (30 minutes supply)	9464.04	11200.41	15.00	406.52	nr	11606.93
15.0 kVA (10 minutes supply)	10315.50	12208.08	10.00	271.01	nr	12479.10
15.0 kVA (30 minutes supply)	12193.68	14430.85	15.00	406.52	nr	14837.38
20.0 kVA (10 minutes supply)	13090.68	15492.43	10.00	271.01	nr	15763.44
20.0 kVA (30 minutes supply)	14012.52	16583.40	15.00	406.52	nr	16989.92
Medium size static; three phase input and output; 10 year battery life; in cubicle						
10.0 kVA (10 minutes supply)	9865.62	11675.67	10.00	271.01	nr	11946.68
10.0 kVA (30 minutes supply)	11393.28	13483.61	15.00	406.52	nr	13890.13
15.0 kVA (10 minutes supply)	10986.18	13001.81	15.00	406.52	nr	13408.34
15.0 kVA (30 minutes supply)	13058.94	15454.86	20.00	542.03	nr	15996.89
20.0 kVA (10 minutes supply)	13529.52	16011.78	20.00	542.03	nr	16553.81
20.0 kVA (30 minutes supply)	14426.52	17073.35	25.00	677.54	nr	17750.89
30.0 kVA (10 minutes supply)	14674.92	17367.33	25.00	677.54	nr	18044.87
30.0 kVA (30 minutes supply)	15709.92	18592.22	30.00	813.04	nr	19405.26
Large size static; three phase input and output; 10 year battery life; in cubicle						
40 kVA (10 minutes supply)	14536.92	17204.01	30.00	813.04	nr	18017.05
40 kVA (30 minutes supply)	19071.60	22570.67	30.00	813.04	nr	23383.71
60 kVA (10 minutes supply)	21276.84	25180.50	35.00	948.55	nr	26129.05
60 kVA (30 minutes supply)	26995.56	31948.44	35.00	948.55	nr	32896.99
100 kVA (10 minutes supply)	24493.62	28987.46	40.00	1084.06	nr	30071.52
200 kVA (10 minutes supply)	42041.70	49755.09	40.00	1084.06	nr	50839.15
300 kVA (10 minutes supply)	62291.82	73720.50	40.00	1084.06	nr	74804.56
400 kVA (10 minutes supply)	82322.52	97426.23	50.00	1355.08	nr	98781.31
500 kVA (10 minutes supply)	100436.40	118863.47	60.00	1626.09	nr	120489.56
600 kVA (10 minutes supply)	115692.30	136918.37	70.00	1897.11	nr	138815.47
800 kVA (10 minutes supply)	150914.04	178602.24	80.00	2168.12	nr	180770.36

V: ELECTRICAL SUPPLY/POWER/LIGHTING SYSTEMS

Item	Net Price £	Material £	Labour hours	Labour £	Unit	Total rate £
Integral diesel rotary; three phase input and output; no break supply; including ventilation and accoustic attenuation, oil day tank and interconnecting pipework						
100 kVA	125085.94	148035.45	100.00	2710.15	nr	**150745.60**
125 kVA	142013.16	168068.32	100.00	2710.15	nr	**170778.47**
150 kVA	152691.98	180706.38	100.00	2710.15	nr	**183416.53**
180 kVA	163617.20	193636.05	100.00	2710.15	nr	**196346.20**
200 kVA	241063.43	285291.34	100.00	2710.15	nr	**288001.49**
250 kVA	248496.06	294087.63	100.00	2710.15	nr	**296797.78**
300 kVA	255685.97	302596.67	120.00	3252.18	nr	**305848.85**
400 kVA	301512.56	356831.07	120.00	3252.18	nr	**360083.25**
500 kVA	331108.23	391856.66	120.00	3252.18	nr	**395108.84**
630 kVA	400561.60	474052.63	140.00	3794.21	nr	**477846.84**
800 kVA	483627.36	572358.47	140.00	3794.21	nr	**576152.68**
1000 kVA	543483.15	643196.00	140.00	3794.21	nr	**646990.21**
1125 kVA	612866.64	725309.29	160.00	4336.24	nr	**729645.53**
1250 kVA	645116.39	763475.89	160.00	4336.24	nr	**767812.13**
1500 kVA	708043.07	837947.73	170.00	4607.26	nr	**842554.98**
1750 kVA	803392.34	950790.73	170.00	4607.26	nr	**955397.98**

V: ELECTRICAL SUPPLY/POWER/LIGHTING SYSTEMS

Item	Net Price £	Material £	Labour hours	Labour £	Unit	Total rate £
V40 : EMERGENCY LIGHTING						
Y71 – LV SWITCHGEAR						
DC central battery systems BS5266 compliant 24/50/110 Volt						
DC supply to luminaires on mains failure; metal cubicle with battery charger, changeover device and battery as integral unit; ICEL 1001 compliant; 10 year design life valve regulated lead acid battery; 24 hour recharge; LCD display & LED indication; ICEL alarm pack; Includes on-site commissioning on 110 Volt systems only						
24 Volt, wall mounted						
300 W maintained, 1 hour	2089.55	2472.92	4.00	108.41	nr	2581.32
635 W maintained, 3 hour	2534.80	2999.86	6.00	162.61	nr	3162.47
470 W non maintained, 1 hour	1862.82	2204.59	4.00	108.41	nr	2313.00
780 W non maintained, 3 hour	2201.74	2605.69	6.00	162.61	nr	2768.30
50 Volt						
935 W maintained, 3 hour	2266.01	2681.76	8.00	216.81	nr	2898.57
1965 W maintained, 3 hour	2786.06	3297.22	8.00	216.81	nr	3514.03
1311 W non maintained, 3 hour	2727.63	3228.07	8.00	216.81	nr	3444.88
2510 W non maintained 3, hour	4165.07	4929.23	8.00	216.81	nr	5146.05
110 Volt						
1603 W maintained, 3 hour	3925.50	4645.71	8.00	216.81	nr	4862.52
4446 W maintained, 3 hour	4482.94	5305.43	10.00	271.01	nr	5576.44
2492 W non maintained, 3 hour	4079.76	4828.27	10.00	271.01	nr	5099.29
5429 W non maintained, 3 hour	6796.87	8043.89	12.00	325.22	nr	8369.11
DC central battery systems; BS EN 50171 compliant; 24/50/110 Volt						
Central power systems						
DC supply to luminaires on mains failure; metal cubicle with battery charger, changeover device and battery as integral unit; 10 year design life valve regulated lead acid battery; 12 hour recharge to 80% of specified duty; low volts discount; LCD display & LED indication; includes on-site commissioning for CPS systems only; battery sized for 'end of life' @ 20°C test pushbutton						
24 Volt, floor standing						
400 W non maintained, 1 hour	1962.17	2322.16	4.00	108.41	nr	2430.57
600 W maintained, 3 hour	2523.12	2986.03	6.00	162.61	nr	3148.64
50 Volt						
2133 W non maintained, 3 hour	4410.49	5219.68	8.00	216.81	nr	5436.50
1900 W maintained, 3 hour	4215.32	4988.71	8.00	216.81	nr	5205.52
110 Volt						
2200 W non maintained, 3 hour	4079.76	4828.27	8.00	216.81	nr	5045.09
4000 W maintained, 3 hour	4553.06	5388.41	12.00	325.22	nr	5713.63

V: ELECTRICAL SUPPLY/POWER/LIGHTING SYSTEMS

Item	Net Price £	Material £	Labour hours	Labour £	Unit	Total rate £
Y73 - LUMINAIRES						
24 Volt/50 Volt/110 volt fluorescent slave luminaires						
For use with DC central battery systems						
Indoor, 8 Watt	35.64	42.18	0.80	21.68	nr	**63.86**
Indoor, exit sign box	44.41	52.56	0.80	21.68	nr	**74.24**
Outdoor, 8 Watt weatherproof	39.73	47.02	0.80	21.68	nr	**68.70**
Conversion module AC/DC	44.41	52.56	0.25	6.78	nr	**59.34**
Self contained; polycarbonate base and diffuser; LED charging light to European sign directive; 3 hour standby						
Non maintained						
Indoor, 8 Watt	39.36	46.58	1.00	27.10	nr	**73.68**
Outdoor, 8 Watt weatherproof, vandal resistant IP65	56.46	66.81	1.00	27.10	nr	**93.92**
Maintained						
Indoor, 8 Watt	57.41	67.94	1.00	27.10	nr	**95.04**
Outdoor, 8 Watt weatherproof, vandal resistant IP65	83.44	98.75	1.00	27.10	nr	**125.85**
Exit signage						
Exit sign; gold effect, pendular including brackets						
Non maintained, 8 Watt	107.12	126.78	1.00	27.10	nr	**153.88**
Maintained, 8 Watt	115.31	136.46	1.00	27.10	nr	**163.56**
Modification kit						
Module & battery for 58W fluorescent modification from mains fitting to emergency; 3 hour standby	36.93	43.71	0.50	13.55	nr	**57.26**
Extra for remote box (when fitting is too small for modification)	18.34	21.70	0.50	13.55	nr	**35.25**
12 Volt low voltage lighting; non maintained; 3 hour standby						
2 x 20 Watt lamp load	127.61	151.03	1.20	32.52	nr	**183.55**
1 x 50 Watt lamp load	139.31	164.86	1.00	27.10	nr	**191.96**
Maintained 3 hour standby						
2 x 20 Watt lamp load	121.24	143.49	1.20	32.52	nr	**176.01**
1 x 50 Watt lamp load	132.33	156.61	1.00	27.10	nr	**183.71**

V: ELECTRICAL SUPPLY/POWER/LIGHTING SYSTEMS

Item	Net Price £	Material £	Labour hours	Labour £	Unit	Total rate £
V40 : EMERGENCY LIGHTING (cont'd)						
Y73 – LUMINAIRES (cont'd)						
Low Power Systems						
DC supply to luminaires on mains failure; metal cubicle with battery charger, changeover device and battery as integral unit; 5 Year design life valve regulated lead acid battery; low volts discount; LED display & LED indication; battery sized for 'end of life' @ 20°C test pushbutton						
24 Volt, floor standing						
300 W non maintained, 1 hour	2089.55	2472.92	4.00	108.41	nr	2581.32
600 W maintained, 3 hour	2312.76	2737.08	6.00	162.61	nr	2899.69
AC static inverter system; BS5266 compliant; one hour standby						
Central system supplying AC power on mains failure to mains luminaires; ICEL 1001 compliant metal cubicle(s) with changeover device, battery charger, battery & static inverter; 10 year design LIfe valve regulated lead acid battery; 24 hour recharge; LED indication and LCD display; pure sinewave output						
One hour						
750 VA, 600 W single phase I/P & O/P	2544.15	3010.92	6.00	162.61	nr	3173.53
3 KVA, 2.55 KW single phase I/P & O/P	4174.42	4940.30	8.00	216.81	nr	5157.11
5 KVA, 4.25 KW single phase I/P & O/P	5511.35	6522.52	10.00	271.01	nr	6793.54
8 KVA, 6.80 KW single phase I/P & O/P	6493.02	7684.29	12.00	325.22	nr	8009.51
10 KVA, 8.5 KW single phase I/P & O/P	8381.56	9919.32	14.00	379.42	nr	10298.74
13 KVA, 11.05 KW single phase I/P & O/P	11213.20	13270.49	16.00	433.62	nr	13704.11
15 KVA, 12.75 KW single phase I/P & O/P	11621.06	13753.17	30.00	813.04	nr	14566.22
20 KVA, 17.0 KW 3 phase I/P & single phase O/P	16050.24	18994.98	40.00	1084.06	nr	20079.04
30 KVA, 25.5 KW 3 phase I/P & O/P	21690.14	25669.63	60.00	1626.09	nr	27295.72
40 KVA, 34.0 KW 3 phase I/P & O/P	27040.23	32001.30	80.00	2168.12	nr	34169.42
50 KVA, 42.5 KW 3 phase I/P & O/P	35266.35	41736.67	90.00	2439.14	nr	44175.80
65 KVA, 55.25 KW 3 phase I/P & O/P	43262.26	51199.58	100.00	2710.15	nr	53909.73
90 KVA, 68.85 KW 3 phase I/P & O/P	57169.19	67658.02	120.00	3252.18	nr	70910.20
120 KVA, 102 KW 3 phase I/P & O/P	63815.30	75523.49	150.00	4065.22	nr	79588.71
Three hour						
750 VA, 600 W single phase I/P & O/P	2977.72	3524.04	6.00	162.61	nr	3686.65
3 KVA, 2.55 KW single phase I/P & O/P	5010.01	5929.19	8.00	216.81	nr	6146.00
5 KVA, 4.25 KW single phase I/P & O/P	7941.98	9399.09	10.00	271.01	nr	9670.11
8 KVA, 6.80 KW single phase I/P & O/P	9880.94	11693.79	12.00	325.22	nr	12019.01
10 KVA, 8.5 KW single phase I/P & O/P	13033.95	15425.29	14.00	379.42	nr	15804.71
13 KVA, 11.05 KW single phase I/P & O/P	16071.27	19019.87	16.00	433.62	nr	19453.50
15 KVA, 12.75 KW single phase I/P & O/P	16334.22	19331.06	30.00	813.04	nr	20144.10
20 KVA, 17.0 KW 3 phase I/P & single phase						

Material Costs/Prices for Measured Works – Electrical Installations

V: ELECTRICAL SUPPLY/POWER/LIGHTING SYSTEMS

Item	Net Price £	Material £	Labour hours	Labour £	Unit	Total rate £
O/P	23782.03	28145.32	40.00	1084.06	nr	29229.38
30 KVA, 25.5 KW 3 phase I/P & O/P	35543.32	42064.45	60.00	1626.09	nr	43690.54
40 KVA, 34.0 KW 3 phase I/P & O/P	42495.62	50292.29	80.00	2168.12	nr	52460.41
50 KVA, 42.5 KW 3 phase I/P & O/P	57316.44	67832.28	90.00	2439.14	nr	70271.42
65 KVA, 55.25 KW 3 phase I/P & O/P	69300.95	82015.59	100.00	2710.15	nr	84725.74
90 KVA, 68.85 KW 3 phase I/P & O/P	82017.02	97064.69	120.00	3252.18	nr	100316.87
120 KVA, 102 KW 3 phase I/P & O/P	101088.23	119634.88	150.00	4065.22	nr	123700.11
AC static inverter system; BS EN 50171 compliant; one hour standby; Low power system (typically wall mounted)						
Central system supplying AC power on mains failure to mains luminaires; metal cubicle(s) with changeover device, battery charger, battery & static inverter; 5 year design life valve regulated lead acid battery; LED indication and LCD display; 12 hour recharge to 80% duty; inverter rated for 120% of load for 100% of duty; battery sized for 'end of life' @ 20°C test pushbutton						
One hour						
300 VA, 240 W single phase I/P & O/P	1145.28	1355.40	3.00	81.30	nr	1436.71
600 VA, 480 W single phase I/P & O/P	1379.01	1632.01	4.00	108.41	nr	1740.42
750 VA, 600 W single phase I/P & O/P	2544.15	3010.92	6.00	162.61	nr	3173.53
Three hour						
150 VA, 120 W single phase I/P & O/P	1262.14	1493.71	3.00	81.30	nr	1575.01
450 VA, 360 W single phase I/P & O/P	1495.87	1770.32	4.00	108.41	nr	1878.73
750 VA, 600 W single phase I/P & O/P	2977.72	3524.04	6.00	162.61	nr	3686.65
AC static inverter system central power system; CPS BS EN 50171 compliant; one hour standby						
Central system supplying AC power on mains failure to mains luminaires; metal cubicle(s) with changeover device, battery charger, battery & static inverter; LED indication and LCD display; pure sinewave output; 10 year design life valve regulated lead acid battery; 12 hour recharge to 80% duty specified; unverter rated for 120% of load for 100% of duty; battery sized for 'end of life' @ 20°C test push button; includes on-site commissioning						
One hour						
750 VA, 600 W single phase I/P & O/P	2848.00	3370.52	6.00	162.61	nr	3533.13
3 KVA, 2.55 KW single phase I/P & O/P	4536.69	5369.04	8.00	216.81	nr	5585.85
5 KVA, 4.25 KW single phase I/P & O/P	5873.63	6951.26	10.00	271.01	nr	7222.28
8 KVA, 6.80 KW single phase I/P & O/P	6855.31	8113.05	12.00	325.22	nr	8438.27
10 KVA, 8.5 KW single phase I/P & O/P	8802.27	10417.22	14.00	379.42	nr	10796.65
13 KVA, 11.05 KW single phase I/P & O/P	11633.91	13768.39	16.00	433.62	nr	14202.01
15 KVA, 12.75 KW single phase I/P & O/P	12041.77	14251.07	30.00	813.04	nr	15064.12
20 KVA, 17.0 KW 3 phase I/P & single phase						

V: ELECTRICAL SUPPLY/POWER/LIGHTING SYSTEMS

Item	Net Price £	Material £	Labour hours	Labour £	Unit	Total rate £
V40 : EMERGENCY LIGHTING (cont'd)						
Y73 – LUMINAIRES (cont'd)						
AC static inverter system (cont'd)						
One hour (cont'd)						
O/P	16587.82	19631.19	40.00	1084.06	nr	**20715.25**
30 KVA, 25.5 KW 3 phase I/P & O/P	22344.59	26444.15	60.00	1626.09	nr	**28070.24**
40 KVA, 34.0 KW 3 phase I/P & O/P	27694.66	32775.80	80.00	2168.12	nr	**34943.92**
50 KVA, 42.5 KW 3 phase I/P & O/P	36066.88	42684.07	90.00	2439.14	nr	**45123.21**
65 KVA, 55.25 KW 3 phase I/P & O/P	44547.77	52720.95	100.00	2710.15	nr	**55431.10**
90 KVA, 68.85 KW 3 phase I/P & O/P	58454.71	69179.39	120.00	3252.18	nr	**72431.57**
120 KVA, 102 KW 3 phase I/P & O/P	65100.81	77044.86	150.00	4065.22	nr	**81110.08**
Three hour						
750 VA, 600 W single phase I/P & O/P	3281.57	3883.64	6.00	162.61	nr	**4046.25**
3 KVA, 2.55 KW single phase I/P & O/P	5372.28	6357.93	8.00	216.81	nr	**6574.75**
5 KVA, 4.25 KW single phase I/P & O/P	8324.30	9851.56	10.00	271.01	nr	**10122.57**
8 KVA, 6.80 KW single phase I/P & O/P	10243.21	12122.54	12.00	325.22	nr	**12447.76**
10 KVA, 8.5 KW single phase I/P & O/P	13454.66	15923.19	14.00	379.42	nr	**16302.61**
13 KVA, 11.05 KW single phase I/P & O/P	16491.99	19517.77	16.00	433.62	nr	**19951.40**
15 KVA, 12.75 KW single phase I/P & O/P	16754.93	19828.96	30.00	813.04	nr	**20642.00**
20 KVA, 17.0 KW 3 phase I/P & single phase O/P	24319.61	28781.53	40.00	1084.06	nr	**29865.59**
30 KVA, 25.5 KW 3 phase I/P & O/P	36197.76	42838.96	60.00	1626.09	nr	**44465.05**
40 KVA, 34.0 KW 3 phase I/P & O/P	43150.07	51066.81	80.00	2168.12	nr	**53234.93**
50 KVA, 42.5 KW 3 phase I/P & O/P	58116.96	68779.68	90.00	2439.14	nr	**71218.81**
65 KVA, 55.25 KW 3 phase I/P & O/P	70586.46	83536.96	100.00	2710.15	nr	**86247.11**
90 KVA, 68.85 KW 3 phase I/P & O/P	87754.54	103854.86	120.00	3252.18	nr	**107107.04**
120 KVA, 102 KW 3 phase I/P & O/P	102373.74	121156.25	150.00	4065.22	nr	**125221.48**

W: COMMUNICATIONS/SECURITY/CONTROL

Item	Net Price £	Material £	Labour hours	Labour £	Unit	Total rate £
W10 : TELECOMMUNICATIONS						
Y61 - CABLES						
Multipair internal telephone cable; BS 6746; loose laid on tray / basket						
0.5 millimetre diameter conductor LSZH insulated and sheathed multipair cables; BT specification CW 1308						
3 pair	0.11	0.13	0.03	0.81	m	0.94
4 pair	0.13	0.15	0.03	0.81	m	0.97
6 pair	0.19	0.22	0.03	0.81	m	1.04
10 pair	0.39	0.46	0.05	1.36	m	1.82
15 pair	0.49	0.58	0.06	1.63	m	2.21
20 pair + 1 wire	0.65	0.77	0.06	1.63	m	2.40
25 pair	0.80	0.95	0.08	2.17	m	3.11
40 pair + earth	1.24	1.47	0.08	2.17	m	3.64
50 pair + earth	1.62	1.92	0.10	2.71	m	4.63
80 pair + earth	2.53	2.99	0.10	2.71	m	5.70
100 pair + earth	3.13	3.70	0.12	3.25	m	6.96
Multipair internal telephone cable; BS 6746; installed in conduit / trunking						
0.5 millimetre diameter conductor LSZH insulated and sheathed multipair cables; BT specification CW 1308						
3 pair	0.11	0.13	0.05	1.36	m	1.49
4 pair	0.13	0.15	0.06	1.63	m	1.78
6 pair	0.19	0.22	0.06	1.63	m	1.85
10 pair	0.39	0.46	0.07	1.90	m	2.36
15 pair	0.49	0.58	0.07	1.90	m	2.48
20 pair + 1 wire	0.65	0.77	0.09	2.44	m	3.21
25 pair	0.80	0.95	0.10	2.71	m	3.66
40 pair + earth	1.24	1.47	0.12	3.25	m	4.72
50 pair + earth	1.62	1.92	0.14	3.79	m	5.71
80 pair + earth	2.53	2.99	0.14	3.79	m	6.79
100 pair + earth	3.13	3.70	0.15	4.07	m	7.77
Low speed data; unshielded twisted pair; solid copper conductors; LSOH sheath; nominal impedance 100 Ohm; Category 3 to ISO IS 1801/EIA/TIA 568B and EN50173/50174 standards to current revisions						
Installed in riser						
25 pair 24AWG	1.52	1.80	0.03	0.81	m	2.61
50 pair 24AWG	3.03	3.59	0.06	1.63	m	5.21
100 pair 24AWG	5.10	6.04	0.10	2.71	m	8.75
Installed below floor						
25 pair 24AWG	1.52	1.80	0.02	0.54	m	2.34
50 pair 24AWG	3.03	3.59	0.05	1.36	m	4.94
100 pair 24AWG	5.10	6.04	0.08	2.17	m	8.20

W: COMMUNICATIONS/SECURITY/CONTROL

Item	Net Price £	Material £	Labour hours	Labour £	Unit	Total rate £
W10 : TELECOMMUNICATIONS (cont'd)						
Y74 - ACCESSORIES						
Telephone outlet: moulded plastic plate with box; fitted and connected; flush or surface mounted						
Single master outlet	6.47	7.63	0.35	9.49	nr	17.11
Single secondary outlet	4.78	5.64	0.35	9.49	nr	15.12
Telephone outlet: bronze or satin chromeplate; with box; fitted and connected; flush or surface mounted						
Single master outlet	10.53	12.42	0.35	9.49	nr	21.90
Single secondary outlet	11.51	13.57	0.35	9.49	nr	23.06
Frames and Box Connections						
Provision and Installation of a Dual Vertical Krone 108A Voice Distribution Frame that can accommodate a total of 138 x Krone 237A Strips	218.90	259.06	1.50	40.65	nr	299.71
Label Frame (traffolyte style)	1.38	1.63	0.27	7.32	nr	8.95
Provision and Installation of a Box Connection 301A Voice Termination Unit that can accommodate a total of 10 x Krone 237A Strips	18.98	22.46	0.25	6.78	nr	29.24
Label Frame (traffolyte style)	0.45	0.53	0.08	2.25	nr	2.78
Provision and Installation of a Box Connection 201 Voice Termination Unit that can accommodate 20 pairs	10.10	11.95	0.17	4.50	nr	16.45
Label Frame (traffolyte style)	0.45	0.53	0.08	2.25	nr	2.78
Terminate, Test and Label Voice Multicore System						
Patch Panels						
Voice; 19" wide fully loaded, finished in black including termination and forming of cables (assuming 2 pairs per port)						
25 port - RJ45 UTP - Krone	47.50	56.21	2.60	70.46	nr	126.68
50 port - RJ45 UTP - Krone	64.75	76.63	4.65	126.02	nr	202.65
900 pair fully loaded Systimax style of frame including forming and termination of 9 x 100 pair cables	535.00	633.16	25.00	677.54	nr	1310.69
Installation and termination of Krone Strip (10 pair block - 237A) including designation label strip	4.25	5.03	0.50	13.55	nr	18.58
Patch panel and Outlet abelling per port (traffolyte style)	0.16	0.19	0.02	0.54	nr	0.73
Provision and installation of a voice jumper, for cross termination on Krone termination Strips	0.05	0.06	0.06	1.63	nr	1.69
CW1308 / Cat 3 cable circuit test per pair	-	-	0.04	1.08	nr	1.08

W: COMMUNICATIONS/SECURITY/CONTROL

Item	Net Price £	Material £	Labour hours	Labour £	Unit	Total rate £
W20 : RADIO/TELEVISION						
Y61 - CABLES						
RADIO						
Radio Frequency Cable; BS 2316 ; PVC sheathed; laid loose						
7/0.41mm tinned copper inner conductor; solid polyethylene dielectric insulation; bare copper wire braid; PVC sheath; 75 ohm impedance Cable	1.10	1.30	0.05	1.36	m	**2.66**
Twin 1/0.58mm copper covered steel solid core wire conductor; solid polyethylene dielectric insulation; barecopper wire braid; PVC sheath; 75 ohm impedance Cable	1.57	1.86	0.05	1.36	m	**3.22**
TELEVISION						
Television aerial cable; coaxial; PVC sheathed; fixed to backgrounds						
General purpose TV aerial downlead; copper stranded inner conductor; cellular polythene insulation; copper braid outer conductor; 75 ohm impedance 7/0.25mm	0.29	0.34	0.06	1.63	m	**1.97**
Low loss TV aerial downlead; solid copper inner conductor; cellular polythene insulation; copper braid outer; conductor; 75 ohm impedance 1/1.12mm	0.47	0.56	0.06	1.63	m	**2.18**
Low loss air spaced; solid copper inner conductor; air spaced polythene insulation; copper braid outer conductor; 75 ohm impedance 1/1.00mm	0.28	0.33	0.06	1.63	m	**1.96**
Satelite aerial downlead; solid copper inner conductor; air spaced polythene insulation; copper tape and braid outer conductor; 75 ohm impedance 1/1.00mm	0.59	0.70	0.06	1.63	m	**2.32**
Satelite TV coaxial; solid copper inner conductor; semi air spaced polyethylene dielectric insulation; plain annealed copper foil and copper braid screen in outer conductor; PVC sheath; 75 ohm impedance 1/1.25mm	0.78	0.92	0.08	2.17	m	**3.09**

W: COMMUNICATIONS/SECURITY/CONTROL

Item	Net Price £	Material £	Labour hours	Labour £	Unit	Total rate £
W20 : RADIO/TELEVISION (cont'd)						
Y61 – CABLES (cont'd)						
TELEVISION (cont'd)						
Satelite TV coaxial; solid copper inner conductor; air spaced polyethylene dielectric insulation; plain annealed copper foil and copper braid screen in outer conductor; PVC sheath; 75 ohm impedance						
1/1.67mm	1.29	1.53	0.09	2.44	m	3.97
Video cable; PVC flame retardant sheath; laid loose						
7/0.1mm silver coated copper covered annealed steel wire conductor; polyethylene dielectric insulation with tin coated copper wire braid; 75 ohm impedance						
Cable	0.74	0.87	0.05	1.36	m	2.23
Y74 - ACCESSORIES						
TV co-axial socket outlet: moulded plastic box; flush or surface mounted						
One way Direct Connection	7.53	8.91	0.35	9.49	nr	18.40
Two way Direct Connection	10.50	12.42	0.35	9.49	nr	21.91
One way Isolated UHF/VHF	13.26	15.69	0.35	9.49	nr	25.18
Two way Isolated UHF/VHF	18.06	21.38	0.35	9.49	nr	30.86

Material Costs/Prices for Measured Works – Electrical Installations

W: COMMUNICATIONS/SECURITY/CONTROL

Item	Net Price £	Material £	Labour hours	Labour £	Unit	Total rate £
W23 : CLOCKS						
Clock timing systems; master and slave units; fixed to background; excluding supports and fixings						
Quartz master clock with solid state digital readout for parallel loop operation; one minute, half minute and one second pulse; maximum of 160 clocks						
Over two loops only	630.00	745.59	4.40	119.25	nr	**864.83**
Power supplies for above, giving 24 hours power reserve						
2 6 Amp hour batteries	183.75	217.46	3.00	81.30	nr	**298.77**
2 15 Amp hour batteries	210.00	248.53	5.00	135.51	nr	**384.04**
Radio receiver to accept BBC Rugby Transmitter MSF signal						
To synchronise time of above Quartz master clock	136.50	161.54	7.04	190.86	nr	**352.40**
Wall clocks for slave (impulse) systems; 24V DC, white dial with black numerals fitted with axispolycarbonate disc; BS 467.7 Class O						
305mm diameter 1 minute impulse	55.17	65.29	1.37	37.13	nr	**102.42**
305mm diameter 1/2 minute impulse	55.17	65.29	1.37	37.13	nr	**102.42**
227mm diameter 1 second impulse	78.75	93.20	1.37	37.13	nr	**130.33**
305mm diameter 1 second impulse	78.75	93.20	1.37	37.13	nr	**130.33**
Quartz battery movement; BS 467.7 Class O; white dial with black numerals and sweep second hand; fitted with axispolycarbonate disc; stove enamel case						
305mm diameter	28.88	34.05	0.77	20.87	nr	**54.92**
Internal wall mounted electric clock; white dial with black numerals; 240v, 50 Hz						
305mm diameter	36.75	43.49	1.00	27.10	nr	**70.59**
458mm diameter	157.50	186.40	1.00	27.10	nr	**213.50**
Matching clock; BS 467.7 Class O; 240V AC, 50/60 Hz mains supply; 12 hour duration; dial with 1-12; IP 66; axispolycarbonate disc; spun metal movement cover; semi flush mount on 6 point fixing bezel						
227mm diameter	210.00	248.53	0.62	16.80	nr	**265.33**
Digital clocks; 240V, 50 hz supply; with/without synchronisation from masterclock;12/24 hour display; stand alone operation; 50mm digits						
Flush - hours/minutes/seconds or minutes/seconds/10th seconds	210.00	248.53	0.57	15.45	nr	**263.98**
Surface - hours/minutes/seconds or minutes/seconds/10th seconds	183.75	465.21	0.57	15.45	nr	**480.65**
Flush - hours/minutes or minutes/seconds	178.50	657.67	0.57	15.45	nr	**673.12**
Surface - hours/minutes or minutes/seconds	152.25	416.23	0.57	15.45	nr	**431.67**

W: COMMUNICATIONS/SECURITY/CONTROL

Item	Net Price £	Material £	Labour hours	Labour £	Unit	Total rate £
W30 : DATA TRANSMISSION						
Cabinets						
Floor standing; suitable for 19" patch panels with glass lockable doors, metal rear doors, side panels, vertical cable management, 2 x 4 way PDU's, 4 way fan, earth bonding kit; installed on raised floor						
600 wide x 800 deep - 18U	655.20	835.60	3.00	85.83	nr	**921.42**
600 wide x 800 deep - 24U	682.50	870.41	3.00	85.83	nr	**956.24**
600 wide x 800 deep - 33U	733.00	934.82	4.00	114.44	nr	**1049.25**
600 wide x 800 deep - 42U	786.25	1002.73	4.00	114.44	nr	**1117.17**
600 wide x 800 deep - 47U	830.00	1058.52	4.00	114.44	nr	**1172.96**
800 wide x 800 deep - 42U	890.00	1135.04	4.00	114.44	nr	**1249.48**
800 wide x 800 deep - 47U	928.25	1183.83	4.00	114.44	nr	**1298.26**
Label cabinet	2.10	2.68	0.25	7.15	nr	**9.83**
Wall mounted; suitable for 19" patch panels with glass lockable doors, side panels, vertical cable management, 2 x 4 way PDU's, 4 way fan, earth bonding kit; fixed to wall						
19 wide x 500 deep - 9U	437.00	557.32	3.00	85.83	nr	**643.15**
19 wide x 500 deep - 12U	447.75	571.03	3.00	85.83	nr	**656.86**
19 wide x 500 deep - 15U	464.00	591.75	3.00	85.83	nr	**677.58**
19 wide x 500 deep - 18U	546.00	696.33	3.00	85.83	nr	**782.16**
19 wide x 500 deep - 21U	568.00	724.39	3.00	85.83	nr	**810.22**
Label cabinet	2.10	2.68	0.25	7.15	nr	**9.83**
Frames						
Floor standing; suitable for 19" patch panels with supports, vertical cable management, earth bonding kit; installed on raised floor						
19 wide x 500 deep - 25U	491.50	626.82	2.50	71.52	nr	**698.35**
19 wide x 500 deep - 39U	600.50	765.84	2.50	71.52	nr	**837.36**
19 wide x 500 deep - 42U	655.25	835.66	2.50	71.52	nr	**907.18**
19 wide x 500 deep - 47U	710.00	905.48	2.50	71.52	nr	**977.01**
Label frame	2.10	2.68	0.25	7.15	nr	**9.83**
Copper Data Cabling						
Unshielded twisted pair; solid copper conductors; PVC insulation; nominal impedance 100 Ohm; Cat 5e to ISO 11801, EIA/TIA 568B and EN 50173/50174 standards to the current revisions						
4 pair 24AWG; nominal outside diameter 5.6mm; installed above ceiling	0.14	0.18	0.02	0.57	m	**0.75**
4 pair 24AWG; nominal outside diameter 5.6mm; installed in riser	0.14	0.18	0.02	0.57	m	**0.75**
4 pair 24AWG; nominal outside diameter 5.6mm; installed below floor	0.14	0.18	0.01	0.29	m	**0.46**
4 pair 24AWG; nominal outside diameter 5.6mm; installed in trunking	0.14	0.18	0.02	0.57	m	**0.75**

W: COMMUNICATIONS/SECURITY/CONTROL

Item	Net Price £	Material £	Labour hours	Labour £	Unit	Total rate £
Unshielded twisted pair; solid copper conductors; LSOH sheathed; nominal impedance 100 Ohm; Cat 5e to ISO 11801, EIA/TIA 568B and EN 50173/50174 standards to the current revisions						
4 pair 24AWG; nominal outside diameter 5.6mm; installed above ceiling	0.21	0.27	0.02	0.57	m	**0.84**
4 pair 24AWG; nominal outside diameter 5.6mm; installed in riser	0.21	0.27	0.02	0.57	m	**0.84**
4 pair 24AWG; nominal outside diameter 5.6mm; installed below floor	0.21	0.27	0.01	0.29	m	**0.55**
4 pair 24AWG; nominal outside diameter 5.6mm; installed in trunking	0.21	0.27	0.02	0.57	m	**0.84**
Unshielded twisted pair; solid copper conductors; PVC insulation; nominal impedance 100 Ohm; Cat 6 to ISO 11801, EIA/TIA 568B and EN 50173/50174 standards to the current revisions						
4 pair 24AWG; nominal outside diameter 5.6mm; installed above ceiling	0.22	0.28	0.02	0.43	m	**0.71**
4 pair 24AWG; nominal outside diameter 5.6mm; installed in riser	0.22	0.28	0.02	0.43	m	**0.71**
4 pair 24AWG; nominal outside diameter 5.6mm; installed below floor	0.22	0.28	0.01	0.23	m	**0.51**
4 pair 24AWG; nominal outside diameter 5.6mm; installed in trunking	0.22	0.28	0.02	0.57	m	**0.85**
Unshielded twisted pair; solid copper conductors; LSOH sheathed; nominal impedance 100 Ohm; Cat 6 to ISO 11801, EIA/TIA 568B and EN 50173/50174 standards to the current revisions						
4 pair 24AWG; nominal outside diameter 5.6mm; installed above ceiling	0.26	0.33	0.02	0.63	m	**0.96**
4 pair 24AWG; nominal outside diameter 5.6mm; installed in riser	0.26	0.33	0.02	0.63	m	**0.96**
4 pair 24AWG; nominal outside diameter 5.6mm; installed below floor	0.26	0.33	0.01	0.31	m	**0.65**
4 pair 24AWG; nominal outside diameter 5.6mm; installed in trunking	0.26	0.33	0.02	0.63	m	**0.96**
Patch panels						
Category 5e; 19" wide fully loaded, finished in black including termination and forming of cables						
24 port - RJ45 UTP - Krone / 110	69.00	88.00	4.75	135.89	nr	**223.89**
48 port - RJ45 UTP - Krone / 110	129.50	165.16	9.35	267.50	nr	**432.65**
Patch panel labelling per port	0.22	0.28	0.02	0.57	nr	**0.85**
Category 6; 19" wide fully loaded, finished in black including termination and forming of cables						
24 port - RJ45 UTP - Krone / 110	132.00	168.34	5.00	143.05	nr	**311.39**
48 port - RJ45 UTP - Krone / 110	240.50	306.72	9.80	280.37	nr	**587.09**
Patch panel labelling per port	0.22	0.28	0.02	0.57	nr	**0.85**

W: COMMUNICATIONS/SECURITY/CONTROL

Item	Net Price £	Material £	Labour hours	Labour £	Unit	Total rate £
W30 : DATA TRANSMISSION (cont'd)						
Work Station						
Category 5e RJ45 data outlet plate and multiway outlet boxes for wall, ceiling and below floor installations including label to ISO 11801 standards						
Wall mounted; fully loaded						
One gang LSOH PVC plate	6.38	8.14	0.15	4.29	nr	12.43
Two gang LSOH PVC plate	9.13	11.64	0.20	5.72	nr	17.37
Four gang LSOH PVC plate	15.62	19.92	0.40	11.44	nr	31.36
One gang satin brass plate	14.50	18.49	0.25	7.15	nr	25.64
Two gang satin brass plate	16.35	20.85	0.33	9.44	nr	30.29
Ceiling mounted; fully loaded						
One gang metal clad plate	7.70	9.82	0.33	9.44	nr	19.26
Two gang metal clad plate	10.45	13.33	0.45	12.87	nr	26.20
Below floor; fully loaded						
Four way outlet box, 5 m length 20mm flexible conduit with glands and starin relief bracket	26.02	33.18	0.80	22.89	nr	56.07
Six way outlet box, 5 m length 25mm flexible conduit with glands and strain relief bracket	33.64	42.90	1.20	34.33	nr	77.23
Eight way outlet box, 5 m length 25mm flexible conduit with glands and strain relief bracket	42.35	54.01	1.40	40.05	nr	94.06
Installation of outlet boxes to desks	-	-	0.40	11.44	nr	11.44
Category 6 RJ45 data outlet plate and multiway outlet boxes for wall, ceiling and below floor installations including label to ISO 11801 standards						
Wall mounted; fully loaded						
One gang LSOH PVC plate	8.14	10.38	0.17	4.72	nr	15.10
Two gang LSOH PVC plate	11.94	15.22	0.22	6.29	nr	21.52
Four gang LSOH PVC plate	21.56	27.50	0.44	12.59	nr	40.08
One gang satin brass plate	13.93	17.77	0.28	7.87	nr	25.63
Two gang satin brass plate	17.43	22.24	0.36	10.39	nr	32.62
Ceiling mounted; fully loaded						
One gang metal clad plate	9.24	11.78	0.36	10.39	nr	22.17
Two gang metal clad plate	13.04	16.62	0.50	14.16	nr	30.79
Below floor; fully loaded						
Four way outlet box, 5 m length 20mm flexible conduit with glands and starin relief bracket	31.89	40.67	0.88	25.18	nr	65.85
Six way outlet box, 5 m length 25mm flexible conduit with glands and strain relief bracket	42.42	54.09	1.32	37.76	nr	91.86
Eight way outlet box, 5 m length 25mm flexible conduit with glands and strain relief bracket	54.03	68.91	1.54	44.06	nr	112.97
Installation of outlet boxes to desks	-	-	0.40	11.44	nr	11.44
Category 5e cable test	-	-	0.10	2.86	nr	2.86
Category 6 cable test	-	-	0.11	3.15	nr	3.15

W: COMMUNICATIONS/SECURITY/CONTROL

Item	Net Price £	Material £	Labour hours	Labour £	Unit	Total rate £
Copper Patch Leads						
Category 5e; straight through booted RJ45 UTP - RJ45 UTP						
Patch lead 1m length	1.60	2.04	0.09	2.57	nr	**4.62**
Patch lead 3m length	1.90	2.42	0.09	2.57	nr	**5.00**
Patch lead 5m length	2.70	3.44	0.10	2.86	nr	**6.30**
Patch lead 7m length	3.95	5.04	0.10	2.86	nr	**7.90**
Category 6; straight through booted RJ45 UTP - RJ45 UTP						
Patch lead 1 m length	4.25	5.42	0.09	2.57	nr	**8.00**
Patch lead 3 m length	6.40	8.16	0.09	2.57	nr	**10.74**
Patch lead 5 m length	7.50	9.56	0.10	2.86	nr	**12.43**
Patch lead 7 m length	9.50	12.12	0.10	2.86	nr	**14.98**
Note 1 - With an intelligent Patching System, add a 25% uplift to the Cat5e and Cat 6 System price. This would be dependant on the System and IMS solution chosen, i.e. iPatch, RIT or iTRACs.						
Note 2 - For a Cat 6 augmented solution (10Gbps capable), add 25% to a Cat 6 System price.						
Fibre Infrastructure						
Fibre optic cable, tight buffered, internal/external application, single mode, LSOH sheathed						
4 core fibre optic cable	1.27	1.62	0.10	2.86	m	**4.48**
8 core fibre optic cable	2.11	2.69	0.10	2.86	m	**5.55**
12 core fibre optic cable	2.62	3.34	0.10	2.86	m	**6.20**
16 core fibre optic cable	3.18	4.06	0.10	2.86	m	**6.92**
24 core fibre optic cable	3.84	4.90	0.10	2.86	m	**7.76**
Fibre optic cable OM1 and OM2, tight buffered, internal/external application, 62.5/125 multimode fibre, LSOH sheathed						
4 core fibre optic cable	1.36	1.73	0.10	2.86	m	**4.60**
8 core fibre optic cable	2.25	2.87	0.10	2.86	m	**5.73**
12 core fibre optic cable	2.86	3.65	0.10	2.86	m	**6.51**
16 core fibre optic cable	3.56	4.54	0.10	2.86	m	**7.40**
24 core fibre optic cable	4.12	5.25	0.10	2.86	m	**8.12**
Fibre optic cable OM3, tight buffered, internal only application, 50/125 multimode fibre, LSOH sheathed						
4 core fibre optic cable	1.68	2.14	0.10	2.86	nr	**5.00**
8 core fibre optic cable	2.62	3.34	0.10	2.86	nr	**6.20**
12 core fibre optic cable	3.80	4.85	0.10	2.86	nr	**7.71**
24 core fibre cable	7.26	9.26	0.10	2.86	nr	**12.12**

W: COMMUNICATIONS/SECURITY/CONTROL

Item	Net Price £	Material £	Labour hours	Labour £	Unit	Total rate £
W30 : DATA TRANSMISSION (cont'd)						
Fibre Infrastructure (cont'd)						
Fibre optic single and multimode connectors and couplers including termination						
ST singlemode booted connector	6.39	8.15	0.25	7.15	nr	15.30
ST multimode booted connector	3.09	3.94	0.25	7.15	nr	11.09
SC simplex singlemode booted connector	9.06	11.55	0.25	7.15	nr	18.71
SC simplex multimode booted connector	3.30	4.21	0.25	7.15	nr	11.36
SC duplex multimode booted connector	6.59	8.40	0.25	7.15	nr	15.56
ST - SC duplex adaptor	16.02	20.43	0.01	0.29	nr	20.72
ST inline bulkhead coupler	3.43	4.37	0.01	0.29	nr	4.66
SC duplex coupler	6.41	8.17	0.01	0.29	nr	8.46
MTRJ small form factor duplex connector	6.69	6.69	0.25	7.15	nr	13.84
LC simplex multimode booted connector	-	-	0.25	7.15	nr	7.15
Singlemode core test per core	-	-	0.20	5.72	nr	5.72
Multimode core test per core	-	-	0.20	5.72	nr	5.72
Fibre; 19" wide fully loaded, labled, alluminium alloy c/w couplers, fibre management and glands (excludes termination of fibre cores)						
8 way ST; fixed drawer	68.78	87.71	0.50	14.30	nr	102.02
16 way ST; fixed drawer	103.16	131.57	0.50	14.30	nr	145.87
24 way ST; fixed drawer	143.32	182.79	0.50	14.30	nr	197.09
8 way ST; sliding drawer	87.50	111.59	0.50	14.30	nr	125.90
16 way ST; sliding drawer	120.25	153.36	0.50	14.30	nr	167.66
24 way ST; sliding drawer	158.25	201.82	0.50	13.55	nr	215.37
8 way (4 duplex) SC; fixed drawer	82.00	104.58	0.50	14.30	nr	118.88
16 way (8 duplex) SC; fixed drawer	120.00	153.04	0.50	14.30	nr	167.34
24 way (12 duplex) SC; fixed drawer	147.50	188.11	0.50	14.30	nr	202.42
8 way (4 duplex) SC; sliding drawer	103.75	132.32	0.50	14.30	nr	146.62
16 way (8 duplex) SC; sliding drawer	142.00	181.10	0.50	14.30	nr	195.40
24 way (12 duplex) SC; sliding drawer	169.25	215.85	0.50	14.30	nr	230.15
8 way (4 duplex) MTRJ; fixed drawer	93.00	118.61	0.50	14.30	nr	132.91
16 way (8 duplex) MTRJ; fixed drawer	142.00	181.10	0.50	14.30	nr	195.40
24 way (12 duplex) MTRJ; fixed drawer	169.25	215.85	0.50	14.30	nr	230.15
8 way (4 duplex) FC/PC; fixed drawer	97.50	124.34	0.50	14.30	nr	138.65
16 way (8 duplex) FC/PC; fixed drawer	149.00	190.02	0.50	14.30	nr	204.33
24 way (12 duplex) FC/PC; fixed drawer	177.75	226.69	0.50	14.30	nr	240.99
Patch panel label per way	0.22	0.28	0.02	0.57	nr	0.85
Fibre Patch Leads						
Single Mode						
Duplex OS1 LC - LC						
Fibre patch lead 1 m length	18.51	23.61	0.08	2.29	nr	25.90
Fibre patch lead 3 m length	20.12	25.66	0.08	2.29	nr	27.95
Fibre patch lead 5 m length	21.14	26.96	0.10	2.86	nr	29.83
Multi Mode						
Duplex 50/125 OM3 MTRJ - MTRJ						
Fibre patch lead 1m length	20.72	26.42	0.08	2.29	nr	28.71
Fibre patch lead 3m length	22.54	28.75	0.08	2.17	nr	30.91
Fibre patch lead 5m length	23.67	30.19	0.10	2.86	nr	33.05
Duplex 50/125 OM3 ST - ST						
Fibre patch lead 1m length	22.19	28.30	0.08	2.29	nr	30.59
Fibre patch lead 3m length	24.15	30.80	0.08	2.29	nr	33.09
Fibre patch lead 5m length	25.37	32.36	0.10	2.86	nr.	35.22

W: COMMUNICATIONS/SECURITY/CONTROL

Item	Net Price £	Material £	Labour hours	Labour £	Unit	Total rate £
Duplex 50/125 OM3 SC - SC						
Fibre patch lead 1m length	19.73	25.16	0.08	2.29	nr	**27.45**
Firbe patch lead 3m length	21.47	27.38	0.08	2.29	nr	**29.67**
Fibre patch lead 5m length	22.55	28.76	0.10	2.86	nr	**31.62**
Duplex 50/125 OM3 LC - LC						
Fibre patch lead 1 m length	24.66	31.45	0.08	2.29	nr	**33.74**
Fibre patch lead 3 m length	26.83	34.22	0.08	2.29	nr	**36.51**
Fibre patch lead 5 m length	28.18	35.94	0.10	2.86	nr	**38.80**

W: COMMUNICATIONS/SECURITY/CONTROL

Item	Net Price £	Material £	Labour hours	Labour £	Unit	Total rate £
W40 : ACCESS CONTROL						
ACCESS CONTROL EQUIPMENT						
Equipment to control the movement of personnel into defined spaces; includes fixing to backgrounds, termination of power and data cables; excludes cable containment and cable installation						
Access control						
Magnetic swipe reader	108.11	140.76	1.50	63.61	nr	204.37
Proximity reader	48.88	63.64	1.50	63.61	nr	127.25
Exit button	10.25	13.35	1.50	63.61	nr	76.95
Exit PIR	37.78	49.19	2.00	84.81	nr	134.00
Emergency break glass double pole	15.97	20.79	1.00	42.41	nr	63.20
Alarm contact flush	1.25	1.63	1.00	42.41	nr	44.03
Alarm contact surface	1.25	1.63	1.00	42.41	nr	44.03
Reader controller 16 door	2580.69	3360.06	4.00	169.62	nr	3529.68
Reader controller 8 door	1631.75	2124.54	4.00	169.62	nr	2294.16
Reader controller 2 door	450.54	586.60	4.00	169.62	nr	756.23
Reader interface	237.24	308.89	1.50	63.61	nr	372.50
Lock power supply 12 volt 3 AMP	51.15	66.60	2.00	84.81	nr	151.41
Lock power supply 24 volt 3 AMP	65.00	84.63	2.00	84.81	nr	169.44
Rechargeable battery 12 volt 7ah	8.48	11.04	0.25	10.60	nr	21.64
Lock Equipment						
Single slimline magnetic lock, monitored	39.76	51.77	2.00	84.81	nr	136.58
Single slimline magnetic lock, unmonitored	36.90	48.04	1.75	74.21	nr	122.25
Double slimline magnetic lock, monitored	79.52	103.53	4.00	169.62	nr	273.16
Double slimline magnetic lock, unmonitored	73.23	95.35	4.50	190.83	nr	286.17
Standard single magnetic lock, monitored	48.91	63.68	1.25	53.01	nr	116.69
Standard single magnetic lock, unmonitored	42.05	54.75	1.00	42.41	nr	97.16
Standard single magnetic lock, double monitored	87.14	113.46	1.50	63.61	nr	177.07
Standard double magnetic lock, monitored	97.72	127.23	1.50	63.61	nr	190.84
Standard double magnetic lock, unmonitored	84.06	109.45	1.25	53.01	nr	162.45
12V electric release fail safe, monitored	37.14	48.36	1.25	53.01	nr	101.36
12V electric release fail secure, monitored	37.14	48.36	1.00	42.41	nr	90.76
Solenoid bolt	57.20	74.47	1.50	63.61	nr	138.08
Electric mortice lock	270.40	352.06	1.00	42.41	nr	394.47

W: COMMUNICATIONS/SECURITY/CONTROL

Item	Net Price £	Material £	Labour hours	Labour £	Unit	Total rate £
W41 : SECURITY DETECTION & ALARM						
SECURITY DETECTION & ALARM EQUIPMENT						
Detection and alarm systems for the protection of property and persons; includes fixing of equipment to backgrounds and termination of power and data cabling; excludes cable containment and cable installation						
Detection, alarm equipment						
Alarm contact flush	4.68	6.09	1.00	42.41	nr	48.50
Alarm contact surface	5.00	6.51	1.00	42.41	nr	48.92
Roller shutter contact	12.74	16.59	1.00	42.41	nr	58.99
Personal attack button	7.08	9.22	1.00	42.41	nr	51.62
Acoustic break glass detectors	19.74	25.70	2.00	84.81	nr	110.51
Vibration detectors	10.48	13.64	2.00	84.81	nr	98.46
12 metre PIR detector	16.56	21.56	1.50	63.61	nr	85.17
15 metre dual detector	20.96	27.29	1.50	63.61	nr	90.90
8 zone alarm panel	128.00	166.66	3.00	127.22	nr	293.87
8-24 zone end station	178.00	231.76	4.00	169.62	nr	401.38
Remote keypad	74.00	96.35	2.00	84.81	nr	181.16
8 zone expansion	44.80	58.33	1.50	63.61	nr	121.94
Final exit set button	4.72	6.15	1.00	42.41	nr	48.55
Self contained external sounder	29.20	38.02	2.00	84.81	nr	122.83
Internal loudspeaker	8.00	10.42	2.00	84.81	nr	95.23
Rechargeable battery 12 volt 7ah (ampere hours)	8.48	11.04	0.25	10.60	nr	21.64
Surveillance Equipment						
Vandal resistant camera, colour	212.50	276.68	3.00	127.22	nr	403.89
External camera, colour	164.45	214.11	5.00	212.03	nr	426.14
Auto dome external, colour	774.40	1008.27	5.00	212.03	nr	1220.30
Auto dome external, colour/monochrome	856.90	1115.68	5.00	212.03	nr	1327.71
Auto dome internal, colour	653.40	850.73	4.00	169.62	nr	1020.35
Auto dome internal colour/monochrome	735.90	958.14	4.00	169.62	nr	1127.77
Mini internal domes	54.45	70.89	3.00	127.22	nr	198.11
Camera switcher, 32 inputs	604.45	786.99	4.00	169.62	nr	956.62
Full function keyboard	219.45	285.72	1.00	42.41	nr	328.13
16 CH multiplexors, simplex	-	-	-	-	nr	-
16 CH multiplexors, duplex	335.50	436.82	2.00	84.81	nr	521.63
16 way DVR, 250 GB harddrive	1470.00	1913.94	3.00	127.22	nr	2041.16
Additional 250 GB harddrive	185.00	722.61	1.00	42.41	nr	765.02
10" colour monitor	153.45	199.79	2.00	84.81	nr	284.60
15" colour monitor	81.95	106.70	2.00	84.81	nr	191.51
17" colour monitor, high resolution	219.45	285.72	2.00	84.81	nr	370.54
21" colour monitor, high resolution	191.45	249.27	2.00	84.81	nr	334.08

W: COMMUNICATIONS/SECURITY/CONTROL

Item	Net Price £	Material £	Labour hours	Labour £	Unit	Total rate £
W50 : FIRE DETECTION AND ALARM						
STANDARD FIRE DETECTION CONTROL PANEL						
Zone control panel; 2 x 12 volt batteries/charge up to 48 hours standby; mild steel case; flush or surface mounting						
1 zone	161.37	190.98	3.00	81.30	nr	**272.28**
2 zone	217.22	257.07	3.51	95.09	nr	**352.17**
4 zone	255.75	302.67	4.00	108.41	nr	**411.08**
8 zone	379.43	449.04	5.00	135.51	nr	**584.55**
12 zone	457.27	541.17	6.00	162.61	nr	**703.77**
16 zone	771.02	912.48	6.00	162.61	nr	**1075.09**
24 zone	1037.59	1227.96	6.00	162.61	nr	**1390.57**
Repeater panels						
8 zone	283.31	335.29	4.00	108.41	nr	**443.69**
EQUIPMENT						
Manual call point units: plastic covered						
Surface mounted						
Call point	9.17	10.85	0.50	13.55	nr	**24.40**
Call point; Weatherproof	84.77	100.32	0.80	21.68	nr	**122.00**
Flush mounted						
Call point	9.17	10.85	0.56	15.18	nr	**26.03**
Call point; Weatherproof	84.77	100.32	0.86	23.32	nr	**123.65**
Detectors						
Smoke, ionisation type with mounting base	31.77	37.60	0.75	20.33	nr	**57.93**
Smoke, optical type with mounting base	31.77	37.60	0.75	20.33	nr	**57.93**
Fixed temperature heat detector with mounting base (60°C)	25.57	30.26	0.75	20.33	nr	**50.59**
Rate of Rise heat detector with mounting base (90°C)	23.96	28.36	0.75	20.33	nr	**48.69**
Duct detector including optical smoke detector and base	203.12	240.39	2.00	54.20	nr	**294.59**
Remote smoke detector LED indicator with base	8.18	9.68	0.50	13.55	nr	**23.23**
Sounders						
6" bell, conduit box	17.69	20.94	0.75	20.33	nr	**41.27**
6" bell, conduit box; weatherproof	-	-	0.75	20.33	nr	**20.33**
Siren; 230V	62.84	74.37	1.25	33.88	nr	**108.25**
Magnetic Door Holder; 230V ; surface fixed	59.26	70.13	1.50	40.69	nr	**110.83**
ADDRESSABLE FIRE DETECTION CONTROL PANEL						
Analogue addressable panel; BS EN54 Part 2 and 4 1998; incorporating 120 addresses per loop (maximum 1-2km length); sounders wired on loop; sealed lead acid integral battery standby providing 48 hour standby; 24 volt DC; mild steel case; surface fixed						
1 loop; 4 x 12 volt batteries	1250.84	1480.33	6.00	162.61	nr	**1642.94**

W: COMMUNICATIONS/SECURITY/CONTROL

Item	Net Price £	Material £	Labour hours	Labour £	Unit	Total rate £
Extra for 1 loop panel						
Loop card	296.82	351.28	1.00	27.10	nr	378.38
Repeater panel	868.39	1027.71	6.00	162.61	nr	1190.32
Network nodes	1486.00	1758.64	6.00	162.61	nr	1921.25
Interface unit; for other systems						
Mains powered	445.00	526.64	1.50	40.65	nr	567.30
Loop powered	195.34	231.18	1.00	27.10	nr	258.28
Single channel I/O	101.50	120.12	1.00	27.10	nr	147.22
Zone module	126.25	149.41	1.50	40.65	nr	190.07
4 loop; 4 x 12 volt batteries; 24 hour standby; 30 minute alarm	2097.00	2481.74	8.00	216.81	nr	2698.55
Extra for 4 loop panel						
Loop card	439.73	520.41	1.00	27.10	nr	547.51
Repeater panel	1245.20	1473.66	6.00	162.61	nr	1636.27
Mimic panel	3057.12	3618.01	5.00	135.51	nr	3753.52
Network nodes	1484.84	1757.26	6.00	162.61	nr	1919.87
Interface unit; for other systems						
Mains powered	445.00	526.64	1.50	40.65	nr	567.30
Loop powered	195.34	231.18	1.00	27.10	nr	258.28
Single channel I/O	101.50	120.12	1.00	27.10	nr	147.22
Zone module	126.25	149.41	1.50	40.65	nr	190.07
Line modules	23.28	27.55	1.00	27.10	nr	54.65
8 loop; 4 x 12 volt batteries; 24 hour standby; 30 minute alarm	4274.84	5059.14	12.00	325.22	nr	5384.36
Extra for 8 loop panel						
Loop card	439.73	520.41	1.00	27.10	nr	547.51
Repeater panel	1245.20	1473.66	6.00	162.61	nr	1636.27
Mimic panel	3057.12	3618.01	5.00	135.51	nr	3753.52
Network nodes	1484.84	1757.26	6.00	162.61	nr	1919.87
Interface unit; for other systems						
Mains powered	445.00	526.64	1.50	40.65	nr	567.30
Loop powered	195.36	231.20	1.00	27.10	nr	258.30
Single channel I/O	42.60	50.42	1.00	27.10	nr	77.52
Zone module	126.25	149.41	1.50	40.65	nr	190.07
Line modules	23.28	27.55	1.00	27.10	nr	54.65
EQUIPMENT						
Manual Call Point						
Surface mounted						
Call point	58.63	69.39	1.00	27.10	nr	96.49
Call point; Weather proof	214.84	254.26	1.25	33.88	nr	288.13
Flush mounted						
Call point	58.63	69.39	1.00	27.10	nr	96.49
Call point; Weather proof	214.84	254.26	1.25	33.88	nr	288.13

W: COMMUNICATIONS/SECURITY/CONTROL

Item	Net Price £	Material £	Labour hours	Labour £	Unit	Total rate £
W50 : FIRE DETECTION AND ALARM (cont'd)						
ADDRESSABLE FIRE DETECTION CONTROL PANEL (cont'd)						
EQUIPMENT (cont'd)						
Detectors						
Smoke, ionisation type with mounting base	59.71	70.67	0.75	20.33	nr	**91.00**
Smoke, optical type with mounting base	59.10	69.94	0.75	20.33	nr	**90.27**
Fixed temperature heat detector with mounting base (60°C)	59.41	70.31	0.75	20.33	nr	**90.64**
Rate of Rise heat detector with mounting base (90°C)	59.41	70.31	0.75	20.33	nr	**90.64**
Duct Detector including optical smoke detector and addressable base	378.58	448.04	2.00	54.20	nr	**502.24**
Beam smoke detector with transmitter and receiver unit	583.64	690.72	2.00	54.20	nr	**744.92**
Zone short circuit isolator	37.53	44.42	0.75	20.33	nr	**64.74**
Plant interface unit	27.65	32.72	0.50	13.55	nr	**46.27**
Sounders						
Xenon flasher, 24 volt, conduit box	63.04	74.61	0.50	13.55	nr	**88.16**
Xenon flasher, 24 volt, conduit box; weatherproof	93.84	111.06	0.50	13.55	nr	**124.61**
6" bell, conduit box	17.69	102.41	0.75	20.33	nr	**122.74**
6" bell, conduit box; weatherproof	51.15	60.53	0.75	20.33	nr	**80.86**
Siren; 24V polarised	63.84	75.55	1.00	27.10	nr	**102.65**
Siren; 240V	62.84	74.37	1.25	33.88	nr	**108.25**
Magnetic Door Holder; 240V ; surface fixed	59.26	70.13	1.50	40.69	nr	**110.83**

W: COMMUNICATIONS/SECURITY/CONTROL

Item	Net Price £	Material £	Labour hours	Labour £	Unit	Total rate £
W51 : EARTHING AND BONDING						
EARTH BAR						
Earth bar; polymer insulators and base mounting; including connections						
Non disconnect link						
6 way	174.03	205.96	0.81	21.95	nr	**227.91**
8 way	191.41	226.53	0.81	21.95	nr	**248.48**
10 way	210.58	249.22	0.81	21.95	nr	**271.17**
Disconnect link						
6 way	195.48	231.35	1.01	27.37	nr	**258.72**
8 way	215.02	254.47	1.01	27.37	nr	**281.84**
10 way	236.52	279.92	1.01	27.37	nr	**307.29**
Soild earth bar; including connections						
150 x 50 x 6 mm	57.71	68.29	1.01	27.37	nr	**95.67**
Extra for earthing						
Disconnecting link						
300 x 50 x 6mm	54.81	64.87	1.16	31.44	nr	**96.31**
500 x 50 x 6mm	60.29	71.35	1.16	31.44	nr	**102.79**
Crimp lugs; including screws and connections to cable						
25 mm	0.65	0.77	0.31	8.40	nr	**9.17**
35 mm	0.95	1.12	0.31	8.40	nr	**9.52**
50 mm	1.11	1.31	0.32	8.67	nr	**9.99**
70 mm	1.81	2.15	0.32	8.67	nr	**10.82**
95 mm	2.19	2.59	0.46	12.47	nr	**15.06**
120 mm	2.33	2.76	1.25	33.88	nr	**36.64**
Earth clamps; connection to pipework						
15mm to 32mm dia	1.23	1.46	0.15	4.07	nr	**5.52**
32mm to 50mm dia	1.53	1.81	0.18	4.88	nr	**6.69**
50mm to 75mm dia	1.81	2.15	0.20	5.42	nr	**7.57**

W: COMMUNICATIONS/SECURITY/CONTROL

Item	Net Price £	Material £	Labour hours	Labour £	Unit	Total rate £
W52 : LIGHTNING PROTECTION						
CONDUCTOR TAPE						
PVC sheathed copper tape						
25 x 3 mm	10.16	12.63	0.30	8.13	m	**20.76**
25 x 6 mm	17.84	22.17	0.30	8.13	m	**30.30**
50 x 6 mm	37.97	47.19	0.30	8.13	m	**55.32**
PVC sheathed copper solid circular conductor						
8mm	6.33	7.49	0.50	13.55	m	**21.04**
Bare copper tape						
20 x 3 mm	7.44	8.80	0.30	8.13	m	**16.93**
25 x 3 mm	7.59	9.43	0.30	8.13	m	**17.56**
25 x 6 mm	15.19	18.88	0.40	10.84	m	**29.72**
50 x 6 mm	30.37	35.94	0.50	13.55	m	**49.49**
Bare copper solid circular conductor						
8mm	5.05	5.98	0.50	13.55	m	**19.53**
Tape fixings; flat; metalic						
PVC sheathed copper						
25 x 3 mm	6.37	7.91	0.33	8.94	nr	**16.86**
25 x 6 mm	6.59	8.19	0.33	8.94	nr	**17.13**
50 x 6 mm	10.86	13.49	0.33	8.94	nr	**22.43**
8mm	7.00	8.69	0.50	13.55	nr	**22.24**
Bare copper						
20 x 3 mm	7.44	9.24	0.30	8.13	nr	**17.37**
25 x 3 mm	7.59	9.43	0.30	8.13	nr	**17.56**
25 x 6 mm	15.19	18.88	0.40	10.84	nr	**29.72**
50 x 6 mm	30.37	35.94	0.50	13.55	nr	**49.49**
8mm	5.05	6.27	0.50	13.55	nr	**19.82**
Tape fixings; flat; non-metalic; PVC sheathed copper						
25 x 3 mm	0.82	1.03	0.30	8.13	nr	**9.16**
Tape fixings; flat; non-metalic; Bare copper						
20 x 3 mm	0.80	1.00	0.30	8.13	nr	**9.13**
25 x 3 mm	0.80	1.00	0.30	8.13	nr	**9.13**
50 x 6 mm	2.06	2.43	0.30	8.13	nr	**10.56**
Puddle flanges; copper						
600 mm long	77.43	91.63	0.93	25.20	nr	**116.84**
AIR RODS						
Pointed air rod fixed to structure; copper 10 mm diameter						
500 mm long	16.29	19.28	1.00	27.10	nr	**46.38**
1000mm long	24.70	29.23	1.50	40.65	nr	**69.88**
Extra for						
Air terminal base	21.07	24.94	0.35	9.49	nr	**34.43**
Strike Pad	29.94	35.43	0.35	9.49	nr	**44.92**

Material Costs/Prices for Measured Works – Electrical Installations

W: COMMUNICATIONS/SECURITY/CONTROL

Item	Net Price £	Material £	Labour hours	Labour £	Unit	Total rate £
16 mm diameter						
500 mm long	23.21	27.47	0.91	24.66	nr	**52.13**
1000mm long	42.41	50.19	1.75	47.43	nr	**97.62**
2000mm long	77.46	91.67	2.50	67.75	nr	**159.43**
Extra for						
Multiple point	45.02	53.29	0.35	9.49	nr	**62.77**
Air terminal base	22.98	27.19	0.35	9.49	nr	**36.68**
Ridge saddle	40.25	47.63	0.35	9.49	nr	**57.12**
Side mounting bracket	40.05	47.40	0.50	13.55	nr	**60.95**
Rod to tape coupling	18.64	22.06	0.50	13.55	nr	**35.61**
Strike Pad	29.94	35.43	0.35	9.49	nr	**44.92**
AIR TERMINALS						
16 mm diameter						
500 mm long	23.21	27.47	0.65	17.62	nr	**45.09**
1000mm long	42.41	50.19	0.78	21.14	nr	**71.33**
2000mm long	77.46	91.67	1.50	40.65	nr	**132.33**
Extra for						
Multiple point	45.02	53.29	0.35	9.49	nr	**62.77**
Flat saddle	22.98	27.19	0.35	9.49	nr	**36.68**
Side bracket	40.05	47.40	0.50	13.55	nr	**60.95**
Rod to cable coupling	18.64	22.06	0.50	13.55	nr	**35.61**
BONDS AND CLAMPS						
Bond to flat surface; copper						
26 mm	4.44	5.25	0.45	12.20	nr	**17.45**
8 mm diameter	17.94	21.23	0.33	8.94	nr	**30.17**
Pipe bond						
26 mm	8.63	10.21	0.45	12.20	nr	**22.40**
8 mm diameter	38.24	45.25	0.33	8.94	nr	**54.20**
Rod to tape clamp						
26 mm	7.21	8.54	0.45	12.20	nr	**20.73**
Square clamp; copper						
25 x 3 mm	7.76	9.19	0.33	8.94	nr	**18.13**
50 x 6 mm	34.44	40.76	0.50	13.55	nr	**54.31**
8 mm diameter	9.15	10.83	0.33	8.94	nr	**19.77**
Test clamp; copper						
26 x 8 mm; oblong	12.11	14.33	0.50	13.55	nr	**27.89**
26 x 8 mm; plate type	33.17	39.26	0.50	13.55	nr	**52.81**
26 x 8 mm; screw down	29.81	35.28	0.50	13.55	nr	**48.83**
Cast in earth points						
2 hole	23.77	28.14	0.75	20.33	nr	**48.46**
4 hole	37.77	44.71	1.00	27.10	nr	**71.81**
Extra for cast in earth points						
Cover plate; 25 x 3 mm	26.69	31.58	0.25	6.78	nr	**38.36**
Cover plate; 8 mm	26.69	31.58	0.25	6.78	nr	**38.36**
Rebar clamp; 8 mm	57.23	67.72	0.25	6.78	nr	**74.50**
Static earth receptacle	129.89	153.72	0.50	13.55	nr	**167.27**

W: COMMUNICATIONS/SECURITY/CONTROL

Item	Net Price £	Material £	Labour hours	Labour £	Unit	Total rate £
W52 : LIGHTNING PROTECTION (cont'd)						
BONDS AND CLAMPS (cont'd)						
Copper braided bonds						
25 x 3 mm						
200 mm hole centres	16.40	19.41	0.33	8.94	nr	**28.35**
400 mm holes centres	24.10	28.52	0.40	10.84	nr	**39.36**
U bolt clamps						
16 mm	9.59	11.35	0.33	8.94	nr	**20.29**
20 mm	10.86	12.86	0.33	8.94	nr	**21.80**
25 mm	13.31	15.75	0.33	8.94	nr	**24.70**
EARTH PITS/MATS						
Earth inspection pit; hand to others for fixing						
Concrete	42.13	49.86	1.00	27.10	nr	**76.96**
Polypropylene	41.81	49.48	1.00	27.10	nr	**76.58**
Extra for						
5 hole copper earth bar; concrete pit	34.00	80.48	0.35	9.49	nr	**89.96**
5 hole earth bar; polypropylene	29.03	34.35	0.35	9.49	nr	**43.84**
Water proof electrode seal						
Single flange	270.93	320.64	0.93	25.20	nr	**345.84**
Double flange	455.88	539.52	0.93	25.20	nr	**564.73**
Earth electrode mat; laid in ground and connected						
Copper tape lattice						
600 x 600 x 3 mm	87.02	102.99	0.93	25.20	nr	**128.19**
900 x 900 x 3 mm	155.88	184.48	0.93	25.20	nr	**209.68**
Copper tape plate						
600 x 600 x 1.5 mm	82.75	97.94	0.93	25.20	nr	**123.14**
600 x 600 x 3 mm	165.52	195.88	0.93	25.20	nr	**221.09**
900 x 900 x 1.5 mm	185.55	219.59	0.93	25.20	nr	**244.79**
900 x 900 x 3 mm	359.99	426.03	0.93	25.20	nr	**451.24**
EARTH RODS						
Solid cored copper earth electrodes driven into ground and connected						
15 mm diameter						
1200 mm long	28.46	33.68	0.93	25.20	nr	**58.89**
Extra for						
Coupling	1.59	1.88	0.06	1.63	nr	**3.50**
Driving stud	1.94	2.29	0.06	1.63	nr	**3.92**
Spike	1.82	2.16	0.06	1.63	nr	**3.79**
Rod Clamp; flat tape	7.21	8.54	0.25	6.78	nr	**15.31**
Rod Clamp; solid conductor	3.51	4.16	0.25	6.78	nr	**10.93**
20 mm diameter						
1200 mm long	51.69	61.17	0.98	26.56	nr	**87.73**

W: COMMUNICATIONS/SECURITY/CONTROL

Item	Net Price £	Material £	Labour hours	Labour £	Unit	Total rate £
Extra for						
Coupling	1.59	1.88	0.06	1.63	nr	3.50
Driving stud	3.21	3.80	0.06	1.63	nr	5.43
Spike	2.96	3.51	0.06	1.63	nr	5.13
Rod Clamp; flat tape	7.21	8.54	0.25	6.78	nr	15.31
Rod Clamp; solid conductor	3.95	4.67	0.25	6.78	nr	11.45
Stainless steel earth electrodes driven into ground and connected						
16 mm diameter						
1200 mm long	66.03	78.14	0.93	25.20	nr	103.34
Extra for						
Coupling	2.06	2.44	0.06	1.63	nr	4.07
Driving head	1.94	2.29	0.06	1.63	nr	3.92
Spike	1.82	2.16	0.06	1.63	nr	3.79
Rod Clamp; flat tape	7.21	8.54	0.25	6.78	nr	15.31
Rod Clamp; solid conductor	3.51	4.16	0.25	6.78	nr	10.93
SURGE PROTECTION						
Single Phase; including connection to equipment						
90 - 150v	288.29	341.18	5.00	135.51	nr	476.69
200 - 280v	288.29	341.18	5.00	135.51	nr	476.69
Three Phase; including connection to equipment						
156 - 260v	569.89	674.45	10.00	271.01	nr	945.46
346 - 484v	569.89	674.45	10.00	271.01	nr	945.46
349 - 484v; remote display	636.92	753.78	10.00	271.01	nr	1024.79
346 - 484v; 60kA	1106.24	1309.20	10.00	271.01	nr	1580.21
346 - 484v; 120kA	2111.91	2499.38	10.00	271.01	nr	2770.40

W: COMMUNICATIONS/SECURITY/CONTROL

Item	Net Price £	Material £	Labour hours	Labour £	Unit	Total rate £
W60 : CENTRAL CONTROL/BUILDING MANAGEMENT						
Equipment						
Switches/sensors; includes fixing in position; electrical work elsewhere. Note - these are normally free issued to the mechanical contractor for fitting. The labour times applied assume the installation has been prepared for the fitting of the component.						
Pressure devices						
Liquid differential pressure sensor	139.76	165.41	0.50	13.55	nr	178.96
Liquid differential pressure switch	76.64	90.71	0.50	13.55	nr	104.26
Air differential pressure transmitter	105.49	124.85	0.50	13.55	nr	138.40
Air differential pressure switch	14.87	17.60	0.50	13.55	nr	31.15
Liquid level switch	53.61	63.44	0.50	13.55	nr	76.99
Static pressure sensor	362.43	428.93	0.50	13.55	nr	442.48
High pressure switch	76.63	90.69	0.50	13.55	nr	104.24
Low pressure switch	76.63	90.69	0.50	13.55	nr	104.24
Water pressure switch	76.63	90.69	0.50	13.55	nr	104.24
Duct averaging temperature sensor	126.23	149.39	1.00	27.10	nr	176.49
Temperature devices						
Return air sensor (fan coils)	6.32	7.48	1.00	27.10	nr	34.59
Frost thermostat	26.14	30.94	0.50	13.55	nr	44.49
Immersion thermostat	54.05	63.96	0.50	13.55	nr	77.51
Temperature high limit	50.54	59.81	0.50	13.55	nr	73.36
Temperature sensor with averaging element	126.23	149.39	0.50	13.55	nr	162.94
Immersion temperature sensor	55.00	65.09	0.50	13.55	nr	78.64
Space temperature sensor	5.41	6.41	1.00	27.10	nr	33.51
Combined space temperature & humidity sensor	129.83	153.65	1.00	27.10	nr	180.76
Outside air temperature sensor	10.82	12.80	2.00	54.20	nr	67.01
Outside air temperature & humidity sensor	149.67	177.13	2.00	54.20	nr	231.33
Duct humidity sensor	139.75	165.39	0.50	13.55	nr	178.94
Space humidity sensor	129.83	153.65	1.00	27.10	nr	180.76
Immersion water flow sensor	129.83	153.65	0.50	13.55	nr	167.21
Rain sensor	184.83	218.74	2.00	54.20	nr	272.95
Wind speed and direction sensor	949.39	1123.57	2.00	54.20	nr	1177.78
Controllers; includes fixing in position; electrical work elsewhere						
Zone						
Fan coil controller	246.44	291.66	2.00	54.20	nr	345.86
VAV controller	246.44	291.66	2.00	54.20	nr	345.86
Plant						
Controller, 96 I/O points (exact configuration is dependent upon the number of I/O boards added)	4769.51	5644.58	0.50	13.55	nr	5658.13
Controller, 48 I/O points (exact configuration is dependent upon the number of I/O boards added)	2551.54	3019.68	0.50	13.55	nr	3033.23
Controller, 32 I/O points (exact configuration is dependent upon the number of I/O boards added)	1812.24	2144.73	0.50	13.55	nr	2158.28

W: COMMUNICATIONS/SECURITY/CONTROL

Item	Net Price £	Material £	Labour hours	Labour £	Unit	Total rate £
Additional Digital Input Boards (12 DI)	568.01	672.22	0.20	5.42	nr	677.64
Additional Digital Output Boards (6 DO)	369.65	437.47	0.20	5.42	nr	442.89
Additional analogue Input Boards (8 AI)	369.65	437.47	0.20	5.42	nr	442.89
Additional analogue Output Boards (8 AO)	369.65	437.47	0.20	5.42	nr	442.89
Outstation Enclosure (fitted in riser with space allowance for controller and network device)	255.86	302.80	5.00	135.51	nr	438.31
Damper actuator; electrical work elsewhere						
Damper actuator 0-10v	76.45	90.48	-	-	nr	90.48
Damper actuator with auxiliary switches	100.98	119.50	-	-	nr	119.50
Frequency inverters: not mounted within MCC; includes fixing in position; electrical work elsewhere						
2.2kW	568.93	673.31	2.00	54.20	nr	727.51
3kW	627.52	742.65	2.00	54.20	nr	796.86
7.5kW	793.42	938.98	2.00	54.20	nr	993.19
11kW	1075.92	1273.32	2.00	54.20	nr	1327.52
15kW	1390.27	1645.35	2.00	54.20	nr	1699.55
18.5kW	1541.75	1824.62	2.50	67.75	nr	1892.37
20kW	1853.71	2193.81	2.50	67.75	nr	2261.56
30kW	2092.64	2476.58	2.50	67.75	nr	2544.33
55kW	4361.08	5161.21	3.00	81.30	nr	5242.52
Miscellaneous; includes fixing in position; electrical work elsewhere						
1kW Thyristor	86.55	102.43	2.00	54.20	nr	156.64
10kW Thyristor	230.81	273.16	2.00	54.20	nr	327.36
Front end and networking; electrical work elsewhere						
PC/monitor	3030.40	3586.39	2.00	54.20	nr	3640.59
Dot matrix printer	549.46	650.26	2.00	54.20	nr	704.47
PC Software	2709.29	3206.37	-	-	nr	3206.37
Lonmaker software	964.11	1141.00	-	-	nr	1141.00
Lonmaker credits	7.15	8.46	-	-	nr	8.46
Network server software	1358.60	1607.86	-	-	nr	1607.86
Router (allows connection to a network)	763.36	903.41	-	-	nr	903.41

ESSENTIAL READING FROM TAYLOR AND FRANCIS

Ethics for the Built Environment

Peter Fewings

Much closer relationships are being formed in the development of the built environment and cultural changes in procurement methods are taking place. A need has grown to re-examine the ethical frameworks required to sustain collaborative trust and transparency. Young professionals are moving around between companies on a frequent basis and take their personal values with them. What can companies do to support their employees?

The book looks at how people develop their personal values and tries to set up a model for making effective ethical decisions.

Selected Contents:

1 Business Ethics
2 Professional Practice
3 The Ethics of Employment
4 Environmental Sustainability
5 Health and Safety Ethics
6 Ethical Relationships
7 Corporate Social Responsibility
CASE STUDIES

2008: 234x156: 384pp
Hb: 978-0-415-42982-5 **£85.00**
Pb: 978-0-415-42983-2 **£29.99**

To Order: Tel: +44 (0) 1235 400524 **Fax:** +44 (0) 1235 400525
or Post: Taylor and Francis Customer Services,
Bookpoint Ltd, Unit T1, 200 Milton Park, Abingdon, Oxon, OX14 4TA UK
Email: book.orders@tandf.co.uk

For a complete listing of all our titles visit:
www.tandf.co.uk

ESSENTIAL READING FROM TAYLOR AND FRANCIS

Project Management Demystified

Third Edition

Geoff Reiss

Concise, practical and entertaining to read, this excellent introduction to project management is an indispensable book for both professionals and students working in or studying project management in business, engineering or the public sector.

Approachable and written in an easy-to-use style, it shows readers how, where and when to use the various project management techniques, demonstrating how to achieve efficient management of human, material and financial resources to make major contributions to projects and be an appreciated and successful project manager.

This new edition contains expanded sections on programme management, portfolio management, and the public sector. An entirely new chapter covers the evaluation, analysis and management of risks and issues. A much expanded section explores the rise and utilisation of methodologies like Prince2.

Contents: Introduction. Setting the Stage. Getting the Words in the Right Order. Nine Steps to a Successful Project. The Scope of the Project and its Objectives. Project Planning. A Fly on the Wall. Resource Management. Progress Monitoring and Control. Advanced Critical-Path Topics. The People Issues. Risk and Issue Management. Terminology

June 2007: 234x156mm: 224 pages
Pb: 978-0-415-42163-8: **£19.99**

To Order: Tel: +44 (0) 1235 400524 **Fax:** +44 (0) 1235 400525
or Post: Taylor and Francis Customer Services,
Bookpoint Ltd, Unit T1, 200 Milton Park, Abingdon, Oxon, OX14 4TA UK
Email: book.orders@tandf.co.uk

For a complete listing of all our titles visit:
www.tandf.co.uk

ESSENTIAL READING FROM TAYLOR AND FRANCIS

Construction Contracts Questions and Answers

David Chappell

Construction law can be a minefield of complications and misunderstandings in which professionals need answers which are pithy and straightforward but also legally rigorous. In *Construction Contracts: Questions and Answers*, specialist in construction law David Chappell answers architects' and builders' common construction contract questions.

Questions range in content and include:

- extensions of time
- liquidated damages
- loss and/or expense
- practical completion
- defects
- valuation
- certificates and payment
- architects' instructions
- adjudication and fees.

Chappell's authoritative and practical advice answers questions ranging from simple queries, such as which date should be put on a contract, through to more complex issues, such as whether the contractor is entitled to take possession of a section of the work even though it is the contractor's fault that possession is not practicable.

In answering genuine questions on construction contracts, Chappell has created an invaluable resource on which not only architects, but also project managers, contractors, QSs, employers and others involved in construction can depend.

2006: 216x138 mm: 240 pages
Pb: 978-0-415-37597-9: **£26.99**

To Order: Tel: +44 (0) 1235 400524 **Fax:** +44 (0) 1235 400525
or Post: Taylor and Francis Customer Services,
Bookpoint Ltd, Unit T1, 200 Milton Park, Abingdon, Oxon, OX14 4TA UK
Email: book.orders@tandf.co.uk

For a complete listing of all our titles visit:
www.tandf.co.uk

PART FOUR

Rates of Wages

Mechanical Installations, *page 603*
Electrical Installations, *page 611*

ESSENTIAL READING FROM TAYLOR AND FRANCIS

Principles of Project and Infrastructure Finance

Willie Tan

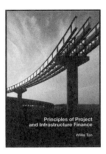

Current books on project finance tend to be non-technical and are either procedural or rely heavily on case studies. In contrast this textbook provides a more analytical perspective, without a loss of pragmatism.

Principles of Project Finance and Infrastructure is written for senior undergraduates, graduate students and practitioners who wish to know how major projects, such as residential and infrastructural developments are financed. The approach is intuitive, yet rigorous, making the book highly readable. Case studies are used to illustrate integration as well as to underscore the pragmatic slant.

Selected Contents: 1. Introduction 2. Time Value of Money 3. Organizations and Projects 4. Corporate Finance I 5. Corporate Finance II 6. Project Development 7. Social Projects 8. Characteristics of Project Finance 9. Risk Management Framework 10. Risk, Insurance, and Bonds 11. Cash Flow Risks 12. Financial Risks 13. Agreements, Contracts, and Guarantees 14. Case Study I: Power Projects 15. Case Study II: Airport Projects 16. Case Study III: Office Projects 17. Case Study IV: Chemical Storage Projects Appendix: Cumulative Standard Normal Distribution

2007: 234x156mm: 296 pages
Hb: 978-0-415-41576-7: **£84.00**
Pb: 978-0-415-41577-4: **£29.99**

To Order: Tel: +44 (0) 1235 400524 **Fax:** +44 (0) 1235 400525
or Post: Taylor and Francis Customer Services,
Bookpoint Ltd, Unit T1, 200 Milton Park, Abingdon, Oxon, OX14 4TA UK
Email: book.orders@tandf.co.uk

For a complete listing of all our titles visit:
www.tandf.co.uk

Mechanical Installations

Rates of Wages

HEATING, VENTILATING, AIR CONDITIONING, PIPING AND DOMESTIC ENGINEERING INDUSTRY

For full details of the wage agreement and the Heating Ventilating Air Conditioning Piping and Domestic Engineering Industry's National Working Rule Agreement, contact:

Heating and Ventilating Contractor's Association
ESCA House,
34 Palace Court,
Bayswater,
London W2 4JG
Telephone: 020 7313 4900
Internet: www.hvca.org.uk

WAGE RATES, ALLOWANCES AND OTHER PROVISIONS

Hourly rates of wages
All districts of the United Kingdom

Main Grades	From 6 October 2008 p/hr
Foreman	14.71
Senior Craftsman (+2nd welding skill)	12.66
Senior Craftsman	12.16
Craftsman (+2nd welding skill)	11.66
Craftsman	11.16
Operative	10.11
Adult Trainee	8.52
Mate (18 and over)	8.52
Mate (16-17)	3.95
Modern Apprentices	
Junior	5.53
Intermediate	7.84
Senior	10.11

Note: Ductwork Erection Operatives are entitled to the same rates and allowances as the parallel Fitter grades shown.

HEATING, VENTILATING, AIR CONDITIONING, PIPING AND DOMESTIC ENGINEERING INDUSTRY

Trainee Rates of Pay

Junior Ductwork Trainees (Probationary)

Age at entry	From 6 October 2008 p/hr
17	5.01
18	5.01
19	5.01
20	5.01

Junior Ductwork Erectors (Year of Training)

	From 6 October 2008		
Age at entry	1 yr p/h	2 yr p/hr	3 yr p/hr
17	6.24	7.76	8.81
18	6.24	7.76	8.81
19	6.24	7.76	8.81
20	6.24	7.76	8.81

Responsibility Allowance (Craftsmen)	From 6 October 2008 p/hr
Second welding skill or supervisory responsibility (one unit)	0.50
Second welding skill and supervisory responsibility (two units)	1.00

Responsibility Allowance (Senior Craftsmen)	From 6 October 2008 p/hr
Second welding skill	0.50
Supervising responsibility	1.00
Second welding skill and supervisory responsibility	1.50

Daily travelling allowance – Scale 2

C: Craftsmen including Installers
M&A: Mates, Apprentices and Adult Trainees

Direct distance from centre to job in miles

		From 6 October 2008	
Over	Not exceeding	C p/hr	M&A p/hr
15	20	2.44	2.10
20	30	6.28	5.44
30	40	9.04	7.82
40	50	11.91	10.19

HEATING, VENTILATING, AIR CONDITIONING, PIPING & DOMESTIC ENGINEERING INDUSTRY

Daily travelling allowance – Scale 1

C: Craftsmen including Installers
M&A: Mates, Apprentices and Adult Trainees

Direct distance from centre to job in miles

		From 6 October 2008	
Over	Not exceeding	C p/hr	M&A p/hr
15	20	9.13	8.79
20	30	12.97	12.12
30	40	15.73	14.51
40	50	18.59	16.88

Weekly Holiday Credit and Welfare Contributions

	From 6 October 2008						
	£ a	£ b	£ c	£ d	£ e	£ f	£ g
Weekly Holiday Credit	70.05	65.12	62.71	60.30	57.99	55.61	53.18
Combined Weekly/Welfare Holiday Credit and Contribution	77.16	72.23	69.82	67.41	65.10	62.72	60.29

	From 6 October 2008					
	£ h	£ i	£ j	£ k	£ l	£ m
Weekly Holiday Credit	48.18	40.60	37.34	26.43	N/A	18.82
Combined Weekly/Welfare Holiday Credit and Contribution	55.29	47.71	44.45	33.54	N/A	25.93

The grades of H&V Operatives entitled to the different rates of Weekly Holiday Credit and Welfare Contribution are as follows:

a
Foreman

b
Senior Craftsman
(RAS & RAW)

c
Senior Craftsman
(RAS)

d
Senior Craftsman
(RAW)

e
Senior Craftsman
Craftsman
(+2 RA)

f
Craftsman
(+ 1RA)

g
Craftsman

h
Installer
Senior Modern Apprentice

i
Adult Trainee
Mate (over 18)

j
Intermediate
Modern Apprentice

k
Junior Modern Apprentice

l
No grade allocated to this Credit Value Category

m
Mate (16-17)

HEATING, VENTILATING, AIR CONDITIONING, PIPING & DOMESTIC ENGINEERING INDUSTRY

	From *6 October 2008*
Daily abnormal conditions money Per day	2.99

	From *6 October 2008*
Lodging allowance Per night	28.60

Explanatory Notes

1. Working Hours

    The normal working week (Monday to Friday) shall be 38 hours.

2. Overtime

    Time worked in excess of 38 hours during the normal working week shall be paid at time and a half until 12 hours have been worked since the actual starting time. Thereafter double time shall be paid until normal starting time the following morning. Weekend overtime shall be paid at time and a half for the first 5 hours worked on a Saturday and at double time thereafter until normal starting time on Monday morning.

PLUMBING MECHANICAL ENGINEERING SERVICES INDUSTRY

PLUMBING MECHANICAL ENGINEERING SERVICES INDUSTRY

The Joint Industry Board for Plumbing Mechanical Engineering Services has agreed a two year wage agreement for 2008 – 2009 with effect from 7 January 2008. Current rates of pay for 2008 will be effective from 1 January 2008.

For full details of this wage agreement and the JIB PMES National Working Rules, contact:

The Joint Industry Board for Plumbing Mechanical Engineering Services in England and Wales
Brook House,
Brook Street,
St Neots,
Huntingdon,
Cambridge PE19 2HW
Telephone: 01480 476925
E-mail: info@jib-pmes.org.uk

WAGE RATES, ALLOWANCES AND OTHER PROVISIONS

EFFECTIVE FROM 7 JANUARY 2008

Basic Rates of Hourly Pay

Applicable in England and Wales

	Hourly rate £
Operatives	
Technical plumber and gas service technician	13.46
Advanced plumber and gas service engineer	12.12
Trained plumber and gas service fitter	10.39
Apprentices	
4th year of training with NVQ level 3	10.05
4th year of training with NVQ level 2	9.10
4th year of training	8.03
3rd year of training with NVQ level 2	7.92
3rd year of training	6.52
2nd year of training	5.77
1st year of training	5.04
Adult Trainees	
3rd 6 months of employment	9.07
2nd 6 months of employment	8.70
1st 6 months of employment	8.11

PLUMBING MECHANICAL ENGINEERING SERVICES INDUSTRY

Major Projects Agreement

Where a job is designated as being a Major Project then the following Major Project Performance Payment hourly rate supplement shall be payable:

Employee Category

	National Payment £	London* Payment £
Technical Plumber and Gas Service Technician	2.20	3.57
Advanced Plumber and Gas Service Engineer	2.20	3.57
Trained Plumber and Gas Service Fitter	2.20	3.57
All 4th year apprentices	1.76	2.86
All 3rd year apprentices	1.32	2.68
2nd year apprentice	1.21	1.96
1st year apprentice	0.88	1.43
All adult trainees	1.76	2.86

* The London Payment Supplement applies only to designated Major Projects that are within the M25 London orbital motorway and are effective from 1 February 2007.

* National payment hourly rates are unchanged for 2007 and will continue to be at the rates shown in Promulgation 138A issued 14 October 2003.

Allowances

Daily travel time allowance plus return fares

All daily travel allowances are to be paid at the daily rate as follows:

Over	Not exceeding	All Operatives	3rd & 4th Year Apprentices	1st & 2nd Year Apprentices
20	30	£3.81	£2.46	£1.52
30	40	£8.88	£5.70	£3.66
40	50	£10.10	£6.06	£3.81

Responsibility/Incentive Pay Allowance

As from Monday 3rd September 2003, Employers may, in consultation with the employees concerned, enhance the basic graded rates of pay by the payment of an additional amount, as per the bands shown below, where it is agreed that their work involves extra responsibility, productivity or flexibility.

Band 1 - an additional rate of £ 0.24 per hour
Band 2 - an additional rate of £ 0.44 per hour
Band 3 - an additional rate of £ 0.64 per hour
Band 4 - an additional rate of £ 0.84 per hour

This allowance forms part of an operative's basic rate of pay and shall be used to calculate premium payments.

Mileage allowance ....£0.40 per mile

Lodging allowance ....£28.35 per night

Subsistence Allowance (London Only) ....£4.37 per night

PLUMBING MECHANICAL ENGINEERING SERVICES INDUSTRY

Plumbers welding supplement
Possession of Gas or Arc Certificate ....£0.27 per hour
Possession of Gas and Arc Certificate ....£0.46 per hour

Weekly Holiday Credit Contributions (35th Issue Stamps Option)

	Public & Annual Gross Value, £	Combined Holiday Credit*, £
Technical Plumber and Gas Service Technician	60.30	63.65
Advance Plumber and Gas Service Engineer	54.20	57.15
Trained Plumber and Gas Service Fitter	46.40	48.85
Adult Trainee	34.80	35.60
Apprentice in last year of training	34.80	35.60
Apprentice 3rd year	25.10	25.80
Apprentice 2nd year	22.10	22.70
Apprentice 1st year	19.30	19.80
Working Principal	25.10	26.70
Ancillary Employee	30.60	32.20

* Public and Annual gross value is the Combined Holiday credit value less JIB administration value.

Explanatory Notes

1. Working Hours

   The normal working week (Monday to Friday) shall be 37½ hours, with 45 hours to be worked in the same period before overtime rates become applicable.

2. Overtime

   Overtime shall be paid at time and a half up to 8.00pm (Monday to Friday) and up to 1.00pm (Saturday). Overtime worked after these times shall be paid at double time.

3. Major Projects Agreement

   Under the Major Projects Agreement the normal working week shall be 38 hours (Monday to Friday) with overtime rates payable for all hours worked in excess of 38 hours in accordance with 2 above. However, it should be noted that the hourly rate supplement shall be paid for each hour worked but does not attract premium time enhancement.

4. Pension

   In addition to their hourly rates of pay, plumbing employees are entitled to inclusion within the Industry Pension Scheme (or one providing equivalent benefits). The current levels of industry scheme contributions are 6½% (employers) and 3¼% (employees).

5. Weekly Holiday Credit Contributions

   There has been a fundamental change in the way the JIB-PMES Holiday Pay Schemes shall apply for the 36th issue, in that 60 credits will be paid over a period of 53 weeks.

   For full details please refer to the 36th issue of the Holiday Credit Values as published by the JIB for Plumbing Mechanical Engineering Services in England and Wales.

ESSENTIAL READING FROM TAYLOR AND FRANCIS

Spon's Irish Construction Price Book

Third Edition

Franklin + Andrews

This new edition of *Spon's Irish Construction Price Book*, edited by Franklin + Andrews, is the only complete and up-to-date source of cost data for this important market.

- All the materials costs, labour rates, labour constants and cost per square metre are based on current conditions in Ireland
- Structured according to the new Agreed Rules of Measurement (second edition)
- 30 pages of Approximate Estimating Rates for quick pricing

This price book is an essential aid to profitable contracting for all those operating in Ireland's booming construction industry.

Franklin + Andrews, Construction Economists, have offices in 100 countries and in-depth experience and expertise in all sectors of the construction industry.

April 2008: 246x174 mm: 510 pages
Hb: 978-0-415-45637-1: **£135.00**

To Order: Tel: +44 (0) 1235 400524 **Fax:** +44 (0) 1235 400525
or Post: Taylor and Francis Customer Services,
Bookpoint Ltd, Unit T1, 200 Milton Park, Abingdon, Oxon, OX14 4TA UK
Email: book.orders@tandf.co.uk

For a complete listing of all our titles visit:
www.tandf.co.uk

Electrical Installations

Rates of Wages

ELECTRICAL CONTRACTING INDUSTRY

For full details of this wage agreement and the Joint Industry Board for the Electrical Contracting Industry's National Working Rules, contact:

The Joint Industry Board for the Electrical Contracting Industry
Kingswood House
47/51 Sidcup Hill,
Sidcup,
Kent DA14 6HP
Telephone : 020 8302 0031
Internet : www.jib.org.uk

WAGES (Graded Operatives)

Rates

Since **7 January 2002** two different wage rates have applied to JIB Graded Operatives working on site, depending on whether the Employer transports them to site or whether they provide their own transport. The two categories are:

Job Employed (Transport Provided)

Payable to an Operative who is transported to and from the job by his Employer. The Operative shall also be entitled to payment for Travel Time, when travelling in his own time, as detailed in the appropriate scale.

Job Employed (Own Transport)

Payable to an Operative who travels by his own means to and from the job. The Operative shall be entitled to payment for Travel Allowance and also Travel Time, when travelling in his own time, as detailed in the appropriate scale.

The JIB rates of wages are set out below:

From and including 7 January 2008, the JIB hourly rates of wages shall be as set out below:

(i) National Standard Rate:

Grade	Transport Provided	Own Transport
Technician (or equivalent specialist grade)	£ 14.02	£ 14.73
Approved Electrician (or equivalent specialist grade)	£ 12.38	£ 13.08
Electrician (or equivalent specialist grade)	£ 11.35	£ 12.06
Senior Graded Electrical Trainee	£ 10.21	£ 10.86
Electrical Improver	£ 10.21	£ 10.86
Labourer	£ 9.01	£ 9.67
Adult Trainee	£ 9.01	£ 9.67

ELECTRICAL CONTRACTING INDUSTRY

Rates of Wages - Electrical Installations

(ii) **London Rate:**

Grade	Transport Provided	Own Transport
Technician (or equivalent specialist grade)	£ 15.70	£ 16.50
Approved Electrician (or equivalent specialist grade)	£ 13.87	£ 14.65
Electrician (or equivalent specialist grade)	£ 12.71	£ 13.51
Senior Graded Electrical Trainee	£ 11.44	£ 12.16
Electrical Improver	£ 11.44	£ 12.16
Labourer	£ 10.09	£ 10.83
Adult Trainee	£ 10.09	£ 10.83

1999 Joint Industry Board Apprentice Training Scheme

From and including 7 January 2008, the JIB hourly rates for Job Employed apprentices shall be:

(i) **National Standard Rates**

	Transport Provided	Own Transport
Stage 1	£ 3.96	£ 4.64
Stage 2	£ 5.84	£ 6.53
Stage 3	£ 8.46	£ 9.16
Stage 4	£ 8.95	£ 9.66

(ii) **London Rate**

	Transport Provided	Own Transport
Stage 1	£ 4.44	£ 5.20
Stage 2	£ 6.54	£ 7.31
Stage 3	£ 9.48	£ 10.26
Stage 4	£ 10.02	£ 10.82

Rates of Wages – Electrical Installations

ELECTRICAL CONTRACTING INDUSTRY

Travelling Time and Travel Allowances

From and including 7 January 2008

Operatives required to start/finish at the normal starting and finishing time on jobs which are 15 miles and over from the shop - in a straight line - receive payment for Travelling Time and where transport is not provided by the Employer, Travel Allowance, as follows:

	Distance	Total Daily Travel Allowance	Total Daily Travelling Time
(a)	**National Standard Rate**		
	Up to 15 miles	Nil	Nil
	Over 15 & up to 20 miles each way	£ 3.13	£ 4.27
	Over 20 & up to 25 miles each way	£ 4.15	£ 5.41
	Over 25 & up to 35 miles each way	£ 5.47	£ 6.60
	Over 35 & up to 55 miles each way	£ 8.72	£ 8.72
	Over 55 & up to 75 miles each way	£ 10.67	£ 10.67

For each additional 10 mile band over 75 miles, additional payment of £ 1.88 for Daily Travel Allowance and £1.88 for Daily Travel Time will be made.

Note: Special arrangements may apply for work in the Merseyside area.

	Distance	Total Daily Travel Allowance	Total Daily Travelling Time
(b)	**London Rate**		
	Up to 15 miles	Nil	Nil
	Over 15 & up to 20 miles each way	£ 3.15	£ 4.58
	Over 20 & up to 25 miles each way	£ 4.18	£ 5.99
	Over 25 & up to 35 miles each way	£ 5.51	£ 7.21
	Over 35 & up to 55 miles each way	£ 8.79	£ 9.73
	Over 55 & up to 75 miles each way	£ 10.75	£ 11.35

For each additional 10 mile band over 75 miles, additional payments of £ 1.89 for Daily Travel Allowance and £1.89 for Daily Travel Time will be made.

Travelling time and travel allowance - Section 8 (Employed permanently at the shop)

Operatives required to start/finish at the normal starting and finishing time on jobs which are 15 miles and over from the shop - in a straight line - receive payment for Travelling Time and where transport is not provided by the Employer, Travel Allowance, as follows:

From and including 7 January 2008

	Distance	Total Daily Travel Allowance	Total Daily Travelling Time
(a)	**National Standard Rate**		
	Up to 15 miles	Nil	Nil
	Over 15 & up to 20 miles each way	£ 3.13	£ 2.51
	Over 20 & up to 25 miles each way	£ 4.15	£ 3.76
	Over 25 & up to 35 miles each way	£ 5.54	£ 5.02
	Over 35 & up to 55 miles each way	£ 8.72	£ 6.28
	Over 55 & up to 75 miles each way	£ 10.67	£ 7.54

For each additional 10 mile band over 75 miles, additional payments of £ 1.88 for Daily Travel Allowance and £1.26 for Daily Travelling Time will be made.

Note: Special arrangements may apply for work in the Merseyside area.

ELECTRICAL CONTRACTING INDUSTRY

Distance	Total Daily Travel Allowance	Total Daily Travelling Time
(b) London Rate		
Up to 15 miles	Nil	Nil
Over 15 & up to 20 miles each way	£ 3.15	£ 2.78
Over 20 & up to 25 miles each way	£ 4.18	£ 4.18
Over 25 & up to 35 miles each way	£ 5.51	£ 5.57
Over 35 & up to 55 miles each way	£ 8.79	£ 6.95
Over 55 & up to 75 miles each way	£ 10.75	£ 8.35

For each additional 10 mile band over 75 miles, additional payments of £ 1.89 for Daily Travel Allowance and £1.27 for Daily Travelling Time will be made.

Lodging Allowances
£29.60 from and including 7 January 2008

Lodgings weekend retention fee, maximum reimbursement
£29.60 from and including 7 January 2008

Annual Holiday Lodging Allowance Retention
Maximum £10.00 per night (£ 70.00 per week) from and including 7 January 2008

Responsibility money

From and including 30 March 1998 the minimum payment increased to 10p per hour and the maximum to £1.00 per hour (no change)

From and including 4 January 1992 responsibility payments are enhanced by overtime and shift premiums where appropriate (no change)

Combined JIB Benefits Stamp Value (from week commencing 1 October 2007)

JIB Grade	Weekly JIB combined credit value £	Holiday Value £
Technician	£ 57.19	£ 44.49
Approved Electrician	£ 51.98	£ 39.28
Electrician	£ 48.70	£ 36.00
Senior Graded Electrical Trainee and Electrical Improver	£ 45.10	£ 32.40
Labourer & Adult Trainee	£ 41.29	£ 28.59

ELECTRICAL CONTRACTING INDUSTRY

Explanatory Notes

1. Working Hours

    The normal working week (Monday to Friday) shall be 37½ hours, with 38 hours to be worked in the same period before overtime rates become applicable.

2. Overtime

    Overtime shall be paid at time and a half for all weekday overtime. Saturday overtime shall be paid at time and a half for the first 6 hours, or up to 3.00pm (whichever comes first). Thereafter double time shall be paid until normal starting time on Monday.

ESSENTIAL READING FROM TAYLOR AND FRANCIS

The ZEDbook
Solutions for a Shrinking World

Bill Dunster

While Zero (fossil) Energy Development (ZED) is generally regarded as a positive innovation, in practice this option is not taken up or fully implemented. A planner, a councillor, a developer, a housing association representative, an architect, a contractor, a building component manufacturer and a government officer would all have different reasons for not espousing ZED standards for their new-build projects, but the result would be the same.

This book tackles the reasons why this happens, systematically dismantling the defence against its adoption. It explains the principles behind ZED fossil Energy Development, and provides a ZED toolkit of methods and case studies to enable construction professionals and policymakers to realize a ZED build as easily as a conventional scheme.

2007: 276x219: 276pp
Pb: 978-0-415-39199-3: **£40.00**

To Order: Tel: +44 (0) 1235 400524 **Fax:** +44 (0) 1235 400525
or Post: Taylor and Francis Customer Services,
Bookpoint Ltd, Unit T1, 200 Milton Park, Abingdon, Oxon, OX14 4TA UK
Email: book.orders@tandf.co.uk

For a complete listing of all our titles visit:
www.tandf.co.uk

PART FIVE

Daywork

Heating and Ventilating Industry, *page 619*
Electrical Industry, *page 622*
Building Industry Plant Hire Costs, *page 625*

When work is carried out in connection with a contract that cannot be valued in any other way, it is usual to assess the value on a cost basis with suitable allowances to cover overheads and profit. The basis of costing is a matter for agreement between the parties concerned but definitions of prime cost for the Heating and Ventilating and Electrical Industries have been published jointly by the Royal Institution of Chartered Surveyors and the appropriate bodies of the industries concerned, for those who wish to use them.

These, together with a schedule of basic plant hire charges are reproduced on the following pages, with the kind permission of the Royal Institution of Chartered Surveyors, who own the copyright.

ESSENTIAL READING FROM TAYLOR AND FRANCIS

Spon's Building Regulations Explained
Eighth Edition

London District Surveyors Association and **John Stephenson**

This fully revised, essential reference takes into account all important aspects of building control including new legislation up to the start of 2008, covering major revisions to Parts A, B, C, F, J, L1A, L1B, L2A, L2B and P and revisions to Part E. Each chapter explains in clear terms the appropriate regulation and any other relevant legislation, before explaining the approved document.

Selected Contents:

1. The Development of Building Control
2. Control of Building Work
3. Application of Building Regulations to Inner London
4. Relaxation of Building Regulations
5. Exempt Buildings and Works
6. Notices and Plans
7. Approved Inspectors
8. Work Undertaken by Public Bodies
9. Approved Document to Support Regulation 7

Chapters 10-22. Approved Documents A to N 23. Other Approved Documents

2009: 297x210: 672pp
Hb: 978-0-415-43067-8 **£80.00**

To Order: Tel: +44 (0) 1235 400524 **Fax:** +44 (0) 1235 400525
or Post: Taylor and Francis Customer Services,
Bookpoint Ltd, Unit T1, 200 Milton Park, Abingdon, Oxon, OX14 4TA UK
Email: book.orders@tandf.co.uk

For a complete listing of all our titles visit:
www.tandf.co.uk

HEATING AND VENTILATING INDUSTRY

DEFINITION OF PRIME COST OF DAYWORK CARRIED OUT UNDER A HEATING, VENTILATING, AIR CONDITIONING, REFRIGERATION, PIPEWORK AND/OR DOMESTIC ENGINEERING CONTRACT (JULY 1980 EDITION)

This Definition of Prime Cost is published by the Royal Institution of Chartered Surveyors and the Heating and Ventilating Contractors Association for convenience, and for use by people who choose to use it. Members of the Heating and Ventilating Contractors Association are not in any way debarred from defining Prime Cost and rendering accounts for work carried out on that basis in any way they choose. Building owners are advised to reach agreement with contractors on the Definition of Prime Cost to be used prior to entering into a contract or sub-contract.

SECTION 1: APPLICATION

1.1 This Definition provides a basis for the valuation of daywork executed under such heating, ventilating, air conditioning, refrigeration, pipework and or domestic engineering contracts as provide for its use.

1.2 It is not applicable in any other circumstances, such as jobbing or other work carried out as a separate or main contract nor in the case of daywork executed after a date of practical completion.

1.3 The terms 'contract' and 'contractor' herein shall be read as 'sub-contract' and 'sub-contractor' as applicable.

SECTION 2: COMPOSITION OF TOTAL CHARGES

2.1 The Prime Cost of daywork comprises the sum of the following costs:
   (a) Labour as defined in Section 3.
   (b) Materials and goods as defined in Section 4.
   (c) Plant as defined in Section 5.

2.2 Incidental costs, overheads and profit as defined in Section 6, as provided in the contract and expressed therein as percentage adjustments, are applicable to each of 2.1 (a)-(c).

SECTION 3: LABOUR

3.1 The standard wage rates, emoluments and expenses referred to below and the standard working hours referred to in 3.2 are those laid down for the time being in the rules or decisions or agreements of the Joint Conciliation Committee of the Heating, Ventilating and Domestic Engineering Industry applicable to the works (or those of such other body as may be appropriate) and to the grade of operative concerned at the time when and the area where the daywork is executed.

3.2 Hourly base rates for labour are computed by dividing the annual prime cost of labour, based upon the standard working hours and as defined in 3.4, by the number of standard working hours per annum. See example.

3.3 The hourly rates computed in accordance with 3.2 shall be applied in respect of the time spent by operatives directly engaged on daywork, including those operating mechanical plant and transport and erecting and dismantling other plant (unless otherwise expressly provided in the contract) and handling and distributing the materials and goods used in the daywork.

3.4 The annual prime cost of labour comprises the following:
   (a) Standard weekly earnings (i.e. the standard working week as determined at the appropriate rate for the operative concerned).
   (b) Any supplemental payments.
   (c) Any guaranteed minimum payments (unless included in Section 6.1(a)-(p)).
   (d) Merit money.
   (e) Differentials or extra payments in respect of skill, responsibility, discomfort, inconvenience or risk (excluding those in respect of supervisory responsibility - see 3.5)
   (f) Payments in respect of public holidays.
   (g) Any amounts which may become payable by the contractor to or in respect of operatives arising from the rules etc. referred to in 3.1 which are not provided for in 3.4 (a)-(f) nor in Section 6.1 (a)-(p).
   (h) Employers contributions to the WELPLAN, the HVACR Welfare and Holiday Scheme or payments in lieu thereof.
   (i) Employers National Insurance contributions as applicable to 3.4 (a)-(h).
   (j) Any contribution, levy or tax imposed by Statute, payable by the contractor in his capacity as an employer.

HEATING AND VENTILATING INDUSTRY

3.5 Differentials or extra payments in respect of supervisory responsibility are excluded from the annual prime cost (see Section 6). The time of principals, staff, foremen, chargehands and the like when working manually is admissible under this Section at the rates for the appropriate grades.

SECTION 4: MATERIALS AND GOODS

4.1 The prime cost of materials and goods obtained specifically for the daywork is the invoice cost after deducting all trade discounts and any portion of cash discounts in excess of 5%.

4.2 The prime cost of all other materials and goods used in the daywork is based upon the current market prices plus any appropriate handling charges.

4.3 The prime cost referred to in 4.1 and 4.2 includes the cost of delivery to site.

4.4 Any Value Added Tax which is treated, or is capable of being treated, as input tax (as defined by the Finance Act 1972, or any re-enactment or amendment thereof or substitution therefore) by the contractor is excluded.

SECTION 5: PLANT

5.1 Unless otherwise stated in the contract, the prime cost of plant comprises the cost of the following:
  (a) use or hire of mechanically-operated plant and transport for the time employed on and/or provided or retained for the daywork;
  (b) use of non-mechanical plant (excluding non-mechanical hand tools) for the time employed on and/or provided or retained for the daywork;
  (c) transport to and from the site and erection and dismantling where applicable.

5.2 The use of non-mechanical hand tools and of erected scaffolding, staging, trestles or the like is excluded (see Section 6), unless specifically retained for the daywork.

SECTION 6: INCIDENTAL COSTS, OVERHEADS AND PROFIT

6.1 The percentage adjustments provided in the contract which are applicable to each of the totals of Sections 3, 4 and 5 comprise the following:
  (a) Head office charges.
  (b) Site staff including site supervision.
  (c) The additional cost of overtime (other than that referred to in 6.2).
  (d) Time lost due to inclement weather.
  (e) The additional cost of bonuses and all other incentive payments in excess of any included in 3.4.
  (f) Apprentices' study time.
  (g) Fares and travelling allowances.
  (h) Country, lodging and periodic allowances.
  (i) Sick pay or insurances in respect thereof, other than as included in 3.4.
  (j) Third party and employers' liability insurance.
  (k) Liability in respect of redundancy payments to employees.
  (l) Employer's National Insurance contributions not included in 3.4.
  (m) Use and maintenance of non-mechanical hand tools.
  (n) Use of erected scaffolding, staging, trestles or the like (but see 5.2).
  (o) Use of tarpaulins, protective clothing, artificial lighting, safety and welfare facilities, storage and the like that may be available on site.
  (p) Any variation to basic rates required by the contractor in cases where the contract provides for the use of a specified schedule of basic plant charges (to the extent that no other provision is made for such variation - see 5.1).
  (q) In the case of a sub-contract which provides that the sub-contractor shall allow a cash discount, such provision as is necessary for the allowance of the prescribed rate of discount.
  (r) All other liabilities and obligations whatsoever not specifically referred to in this Section nor chargeable under any other Section.
  (s) Profit.

6.2 The additional cost of overtime where specifically ordered by the Architect/Supervising Officer shall only be chargeable in the terms of a prior written agreement between the parties.

Daywork

HEATING AND VENTILATING INDUSTRY

MECHANICAL INSTALLATIONS

Calculation of Hourly Base Rate of Labour for Typical Main Grades applicable from 6 October 2008, refer to notes within Section Three – Rates of Wages.

	FOREMAN	SENIOR CRAFTSMAN (+ 2nd Welding Skill)	SENIOR CRAFTSMAN	CRAFTSMAN	INSTALLER	MATE OVER 18
Hourly Rate from 6 October 2008	14.71	12.66	12.16	11.16	10.11	8.52
Annual standard earnings excluding all holidays, 45.8 weeks x 38 hours	25,601.28	22,033.46	21,163.26	19,422.86	17,595.44	14,828.21
Employers national insurance contributions from 6 April 2008	2,631.78	2,175.10	2,063.72	1,840.95	1,607.04	1,215.83
Weekly holiday credit and welfare contributions (52 weeks) from 6 October 2008	4,012.32	3,505.32	3,385.20	3,135.08	2,875.08	2,480.92
Annual prime cost of labour	32,068.59	27,561.01	26,461.38	24,262.65	21,954.32	18,457.96
Hourly base rate	18.43	15.84	15.20	13.94	12.61	10.61

Notes:

(1) Annual industry holiday (4.6 weeks x 38 hours) and public holidays (1.6 weeks x 38 hours) are paid through weekly holiday credit and welfare stamp scheme.
(2) Where applicable, Merit money and other variables (e.g. daily abnormal conditions money), which attract Employer's National Insurance contribution, should be included.
(3) Contractors in Northern Ireland should add the appropriate amount of CITB Levy to the annual prime cost of labour prior to calculating the hourly base rate.
(4) Hourly rate based on 1,740.40 hours per annum and calculated as follows

```
52 Weeks @ 38 hrs/wk                              =              1,976.00
Less
Public Holiday = 8/5 = 1.6 weeks @ 38 hrs/wk      =    60.80
Annual holidays = 4.6 weeks @ 38 hrs/wk           =   174.80       235.60
                                                                --------
Hours                                             =              1,740.40
                                                                --------
```

(5) For calculation of Holiday Credits and ENI refer to detailed labour rate evaluation
(6) National Insurance contributions are those effective from 6 April 2008.
(7) Weekly holiday credit/welfare stamp values are those effective from 6 October 2008.
(8) Hourly rates of wages are those effective from 6 October 2008.

ELECTRICAL INDUSTRY

DEFINITION OF PRIME COST OF DAYWORK CARRIED OUT UNDER AN ELECTRICAL CONTRACT (MARCH 1981 EDITION)

This Definition of Prime Cost is published by The Royal Institution of Chartered Surveyors and The Electrical Contractors' Associations for convenience and for use by people who choose to use it. Members of The Electrical Contractors' Association are not in any way debarred from defining Prime Cost and rendering accounts for work carried out on that basis in any way they choose. Building owners are advised to reach agreement with contractors on the Definition of Prime Cost to be used prior to entering into a contract or sub-contract.

SECTION 1: APPLICATION

1.1 This Definition provides a basis for the valuation of daywork executed under such electrical contracts as provide for its use.
1.2 It is not applicable in any other circumstances, such as jobbing, or other work carried out as a separate or main contract, nor in the case of daywork executed after the date of practical completion.
1.3 The terms 'contract' and 'contractor' herein shall be read as 'sub-contract' and 'sub-contractor' as the context may require.

SECTION 2: COMPOSITION OF TOTAL CHARGES

2.1 The Prime Cost of daywork comprises the sum of the following costs:
    (a) Labour as defined in Section 3.
    (b) Materials and goods as defined in Section 4.
    (c) Plant as defined in Section 5.
2.2 Incidental costs, overheads and profit as defined in Section 6, as provided in the contract and expressed therein as percentage adjustments, are applicable to each of 2.1 (a)-(c).

SECTION 3: LABOUR

3.1 The standard wage rates, emoluments and expenses referred to below and the standard working hours referred to in 3.2 are those laid down for the time being in the rules and determinations or decisions of the Joint Industry Board or the Scottish Joint Industry Board for the Electrical Contracting Industry (or those of such other body as may be appropriate) applicable to the works and relating to the grade of operative concerned at the time when and in the area where daywork is executed.
3.2 Hourly base rates for labour are computed by dividing the annual prime cost of labour, based upon the standard working hours and as defined in 3.4 by the number of standard working hours per annum. See examples.
3.3 The hourly rates computed in accordance with 3.2 shall be applied in respect of the time spent by operatives directly engaged on daywork, including those operating mechanical plant and transport and erecting and dismantling other plant (unless otherwise expressly provided in the contract) and handling and distributing the materials and goods used in the daywork.
3.4 The annual prime cost of labour comprises the following:
    (a) Standard weekly earnings (i.e. the standard working week as determined at the appropriate rate for the operative concerned).
    (b) Payments in respect of public holidays.
    (c) Any amounts which may become payable by the Contractor to or in respect of operatives arising from operation of the rules etc. referred to in 3.1 which are not provided for in 3.4(a) and (b) nor in Section 6.
    (d) Employer's National Insurance Contributions as applicable to 3.4 (a)-(c).
    (e) Employer's contributions to the Joint Industry Board Combined Benefits Scheme or Scottish Joint Industry Board Holiday and Welfare Stamp Scheme, and holiday payments made to apprentices in compliance with the Joint Industry Board National Working Rules and Industrial Determinations as an employer.
    (f) Any contribution, levy or tax imposed by Statute, payable by the Contractor in his capacity as an employer.
3.5 Differentials or extra payments in respect of supervisory responsibility are excluded from the annual prime cost (see Section 6). The time of principals and similar categories, when working manually, is admissible under this Section at the rates for the appropriate grades.

ELECTRICAL INDUSTRY

SECTION 4: MATERIALS AND GOODS

4.1 The prime cost of materials and goods obtained specifically for the daywork is the invoice cost after deducting all trade discounts and any portion of cash discounts in excess of 5%.

4.2 The prime cost of all other materials and goods used in the daywork is based upon the current market prices plus any appropriate handling charges.

4.3 The prime cost referred to in 4.1 and 4.2 includes the cost of delivery to site.

4.4 Any Value Added Tax which is treated, or is capable of being treated, as input tax (as defined by the Finance Act 1972, or any re-enactment or amendment thereof or substitution therefore) by the Contractor is excluded.

SECTION 5: PLANT

5.1 Unless otherwise stated in the contract, the prime cost of plant comprises the cost of the following:
 (a) Use or hire of mechanically-operated plant and transport for the time employed on and/or provided or retained for the daywork;
 (b) Use of non-mechanical plant (excluding non-mechanical hand tools) for the time employed on and/or provided or retained for the daywork;
 (c) Transport to and from the site and erection and dismantling where applicable.

5.2 The use of non-mechanical hand tools and of erected scaffolding, staging, trestles or the likes is excluded (see Section 6), unless specifically retained for daywork.

5.3 Note: Where hired or other plant is operated by the Electrical Contractor's operatives, such time is to be included under Section 3 unless otherwise provided in the contract.

SECTION 6: INCIDENTAL COSTS, OVERHEADS AND PROFIT

6.1 The percentage adjustments provided in the contract which are applicable to each of the totals of Sections 3, 4 and 5, compromise the following:
 (a) Head Office charges.
 (b) Site staff including site supervision.
 (c) The additional cost of overtime (other than that referred to in 6.2).
 (d) Time lost due to inclement weather.
 (e) The additional cost of bonuses and other incentive payments.
 (f) Apprentices' study time.
 (g) Travelling time and fares.
 (h) Country and lodging allowances.
 (i) Sick pay or insurance in lieu thereof, in respect of apprentices.
 (j) Third party and employers' liability insurance.
 (k) Liability in respect of redundancy payments to employees.
 (l) Employers' National Insurance Contributions not included in 3.4.
 (m) Use and maintenance of non-mechanical hand tools.
 (n) Use of erected scaffolding, staging, trestles or the like (but see 5.2.).
 (o) Use of tarpaulins, protective clothing, artificial lighting, safety and welfare facilities, storage and the like that may be available on site.
 (p) Any variation to basic rates required by the Contractor in cases where the contract provides for the use of a specified schedule of basic plant charges (to the extent that no other provision is made for such variation - see 5.1).
 (q) All other liabilities and obligations whatsoever not specifically referred to in this Section nor chargeable under any other Section.
 (r) Profit.
 (s) In the case of a sub-contract which provides that the sub-contractor shall allow a cash discount, such provision as is necessary for the allowance of the prescribed rate of discount.

6.2 The additional cost of overtime where specifically ordered by the Architect/Supervising Officer shall only be chargeable in the terms of a prior written agreement between the parties.

ELECTRICAL INDUSTRY

ELECTRICAL INSTALLATIONS

Calculation of Hourly Base Rate of Labour for Typical Main Grades applicable from 7 January 2008

	TECHNICIAN	APPROVED ELECTRICIAN	ELECTRICIAN	LABOURER
Hourly Rate from 7 January 2008 (London Rates)	16.50	14.65	13.51	10.83
Annual standard earnings excluding all holidays, 46 weeks x 37.5 hours	28,462.50	25,271.25	23,304.75	18,681.75
Employers national insurance contributions from 6 April 2008	2,998.02	2,589.54	2,337.83	1,746.08
JIB Combined benefits from 25 September 2006	2,973.88	2,702.96	2,532.40	2,147.08
Holiday top up funding	1,398.80	1,253.72	1,167.92	950.04
Annual prime cost of labour	35,833.20	31,817.47	29,342.90	23,524.95
Hourly base rate	20.77	18.44	17.01	13.64

Notes:

(1) Annual industry holiday (4.4 weeks x 37.5 hours) and public holidays (1.6 weeks x 37.5 hours)
(2) It should be noted that all labour costs incurred by the Contractor in his capacity as an Employer, other than those contained in the hourly rate above, must be taken into account under Section 6.
(3) Public Holidays are paid through weekly holiday credit and welfare stamp scheme.
(4) Contractors in Northern Ireland should add the appropriate amount of CITB Levy to the annual prime cost of labour prior to calculating the hourly base rate.
(5) Hourly rate based on 1,725 hours per annum and calculated as follows:

```
52 Weeks @ 37.5 hrs/wk                              =                            1,950.00
Less
Public Holiday = 8/5 = 1.6 weeks @ 37.5 hrs/wk      =           60.00
Annual holidays = 4.4 weeks @ 37.5 hrs/wk           =          165.00              225.00
                                                                                -------
Hours                                               =                            1,725.00
                                                                                -------
```

(6) For calculation of holiday credits and ENI refer to detailed labour rate evaluation.
(7) Hourly wage rates are those effective from 7 January 2008.
(8) National Insurance contributions are those effective from 6 April 2008.
(9) JIB Combined Benefits Values are those effective from 1 October 2007.

BUILDING INDUSTRY PLANT HIRE COSTS

SCHEDULE OF BASIC PLANT CHARGES (MAY 2001)

This Schedule is published by the Royal Institution of Chartered Surveyors and is for use in connection with Dayworks under a Building Contract.

EXPLANATORY NOTES

1. The rates in the Schedule are intended to apply solely to daywork carried out under and incidental to a Building Contract. They are NOT intended to apply to:
   (i) jobbing or any other work carried out as a main or separate contract; or
   (ii) work carried out after the date of commencement of the Defects Liability Period.

2. The rates apply to plant and machinery already on site, whether hired or owned by the Contractor.

3. The rates, unless otherwise stated, include the cost of fuel and power of every description, lubricating oils, grease, maintenance, sharpening of tools, replacement of spare parts, all consumable stores and for licences and insurances applicable to items of plant.

4. The rates, unless otherwise stated, do not include the costs of drivers and attendants (unless otherwise stated).

5. The rates in the Schedule are base costs and may be subject to an overall adjustment for price movement, overheads and profit, quoted by the Contractor prior to the placing of the Contract.

6. The rates should be applied to the time during which the plant is actually engaged in daywork.

7. Whether or not plant is chargeable on daywork depends on the daywork agreement in use and the inclusion of an item of plant in this schedule does not necessarily indicate that item is chargeable.

8. Rates for plant not included in the Schedule or which is not already on site and is specifically provided or hired for daywork shall be settled at prices which are reasonably related to the rates in the Schedule having regard to any overall adjustment quoted by the Contractor in the Conditions of Contract.

NOTE: All rates in the schedule were calculated during the first quarter of 2001.

Daywork

BUILDING INDUSTRY PLANT HIRE COSTS

MECHANICAL PLANT AND TOOLS

Item of Plant	Size/Rating	Unit	Rate/hr
PUMPS			
Mobile Pumps			
Including pump hoses, valves and strainers etc.			
Diaphragm	50mm dia.	Each	0.87
Diaphragm	76mm dia.	Each	1.29
Submersible	50mm dia.	Each	1.18
Induced flow	50mm dia.	Each	1.54
Induced flow	76mm dia.	Each	2.05
Centrifugal, self priming	50mm dia.	Each	1.96
Centrifugal, self priming	102mm dia.	Each	2.52
Centrifugal, self priming	152mm dia.	Each	3.87
SCAFFOLDING, SHORING, FENCING			
Complete Scaffolding			
Mobile working towers, single width	1.8m x 0.8m base x 7m high	Each	2.00
Mobile working towers, single width	1.8m x 0.8m base x 9m high	Each	2.80
Mobile working towers, double width	1.8m x 1.4m base x 7m high	Each	2.15
Mobile working towers, double width	1.8m x 1.4m base x 15m high	Each	5.10
Chimney scaffold, single unit		Each	1.79
Chimney scaffold, twin unit		Each	2.05
Chimney scaffold, four unit		Each	3.59
Trestles			
Trestle, adjustable	Any height	Pair	0.10
Trestle, painters	1.8m high	Pair	0.21
Trestle, Painters	2.4m high	Pair	0.26
Shoring, Planking and Strutting			
'Acrow' adjustable prop	Sizes up to 4.9m (open)	Each	0.10
'Strong boy' support attachment		Each	0.15
Adjustable trench struts	Sizes up to 1.67m (open)	Each	0.10
Trench sheet		Metre	0.01
Backhoe trench box		Each	1.00
Temporary Fencing			
Including block and coupler			
Site fencing steel grid panel	3.5m x 2.0m	Each	0.08
Anti-climb site steel grid fence panel	3.5m x 2.0m	Each	0.08
LIFTING APPLIANCES AND CONVEYORS			
Cranes			
Mobile Cranes			
Rates are inclusive of drivers			
Lorry mounted, telescopic jib			
Two wheel drive	6 tonnes	Each	24.40
Two wheel drive	7 tonnes	Each	25.00
Two wheel drive	8 tonnes	Each	25.62
Two wheel drive	10 tonnes	Each	26.90
Two wheel drive	12 tonnes	Each	28.25
Two wheel drive	15 tonnes	Each	29.66
Two wheel drive	18 tonnes	Each	31.14
Two wheel drive	20 tonnes	Each	32.70
Two wheel drive	25 tonnes	Each	34.33

BUILDING INDUSTRY PLANT HIRE COSTS

MECHANICAL PLANT AND TOOLS

Item of Plant	Size/Rating	Unit	Rate/hr
Four wheel drive	10 tonnes	Each	27.44
Four wheel drive	12 tonnes	Each	28.81
Four wheel drive	15 tonnes	Each	30.25
Four wheel drive	20 tonnes	Each	33.35
Four wheel drive	25 tonnes	Each	35.19
Four wheel drive	30 tonnes	Each	37.12
Four wheel drive	45 tonnes	Each	39.16
Four wheel drive	50 tonnes	Each	41.32

Track-mounted tower crane
Rates inclusive of driver
Note : Capacity equals maximum lift in tonnes times maximum radius at which it can be lifted

	Capacity (metre/tonnes) up to	Height under hook above ground (m) up to		
Tower crane	10	17	Each	7.99
Tower crane	15	18	Each	8.59
Tower crane	20	20	Each	9.18
Tower crane	25	22	Each	11.56
Tower crane	30	22	Each	13.78
Tower crane	40	22	Each	18.09
Tower crane	50	22	Each	22.20
Tower crane	60	22	Each	24.32
Tower crane	70	22	Each	23.00
Tower crane	80	22	Each	25.91
Tower crane	110	22	Each	26.45
Tower crane	125	30	Each	29.38
Tower crane	150	30	Each	32.35

Static tower cranes
Rates inclusive of driver
To be charged at 90% of the above rates for track mounted tower cranes

Crane Equipment			
Muck tipping skip	Up to 0.25m³	Each	0.56
Muck tipping skip	0.5m³	Each	0.67
Muck tipping skip	0.75m³	Each	0.82
Muck tipping skip	1.0m³	Each	1.03
Muck tipping skip	1.5m³	Each	1.18
Muck tipping skip	2.0m³	Each	1.38
Mortar skips	up to 0.38m³	Each	0.41
Boat skips	1.0m³	Each	1.08
Boat skips	1.5m³	Each	1.33
Boat skips	2.0m³	Each	1.59
Concrete skips, hand levered	0.5m³	Each	1.00
Concrete skips, hand levered	0.75m³	Each	1.10
Concrete skips, hand levered	1.0m³	Each	1.25
Concrete skips, hand levered	1.5m³	Each	1.50
Concrete skips, hand levered	2.0m³	Each	1.65

BUILDING INDUSTRY PLANT HIRE COSTS

MECHANICAL PLANT AND TOOLS

Item of Plant	Size/Rating		Unit	Rate/hr
Crane Equipment (cont'd)				
Concrete skips, geared	0.5m^3		Each	1.30
Concrete skips, geared	0.75m^3		Each	1.40
Concrete skips, geared	1.0m^3		Each	1.55
Concrete skips, geared	1.5m^3		Each	1.80
Concrete skips, geared	2.0m^3		Each	2.05
Hoists				
Scaffold hoists	200kg		Each	1.92
Rack and pinion (goods only)	500kg		Each	3.31
Rack and pinion (goods only)	1100kg		Each	4.28
Rack and pinion goods and passenger	15 person, 1200kg		Each	5.62
Wheelbarrow chain sling				0.31
Conveyors				
<u>Belt conveyors</u>				
Conveyor	7.5m long x 400mm wide		Each	6.41
Miniveyor, control box and loading hopper	3m unit		Each	3.59
<u>Other Conveying Equipment</u>				
Wheelbarrow			Each	0.21
Hydraulic superlift			Each	2.95
Pavac slab lifter			Each	1.03
Hand pad and hose attachment			Each	0.26
Lifting Trucks				
Fork lift, two wheel drive	Payload	Max Lift		
Fork lift, two wheel drive	1100kg	up to 3.0m	Each	4.87
Fork lift, two wheel drive	2540kg	up to 3.7m	Each	5.12
Fork lift, four wheel drive	1524kg	up to 6.0m	Each	6.04
Fork lift, four wheel drive	2600kg	up to 5.4m	Each	7.69
Lifting Platforms				
Hydraulic platform (Cherry picker)	7.5m		Each	4.23
Hydraulic platform (Cherry picker)	13m		Each	9.23
Scissors lift	7.8m		Each	7.56
Telescopic handlers	7m, 2 tonne		Each	7.18
Telescopic handlers	13m, 3 tonne		Each	8.72
Lifting and Jacking Gear				
Pipe winch including gantry	1 tonne		Sets	1.92
Pipe winch including gantry	3 tonnes		Sets	3.21
Chain block	1 tonne		Each	0.45
Chain block	2 tonnes		Each	0.71
Chain block	5 tonnes		Each	1.22
Pull lift (Tirfor winch)	1 tonne		Each	0.64
Pull lift (Tirfor winch)	1.6 tonnes		Each	0.90
Pull lift (Tirfor winch)	3.2 tonnes		Each	1.15
Brother or chain slings, two legs	not exceeding 4.2 tonnes		Set	0.35
Brother or chain slings, two legs	not exceeding 7.5 tonnes		Set	0.45
Brother or chain slings, four legs	not exceeding 3.1 tonnes		Set	0.41
Brother or chain slings, four legs	not exceeding 11.2 tonnes		Set	1.28

BUILDING INDUSTRY PLANT HIRE COSTS

MECHANICAL PLANT AND TOOLS

Item of Plant	Size/Rating	Unit	Rate/hr
CONSTRUCTION VEHICLES			
Lorries			
Plated lorries			
Rates are inclusive of driver			
Platform lorries	7.5 tonnes	Each	19.00
Platform lorries	17 tonnes	Each	21.00
Platform lorries	24 tonnes	Each	26.00
Platform lorries with winch and skids	7.5 tonnes	Each	21.40
Platform lorries with crane	17 tonnes	Each	27.50
Platform lorries with crane	24 tonnes	Each	32.10
Tipper Lorries			
Rates are inclusive of driver			
Tipper lorries	15/17 tonnes	Each	19.50
Tipper lorries	24 tonnes	Each	21.40
Tipper lorries	30 tonnes	Each	27.10
Dumpers			
Site use only (excluding tax, insurance and extra Cost of DERV etc. when operating on highway)			
	Makers capacity		
Two wheel drive	0.8 tonnes	Each	1.20
Two wheel drive	1 tonne	Each	1.30
Two wheel drive	1.2 tonnes	Each	1.60
Four wheel drive	2 tonnes	Each	2.50
Four wheel drive	3 tonnes	Each	3.00
Four wheel drive	4 tonnes	Each	3.50
Four wheel drive	5 tonnes	Each	4.00
Four wheel drive	6 tonnes	Each	4.50
Dumper Trucks			
Rates are inclusive of drivers			
Dumper trucks	10/13 tonnes	Each	20.00
Dumper trucks	18/20 tonnes	Each	20.40
Dumper trucks	22/25 tonnes	Each	26.30
Dumper trucks	35/40 tonnes	Each	36.60
Tractors			
<u>Agricultural Type</u>			
Wheeled, rubber-clad tyred			
Light	48 h.p.	Each	4.65
Heavy	65 h.p.	Each	5.15
<u>Crawler Tractors</u>			
With bull or angle dozer	80/90 h.p.	Each	21.40
With bull or angle dozer	115/130 h.p.	Each	25.10
With bull or angle dozer	130/150 h.p.	Each	26.00
With bull or angle dozer	155/175 h.p.	Each	27.74
With bull or angle dozer	210/230 h.p.	Each	28.00
With bull or angle dozer	300/340 h.p.	Each	31.10
With bull or angle dozer	400/440 h.p.	Each	46.90
With loading shovel	0.8m³	Each	25.00
With loading shovel	1.0m³	Each	28.00
With loading shovel	1.2m³	Each	32.00
With loading shovel	1.4m³	Each	36.00
With loading shovel	1.8m³	Each	45.00

BUILDING INDUSTRY PLANT HIRE COSTS

MECHANICAL PLANT AND TOOLS

Item of Plant	Size/Rating	Unit	Rate/hr
Light Vans			
Ford Escort or the like		Each	4.74
Ford Transit or the like	1.0 tonnes	Each	6.79
Luton Box Van or the like	1.8 tonnes	Each	8.33
Water/Fuel Storage			
Mobile water container	110 litres	Each	0.28
Water bowser	1100 litres	Each	0.55
Water bowser	3000 litres	Each	0.74
Mobile fuel container	110 litres	Each	0.28
Fuel bowser	1100 litres	Each	0.65
Fuel bowser	3000 litres	Each	1.02
EXCAVATORS AND LOADERS			
Excavators			
Wheeled, hydraulic	7/10 tonnes	Each	12.00
Wheeled, hydraulic	11/13 tonnes	Each	12.70
Wheeled, hydraulic	15/16 tonnes	Each	14.80
Wheeled, hydraulic	17/18 tonnes	Each	16.70
Wheeled, hydraulic	20/23 tonnes	Each	16.70
Crawler, hydraulic	12/14 tonnes	Each	12.00
Crawler, hydraulic	15/17.5 tonnes	Each	14.00
Crawler, hydraulic	20/23 tonnes	Each	16.00
Crawler, hydraulic	25/30 tonnes	Each	21.00
Crawler, hydraulic	30/35 tonnes	Each	30.00
Mini excavators	1000/1500kg	Each	4.50
Mini excavators	2150/2400kg	Each	5.50
Mini excavators	2700/3500kg	Each	6.50
Mini excavators	3500/4500kg	Each	8.50
Mini excavators	4500/6000kg	Each	9.50
Loaders			
Wheeled skip loader		Each	4.50
Shovel loaders, four wheel drive	1.6m^3	Each	12.00
Shovel loaders, four wheel drive	2.4m^3	Each	19.00
Shovel loaders, four wheel drive	3.6m^3	Each	22.00
Shovel loaders, four wheel drive	4.4m^3	Each	23.00
Shovel loaders, crawlers	0.8m^3	Each	11.00
Shovel loaders, crawlers	1.2m^3	Each	14.00
Shovel loaders, crawlers	1.6m^3	Each	16.00
Shovel loaders, crawlers	2m^3	Each	17.00
Skid steer loaders wheeled	300/400kg payload	Each	6.00
Excavator Loaders			
Wheeled tractor type with back-hoe excavator			
Four wheel drive	2.5/3.5 tonnes	Each	7.00
Four wheel drive, 2 wheel steer	7/8 tonnes	Each	9.00
Four wheel drive, 4 wheel steer	7/8 tonnes	Each	10.00
Crawler, hydraulic	12 tonnes	Each	20.00
Crawler, hydraulic	20 tonnes	Each	16.00
Crawler, hydraulic	30 tonnes	Each	35.00
Crawler, hydraulic	40 tonnes	Each	38.00

Daywork

BUILDING INDUSTRY PLANT HIRE COSTS

MECHANICAL PLANT AND TOOLS

Item of Plant	Size/Rating	Unit	Rate/hr
COMPACTION EQUIPMENT			
Rollers			
Vibrating roller	368kg - 420kg	Each	1.68
Single roller	533kg	Each	1.92
Single roller	750kg	Each	2.41
Vibrating roller	368kg - 420kg	Each	1.68
Single roller	533kg	Each	1.92
Single roller	750kg	Each	2.41
Twin roller	698kg	Each	1.93
Twin roller	851kg	Each	2.41
Twin roller with seat end steering wheel	1067kg	Each	3.03
Twin roller with seat end steering wheel	1397kg	Each	3.17
Pavement rollers	3 - 4 tonnes dead weight	Each	3.18
Pavement rollers	4 - 6 tonnes	Each	4.13
Pavement rollers	6 - 10 tonnes	Each	4.84
Rammers			
Tamper rammer 2 stroke-petrol	225mm - 275mm	Each	1.59
Soil Compactors			
Plate compactor	375mm - 400mm	Each	1.20
Plate compactor rubber pad	375mm - 1400mm	Each	0.33
Plate compactor reversible plate - petrol	400mm	Each	2.20
CONCRETE EQUIPMENT			
Concrete/Mortar Mixers			
Open drum without hopper	0.09/0/06m^3	Each	0.62
Open drum without hopper	0.12/0.09m^3	Each	0.68
Open drum without hopper	0.15/0.10m^3	Each	0.72
Open drum with hopper	0.20/0.15m^3	Each	0.80
Concrete/Mortar Transport Equipment			
Concrete pump including hose, valve and couplers			
Lorry mounted concrete pump	23m max. distance	Each	36.00
Lorry mounted concrete pump	50m max. distance	Each	46.00
Concrete Equipment			
Vibrator, poker, petrol type	up to 75mm dia.	Each	1.62
Air vibrator (excluding compressor and hose)	up to 75mm dia.	Each	0.79
Extra poker heads	5m	Each	0.77
Vibrating screed unit with beam	3m - 5m	Each	1.77
Vibrating screed unit with adjustable beam	725mm - 900mm	Each	2.18
Power float		Each	1.72
Power grouter		Each	0.92
TESTING EQUIPMENT			
Pipe Testing Equipment			
Pressure testing pump, electric		Sets	1.87
Pipe pressure testing equipment, hydraulic		Sets	2.46
Pressure test pump		Sets	0.64

BUILDING INDUSTRY PLANT HIRE COSTS

MECHANICAL PLANT AND TOOLS

Item of Plant	Size/Rating	Unit	Rate/hr
SITE ACCOMMODATION AND TEMPORARY SERVICES			
Heating Equipment			
Space heaters - propane		Each	0.77
Space heaters - propane/electric	80,000Btu/hr	Each	1.56
Space heaters - propane/electric	125,000Btu/hr	Each	1.79
Space heaters, propane	250,000Btu/hr	Each	1.33
Space heaters, propane	125,000Btu/hr	Each	1.64
Cabinet heaters	260,000Btu/hr	Each	0.41
Cabinet heater catalytic		Each	0.46
Electric halogen heaters		Each	1.28
Ceramic heaters		Each	0.79
Fan heaters	3kW	Each	0.41
Cooling Fan	3kW	Each	1.15
Mobile cooling unit - small		Each	1.38
Mobile cooling unit - large		Each	1.54
Air conditioning unit		Each	2.62
Site Lighting and Equipment			
Tripod floodlight	500W		
Tripod floodlight	1000W	Each	0.36
Towable floodlight	4 x 1000W	Each	0.34
Hand held floodlight	500W	Each	2.00
		Each	0.22
Rechargeable light		Each	0.62
Inspection light		Each	0.15
Plasterers light		Each	0.56
Lighting mast		Each	0.92
Festoon light string	33m	Each	0.31
Site Electrical Equipment			
Extension leads	240V/14m	Each	0.20
Extension leafs	110V/14m	Each	0.20
Cable reel	25m 110V/240V	Each	0.28
Cable reel	50m 110V/240V	Each	0.33
4 way junction box	110V	Each	0.17
Power Generating Units			
Generator - petrol	2kVA	Each	1.08
Generator - silenced petrol	2kVA	Each	1.54
Generator - petrol	3VA	Each	1.38
Generator - diesel	5kVA	Each	1.92
Generator - silenced diesel	8kVA	Each	3.59
Generator - silenced diesel	1.5kVA	Each	7.69
Tail adaptor	240V	Each	0.20
Transformers			
Transformer	3kVA	Each	0.36
Transformer	5kVA	Each	0.51
Transformer	7.5kVA	Each	0.82
Transformer	10kVA	Each	0.87
Rubbish Collection and Disposal Equipment			
Rubbish Chutes			
Standard plastic module	1m section	Each	0.18
Steel liner insert		Each	0.26
Steel top hopper		Each	0.20
Plastic side entry hopper		Each	0.20
Plastic side entry hopper liner		Each	0.20

BUILDING INDUSTRY PLANT HIRE COSTS

MECHANICAL PLANT AND TOOLS

Item of Plant	Size/Rating	Unit	Rate/hr
SITE EQUIPMENT			
Welding Equipment			
<u>Arc-(Electric) Complete with Leads</u>			
Welder generator - petrol	200 amp	Each	2.26
Welder generator - diesel	300/350 amp	Each	3.33
Welder generator - diesel	400 amp	Each	4.74
Extra welding lead sets		Each	0.29
<u>Gas-Oxy Welder</u>			
Welding and cutting set (including oxygen and acetylene, excluding underwater equipment and thermic boring)			
Small		Each	1.41
Large		Each	2.00
Mig welder		Each	1.00
Fume extractor		Each	0.92
Road Works Equipment			
Traffic lights, mains/generator	2-way	Set	4.01
Traffic lights, mains/generator	3-way	Set	7.92
Traffic lights, mains/generator	4-way	Set	9.81
Traffic lights, mains/generator - trailer mounted	2-way	Set	3.98
Flashing lights		Each	0.20
Road safety cone	450mm	10	0.26
Safety cone	750mm	10	0.38
Safety barrier plank	1.25m	Each	0.03
Safety barrier plank	2m	Each	0.04
Road sign		Each	0.26
DPC Equipment			
Damp proofing injection machine		Each	1.49
Cleaning Equipment			
Vacuum cleaner (industrial wet) single motor		Each	0.62
Vacuum cleaner (industrial wet) twin motor		Each	1.23
Vacuum cleaner (industrial wet) triple motor		Each	1.44
Vacuum cleaner (industrial wet) back pack		Each	0.97
Pressure washer, light duty, electric	1450 PSI	Each	0.97
Pressure washer, heavy duty, diesel	2500 PSI	Each	2.69
Cold pressure washer, electric		Each	1.79
Hot pressure washer, petrol		Each	2.92
Cold pressure washer, petrol		Each	2.00
Sandblast attachment to last washer		Each	0.54
Drain cleaning attachment to last washer		Each	0.31
Surface Preparation Equipment			
Rotavators	5 h.p.	Each	1.67
Scabbler, up to three heads		Each	1.15
Scabbler, pole		Each	1.50
Scabbler, multi-headed floor		Each	4.00
Floor preparation machine		Each	2.82

BUILDING INDUSTRY PLANT HIRE COSTS

MECHANICAL PLANT AND TOOLS

Item of Plant	Size/Rating	Unit	Rate/hr
Compressors and Equipment			
Portable Compressors			
Compressors - electric	0.23m³/min	Each	1.59
Compressors - petrol	0.28m³/min	Each	1.74
Compressors - petrol	0.71m³/min	Each	2.00
Compressors - diesel	up to 2.83m³/min	Each	1.24
Compressors - diesel	up to 3.68m³/min	Each	1.49
Compressors - diesel	up to 4.25m³/min	Each	1.60
Compressors - diesel	up to 4.81m³/min	Each	1.92
Compressors - diesel	up to 7.64m³/min	Each	3.08
Compressors - diesel	up to 11.32m³/min	Each	4.23
Compressors - diesel	up to 18.40m³/min	Each	5.73
Mobile Compressors			
Lorry mounted compressors	2.86-4.24m³/min	Each	12.50
(machine plus lorry only)			
Tractor mounted compressors	2.86-3.40m³/min	Each	13.50
(machine plus rubber tyred tractor)			
Accessories (Pneumatic Tools)			
(with and including up to 15m of air hose)			
Demolition pick		Each	1.03
Breakers (with six steels) light	up to 150kg	Each	0.79
Breakers (with six steels) medium	295kg	Each	1.08
Breakers (with six steels) heavy	386kg	Each	1.44
Rock drill (for use with compressor) hand held		Each	0.90
Additional hoses	15m	Each	0.16
Muffler, tool silencer		Each	0.14
Breakers			
Demolition hammer drill, heavy duty, electric		Each	1.00
Road breaker, electric		Each	1.65
Road breaker, 2 stroke, petrol		Each	2.05
Hydraulic breaker unit, light duty, petrol		Each	2.05
Hydraulic breaker unit, heavy duty, petrol		Each	2.60
Hydraulic breaker unit, heavy duty, diesel		Each	2.95
Quarrying and Tooling Equipment			
Block and stone splitter, hydraulic	600mm x 600mm	Each	1.35
Block and slab splitter, manual		Each	1.10
Steel Reinforcement Equipment			
Bar bending machine - manual	up to 13mm dia. rods	Each	0.90
Bar bending machine - manual	up to 20mm dia. rods	Each	1.28
Bar shearing machine - electric	up to 38mm dia. rods	Each	2.82
Bar shearing machine - electric	up to 40mm dia. rods	Each	3.85
Bar cropper machine - electric	up to 13mm dia. rods	Each	1.54
Bar cropper machine - electric	up to 20mm dia. rods	Each	2.05
Bar cropper machine - electric	up to 40mm dia. rods	Each	2.82
Bar cropper machine - 3 phase	up to 40mm dia. rods	Each	3.85
Dehumidifiers			
110/240v Water	68 litres extraction per 24 hrs	Each	1.28
110/240v Water	90 litres extraction per 24 hrs	Each	1.85

BUILDING INDUSTRY PLANT HIRE COSTS

MECHANICAL PLANT AND TOOLS

Item of Plant	Size/Rating	Unit	Rate/hr
Compressors and Equipment (cont'd)			
SMALL TOOLS			
Saws			
Masonry bench saw	350mm - 500mm dia.	Each	2.80
Floor saw	350mm dia., 125mm max. cut	Each	1.90
Floor saw	450mm dia., 150mm max. cut	Each	2.60
Floor saw, reversible	Max. cut 300mm	Each	13.00
Chop/cut off saw, electric	350mm dia.	Each	1.33
Circular saw, electric	230mm dia.	Each	0.60
Tyrannosaw		Each	1.20
Reciprocating saw		Each	0.60
Door trimmer		Each	0.90
Chainsaw, petrol	500mm	Each	2.13
Full chainsaw safety kit		Each	0.50
Worktop jig		Each	0.60
Pipework Equipment			
Pipe bender	15mm - 22mm	Each	0.33
Pipe bender, hydraulic	50mm	Each	0.60
Pipe bender, electric	50mm - 150mm dia.	Each	1.35
Pipe cutter, hydraulic		Each	1.84
Tripod pipe vice		Set	0.40
Ratchet threader	12mm - 32mm	Each	0.55
Pipe threading machine, electric	12mm - 75mm	Each	2.40
Pipe threading machine, electric	12mm - 100mm	Each	3.00
Impact wrench, electric		Each	0.54
Impact wrench, two stroke, petrol		Each	4.49
Impact wrench, heavy duty, electric		Each	1.13
Plumber's furnace, calor gas or similar		Each	2.16
Hand-held Drills and Equipment			
Impact or hammer drill	up to 25mm dia.	Each	0.50
Impact or hammer drill	35mm dia.	Each	0.90
Angle head drills		Each	0.70
Stirrer, mixer drills		Each	0.70
Paint, Insulation Application Equipment			
Airless spray unit		Each	4.20
Portaspray unit		Each	1.65
HVLP turbine spray unit		Each	1.65
Compressor and spray gun		Each	2.20
Other Handtools			
Screwing machine	13mm - 50mm dia.	Each	0.77
Screwing machine	25mm - 100mm dia.	Each	1.57
Staple gun		Each	0.33
Air nail gun	110V	Each	3.33
Cartridge hammer		Each	1.00
Tongue and groove nailer complete with mallet		Each	0.93
Chasing machine	152mm	Each	1.72
Chasing machine	76mm - 203mm	Each	5.99
Floor grinder		Each	3.00
Floor plane		Each	3.67
Diamond concrete planer		Each	2.05
Autofeed screwdriver, electric		Each	1.13
Laminate trimmer		Each	0.64

BUILDING INDUSTRY PLANT HIRE COSTS

MECHANICAL PLANT AND TOOLS

Item of Plant	Size/Rating	Unit	Rate/hr
Biscuit jointer		Each	0.87
Random orbital sander		Each	0.72
Floor sander		Each	1.33
Palm, delta, flap or belt sander	300mm	Each	0.38
Saw cutter, 2 stroke, petrol	up to 225mm	Each	1.26
Grinder, angle or cutter	300mm	Each	0.60
Grinder, angle or cutter		Each	1.10
Mortar raking tool attachment	325mm	Each	0.15
Floor/polisher scrubber		Each	1.03
Floor tile stripper		Each	1.74
Wallpaper stripper, electric		Each	0.56
Electric scraper		Each	0.51
Hot air paint stripper	All sizes	Each	0.38
Electric diamond tile cutter		Each	1.38
Hand tile cutter		Each	0.36
Electric needle gun		Each	1.08
Needle chipping gun	1.2m wide	Each	0.72
Pedestrian floor sweeper		Each	0.87

Tables and Memoranda

Conversion Tables, *page 638*
Formulae, *page 640*
Fractions, Decimals and Millimetre Equivalents, *page 641*
Imperial Standard Wire Gauge, *page 642*
Water Pressure Due to Height, *page 643*
Table of Weights for Steelwork, *page 644*
Dimensions and Weights of Copper Pipes, *page 649*
Dimensions of Stainless Steel Pipes, *page 650*
Dimensions of Steel Pipes, *page 651*
Approximate Metres per Tonne of Tubes, *page 652*
Flange Dimension Chart, *page 653*
Minimum Distances Between Supports/Fixings, *page 655*
Litres of Water Storage Required per Person per Building Type, *page 656*
Recommended Air Conditioning Design Loads, *page 656*
Capacity and Dimensions of Galvanised Mild Steel Cisterns, *page 657*
Capacity of Cold Water Polypropylene Storage Cisterns, *page 657*
Minimum Insulation Thickness to Protect Against Frost, *page 658*
Insulation Thickness for Chilled Water Supplies to Prevent Condensation, *page 658*
Insulation Thickness for Non-Domestic Heating Installations to Control Heat Loss, *page 659*

CONVERSION TABLES

LENGTH

Millimetre	(mm)	1 in	=	25.4	mm	: 1 mm	=	0.0394	in
Metre	(m)	1 ft	=	0.3048	m	: 1 m	=	3.2808	ft
Kilometre	(km)	1 yd	=	0.9144	m	: 1 m	=	1.0936	yd
Kilometre	(km)	1 mile	=	1.6093	km	: 1 km	=	0.6214	mile

NOTE :

1 cm	=	10	mm	1 ft	=	12	in
1 m	=	100	cm	1 yd	=	3	ft
1 km	=	1000	m	1 mile	=	1760	yd

AREA

Square Millimetre	(mm^2)	1 in^2	=	645.2	mm^2	: 1 mm^2	=	0.0016	in^2
Square Centimetre	(cm^2)	1 in^2	=	6.4516	cm^2	: 1 cm^2	=	0.1550	in^2
Square Metre	(m^2)	1 ft^2	=	0.0929	m^2	: 1 m^2	=	10.764	ft^2
Square Metre	(m^2)	1 yd^2	=	0.8361	m^2	: 1 m^2	=	1.1960	yd^2
Square Kilometre	(km^2)	1 $mile^2$	=	2.590	km^2	: 1 km^2	=	0.3861	$mile^2$
Hectare	(ha)	1 acre	=	0.405	ha	: 1 ha	=	2.471	acre

NOTE :

1 cm^2	=	100	mm^2	1 ft^2	=	144	in^2
1 m^2	=	10000	cm^2	1 yd^2	=	9	ft^2
1 km^2	=	100	ha	1 $mile^2$	=	640	acre
				1 acre	=	4840	yd^2

VOLUME

Cubic Centimetre	(cm^3)	1 cm^3	=	0.0610	in^3	: 1 in^3	=	16.387	cm^3
Cubic Decimetre	(dm^3)	1 dm^3	=	0.0353	ft^3	: 1 ft^3	=	28.329	dm^3
Cubic Metre	(m^3)	1 m^3	=	35.315	ft^3	: 1 ft^3	=	0.0283	m^3
Cubic Metre	(m^3)	1 m^3	=	1.3080	yd^3	: 1 yd^3	=	0.7646	m^3
Litre	(L)	1 L	=	1.76	pint	: 1 pint	=	0.5683	L
Litre	(L)	1 L	=	2.113	US pt	: 1 pint	=	0.4733	US L
Litre	(L)	1L	=	0.220	gal	1 gal	=	4.546	L

NOTE :

1 dm^3	=	1000	cm^3	1 ft^3	=	1728	in^3
1 m^3	=	1000	dm^3	1 yd^3	=	27	ft^3
1 L	=	1	dm^3	1 pint	=	20	fl oz
1 HL	=	100	L	1 gal	=	8	pints

MASS

Milligram	(mg)	1 mg	=	0.0154	grain	: 1 grain	=	64.935	mg
Gram	(g)	1 g	=	0.0353	oz	: 1 oz	=	28.35	g
Kilogram	(kg)	1 kg	=	2.2046	lb	: 1 lb	=	0.4536	kg
Kilogram	(kg)	1 kg	=	0.020	cwt	: 1 cwt	=	50.802	kg
Tonne	(t)	1 t	=	0.9842	ton	: 1 ton	=	1.016	t

NOTE :

1 g	=	1000	mg	1 oz	=	437.5	grains
1 kg	=	1000	g	1 lb	=	16	oz
1 t	=	1000	kg	1 stone	=	14	lb
				1 cwt	=	112	lb
				1 ton	=	20	cwt

CONVERSION TABLES

FORCE

Newton	(N)	1 lb f	=	4.448	N	: 1 kg f	=	9.807	N
Kilonewton	(kN)	1 lb f	=	0.004448	kN	: 1 ton f	=	9.964	kN
Meganewton	(MN)	100 ton f	=	0.9964	MN				

POWER

Kilowatt	(kW)	1 kW	=	1.310	HP	: 1 HP	=	0.746	kW

PRESSURE AND STRESS

Kilonewton		1 lb f/in^2	=	6.895	kN/m^2
per square metre	(kN/m^2)	1 bar	=	100	kN/m^2
		1 ton f/ft^2	=	107.3	kN/m^2
		1 kg f/cm^2	=	98.07	kN/m^2
		1 lb f/ft^2	=	0.0479	kN/m^2

TEMPERATURE

Degrees   °C = 5/9 (°F − 32°)   °F = 9/5 (°C + 32°)

FORMULAE

Two dimensional figures

Figure	Area
Triangle	$0.5 \times$ base $\times$ height, or $\sqrt{s(s-a)(s-b)(s-c)}$ where $s = 0.5 \times$ the sum of the three sides and a, b and c are the lengths of the three sides, or $a^2 = b^2 + c^2 - 2 \times bc \times \cos A$ where A is the angle opposite side a
Hexagon	$2.6 \times (\text{side})^2$
Octagon	$4.83 \times (\text{side})^2$
Trapezoid	height $\times$ 0.5 (base + top)
Circle	$3.142 \times \text{radius}^2$ or $0.7854 \times \text{diameter}^2$ (circumference = $2 \times 3.142 \times$ radius or $3.142 \times$ diameter)
Sector of a circle	$0.5 \times$ length of arc $\times$ radius
Segment of a circle	area of sector - area of triangle
Ellipse of a circle	$3.142 \times AB$ (where $A = 0.5 \times$ height and $B = 0.5 \times$ length)
Spandrel	$3/14 \times \text{radius}^2$

Three dimensional figure

Figure	Volume	Surface Area
Prism x height	Area of base $\times$ height	circumference of base
Cube	(length of side) cubed	$6 \times (\text{length of side})^2$
Cylinder	$3.142 \times \text{radius}^2 \times$ height	$2 \times 3.142 \times$ radius $\times$ (height - radius)
Sphere	$4/3 \times 3.142 \times \text{radius}^3$	$4 \times 3.142 \times \text{radius}^2$
Segment of a sphere	$[(3.142 \times h)/6] \times (3 \times r^2 + h^2)$	$(2 \times 3.142 \times r \times h)$
Pyramid	$1/3 \times$ (area of base $\times$ height)	$0.5 \times$ circumference of base $\times$ slant height

FRACTIONS, DECIMALS AND MILLIMETRE EQUIVALENTS

Fractions	Decimals	mm	Fractions	Decimals	mm
1/64	0.015625	0.396875	33/64	0.515625	13.096875
1/32	0.03125	0.79375	17/32	0.53125	13.49375
3/64	0.046875	1.190625	35/64	0.546875	13.890625
1/16	0.0625	1.5875	9/16	0.5625	14.2875
5/64	0.078125	1.984375	37/64	0.578125	14.684375
3/32	0.09375	2.38125	19/32	0.59375	15.08125
7/64	0.109375	2.778125	39/64	0.609375	15.478125
1/8	0.125	3.175	5/8	0.625	15.875
9/64	0.140625	3.571875	41/64	0.640625	16.271875
5/32	0.15625	3.96875	21/32	0.65625	16.66875
11/64	0.171875	4.365625	43/64	0.671875	17.065625
3/16	0.1875	4.7625	11/16	0.6875	17.4625
13/64	0.203125	5.159375	45/64	0.703125	17.859375
7/32	0.21875	5.55625	23/32	0.71875	18.25625
15/64	0.234375	5.953125	47/64	0.734375	18.653125
1/4	0.25	6.35	3/4	0.75	19.05
17/64	0.265625	6.746875	49/64	0.765625	19.446875
9/32	0.28125	7.14375	25/32	0.78125	19.84375
19/64	0.296875	7.540625	51/64	0.796875	20.240625
5/16	0.3125	7.9375	13/16	0.8125	20.6375
21/64	0.328125	8.334375	53/64	0.828125	21.034375
11/32	0.34375	8.73125	27/32	0.84375	21.43125
23/64	0.359375	9.128125	55/64	0.859375	21.828125
3/8	0.375	9.525	7/8	0.875	22.225
25/64	0.390625	9.921875	57/64	0.890625	22.621875
13/32	0.40625	10.31875	29/32	0.90625	23.01875
27/64	0.421875	10.71563	59/64	0.921875	23.415625
7/16	0.4375	11.1125	15/16	0.9375	23.8125
29/64	0.453125	11.50938	61/64	0.953125	24.209375
15/32	0.46875	11.90625	31/32	0.96875	24.60625
31/64	0.484375	12.30313	63/64	0.984375	25.003125
1/2	0.5	12.7	1.0	1	25.4

IMPERIAL STANDARD WIRE GAUGE (SWG)

SWG No	Diameter inches	mm	SWG No	Diameter inches	mm
7/0	0.5	12.7	23	0.024	0.61
6/0	0.464	11.79	24	0.022	0.559
5/0	0.432	10.97	25	0.02	0.508
4/0	0.4	10.16	26	0.018	0.457
3/0	0.372	9.45	27	0.0164	0.417
2/0	0.348	8.84	28	0.0148	0.376
1/0	0.324	8.23	29	0.0136	0.345
1	0.3	7.62	30	0.0124	0.315
2	0.276	7.01	31	0.0116	0.295
3	0.252	6.4	32	0.0108	0.274
4	0.232	5.89	33	0.01	0.254
5	0.212	5.38	34	0.009	0.234
6	0.192	4.88	35	0.008	0.213
7	0.176	4.47	36	0.008	0.193
8	0.16	4.06	37	0.007	0.173
9	0.144	3.66	38	0.006	0.152
10	0.128	3.25	39	0.005	0.132
11	0.116	2.95	40	0.005	0.122
12	0.104	2.64	41	0.004	0.112
13	0.092	2.34	42	0.004	0.102
14	0.08	2.03	43	0.004	0.091
15	0.072	1.83	44	0.003	0.081
16	0.064	1.63	45	0.003	0.071
17	0.056	1.42	46	0.002	0.061
18	0.048	1.22	47	0.002	0.051
19	0.04	1.016	48	0.002	0.041
20	0.036	0.914	49	0.001	0.031
21	0.032	0.813	50	0.001	0.025
22	0.028	0.711			

WATER PRESSURE DUE TO HEIGHT

Imperial

Head Feet	Pressure lb/in²	Head Feet	Pressure lb/in²
1	0.43	70	30.35
5	2.17	75	32.51
10	4.34	80	34.68
15	6.5	85	36.85
20	8.67	90	39.02
25	10.84	95	41.18
30	13.01	100	43.35
35	15.17	105	45.52
40	17.34	110	47.69
45	19.51	120	52.02
50	21.68	130	56.36
55	23.84	140	60.69
60	26.01	150	65.03
65	28.18		

Metric

Head m	Pressure bar	Head m	Pressure bar
0.5	0.049	18.0	1.766
1.0	0.098	19.0	1.864
1.5	0.147	20.0	1.962
2.0	0.196	21.0	2.06
3.0	0.294	22.0	2.158
4.0	0.392	23.0	2.256
5.0	0.491	24.0	2.354
6.0	0.589	25.0	2.453
7.0	0.687	26.0	2.551
8.0	0.785	27.0	2.649
9.0	0.883	28.0	2.747
10.0	0.981	29.0	2.845
11.0	1.079	30.0	2.943
12.0	1.177	32.5	3.188
13.0	1.275	35.0	3.434
14.0	1.373	37.5	3.679
15.0	1.472	40.0	3.924
16.0	1.57	42.5	4.169
17.0	1.668	45.0	4.415

1 bar	=	14.5038 lbf/in²
1 lbf/in²	=	0.06895 bar
1 metre	=	3.2808 ft or 39.3701 in
1 foot	=	0.3048 metres
1 in wg	=	2.5 mbar (249.1 N/m²)

TABLE OF WEIGHTS FOR STEELWORK

Mild Steel Bar

Diameter (mm)	Weight (kg/m)	Diameter (mm)	Weight (kg/m)
6	0.22	20	2.47
10	0.62	25	3.85
12	0.89	30	5.55
16	1.58	32	6.31

Mild Steel Flat

Size (mm)	Weight (kg/m)	Size (mm)	Weight (kg/m)
15 x 3	0.36	15 x 5	0.59
20 x 3	0.47	20 x 5	0.79
25 x 3	0.59	25 x 5	0.98
30 x 3	0.71	30 x 5	1.18
40 x 3	0.94	40 x 5	1.57
45 x 3	1.06	45 x 5	1.77
50 x 3	1.18	50 x 5	1.96
20 x 6	0.94	20 x 8	1.26
25 x 6	1.18	25 x 8	1.57
30 x 6	1.41	30 x 8	1.88
40 x 6	1.88	40 x 8	2.51
45 x 6	2.12	45 x 8	2.83
50 x 6	2.36	50 x 8	3.14
55 x 6	2.60	55 x 8	3.45
60 x 6	2.83	60 x 8	3.77
65 x 6	3.06	65 x 8	4.08
70 x 6	3.30	70 x 8	4.40
75 x 6	3.53	75 x 8	4.71
100 x 6	4.71	100 x 8	6.28
20 x 10	1.57	20 x 12	1.88
25 x 10	1.96	25 x 12	2.36
30 x 10	2.36	30 x 12	2.83
40 x 10	3.14	40 x 12	3.77
45 x 10	3.53	45 x 12	4.24
50 x 10	3.93	50 x 12	4.71
55 x 10	4.32	55 x 12	5.12
60 x 10	4.71	60 x 12	5.65
65 x 10	5.10	65 x 12	6.12
70 x 10	5.50	70 x 12	6.59
75 x 10	5.89	75 x 12	7.07
100 x 10	7.85	100 x 12	9.42

TABLE OF WEIGHTS FOR STEELWORK

Mild Steel Equal Angle

Size (mm)	Weight (kg/m)	Size (mm)	Weight (kg/m)
13 x 13 x 3	0.56	60 x 60 x 10	8.69
20 x 20 x 3	0.88	70 x 70 x 10	10.30
25 x 25 x 3	1.11	75 x 75 x 10	11.05
30 x 30 x 3	1.36	80 x 80 x 10	11.90
40 x 40 x 3	1.82	90 x 90 x 10	13.40
45 x 45 x 3	2.06	100 x 100 x 10	15.00
50 x 50 x 3	2.30	120 x 120 x 10	18.20
		150 x 156 x 10	23.00
30 x 30 x 6	2.56	75 x 75 x 12	13.07
40 x 40 x 6	3.52	80 x 80 x 12	14.00
45 x 45 x 6	4.00	90 x 90 x 12	15.90
50 x 50 x 6	4.47	100 x 120 x 12	21.60
60 x 60 x 6	5.42	120 x 120 x 12	21.90
70 x 70 x 6	6.38	150 x 150 x 12	27.30
75 x 75 x 6	6.82	200 x 200 x 12	36.74
80 x 80 x 6	7.34		
90 x 90 x 6	8.30		
40 x 40 x 8	4.55		
50 x 50 x 8	5.82		
60 x 60 x 8	7.09		
70 x 70 x 8	8.36		
75 x 75 x 8	8.96		
80 x 80 x 8	9.63		
90 x 90 x 8	10.90		
100 x 100 x 8	12.20		
120 x 120 x 8	14.70		

Mild Steel Unequal Angle

Size (mm)	Weight (kg/m)	Size (mm)	Weight (kg/m)
40 x 25 x 6	2.79	100 x 65 x 10	12.30
50 x 40 x 6	4.24	100 x 75 x 10	13.00
60 x 30 x 6	3.99	125 x 75 x 10	15.00
65 x 50 x 6	5.16	150 x 75 x 10	17.00
75 x 50 x 6	5.65	150 x 90 x 10	18.20
80 x 60 x 6	6.37	200 x 100 x 10	23.00
125 x 75 x 6	9.18		
75 x 50 x 8	7.39	100 x 75 x 12	15.40
80 x 60 x 8	8.34	125 x 75 x 12	17.80
100 x 65 x 8	9.94	150 x 75 x 12	20.20
100 x 75 x 8	10.60	150 x 90 x 12	21.60
125 x 75 x 8	12.20	200 x 100 x 12	27.30
137 x 102 x 8	14.88	200 x 150 x 12	32.00

TABLE OF WEIGHTS FOR STEELWORK

Rolled Steel Channels

Size (mm)	Weight (kg/m)	Size (mm)	Weight (kg/m)
32 x 27	2.80	178 x 76	20.84
38 x 19	2.49	178 x 79	26.81
51 x 25	4.46	203 x 76	23.82
51 x 38	5.80	203 x 89	29.78
64 x 25	6.70	229 x 76	26.06
76 x 38	7.46	229 x 89	32.76
76 x 51	9.45	254 x 76	28.29
102 x 51	10.42	254 x 89	35.74
127 x 64	14.90	305 x 89	41.67
152 x 76	17.88	305 x 102	46.18
152 x 89	23.84	381 x 102	55.10

Rolled Steel Joists

Size (mm)	Weight (kg/m)	Size (mm)	Weight (kg/m)
76 x 38	6.25	152 x 76	17.86
76 x 76	12.65	152 x 89	17.09
102 x 44	7.44	152 x 127	37.20
102 x 64	9.65	178 x 102	21.54
102 x 102	23.06	203 x 102	25.33
127 x 76	13.36	203 x 152	52.03
127 x 114	26.78	254 x 114	37.20
127 x 114	29.76	254 x 203	81.84
		305 x 203	96.72

Universal Columns

Size (mm)	Weight (kg/m)	Size (mm)	Weight (kg/m)
152 x 152	23.00	254 x 254	89.00
152 x 152	30.00	254 x 254	107.00
152 x 152	37.00	254 x 254	132.00
203 x 203	46.00	254 x 254	167.00
203 x 203	52.00	305 x 305	97.00
203 x 203	60.00	305 x 305	118.00
203 x 203	71.00	305 x 305	137.00
203 x 203	86.00	305 x 305	158.00
254 x 254	73.00	305 x 305	198.00

TABLE OF WEIGHTS FOR STEELWORK

Universal Beams

Size (mm)	Weight (kg/m)	Size (mm)	Weight (kg/m)
203 x 133	25.00	305 x 127	48.00
203 x 133	30.00	305 x 165	40.00
254 x 102	22.00	305 x 165	46.00
254 x 102	25.00	305 x 165	54.00
254 x 102	28.00	356 x 127	33.00
254 x 146	31.00	356 x 127	39.00
254 x 146	37.00	356 x 171	45.00
254 x 146	43.00	356 x 171	51.00
305 x 102	25.00	356 x 171	57.00
305 x 102	28.00	356 x 171	67.00
305 x 102	33.00	381 x 152	52.00
305 x 127	37.00	381 x 152	60.00
305 x 127	42.00	381 x 152	67.00

Circular Hollow Sections

Size (mm)	Weight (kg/m)	Size (mm)	Weight (kg/m)
21.3 x 3.2	1.43	76.1 x 3.2	5.75
26.9 x 3.2	1.87	76.1 x 4.0	7.11
33.7 x 2.6	1.99	76.1 x 5.0	8.77
33.7 x 3.2	2.41	88.9 x 3.2	6.76
33.7 x 4.0	2.93	88.9 x 4.0	8.36
42.4 x 2.6	2.55	88.9 x 5.0	10.30
42.4 x 3.2	3.09	114.3 x 3.6	9.83
42.4 x 4.0	3.79	114.3 x 5.0	13.50
48.3 x 3.2	3.56	114.3 x 6.3	16.80
48.3 x 4.0	4.37	139.7 x 5.0	16.60
48.3 x 5.0	5.34	139.7 x 6.3	20.70
60.3 x 3.2	4.51	139.7 x 8.0	26.00
60.3 x 4.0	5.55	139.7 x 10.0	32.00
60.3 x 5.0	6.82	168.3 x 5.0	20.10

TABLE OF WEIGHTS FOR STEELWORK

Square Hollow Sections

Size (mm)	Weight (kg/m)	Size (mm)	Weight (kg/m)
20 x 20 x 2.0	1.12	90 x 90 x 3.6	9.72
20 x 20 x 2.6	1.39	90 x 90 x 5.0	13.30
30 x 30 x 2.6	2.21	90 x 90 x 6.3	16.40
30 x 30 x 3.2	2.65	100 x 100 x 4.0	12.00
40 x 40 x 2.6	3.03	100 x 100 x 5.0	14.80
40 x 40 x 3.2	3.66	100 x 100 x 6.3	18.40
40 x 40 x 4.0	4.46	100 x 100 x 8.0	22.90
50 x 50 x 3.2	4.66	100 x 100 x 10.0	27.90
50 x 50 x 4.0	5.72	120 x 120 x 5.0	18.00
50 x 50 x 5.0	6.97	120 x 120 x 6.3	22.30
60 x 60 x 3.2	5.67	120 x 120 x 8.0	27.90
60 x 60 x 4.0	6.97	120 x 120 x 10.0	34.20
60 x 60 x 5.0	8.54	150 x 150 x 5.0	22.70
70 x 70 x 3.2	7.46	150 x 150 x 6.3	28.30
70 x 70 x 5.0	10.10	150 x 150 x 8.0	35.40
80 x 80 x 3.6	8.59	150 x 150 x 10.0	43.60
80 x 80 x 5.0	11.70		
80 x 80 x 6.3	14.40		

Rectangular Hollow Sections

Size (mm)	Weight (kg/m)	Size (mm)	Weight (kg/m)
50 x 30 x 2.6	3.03	120 x 80 x 5.0	14.80
50 x 30 x 3.2	3.66	120 x 80 x 6.3	18.40
60 x 40 x 3.2	4.66	120 x 80 x 8.0	22.90
60 x 40 x 4.0	5.72	120 x 80 x 10.0	27.90
80 x 40 x 3.2	5.67	150 x 100 x 5.0	18.70
80 x 40 x 4.0	6.97	150 x 100 x 6.3	23.30
90 x 50 x 3.6	7.46	150 x 100 x 8.0	29.10
90 x 50 x 5.0	10.10	150 x 100 x 10.0	35.70
100 x 50 x 3.2	7.18	160 x 80 x 5.0	18.00
100 x 50 x 4.0	8.86	160 x 80 x 6.3	22.30
100 x 50 x 5.0	10.90	160 x 80 x 8.0	27.90
100 x 60 x 3.6	8.59	160 x 80 x 10.0	34.20
100 x 60 x 5.0	11.70	200 x 100 x 5.0	22.70
100 x 60 x 6.3	14.40	200 x 100 x 6.3	28.30
120 x 60 x 3.6	9.72	200 x 100 x 8.0	35.40
120 x 60 x 5.0	13.30	200 x 100 x 10.0	43.60
120 x 60 x 6.3	16.40		

DIMENSIONS AND WEIGHTS OF COPPER PIPES TO BSEN 1057, BSEN 12499, BSEN 14251

Outside Diameter (mm)	Internal Diameter (mm)	Weight per Metre (kg)	Internal Diameter (mm)	Weight per Metre (kg)	Internal Diameter (mm)	Weight per Metre (kg)
	Formerly Table X		Formerly Table Y		Formerly Table Z	
6	4.80	0.0911	4.40	0.1170	5.00	0.0774
8	6.80	0.1246	6.40	0.1617	7.00	0.1054
10	8.80	0.1580	8.40	0.2064	9.00	0.1334
12	10.80	0.1914	10.40	0.2511	11.00	0.1612
15	13.60	0.2796	13.00	0.3923	14.00	0.2031
18	16.40	0.3852	16.00	0.4760	16.80	0.2918
22	20.22	0.5308	19.62	0.6974	20.82	0.3589
28	26.22	0.6814	25.62	0.8985	26.82	0.4594
35	32.63	1.1334	32.03	1.4085	33.63	0.6701
42	39.63	1.3675	39.03	1.6996	40.43	0.9216
54	51.63	1.7691	50.03	2.9052	52.23	1.3343
76.1	73.22	3.1287	72.22	4.1437	73.82	2.5131
108	105.12	4.4666	103.12	7.3745	105.72	3.5834
133	130.38	5.5151	--	--	130.38	5.5151
159	155.38	8.7795	--	--	156.38	6.6056

DIMENSIONS OF STAINLESS STEEL PIPES TO BS 4127

Outside Diameter (mm)	Maximum Outside Diameter (mm)	Minimum Outside Diameter (mm)	Wall Thickness (mm)	Working Pressure (bar)
6	6.045	5.940	0.6	330
8	8.045	7.940	0.6	260
10	10.045	9.940	0.6	210
12	12.045	11.940	0.6	170
15	15.045	14.940	0.6	140
18	18.045	17.940	0.7	135
22	22.055	21.950	0.7	110
28	28.055	27.950	0.8	121
35	35.070	34.965	1.0	100
42	42.070	41.965	1.1	91
54	54.090	53.940	1.2	77

DIMENSIONS OF STEEL PIPES TO BS 1387

Nominal Size	Approx. Outside Diameter	Outside Diameter				Thickness		
		Light		Medium & Heavy		Light	Medium	Heavy
		Max.	Min.	Max.	Min.			
mm	mm	mm	mm	mm	mm	mm	mm	mm
6	10.20	10.10	9.70	10.40	9.80	1.80	2.00	2.65
8	13.50	13.60	13.20	13.90	13.30	1.80	2.35	2.90
10	17.20	17.10	16.70	17.40	16.80	1.80	2.35	2.90
15	21.30	21.40	21.00	21.70	21.10	2.00	2.65	3.25
20	26.90	26.90	26.40	27.20	26.60	2.35	2.65	3.25
25	33.70	33.80	33.20	34.20	33.40	2.65	3.25	4.05
32	42.40	42.50	41.90	42.90	42.10	2.65	3.25	4.05
40	48.30	48.40	47.80	48.80	48.00	2.90	3.25	4.05
50	60.30	60.20	59.60	60.80	59.80	2.90	3.65	4.50
65	76.10	76.00	75.20	76.60	75.40	3.25	3.65	4.50
80	88.90	88.70	87.90	89.50	88.10	3.25	4.05	4.85
100	114.30	113.90	113.00	114.90	113.30	3.65	4.50	5.40
125	139.70	--	--	140.60	138.70	--	4.85	5.40
150	165.1*	--	--	166.10	164.10	--	4.85	5.40

* 165.1mm (6.5in) outside diameter is not generally recommended except where screwing to BS 21 is necessary.

All dimensions are in accordance with ISO R65 except approximate outside diameters which are in accordance with ISO R64.

Light quality is equivalent to ISO R65 Light Series II.

APPROXIMATE METRES PER TONNE OF TUBES TO BS 1387

Nom. Size mm	BLACK						GALVANISED					
	Plain/screwed ends			Screwed & socketed			Plain/screwed ends			Screwed & socketed		
	L m	M m	H m	L m	M m	H m	L m	M m	H m	L m	M m	H m
6	2765	2461	2030	2743	2443	2018	2604	2333	1948	2584	2317	1937
8	1936	1538	1300	1920	1527	1292	1826	1467	1254	1811	1458	1247
10	1483	1173	979	1471	1165	974	1400	1120	944	1386	1113	939
15	1050	817	688	1040	811	684	996	785	665	987	779	661
20	712	634	529	704	628	525	679	609	512	673	603	508
25	498	410	336	494	407	334	478	396	327	474	394	325
32	388	319	260	384	316	259	373	308	254	369	305	252
40	307	277	226	303	273	223	296	268	220	292	264	217
50	244	196	162	239	194	160	235	191	158	231	188	157
65	172	153	127	169	151	125	167	149	124	163	146	122
80	147	118	99	143	116	98	142	115	97	139	113	96
100	101	82	69	98	81	68	98	81	68	95	79	67
125	--	62	56	--	60	55	--	60	55	--	59	54
150	--	52	47	--	50	46	--	51	46	--	49	45

The figures for `plain or screwed ends' apply also to tubes to BS 1775 of equivalent size and thickness.

Key

L – Light
M – Medium
H – Heavy

FLANGE DIMENSION CHART TO BS 4504 & BS 10

Normal Pressure Rating (PN 6) 6 Bar

Nom. Size	Flange Outside Diam.	Table 6/2 Forged Welding Neck	Table 6/3 Plate Slip on	Table 6/4 Forged Bossed Screwed	Table 6/5 Forged Bossed Slip on	Table 6/8 Plate Blank	Raised Face Diam.	Raised Face T'ness	Nr. Bolt Hole	Size of Bolt
15	80	12	12	12	12	12	40	2	4	M10 x 40
20	90	14	14	14	14	14	50	2	4	M10 x 45
25	100	14	14	14	14	14	60	2	4	M10 x 45
32	120	14	16	14	14	14	70	2	4	M12 x 45
40	130	14	16	14	14	14	80	3	4	M12 x 45
50	140	14	16	14	14	14	90	3	4	M12 x 45
65	160	14	16	14	14	14	110	3	4	M12 x 45
80	190	16	18	16	16	16	128	3	4	M16 x 55
100	210	16	18	16	16	16	148	3	4	M16 x 55
125	240	18	20	18	18	18	178	3	8	M16 x 60
150	265	18	20	18	18	18	202	3	8	M16 x 60
200	320	20	22	--	20	20	258	3	8	M16 x 60
250	375	22	24	--	22	22	312	3	12	M16 x 65
300	440	22	24	--	22	22	365	4	12	M20 x 70

FLANGE DIMENSION CHART TO BS 4504 & BS 10

Normal Pressure Rating (PN 16) 16 Bar

Nom. Size	Flange Outside Diam.	Table 6/2 Forged Welding Neck	Table 6/3 Plate Slip on	Table 6/4 Forged Bossed Screwed	Table 6/5 Forged Bossed Slip on	Table 6/8 Plate Blank	Raised Face Diam.	Raised Face T'ness	Nr. Bolt Hole	Size of Bolt
15	95	14	14	14	14	14	45	2	4	M12 x 45
20	105	16	16	16	16	16	58	2	4	M12 x 50
25	115	16	16	16	16	16	68	2	4	M12 x 50
32	140	16	16	16	16	16	78	2	4	M16 x 55
40	150	16	16	16	16	16	88	3	4	M16 x 55
50	165	18	18	18	18	18	102	3	4	M16 x 60
65	185	18	18	18	18	18	122	3	4	M16 x 60
80	200	20	20	20	20	20	138	3	8	M16 x 60
100	220	20	20	20	20	20	158	3	8	M16 x 65
125	250	22	22	22	22	22	188	3	8	M16 x 70
150	285	22	22	22	22	22	212	3	8	M20 x 70
200	340	24	24	--	24	24	268	3	12	M20 x 75
250	405	26	26	--	26	26	320	3	12	M24 x 90
300	460	28	28	--	28	28	378	4	12	M24 x 90

MINIMUM DISTANCES BETWEEN SUPPORTS/FIXINGS

Material	BS Nominal Pipe Size		Pipes - Vertical	Pipes - Horizontal on to low gradients
	inch	mm	support distance in metres	support distance in metres
Copper	0.50	15.00	1.90	1.30
	0.75	22.00	2.50	1.90
	1.00	28.00	2.50	1.90
	1.25	35.00	2.80	2.50
	1.50	42.00	2.80	2.50
	2.00	54.00	3.90	2.50
	2.50	67.00	3.90	2.80
	3.00	76.10	3.90	2.80
	4.00	108.00	3.90	2.80
	5.00	133.00	3.90	2.80
	6.00	159.00	3.90	2.80
muPVC	1.25	32.00	1.20	0.50
	1.50	40.00	1.20	0.50
	2.00	50.00	1.20	0.60
Polypropylene	1.25	32.00	1.20	0.50
	1.50	40.00	1.20	0.50
uPVC	--	82.40	1.20	0.50
	--	110.00	1.80	0.90
	--	160.00	1.80	1.20
Steel	0.50	15.00	2.40	1.80
	0.75	20.00	3.00	2.40
	1.00	25.00	3.00	2.40
	1.25	32.00	3.00	2.40
	1.50	40.00	3.70	2.40
	2.00	50.00	3.70	2.40
	2.50	65.00	4.60	3.00
	3.00	80.40	4.60	3.00
	4.00	100.00	4.60	3.00
	5.00	125.00	5.50	3.70
	6.00	150.00	5.50	4.50
	8.00	200.00	8.50	6.00
	10.00	250.00	9.00	6.50
	12.00	300.00	10.00	7.00
	16.00	400.00	10.00	8.25

LITRES OF WATER STORAGE REQUIRED PER PERSON PER BUILDING TYPE

Type of Building	Storage litres
Houses and flats (up to 4 bedrooms)	120/bedroom
Houses and flats (more than 4 bedrooms)	100/bedroom
Hostels	90/bed
Hotels	200/bed
Nurses homes and medical quarters	120/bed
Offices with canteen	45/person
Offices without canteen	40/person
Restaurants	7/meal
Boarding schools	90/person
Day schools - Primary	15/person
Day schools - Secondary	20/person

RECOMMENDED AIR CONDITIONING DESIGN LOADS

Building Type	Design Loading
Computer rooms	500 W/m² of floor area
Restaurants	150 W/m² of floor area
Banks (main area)	100 W/m² of floor area
Supermarkets	25 W/m² of floor area
Large Office Block (exterior zone)	100 W/m² of floor area
Large Office Block (interior zone)	80 W/m² of floor area
Small Office Block (interior zone)	80 W/m² of floor area

CAPACITY AND DIMENSIONS OF GALVANISED MILD STEEL CISTERNS – BS 417

Capacity (litres)	BS type (SCM)	Dimensions		
		Length (mm)	Width (mm)	Depth (mm)
18	45	457	305	305
36	70	610	305	371
54	90	610	406	371
68	110	610	432	432
86	135	610	457	482
114	180	686	508	508
159	230	736	559	559
191	270	762	584	610
227	320	914	610	584
264	360	914	660	610
327	450/1	1220	610	610
336	450/2	965	686	686
423	570	965	762	787
491	680	1090	864	736
709	910	1070	889	889

CAPACITY OF COLD WATER POLYPROPYLENE STORAGE CISTERNS - BS 4213

Capacity (litres)	BS type (PC)	Maximum Height mm
18	4	310
36	8	380
68	15	430
91	20	510
114	25	530
182	40	610
227	50	660
273	60	660
318	70	660
455	100	760

MINIMUM INSULATION THICKNESS TO PROTECT AGAINST FREEZING FOR DOMESTIC COLD WATER SYSTEMS (8 Hour Evaluation Period)

Pipe size (mm)	Insulation thickness (mm)					
	Condition 1			Condition 2		
	$\lambda = 0.020$	$\lambda = 0.030$	$\lambda = 0.040$	$\lambda = 0.020$	$\lambda = 0.030$	$\lambda = 0.040$
Copper pipes						
15	11	20	34	12	23	41
22	6	9	13	6	10	15
28	4	6	9	4	7	10
35	3	5	7	4	5	7
42	3	4	5	8	4	6
54	2	3	4	2	3	4
76	2	2	3	2	2	3
Steel pipes						
15	9	15	24	10	18	29
20	6	9	13	6	10	15
25	4	7	9	5	7	10
32	3	5	6	3	5	7
40	3	4	5	3	4	6
50	2	3	4	2	3	4
65	2	2	3	2	3	3

Condition 1 : water temperature 7°C; ambient temperature –6°C; evaluation period 8 h; permitted ice formation 50%; normal installation i.e. inside the building and inside the envelope of the structural insulation

Condition 2 : water temperature 2°C; ambient temperature –6°C; evaluation period 8 h; permitted ice formation 50%; extreme installation, i.e. inside the building but outside the envelope of the structural insulation

λ = thermal conductivity [W/(mK)]

INSULATION THICKNESS FOR CHILLED AND COLD WATER SUPPLIES TO PREVENT CONDENSATION
On A Low Emissivity Outer Surface (0.05, i.e. Bright Reinforced Aluminium Foil) With An Ambient Temperature Of +25°C And A Relative Humidity Of 80%

Steel pipe size (mm)	$t = +10$			$t = +5$			$t = 0$		
	Insulation thickness (mm)			Insulation thickness (mm)			Insulation thickness (mm)		
	$\lambda = 0.030$	$\lambda = 0.040$	$\lambda = 0.050$	$\lambda = 0.030$	$\lambda = 0.040$	$\lambda = 0.050$	$\lambda = 0.030$	$\lambda = 0.040$	$\lambda = 0.050$
15	16	20	25	22	28	34	28	36	43
25	18	24	29	25	32	39	32	41	50
50	22	28	34	30	39	47	38	49	60
100	26	34	41	36	47	57	46	60	73
150	29	38	46	40	52	64	51	67	82
250	33	43	53	46	60	74	59	77	94
Flat surfaces	39	52	65	56	75	93	73	97	122

t = temperature of contents (°C)
λ = thermal conductivity at mean temperature of insulation [W/(mK)]

INSULATION THICKNESS FOR NON-DOMESTIC HEATING INSTALLATIONS TO CONTROL HEAT LOSS

Steel pipe size (mm)	$t = 75$ Insulation thickness (mm)			$t = 100$ Insulation thickness (mm)			$t = 150$ Insulation thickness (mm)		
	$\lambda = 0.030$	$\lambda = 0.040$	$\lambda = 0.050$	$\lambda = 0.030$	$\lambda = 0.040$	$\lambda = 0.050$	$\lambda = 0.030$	$\lambda = 0.040$	$\lambda = 0.050$
10	18	32	55	20	36	62	23	44	77
15	19	34	56	21	38	64	26	47	80
20	21	36	57	23	40	65	28	50	83
25	23	38	58	26	43	68	31	53	85
32	24	39	59	28	45	69	33	55	87
40	25	40	60	29	47	70	35	57	88
50	27	42	61	31	49	72	37	59	90
65	29	43	62	33	51	74	40	63	92
80	30	44	62	35	52	75	42	65	94
100	31	46	63	37	54	76	45	68	96
150	33	48	64	40	57	77	50	73	100
200	35	49	65	42	59	79	53	76	103
250	36	50	66	43	61	80	55	78	105

t = hot face temperature (°C)
λ = thermal conductivity at mean temperature of insulation [W/(mK)]

ESSENTIAL READING FROM TAYLOR AND FRANCIS

Spon's Estimating Costs Guide to Plumbing and Heating
Unit Rates and Project Costs

Fourth Edition

Bryan Spain

Do you work on jobs between £50 and £50,000? - Then this book is for you.

All the cost data you need to keep your estimating accurate, competitive and profitable.

Specially written for contractors and small businesses carrying out small works, *Spon's Estimating Cost Guide to Plumbing and Heating* contains accurate information on thousands of rates each broken down to labour, material overheads and profit.

The first book to include typical project costs for:

- rainwater goods installations
- bathrooms
- external waste systems
- central heating systems
- hot and cold water systems.

July 2008: 216x138: 264pp
Pb: 978-0-415-46905-0: **£29.99**

To Order: Tel: +44 (0) 1235 400524 **Fax:** +44 (0) 1235 400525
or Post: Taylor and Francis Customer Services,
Bookpoint Ltd, Unit T1, 200 Milton Park, Abingdon, Oxon, OX14 4TA UK
Email: book.orders@tandf.co.uk

**For a complete listing of all our titles visit:
www.tandf.co.uk**

Index

Note: Italics Represents Typical Engineering Drawings

Above Ground Drainage
    cast iron
        EDPM rubber gasket   163
        nitrile rubber gasket   159
    plastic
        ABS waste   148
        MuPVC waste   145
        polypropylene waste   150
        PVC-u overflow   145
        PVC-u soil   153
ABS pipe & solvent weld fittings   195
Access Control Equipment   42, *76*, 588
Acoustic Housings (for generators)   482
Adaptable Boxes
    PVC   489
    steel   501
Air Circuit Breakers   544
Air Conditioning Design Loads   656
Air Cooled Condensers   348
Air Curtains   473
Air Handling Units   75
Airport Terminal   90
All Air Systems (FGU)   53, 78
All-In-Rates   74
    above ground drainage   74
    access control   76
    air handling units   75
    CCTV   76
    chilled water   74
    data cabling   77
    ductwork   75
    electrical works in connection with mechanical (EWIC)   82
    external lighting   83
    extract fans   75
    fire alarms   82–83
    generators   80
    heat rejection   74
    heat source   74
    hosereels/dry risers   76
    HV switchgear   80
    lifts and escalators   84–89
    lighting   82
    LV switchgear   80
    pipework   77–79
    pumps   75
    ring main units   80
    small power   81
    sprinklers   76

    substation (packaged)   80
    transformers   80
    UPS   80
    water installations   74
Approximate Estimating
    cost indices   61
    directions   59
    stage A feasibility costs   62
    stage C elemental rates   67
Armoured Cable
    SWA & AWA XLPE PVC
        clipped   505
        in trench   503
        terminations   507
    SWA XLPE LSOH (LSF)
        clipped   511
        in trench   509
        terminations   513
Attenuators   432
Automatic Air Vents   294
Automatic Power Factor Correction   546
Automatic Transfer Switches   547

Batten Lampholders   555
Boards and Panels   548
Boilers
    all-in-rates   74
    atmospheric   247
    condensing   247
    domestic   243
    forced draft
        cast iron sectional   244
        steel shell   246
Building Management/Control equipment   598
Building Management Installations   47, *48*, 114
Busbar
    chamber   553
    lighting   530
    main   528
    rising   529
    under-floor   531
Business Parks   95

Cable Basket   542
Cable Tray
    galvanised steel   535
    GRP   341
Cables
    high voltage   483

Cables (cont'd)
    low voltage 503
    armoured 504
    earth cable 516
    flexible 518
    fire rated 518
    modular 524
    un-armoured 515
Calorifiers
    non storage steam 342
    non storage water 303
    storage 224
Cast Resin Transformers 486
CCTV Systems 44, 76
Ceiling Roses 561
Chilled Beams 50, 356
Chilled Ceiling 49
Chilled Water Pipework 352
Chillers
    absorption 346
    air cooled 344
    all-in-rates 74
    water cooled 345
Circuit Breakers
    HV 485
    LV moulded ACB 545
Clock Systems 581
Combined Heat & Power (CHP) 482
Connection Units 565
Consumer Units 551
Contactor Relays 553
Containment
    basket 542
    cable tray galvanised ms 534
    cable tray GRP 541
    conduit 488
    ladder rack 532
Control Components
    FCU DDC control pack 472
    Immersion thermostats 312
    room thermostats 311
    surface thermostats 311
    thermostatic radiator valves 311
Conversion Tables 638
Cooker Control Units 566
Cooling Towers 349
Cooling Units
    local 358
Copper Pipework (Cold Water) 169
    brazed (silver) fittings 180
    brazed (silver) flanges 182
    capillary fittings 169
    capillary fittings heavy duty 177
    compression fittings 178
    compression fittings DZR 180
    pressfit mechanical joints 185
Copper Pipework (CHW) 353
Cost Indices 60
Cylinders
    copper; direct 226
    copper; indirect 225
    storage 224

Dampers
    fire 424
    smoke 425
    volume control 423
Data Centre 112
Data Transmission Equipment/Cabling 77, 582
Day Care Units 63
Daywork
    Electrical industry rates 624
    H&V industry rates 621
    Plant hire 625
Deaerators 296
Department Stores 64
Diffusers
    circular and rectangular 439
    floor swirl 441
    perforated 441
    slot 440
Dimension of Copper and Stainless Steel Pipes 649, 650
Dimension of Steel Pipes 651
Dimmer Switches 560
Dirt Separators 297
Displacement System 48
Distribution Boards 549
Distribution Centres 62, 111
Dry Risers 76
Ductwork
    all-in-rates 74
    access doors 399
    approximate estimating 53
    circular ductwork, class B 360
    circular polypropylene 371
    circular PVC ductwork 369
    dampers
        fire 424
        smoke 425
        volume control 423
    fire rated ductwork
        duct, 2 hr protection 445
        duct, 4 hr protection 458
        slab, mineral wool 445
    flat oval ductwork 364
    flexible ductwork 368
    rectangular ductwork, class B 360
    rectangular ductwork, class C 400

Earth Bars 593
Earth Pits 596
Earth Rods 596
Electromagnetic Water Conditioner 220
Elemental Rates 67
Emergency Lighting Central Battery Systems 573
Emergency Lighting Luminaires 572
Energy Meters
    CHW 357
    LTHW 310

Engineering Details 31
Engineering Features 1
Entertainment and Recreation 63
Expansion Joints 285
External Lighting 83, 557

Factories 62
Fan Coil Units 47, 469
Fans
   all-in-rates 74
   axial flow fans 427
   bifurcated 427
   centrifugal 429
   in-line centrifugal 429
   kitchen fans 429
   multivent fans 428
   roof fans 428
   toilet fans – domestic 429
   twin extract 428
Filters, air
   bag 430
   panel 431
Fire Detection and Alarm Equipment 41, 82, 590
Fire Extinguishers 241
Fire Hydrants 241
Fire Resistant Cables 509, 520
Flange Dimension Chart 653
Flexible Cable 516
Flexible Conduit 488
Floor Service Outlet Box 564
Floor Trunking 489
Flue Systems
   domestic 248
   domestic/small commercial 254
Formulae 640
Fractions, Decimals & Millimetre Equivalents 641
Fuel Cells 19
Fused Switches Heavy Duty 552

Galvanised Steel Pipework 201
   fittings BS 1740 206
   fittings Heavy Steel BS 1387 205
   flanges screwed 204
   malleable iron fittings BS 143 207
Gas Booster Sets 231
Gas Pipework
   copper (plastic coated) 230
      capillary fittings 231
   MDPE (Yellow) 228
      compression fittings 228
      electrofusion fittings 228
   steel screwed 229
   steel welded 230
Gauges 295
Generating Sets
   standby diesel 40, 80
Grilles
   air diffusion 438
   exhaust 437

supply 435
Ground Water Cooling 13
Gutters
   cast iron 132
   PVC-u 125

Heat Emitters
   air curtains 474
   fan convectors 306
   perimeter heaters 303
   radiant strip heaters 303
   radiators
      flat panel 305
      pressed steel 304
   trench heating 308
   under floor heating 309
   unit heaters 343
Heat Exchangers
   CHW 354
   LTHW 302
Heat Rejection 48, 347
   absorption 346
   air cooled condensers 348
   all-in-rates 74
   cooling towers 349
   dry air liquid coolers 347
Hose Reels 76
Hospitals 63, 103
Hotels 66, 71, 72, 101
HRC Fused Distribution Boards 38, 485
HV Intakes 36, 80
HV Switchgear 486

Imperial Standard Wire Gauge 642
Industrial Sockets/Plugs 567
Intruder Detection 43, *589*
Isolating Switches 552

Junction Boxes
   Weatherproof 554

Labour Rates
   ductwork 122
   electrical 478, 612
   mechanical 120, 603
Ladder Rack 532
Leak Detection 356
Leisure Centres 64
Lifts and Escalators 84–89
Lighting
   all-in-rates 82
Lighting Busbar 530
Lighting Control 39, 561
Lighting Switches 559
Lighting Track 558
Lightning Protection Equipment 594
Louvres
   acoustic 442
   weather 443

Low Temp' Hot Water Pipework
   Mech' grooved steel pipe 279
      fittings 279
   plastic (polypropylene) pipe 280
      fittings 280
   pressfit carbon steel pipe 278
      fittings 278
   screwed BS 1387 steel pipe 258
      fittings 259
      flanges 261
   welded BS 1387 steel pipe 268
      fittings 272
      flanges 270
   welded carbon steel 275
      fittings 279
      flanges 277
Luminaires
   ceiling recessed asymmetric 556
   downlighters 557
   emergency 572
   external 557
   high/low bay 557
   linear fluorescent 557
   modular recessed 535
   surface fluorescent 555
   suspended fluorescent 556
   wall mounted fluorescent 556
LV Intakes 35
LV Switchboards 544
LV Switchgear 574
   all-in-rates 80

MCB Distribution Boards 549
MCCB Panel Boards 548
Medium Density Polyethylene
   MDPE (Blue) 191
      compression fittings 193
      electrofusion fittings 192
   MDPE (Yellow) 228
      compression fittings 228
      electrofusion fittings 228
Midel-Filled Transformers 486
Mineral Insulated Cables 518
Miniature Circuit Breakers 549
Minimum Distance between Supports 655
Modular Wiring 524

Offices 62, 63, 67-70
Oil-Filled Transformers 486

Part L Building Regulations 3
Performing Arts Centre 64, 94
Pipe Freezing 310
Pipe in Pipe (Gas) 229
Pipes
   ABS
      waste 148
      water 195
   all-in-rates 74

   black steel
      screwed chilled water 352
      screwed LTHW 258
      welded chilled water 352
      welded LTHW 268
   carbon steel 275
   cast iron
      rainwater dry joints 141
      soil EPDM rubber joints 163
      soil nitrile rubber joints 159
   copper
      chilled water 353
      cold water 169
      hot water 224
   galvanised steel 201
   MDPE
      blue 191
      yellow 228
   mechanical grooved 279
   MuPVC Waste
      solvent joints 145
   Polypropylene for LTHW
      thermally fused joints 280
   Polypropylene Waste
      push fit 150
   Pressfit
      carbon steel 278
      copper 186
      stainless steel 190
   PVC-C for cold water 200
   PVC-U
      overflow 125
      rainwater
      – dry push fit joints 128
      – solvent welded joints 131
      waste, solvent joints 145
      water 200
   Stainless Steel 187
Pipework Supports, minimum distance 655
Power Posts, Poles, Pillars 565
Pressurisation Units
   CHW 355
   LTHW 296
Pumps
   accelerator 301
   all-in-rates 75
   belt drive – CHW & LTHW 298
   circulator 301
   close coupled 299
   glandless – domestic 300
   pressurisation – cold water 217
   sump 217
   twin head belt drive 298
   variable speed 300
Push Button Stations 553
PVC Trunking 501
   PVC-C 200
      solvent weld fittings 200
   PVC-u (cold water) 197
      solvent weld fittings 197

Index

Rainwater Systems
    cast iron 132
    PVC-u 125
Rates of Wages
    Electrical 611
    Plumbing 607
    Mechanical 603
Regional Variations 61
Renewable Energy Options 5
Residential Buildings 73, 106
Residual Current Circuit Breakers 550
Retail 64, *83*, 110
Reverse Cycle Heat Pump 54
Ring Main Units (RMU's) 80
Rising Main Busbars 529

Schools 105
Scientific Buildings 65
Security Detection and Alarm Equipment 589
Shaver Sockets 566
Shopping Malls 64, 92
Small Power
    all-in-rates 81
Socket Outlets 563
Split Systems for Heating & Cooling 358
Sports Hall 64, 100
Sprinkler Alarms 238
Sprinkler Heads 238
Sprinkler Pipework & fittings
    screwed 201
    victaulic 237
    welded 235
Sprinkler Systems
    all-in-rates 76
Sprinkler Tanks 239
Sprinkler Valves 238
Stadium 102
Stainless Steel Pipework 187
    capillary fittings 187
    pressfit fittings 185
Standby Generators 40, 481
Steam Generators 257
Steam pressure reducing valves 341
Steam sight glasses 340
Steam strainers 339
Steam Traps
    cast iron 339
    stainless steel 339
Structured Cabling (Data) 38, 582
Supermarkets 65, 110
Surge Protection 597
Surveillance Equipment 44, *76*, 589
Switch Disconnectors 552, 566
Synchronisation Panels (for generators) 481

Tables and Memoranda 637
Tanks and Cisterns
    cold water storage tanks
        sectional GRP 219
        sectional steel 218

fuel oil storage tanks 232
moulded glass fibre cisterns 218
polypropylene cisterns 218
sprinkler tanks
    sectional GRP 240
    sectional steel 239
Telecommunications Cables 577
Telephone Outlets 578
Television and Radio Aerial Cable 579
Television Co-Axial Socket Outlet 580
Thermal Insulation
    ductwork – all types/finishes 444
    pipework – closed cell 222
    pipework – mineral fibre
        concealed 313
        external 329
        plantroom 319
Thermal Insulation Thickness Tables 658
Thermostats 311
Trace Heating System 354
Transformers 80
Tripping Batteries 485
Trunking
    dado/skirting 501
    galvanised steel 494
    floor 494
    PVC 495
    PVC bench 498
    PVC miniature 497
    Under-floor 498

Ultra Violet Water Sterilising Unit 220
Un-Armoured Cable 515
Under-floor Busbar 531
Uninterruptible Power Supply 570

Vacuum Circuit Breakers 35, 485
Valves
    Automatic Air Vent 294
    ball float valves 214
    ball valves 287
    check valves 214, 290
    check valves – DZR 214
    cocks 214, 294
    commissioning valves 290
    control valves 293
    gate valves – DZR 214, 286
    globe valves 291
    isolating 214, 286
    pressure reducing 341
    radiator valves 295
    regulators 214, 292
    safety and relief valves 293, 341
    stopcocks 214, 294
    strainers 292, 339
VAV – Variable Air Vol' System
    Terminal Units 51, 468
VRV – Variable Refrigerant Vol' System 52
Voice Cabling 591

Warehouses 62
Water Conditioner – Electromagnetic 220
Water Heater (Electric) 227
Water Installations
    all-in-rates 74
Water Pressure Due to Height 643

Water Softener 220
Water Storage Capacities/dimensions 657
Water Storage Requirements 656
Weights of Copper Pipework 649
Weights of Steel Pipework 652
Weights of Steelwork 644

Software and eBook Single-User Licence Agreement

We welcome you as a user of this Taylor & Francis Software and eBook and hope that you find it a useful and valuable tool. Please read this document carefully. **This is a legal agreement** between you (hereinafter referred to as the "Licensee") and Taylor and Francis Books Ltd. (the "Publisher"), which defines the terms under which you may use the Product. **By breaking the seal and opening the document inside the back cover of the book containing the access code you agree to these terms and conditions outlined herein. If you do not agree to these terms you must return the Product to your supplier intact, with the seal on the document unbroken.**

1. **Definition of the Product**
   The product which is the subject of this Agreement, *Spon's Mechanical and Electrical Services Price Book 2009* Software and eBook (the "Product") consists of:
   1.1 Underlying data comprised in the product (the "Data")
   1.2 A compilation of the Data (the "Database")
   1.3 Software (the "Software") for accessing and using the Database
   1.4 An electronic book containing the data in the price book (the "eBook")

2. **Commencement and licence**
   2.1 This Agreement commences upon the breaking open of the document containing the access code by the Licensee (the "Commencement Date").
   2.2 This is a licence agreement (the "Agreement") for the use of the Product by the Licensee, and not an agreement for sale.
   2.3 The Publisher licenses the Licensee on a non-exclusive and non-transferable basis to use the Product on condition that the Licensee complies with this Agreement. The Licensee acknowledges that it is only permitted to use the Product in accordance with this Agreement.

3. **Multiple use**
   For more than one user or for a wide area network or consortium, use is only permissible with the purchase from the Publisher of a multiple-user licence and adherence to the terms and conditions of that licence.

4. **Installation and Use**
   4.1 The Licensee may provide access to the Product for individual study in the following manner: The Licensee may install the Product on a secure local area network on a single site for use by one user.
   4.2 The Licensee shall be responsible for installing the Product and for the effectiveness of such installation.
   4.3 Text from the Product may be incorporated in a coursepack. Such use is only permissible with the express permission of the Publisher in writing and requires the payment of the appropriate fee as specified by the Publisher and signature of a separate licence agreement.
   4.4 The Product is a free addition to the book and no technical support will be provided.

5. **Permitted Activities**
   5.1 The Licensee shall be entitled:
      5.1.1 to use the Product for its own internal purposes;
      5.1.2 to download onto electronic, magnetic, optical or similar storage medium reasonable portions of the Database provided that the purpose of the Licensee is to undertake internal research or study and provided that such storage is temporary;
   5.2 The Licensee acknowledges that its rights to use the Product are strictly set out in this Agreement, and all other uses (whether expressly mentioned in Clause 6 below or not) are prohibited.

6. **Prohibited Activities**
   The following are prohibited without the express permission of the Publisher:
   6.1 The commercial exploitation of any part of the Product.
   6.2 The rental, loan, (free or for money or money's worth) or hire purchase of this product, save with the express consent of the Publisher.
   6.3 Any activity which raises the reasonable prospect of impeding the Publisher's ability or opportunities to market the Product.
   6.4 Any networking, physical or electronic distribution or dissemination of the product save as expressly permitted by this Agreement.
   6.5 Any reverse engineering, decompilation, disassembly or other alteration of the Product save in accordance with applicable national laws.
   6.6 The right to create any derivative product or service from the Product save as expressly provided for in this Agreement.
   6.7 Any alteration, amendment, modification or deletion from the Product, whether for the purposes of error correction or otherwise.

7. General Responsibilities of the License

7.1 The Licensee will take all reasonable steps to ensure that the Product is used in accordance with the terms and conditions of this Agreement.

7.2 The Licensee acknowledges that damages may not be a sufficient remedy for the Publisher in the event of breach of this Agreement by the Licensee, and that an injunction may be appropriate.

7.3 The Licensee undertakes to keep the Product safe and to use its best endeavours to ensure that the product does not fall into the hands of third parties, whether as a result of theft or otherwise.

7.4 Where information of a confidential nature relating to the product of the business affairs of the Publisher comes into the possession of the Licensee pursuant to this Agreement (or otherwise), the Licensee agrees to use such information solely for the purposes of this Agreement, and under no circumstances to disclose any element of the information to any third party save strictly as permitted under this Agreement. For the avoidance of doubt, the Licensee's obligations under this sub-clause 7.4 shall survive the termination of this Agreement.

8. Warrant and Liability

8.1 The Publisher warrants that it has the authority to enter into this agreement and that it has secured all rights and permissions necessary to enable the Licensee to use the Product in accordance with this Agreement.

8.2 The Publisher warrants that the Product as supplied on the Commencement Date shall be free of defects in materials and workmanship, and undertakes to replace any defective Product within 28 days of notice of such defect being received provided such notice is received within 30 days of such supply. As an alternative to replacement, the Publisher agrees fully to refund the Licensee in such circumstances, if the Licensee so requests, provided that the Licensee returns this copy of *Spon's Mechanical and Electrical Services Price Book 2009* to the Publisher. The provisions of this sub-clause 8.2 do not apply where the defect results from an accident or from misuse of the product by the Licensee.

8.3 Sub-clause 8.2 sets out the sole and exclusive remedy of the Licensee in relation to defects in the Product.

8.4 The Publisher and the Licensee acknowledge that the Publisher supplies the Product on an "as is" basis. The Publisher gives no warranties:
    8.4.1 that the Product satisfies the individual requirements of the Licensee; or
    8.4.2 that the Product is otherwise fit for the Licensee's purpose; or
    8.4.3 that the Data are accurate or complete or free of errors or omissions; or
    8.4.4 that the Product is compatible with the Licensee's hardware equipment and software operating environment.

8.5 The Publisher hereby disclaims all warranties and conditions, express or implied, which are not stated above.

8.6 Nothing in this Clause 8 limits the Publisher's liability to the Licensee in the event of death or personal injury resulting from the Publisher's negligence.

8.7 The Publisher hereby excludes liability for loss of revenue, reputation, business, profits, or for indirect or consequential losses, irrespective of whether the Publisher was advised by the Licensee of the potential of such losses.

8.8 The Licensee acknowledges the merit of independently verifying Data prior to taking any decisions of material significance (commercial or otherwise) based on such data. It is agreed that the Publisher shall not be liable for any losses which result from the Licensee placing reliance on the Data or on the Database, under any circumstances.

8.9 Subject to sub-clause 8.6 above, the Publisher's liability under this Agreement shall be limited to the purchase price.

9. Intellectual Property Rights

9.1 Nothing in this Agreement affects the ownership of copyright or other intellectual property rights in the Data, the Database of the Software.

9.2 The Licensee agrees to display the Publishers' copyright notice in the manner described in the Product.

9.3 The Licensee hereby agrees to abide by copyright and similar notice requirements required by the Publisher, details of which are as follows:
"© 2009 Taylor & Francis. All rights reserved. All materials in *Spon's Mechanical and Electrical Services Price Book 2009* are copyright protected. All rights reserved. No such materials may be used, displayed, modified, adapted, distributed, transmitted, transferred, published or otherwise reproduced in any form or by any means now or hereafter developed other than strictly in accordance with the terms of the licence agreement enclosed with *Spon's Mechanical and Electrical Services Price Book 2009*. However, text and images may be printed and copied for research and private study within the preset program limitations. Please note the copyright notice above, and that any text or images printed or copied must credit the source."

9.4 This Product contains material proprietary to and copyedited by the Publisher and others. Except for the licence granted herein, all rights, title and interest in the Product, in all languages, formats and media throughout the world, including copyrights therein, are and remain the property of the Publisher or other copyright holders identified in the Product.

10. **Non-assignment**

This Agreement and the licence contained within it may not be assigned to any other person or entity without the written consent of the Publisher.

11. **Termination and Consequences of Termination.**

11.1 The Publisher shall have the right to terminate this Agreement if:

11.1.1 the Licensee is in material breach of this Agreement and fails to remedy such breach (where capable of remedy) within 14 days of a written notice from the Publisher requiring it to do so; or

11.1.2 the Licensee becomes insolvent, becomes subject to receivership, liquidation or similar external administration; or

11.1.3 the Licensee ceases to operate in business.

11.2 The Licensee shall have the right to terminate this Agreement for any reason upon two month's written notice. The Licensee shall not be entitled to any refund for payments made under this Agreement prior to termination under this sub-clause 11.2.

11.3 Termination by either of the parties is without prejudice to any other rights or remedies under the general law to which they may be entitled, or which survive such termination (including rights of the Publisher under sub-clause 7.4 above).

11.4 Upon termination of this Agreement, or expiry of its terms, the Licensee must destroy all copies and any back up copies of the product or part thereof.

12. **General**

12.1 **Compliance with export provisions**

The Publisher hereby agrees to comply fully with all relevant export laws and regulations of the United Kingdom to ensure that the Product is not exported, directly or indirectly, in violation of English law.

12.2 **Force majeure**

The parties accept no responsibility for breaches of this Agreement occurring as a result of circumstances beyond their control.

12.3 **No waiver**

Any failure or delay by either party to exercise or enforce any right conferred by this Agreement shall not be deemed to be a waiver of such right.

12.4 **Entire agreement**

This Agreement represents the entire agreement between the Publisher and the Licensee concerning the Product. The terms of this Agreement supersede all prior purchase orders, written terms and conditions, written or verbal representations, advertising or statements relating in any way to the Product.

12.5 **Severability**

If any provision of this Agreement is found to be invalid or unenforceable by a court of law of competent jurisdiction, such a finding shall not affect the other provisions of this Agreement and all provisions of this Agreement unaffected by such a finding shall remain in full force and effect.

12.6 **Variations**

This agreement may only be varied in writing by means of variation signed in writing by both parties.

12.7 **Notices**

All notices to be delivered to: Spon's Price Books, Taylor & Francis Books Ltd., 2 Park Square, Milton Park, Abingdon, Oxfordshire, OX14 4RN, UK.

12.8 **Governing law**

This Agreement is governed by English law and the parties hereby agree that any dispute arising under this Agreement shall be subject to the jurisdiction of the English courts.

If you have any queries about the terms of this licence, please contact:

Spon's Price Books
Taylor & Francis Books Ltd.
2 Park Square, Milton Park, Abingdon, Oxfordshire, OX14 4RN
Tel: +44 (0) 20 7017 6672
Fax: +44 (0) 20 7017 6702
www.tandfbuiltenvironment.com/

Software Installation and Use Instructions

System requirements

Minimum

- Pentium processor
- 256 MB of RAM
- 20 MB available hard disk space
- Microsoft Windows 98/2000/NT/ME/XP/Vista
- SVGA screen
- Internet connection

Recommended

- Intel 466 MHz processor
- 512 MB of RAM (1,024MB for Vista)
- 100 MB available hard disk space
- Microsoft Windows XP/Vista
- XVGA screen
- Broadband Internet connection

Microsoft® is a registered trademark and Windows™ is a trademark of the Microsoft Corporation.

Installation

Spon's Mechanical and Electrical Services Price Book 2009 Electronic Version is supplied solely by internet download. No CD-ROM is supplied.

In your internet browser type in www.ebookstore.tandf.co.uk/supplements/pricebook and follow the instructions on screen. Then type in the unique access code which is sealed inside the back cover of this book.

If the access code is successfully validated, a web page with details of the available download content will be displayed.

Please note: you will only be allowed one download of these files and onto one computer.

Click on the download links to download the file. A folder called *PriceBook* will be added to your desktop, which will need to be unzipped. Then click on the application called *install*.

Use

- The installation process will create a folder containing the price book program links as well as a program icon on your desktop.
- Double click the icon (from the folder or desktop) installed by the Setup program.
- Follow the instructions on screen.

Technical Support

Support for the installation is provided on
http://www.ebookstore.tandf.co.uk/html/helpdesk.asp

The *Electronic Version* is a free addition to the book. For help with the running of the software please visit www.pricebooks.co.uk

---

All materials in *Spon's Mechanical and Electrical Services Price Book 2009 Electronic Version* are copyright protected. No such materials may be used, displayed, modified, adapted, distributed, transmitted, transferred, published or otherwise reproduced in any form or by any means now or hereafter developed other than strictly in accordance with the terms of the above licence agreement.

The software used in *Spon's Mechanical and Electrical Services Price Book 2009 Electronic Version* is furnished under a single user licence agreement. The software may be used only in accordance with the terms of the licence agreement, unless the additional multi-user licence agreement is purchased from Taylor & Francis, 2 Park Square, Milton Park, Abingdon, Oxon OX14 4RN Tel: +44 (0) 20 7017 6000

© COPYRIGHT ALL RIGHTS RESERVED

Multiple-user use of the Spon Press Software and eBook

To buy a licence to install your Spon Press Price Book Software and eBook on a secure local area network or a wide area network, and for the supply of network key files, for an agreed number of users please contact:

Spon's Price Books
Taylor & Francis Books Ltd.
2 Park Square, Milton Park, Abingdon, Oxfordshire, OX14 4RN
Tel: +44 (0) 207 017 6672
Fax: +44 (0) 207 017 6072
www.pricebooks.co.uk

Number of users	Licence cost
2–5	£450
6–10	£915
11–20	£1400
21–30	£2150
31–50	£4200
51–75	£5900
76–100	£7100
Over 100	Please contact Spon for details